International Conference on Innovative Ideas in Science (IIS 2016)

IOP Conference Series: Materials Science and Engineering
Volume 200

Baia Mare, Romania
10 – 11 November 2016

ISBN: 978-1-5108-4096-6
ISSN: 1757-8981

Printed from e-media with permission by:

Curran Associates, Inc.
57 Morehouse Lane
Red Hook, NY 12571

CURRAN ASSOCIATES INC. proceedings.com

Some format issues inherent in the e-media version may also appear in this print version.

Printed by Curran Associates, Inc. (2017)

For permission requests, please contact the Institute of Physics
at the address below.

Institute of Physics
Dirac House, Temple Back
Bristol BS1 6BE UK

Phone: 44 1 17 929 7481
Fax: 44 1 17 920 0979

techtracking@iop.org

Additional copies of this publication are available from:

Curran Associates, Inc.
57 Morehouse Lane
Red Hook, NY 12571 USA
Phone: 845-758-0400
Fax: 845-758-2633
Email: curran@proceedings.com
Web: www.proceedings.com

Table of contents

Volume 200

Innovative Ideas in Science 2016

10–11 November 2016, Baia Mare, Romania

Accepted papers received: 9 May 2017
Published online: 25 May 2017

Preface

Papers

IOP Conf. Series: Materials Science and Engineering **200** (2017) 011001 doi:10.1088/1757-899X/200/1/011001

Technical University of Cluj-Napoca, North University Centre of Baia Mare
Faculty of Engineering, Department of Electrical Engineering, Electronics and Computers

University of Business Engineering and Management Banja Luka

INNOVATIVE IDEAS IN SCIENCE
IIS2016

Baia Mare, Romania, November 10-11, 2016

The aim of the International Conference on *Innovative Ideas in Science IIS2016* is to provide an opportunity for academics, practitioners and researchers to debate new achievements or concepts, approaches and innovative practices within the continuously progressing world.

This conference aims at including different fields of science under the general purpose of developing future interdisciplinary works and projects. Topics of the conference covers a comprehensive spectrum of issues from *Engineering and Technology, Computer Sciences, Environmental Science, Economical Sciences, Fundamental Sciences, Medical and Health Sciences, Social Sciences, Humanities, etc.*

On the basis of more than 120 works presented at the conference, 69 selected papers have been included in this volume of *IOP Conference Series: Materials Science and Engineering.*

CONFERENCE COORDINATORS

Cristian BARZ, Technical University of Cluj-Napoca, North University Center of Baia Mare, Romania
Vesna RODIC, University PIM Republic of Srpska, Bosnia and Herzegovina

ORGANIZING COMMITTEE

Biljana PETROVIC, University of Business Engineering and Management – Banja Luka, Bosnia and Herzegovina
Ilija J. DZOMBIC, University of Business Engineering and Management – Banja Luka, Bosnia and Herzegovina
Ranko MOJIC, University of Business Engineering and Management – Banja Luka, Bosnia and Herzegovina
Ognjen PANTELIUC, University of Business Engineering and Management – Banja Luka, Bosnia and Herzegovina
Veljko VUKOVIC, University of Business Engineering and Management – Banja Luka, Bosnia and Herzegovina
Radenko DJURICA, University of Business Engineering and Management – Banja Luka, Bosnia and Herzegovina
Mladen IVIC, University of Business Engineering and Management – Banja Luka, Bosnia and Herzegovina
Mircea HORGOŞ, Technical University of Cluj-Napoca, North University Center of Baia Mare, Romania
Anamaria DĂSCĂLESCU, Technical University of Cluj-Napoca, North University Center of Baia Mare, Romania
Zoltan ERDEI, Technical University of Cluj-Napoca, North University Center of Baia Mare, Romania
Adrian PETROVAN, Technical University of Cluj-Napoca, North University Center of Baia Mare, Romania

We thank the authors for their contributions and we would also like to express our gratitude everyone who contributed to the success of this conference.

Ludovic Dan Lemle
Editor of the Proceedings of IIS2016

IOP Conf. Series: Materials Science and Engineering **200** (2017) 011002 doi:10.1088/1757-899X/200/1/011002

Peer review statement

All papers published in this volume of *IOP Conference Series: Materials Science and Engineering* have been peer reviewed through processes administered by the proceedings Editors. Reviews were conducted by expert referees to the professional and scientific standards expected of a proceedings journal published by IOP Publishing.

IOP Conf. Series: Materials Science and Engineering **200** (2017) 012001 doi:10.1088/1757-899X/200/1/012001

Comparative analysis of the electroseparation results for the Aghires quartz sand in order to enrich it, using two types of active electrodes

C Plesa, R Morar and M L Vadan

Tehnical University of Cluj Napoca, Electrotechnics and Measurements Department, Baritiu str., no. 26, Cluj Napoca, Romania

E-mail: calin.plesa64@yahoo.com

Abstract. The paper presents the laboratory researches results regarding the Aghires quarts sand electroseparation, in two configurations for the ILES laboratory installation. The first studied configuration used a performant electrode (multi needles) from the endowment of the Intense Electric Field Laboratory from UTCN, while in the second analyzed configuration, the authors use an original multithreaded electrode, for which they have obtained an OSIM patent.
The sand sample preparation for the electroseparation was made according to an original technological flow, proposed by the authors. The comparative analysis of the electroseparation results for the two configurations were made with the help of the statistical analysis program MODDE.5.
The comparative study reveals the fact that from the quantitative point of view the multi needles electrode type is preferred, while, in the case in which a superior enrichment is needed for the quartz sand, the multithreaded electrode is preferred. The enriched through electroseparation quartz sand, respectively with low content of Fe_2O_3, is an essential raw material in the porcelain industry.

1. Introduction

The Aghires quartz sand represents an important raw material in the processing porcelain industry [1]. One of the quality and aspect threatening factors for the porcelain objects, is the content of metallic oxides, especially of Fe_2O_3 .

Considering, according to the physicochemical analysis of the sand samples taken from the dump, that its main components are SiO_2-94,73%; Fe_2O_3 -0,34%; raising the SiO_2 content and decreasing the Fe_2O_3 content to minimum values was imposed.

Since the two essential elements of the quartz sand have different electrical conductivities, the authors proposed the use of the charging with electrical charge principle, through the corona phenomenon, of the granular mixture, followed by an electroseparation process [2], [3], after which a separation in three fractions is performed, namely: nonconductive fraction(NC)-SiO_2-; mixed fraction(M); conductive fraction(C) Fe_2O_3 and others.

The essential element of the electroseparation installation is considered to be the active electrode, fed in high voltage, producing a corona discharge field between the active electrode and the grounded rotating cylindrical electrode [4].

IOP Conf. Series: Materials Science and Engineering **200** (2017) 012001 doi:10.1088/1757-899X/200/1/012001

2. Constructive aspect presentation of the two types of electrodes

The multineedles electrode from Figure 1 has as active elements a number of 150 metallic needles with the length of 28mm and a section of $2mm^2$, placed on 15 rows of 10 needles each on a cylindrical surface from textolit with the thickness of 0.6mm. The radius of this surface can be adjusted with the help of M4 screws screwed in a plexiglas holder with the dimensions of 160x100mm, which can be fixed in the electrode holder with the help of a bridle from single pickled sheet.

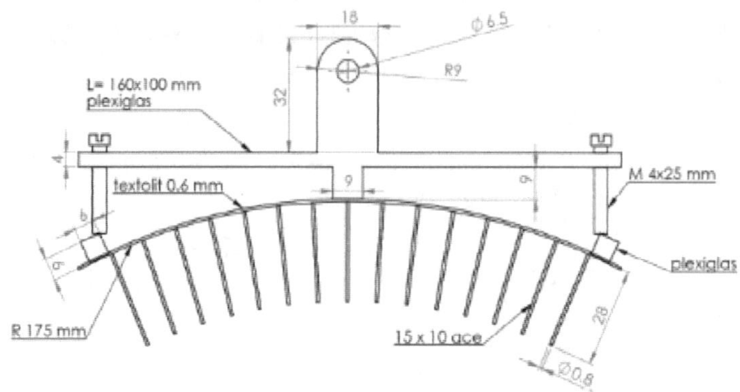

Figure 1. Multineedles electrode

The active elements of the multithreaded electrode are three conductive wolfram wires, with the thickness of 0.3mm. These are parallel between them and are attached on the threaded ends of some steel cylindrical support. The cylindrical support has an exterior diameter of 10 mm, on a length of 24mm, having at one end a threaded part with M4 in a length of 6mm (total 30mm), and at the other end a clogged threaded hole with M4, on a depth of 8 mm. At the base of the threaded part through holes were made perpendicular to the cylindrical support shaft, through which the two ends of the wolfram conductor are inserted, are prestressed, after which is fixed and stretched with the help of some threaded plugs from an isolating material, namely teflon (Figure 2).

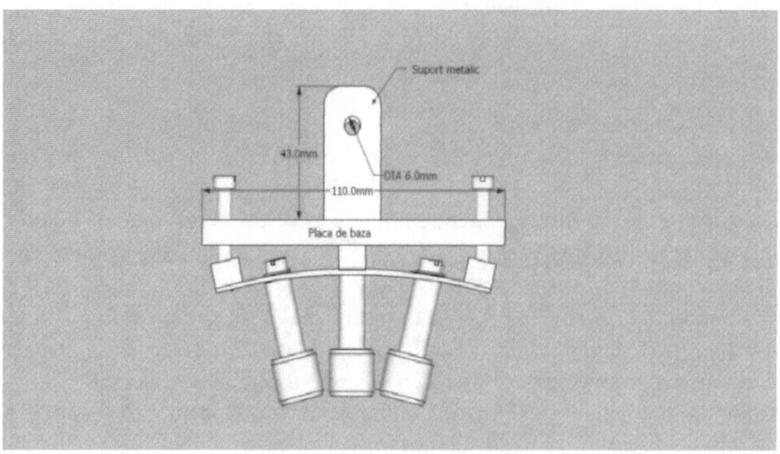

Figure 2. Frontal view of the multithreaded electrode

3. Technological process proposed by the authors in order to prepare the quartz sand samples for the electroseparation

The technological process proposed for the sand preparation for the electroseparation process is presented in Figure 3.

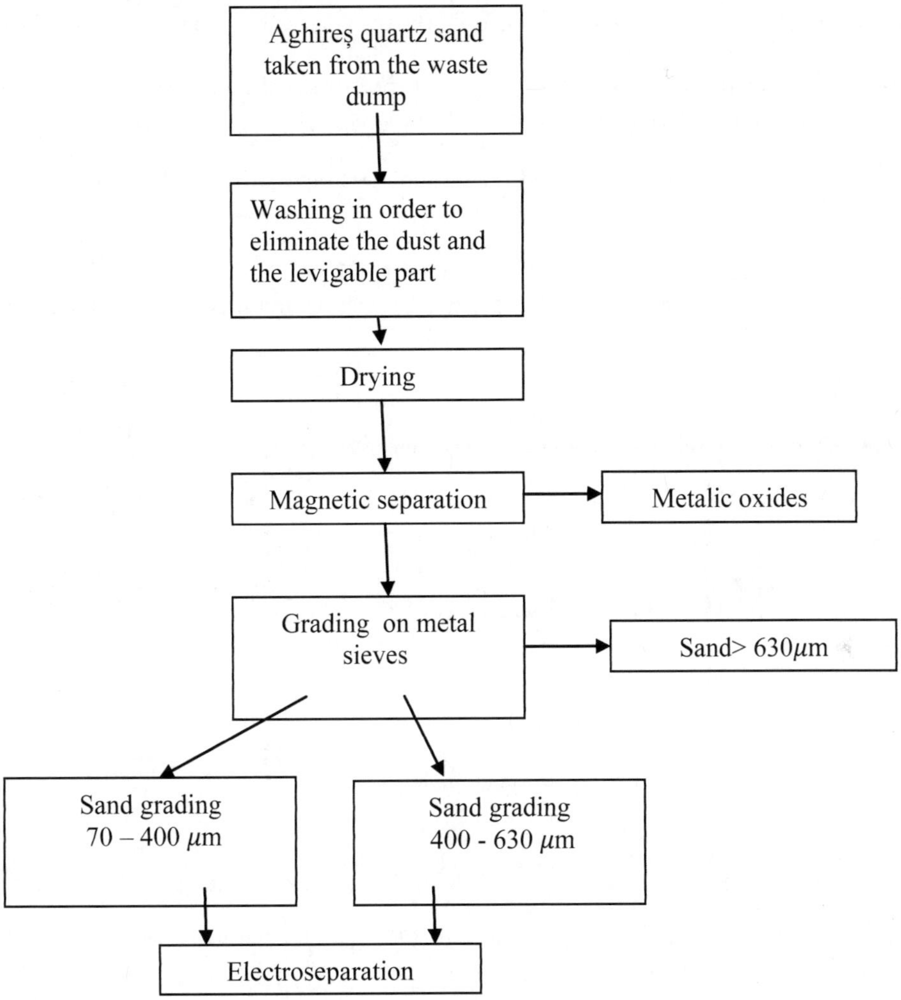

Figure 3. Technological process proposed by the authors in order to prepare the quartz sand samples for the electroseparation

4. Main factors which influence the quartz sand electroseparation

Following the preliminary laboratory researches, by varying some factors that influence the electroseparation [5] and by performing a large number of measurements, it was determined that the main influencing factors in the Aghires quartz sand electroseparation are [6]:

- the configuration and the active electrodes type;
- the supply voltage of the active electrodes U, variable between (15-40) [kV];
- speed of the rotating cylinder n, variable between (20-60) [rpm];
- placement angle to the vertical of the active electrodes α, variable between (45-70) [°].

It has to be mentioned that throughout the entire experimental plan two parameters were maintained constant: the distance S between the active electrode and the rotating cylinder surface, and also the granular mixture flow introduced by the vibrotransportor in the active area of the electroseparator.

The electroseparated sand was collected and measured (weighed) in three fractions: m 1-nonconductive material (boxes 1-3); m2- mixt material (boxes 4-7); m3- conductive material (boxes 8-10).

The used material as raw material in the production process is the one from the m 1 fraction, requiring a high concentration of SiO_2 and low concentration of Fe_2O_3 .

5. Statistical analysis of the results of the quartz sand electroseparation in the two considered configurations, using MODDE.5 software program

5.1. The principle of the statistical analysis program MODDE.5

According to the program MODDE.5 some process with m factors and n answers is considered, as in Figure 4.

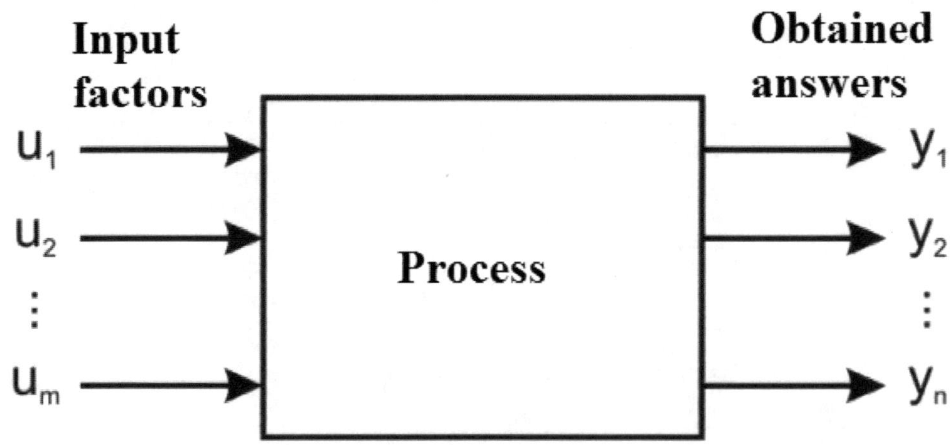

Figure 4. Symbolic representation of a process with u inputs and y outputs

The y answers may be expressed as polynomial functions with the variables ui, i=1, 2,…, m of the following form :

$$y = c_0 + \sum c_i u_i + \sum c_{ij} u_i u_j + \sum c_{ij} u_i^2 \quad , \text{i, j} = 1, 2, \ldots, m \tag{1}$$

The MODDE.5 program allows the graphical representation of the output functions as equipotential surfaces, marked with the resulting values on the contour [7].

5.2. The configuration with multineedle electrode and the statistical analysis of the electroseparation results

The configuration for the multineedle electrode can be observed in Figure 5, while in Table 1 the measurements for this configuration are presented.

Figure 5. Configuration with corona multineedle electrode for the ILES installation

Table 1. Measurements in the multineedle electrode configuration – grading 70-400 μm.

Crt. No.	U [kV]	n [rot/min]	α1 [°]	m1 [g]	m2 [g]	m3 [g]
1	20	20	50	85.24	13.86	0.85
2	20	60	50	92.92	5.82	0.52
3	40	20	50	94.07	5.22	0.08
4	40	60	50	95.54	3.82	0.04
5	20	20	70	87.02	12	0.62
6	20	60	70	91.1	7.84	0.26
7	40	20	70	96.3	3.04	0.18
8	40	60	70	96.3	3.14	0.26
9	30	20	60	92.28	7.21	0.21
10	30	60	60	93.15	6.12	0.17
11	20	40	60	93.42	6.25	0.42
12	40	40	60	96.34	2.63	0.14
13	30	40	50	96.16	3.18	0.11
14	30	40	70	96.76	2.27	0.12
15	30	40	60	96.69	2.14	0.39
16	30	40	60	96.62	2.52	0.42
17	30	40	60	96.56	2.68	0.36

In Figure 6 the allure of the response functions for the functions m1 and m2 can be observed. In the case of m1, the maximum obtained value is in the area represented in red, delimited by the border labeled ''98.8'', for a speed of approximately 40 rev/min and a supply voltage of approximately 40kV.

For m2, the same area includes the minimum obtained value, while the maximum appears around the values of 20 rev/min and 20kV.

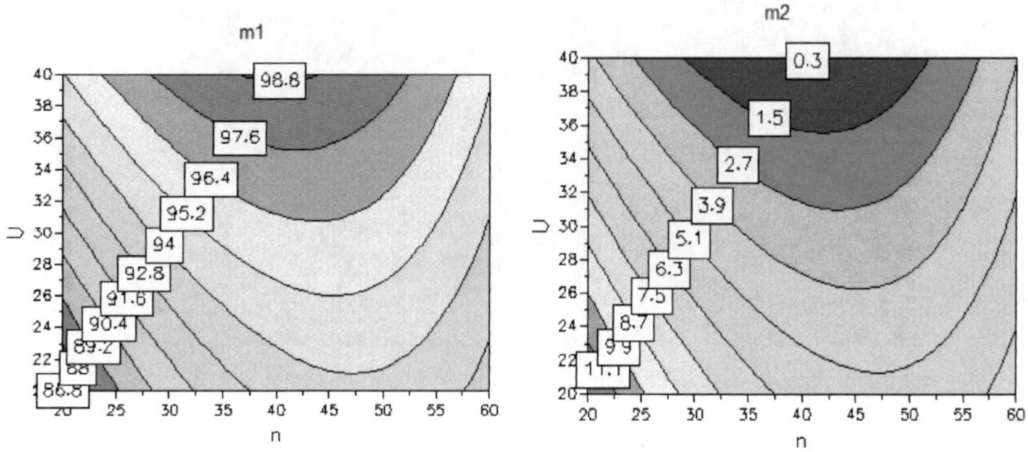

Figure 6. m_1 and m_2 masses predicted for the experimental plan in the case of the multineedles electrode configuration and grading of 70-400μm, represented as equipotential surfaces

For the representation of the m3 mass, we only have one factor, namely the supply voltage of the active electrode. The representation from Figure 7 was chosen, where the program adds the values of a confidence interval, represented by the red and blue curves.

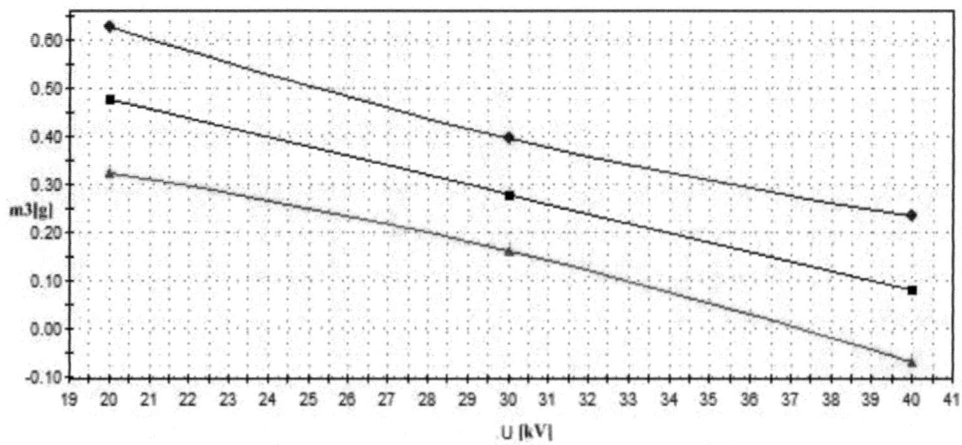

Figure 7. m_3 mass predicted depending on the U voltage at confidence intervals of 95%

After a number of iterations, the MODDE.5 program optimizer indicates the values outlined in black, as an optimum operating point (Figure 8).

IOP Conf. Series: Materials Science and Engineering **200** (2017) 012001 doi:10.1088/1757-899X/200/1/012001

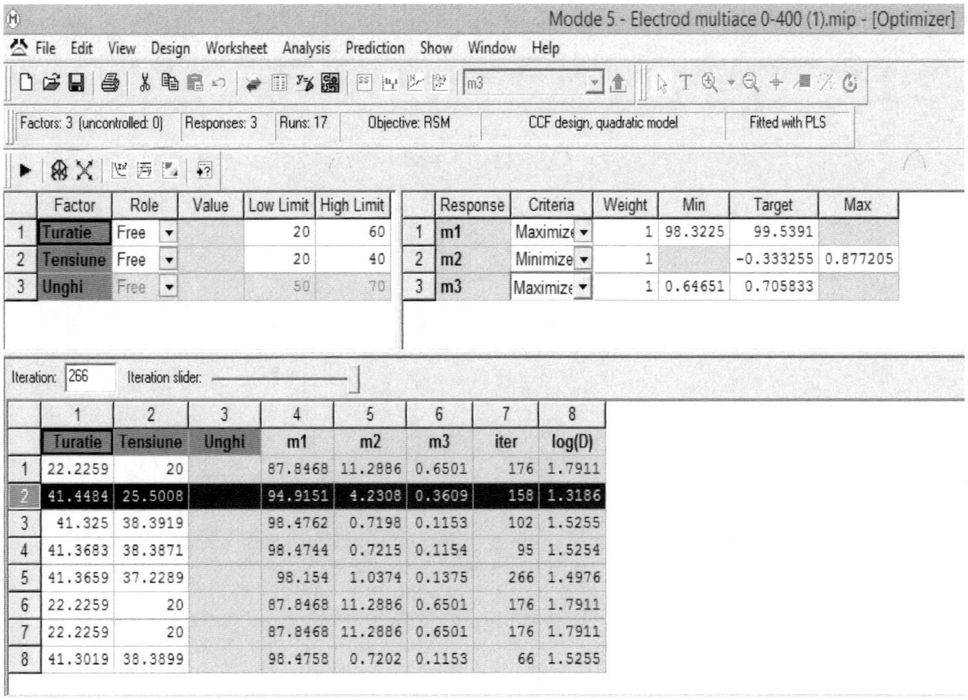

	Factor	Role	Value	Low Limit	High Limit
1	Turatie	Free		20	60
2	Tensiune	Free		20	40
3	Unghi	Free		50	70

	Response	Criteria	Weight	Min	Target	Max
1	m1	Maximize	1	98.3225	99.5391	
2	m2	Minimize	1		-0.333255	0.877205
3	m3	Maximize	1	0.64651	0.705833	

	1	2	3	4	5	6	7	8
	Turatie	Tensiune	Unghi	m1	m2	m3	iter	log(D)
1	22.2259	20		87.8468	11.2886	0.6501	176	1.7911
2	41.4484	25.5008		94.9151	4.2308	0.3609	158	1.3186
3	41.325	38.3919		98.4762	0.7198	0.1153	102	1.5255
4	41.3683	38.3871		98.4744	0.7215	0.1154	95	1.5254
5	41.3659	37.2289		98.154	1.0374	0.1375	266	1.4976
6	22.2259	20		87.8468	11.2886	0.6501	176	1.7911
7	22.2259	20		87.8468	11.2886	0.6501	176	1.7911
8	41.3019	38.3899		98.4758	0.7202	0.1153	66	1.5255

Figure 8. Optimum operating point calculated for the convergence criteria maximum m_1, m_3, respectively minimum m_2

5.3. The configuration with multithreaded electrode and the statistical analysis of the electroseparation results

The configuration for the multithreaded electrode can be observed in Figure 9, while in Table 2 the measurements for this configuration are presented.

Figure 9. Configuration with corona multithreaded electrode for the ILES installation

Table 2. Measurements in the multithreaded electrode configuration – grading 70-400 μm.

Crt. No.	U [kV]	n [rot/min]	α_1 [°]	m_1[g]	m_2 [g]	m_3 [g]
1	15	20	45	1.15	97.87	0.62
2	15	60	45	1.69	97.64	0.44
3	35	20	45	51.41	45.43	3.98
4	35	60	45	78.47	18.28	2.89
5	15	20	65	1.2	98.13	0.11
6	15	60	65	0.76	98.59	0.18
7	35	20	65	54.8	41.12	4.43
8	35	60	65	69.21	29.47	1.24
9	25	20	55	2.79	72.89	23.61
10	25	60	55	5.67	65.26	28.12
11	15	40	55	7.89	90.65	1.32
12	35	40	55	60.84	36.22	2.15
13	25	40	45	4.62	71.27	23.86
14	25	40	65	5.55	74.21	19.89
15	25	40	55	5.23	73.16	29.58
16	25	40	55	5.14	72.82	21.86
17	25	40	55	5.48	73.11	21.37

In Figure 10 the allure of the output functions can be seen. In the case of m1, the maximum value at a speed of 50÷60 rev/min and a voltage of the multithreaded electrode over 34 kV. For m2, this is the area with minumum values, the maximum values appearing for a voltage of 15÷18kV, regardless of the speed. The m3 value is maximum for a voltage value of 25÷26kV, for all speed range.

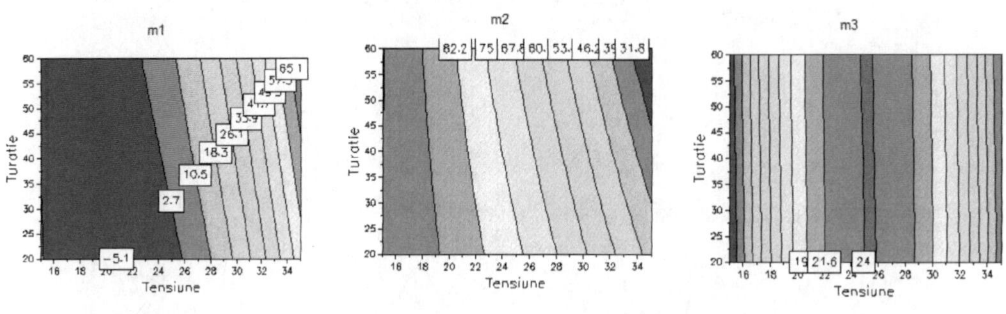

Figure 10. m_1, m_2 and m_3 masses predicted for the experimental plan in the case of the multithreaded electrode configuration and grading of 70-400μm, represented as equipotential surfaces

After a few iterations, the MODDE.5 program optimizer indicates the values outlined in black, as an optimum operating point for this case as it can be seen in Figure 11.

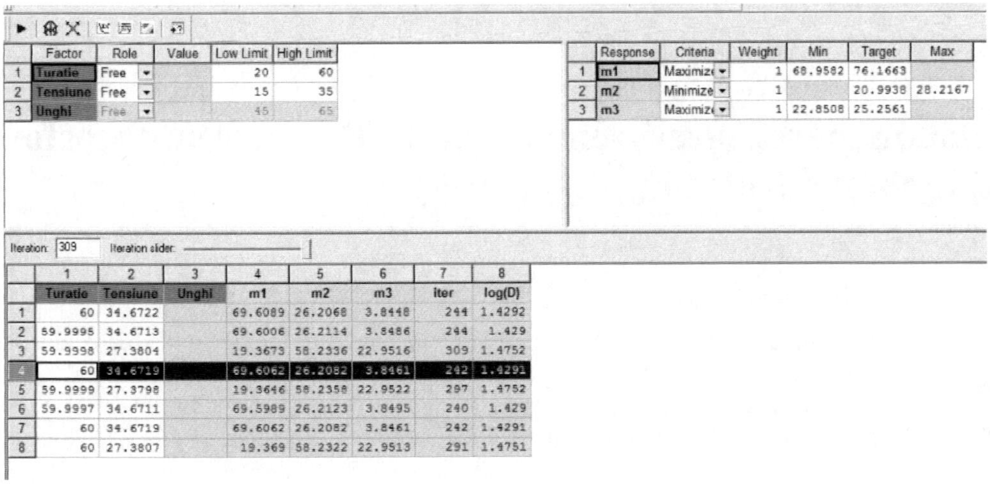

Figure 11. Optimum operating point calculated for the convergence criteria maximum m_1, m_3, respectively minimum m_2

6. Conclusions

Aghires quartz sand electroseparation ensures its enrichment, through the increase of SiO_2 content and decrease of the Fe_2O_3 content in the useful fraction m 1.

Using MODDE.5 program assured the electroseparation result analysis for a relatively small number of experiments.

From the quantitative point of view, the best results for the m1 fraction were obtained for the multineedle electrode configuration, with a Fe_2O_3 content of 0,17%. the configuration with the original multithreaded corona electrode, ensures a superior quality electroseparation, thus obtaining a smaller amount in the useful fraction m1, but with a higher purity (Fe_2O_3 -0,15%).

The original multithreaded electrode, proposed by the authors, won the OSIM patent no.201500033/19/01/2015. Also, the multithreaded corona electrode, was appreciated at the International Exhibition of Inventions, PRO INVENT, edition XIII 2015, Cluj Napoca, Romania, with excellence diploma and gold medal.

References

[1] Brana V, Avramescu C and Calugaru I 1986 *Substante minerale nemetalifere*, EdituraTehnica, Bucuresti, Romania

[2] Knoll F S and Taylor J B 1985 Advances in Electrostatic Separation, *Minerals and Metallurgical Processing* **2** 106-114

[3] Samuilă A and Dăscălescu L 1999 *Comportarea materialelor granulare în câmp electric,* Editura Mediamira, Cluj-Napoca, Romania

[4] Dascalescu L, Iuga Al and Morar R 1999 *Electroseparerea materialelor granulare*, in: Electrostatica, Vol.1 (FTanasescu F T, Bologa M, Cramaciuc R Eds.), Editura Academiei Romane, Bucuresti, Romania, pp 86-116

[5] Morar R, Iuga A, Dascalescu L and Samuila A 1993 Factors which influence the insulation/metal electroseparation, *Journal of Electrostatics* **30** 403-412

[6] Neamţu V 1998 *Electroseparerea materialelor granulare*, Editura Casa Cărţii de Stiinţă, Cluj-Napoca, Romania

[7] ***Umetrics AB 1999 *MODDE 5.0 User guide and tutorial*, Umetrics, Umea, Sweden

IOP Conf. Series: Materials Science and Engineering **200** (2017) 012002 doi:10.1088/1757-899X/200/1/012002

Partial discharge measurements on 110kV current transformers. Setting the control value. Case study

C Dan[1] and R Morar[2]

[1]Transilvania Sud Power Distribution Company, Subsidiary Sibiu, 1-7, Uzinei street, 550253, Sibiu, Romania
[2]Technical University Cluj Napoca, Romania, Department of Electrical Engineering, 26-28, George Bariţiu street, 400027, Cluj-Napoca, Romania

E-mail: cristian.dan@eletricats.ro

Abstract. The case study presents a series of partial discharge measurements, reflecting the state of insulation of 110kV CURRENT TRANSFORMERS located in Sibiu county substations. Measurements were performed based on electrical method, using MPD600: an acquisition and analysis toolkit for detecting, recording, and analyzing partial discharges. MPD600 consists of one acquisition unit, an optical interface and a computer with dedicated software. The system allows measurements of partial discharge on site, even in presence of strong electromagnetic interferences because it provides synchronous acquisition from all measurement points. Therefore, measurements, with the ability to be calibrated, do render:

- a value subject to interpretation according to IEC 61869-1:2007 + IEC 61869-2:2012 + IEC 61869-3:2011 + IEC 61869-5:2011 and IEC 60270: 2000;
- the possibility to determine the quantitative limit of PD (a certain control value) to which the equipment can be operated safely and repaired with minimal costs (relative to the high costs implied by eliminating the consequences of a failure) identified empirically (process in which the instrument transformer subjected to the tests was completely destroyed).

1. Introduction

Electrical utility industry all over the world faces a tough challenge with ageing of the power grid components population as failure of these assets may cause interruption of power supply and also revenue losses. The insulation condition represents an essential cast for reliability in operating high voltage equipment. Considering this, it is very important to identify the healthy units in the ageing equipment' population as not extending service to these units will result in substantial cost savings for the power company [1].

Modern technologies and developments in signal acquisition and analysis techniques do provide new tools for diagnostic of electrical power equipment.

2. Instrument transformers

Per se, instrument transformers are non – rotating electromagnetic machines, powering electrical equipment falling into a broad category of metering and protection devices. Judging by their manufacturing technology, they have some common features with other non-rotating electromagnetic machines (power transformers), still their functional properties resembling more to those of electrical apparatus.

The order of magnitude of voltage and current in a power system is very diverse. Adapting both voltage or current metering systems and protective devices to this wide range of values is neither technically nor economically justified. Therefore, these devices are connected to the electrical circuitry indirectly, via current and voltage transformers – instrument transformers whose construction and operation is tailored specifically for this purpose. The instrument transformers possessing a primary winding and a certain number of secondary windings, performing an isolating function (isolating the utilization current or voltage from the supply current/voltage to ensure safety for both, operator and in use end device), change currents and voltages from one magnitude to another, suitable for power metering and protection devices.

Instrument transformers, in general acceptation [2]:

- transform current and voltage to standardized magnitude appropriate for supplying current and voltage windings of the metering and protection equipment;
- isolate high voltage circuit (the grid) from the low voltage circuit (metering and / or protection), thus removing meters and protection devices from the area influenced by the strong electric and magnetic fields of power grid circuitry, eliminating disruptive action that these fields may have on measurement accuracy and correct operation of the equipment;
- allow rational concentration of metering and protection devices in the control room right outside the substation, enabling thereby direct both command and monitoring of the entire grid.

For each instrument transformer, insulation issues are fundamental and do determine a priori the possible limits of their functional properties [3].

In a CIGRE survey also the used maintenance techniques prior to instrument transformers' failure are listed [4] in table 1.

Table 1. Maintenance strategy used prior to failure from 3004 reported failures [4]

Maintenance strategy	How often used
Regular visual inspection	95%
Check of oil level and/or pressure gauge	61%
Secondary voltage monitoring for CVTs	15%
Insulation resistance checks	11%
DGA and/or moisture of oil	7%
Thermovision inspection	4%
DF measurement at mains frequency	2%
PD measurement	***1%***

3. Partial discharge measurement

Partial Discharge (PD) measurement is a worldwide accepted tool for quality control of high voltage equipment [5]. PD indicates partial loss of the insulating capacity and is, thereby, considered a measure of electrical ageing of any insulation system [6].

The classical measuring circuit, according to IEC60270: High voltage test techniques – Partial discharge measurements, requires a coupling capacitor for the connection of the PD measurement instrument and works at a measuring frequency from several hundred kHz up to one MHz [7], detecting therefore, the line based pulse propagation. Beside these methods, there are also non-electrical PD measurements (e.g. acoustic), used above all for localizing partial discharges. Nevertheless, electrical PD measurement is being by far the most used technique standardized. As a general rule, successful diagnosis requires distinguishing internal partial discharges from any external interference, particularly for onsite applications. Excellent measurement and interference suppression algorithms have therefore been developed for conventional electrical measurement [7].

3.1. *Interference – free measurement setup*

Often lacking screening, PD signals are cvasi every time superposed by various noise pulses, a fact that adds difficulty to PD data analysis for both human and software expert systems. Thus, handling disturbances is one of the main challenges when the goal is to accurately measure PD. The effects of external interference in the measuring circuit are minimized through electrical isolation between the measuring system and the control unit.

3.2. *Frequency – selective, narrow band and wide – band measurement*

The measuring systems used today extend the measuring range historically limited to ~ 1MHz, as far as 20MHz. Within this wide frequency range, the user can select a band that has only marginal superimposed interference or none at all.

3.3. *Synchronous multi – location PD measurement*

PD measurements are often performed under noisy conditions. By using synchronous multi-channel PD acquisition it will be possible to gain de-noised PD data from separated PD sources. The PD signal is superposed by stochastic noise pulses or even multiple PD sources, leading to a complex phase-resolved PD pattern that is not easy to analyze. Conventional frequency filters are not able to eliminate these pulse-shaped disturbances [6]. The superposition of multiple PD sources and noise will, nevertheless, raise laboriousness for PD experts and automated computer expert systems. Some well-known evaluation techniques as pulse-sequence analyses would even fail with non-correlated PD pulses to be compared [7]. Separating noise from PD and then separating multiple PD sources must be the starting point onto obtaining free-noise PD data.

3.4. *3PARD evaluation procedures*

A new field of evaluation methods is opened by fully synchronous multi-channel PD acquisition in order to gain more reliable measuring results combined with effective noise suppression. Therefore the 3-Phase-Amplitude-Relation-Diagram (3PARD) was introduced as a new powerful analysis tool [3] to distinguish between PD sources and noise pulses when measuring 3-phase high voltage equipment [7]. Through suitable superimposition of the amplitudes, the traditional phase – resolved PD (PRPD) pattern can be broken down into individual components, allowing individual sources to be singled out.

3.5. *Analysis of the frequency spectrums*

With these processes, it is not the amplitudes measured at various locations that are superimposed, but rather the spectral components measured selectively but still synchronously at one single location. The frequently noisy PD patterns can also be broken down into individual sources using these methods.

IOP Conf. Series: Materials Science and Engineering **200** (2017) 012002 doi:10.1088/1757-899X/200/1/012002

4. PD Measuring System

The MPD600 as a fully digital PD measuring system is capable of performing synchronous multi-channel PD measurements [8]. It consists of one or more acquisition units (Figure 1), an optical interface (FO bus-controller) and a PC including the measuring software [9]. The PD signals are filtered, amplified and digitized. A frequency band can be chosen to avoid continuous-wave disturbances and to reach a high signal-to-noise-ratio (SNR) even under noisy conditions on site.

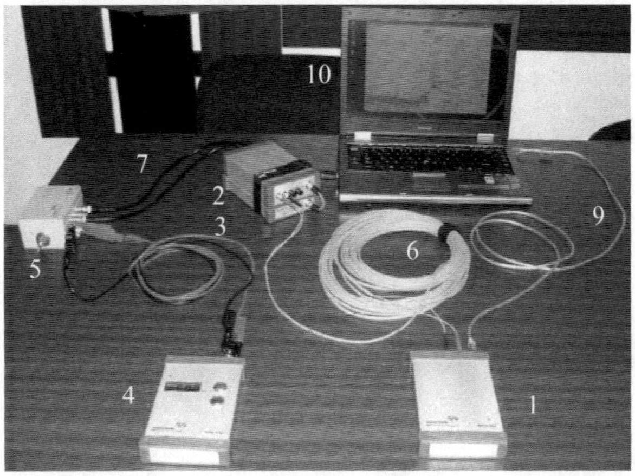

Figure 1. Components of a MPD system:
optic-USB converter (1), PD acquisition unit DP (2), battery pack (lithium-ion) (3), calibration unit CAL542 (4), measuring impedance (5), fiber optical cabling (6), Ethernet/BNC cables (7), battery cable (8), USB cable (9), notebook (10)

For multi-channel PD measurements several acquisition units can be connected to one distributed PD system [7] while a maximum number of 960 units can be operated in a fully synchronized mode.

5. Case Study

The case study of the current transformer, type **CESU[1] 110kV (925011/1979)**, operating in a 110/20kV transformer power substation, in a 110kV line bay, identified ever since 11 October 2000 (back then by gas chromatography) with potential failure: *high-intensity partial discharges, active failure*!

Temporal evolution of the technical condition both in terms of chromatographic analysis (for in the insulating oil dissolved gases) and PD measurement results is presented in full in attached Annex 1.

[1] Encoding current transformers denomination:
C – current transformer;
E – exterior / outdoor mounting;
S – support;
U – insulating environment, **u**lei (oil);
m – embodiment with metal compensator;
o – construction type (monospiral, with bar crossing and median round flange fixing);
110(123) – rated insulation voltage [kV];
after these groups of letters and numbers, the rated ratio is printed as: rated primary current / rated secondary current.

Electrical measurements of partial discharge, for this particular current transformer, on site (power substation) were made (starting in July 2007) according to the provisions of standard EN 61869-1:2007 „*Instrument transformers. General requirements*", applying, assimilated to it, *A*[2] partial discharge test procedure, using as test voltage the very rated 110kV distribution grid voltage [2].

Considering paramount maintaining the same operation conditions for the instrument transformers and precisely to the purpose of an increased reproductibility of measurement results, following the guidelines of standard EN 61869-1:2007, we knowingly used the rated voltage of the 110kV power distribution grid to perform PD measurements.

One of the main benefits of the PD measuring system used is that the PD acquisition units can be placed very close to the transformer. This minimizes the required length of electrical connecting cables and allows PD signal acquisition with high bandwidth, while minimizing electromagnetic interference.

Using the built-in frequency sweep function, a measuring frequency of 1MHz (1.5MHz bandwidth) was selected. Results of the PD measurements on CT (110kV CESU (925011/1979)), are shown in the following Figures 2 and 3.

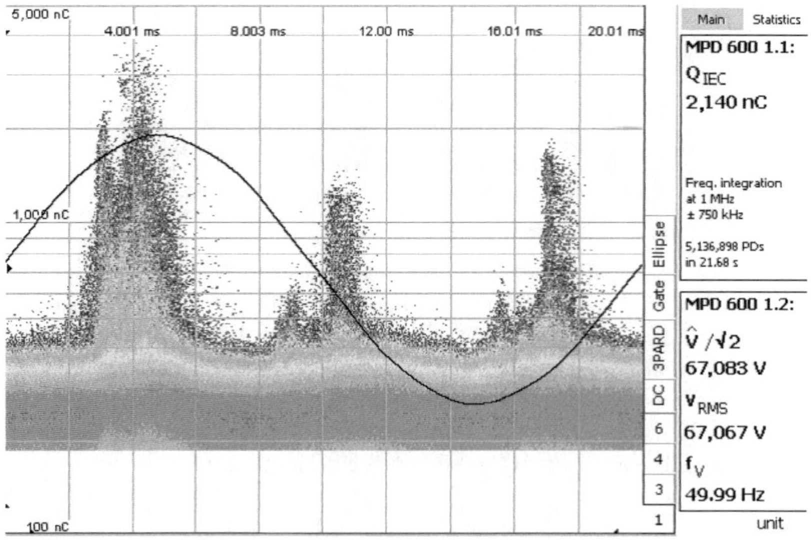

Figure 2. Electrical PD measurement according to IEC 60270, center frequency: 1MHz, bandwidth: 1.5MHz, all PD data

As usual for conventional on-line PD measurements, the diagram shows a notable 3-phase corona cross-talk from the substation and the near overhead transmission lines.

The external noise was successfully gated with the information from one second PD acquisition unit. Figure 3 shows the corrected PD pattern recorded.

[2] Partial discharge test voltage is reached while decreasing rated power-frequency withstand voltage.

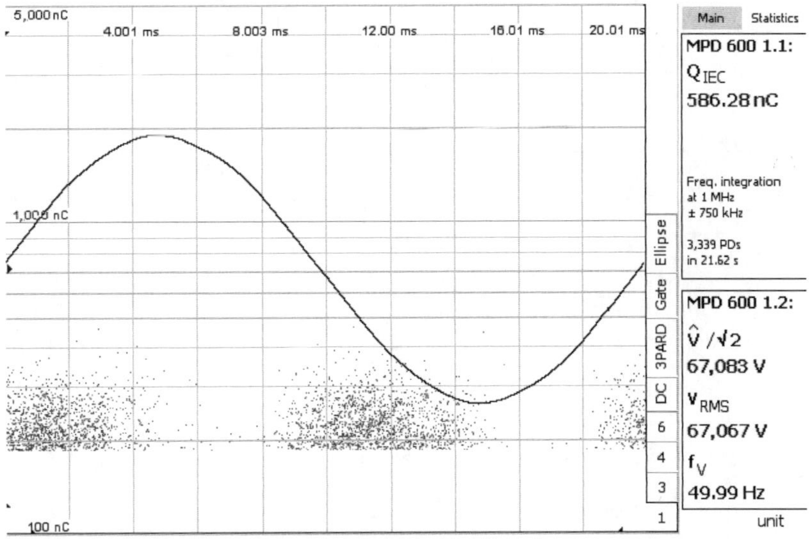

Figure 3. Gated electrical PD measurement, center frequency: 1MHz, bandwidth: 1.5MHz

This significant increase in PD level was measured in August 27, 2011, about two weeks before the complete destruction of insulation: 09.07.2011 (Figure 4).
This PD level value, exceeding from a far the permissible PD level [10], according to IEC 61869-2:2012[3] (10pC), *was identified empirically* (process in which the instrument transformer subjected to the tests was completely destroyed).

Figure 4. Current transformer, CESU (925011/1979), at 07.09.2011

Judging by the temporal evolution of the PD level, measured for 5 years (2007-2011) and considering the maximum slope segment (specific time intervals of 1 year) represented graphically in figure 5, it is necessary to use a safety factor of 17.95% of the maximum measured value $(586{,}285pC)$.

[3] Which applies only to newly manufactured current transformers.

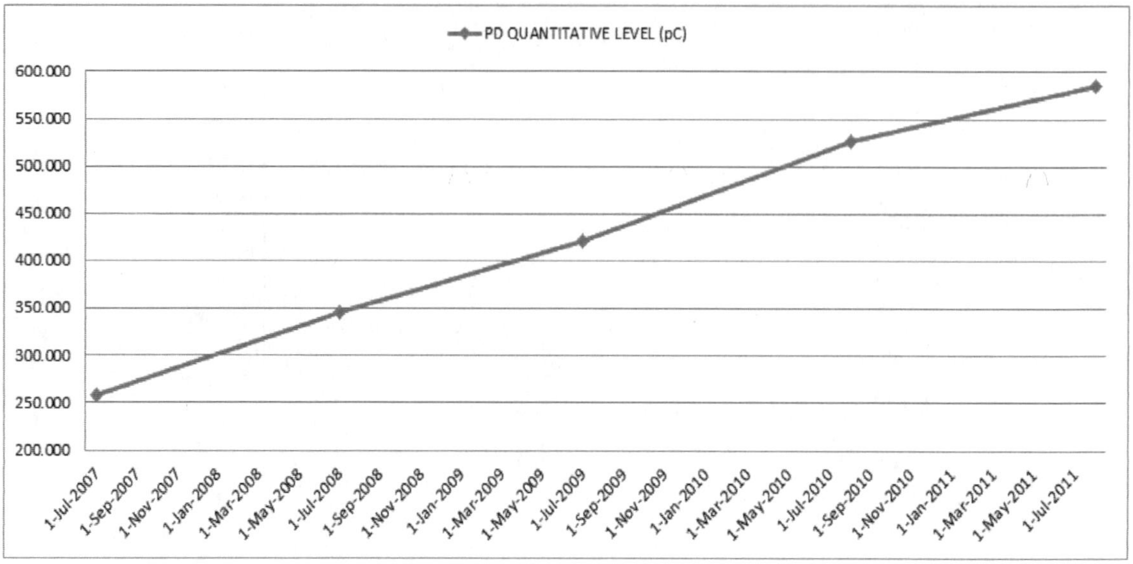

Figure 5. PD level quantitative development for current transformer, CESU (925011/1979)

6. Conclusion

The quantitative limit of PD to which the equipment can be operated safely and still be repaired with minimal costs (relative to the high costs implied by eliminating the consequences of a failure) is $480nC$.

References

[1] Talib M A, Miswan N and Supramaniam G K 2010 *Condition Assessment of HV Equipments in Tenaga Nasional Berhad, Malaysia*, Instrument Transformer Measurement Forum, Salzburg, Austria

[2] ***IEC 61869-1:2007 (cancels and replaces first edition of IEC 60044-1 (1996), its Amendment 1 (2000) and Amendment 2 (2002), *Instrument transformers – General requirements*

[3] Plath K D, Plath R, Emanuel H and Kalkner W 2002 *Three-phase synchronous partial discharge measurements on transformers on site and in the laboratory*, ETG Conference on Diagnostic, Berlin, Germany

[4] ***CIGRÉ Study Committee A3, Technical Brochure 394/2009 *State of the Art of Instrument Transformers*, Paris

[5] Smit J J 2005 *Trends in PD-diagnostics for Asset Management of Aging HV Infrastructures*, 14th International Symposium on High Voltage Engineering, Beijing, China, August 25-29

[6] Montanari J C 2006 *Insulation diagnosis of high voltage apparatus by partial discharge investigation*, Liu-Yeda Memorial Lecture, in Proceedings IEEE ICPADM, Bali, Indonesia, June 26-30

[7] Kraetge A, Rethmeier K, Krüger M and Winter P 2009 *Synchronous Multi-Channel PD Measurements and the Benefits for PD Analyses*, OMICRON Training, Klaus, Austria

[8] ***2014 *MPD 600: High-end measurement and analysis system for partial discharges*, www.omicron.at | www.omicronusa.com, OMICRON L2317

[9] Holle R, Plath R, Schon K and Lucas W 2006 *Type testing a digital PD measuring system according to IEC 60270*, ETG Conference on Diagnostic, Kassel, Germany

[10] ***IEC 60270:2000 (SR EN 60270:2003) *Tehnici de încercare la înaltă tensiune – Măsurarea descărcărilor parțiale*, versiune română a Standardului European IEC60270:2000 (standard tradus de ASRO, publicat cu permisiunea CENELEC)

ANNEX 1
RESULTS for the **110kV CURRENT TRANSFORMER in substation A. Vlaicu, bay Agnita, phase S**
of the ANALYSIS of GASES DISSOLVED in ELECTRICAL INSULATING OIL by GAS CHROMATOGRAPHY / **2000 – 2011**
and its AMOUNT OF PARTIAL DISCHARGES (pC) / **2007 – 2011**

NR.	EQUIPMENT	TYPE	PRODUCER	SERIAL NUMBER	LOCATION	FLUID TYPE	MANUFACTURING YEAR	MODEL	NOMINAL kV	RESPIRATION	SAMPLE DATE	DGRESULT / KEY GAS LIMITS H_2	CH_4	C_2H_6	C_2H_4	C_2H_2	CO	CO_2	TOTALGAS	EQUIPMENT CONDITION	GAS CHROMATOGRAPHY RESOLUTION	NEXT DGA / TIME LAPSE	NEXT DGA DATE	EVALUATION STANDARD	PD AMOUNT (pC)
												150	120	130	10	5	1.100	4.000	5.515						
1	CT-A. Vlaicu-Agnita-S	iTRN	EPC	925011	110/20kV Aurel Vlaicu substation	OIL	1979	CESU	110	FREE	11-Oct-2000	1.452	94	79	2	2	99	2.124	3.852	4	High-intensity partial discharge. Active fault	PD / 1 MONTH	10-Nov-2000	IEEE DG	
2	CT-A. Vlaicu-Agnita-S	iTRN	EPC	925011	110/20kV Aurel Vlaicu substation	OIL	1979	CESU	110	FREE	10-Nov-2000	1.493	108	81	5	4	93	1.931	3.715	4	High-intensity partial discharge. Active fault	PD / 1 MONTH	10-Dec-2000	IEEE DG	
3	CT-A. Vlaicu-Agnita-S	iTRN	EPC	925011	110/20kV Aurel Vlaicu substation	OIL	1979	CESU	110	FREE	10-Dec-2000	1.565	129	88	11	9	85	1.792	3.679	4	High-intensity partial discharge. Active fault	PD / 1 MONTH	9-Jan-2001	IEEE DG	
4	CT-A. Vlaicu-Agnita-S	iTRN	EPC	925011	110/20kV Aurel Vlaicu substation	OIL	1979	CESU	110	FREE	9-Jan-2001	1.623	149	93	13	20	78	1.558	3.535	4	High-intensity partial discharge. Active fault	PD / 3 MONTHS	9-Apr-2001	IEEE DG	
5	CT-A. Vlaicu-Agnita-S	iTRN	EPC	925011	110/20kV Aurel Vlaicu substation	OIL	1979	CESU	110	FREE	9-Apr-2001	1.732	192	99	18	29	60	589	2.719	4	High-intensity partial discharge. Active fault	PD / 6 MONTHS	6-Oct-2001	IEEE DG	
6	CT-A. Vlaicu-Agnita-S	iTRN	EPC	925011	110/20kV Aurel Vlaicu substation	OIL	1979	CESU	110	FREE	13-Nov-2001	2.004	271	108	25	36	104	786	3.334	4	High-intensity partial discharge. Active fault	PD / 6 MONTHS	15-May-2002	IEEE DG	
7	CT-A. Vlaicu-Agnita-S	iTRN	EPC	925011	110/20kV Aurel Vlaicu substation	OIL	1979	CESU	110	FREE	15-May-2002	2.262	393	118	29	50	147	1.113	4.111	4	High-intensity partial discharge. Active fault	PD / 6 MONTHS	14-Nov-2002	IEEE DG	
8	CT-A. Vlaicu-Agnita-S	iTRN	EPC	925011	110/20kV Aurel Vlaicu substation	OIL	1979	CESU	110	FREE	14-Nov-2002	2.473	476	139	35	69	237	1.570	4.999	4	High-intensity partial discharge. Active fault	PD / 6 MONTHS	16-May-2003	IEEE DG	
9	CT-A. Vlaicu-Agnita-S	iTRN	EPC	925011	110/20kV Aurel Vlaicu substation	OIL	1979	CESU	110	FREE	1-Jun-2003	2.749	624	151	39	75	295	2.224	6.158	4	High-intensity partial discharge. Active fault	PD / 12 MONTHS	2-Jun-2004	IEEE DG	
10	CT-A. Vlaicu-Agnita-S	iTRN	EPC	925011	110/20kV Aurel Vlaicu substation	OIL	1979	CESU	110	FREE	2-Jun-2004	3.460	687	173	51	93	422	2.917	7.802	4	High-intensity partial discharge. Active fault	PD / 12 MONTHS	2-Jun-2005	IEEE DG	
11	CT-A. Vlaicu-Agnita-S	iTRN	EPC	925011	110/20kV Aurel Vlaicu substation	OIL	1979	CESU	110	FREE	15-Jul-2005	4.183	789	189	60	123	575	4.043	9.962	4	High-intensity partial discharge. Active fault	PD / 12 MONTHS	15-Jul-2006	IEEE DG	
12	CT-A. Vlaicu-Agnita-S	iTRN	EPC	925011	110/20kV Aurel Vlaicu substation	OIL	1979	CESU	110	FREE	15-Jul-2006	6.416	1.152	299	114	351	569	3.745	12.646	4	High-intensity partial discharge. Active fault	PD / 12 MONTHS	17-Jul-2007	IEEE DG	
13	CT-A. Vlaicu-Agnita-S	iTRN	EPC	925011	110/20kV Aurel Vlaicu substation	OIL	1979	CESU	110	FREE	17-Jul-2007	9.258	1.406	350	128	669	554	3.542	15.907	4	High-intensity partial discharge. Active fault	PD / 12 MONTHS	20-Jul-2008	IEEE DG	258.260
14	CT-A. Vlaicu-Agnita-S	iTRN	EPC	925011	110/20kV Aurel Vlaicu substation	OIL	1979	CESU	110	FREE	27-Jul-2008	11.811	1.663	444	165	1.001	541	3.301	18.926	4	High-intensity partial discharge. Active fault	PD / 12 MONTHS	31-Jul-2009	IEEE DG	345.763
15	CT-A. Vlaicu-Agnita-S	iTRN	EPC	925011	110/20kV Aurel Vlaicu substation	OIL	1979	CESU	110	FREE	31-Jul-2009	14.335	1.879	507	189	1.514	490	3.293	22.207	4	High-intensity partial discharge. Active fault	PD / 12 MONTHS	3-Aug-2010	IEEE DG	420.562
16	CT-A. Vlaicu-Agnita-S	iTRN	EPC	925011	110/20kV Aurel Vlaicu substation	OIL	1979	CESU	110	FREE	13-Aug-2010	17.911	2.138	603	236	2.064	434	3.225	26.610	4	High-intensity partial discharge. Active fault	PD / 12 MONTHS	21-Aug-2011	IEEE DG	526.099
17	CT-A. Vlaicu-Agnita-S	iTRN	EPC	925011	110/20kV Aurel Vlaicu substation	OIL	1979	CESU	110	FREE	27-Aug-2011	19.520	2.179	687	293	2.418	384	3.173	28.654	4	High-intensity partial discharge. Active fault	PD / 12 MONTHS	28-Aug-2012	IEEE DG	586.285

Apparent charge – hydrogen H_2 (in insulating oil dissolved gas) dependancy for 110kV CT, Aurel Vlaicu substation, bay AGNITA, phase S, 2007 - 2011

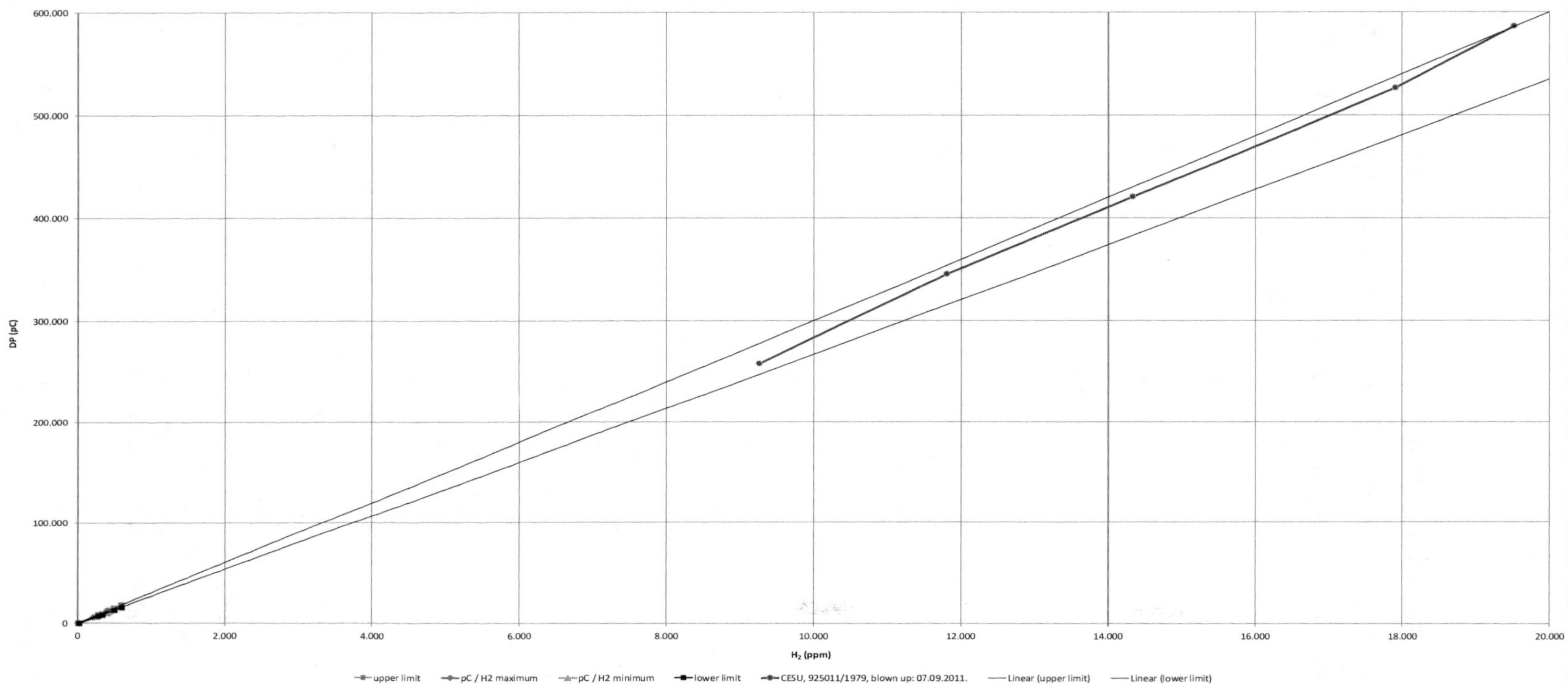

Determining insulation condition of 110kV instrument transformers. Linking PD measurement results from both gas chromatography and electrical method

C Dan[1] and R Morar[2]

[1]Transilvania Sud Power Distribution Company, Subsidiary Sibiu, 1-7, Uzinei street, 550253, Sibiu, Romania
[2]Technical University Cluj Napoca, Romania, Department of Electrical Engineering, 26-28, George Barițiu street, 400027, Cluj-Napoca, Romania

E-mail: cristian.dan@eletricats.ro

Abstract. Working methods for on site testing of insulations:
1. Gas chromatography (using the TFGA-P200 chromatographer);
2. Electrical measurements of partial discharge levels using the digital detection, recording, analysis and partial discharge acquisition system, MPD600.
First performed, between 2000–2015, were the chromatographic analyses concerning electrical insulating environments of:
A. 102 current transformers, 110kV. Items in operation, functioning in 110/20kV substations.
B. 38 voltage transformers, 110kV also in operation, functioning in 110/20kV substations.
Then, electrical measurements of partial discharge inside instrument transformers, on site (power substations) were made (starting in the year 2009, over a 7-year period, collecting data until the year 2015) according to the provisions of standard EN 61869-1:2007 „Instrument transformers. General requirements", applying, assimilated to it, type A partial discharge test procedure, using as test voltage the very rated 110kV distribution grid voltage.
Given the results of two parallel measurements, containing:
I. to this type of failure specific gas amount (H_2) and
II. the quantitative partial discharge' level,
establishing a clear dependence between the quantity of partial discharges and the type and amount of in oil dissolved gases inside equipments affected by this type of defect: partial discharges, was expected.
Of the „population" of instrument transformers subject of the two parallel measurements, ***the dependency*** between Q_{IEC} (apparent charge) and (H_2) (hydrogen, gas amount dissolved within their insulating environment) ***represents a finite assemblage*** situated between the two limits developed on an empirical basis.

1. Partial discharges. General considerations. Damaging effects on insulation systems

Insulating materials used in manufacturing processes of high voltage electrical equipment, do come sometimes with small local defects consisting of gaseous inclusions or voids (for solid insulation), gas bubbles/cavities (in liquid dielectrics) and metallic / non-metallic particles (metal grit / dust or lint) in quite all insulation systems either solid, liquid or gaseous.

Inside electrical equipment during operation[1], at certain amplitude of the electric field, local and nondestructive discharges occur, limited to a small portion of the dielectric, especially within local faults as described above.

PARTIAL DISCHARGES – PD result in an impulse current occurring within the dielectric providing to its external circuit, impulses, which although engaging low energy levels, are leading to progressive degradation of insulating material properties.

Without going into specific technical details, we state that PDs do modify the insulating materials and depending on the repetition rate and intensity of the process, sooner or later, destruction of the dielectric will be imminent.

Partial discharges stand out by the following features:

- transferred charge of low value, expressed in picocoulombs (pC);
- very short pulse duration $\left(10^{-9}...10^{-6}s\right)$;
- discharge does not occur throughout the cavity volume, but PARTIALLY;
- for the homogeneous, but also for the non-homogeneous electrical field, PD geometry is not identical to that of electric field that produces them.

Each discharge dissipates a small amount of energy which shall be located inside the cavity. Under these conditions, PD (by its generated pulses) are source of noise within power circuits, of dielectric losses and are always accompanied by emission of sound, heat and decomposition of affected insulating material into gaseous compounds [1]. These events point out multiple methods of detection, measurement and analysis of the PD phenomena.

The effects described above are capable of causing insulation degradation next to the cavities and therefore, if not detected and stopped on time, mechanical erosion of the insulation commencing from void walls will evolve in the mass of the healthy material, fundamentally shortening the life of that insulation system exhibiting such defects.

2. Two methods for detection and measurement of partial discharges

For the ongoing research, taking into account the studied objects – *high voltage instrument transformers, 110kV* – we chose a combination of *two methods for detection, measurement* of partial discharges, namely:

1. non-electrical, chemical: gas chromatography;
2. electrical, using pulses amplification measured from a network element (test object).

2.1. Non-electrical method. Gas chromatography

Due to internal partial discharges (local penetrations of insulation, possessing a high density of ionization or appearance of low energy arcing), decomposition of the insulating materials and formation of various combustible and noncombustible gases will occur; these gas amounts can be detected and measured through gas chromatography. By brief overview of characteristics and performance of gas chromatography, the applicability and effectiveness of this exceptional analytical technique stand out plainly. Very reproductive results recommend the use of gas chromatography. Gas analysis is not only a science but an art subject to variability arising from the vastness of electrical equipment operation and maintenance experience. It is now one of the most used methods for determining and then assessing electrical equipment internal faults [2].

Two are the main components of the complex insulation systems of electrical equipment in service that are affected by PD, namely:

① oil – the insulating liquid environment;
② paper – solid insulation.

[1] This is the framework within the analysis of partial discharge effects on the state of insulation of high voltage instrument transformers has been developed.

Gas chromatography[2] is an analytical technique used for separation of sample components between two **phases: a mobile phase (the carrier gas, in this case, helium** (He) **and a stationary phase (the column packing or coating)**.

Separation begins when a gaseous sample (which diffuses through molecular agitation in a free volume well proportioned compared to the volume of the oil in which they were dissolved) fills the sample loop of the **injector**. The sample is then injected from the loop onto the column. Actual separation occurs inside the column. As each component emerges from the column, separately, (seven (7) gases, key gases: hydrogen(H_2), methane(CH_4), carbon monoxide(CO), ethylene(C_2H_4), ethane(C_2H_6), acetylene(C_2H_2), which are combustible gases and carbon dioxide(CO_2), non – combustible gas) they pass through the **detector**, one at a time. The detector response versus time plot will be a *peak*. This chain of peaks is **the end result** – **the chromatogram** – with accurate data of the gas chromatography [3].

Computer systems can serial communicate with gas analyzers (Figure 1). This feature of the device allows the computer to control the entire chromatographic process.

It requires a small amount of oil, namely $100ml$ (without air inclusions or other impurities), which will be extracted directly from the electrical equipment in a special container – syringe – (Figure 2) and then immediately airtight sealed.

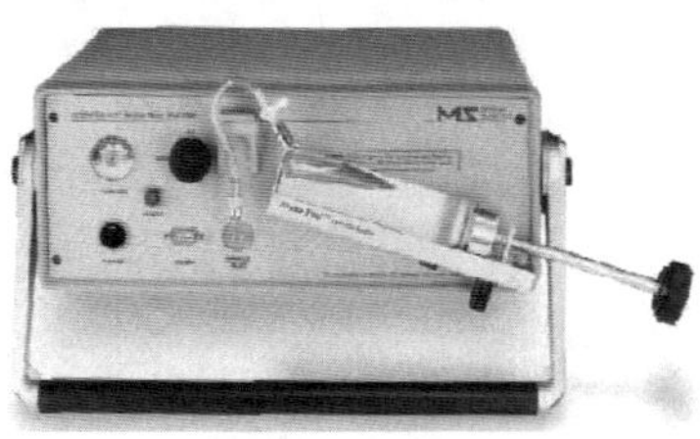

Figure 1. Portable gas chromatograph TFGA – P200

Figure 2. Special container, syringe. for insulating oil sampling

The dissolved in oil gas analysis itself follows (120 sec.). Gases that due to the strong agitation will elute from oil into a free space ($10ml$ of oil will be evacuated to allow this method of gases diffusion – molecular agitation – be applied), from where they will be sampled in by the gas analyzer.

2.2. Electrical method. Pulses amplification measured from a network element

It should be specified that the pulse shape that reaches the measuring instrument depends on the pulse involved at the site of the discharge inside the dielectric, on the geometry of the insulation system and on the characteristics of the external measuring circuit. This pulse will always be different from the one that emerges at the very site of the discharge.

[2] It dwells on the use of the portable chromatographer TFGA – P200 produced by Morgan Schaffer (Figure 1).

The *SR EN 60270:2003,* standard: *High voltage test techniques – Partial discharge measurements* [4], the Romanian version of the English text for the European standard EN60270:2001/ IEC60270:2000):

- *defines the used terms and the quantities to be measured;*
- *describes test and measuring circuits which may be used;*
- *defines analogue and digital measuring methods required for common applications;*
- *specifies methods for calibration and requirements of instruments used for calibration;*
- *gives guidance on test procedures and assistance concerning the discrimination of partial discharges from external interference.*

The difficulty in measuring partial discharges is that the pulses can not be measured directly at the site where they occur but only after passing through the test object detection circuit (direct, electromagnetic, etc.).

To measure PD inside the insulation of high voltage instrument transformers, we'll use the method of partial discharges measurement by using broadband amplifiers, trying to determine the quantitative limit $PD(pC)$, to which the electric element can operate under maximum operating safety.

The measurements of partial discharges to determine the insulation state of 110kV high voltage instrument transformers starting in the year 2009 were achieved using the digital system, MPD600 (**M**easuring **P**artial **D**ischarge System) for detecting, recording and analyzing partial discharge events.

Measurements by the ability to be calibrated have resulted in: a quantity subject to interpretation according to the standard SR EN 60270:2003: **apparent charge**, $Q_{IEC}(pC)$.

Displaying individual output voltage pulses on an oscilloscope screen enables both recognition of partial discharges origin and their separation from any interference / background noises.

Thus, the PD apparent charge can be visualized either on a time base (depending on test voltage) or on a sinusoidal time base synchronized with the test voltage' frequency or on an elliptical time base crossed synchronous with the frequency of the test voltage [5].

According to recommendations of standard: *SR EN 60270:2003* for measuring partial discharges, the procedure:

Determining partial discharges amplitude at a specified test **voltage** - is the subject addressed in this paper' measurements and for which:

the partial discharges' amplitude (expressed in terms specified) is measured at the nominal voltage of the power grid, voltage value much higher than the alleged partial discharges inception voltage.

The test object will be brought to the specified voltage, voltage to be maintained throughout the range of measurement performing. DP associated characteristics will be recorded throughout the whole period of measurement (DP activity (Q_{IEC}) was measured over a preset, 2 minutes, period of time and the PD pulses recorded $\left(\frac{k_{PDs}}{s}\right)$ were those just above the threshold / offset of: *1pC*).

3. Instrument transformers. General considerations. Measuring circuits

To safe operate a power system, appliances capable of measuring electrical quantities (current, voltage, power, energy, frequency, power factor ($\cos\varphi$), etc.) are indispensable and also those that ensure correct functioning of the network and / or do limit proportions of possible damage to it. First are called *meters*, and other, *protective devices*.

Instrument transformers are non – rotating electromagnetic machines, powering electrical equipment falling into a broad category of metering and protection devices. In this regard, judging by their manufacturing technologies, they have some common features with other non-rotating electromagnetic machines (power transformers), but their functional properties do resemble more to those of electrical apparatus.

High voltage – 110kV – instrument transformers, part of the survey:

A. current transformers (102 elements);

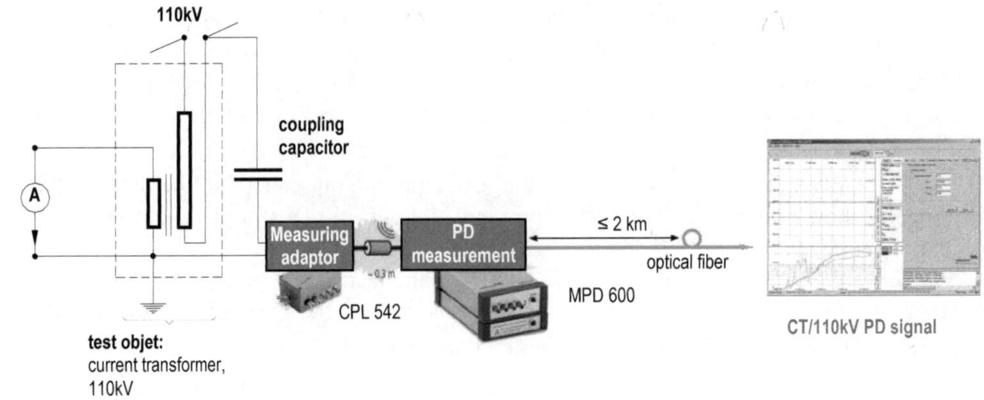

Figure 3. Test circuit for partial discharge measurement on current transformers

B. voltage transformers (38 elements);

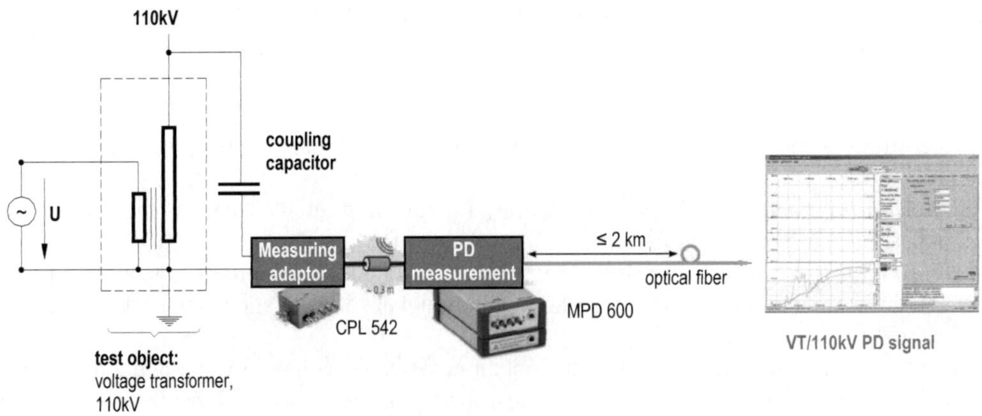

Figure 4. Test circuit[3] for partial discharge measurement on voltage transformers

[3] Using a high-voltage coupling capacitor: 112kV (1,2nF).

IOP Conf. Series: Materials Science and Engineering **200** (2017) 012003 doi:10.1088/1757-899X/200/1/012003

4. Linking high voltage instrument transformers insulation PD measurement results from both methods: non-electrical and electrical

Working methods for on site testing of liquid insulations (insulating oils) were:

1. Gas chromatography for the electrical insulating environment (using the TFGA-P200 chromatographer manufactured by Morgan Schaffer);
2. Electrical measurements of partial discharge levels using the digital detection, recording, analysis and partial discharge acquisition system, MPD600 (Measuring Partial Discharge System) produced by Mtronix Precision Measuring Instruments.

First we performed, between 2000 – 2015, chromatographic analyses (2,545) concerning the electrical insulating environments of:

A. **102 current transformers, 110kV**. Items (indigenous production of various types) in operation, functioning in 110/20kV substations.

B. **38 voltage transformers, 110kV**. Devices (all indigenous production, three constructive types) in operation, functioning in 110/20kV substations, mounted in 110kV OHL bays or in 110kV metering/measure bays.

Then, electrical measurements of partial discharge ***(1,125 individual measurements****), inside instrument transformers (figure 3 and 4), on site (power substations)* were made (starting in 2009, over a **7-year period**, collecting data **until 2015**) according to the provisions of standard EN 61869-1:2007 „*Instrument transformers. General requirements*", applying, assimilated to it, *A*[4] partial discharge test procedure, using as test voltage the very rated 110kV distribution grid voltage [6].

Considering paramount maintaining the same operation conditions for the instrument transformers and precisely to the purpose of an increased reproductibility of measurement results (of gas chromatography and electric measurements), following the guidelines of standard EN 61869-1:2007, we knowingly used the rated voltage of the 110kV power distribution grid to perform PD measurements [6].

Given the results of two parallel measurements, containing (***3,760 measurements***):

I. to this type of failure specific gas amount (H_2), as results of gas chromatographies performed on the insulating environment of high voltage instrument transformers (current and voltage) 110kV and

II. the quantitative partial discharge' level inside the same high voltage (110kV) instrument transformers,

establishing a certain dependence between the quantity of PD and the type and amount of in oil dissolved gases inside equipments affected by this type of defect: partial discharges, was expected.

5. Conclusion

Of the „population" of instrument transformers subject to the two parallel measurements, ***the dependency between*** Q_{IEC} ***(apparent charge)*** and H_2 ***(hydrogen, gas amount dissolved within their insulating environment) represents a finite assemblage situated between the two limits developed on an empirical basis*** (figure 5):

- upper limit: $\dfrac{Q_{IEC}(pC)}{H_2(ppm)}$, assimilated, $y = 19\sqrt{2}\cdot x - 230;$ (1)

- lower limit: $\dfrac{Q_{IEC}(pC)}{H_2(ppm)}$, assimilated, $y = \dfrac{52}{\sqrt{3}}\cdot x - 521,$ (2)

[4] Partial discharge test voltage is reached while decreasing rated power-frequency withstand voltage.

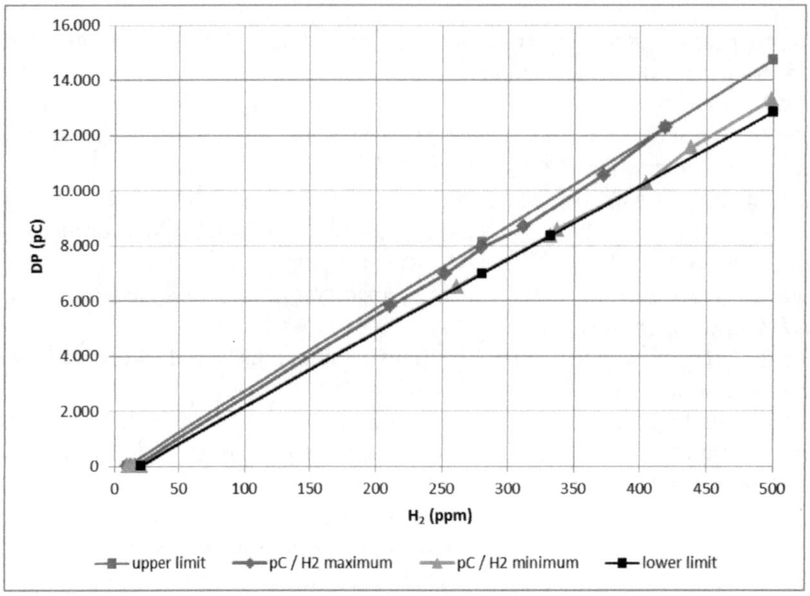

Figure 5. Dependency between Q_{IEC} (apparent charge) and H_2

The two lowest amounts of gas H_2 (hydrogen) dissolved within the insulating environment of those instrument transformers are the maximum quantitive values that can exist (threshold) in the insulating oil of an electrical equipment without it ever functioning [7].

References

[1] Munteanu T 1998 *Descărcări electrice parţiale în materiale electroizolante*, Editura Tehnică, Bucureşti, Romania

[2] ***ANSI/IEEE C57-104:2008 *IEEE Guide for the Interpretation of Gases Generated in Oil – Imersed Transformers*

[3] ***2000 *Transformer Fault Gas Analyzer TFGA P-200, Incipient Fault Detection – Operation Notes*, version 2.1.3, Morgan Schaffer Systems, Montreal, Canada

[4] ***IEC 60270:2000 (SR EN 60270:2003) *Tehnici de încercare la înaltă tensiune – Măsurarea descărcărilor parţiale*, versiune română a Standardului European IEC60270:2000 (standard tradus de ASRO, publicat cu permisiunea CENELEC)

[5] ***2014 *MPD 600: High-end measurement and analysis system for partial discharges*, www.omicron.at | www.omicronusa.com, OMICRON L2317

[6] ***IEC 61869-1:2007 *Instrument transformers. General requirements*, International Electrotechnical Commission

[7] Montanari J C 2006 *Insulation diagnosis of high voltage apparatus by partial discharge investigation*, Liu-Yeda Memorial Lecture, in Proceedings IEEE ICPADM, Bali, Indonesia, June 26-30

IOP Conf. Series: Materials Science and Engineering **200** (2017) 012004 doi:10.1088/1757-899X/200/1/012004

Research on the impact of LMX leadership theory on mutual trust and organisational commitment of employees in Bosnia and Herzegovina

E Strukan and M Nikolić

University of Novi Sad, Technical Faculty "Mihajlo Pupin", Serbia

E-mail: edinstrukan@gmail.com

Abstract: The paper presents the theoretical foundations of leadership based on the LMX exchange leader-member theory which essentially involves leadership process where a continuous and creative interaction between leaders and followers is at the centre of attention, during which the leader, among other things, affects the degree of mutual trust and organisational commitment and the quality of relationships in his/her organisation, which actually has a direct impact on organisational performances, effectiveness of the organisation's business and its market positioning.

Also, the paper presents the results of research conducted in organisations in Bosnia and Herzegovina, which undoubtedly point to the fact that the dimensions of leadership based on the LMX exchange leader-member theory, is strongly correlated with the dimensions of mutual trust and organisational commitment, and to have a significant positive impact on them, and therefore on organisational performances and effectiveness of business organisations that were included in this study.

1. Introduction and theoretical basis

Leadership as a term has first appeared in non-economic social theories. Regardless of which leadership it is about, political or business, *Gardner* has defined six characteristics of leadership: *story* (a leader must have a central story or a message to a large heterogeneous groups), *audience* (listeners to whom a message is transmitted), *organisation* (necessary institutions through which the leadership messages are forwarded to the audience), *the embodiment* (creator of the message should not be in contradiction with its message), *address* (can be directly to the audience as the political leaders do it, or indirectly through a certain product, as business leaders do), *skills* (a high degree of expert knowledge in their work or credibility in their life) [1]. Since the last years the business leadership is in the focus of scientific interest, the works will, in the context of the application of leadership competencies in management, deal precisely with business leadership, that in the shortest may be defined as *conducting* the entire organisation or one of its organisational units.

Despite the large number of different approaches to leadership that are recently studied using various quantitative and qualitative methods, leadership in the research literature is generally defined as a complex and multidimensional process that can significantly improve organisational performances regardless of the sector in which the organisations operate, what is its size, what kind of structure, etc. In this context, there are many theories that explain the complexity of the leadership process [2-5].

Bass suggests that some definitions observe leadership as *a centre of group processes* with a leader at the centre who creates, initiates, and through the process implements positive change. In the second group of definitions of leadership are those whose leadership are viewed from *the aspect of personality*, defining it as a combination of special features and characteristics that an individual (manager) who controls certain processes or leads a certain group of people, possesses. By the same author, there are those definitions of leadership that are observed in *terms of skills*, which emphasise certain knowledge and skills that provide effective leadership and achieving better individual and organisational performance for managers, and consequently improve operating results [2].

Essentially, regardless of the many definitions, as a key component of leadership the following may be set up: (a) leadership is a process, (b) through leadership the influence is achieved, (c) leadership occurs in the context of the group, (d) leadership is assumed to achieve objectives [6].In this context, the most appropriate definition of leadership is that leadership is a process in which the individual realises the impact on the group to achieve a common goal [6].

Some authors emphasise that in the process of leadership leaders and followers are involved together, because for that process to be successful, leaders need followers and vice versa, followers need leaders [7-10], despite this strong correlation, leaders are still those who create and promote the establishment of relation, good conditions for communication as a key prerequisite for effective leadership and take care of the maintenance of these relations. The leader knows that all employees want to develop as a person through their work, to develop and use their talents and skills, to achieve operating results and achievements that will be recognised [11], while the job does not only involve job performance, but also active interaction with colleagues, managers, respect of the rules and policies of the organisation, achieving standards of work performance, etc.

If the task of the leaders would be showed as a quite a simplified method, it might have been said that it is reflected in two key things, situation assessment and taking action. However, in the business world there is a clear distinction between leaders and managers.

So, managers and leaders are very different [12]. The leader is the person who takes responsibility (planning and implementing the vision he/she believes in) and that coordinates the activities of people in their efforts to achieve the goal [13]. According to most experts in this field, the leadership has a more important role than the standard managerial control, simply because the conduct is the static form of determining the direction of movement of the organisation, and leadership necessarily has a dynamic form of movement toward a goal that can be defined as the dynamic change in order to achieve greater market competitiveness [14].

Also, skilful leaders can properly predict future needs and empower others to share and implement the vision [15]. *Benis* states that effective leaders have four common characteristics: provide guidance and set objectives and targets, create trust in the environment, take the risk and communicate effectively [16].

It is noteworthy that despite the great importance and expansion of the study of leadership as a science in the last three decades in the countries of the region (Western Balkan), conducted research on the topic of leadership is very rare, which is why this paper will focus on leadership as a process in which leader managers generate specific interaction with employees and thus transmit to them all the necessary knowledge, motivating them to achieve better business results.

Bearing in mind the variety of models and styles of leadership, four basic types of leader behaviour can be identified: supportive, participatory, directive and charismatic, whereby the effectiveness of each type is conditioned by situational factors of nature, mainly by the characteristics of tasks, organisation and employees [17].

1.1. From traditional managerial control to management and effective leadership

The essence of effective leadership, but also of management, is the relationship to the people as the holders of all business processes in the organisation. It is therefore essential that modern managers who aspire to become leaders, place precisely this resource (people) in the centre of their activities and to strive to keep away from the traditional approach to human resources as much as possible, which is based on strict rules, procedures and controls, and to creatively develop modern guidance.

Leadership is a process of directing the behaviour of others toward the realisation of a goal, which implies that the central idea is keeping things done through people [18]. For the successful conduct, an important prerequisite is to exercise a certain influence on others, which is the main item that determines how to be a successful manager [19]. Today's guidance, thanks to incredible opportunities that are achieved due to improvements in communication technology and the expansion of international business, is more important than ever before [20]. To lead is not the same as to manage. Although some managers are leaders, and some leaders are managers, leadership and management are not the same [21], because owning management skills is no longer enough for the survival of managers in the business world. Namely, management consists of a rational assessment of the situation and of systematic choosing the goals and objectives (what to do), systematic strategy development to achieve these objectives, the allocation of the necessary resources, the rational design, organisation, directing and controlling the activities required to achieve a chosen purpose, and finally, motivating and rewarding people to do the job well [22].

In fact, by the combination of management and leadership, the ability of calculated and logical access to organisational processes (management), and genuine concern for workers as human beings (leadership) are sought [23]. Successful leaders generally have the following characteristics: they are intelligent, including the ability to assess and expression, they are communicative, emotionally mature and stable, and they are persistent, motivated for constant achievements, ambitious, yet modest, etc.

Modern theorists of management and managers are more likely to say that people are not born leaders, but are being trained, or in other words, leaders are made, not born [24]. A crucial factor in the successful conduct is the leadership style, which actually involves the behaviour of leaders while directing members of the organisation in the right direction [25]. In the past few decades, five leadership styles that match the increasingly dynamic global business environment appeared [26]:

(1) *Transformational leadership,* which encourages the success of the organisation constantly creating positive change and strongly influencing the belief of its followers in what an organisation actually needs to be, and their system of values, such as justice and fairness [27].
(2) *Training* is a leadership style by which manager gives instructions to his/her subordinates on how to overcome specific organisational challenges that they need to deal with [28]. *Super-leading* implies teaching followers how to think independently and act constructively and independently [29]. The goal of super-leading is to create followers who need a little guidance, or who need minimal attention in the process of performing work.
(3) *Servant leaders*, is the approach (style) in which the leader considers that his/her primary role is helping subordinates looking to gain their personal needs, aspirations and interests [30]. Generally speaking, the servant leaders think that their main task is to take care of human resources of the organisation [31].
(4) *Entrepreneurial leadership* is based on the view that the leader is an entrepreneur. These leaders are behaving as if their role in the organisation is the key role, and not a side, ready to take risks, approach to any error as it is very important and not insignificant, and as such will neutralise the normal functioning of the organisation [32].

Most researchers agree that the two concepts (management and leadership) nevertheless differ, because the prevailing management function is to provide order and consistency in the organisation, while the primary function of leadership is to induce positive changes and shifts [33]. By the same author, management is engaged in establishing peace and stability, and leadership is in search of adaptive and structural changes. On the other hand, if the organisation has a strong leadership without management, the outcome may be insignificant and meaningless, that is, misguided change for the sake of change [33]. To be effective, organisations must have competent management and skilled leaders, because the managers manage, and leaders influence the people taking them through the execution of their tasks.

The next chapter shows, according to the authors, one of the most interesting theories of leadership through which the influence on followers is exerted.

1.2. LMX leadership theory of measuring the quality of the leader-member relations

LMX theory (Leader-member exchange theory) is among the most widely researched theories of leadership [34]. LMX theory measures the quality of the relationship and trust between leaders - member

[35]. In the leader-memberexchangetheory (LMX) leadership is defined as a process in which centre is *the interaction* between leaders and followers. This theory was first presented by *Dansereau, Graen* and *The Hague* [36], *Graen* and *Cashman [*37] and *Graen* [38].After it was first released, it has undergone several revisions and it remains equally interesting for researchers who study the process of leadership. In this theory, special attention is paid to what was the connection between the quality of the exchange in relation to the leader-member and the positive results for leaders, followers (subordinate employees), groups and organisations in general [39].The researchers have found that high-quality leader-member exchange results in smaller departure of employees, higher grade of performance, desirable work assignments, better (positive) attitude towards work, greater attention and support of leaders, prominent participation and faster career progression over a period of 25 years [39-40].

High quality of LMX implies mutual support, mutual trust, respect, exchange of formal and informal rewards, loyalty, connectivity and sense of obligation to their superior [41], [39]. High quality of LMX is an upgrade which goes beyond what is stipulated in the employment contract. Low quality of LMX is characterised by lack of trust and support, poor relationships, low trust and organisational commitment. It is purely an economic relationship that progresses no more than it is stipulated in the employment contract [42]. High quality of LMX brings many benefits for the organisation, and therefore it is good that the leaders have a high-quality relationship with a subordinate [43].

Because communication is a "bloodstream" of the organisation, communication skills are important competency for leaders, which is why it is advisable to constantly develop them in formal and informal communication which is a network that consists of a complete system of full flow of various information, orders, wishes and hints [44].Leaders who succeed to develop their communication skills, among other things, take care of their subordinates easier and help them to do their jobs following their success, but through quality communication helps them to develop personally and professionally.

Numerous studies show that there is a strong, positive relationship between effective internal communications and business results and other organisational outcomes. For example, according to *Garcia-Morales, Matias-Reche, Verdu-Jover* [45], internal communication affects the technological proactively, organisational learning and organisational innovation. References *Andersen, Segars* [46], *Yates* [47], show that the improvement of internal communication provides better financial results of the organisation. Similar to the above, a significant number of references confirms the influence of internal (as the most important leadership communication) on job satisfaction *Kang* [48], *Schweitzer* [49], which is certainly in close relation with the dimensions of mutual trust and organisational commitment. Open communication in the workplace increases productivity and impacts on employee satisfaction [50]. According to *Beebe, Blaylock & Sweetser* [51], good communication leads to an increase in satisfaction and motivation of employees. Reference *DiFonzo & Bordie* [52] points to the feedback; job dissatisfaction often has an impact on internal communication, and so on all the organisational performances including mutual trust and organisational commitment. Finally, a leader knows that organisational communication is directly related to the objectives, functions and structure of the organisation of people, and the organisation's success largely depends on the effectiveness of organisational communication [53].

2. Methodological item of research and used instruments

The questionnaire used for this study is designed to measure reliably and validly, the quality of labour relations and exchanges between leaders and followers, but is also designed so that leaders can use it for self-evaluation, as well as a good tool for training and developing their own leadership skills. Also, it offers answers to questions about the extent, scope and intensity of mutual trust and organisational commitment of employees. Thus designed questionnaire had the purpose to enable the measurement of the above parameters and their placement in mutual correlation, in order to determine the extent and intensity of the impact of LMX theory of leadership on mutual trust and organisational commitment of employees in organisations in Bosnia and Herzegovina covered by this study.

Leader-member exchange (LMX). In this study, to measure the quality of LMX, a LMX-7questionnaire was applied, an instrument used to measure the degree of respect, trust and obligation between leader - member. This instrument has seven items, and when assessing the respondents used a five-point Likert scales [39]. At the same time, LMX-7 is an instrument that is most commonly used in the LMX quality researches [54].

Mutual trust at work. Instrument Interpersonal trust at work [55] has 12 items and measures the faith in the purposes and confidence in the actions of colleagues and management through 4 dimensions.

Organisational commitment was measured by instrument of authors *Cook and Wall* [55]. The instrument has 9 items and measures the organisational identification, engagement and loyalty of employees.

Wishing to determine the individual effects of each item separately, as well as their mutual correlation with the intention of obtaining accurate results on the intensity and scope of influence of each item of LMX theory to each item of mutual trust and organisational commitment, the research has shown that this form and related items are not grouped in dimensions as it is normally done on a research that use those instruments.

2.1. Objectives, tasks and hypothesis of the research

The main reason why this research has been conducted was to obtain the relevant indicators and determine the extent to which leadership based on the LMX theory of developing better relations between leaders and subordinates have a positive impact on mutual trust and organisational commitment of all actors within the working processes.

The main objective of the study was to, by adequate analysis done on the basis of measurements, determine the extent to which the concept of LMX theory of leadership has an impact on the performances and effectiveness of business processes in organisations in Bosnia and Herzegovina.

The main task of the research was to, on the basis of determined values obtained from the questionnaire that was used during the research; establish and present to managers in which direction they need to go in the future, in order to gain additional competencies useful for achieving outstanding results in the process of creating better relations with their subordinates.

Research hypothesis. Taking into account theoretical foundations presented in the work and other researches in this area, it has been hypothesised that there is a positive correlation between the dimensions of leadership based on the LMX theory and mutual trust, and organisational commitment of employees. So, LMX has a significant positive impact on mutual trust at work and organisational commitment in organisations in Bosnia and Herzegovina.

2.2. Method and mode of work

This study was conducted in 103 business organisations from the private and public sector in Bosnia and Herzegovina during 2016.The study included 334 employees and mid-level managers who are directly related to the implementation of business processes. Research was preceded by the introductory explanation to managers about the purpose and significance of the research. To conduct the survey a method of scientific research, surveys and interviewing techniques, as well as an instrument in the form of questionnaires were used. The survey is anonymous, and the questionnaire, in addition to general information (gender, age, education degree, field of education, national origin, organisation and ownership structure), consisted of 19 questions. The questions covered a wide range of leadership based on the LMX theory, mutual trust and organisational commitment.

3. Research results

The research was conducted on a sample of N = 334 respondents.184 (55.10%) were male and 150 (44.90%) were women. According to age, the subjects were divided into three age groups: younger managers (under 35 years of age), and there were 127 (38.02%) of them, managers of middle age (35 to 50), there were 124 (37.13%) of them, and managers of the older age (over 50 years) and there were 83 (24.85%) of them. Education degree of examinees is different, but most of the managers are with higher education, university education first cycle. There were 44 respondents (13.17%) with secondary school, 55 (16.47%) with higher education, 198 (59.28%) with university education (university education first cycle), 24 (7.19%) masters (second cycle university education), and 13 (3.89%) Ph Ds. When it comes to fields of education, 97 (29.04%) of the respondents were educated in the field of natural sciences, 144 (43.11%) in the social sciences, and 93 (27.85%) in other areas of education. The structure of the organisation in which the subjects are engaged is also different. Of the 103 included organisations, 86 (83.50%) of organisations have Bosnian national origin, 17 (16.50%) have the other (mainly EU origin). 48 (46.60%) organisations have a private ownership structure, and 55 (53.40%) are under state ownership structure.296 (88.62%) of respondents are employed in organisations of Bosnian national origin, 38 (11.38%) in the organisations of the European (EU) national origin. 185 (55.39%) of respondents are employed in organisations from the public sector, and 149 (44.61%) in the private sector organisations.

Descriptive statistics for the dimensions of leadership based on the LMX theory at the centre of which is the interaction between leaders and subordinates, during which special attention is paid to the quality of relationships (LMQ questionnaire), and the dimensions of mutual trust and organisational commitment, are shown in Table 1. In the same Table, among other things, are given the names of dimensions and a short name for each dimension, the mean value and standard deviation of dimensions. The independent variable is the level of LMX leadership shown in the table as short codes L1 to L7 (LMX short questionnaire consisting of 7 questions with possible responses ranging from 1 "very bad" to 5 "excellent"). The dependent variables are the dimensions or the achieved level of mutual trust, shown in the table as short codes of IT1 to IT12 (a questionnaire with 12 questions and ratings ranging from 1 "strongly disagree" to 7 "strongly agree"), and the dimensions or the degree of organisational commitment, shown in the tables short codes OCM1 to OCM9 (questionnaire of 9 questions and answers in the range from 1 "strongly disagree" to 7 "strongly agree").

Table 1.Descriptive statistics

Dimensions	Short name	Min.	Max.	Arithmetic mean	Standard deviation
The level of information about satisfaction or dissatisfaction of the manager with staff performance.	L1	1	5	3.5239	1.03004
The level of understanding the problems and needs of the working staff from the manager.	L2	1	5	3.2275	1.07762
The level of feeling that the manager understands the potential of employees.	L3	1	5	3.3053	1.11615
The level of readiness to use the power of the managers in the process of helping to solve labour problems during the work.	L4	1	5	3.2485	1.13948
The level of readiness of managers to "get out of the pulley" employee if he/she really needs help, even to their own detriment.	L5	1	5	2.7095	1.20911
The level of trust of employees in the decision of the manager, in order to defend them	L6	1	5	3.2874	1.04322
Working relations' efficiency of managers and employees	L7	1	5	3.3982	1.02526
Sincere respect of opinions of employees by the management organisation.	IT1	1	7	4.0748	1.80017
Opinion of employees on the good future of the organisation and operation of managers.	IT2	1	7	4.1676	1.90195
The level of mutual assistance of colleagues in the event of difficulties in the workplace.	IT3	1	7	4.6766	1.57025
The level of belief that management has	IT4	1	7	4.2365	1.82928

made good decisions for the future of the organisation.					
The level of belief that colleagues (people working with them) would help if necessary.	IT5	1	7	4.7335	1.54344
The level of belief that the leadership of the organisation is effectively doing its job.	IT6	1	7	4.1556	1.85665
The level of solid belief that the organisation will always be fair to employees.	IT7	1	7	3.4251	2.01766
The level of reliance on its collaborators and the belief that most associates would do what is required of them.	IT8	1	7	4.6946	1.46105
The level of full confidence in the abilities and skills of its associates.	IT9	1	7	4.6946	1.48551
The level of belief that most associates would do their job, even when supervisors are not nearby.	IT10	1	7	4.5688	1.59382
The reliance on its collaborators to conscientiously do their job.	IT11	1	7	4.6616	1.57187
The level of belief that the leadership of the organisation does not deceive its employees.	IT12	1	7	4.3892	1.80539
The level of pride that they are employed and work for their organisation.	OCM1	1	7	4.6047	1.82560
The level of desire to work in their organisation forever.	OCM2	1	7	4.2305	2.06591
The level of desire to stand out to help their organisation.	OCM3	1	7	5.3562	1.50534
The level of commitment to the organisation (even if my organisation does not stand well financially, I would not have gone to another organisation).	OCM4	1	7	3.2874	2.00180
The level of sense of belonging to the organisation (I feel I am part of the organisation).	OCM5	1	7	4.4491	1.75755
The level of giving its maximum in the workplace (always give my best at the workplace, not only for myself but also for the organisation).	OCM6	1	7	5.4940	1.50223
The level of commitment to the organisation in terms of benefits (If other organisations offered me more money; it would not have led me to seriously think about changing jobs).	OCM7	1	7	3.4131	2.09316
The level of confidence in the good future of their organisation (I would recommend to my friend to get a job in my organisation).	OCM8	1	7	4.4221	1.73938
The level of satisfaction if the work contributes to the well-being of the organisation.	OCM9	1	7	5.2724	1.60910
Valid N = 334					

Table 2 shows the results of correlation analysis: correlation dimension of leadership based on the LMX theory with cantered quality of the relationship that is being implemented through the active interaction between leaders and subordinates (as the independent variable), with dimensions of mutual trust and organisational commitment as organisational outcomes of this process (as dependent variables). Results relate to the total sample of N = 334 respondents and Pearson correlation was used. Statistically significant correlations are marked as follows: *p<0,05, **p<0,01.

Table 2. Pearson's correlation coefficient between the dimensions of LMX leadership and dimension of mutual trust (IT) and organisational commitment (OCM)

	IT1	IT2	IT3	IT4	IT5	IT6	IT7	IT8	IT9	IT10	IT11	IT12
L1	.463**	.378**	.259**	.446**	.317**	.468**	.392**	.294**	.242**	.118*	.160**	.428**
L2	.649**	.607**	.358**	.609**	.401**	.649**	.535**	.372**	.340**	.206**	.237**	.612**
L3	.679**	.622**	.375**	.591**	.433**	.602**	.566**	.391**	.372**	.273**	.297**	.594**
L4	.574**	.517**	.409**	.542**	.425**	.585**	.550**	.331**	.284**	.206**	.248**	.524**
L5	.683**	.578**	.436**	.585**	.488**	.642**	.660**	.436**	.378**	.331**	.337**	.591**
L6	.679**	.649**	.411**	.676**	.458**	.696**	.625**	.412**	.386**	.259**	.336**	.663**
L7	.628**	.569**	.377**	.613**	.434**	.627**	.615**	.376**	.330**	.217**	.293**	.606**

	OCM1	OCM2	OCM3	OCM4	OCM5	OCM6	OCM7	OCM8	OCM9
L1	.307**	.203**	.257**	.263**	.296**	.257**	.291**	.282**	.193**
L2	.510**	.421**	.379**	.384**	.491**	.331**	.362**	.476**	.291**
L3	.530**	.464**	.373**	.442**	.507**	.284**	.457**	.495**	.402**
L4	.433**	.397**	.321**	.404**	.424**	.253**	.440**	.465**	.335**
L5	.559**	.496**	.387**	.461**	.528**	.314**	.475**	.494**	.382**
L6	.571**	.479**	.470**	.495**	.563**	.407**	.424**	.515**	.366**
L7	.524**	.467**	.433**	.453**	.520**	.443**	.479**	.503**	.409**

4. Discussion

As demonstrated by the data presented in Table 1, which refer to the dimensions of leadership (LMX questionnaire), it is evident that the mean of the maximum dimension L1 has a level of knowledge about satisfaction or dissatisfaction of the manager with the work of employees, with an average of 3.5239. Other dimensions of leadership from this questionnaire have a uniform and relatively high average arithmetic mean as well, which suggests that managers in organisations that were included in this study, have quite a high level of qualities, skills and techniques, and that they use LMX leadership that prefers a strong interaction with subordinates and constantly improve the quality of relationships with them, which of course results in increasing the level of mutual trust and organisational commitment as the final outcome of the leadership process. This way they can adequately respond to the constant changes in the business environment, thereby increasing individual and organisational performances, and thus the competitiveness of its business organisations.

When it comes to dimensions of LMX leadership, it is evident that the mean of the smallest dimension of L5 has a level of readiness of managers to "get out of the pulley" employee if he/she really needed help, even to their own detriment, with an average of 2.7095,which suggests that managers included in this study should work on self-evaluation to additional education and training to improve their personal effectiveness in terms of the development of this dimension, which is in direct mutual-effect relationship with the dimensions of mutual trust, and that could seriously affect the motivation and job satisfaction, and reflect on the organisational commitment and the effectiveness of employees' performances.

As for the dimensions of mutual trust (IT) from the questionnaire on measuring the level of mutual trust, it is evident that the highest arithmetic mean has IT5, level of belief that colleagues (people working with them) would help if needed with an average 4.7335which leads to the conclusion that the level of mutual trust between people who work together (co-workers, members of a team), is at the very high level, which can be understood as the result of a high level of quality of relationships within the team.

In this context it is important to note that the average arithmetic means of the other dimensions of mutual trust are balanced and at a relatively high level, except dimension IT7 (level of solid belief that the organisation will always be fair to employees), which has the lowest mean with the average of 3.4251which is somewhat logical given the fact that the area in which the research was conducted (Bosnia and Herzegovina, the Western Balkans), is still in the transition process, and that the concept of market operations still has not been fully implemented, both in the state nor in private sector, which is why employees cannot be firmly convinced that in their organisation in the near future, there will not be some transitional changes that would inevitably bring the uncertainty of employees.

Finally, it is evident that the mean dimension of organisational commitment is quite high and balanced suggesting that organisational commitment is at a relatively high level. The highest mean of the dimension has OCM6 (level of giving its maximum in the workplace) with an average arithmetic mean 5.4640, and the smallest mean has the dimension OCM4 (level of commitment to the organisation even if the organisation does not make good financial standing), with an average arithmetic mean 3.2874.

Table 2 shows a statistically significant correlation dimensions of LMX leadership and dimensions of mutual trust and organisational commitment, thus confirming the hypothesis of the existence of the mentioned correlation which positively affects the effectiveness of business and organisational performance. Of dimensions of leadership, almost all dimensions have a strong correlation, which is actually logical, because leaders who aspire to provide good long-term individual and overall organisational performance frothier organisation, place strong interaction processes in the centre of their attention, whose main task is to increase the quality of relations and thus mutual trust and organisational commitment of employees, emphasising their long-term positive impact for the organisation in the future.

Of dimensions of mutual trust (IT), a statistically significant and quite a balanced correlation have IT1 (sincere appreciation of employees' opinions by the management of organisation), IT2 (opinion of employees on the good future of the organisation and work of the managers), and IT6 (level of belief that the leadership of the organisation is effectively doing its job), as opposed to, for example IT9 (level of full confidence in the abilities and skills of their associates) and IT10 (level of belief that most associates would do their job, even when supervisors are not nearby), who also have quite a balanced correlation, but with lower intensity, which is understandable because these dimensions of mutual trust, in addition to human resources, which are the focus of leaders, are largely influenced by other factors of the business environment, first of all unregulated market system of business, but also, generally speaking, unregulated society in which organisations covered by this research do business, which may result in irregular income of financial benefits (salaries), low standard of living, etc..

Also, from the obtained results shown in Table 2, it can be concluded that there is a positive correlation, which is statistically significant at the level $^{**}p<0.01$ between dimensions of LMX leadership and dimension of mutual trust, particularly between:L3 (level of feeling that the manager understands the potential of employees) and IT1 (sincere appreciation of employees' opinions by the management organisation);then between L5 (level of readiness of managers to "get out of the pulley" employee if he/she really needed help, even to their own detriment) and IT1 (sincere appreciation of employees' opinions by the management organisation); as well as between L5 (level of readiness of managers to "get out of the pulley" employee if he/she really needed help, even to their own detriment) and IT7 (level of solid belief that the organisation will always be fair to employees), and betweenL6 (level of confidence of employees in the decision of the managers, in order to defend them) and OCM1 (level of pride in what they do for their organisation), between L6 (level of confidence of employees in the decision of the managers, in order to defend them) and OCM5 (level of sense of belonging to the organisation).

On the basis of the above conclusions based on statistical correlations, it can be concluded that the dimensions of LMX leadership theory, based on continuous development of the quality of the relationship between leaders and followers, with continuous interaction and adequate channels of communication (skills and techniques) and good motivation, have a significant impact on dimensions of mutual trust and organisational commitment. On the other hand, these dimensions significantly affect the individual and organisational performances, the effectiveness of business processes and thus on the overall market position of the organisation.

Almost all correlations are statistically significant at the level $^{**}p < 0.01$.

5. Conclusion

The research confirmed the hypothesis that leadership based on LMX leader-member exchange theory has a strong and positive impact on interpersonal relations and organisational commitment in organisations that were included in this study. The article summarises the theoretical foundations as well as the results of this study and confirms previously known fact that the application of LMX leadership theory positively affects the results of the organisation, especially on performance, organisational commitment, the atmosphere at work, innovations, belonging to the organisation, giving jurisdiction, career advancement, and many other important organisational variables.

LMX theory warns that managers need to be aware of the way they treat their subordinates, and the desirability of developing a higher level of sensitivity to the needs of subordinates, as well as the need of fairenablingfor all employees to be adequately involved in the working process of their organisation in the most appropriate way. Also, the message of this theory is that managers need to continuously build relationships of trust and partnership with their subordinates.

So, this theory allows managers to assess their own leadership in terms of the quality of relationships with subordinates and allow them to be more responsive in working processes and thus enhance the effectiveness of their work units by building strong relationships with their subordinates. In this way, overall organisational performances are directly improving, which is why as a general conclusion of this study a statement arises on the need for continuous application of the elements of this leadership theory based on the process of interaction between leaders and subordinates, if they want to achieve better performance for their organisation, or generally better market position in an increasingly competitive and dynamic business environment.

References

[1] Gardner H 1996 *Leading Minds: An anatomy of leadership,* Free Press, New York, USA
[2] Bass B M 1990 Bass and and Stogdill's handbook of leadership: *A survey of theory and research,* Free Press, New York, USA
[3] Rost J C 1991 *Leadership for the twenty-first century*, Preager, New York, USA
[4] Hickman G R (Ed.) 1998 *Leading organisations: Perspectives for a new era,* Thousand Oaks, CA:Sage
[5] Mumford M D 2006 Pathways to outstanding leadership*: A comparative analysis of charismatic ideological,and pragmatic leaders,* Mahwah, Nj: Lawrence Erlbaum
[6] Northouse P G 2008 *Leadership, Theory and Practice*, Fourth edition, p. 2. Data status, Beograd, Serbia
[7] Burns J M 1978 *Leadership*, Harper & Row, New York, USA
[8] Heller T & Van Til J 1983 Leadership andfollowership: Some summary propositions, *Journal of Applied Behavioral Science* **18** 405-414
[9] Hollander E P 1992 Leadership, followership, self, and others, *Leadereship Quarterly* **3**(1) 43-54
[10] Jago A G 1982 Leadership: Perspectives in theory and research, *Management science* **28**(3) 315-336
[11] Davis R V 1992 *Job satisfaction. In: Jones LK (ed.),* Encyclopaedia of career change and work issues (PP. 142-143), The Oryx press, Phoenix, USA
[12] Zaleznik A 2004 *Managers and leadres: are they different,* Harvard Busines Review, USA
[13] Bass B M 1991 *Staglills handhode of leadership,* Free Press, New York, USA
[14] Collins J 2007 *Od dobrog do izvrsnog,* Binoza press, Zagreb, Croatia
[15] Kelley R C, Thornton B and Daugherty R 2005 Relationships between measures of leadership and school climate, *Education* **126** 101: 17-25
[16] Bennis W and Nanus B 1985 *Leadership,* Harper and Row Publishers, New York, USA
[17] Howell J P and Costley D L 2001*Understanding Behaviors For Effective Leadership*, Prentice Hall, New Jersey, USA

[18] Goldman E 1998 The Significance of Leadership Style, *Educational Leadership* **55** (7) 20-21
[19] Torrington D and Weightman J 1987 Middle Management Work, *Journal of General Management* **13** 74-89
[20] Nadler D and Tushman M L 1990 Beyond the Charismatic Leader: Leadership and Organisational Change, *California Management Review* **32** 77-97
[21] Zaleznik A 1989 Executives and Organisations: Real Work, *Harvard Business Review* January 57-64
[22] Levitt T 1976 *Management and the Post-Industrial Society,* Public Interest 73
[23] Townsend P L and Gebhardt J E 1989 We Have Lots of Managers, We Need Leaders, *Journal For Quality and Parcitipation* September 18-20
[24] Sessa V 2003 Creating Leaderful Organisations: How to Bring Out Leadership in Everyone, *Personnel Psyhology* **56** 762
[25] Ribelin P 2003 Retention Reflects Leadership Style, *Nursing Management* **34** 18
[26] DuBrin A J 1996 Reengineering Survival Guide. *Cincinnati, OH: Thomson Executive Press* 115-29
[27] Kuhnert K W and Lewis P 1987 Transactional and Transformational Leadership: A Constructive/Developmental Analysis, *Academy of Management Review* October 648-657
[28] Blades B 2001 Great Coaching Can Increase Revenue, *Arizona Business Gazette* January **5**
[29] Manz C C 1991 Helping Yourself and Others to Master Self Leadership, *Supervisory Management* 19-38
[30] Greenleaf R K 1977 *Servant Leadership: A Journey into the Nature of Legitimate Power and Greatness,* Mahwah, Pualist Press. New Jerey, USA
[31] Sendjaya S and Sarros J C 2002 Servant Leadership: Its Origin, Development, and Application in Organisations, *Journal of Leadership and Organisational Studies* **9**(2) 57-64
[32] McCarthy D J, Puffer S M and Naumov A I 1997 Case study – Olga Kirova: A Rusian Enterpreneur's Quality Leadership, *International Journal of Organisational analysis* **5**(3) 267-90
[33] Kotter J P 1990 *A force for change: How leadership differs from management,* Free Press, New York, USA
[34] Goertzen B J and Fritz S M 2004 Does sex of dyad member really matter?: A review of leader-member exchange, *Journal of Leadership Education* **3**(2) 3-18
[35] Seabright M A, Leventhal D A and Fichman M 1992 Role of individual attachments to the dissolution of inter-organisational relationships, *Academy of Management Journal* **35** 22-60
[36] Danserau F, Graen G B and Haga W 1975 A vertical dyad linkage approach to leadership in formal organisations, *Organisational Behavior and Human Performance* **13** 46-78
[37] Graen G B and Cashman J 1975 *A role-making model leadership in formal organisations: A developmental approach,* in J G Hunt & L L Larson (Eds.), Leadership frontiers (pp 143-166), Kent, OH: Kent State University Press, USA
[38] Graen G B 1976 *Role-making process within complex organisations,* in M D Dunnette (Ed.), Handbook of industrial and organisational psychology (pp 1202-1245). Chicago: Rand McNally
[39] Graen G B and Uhl-Bien M 1995 Relationship-based approach to leadership: Development of leader-member exchange (LMX) theory of leadership over 25 yares: Applyng a multi-level, multi-domain perspective, *Leadership Quarterly* **6**(2) 219-247
[40] Liden R C, Wayne S J and Stilwell D 1993 A longitudinal study on the early development of leader-member exchange, *Journal of Applied Psychology* **78** 662-674
[41] Dienesch R M and Liden R C 1986 Leader-member exchange model of leadership. A critique and further development, *Academy of Management Review* **11**(3) 618-634
[42] Sparoowe R T and Liden R C 1997 Process and structure in leader-member exchange, *Academy of Management review* **22** 522-552
[43] Graen G B and Uhl-Bien M 1991 The transformation of professionals into self-managing and partially self-designing contributions: Toward a theory of leader-making, *Journal of Management Systems* **3**(3) 33-48

[44] Fox R 2006 *Business Communication,* Public Open University, Zagreb, Croatia
[45] Garcia-Morales V J, Matias-Reche F and Verdu-Jover A J 2011 Influence of Internal Communication on Technological Proactivity. Organisational Learning, and Organisational Innovation in the Pharmaceutical Sector, *Journal of Communication* **61** 150-177
[46] Andersen T J and Segars A H 2001 The impact of IT on decision structure and firm performance: evidence from the textile and apparel industry, *Information & Management*
[47] Yates K 2006 Internal communication effectiveness enhances bottom-line results, *Journal of Organisational excellence* **25**(3) 71-79
[48] Kang J A 2010 Ethical conflict and job satisfaction of public relation practitioners, *Public Relations Review* **36**(2) 152-156
[49] Schweitzer J C 1989 The role of formal communication in employee job satisfaction, *Public Relations Review* **15**(3) 52
[50] Irwin H and More E 1994 *Managing corporate communications,* Allen & Unwin Pty Ltd, Sidney, Australia
[51] Beebe A, Blaylock A and Sweetser K D 2009 Job satisfaction in public relations internships, *Public Relations Review* **35**(2) 156-158
[52] DiFonzo N and Bordia P 2000 How top PR professionals handle hearsay: corporate rumors, their effects, andstrategies to manage them, *Public Relations review* **26**(2) 173-190
[53] Visser B 2000 Organisational Communication Structure and Performance, *Journal of Economic Behavior Organisation* **42**(2) 231-252
[54] Hughes L W, Avey J B and Nixon D R 2010 Relationships between leadership and followers' quitting intentions and job search behaviors, *Journal of Leadership and Organisational Studies* **17**(4) 351-362
[55] Cook J and Wall T 1980 New Work Attitude Measures of Trust, Organisational Commitment and Personal Need Non-Fulfilment, *Journal of Occupational Psychology* **53** 39-52

Estimation of coefficient of rolling friction by the evolvent pendulum method

S Alaci[1], F C Ciornei[1], A Ciogole[1] and M C Ciornei[2]

[1]"Stefan cel Mare" University of Suceava, Mechanics and Technologies Department, Str. Universitatii no.13, 720229, Suceava, Romania
[2]"Carol Davila" University of Medicine and Pharmacy, Department 2 Physiology I, Bd. ErolilorSanitari no.8, Bucuresti, Romania

E-mail: alaci@fim.usv.ro

Abstract. The paper presents a method for finding the coefficient of rolling friction using an evolvent pendulum. The pendulum consists in a fixed cylindrical body and a mobile body presenting a plane surface in contact with a cylindrical surface. The mobile body is placed over the fixed one in an equilibrium state; after applying a small impulse, the mobile body oscillates. The motion of the body is video recorded and afterwards the movie is analyzed by frames and the decrease with time of angular amplitude of the pendulum is found. The equation of motion is established for oscillations of the mobile body. The equation of motion, differential nonlinear, is integrated by Runge-Kutta method. Imposing the same damping both to model's solution and to theoretical model, the value of coefficient of rolling friction is obtained. The last part of the paper presents results for actual pairs of materials. The main advantage of the method is the fact that the dimensions of contact regions are small, of order a few millimeters, and thus is substantially reduced the possibility of variation of mechanical characteristic for the two surfaces.

1. Introduction

Three types of friction may occur between two bodies: sliding, spinning and rolling, the last happening only in higher pairs when the contact between the boundary surfaces of the contacting bodies is a concentrated (Hertzian) contact. In order to characterize the rolling friction, finding the rolling friction torque is required, [1-4]. For the general case this is an extremely difficult task due to complex motion from higher pair, on one side, and to the surfaces shape of contacting bodies, on the other side. To overcome these difficulties, bodies with simple geometries are involved, between which plane-parallel motion exists. The most usual pair of surfaces is the cylinder and plane pair [5]. When between the two bodies the plane-parallel motion is present, the motion can be studied in a plane normal to the axis of rotation, as the relative motion between a circle and a straight line. The higher pairs are irreversible. Thus, if the straight line is maintained immobile and the circle is mobile, the points from the circle describe a family of cycloids, and the pendulum is a cycloid pendulum [6].The method is applicable for studying materials with significant gradients of elastic characteristics, when regions implied in tests require small dimensions. A further example refers to biomechanics, for finding the tribological characteristics of biological tissues with complex geometry, like teeth or bones [7-10].

2. Pendulum dynamic equation

The contact between a circle and a straight line is considered. Maintaining fixed the circle, a basis circle C_b, and obliging the straight line Δ to roll without sliding over the fixed circle, Figure 1, the points from the straight line will describe a family of evolvents (common-the point on the circle, extended-the point outside the circle and shortened –the points inside the circle [11]). Based on Figure 1, the equation of motion of the pendulum is deduced under the hypothesis of pure rolling and proportionality between normal force $\boldsymbol{N}$ and the moment of rolling friction M_r.

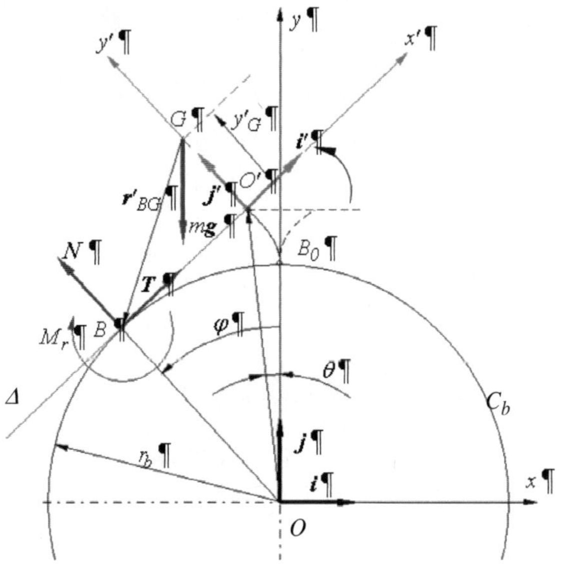

Figure 1. Principle of evolvent pendulum

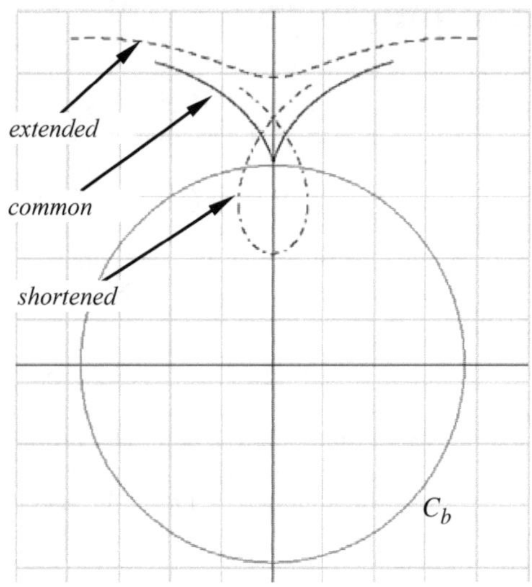

Figure 2. Common, extended and shortened involute generated using the same basic circle

It is considered that the contact between the straight line and the circle is initially in B_0 point. At a certain time, after the straight line rolled in trigonometric direction, the contact point is in B, whose position is précised by the polar angle φ. A ground frame of reference is chosen with the origin in the centre of the base circle and a moving coordinate system, attached to the straight line, with the origin in the initial contact point, the current position of the moving system is given by the polar radius $\overline{OO'}$ and polar angle φ. The pure rolling condition between the straight line Δ and the base circle C_b is obtained by perceiving that the length of the arch B_0B should be equal to the length of segment, BO' mathematically expressed as:

$$r_b\varphi = r_b \tan(\varphi - \theta) \tag{1}$$

Additionally:

$$\rho = r_b / \cos(\varphi - \theta) \tag{2}$$

Considering that the pendulum is symmetric, the centre of mass G will be positioned on the axis of symmetry and having in the attached reference frame the coordinates $(0, y'_G)$. The vector relation:

$$\boldsymbol{r}_G = \overline{OO'} + y'_G \boldsymbol{j}' \tag{3}$$

is projected on the axis of stationary coordinate system. The coordinates of the centre of mass of the pendulum are:

$$\begin{cases} x_G = -(y'_G + r_b)\sin\varphi + r_b\varphi\cos\varphi \\ y_G = -(y'_G + r_b)\cos\varphi + r_b\varphi\sin\varphi \end{cases} \tag{4}$$

The equations (4) describe the common, $y'_G = 0$, extended $y'_G > 0$ and shortened $y'_G < 0$ involute, as in Figure 2.

To figure out the equations of motion, the mass of the pendulum is symbolized by M and the moment of inertia with respect to a centric axis normal to the plane of motion is denoted by J_z. The only external force action upon the pendulum is its own weight. In the contact point there are acting:

- The normal reaction $\boldsymbol{N}$, on the direction of radius of contact point;
- The tangential reaction $\boldsymbol{T}$, tangent in the contact point;
- The rolling friction torque, $\boldsymbol{M_r}$ normal to the plane of motion and with the sense opposite to angular velocity $\omega = \dot{\varphi}$ of the pendulum;

The equations of motion of the pendulum are represented by the theorem of motion of the centre of mass:

$$M\boldsymbol{a}_G = \boldsymbol{G} + \boldsymbol{N} + \boldsymbol{T} \tag{5}$$

and the moment of momentum theorem written with respect to the centre of mass:

$$J_z\ddot{\varphi}\boldsymbol{k} = M_r\boldsymbol{k} + \boldsymbol{r}_{BG} \times (\boldsymbol{N} + \boldsymbol{T}) \tag{6}$$

The system of scalar equations generated by equations (5) and (6) is:

$$\begin{cases} M\ddot{x}_G = -N\sin\varphi + T\cos\varphi \\ M\ddot{y}_G = N\cos\varphi + T\sin\varphi - Mg \\ J_z\ddot{\varphi} = -Nr_b\varphi + Ty'_G - M_r \end{cases} \tag{7}$$

Under the hypothesis of pure rolling, the position of centre of mass of the pendulum is fully determined by the φ angle, equation (4), the unknowns of the system being the magnitude of normal reaction $\boldsymbol{N}$, the magnitude of tangential reaction $\boldsymbol{T}$, the magnitude of moment of friction M_r and the angle φ. An extra equation is required in order to make a compatible system (3). It is assumed that the friction torque is proportional to the normal force $\boldsymbol{N}$ by means of a coefficient of proportionality s_r (rolling friction coefficient):

$$M_r = -sN\,\mathrm{sgn}(\dot{\varphi}) \tag{8}$$

The differential equation of motion is obtained from equations (4), (7) and (8):

$$\ddot{\varphi} = -M\frac{\left[\left(1 - \frac{y'_G}{r_b}\right)\frac{s_r}{r_b}\mathrm{sgn}\,\dot{\varphi} + \varphi\right]r_b^2}{J_z + \left[\dot{\varphi}^2 + \frac{s_r}{r_b}\mathrm{sgn}\,\dot{\varphi} + \frac{y'^2_G}{r_b^2}\right]Mr_b^2}\dot{\varphi}^2 - Mg\frac{\left(\frac{s_r}{r_b}\mathrm{sgn}\,\dot{\varphi} + \varphi\right)\cos\varphi - \frac{y'_G}{r_b}\sin\varphi}{J_z + \left[\dot{\varphi}^2 + \frac{s_r}{r_b}\mathrm{sgn}\,\dot{\varphi} + \frac{y'^2_G}{r_b^2}\right]Mr} \tag{9}$$

With the known law of motion $\varphi = \varphi(t)$, the other two unknowns are found using the relations:

$$\begin{aligned} N &= M\,\ddot{y}_G\cos\varphi - M\ddot{x}_G\sin\varphi + Mg\cos\varphi \\ T &= M\,\ddot{y}_G\sin\varphi + M\ddot{x}_G\cos\varphi + Mg\sin\varphi \end{aligned} \tag{10}$$

The relations (10) are necessary to validate the condition:

IOP Conf. Series: Materials Science and Engineering **200** (2017) 012005 doi:10.1088/1757-899X/200/1/012005

$$-\mu N < T < \mu N \tag{11}$$

condition necessary for pure rolling existence.

3. Experimental device: description and function

The experimental set-up consists of a base 1 that allows for positioning the cylindrical body 2 which materialises the base circle. On top of it, a rod 3 is placed. This rod that materializes the mobile straight line has square cross-section. In order to be able to modify the position of the centre of mass y'_G of the pendulum, two threaded shafts 4 are attached at the ends of the rod, with identical bodies supported. To mention the position of the rod is sufficient to specify the angle that the rod makes with horizontal line. To avoid the occurrence of supplementary forces in the system, a small mirror 5 was fixed on top of the rod (Figure 4.a), inclined at and angle $\gamma = 45^o$ with respect to the rod, the theoretical point M where the mirror is fixed being situated at a distance $h = 30mm$ from the initial contact point, Figure 4b.

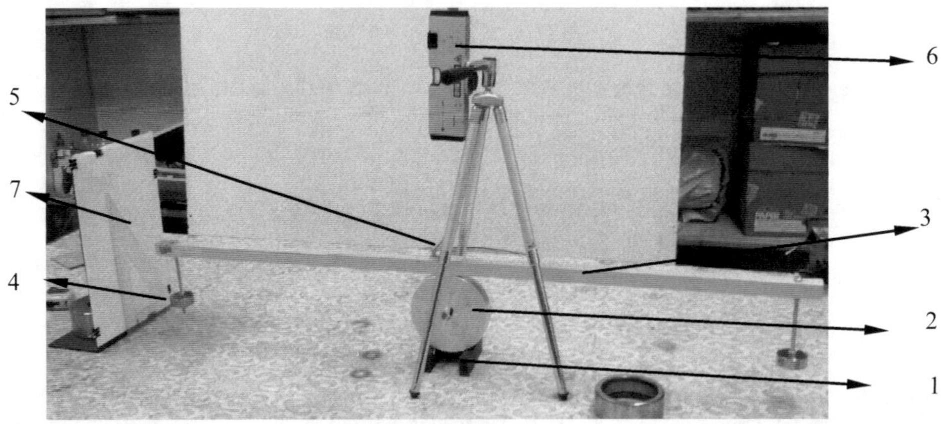

Figure 3. Experimental set-up

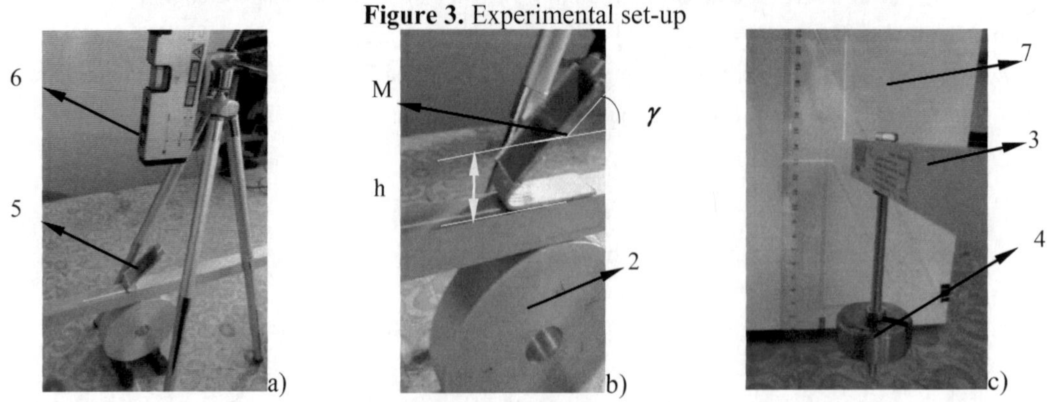

Figure 4. Details of experimental set-up

A laser beam, produced by a vertical laser level generator 6 is reflected by the mirror and the image is formed in the point E on a screen 7 (Figure 4.c) placed at d distance from the vertical in the initial contact point. The coordinates of mounting point M in the stationary frame are found using relations (4), where y'_G must be replaced by h. It is shown that the ordinate of image point from the screen is:

$$y_E(\varphi) = \frac{x_M(\varphi)}{\tan(\varphi+\gamma)} + y_M - \frac{d}{\tan[2(\varphi+\gamma)]} \tag{12}$$

The pendulum was modelled in a specialised software, Figure 5, in order to find the parameters J_z and y'_G. When the pendulum oscillates, the laser fascicle reflected by the mirror will produce a mobile spot on the screen, whose motion is video captured, Figure 6.

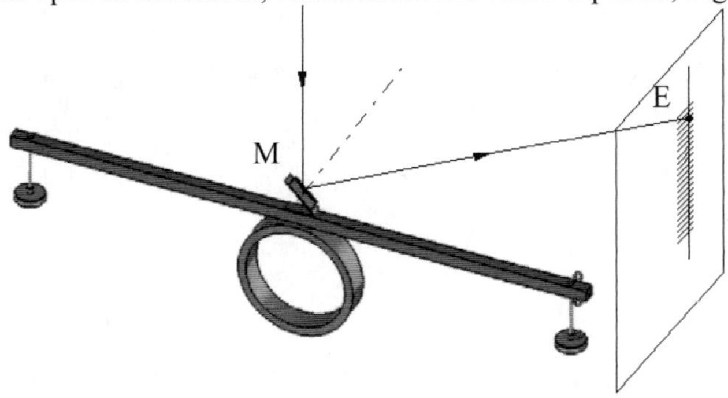

Figure 5. CAD modeling of pendulum for precise inertial characteristics finding (position of center of mass and moment of inertia)

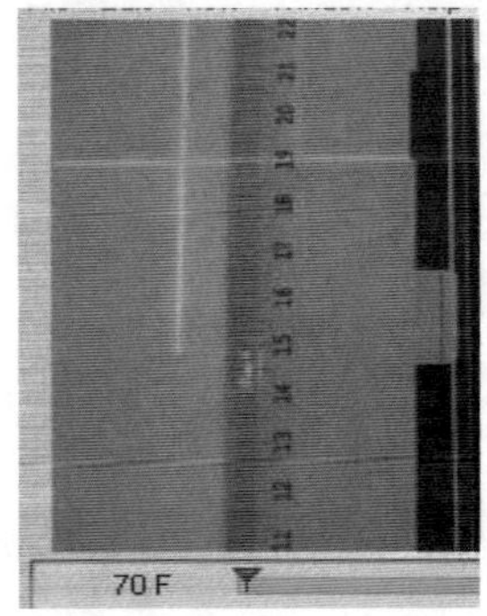

Figure 6. Movie analysis frame by frame

4. Experimental results. Discussions

In order to find the rolling friction coefficient between two materials, two bodies are necessary: a cylinder made from the first material and a small prism made from the second material. The prism is interposed between the cylinder and the plane inferior surface of the pendulum rod.

Figure 7. Bodies from different materials used in experiments

The pendulum is set into motion and let to oscillate free. Using a video camera, the motion of the reflected laser beam on the screen is filmed. The movie is afterwards analyzed frame by frame and the instants of maximum amplitude are identified. Using the relation (12), the angular amplitudes of the pendulum are found. Thus it is obtained the experimental amplitude decrease of involute pendulum. The differential equation of the pendulum is integrated, applying the Runge-Kuta methodology [12] with the initial conditions for the equation: $\varphi = \varphi_0$, $\dot{\varphi} = 0$ (where φ_0 is one of the experimental amplitudes). The obtained theoretical and experimental results are plotted, as in Figure 8.

By suitable choice of coefficient s_r in the theoretical model, the experimental data and theoretical ones will correspond. In figure 9 there are represented the experimental results and the theoretical interpolation curve. The interpolation was made by variation of the value of coefficient of rolling friction. For the obtained value of coefficient of rolling friction $s_r = 3.5\,\mu m$ (steel-steel pair), applying

IOP Conf. Series: Materials Science and Engineering **200** (2017) 012005 doi:10.1088/1757-899X/200/1/012005

equation (10), the forces N and T were found and it was validated the pure rolling condition (11). The time variation of the ratio of reaction forces T/N is represented in Figure 10 and it is noticed that this ratio is much smaller than the coefficient of sliding friction (for steel-steel, about 0.4 for dry sliding [13], [14]).

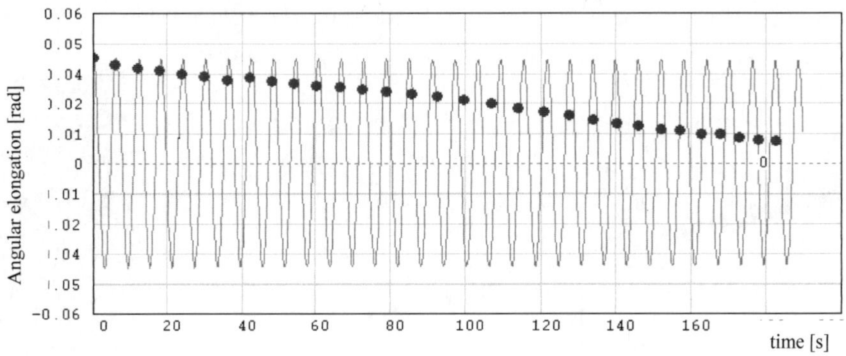

Figure 8. Unfitted experimental results and a solution of the proposed model

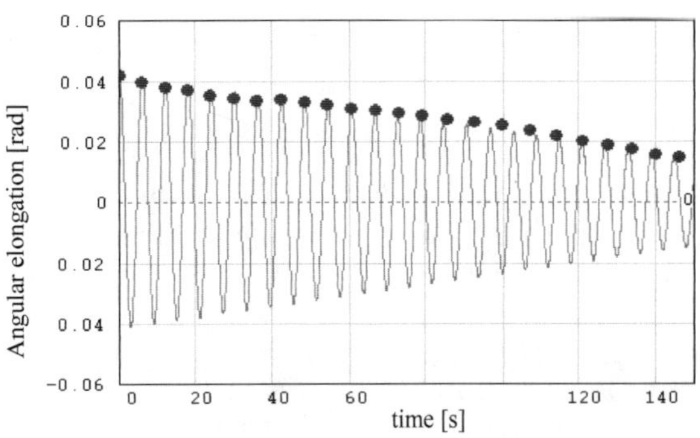

Figure 9. Fitted experimental data and theoretical model for steel–steel pair

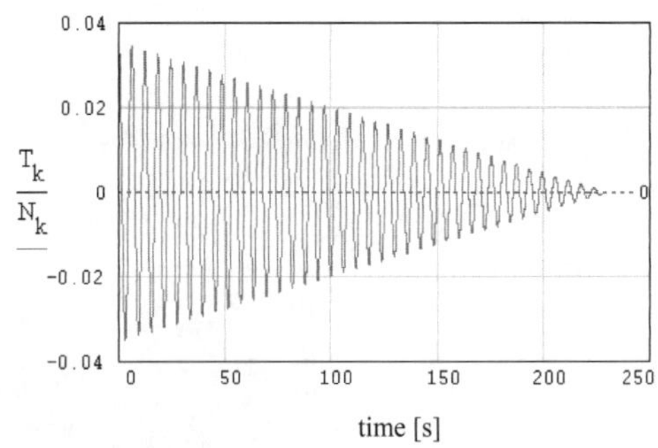

Figure 10. Pure rolling condition test for a steel–steel pair

For the pair rubber plate-steel cylinder, the results are analyzed similarly, as presented in Figures 11 and 12, with the mention that the value found for the coefficient of rolling friction is $s_r = 42\,\mu m$.

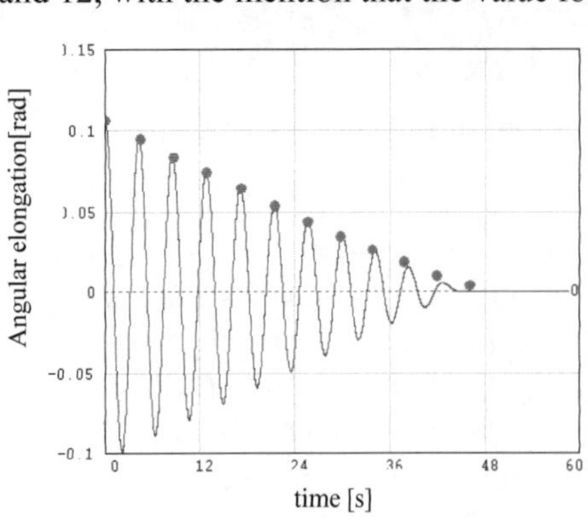

Figure 11. Fitted experimental data and theoretical model for steel–rubber pair

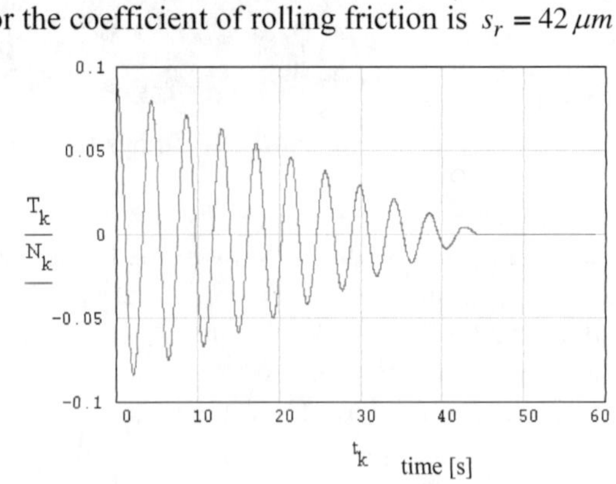

Figure 12. Pure rolling condition test for a steel–rubber pair

5. Conclusions

The paper presents a method for finding the coefficient of rolling friction using the principle of involute pendulum. The pendulum consists of a prismatic rod in contact with an immobile cylinder.

The oscillation non-linear differential equation of the pendulum is deduced for pure rolling condition and then integrated using a numerical procedure. For a chosen pair of materials to be tested, one has to be an immobile base cylinder and the other one, a plate attached to the pendulum rod.

The motion of the pendulum was analyzed via a non-contact method using a laser beam reflected by the rod and the equation of motion of the reflected spot was deduced. The experimental amplitudes of the rod, decreasing in time, were compared to the results generated by the theoretical model. The coincidence between experimental data and theoretical amplitudes was obtained by optimum choice of coefficient of rolling friction. After establishing the coefficient of rolling friction, the pure rolling condition was verified and the validation of the method is therefore ensured.

Two experimental cases are presented, for steel-steel and rubber-steel pair of materials. The values of the coefficient of rolling friction obtained are in good agreement to the ones from technical literature.

The main advantage of the method is the fact that the regions implied in tests have small dimensions and so it can be used for analyzing materials with significant gradients of elastic characteristics. Another advantage refers to the studies from biomechanics: for finding the tribological characteristics of biological tissues with complex geometry and/or reduced dimensions, these will play the part of the cylinder and the moving part (pendulum) will have a simpler geometry.

References

[1] Reynolds O 1876 On Rolling-Friction, *Phil. Trans. Royal. Soc. London* **166** 155-174
[2] Dowson D 1979 *History of tribology,* Longman
[3] Cherepanov G P 2014 Theory of rolling: solution of the Coulomb problem, *Journal of Applied Mechanics and Technical Physics* **55**(1) 182–189
[4] Stolarski T A and Tobe S 2001 *Rolling friction,* Tribology in Practice Series, Wiley
[5] Poschel T, Schwager T and Brilliantov N V 1999 Rolling friction of a hard cylinder on a viscous plane, *The European Physical Journal B* **10** 169-174
[6] Legeza V P 2012 Cycloidal Pendulum with a Rolling Cylinder, *Mechanics of Solids* **47**(4) 380–384
[7] Dowson D 1998 *Advances in medical tribology,* Mechanical Engineering Publications
[8] Zhongmin Jin 2014 *Computational Modelling of Biomechanics and Biotribology in the Musculoskeletal System,* Biomaterials and Tissues, Woodhead Publishing Series in Biomaterials **81**
[9] Alaci S, Ciornei M C, Filote C, Ciornei F C and Gradinariu M C 2016 Considerations upon applying tripodic coupling in artificial hip joint, *IOP Conf. Ser.: Mater. Sci. Eng.* **147** 012074
[10] Alaci S, Cerlincă D A, Ciornei F C, Filote C and Frunză G 2015 Experimental Highlight of Hysteresis Phenomenon in Rolling Contact, *Journal of Physics: Conference Series* **585** 012010 1-9
[11] Litvin F L and Fuentes A 2004 *Gear Geometry and Applied Theory,* 2nd Ed, Cambridge University Press
[12] Gear C W 1971 *Numerical Initial Value Problems in Ordinary Differential Equations,* Englewood Cliffs, NJ:Prentice-Hall
[13] Alaci S, Ciornei F C, Filote C, Romanu I C and Nastaca C P 2016 Determination of the coefficient of friction using spinning motion, *IOP Conf. Ser.: Mater. Sci. Eng.* **147** 012024
[14] Davis J R 2010 *Concise Metals Engineering Data Book,* 5th Ed, ASM International

IOP Conf. Series: Materials Science and Engineering **200** (2017) 012006 doi:10.1088/1757-899X/200/1/012006

Valuation of coefficient of rolling friction by the inclined plane method

F C Ciornei[1], S Alaci[1], V I Ciogole[1] and M C Ciornei[2]

[1]"Stefan cel Mare" University of Suceava, Mechanics and Technologies Department, Str. Universitatii no.13, 720229, Suceava, Romania
[2]"Carol Davila" University of Medicine and Pharmacy, Department 2 Physiology I, Bd. ErolilorSanitari no.8, Bucuresti, Romania

E-mail: florina@fim.usv.ro

Abstract. A major objective of tribological researches is characterisation of rolling friction, due to various cases encountered in classical engineering applications, like gear transmissions and cam mechanisms or more recent examples met in bioengineering and biomedical devices. A characteristic of these examples consists in reduced dimensions of the contact zones, theoretically zero, the relative motion occurring between the contact points being either sliding or rolling. A characteristic parameter for the rolling motion is the coefficient of rolling friction. The paper proposes a method for estimation of coefficient of rolling friction by studying the motion of a body of revolution on an inclined plane. Assuming the hypothesis that moment of rolling friction is proportional to the normal reaction force, the law of motion for the body on the inclined plane is found under the premise of pure rolling. It is reached the conclusion that there is an uniformly accelerated motion, and thus for a known plane slope, it is sufficient to find the time during which the body runs a certain distance, starting from motionless situation. To obtain accurate results assumes finding precisely the time of motion. The coefficient of rolling friction was estimated for several slopes of the inclined plane and it is concluded that with increased slope, the values of coefficient of rolling friction increase, fact that suggest that the proportionality between the rolling torque and normal load is valid only for domains of limited variations of normal load.

1. Introduction

The interaction between two elements of a system can be made either by conform contact, when the bodies get contact on extended surfaces, or by nonconform contact, when the boundary surfaces coming into contact are different and the contact is established on set of points of null area (Hertzian contacts). Unlike conform contact, when between the points of contact region always exists relative velocity and thus sliding is present, for nonconform contacts, besides the sliding possibility there also exists the possibility of pure rolling - which occurs when between the theoretical contact points there is no relative motion. Characterizing friction in the case of pure rolling assumes establishing the moment of rolling. In dynamics literature, rolling friction torque is characterized assuming the proportionality between the friction torque and normal reaction force, the proportionality factor being the coefficient of rolling friction. Recent researches based on theory of elasticity lead to more complex dependencies [1-3].

Bioengineering research require tribological characterization of actual materials, [4], [5] which exhibit elastic or viscoelastic properties - like biological tissues or for the ones designed to replace these, leading to energy losses due to internal friction phenomena [6], [7]. For pure rolling case, there are not losses due to friction [8] for rigid bodies; but for real materials like biological tissues, rolling is accompanied by hysteresis losses [9].

2. Theoretical considerations

In the present paper is considered the motion of a bearing ball on an inclined plane, assuming the proportionality between friction torque and normal force, Figure 1. The axis is chosen as the slope of the inclined plane and the motion of the ball is considered as plane motion, the position parameters of the ball are the displacement of the centre of mass of the ball $\boldsymbol{x}$ and the rotation angle about an axis normal to the plane of motion, φ. The external force acting upon the ball is the force of gravity $\boldsymbol{G}$. In the theoretical contact point C between the ball and the inclined plane the normal reaction force, N the friction force (parallel to the plane) T and the rolling friction torque M_r will occur.

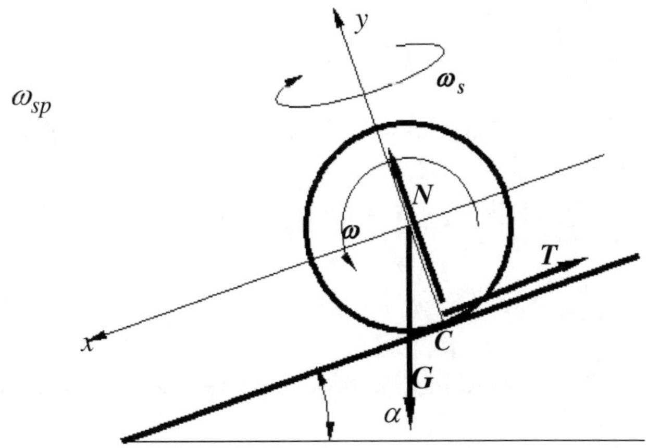

Figure 1. Motion of ball on an inclined plane

The equations of motion are represented by the center of mass theorem and the moment of momentum theorem, [10], written with respect to the centre of mass:

$$\begin{cases} m\ddot{x} = mg\sin\alpha - T \\ 0 = N - mg\cos\alpha \\ J_z\ddot{\varphi} = Tr - M_r \end{cases} \tag{1}$$

For a sphere made by homogenous material, the moment of inertia $\boldsymbol{J_z}$ with respect to the centre of mass, from equation (1), is:

$$J_z = \frac{2}{5}mr^2 \tag{2}$$

where m and r are the mass and the radius of the ball, respectively. The unknown of the system are x, φ, N, T and M_r. To ensure compatibility, two additional equations are required. The first one is a constitutive relation that accepts the proportionality between the rolling friction torque and the magnitude of normal compression force, the proportionality factor being the coefficient of rolling friction s_r. The friction torque should oppose to the rotation tendency of the ball:

$$M_r = s_r N \tag{3}$$

The second equation depends on the type of friction thought to occur in point C. If sliding exists between the points contacting in C, the friction force is known and found using the Coulomb law that considers the coefficient of dynamic friction μ_d:

$$T = \mu_d N \tag{4}$$

In the case when pure rolling is supposed to be present, then, the point C should be the instant centre of rotation (instantaneous velocity center) with permanent zero velocity. A consequence of this fact is the simple relation between the velocity of the centre of mass of the ball, $\dot{x}$ and the angular velocity $\omega = \dot{\varphi}$, explicitly:

$$\dot{x} = \dot{\varphi} r \tag{5}$$

In this case, the size of friction force is an unknown of the problem, the fulfillment of the following being required:

$$|T| < \mu_s N \tag{6}$$

were μ_s is the coefficient of static friction for the limiting case. The equations (1), (3) and (5) allow for finding the acceleration of the centre of mass of the ball:

$$a = \ddot{x} = g\cos\alpha\left(\tan\alpha - \frac{s_r}{r}\right) / \left(1 + \frac{J_z}{mr^2}\right). \tag{7}$$

3. Experimental device

The equation (7) suggests a straightforward method for establishing the coefficient of rolling friction, similar to the method for finding the coefficient of sliding friction. Practically, the ball is set at rest on the plane, initially in horizontal position, and afterwards the plane is slowly inclined until the instant the ball starts to move. Quite prior to this instant, the ball was immobile at the limit; thus, form equation (7) it results directly:

$$s_r = r\tan\alpha \tag{8}$$

The weakness of the method consist in the fact that for common values of coefficient of rolling friction $s_r = (10^{-5} \div 10^{-4})m$ and a radius of the bearing ball $r = 0.02m$, based on equation (8), it results that the angle needed for the start of ball motion is:

$$\alpha = 0.038^o \div 0.382^o \tag{9}$$

Consequently, accurate instruments for angle measurements are essential. Furthermore, the precision of achieving the horizontal position of the plane is directly affecting the value of coefficient of rolling friction. The present paper proposes the valuation of coefficient of rolling friction based on equation (7) by determining the acceleration with which the centre of mass of the ball moves.

A problem arising is how to eliminate the spin motion that the ball may possibly perform; in this case, the motion of the ball is no longer a planar one. To this purpose, instead of a single ball, two identical balls, connected via an aluminum rod are used, as shown in Figure 2. Another difficulty to be overcome is finding accurately the time of motion of the ball between two limits. Thus, in the launching position, the ball closes an electrical circuit. At the base of the plane there are two regions where on the surface of inclined plane, two conductive films are placed. When the rolling body reaches the base of the inclined plane, another electrical circuit closes. In this manner the instants of starting – opening of superior circuit, and ending –closing the second circuit, are precisely found. With known distance of running, d, under the hypothesis of constant friction torque, the motion of the body is uniformly accelerated with the acceleration $a = 2d/t^2$.

In Figure 3 is presented the experimental test-rig. Each electrical circuit also contains a light bulb, placed on the inclined plane which is made from a plate of glass 10 mm thick. At the midst of the inclined plane, on the inferior side, a wire was placed, at certain distance from launching line. Thus, the hypothesis of constant acceleration was verified, Figure 4.

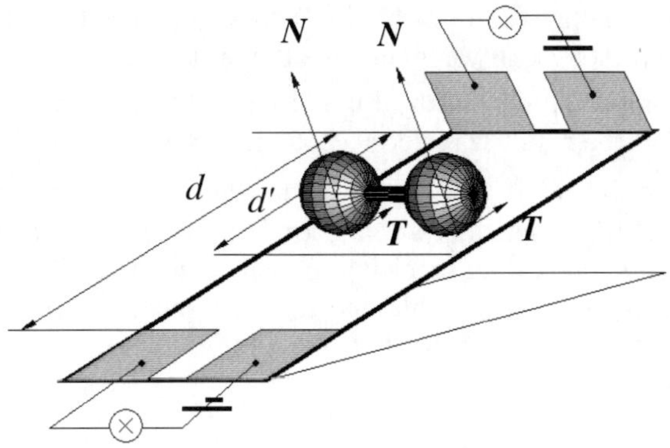

Figure 2. Principle of method

An experimental methodology implies video-capturing the region where the two light bulbs are placed. By analyzing the movie frame by frame, the time can be found. For a series of values of plane slope, three launches were made and then finding the accelerations and coefficients of rolling friction.

Figure 3. Experimental set-up before launching instant

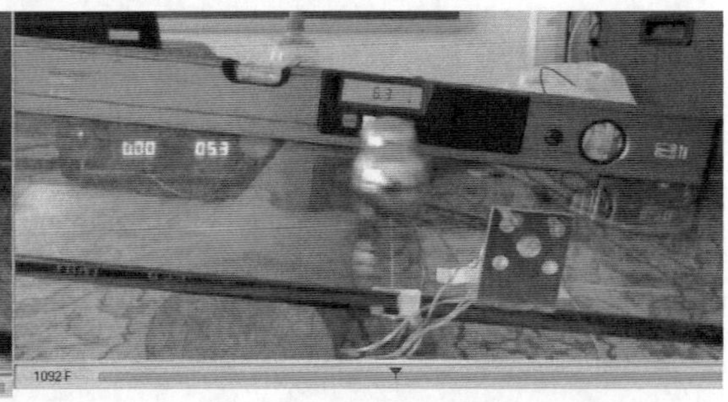

Figure 4. Identification of instants when the moving body passes above the wire position

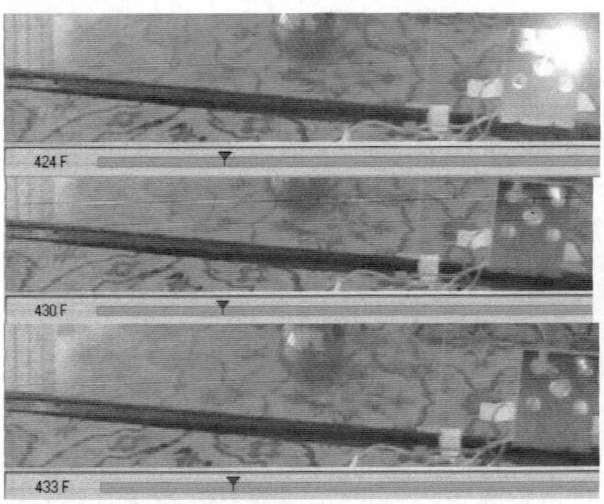

Figure 5. Final instant determined by optical method

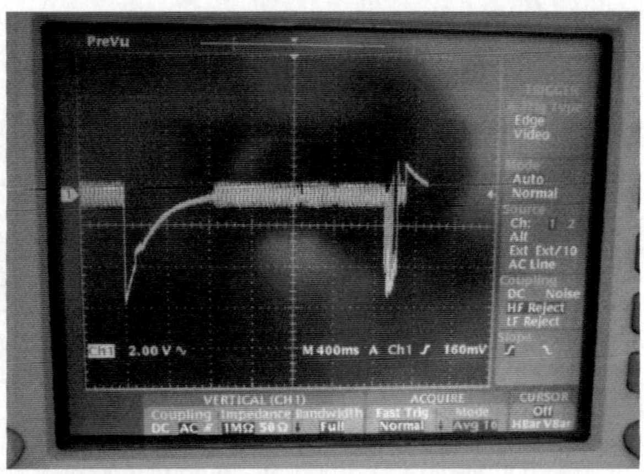

Figure 6. Identification of initial and final instants of the motion

An essential aspect concerns the precision of identifying the initial and final instants of the motion. The variation of current in the electric circuit was evidenced using an electronic oscilloscope, Figure 6. It can be noticed that the instant of contact opening is very well identifiable while the instant corresponding to passing the imposed distance is difficult to spot, either by oscilloscope or by optical system. In Figure 5 there are presented three frames from the movie, correlated to the vicinity of final instant. It can be observed that the light of the bulb faints during an interval of approximately ten frames, ($\approx 10 \cdot (1/120)s = 0.08s$). It was made an attempt to mount a block at the base of the inclined plane for obtaining an impact at the finish of the motion (which is an identical situation with the start, but in reverse sense) but the result was not acceptable since after the collision the contact opens and the circuit is closed only for a period equal to the duration of collision, $\approx 10^{-4} s$, duration imperceptible, [11], [12], both optically and electronically. In order to reduce and limit the effect of time measurement errors upon the precision of the results, it is required that the duration of experiments should be as great as possible and thus the absolute error, constant in this case, would affect as little as possible the relative error. It is required the construction of a test rig with greater length and smaller slope.

4. Experimental results and discussions

The values of the durations necessary for covering the two distances are presented in Figure 7, the points corresponding to the same slope of the inclined plane being practically overlapped. The repeatability of the results is shown in Figure 8; the accelerations obtained for different angles are the same for the three launchings under similar conditions and in the plot, the points overlie.

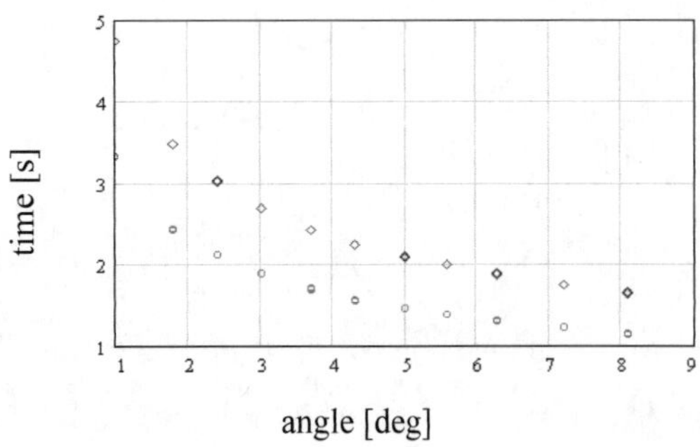

Figure 7. Coincidence of times corresponding to repeated tests

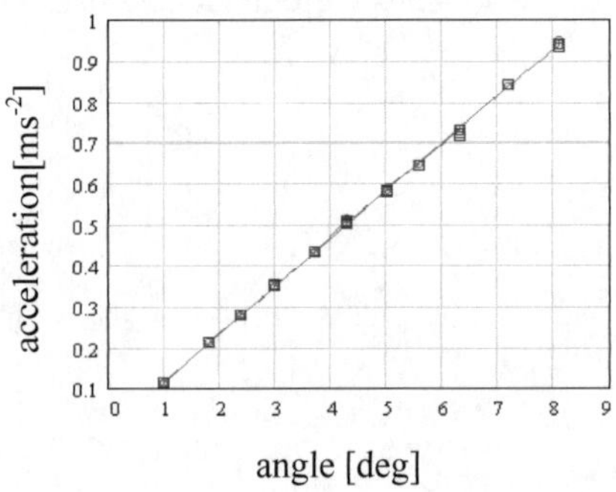

Figure 8. Validation of acceleration steadiness and repeatability of results

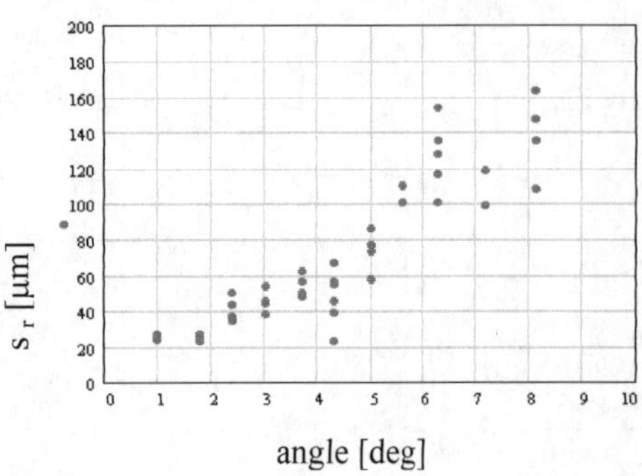

Figure 9. Dependence of experimental coefficient of rolling friction s_r upon slope in Cartesian coordinates

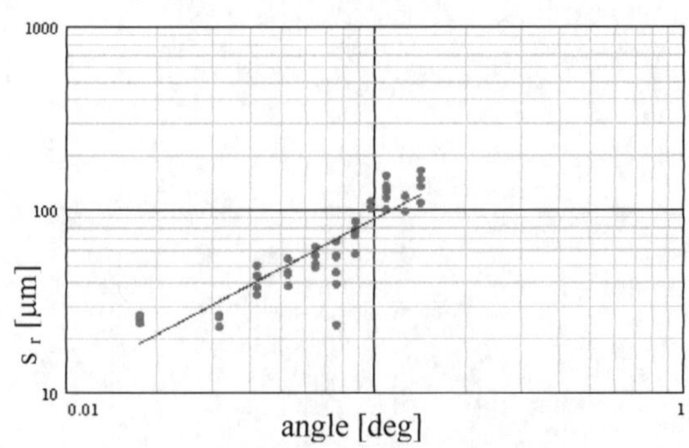

Figure 10. Dependence of experimental coefficient of rolling friction s_r upon slope in logarithmic coordinates

In Figure 9 there are presented the values the experimental dependencies of coefficient of rolling friction on the slope angle of the inclined plane and in Figure 10, using logarithmic coordinates, the data are interpolated by a line.

Two major conclusions result from these plots:

- it is infirmed the hypothesis of constant coefficient of rolling friction: the obtained values of s_r are increasing with the increasing slope of the plane;
- the experimental values of s_r are within the range $[24 \div 163 \mu m]$, values greater than the ones given in technical literature, [13], [14].

5. Conclusions

The paper presents a method and device proposed for evaluation of coefficient of rolling friction. The principle of the method consist in finding the acceleration, assumed constant, of a ball rolling over an inclined plane. Several difficulties had to be surpassed:

- elimination of spin motion, that was made using two identical balls connected by a rod;
- precise identification of the instants of start and end of motion; to this purpose, the mobile balls were introduced into electrical circuits;
- validation of hypothesis of constant acceleration: it was verified by measuring the periods in which two different distances were passed.

Though the time measurement errors influence considerably the accuracy of the experiments, the values of coefficient of rolling friction found by the present method are in good agreement with the ones given in literature. The main aspect underlined is the dependency of coefficient of rolling friction upon the slope of the inclined plane, in contradiction to a series of reference works where the hypothesis of constant value of s_r is stated.

References

[1] Cherepanov G P 2014 Theory of rolling: solution of the Coulomb problem, *Journal of Applied Mechanics and Technical Physics* **55**(1) 182–189

[2] Poschel T, Schwager T and Brilliantov N V 1999 Rolling friction of a hard cylinder on a viscous plane, *The European Physical Journal B* **10** 169-174

[3] Dzhilavdari Z and Riznookaya N N 2007 Features of Small Oscillations of Pendulum on Elastic Surface with Hysteresis Friction, *Journal of Friction and Wear* **28** (2) 145–153

[4] Mow V C, Ratcliffe A and Woo S L Y 1990 *Biomechanics of Diarthrodial Joints,* Vol. 1 Springer

[5] Martin R B, Burr D B, Sharkey N A and Fyhrie D P 2015 *Skeletal Tissue Mechanics*, 2Ed Springer

[6] Alaci S, Ciornei M C, Ciornei F C, Filote C and Amarandei D 2016 Method and device for dynamic modelling of rubbery materials applied to human soft tissues. Part I: determination of mechanical characteristics and dynamic model proposal, *IOP Conf. Ser.: Mater. Sci. Eng.* **161** 012056

[7] Alaci S, Ciornei M C, Ciornei F C, Filote C and Românu I C 2016 Method and device for dynamic modelling of rubbery materials applied to human soft tissues. Part II: device and experimental results, *IOP Conf. Ser.: Mater. Sci. Eng.* **161** 012057

[8] Voinea R, Voiculescu D and Ceausu V 1983 *Mecanica,* EDP Bucuresti

[9] Alaci S, Cerlincă D A, Ciornei F C, Filote C and Frunză G 2015 Experimental Highlight of Hysteresis Phenomenon in Rolling Contact, *J of Physics: Conf Series* **585** 012010 1-9

[10] Mangeron D and Irimiciuc N 1978 *Mecanica rigdelor cu aplicatii in inginerie. Volumul I. Mecanica rigidului,* Ed. Tehnica Bucuresti

[11] Alaci S, Ciornei F C and Filote C 2016 Some remarks upon collision between a dropping ball and a rotating disk, *Journal of the Balkan Tribological Association* **22**(2) 1560–1579

[12] Alaci S, Cerlincă D A, Ciornei F C, Filote C and Frunză G 2015 Method of Integration for Equation of Two Balls in Dumped Collision, *J of Physics: Conf Series* **585** 012008

[13] Oberg E 2012 *Machinery's Hanbook*, 29th Ed Industrial Press
[14] Stolarski T A and Tobe S 2001 *Rolling friction,* Tribology in Practice Series, Wiley

IOP Conf. Series: Materials Science and Engineering **200** (2017) 012007 doi:10.1088/1757-899X/200/1/012007

Establish the CNC machining strategy in relation with geometric complexity of the parts made from aluminum alloy extruded profile

Ș A Moldovan, M Cosma and V Năsui

Technical University of Cluj-Napoca - North University Center of Baia Mare, Romania

E-mail: moldovan.stefanadrian@gmail.com

Abstract. In this paper we present a technological problem encountered in the machining accuracy of the parts for aerospace made of aluminum alloy extruded profile with length up to 10 meters. Those parts have very tight tolerances and on milling process appear several factors that influence the repeatability of machining processes. Several factors must be considered when developing the machining process for a specific part, including: establishing the machining strategy in relation with piece geometric complexity, analysis of machined parts through coordinate measuring machine and statistical analysis, to determinate the proper machining strategy for obtaining parts in tolerance. Through several tests and recording all dimensions changes during the milling process, will be modified the machining strategy. By analysing the machining strategy at different lengths of extrusions and records of dimensions fluctuations along the processing chain has been created a proper machine strategy which will obtain a repeatability of the machining process.

1. Introduction

The application methodology for this research is to solve a technological problem encountered in the machining accuracy of the parts for aerospace industry made of aluminum alloy extruded profiles with length up to 10000 mm (Figure 1).

Those parts have very tight tolerances and on milling process appear several factors that influence the repeatability of machining processes [1].

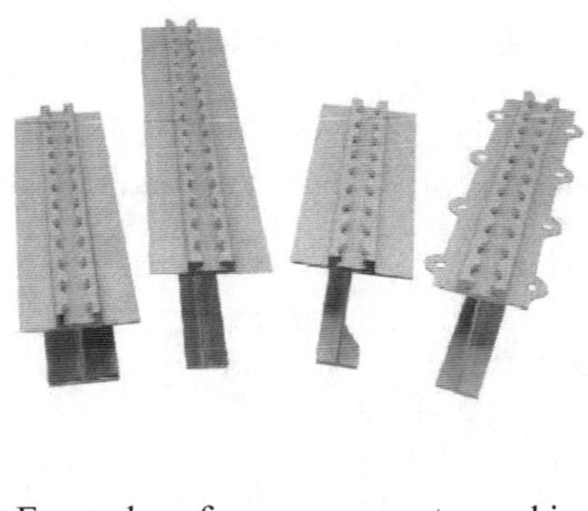

Figure 1. Examples of aerospace parts used in researches

In this research is an analysis of the machine-tools system [2] and the main contributions are: measurements and collecting geometrical deviation values from machined parts on X-axis direction, creating a solution based on CNC machining strategy to solve the deviations problem and achievement the experimental tests.

2. The main elements that are part of the machining process

2.1. CNC machine definition

The machining process is performed in a normal processing unit, without climate control and the CNC machine used is Handtmann PBZ NT (Figure 2).

Figure 2. Handtmann PBZ NT

The PBZ NT is a profile machining center features impressive maximum flexibility and great cost effectiveness in machining aluminum and steel.

A flexible clamping system for machining different profile cross sections and up to 90 tool slots allows for a wide range of applications up to 25 meters in length. An integrated sawing unit for spatial cuts completes the highly efficient PBZ NT machine concept [3].

2.2. Aluminum alloy extrusion

The research was conducted on extruded aluminum alloys work pieces used in aerospace industry, with a wide range of alloys and treatments (Figure 3).

With thin walls and the section represents a high resistance to shock and breakage, develop from advanced alloys tailored for specific applications to help find the right balance of strength, damage tolerance, and corrosion resistance.

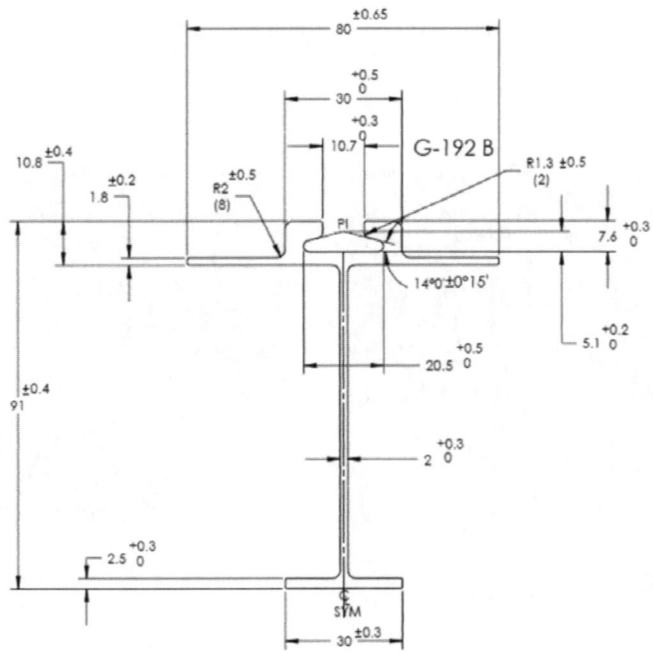

Figure 3. Extrusion tolerance

2.3. Parts complexity

Manufacturing process of those parts are define through: drilling holes with 19.7 diameter on middle side, holes 7.3 diameter on both sides of the raw material (Figure 4) and additional milling pockets from the front side (Figure 5). All holes have tolerance ±0.2 up to 10000 mm length. On the CNC machine tests we use extrusion with 4000 mm length [4].

Figure 4. Drilling top Holes

Figure 5. Front pockets

2.4. Problem detection

All machined parts specific for aerospace industry needs to be validated by CMM (coordinate measuring machine), placed in a room with temperature control environment.

These axes are orthogonal to each other in a typical three-dimensional coordinate system. Each axis has a scale system that indicates the location of that axis. The machine will read the input from the touch probe, as directed by the operator or programmer. The machine then uses the X, Y, Z coordinates of each of these points to determine size and position with micrometer precision typically [5].

Metris is an high quality and innovative metrology solutions to a wide range of industries (Figure 6). The geometrical deviation was determined based on holes position (Figure 7).

Figure 6. Metris CMM, 3 "bridge" used for parts control

Figure 7. Measuring holes position

3. CNC machining strategy

3.1. First machining approach

The CNC machining strategy was determinate in following way (Figure 8):

- First drilling the TOP holes, 19.7 mm diameter and 7.3 mm diameter.
- Second operation will machine the front pockets.

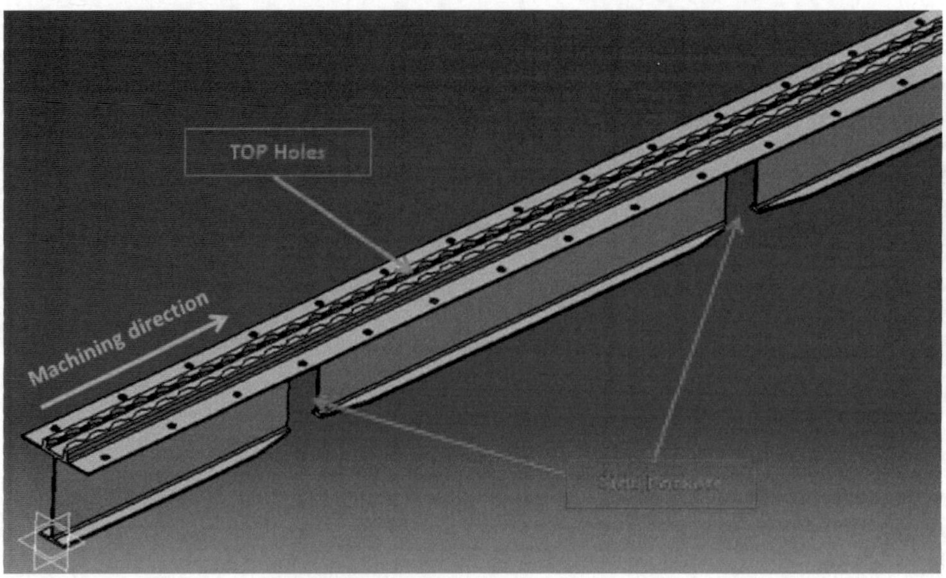

Figure 8. First machining approach

After processing six pieces with the first machining approach, they were measured in CMM, on Table 1 are represented the measurement results [6].

Table 1. Measurement results from first machining approach

X- axis position	X- axis value position (mm)	P1	P2	P3	p4	P5	P6
1	0	0	0,021	0	0,048	0	0,006
2	76	0,009	0,039	0,07	0,116	0,061	0,08
3	152	0,042	0,07	0,099	0,142	0,085	0,099
4	228	0,016	0,022	0,089	0,118	0,063	0,077
5	304	0,012	0,004	0,088	0,104	0,049	0,067
6	380	0,021	0,009	0,083	0,101	0,061	0,065
7	456	0,021	0,007	0,084	0,095	0,065	0,056
8	532	-0,025	-0,019	-0,099	-0,098	-0,064	-0,056
9	608	-0,064	-0,048	-0,128	-0,127	-0,095	-0,08
10	684	-0,031	-0,016	-0,092	-0,105	-0,057	-0,052
....	....	....	....	....	....	....	....
40	2964	-0,256	-0,263	-0,315	-0,349	-0,298	-0,378
41	3040	-0,22	-0,232	-0,291	-0,318	-0,274	-0,344
42	3116	-0,181	-0,207	-0,25	-0,284	-0,241	-0,328
43	3192	-0,18	-0,216	-0,245	-0,286	-0,238	-0,326
44	3268	-0,217	-0,235	-0,272	-0,32	-0,256	-0,338
45	3344	-0,22	-0,229	-0,279	-0,308	-0,253	-0,318
46	3420	-0,246	-0,255	-0,317	-0,34	-0,286	-0,343
47	3496	-0,248	-0,238	-0,318	-0,338	-0,284	-0,285
48	3572	-0,241	-0,242	-0,268	-0,292	-0,251	-0,298
49	3648	-0,235	-0,25	-0,244	-0,268	-0,253	-0,291
50	3724	-0,248	-0,255	-0,249	-0,27	-0,261	-0,283
51	3800	-0,262	-0,261	-0,244	-0,258	-0,26	-0,275
52	3876	-0,258	-0,267	-0,259	-0,245	-0,26	-0,268
53	3952	-0,261	-0,273	-0,268	-0,235	-0,259	-0,26

It is visible an improvement on the holes positions, there are in tolerance (Figure 9).

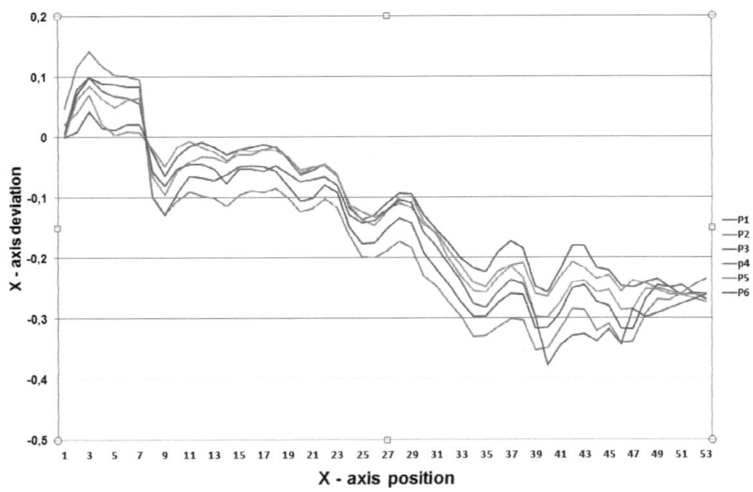

Figure 9. X-axis deviation chart from first machining approach

If the side pocket are machined first the residual stress from extrusion is eliminated and holes drilled after are in tolerance.

3.2. Second machining approach

The CNC machining strategy was changed with a new approach made in following way:

- First milling the front pockets.
- Second drilling the TOP holes, 19.7 mm diameter and 7.3 mm diameter.

Based on the next six pieces machined, we have the following CMM measurement results (Table 2).

Table 2. Measurement results from second machining approach

X- axis position	X- axis value position (mm)	P7	P8	P9	P10	P11	P12
1	0	0	0,019	0	-0.047	0,034	0,036
2	76	0,004	0,015	-0,007	0.010	0,035	0,027
3	152	0,028	0,019	-0,033	0.019	-0,009	0,017
4	228	0,02	0,018	-0,015	0.001	0	0,023
5	304	0,038	0,016	-0,039	-0.007	0,017	0,039
6	380	0,003	0,011	-0,045	-0.010	0,015	0,027
7	456	0,019	0,019	-0,038	-0.013	0,023	0,03
8	532	0,059	0,006	-0,002	-0.001	0,014	0,044
9	608	0,031	0,013	-0,007	0.035	0,005	0,022
10	684	0,059	0,018	0,033	0.011	0,011	0,026
………..	………..	………..	………..	………..	………..	………..	………..
40	2964	0,037	0,048	-0,024	0,046	0,007	0,026
41	3040	0,058	0,048	-0,004	0,013	0	0,012
42	3116	0,027	0,055	-0,009	0,014	0,015	0,006
43	3192	0,027	0,055	-0,004	-0,04	0,009	0,018
44	3268	0,073	0,052	0,018	-0,004	0,034	0,04
45	3344	0,06	0,052	0,003	-0,006	0,027	0,046
46	3420	0,032	0,062	-0,02	-0,033	0,032	0,041
47	3496	-0,009	0,069	-0,002	0,029	0,011	0,04
48	3572	0,034	0,061	-0,004	-0,008	0,02	0,036
49	3648	0,021	0,069	-0,003	0,036	0,023	0,038
50	3724	0,048	0,061	0,011	0,06	0,024	0,034
51	3800	0,008	0,07	-0,02	-0,016	0,069	0,048
52	3876	0,008	0,073	-0,039	-0,004	0,052	0,017
53	3952	0,031	0,069	-0,049	-0,008	0,063	0,008

It is visible an improvement on the holes positions, there are in tolerance. If the side pocket are machined first the residual stress from extrusion is eliminated and holes drilled after are in tolerance (Figure 10).

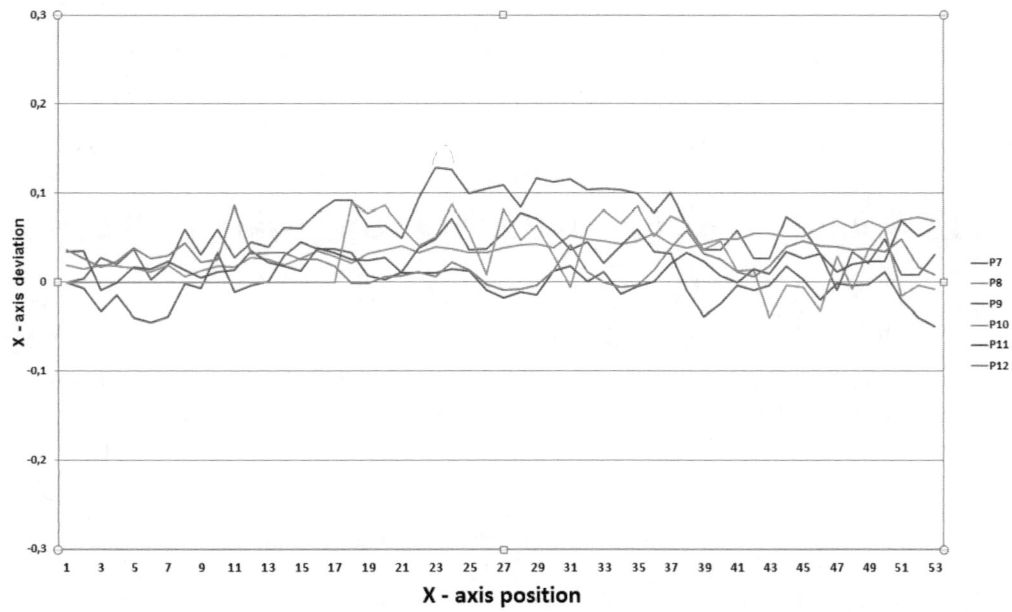

Figure 10. X-axis deviation chart from second machining approach

4. Conclusions

Experimental research bring a solution using different approach on CNC machining strategy, which is a major concern for processing repeatability of specific parts from aerospace industry. This research methodology improve the machining process control.

The original contribution for this research is: creating a solution based on CNC machining strategy to solve the deviations problem.

The experimental research conducted with the support of the machining department of SC Universal Alloy Corporation Europe SRL, Dumbrăviţa, (Research contract Nr.3657/09.02.2016).

References

[1] Năsui V 2000 *Bazele cercetării experimentale*, Editura Universităţii de Nord, Baia Mare
[2] Daraba D 2015 Analysis of the Elastic System of Machine-Tools before Remanufacturing, *Scientific Bulletin Serie C* **XXIX** (Fascicle Mechanics, Tribology, Machine Manufacturing Technology) 26-28
[3] ***2010 *Training Guide PBZ NT Ver.* 2010
[4] ***2016 https://www.handtmann.de/en/machining-centres/products/pbz-profile-machining-centres/pbz-nt
[5] ***2016 *Metris Bridge CMM User Guide*
[6] ***2016 http://www.universalalloy.com/Manufacturing-Capabilities.html

IOP Conf. Series: Materials Science and Engineering **200** (2017) 012008 doi:10.1088/1757-899X/200/1/012008

LabVIEW application for motion tracking using USB camera

R Rob, G O Tirian and M Panoiu

Politehnica University of Timisoara, Department of Electrical Engineering and Industrial Informatics, 5 Revolution Street, Hunedoara, 331128, Romania

E-mail: raluca.rob@fih.upt.ro

Abstract. The technical state of the contact line and also the additional equipment in electric rail transport is very important for realizing the repairing and maintenance of the contact line. During its functioning, the pantograph motion must stay in standard limits. Present paper proposes a LabVIEW application which is able to track in real time the motion of a laboratory pantograph and also to acquire the tracking images. An USB webcam connected to a computer acquires the desired images. The laboratory pantograph contains an automatic system which simulates the real motion. The tracking parameters are the horizontally motion (zigzag) and the vertically motion which can be studied in separate diagrams. The LabVIEW application requires appropriate tool-kits for vision development. Therefore the paper describes the subroutines that are especially programmed for real-time image acquisition and also for data processing.

1. Introduction

This paper presents the technologies for optical signal processing in motion tracking using an USB camera. For experimentally validation a laboratory pantograph stand is used. The laboratory pantograph is made on 1:2 scales and it is controlled by a pneumatic system. By functioning of this mechanism, the pantograph-contact wire assembly describes a trajectory in horizontal plan (zigzag) and another trajectory in vertical plan [1], [2]. A LabVIEW application is accomplished for optical tracking the pantograph motion and for acquiring the trajectory coordinates for statistics. Other paragraphs are indented (BodytextIndented style).

2. Principle formulation

In order to acquire images in real time, the system composed by USB camera and the LabVIEW application has to work properly. The minimum requirements for the USB camera are the followings: 1280x720 pixels, refresh rate 25 fps, autofocus system. Figure 1 presents the block diagram for the proposed system. The laboratory pantograph motion in vertically and horizontally plans are tracked by the USB camera. The Figure 2 depicts the acquiring images flow.

Configuring the image acquisition system implies to determine the minimum feature size in physical units such as millimeters and to determine the number of pixels to allocate to the minimum-sized feature. A very important thing is to arrange the lighting system in order to obtain high contrast for the features of interest and to suppress distracting background features that complicate subsequent software processing.

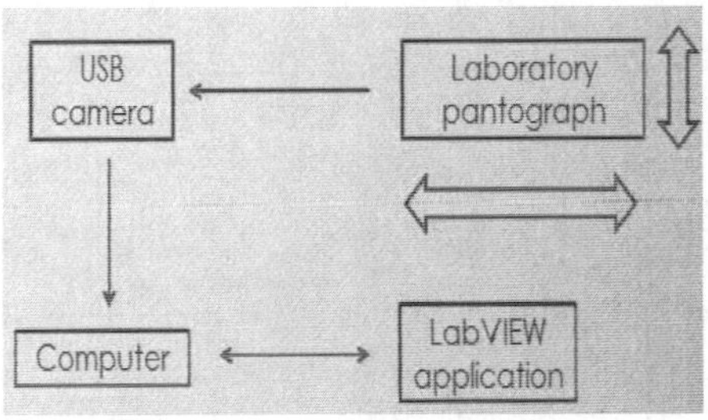

Figure 1. Block diagram for the proposed system

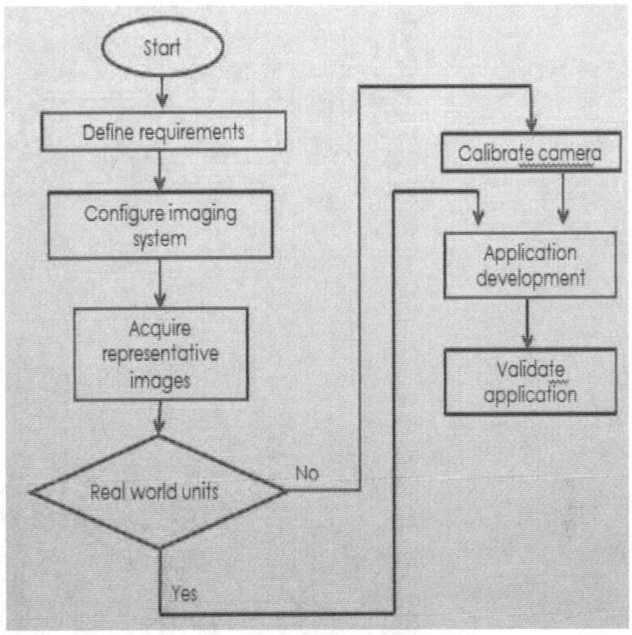

Figure 2. Acquiring images flow

Present application is accomplished using RGB colour images, but for a better efficiency, the grayscale images are preferred. Colour images tend to slow down the display, especially at higher resolutions.

The proposed solution for optically tracking the pantograph motion presumes to design a LabVIEW application which uses the Colour pattern matching technology for checking the presence of a template in the entire colour image or in a region of interest [3], [4].

Two colour markers are proposed to be programmed on the pantograph image: the green marker for tracking the horizontally motion (zigzag) and the orange marker for tracking the vertically motion (Figure 3). These markers will have the same motion as the contact line and the pantograph. The application is able to save into .txt files the coordinates of the motion trajectories.

For acquiring imaging in real-time and also for inline processing, the special toolkits are necessary: *Vision Acquisition Software* and *Vision Development Module Run-Time.*

3. Description of the application for optically tracking

The logic scheme for the proposed tracking method is presented in Figure 4.

Vision Acquisition Software once installed, the main application is able to call the *Vision Aquisition.vi* subroutine which can acquire, save and display in real time the images from the camera connected to the computer.

The configuration of this subroutine (Figure 5) presumes passing of some steps in which the acquisition method is selected: finite or continuous, with in-line or post-processing the images. The configuration implies the selection of the parameters along the image: resolution (implies the refresh rate), brightness, contrast, sharpness, gamma, etc.

These elements can be introduced into the main program as variables, so they can be modified in real-time.

Vision Assistant.vi subroutine permits to process the images acquired by the camera, using a large variety of programming technologies. For measuring the vertically and horizontally motion, the *Colour pattern matching* technology is used.

Figure 3. Proposed solution using *Colour pattern matching* technology

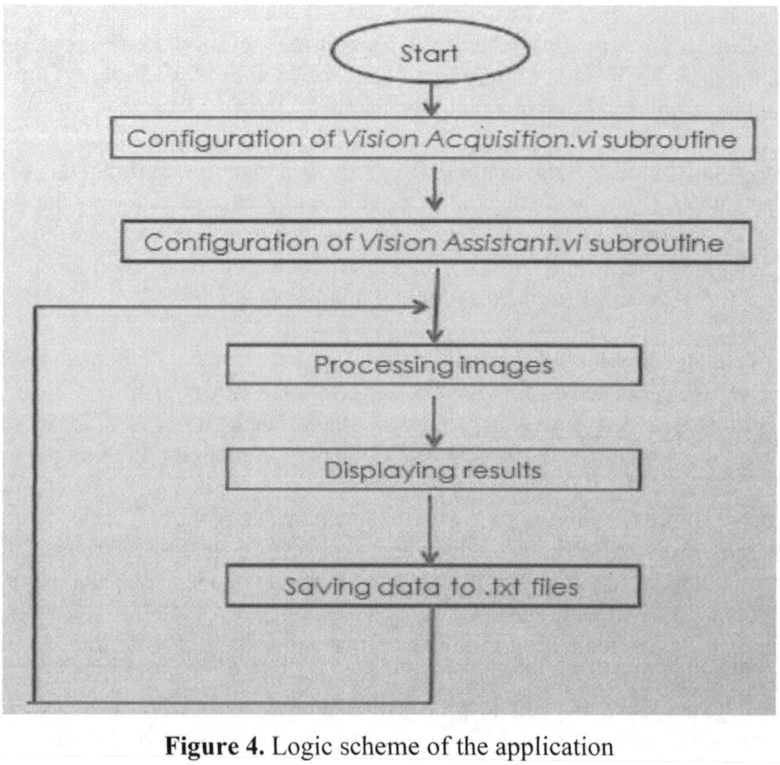

Figure 4. Logic scheme of the application

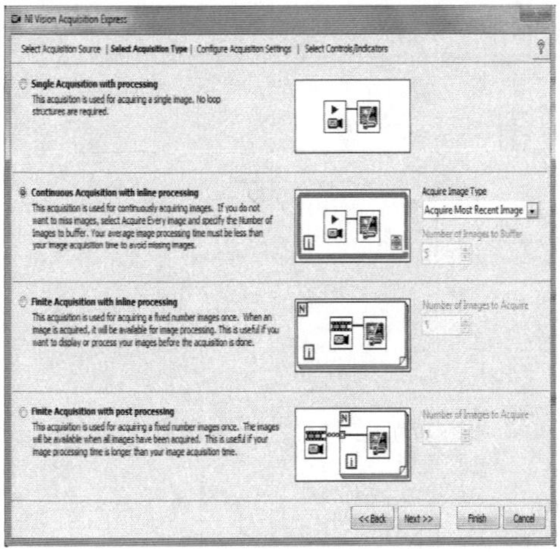

Figure 5. Configuration of *Vision Acquisition.vi*

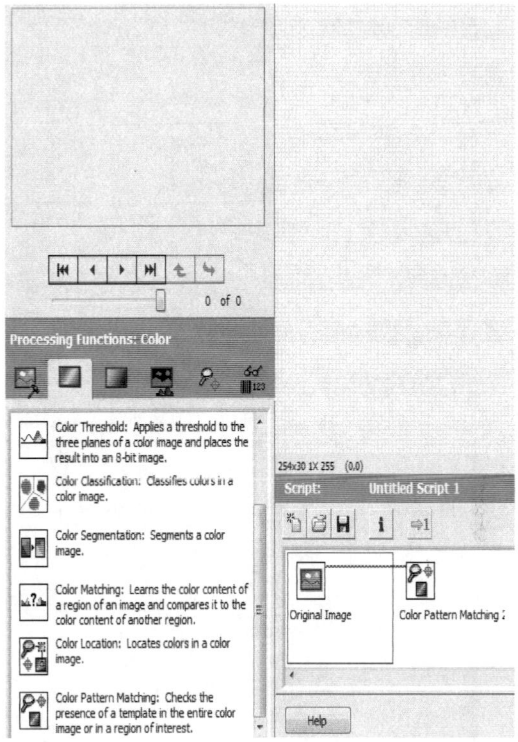

Figure 6. Configuration of *Vision Assistant.vi*

The application will follow the motion of a colour pattern, saving into a .txt file the coordinates of the chosen pattern centre.

These coordinates are saved in order to be used for a statistical purpose. The refresh rate is essential, because it will determine the reconstruction of the elements trajectories [5], [6]. For an image with continuity effect, a refresh rate of minimum 25 fps is necessary.

The attributes which can be modified using this technology are the followings: the number of matches, minimum score, colour sensitivity, searching strategy, the possibility of searching rotated patterns, etc. (Figure 6).

The source code of the main application is presented in Figure 7. The continuous acquisition with inline processing is permitted using a *While* loop. Due to the fact that two elements must be tracked (using green and orange markers), a series of two *Vision Assistant.vi* subroutines must be configured. The information containing the templates which must be optically tracked flows into a cluster form. The templates will be followed at each iteration of the main structure by two markers (green and orange). The coordinates of the markers centres are saved into .txt documents. The subroutines which make possible the saving information into .txt files are presented in Figure 8.

The coordinates of the markers centres form a vector which is written into a text file together with the saving folder path and the file name.

A series of subroutines: *Create File.vi, Write to Text File.vi, Close File.vi* and *Simple Error Handler.vi* serve to create and save the trajectory information into .txt file.

In order to accomplish a graphical variation of the vertically and horizontally motion of the pantograph, another LabVIEW application is designed.

The source code of this application is presented in Figure 9. Figure 10 presents the variation of the pantograph motion in vertically and also in the horizontally plan.

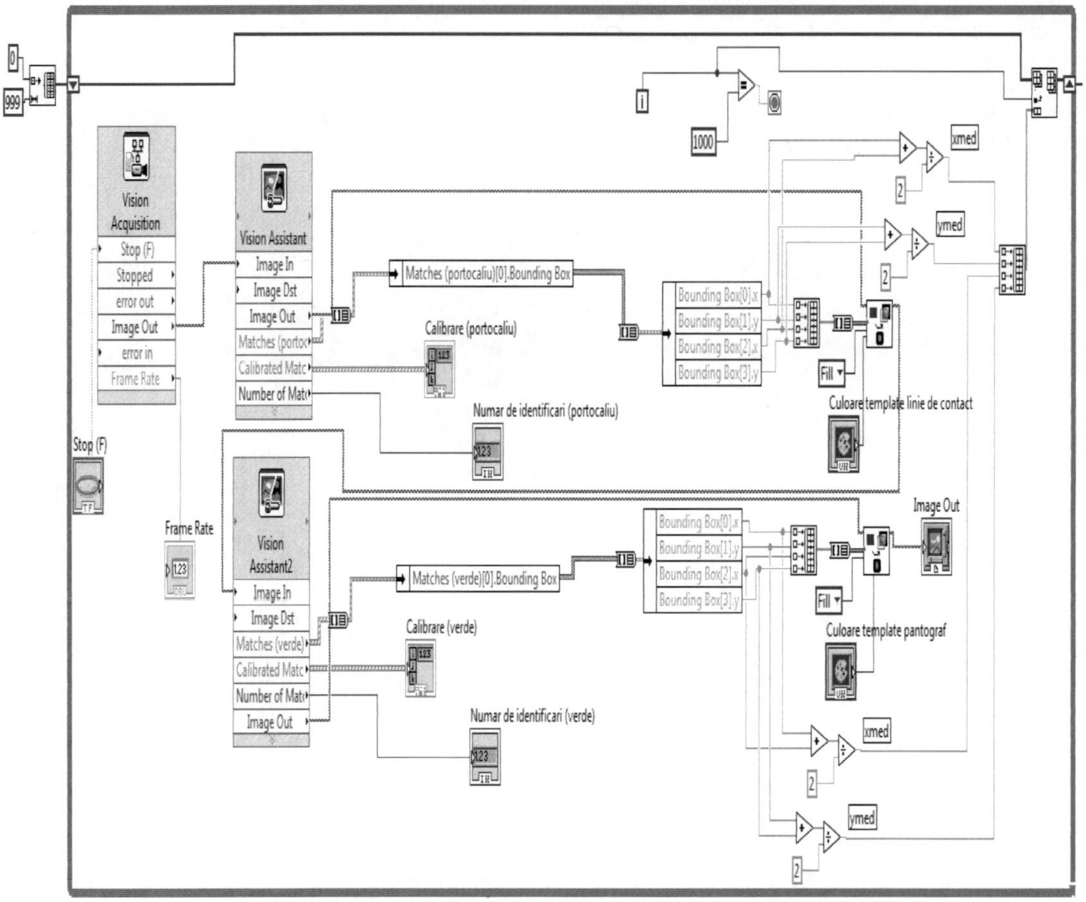

Figure 7. Source code of the main application

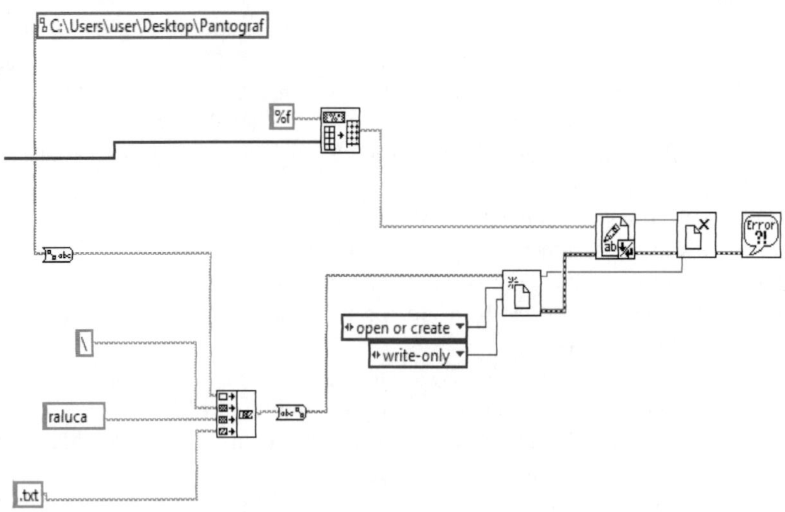

Figure 8. Source code for the saving information into .txt files

IOP Conf. Series: Materials Science and Engineering **200** (2017) 012008 doi:10.1088/1757-899X/200/1/012008

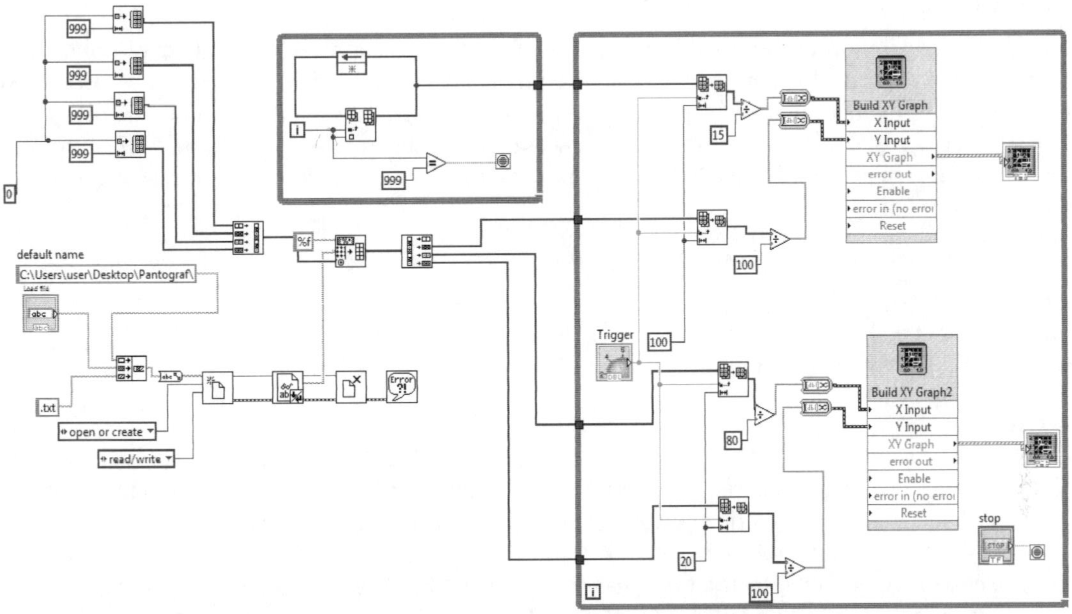

Figure 9. Source code of the coordinate's variation application

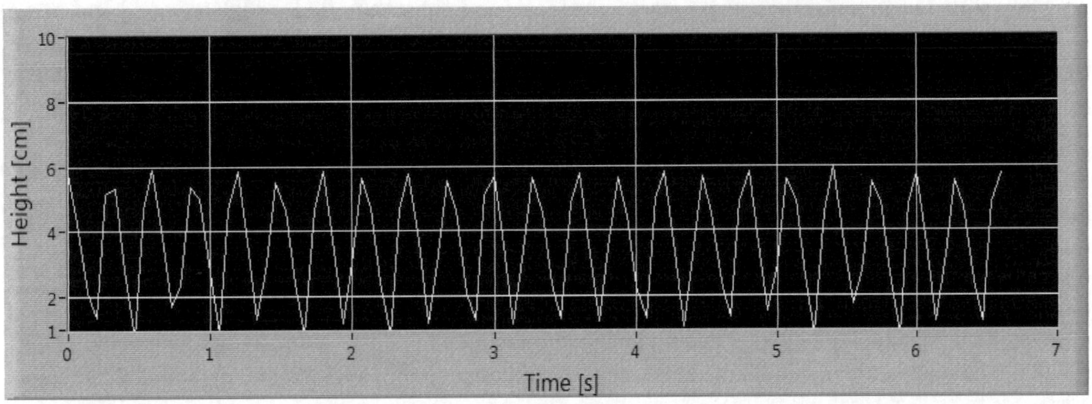

(a)

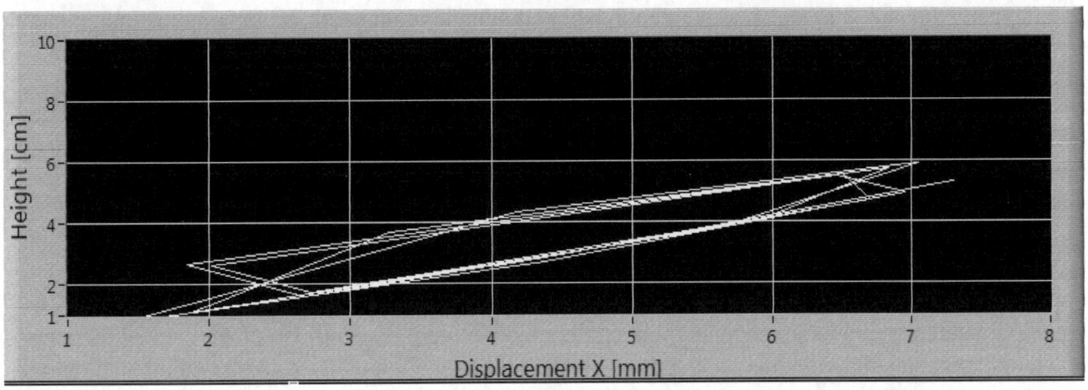

(b)

Figure 10. (a) Pantograph motion (vertically plan); (b) Contact wire motion (horizontally plan)

4. Conclusions

The optically tracking object represents an actual method for measuring distances between objects in motion without requiring of any other sensing devices. From Figure 10 (a) it can be observed an amplitude of approximate 4cm in pantograph vertically motion. The zigzag motion (b) follows an elliptical trajectory. It can also be observed the low quality of this trajectory due to the camera refresh rate (15 fps). The represented trajectory can be improved by using a high definition imaging from an appropriated camera.

Acknowledgement
This work was supported by a grant of the Romanian National Authority for Scientific Research and Innovation, CNCS – UEFISCDI, project number PN-II-RU-TE-2014-4-1788.

References

[1] Rusu-Anghel S, Miklos C, Averseng J and Demian D 2011 *Control System for Catenary-Pantograph Dynamic Interaction Force*, Railway infrastructure development for an efficient&green transport, Railway Pro Science&Technology, pp 36-40

[2] Rusu-Anghel S and Gherman L 2011 Using the Matlab-Simulink simulation for operating active filters ordered by digital signal processor (DSP), *Annals of the Faculty of Engineering Hunedoara* **IX**(4) 225-230

[3] Klinger T 2003 *Image processing with LabVIEW and IMAQ Vision*, Prentice Hall

[4] Smith L N and Smith M 2005 Automatic machine vision calibration using statistical and neural network methods, *Vision and Image Computing* **23** 887-899

[5] Posada-Gómez R, Sandoval-González O and Martínez Sibaja A *Digital Image Processing Using LabView,* Practical Applications and Solutions Using LabVIEW™ Software, InTech, pp 297-316

[6] Cuntan C D, Baciu I and Osaci M 2015 Studies on the Necessity to Integrate the FPGA (Field Programmable Gate Array) Circuits in the Digital Electronics Lab Didactic Activity, *International Journal of Modern Education and Computer Science* **7**(6) 9-15

Composites materials for friction and braking application

A L Crăciun, C Pinca-Bretotean, C Birtok-Băneasă and A Josan

Politehnica University of Timisoara, Department of Engineering and Management, 5 Revolution Street, Hunedoara, 331128, Romania

E-mail: camelia.bretotean@fih.upt.ro

Abstract. The brake pads are an important component in the braking system of automotive. Materials used for brake pads should have stable and reliable frictional and wear properties under varying conditions of load, velocity, temperature and high durability. These factors must be satisfied simultaneously which makes it difficult to select effective brake pads material. The paper presents the results of the study for characterisation of the friction product used for automotive brake pads. In the study it was developed four frictional composites by using different percentages of coconut fibres (0%, 5%, 10%, 15%) reinforcement in aluminium matrix. The new composites tested in the laboratory, modelling appropriate percentage ratio between matrix and reinforcement volume and can be obtained with low density, high hardness properties, good thermal stability, higher ability to hold the compressive force and have a stable friction coefficient. These characteristics make them useful in automotive industry.

1. Introduction

Braking is the process of partially or completely decrease traveling speed of a vehicle. During braking, a part of the kinetic energy gained by the vehicle turns into heat through friction. Other part is consumed for overcoming resistance to rolling and air resistance, which opposing the movement. The large amount of heat that is generated during braking contributes to the worsening of the vehicle braking qualities, hastens wear and friction linings and other brake system components. In recent years, the continued growth of the dynamic qualities of the vehicle and traffic, stressed the importance of braking systems, [1].

A braking system must have the following qualities: effectiveness, stability, fidelity and comfort. The effectiveness of the braking system ensures enhancement of the performance speed of the vehicles. A high efficiency of the braking system leads to obtaining higher speeds, in which case the vehicle's performances are elevated, [2]. Stability of the braking system, involves conservation of braking qualities of the vehicle. This is achieved if the braking force applied is maintained constantly in all operating conditions, [2]. In the case of brakes with friction, this depends on the stability of the friction coefficient of the brake pads. In normal use, the brake thermal regime should not exceed 300^0C, to ensure this constant friction. In this scope, the brake pads must ensure evacuation of heat that occurs during braking, because in the event of excessive heating, noises and vibrations, the structure of material is change and the friction coefficient is reduced. All these leading to lower efficiency and decrease traffic safety, [3]. This is the reason why special attention is paid to the design and manufacture of braking systems components. The brake pads are an important component in the braking system of automotive. These are of different types, suitable for different types of automotive and engines. Brake pads are designed for: friction stability, durability, minimization of noise and vibration. The typology of the brake pads depends on the material which they are made. Materials used for brake pads should have stable and reliable frictional and wear properties under varying conditions of load, velocity, temperature and high durability, [4]. These factors must be satisfied simultaneously which makes it difficult to select effective brake pads material, [5].

Range of materials used in the manufacture of brake pads varies from asbestos formulations to semi metallic, organic and ceramics. In the past, asbestos was a favorite material in the manufacture of brake pads because it has a good ability to dissipate heat, but it has proved to be harmful to human health, [2], [3]. Currently, the brake pads are made from mineral fibers, aramid fibers, cellulose, glass fiber, steel fibers and copper, [6], [7]. Each material has its advantages and disadvantages related to: environmental conditions, wear, noise, braking capacity, but the final properties of each type of brake pads are influenced by the abrasive material. In technical literature are some types of brake pads, depending on the type of material that comes into contact with the brake disc: semi metallic, non-asbestos organic, low-carbon non-asbestos organic and ceramic.

Semi-metallic brake pads have metal content in percentage of 30-65%, organic compounds and resins for connection. They have a high thermal conductivity and high durability, but presents a loud noise and producing a high wear of the brake discs. Non-asbestos organic brake pads are composed of fiberglass, carbon, rubber, Kevlar and high temperature resins. They are quieter, but wear faster and produce more dust. Non-asbestos organic brake pads with a low metal content are made with a low metal content, to provide 10-30%, better heat transfer and better braking. They produce a large amount of dust and high levels of noises. Ceramic brake pads are made of ceramic fibre, non-ferrous materials, bonding agents and sometimes small amounts of metal. They have a high heat transfer capacity, high durability, produce less dust, have low noise and reduced wear of the brake disc. They are only used in light vehicles but not recommended for medium and heavy vehicles. On the other side, their price is higher than other brake pads.

Regarding the manufacture of brake pads, literature indicates that there are government rules that restrict the use of materials in the manufacture of brake pads, [2]. Recently, more research has been focused on the non-asbestos organic fibre reinforced metallic friction composites, [3].

The aim of this paper is to develop new organic material with low metal content for automotive brake pads without any harmful effect. In this sense, were prepared four different laboratory formulations with varying percentage of coconut fiber, friction modifiers, abrasive material and solid lubricant using powder metallurgy technique. The properties examined are density, porosity, hardness, mechanical properties, loss of weight, microstructural analysis and friction coefficient.

2. Technical requirements

2.1. Experimental procedures for formulation the samples

The new composite for brake pads was developed through the process beginning with the selection of raw materials, weighing, mixing, compacting and sintering, [8].

There are four formulations with different composition of coconut fibre content. The fabrication of composites containing seven ingredients was based on keeping the composition of five ingredients (around 75%) constant and varying two ingredients, namely aluminium and coconut fibre (around 25 %), Table 1, [8], [9].

In establish of the recipes it was considered the study realized in paper [2] referring to automotive brake pad formulations. Similar recipes were used by authors in paper [3]. In present paper, the difference lies of: adding hexametiltetramine, various proportions of components and in replacement of aluminium oxide with titanium oxide. We considered this approach because a small change of the composites recipe and manufacturing technology can improve the characteristics of the final product.

All materials were prepared in powder form, [3]. Graphite was added to minimize fracture in the mixing process. In this investigation the coconut fibre was used as a filler material, [3]. The method used in fabrication of new developed composite was powder metallurgy technique. The prime reason for using the powder metallurgy process is the possibility of obtaining uniform parts and reducing tedious and expensive machining processes, [3].

Table 1. The recipes used in the product of composite brake pads material, [8], [9]

Samples	Aluminium (%)	Hexametil tetramine (%)	Graphite (%)	Zirconium oxide (%)	Silicon carbide (%)	Titanium oxide (%)	Phenolic resin (%)	Coconut fibre (%)
C1	25	6	4	2	10	13	40	0
C2	20	6	4	2	10	13	40	5
C3	15	6	4	2	10	13	40	10
C4	10	6	4	2	10	13	40	15

The ingredients were mixed in a drum mixing machine, type PM 100, to ensure the macroscopic homogeneity, using a chopper speed of 1500 rpm. First, aluminium and titanium oxide were churned together for 6 minutes. Zirconium oxide and silicon carbide were switched on at 120 rpm. After this, graphite and coconut fibre were added and mixed for 16 minutes. Then phenolic resin was added and the mixer was run for 5 minutes. The mixing sequence and time of mixing of each component led to proper uniformity in the mixture. If mixing time is low, proper homogeneity cannot be achieved. If it is too high, it does not improve the homogeneity further. Then, the materials were introduced in a mould presented in Figure 1. The settlement order in the mould is: an aluminium foil layer at the bottom of the mould, a layer of graphite on the aluminium foil, introduce the mixture into the mould, the piston is wrapped in aluminium foil, after that the mould with the mixture were pressed at the temperature room, finally the mould is insert in the oven to the temperature of 200^0 C, [10].

Figure 1. The mould for making samples, [10]

Figure 2. Finished composite samples with variable addition of coconut fibre

After 15 minutes, a compaction was performed using a hydraulic press machine under a force of 0.5 KN. After that, the material was sintered in an oven at a temperature of 200°C for an hours, after which cooling was carried out in air for 12 hours. Good consistency of the sample is due by heating-cooling regime. Extracting the samples from the mould was done without sticking problems, which is solved by using aluminium foil layer graphite. All samples are appropriate in appearance, shape and consistency. The surfaces of the samples were then polished with a grinding wheel to attain the desired thickness and smooth surface. The finished composite samples were shown in Figure 2.

In order to characterize the new composite material, were achieved tests to determine density, porosity, hardness, thermo-gravimetric analysis, mechanical testing, and microstructure. The samples for each test were carried out by cutting in concordance with actual standards.

2.2. Characterization of the brake pad material

A density measurement test has been carried out on a laboratory scale to examine the density of the new material obtained. The true density was determined by weighing each sample on a digital weighing machine and measuring their volume by liquid displacement method. The specific gravity formula is, [3]:

$$\rho = \frac{m}{V} \tag{1}$$

The results shown in Table 2 and represents the average density of five readings for each formulation.

Table 2. Density, porosity and hardness for new developed brake pad materials

Samples	Density (g cm^{-3})	Porosity (%)	Hardness HRS
C1	2.59	21.23	55.7
C2	2.42	15.63	69.3
C3	2.38	16.60	62.6
C4	2.29	23.43	41.7

Composite C4 with 15 % coconut fibre content shows lower density compared with other three formulations. The density of composites which have more coconut fibre shows lower density than formulation of C1 without coconut fibre. The high density of C1 is explained by the greater amount of aluminium in the recipe. Comparing the densities of the materials obtained in this paper with the paper [3] it was observed higher densities. This is explained by the fact that the chosen recipe consists of titanium oxide which is denser than aluminium oxide. On the other hand, the densities obtained are much lower than in the case of metallic and semi-metalic pads.

The function of porosity is to absorb energy and heat. Theoretically, lower porosity will result in higher friction coefficient and wear rate, [3]. The paper [11] presented that increasing porosity by more than 10% could reduce the brake noise. On the other side, brake pad should have a certain amount of porosity to minimize the effect of water and oil on the friction coefficient, [3]. Porosity test was performed in accordance with Japanese Standard JIS D 4418: 1996, [3]. The specimen was cut to dimensions: 25 mm x 25 mm x 7 mm, then left in a desiccator for 24 hours at room temperature and cooled to room temperature in desiccator, Figure 3. The sample was weighed to the nearest 1 mg before test sample placed in the test oil in the container and keeps at 90 ± 10 °C for 8 hours. The test sample was left in the oil container for 12 hours until the oil cools to the room temperature and then withdrawn from the oil container, [3], Figure 4.

Figure 3. The test samples left in a desiccator

Figure 4. The test samples immersed in the oil container

Finally, the sample was rolled on a piece of cloth to remove oil from the test sample. The sample was weighed again to the nearest 1 mg, [3]. Table 2 shows the porosity test results for all formulations of brake pad materials. From the porosity results it can be seen that two break disc formulations, such as C2 and C3, shown lower percentage of porosity compared to other two formulations. Same conclusions were obtained in the paper [3], the difference consisting on values of porosity. In the present paper were obtained higher values of porosity which is explained by the different manufacturing technology of composite materials.

The hardness test was conducted on a Rockwell unit PH-C-01 /02 in accordance with EN ISO 6508-1 standard: 2002. The hardness measurements were conducted under test load of 980.6 N, the steel ball diameter is 1.58 mm, using scale B as stipulated in the standard. The measurements were performed at a distance of 13 mm. Drive speed of the load was 0.8 m/s and the holding time was 10 s. The hardness of the sample represented the arithmetic mean of the readings five indentations, [8].

Table 2 shown the hardness of the four formulations brake pads material. It can be seen that the hardness value of C2 is the highest of all and the hardness value of C4 show lower hardness value, which is explaine by the ductile nature of the material. Formulation C2 composite has the highest hardness value of 69.3 HRS because of having more aluminium content used as a matrix. Varying content of aluminium and coconut fibre will exhibit different hardness value of the material. Compared to work [3], hardness achieved are higher due hexametiltatramine, which has the role to transform the phenolic resin which is a thermoplastic material into thermosets.

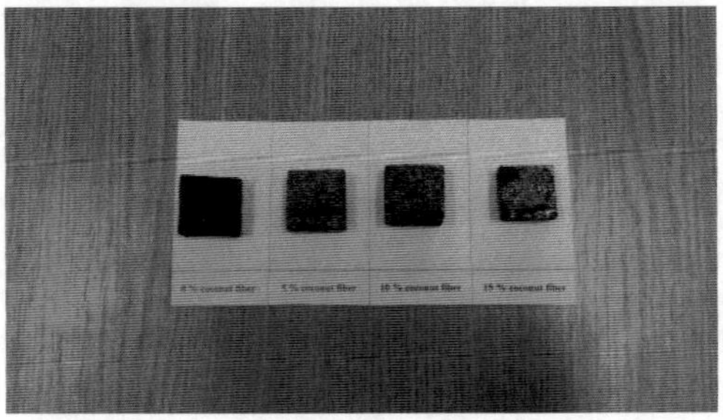

Figure 5. A complete set of samples for compression test

Figure 6. Sample C3 during compression tests

The compressive strength was done using a universal testing machine type Zwick/Roll Z005 with a corresponding software name Test-Xpert II, that retrieves data from experimental testing machine and process them statistically. The dimensions of the samples are: 25mm x 25mm x 7 mm. Figure 5 shows a complete set of samples to achieve compression test. The samples were subjected to compressive force, loaded continuously until failure occurred. The load at which failure occurred was then recorded, Figure 6. Table 3 shows the compressive strength for new composite materials developed.

Table 3. Compressive strength for new composite materials developed

Samples	Mechanical strength	
	Compression load at tensile strength (N)	Compression strength (MPa)
C1	1260	21.45
C2	2710	41.56
C3	2610	39.72
C4	2940	47.67

The compressive strength showed the formulation of C4 exhibited higher strength to withstand the load application and higher ability to hold the compressive force. The ultimate strength of the formulation C4 is corresponds to the stress of 47.67 MPa. The sample starts to break at stress of 45.21 MPa.

The four formulations for brake pads are subject to thermo-gravimetric analysis. Thermo-gravimetric analysis reveals the weight loss when the composite materials were exposed to 50–300^0 C, [6]. Thermo-gravimetric analysis of the developed composites is carried out in a TLB 180 series type machine at a heating rate of 15^0 C/min. The sample is weighed accurately with a balance of fourth decimal. Figure 7 present the procedure of weighing of the samples utilized in thermo-gravimetric analysis. Each sample was placed in a crucible that is inserted into the furnace maintained at 250 to 300^0 C and soaked for two hours. Then the crucible shall be removed from the furnace, cooled in a dessicator and weighed, [6].

Figure 7. A complete set of samples for compression test

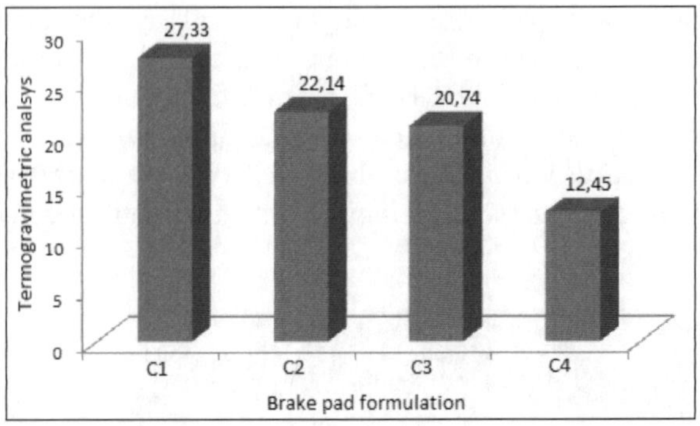

Figure 8. Termo-gravimetric analysis results

The weight loss % of composite material during exposure was calculated with relation (2), [6]:

$$\Delta M = \frac{M_2 - M_3}{M_2 - M_1} \cdot 100 \tag{2}$$

In relation (2) M_1 represents the weight of empty crucible, [g]; M_2 is the weight of crucible and the sample, [g]; and M_3 is weight of crucible and the sample after ignition, [g].

Figure 8 shows thermo-gravimetric analysis results for the developed composites which reveals the weight loss % when the samples were exposed to 50–300^0 C.

Loss by weight % is as follows: C4 < C3 < C2 < C1. Hence C4 has better thermal stability than the other composites. We can observe that the higher of coconut fibre content offer to composite materials higher thermal stability.

For morphology study scanning electron microscopy SEM was used to observe and investigate the distribution of microstructural features along with elemental constituents of new composite reinforced with various content of coconut fibre, Figures 9.

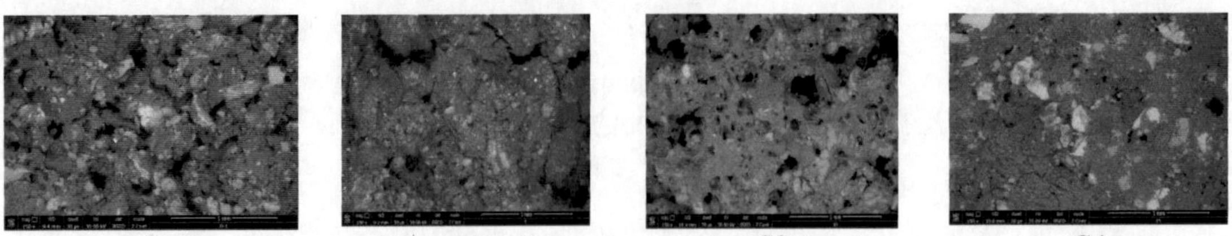

C1 C2 C3 C4

Figure 9. SEM microstructure for composite samples with variable addition of coconut fibre

The resin binder in dark region along with aluminium distribution in white region can be saw in microstructure for composite C1.This formulation showed that aluminium with graphite and zirconium oxide content more aluminium element. The graphite element in grey regions also can be visualized in Figure 9.

The microstructures of C2 composite showed the homogeneous distribution of abrasive, solid lubricant, binder and friction modifier in the aluminium matrix. Higher amount of aluminium is present in C4 due to more coconut fibre in the formulation. The other three structures showed the heterogeneously distribution of elements which explain by different weight of elements. Also the sintering process of materials at different temperatures contribute to change the structure.

To determine the friction coefficient of the composites developed was used Tribometer TR-20 Micro, Figure 10. This equipment allows the study of the wear behaviour of materials by pin-on-disk method. This is computerized and programmable to study friction against speed, load, temperature and wear. The device enables mounting of circular or rectangular samples which moves a steel ball with a diameter of 6 mm. In order to achieve tests were used samples with dimensions 25x25x10 mm. Such tests were carried out on four samples of each composite recipe.

Figure 10. Tribometer TR 20 Micro

The results shown in this paper are the average of four tests per sample. In conducting the experimental determinations, test parameters were: the loading force F=20N, test range R=7,5 mm, ball speed n=150 rot/ min and time test t = 60 min. The distribution of load will also affect the friction level. The initial running of a new sample is associated with a slow increase of the friction coefficient. The results of various composites named C1-C4 are shown below in Figures 11 to 14.

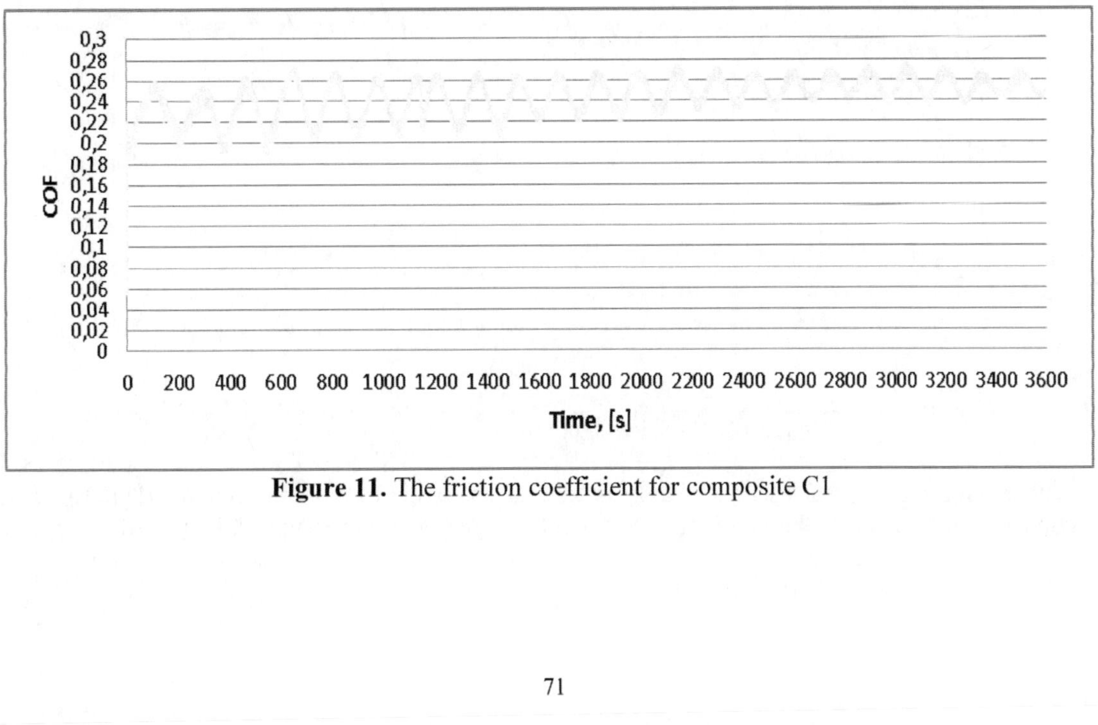

Figure 11. The friction coefficient for composite C1

IOP Conf. Series: Materials Science and Engineering **200** (2017) 012009 doi:10.1088/1757-899X/200/1/012009

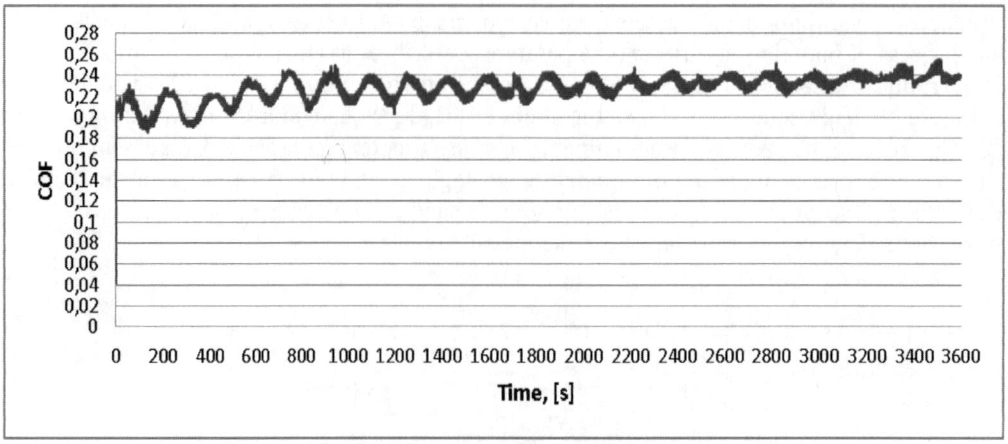

Figure 12. The friction coefficient for composite C2

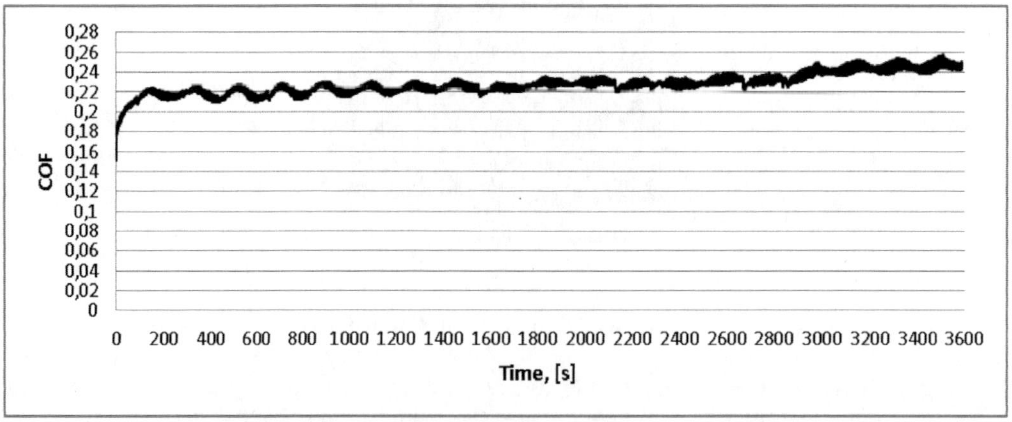

Figure 13. The friction coefficient for composite C3

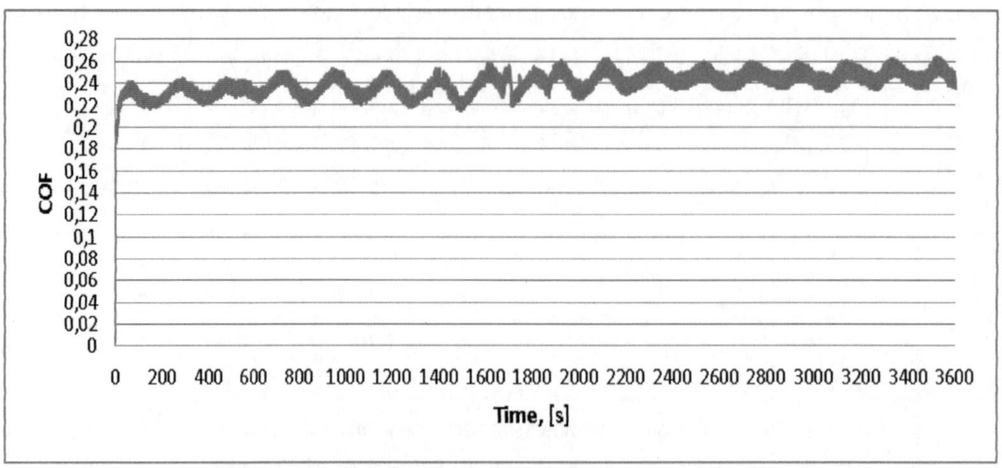

Figure 14. The friction coefficient for composite C4

The initial running of each sample is associated with a slow increase of the friction coefficient. The friction coefficient, touch the value of 0.26 for all samples. For the sample C1 and C2 there is a fluctuation of friction coefficient, and to the other two, after a certain trial, the friction coefficient is stable. From Figure 13, the friction coefficient for C4, goes as high as 0.26 which is attributed to the good thermal stability caused due to lesser amount of weight loss as found from the loss of ignition test. Larger amount of coconut fibre in brake composite does not only provide high and stable friction coefficient, but also strengthens composite body. More content of coconut fibre in the brake composite C4 causes more fibres to appear on the friction surface. This made the true contact area between the fibres and steel ball, so a higher friction coefficient is achieved.

In general, the brake pads with high coefficient of friction provides better braking with a low effort brake pedal, but tend to lose efficiency at high temperatures, increasing the braking distance. The brake pads with a constant low friction coefficient and do not lose braking efficiency at high temperatures, but they need a force of the brake pedal higher, [2].

3. Conclusions

Based on the results obtained in this paper the following conclusions can be made:

- Composite C4 has lower density, but higher porosity compared to other formulations;
- C2 and C3 composites even have a greater density than C4, but they have less of porosity;
- Formulation C2 has the highest hardness value of 69,3 HRS;
- Composite C4 exhibited higher strength to withstand the load application and higher ability to hold the compressive force;
- Composite C4 is more efficient than other compositions since it has good thermal stability. A good thermal stability have composites C2 and C3;
- From the SEM study it can be see that microstructures of C2 composite showed the homogeneous distribution and this composite modelling appropriate percentage between matrix and reinforcement volume;
- Composites C3 and C4 is more efficient than other compositions since it has stabile friction coefficient. It can be concluded that C3 and C4 showed almost similar properties from the four formulations and it could be utilize in the manufacturing of brake pads.
- The higher coconut fibre content, led to the better friction performance in the pin on disc test procedure.

Future research directions proposed prescription in the recipe to increase the friction coefficient. The study led that natural coconut fibre is a potential candidate filler material for the light and medium weight vehicles brake pads.

References

[1] Amaguci Y and Ahodas J 1991 Nob-asbestos, Non-Metallic and Non-Glass Brake Pad Composite, *Automotive Engineering* **99** 12

[2] Blau P J 2001 *Compositions, Functions, and Testing of Friction Brake Materials and Their Additives*, Metals and Ceramics Division Oak Ridge

[3] Maleque M A, Atiqua A, Talib R J and Zahurin H 2012 New natural fibre reinforced aluminium composite for automotive brake pad, *International Journal of Mechanical Engineering* **7**(2) 166-170

[4] Park J H, Chung J O and Kim H R 2010 Friction characteristics of brake pads with aramid fiber and acrylic fiber, *Industrial Lubrication and Tribology* **62**(2) 91–98

[5] Liew N 2009 On the Dry and Wet Sliding Performance of Potentially New Friction Brake Pad Materials for Automotive Industry, *Wear* **266** 275-287

[6] Vijay R , Jees J M , Saibalaji M A and Thiyagarajan V 2013 Optimization of Tribological Properties of Nonasbestos Brake Pad Material by Using Steel Wool, *Advances in Tribology* 165859

[7] O sterle W, Griepentrog M, Gross T and Urban I 2001 Chemical and microstructural changes induced by friction and wear of brakes, *Wear* **250-251**(2) 1469–1476

[8] Crăciun A L, Heput T and Pinca-Bretotean C 2016 Formulation of materials with natural fibre

for brake system components, *International Journal of Engineering* **XVII**(3) 17-20
[9] Crăciun AL, Pinca-Bretotean C, Uțu D and Josan A 2016 Tribological properties of nonasbestos brake pad material by using coconut fiber, *IOP Conf.Ser.:Mater. Sci.Eng.* **163** 012014
[10] Pascu LV 2014 *Research on improving the quality of brake shoes intended for rolling stock*, Doctoral Thesis, University Politehnica Timisoara, Romania
[11] Sasaki Y 1995 *Development Philosophy of Friction Materials for Automobile Disc Brakes*, 8th International Pacific Conference on Automobile Engineering, pp 407-412

IOP Conf. Series: Materials Science and Engineering **200** (2017) 012010 doi:10.1088/1757-899X/200/1/012010

Research on the influence of moulding-casting technology on the quality of castings

A Josan, C Pinca Bretotean, S Rațiu, E Ardelean and M Ardelean

Politehnica University of Timisoara, Department of Engineering and Management, 5 Revolution Street, Hunedoara, 331128, Romania

E-mail: ana.josan@fih.upt.ro

Abstract. The quality of castings has a particularly role in the Romanian foundries. In this context, quality assurance is the overall objective of the foundries. The paper presents the critical analysis performed on moulding-casting technology of the type Lifting mechanism. This casting is a subset of the lifting and rotating mechanism of the furnace vault. The casting analysed is a medium size, with weight of 114 kg. The current moulding-casting technology involves moulding into three mould-parts leading to the occurrence of defects (decentering of the core, displacement of the lower mould and the middle mould and occurrence of burrs in area separated. Thus, to reduce the percentage of defects registered in industrial practice is necessary to change the moulding-casting technology. This requires the use of two mould-parts, re-dimensioning of the core and the core box and dimensioning of the runner network. The adoption of these changes in industrial practice has direct implications on the cost of casting and foundry costs default.

1. Introduction

In the modern technique, the technology is the predominant factor of progress, determining in most cases the qualitative leaps in manufacturing or basic condition of the technical performance [1]. Thus, the knowledge of technology is required both for manufacturing sector and research and development sector [1].

The casting technology of parts, respectively the technological process, has accurately scientific basis and should be projected, in all its stages, before the start of manufacture (from the achieving of moulds and cores, passing on to elaborating of the liquid alloy, the treating of alloys in the liquid phase, the pouring and solidification of the alloy into the mould and finishing with the cooling, cleaning and thermal treatment applied).

The technology must find always the best solutions of technically point of view (to fit the product in techical conditions imposed) but also the most economcal. This is reflected in the correlation between product cost, wich must be competitive or, in the absence of competition to be economical and profitable [1].

Steel castings represent 7-12% of the total quantity of castings [1]. They have applications across the economy and especially where high mechanical properties are required, at machines subjected to strong shocks and to wear stress, to high temperatures anf of chemical substances attack. In this context, the analysis of fabrication technology of cast steel castings has a important role in ensuring the quality of castings.

IOP Conf. Series: Materials Science and Engineering **200** (2017) 012010 doi:10.1088/1757-899X/200/1/012010

Making of the moulds can be done by several methods related to the multitude of factors that influence the manufacturing of castings: moulding in two mould part, moulding in three mould part, moulding into the foundry ground, mechanized moulding etc.

The defects which appear to castings are the result of causes which act unilaterally and combined during the technological process of castings, starting the design stage and ending with their delivery [2].

Thus, a particulary important rol it has the knowledge the formation mechanism of defects and the possible causes that they generate, respectively their measures for preventing.

2. The current moulding-casting technology for casting analyzed

The casting analysed is cast of steel 340-550 W (according to SR ISO EN 3755-95) and is part of the *Lifting mechanism* of an industrial furnace vault.

In according to the SR ISO 3755-95, for cast steel 340-550W, the chemical composition must be within the following limits (Table 1) [3]:

Table 1. The chemical composition for cast steel 340-550W

Chemical composition, %				
C	Mn	Si	S_{max}	P_{max}
0.40...0.50	0.40...0.80	0.25...0.50	0.040	0.040

The mechanical properties at ambient temperature, on samples of 28 mm thich, for cast steel 340-550W, according to SR ISO 3755-95, are presented in Table 2 [3].

Table 2. Mechanical characteristics for cast steel 340-550W

Grades	Yield point, $R_{p0,2}$ min, MPa	Tensile strenght, R_m, min, MPa	Breaking elongation, A min, %	Choosing, according to the order	
				Necking, Z min, %	Absorbed energy at breaking, KV min, J
340-550W	340	590	12	18	20

This steel has relatively good casting properties and favorable mechanical properties, especially after a thermal treatment performed correctly [3-7].

The steels for castings were elaborated in an electric arc furnace type DSN-3, with basic lining and a nominal capacity of 3000 kg. Elaboration of batches was placed in 200…300min/batch [6],[7].

Critical analysis of moulding-casting technology of this type of castings lead to the idea that the rejected castings percentage registered in industrial practice (13%) is too high for the company costs. In this context was analyzed the current moulding-casting technology of this type of casting starting from the finished piece design (Figure 1) [6-10].

The moulding technology for this type of casting involves placing two separation plane (Figure 2). Thus, the casting is performed in three mould jacket [1],[6], [7], [11].

IOP Conf. Series: Materials Science and Engineering **200** (2017) 012010 doi:10.1088/1757-899X/200/1/012010

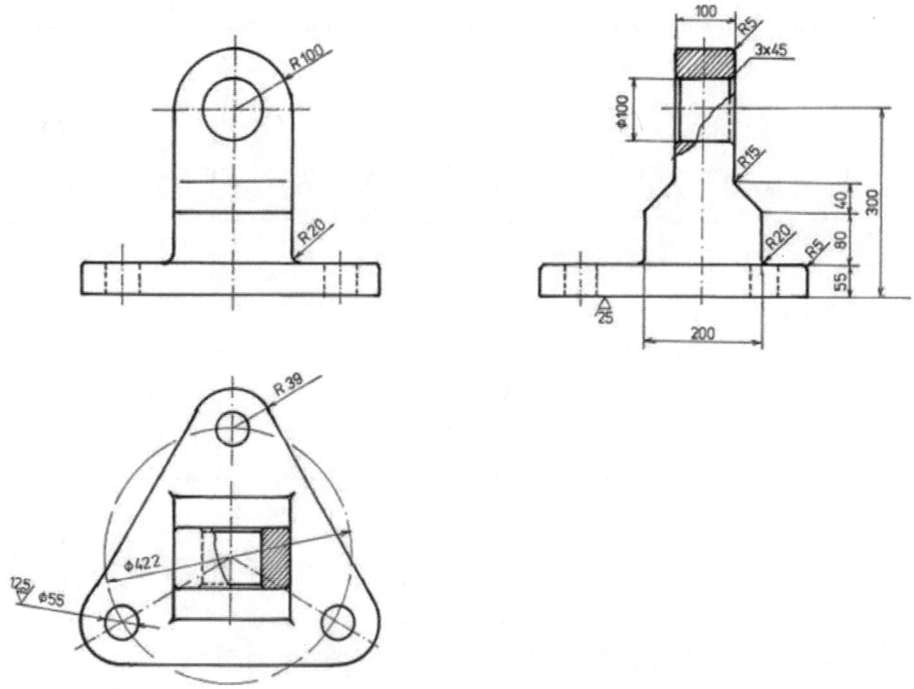

Figure 1. The finished piece design

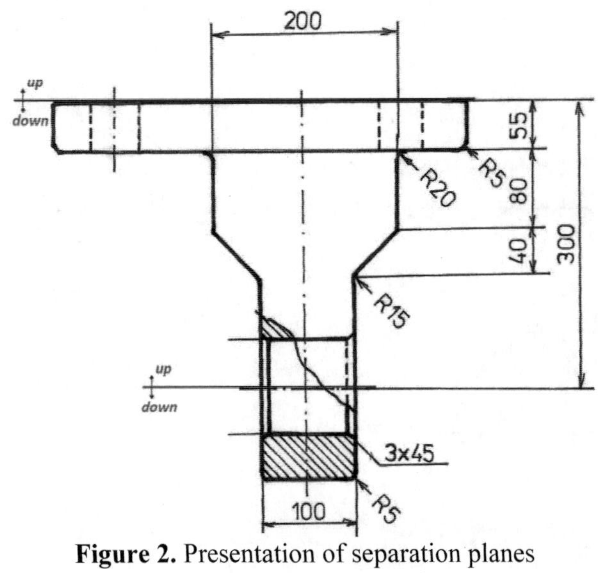

Figure 2. Presentation of separation planes

After etablishing the separation planes the moulding-casting technology involves the following steps [1], [6-9]:

1. Establishing the processing allowance of casting and constructive inclinations in order to facilitate the demoulding of the pattern (Figure 3)
2. Establishing the configuration and size of the wooden pattern (Figure 4)
3. Placing the core in the lower mould part (Figure 5)
4. Dimensioning of center riser and realization the wooden pattern for this (Figure 6)

5. Realization of technological design (Figure 7)
6. Establishing the dimensions of mould jackets and casting of the liquid alloy (Figure 8).

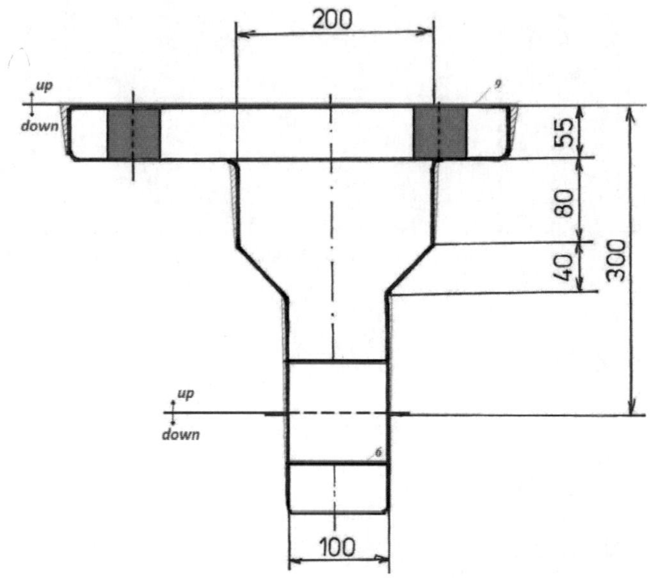

Figure 3. The processing allowance of casting

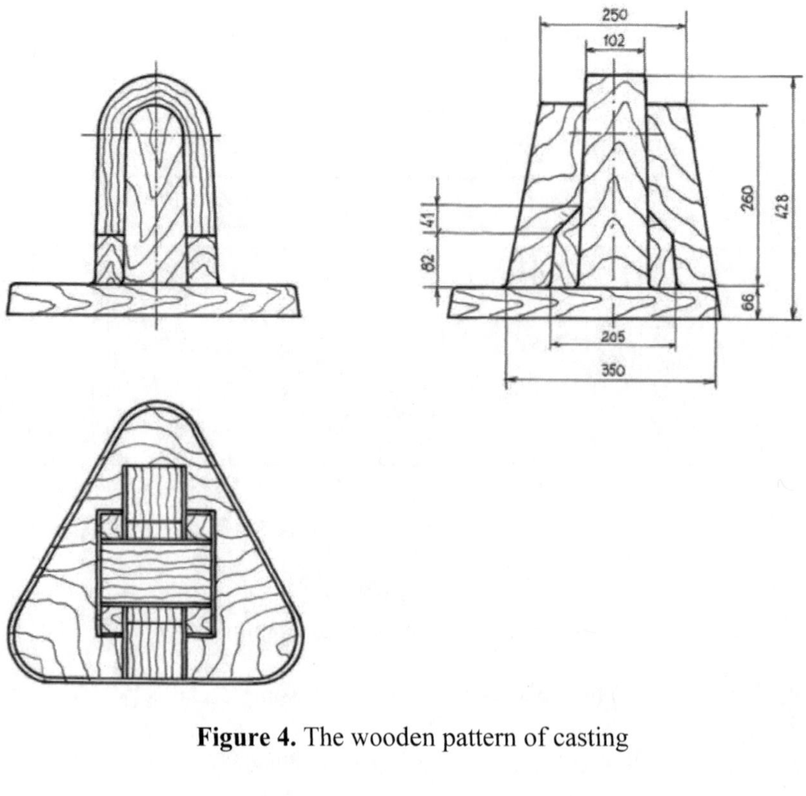

Figure 4. The wooden pattern of casting

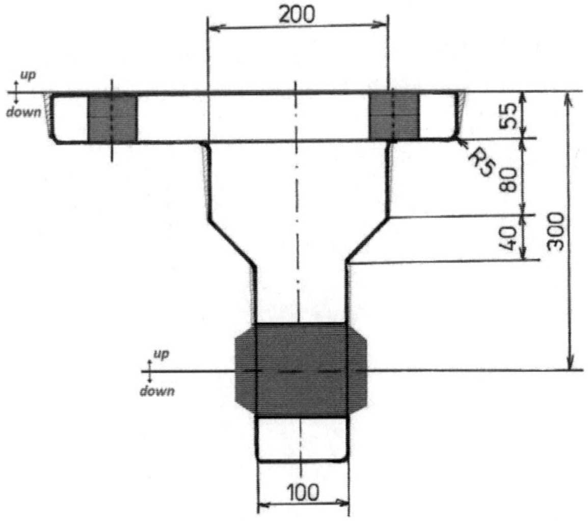

Figure 5. Placing the core on technological design

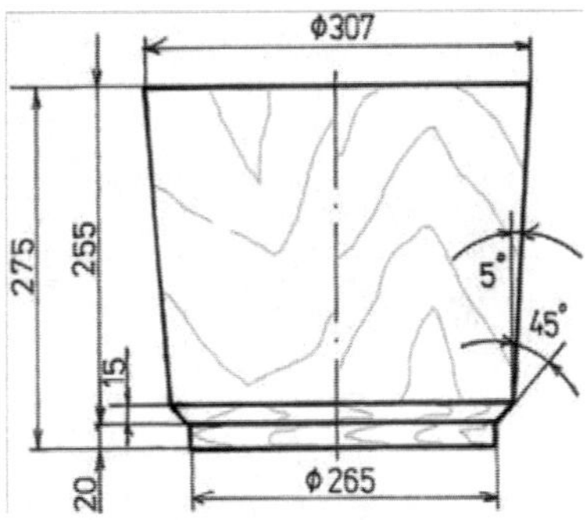

Figure 6. The center riser pattern (made of wood)

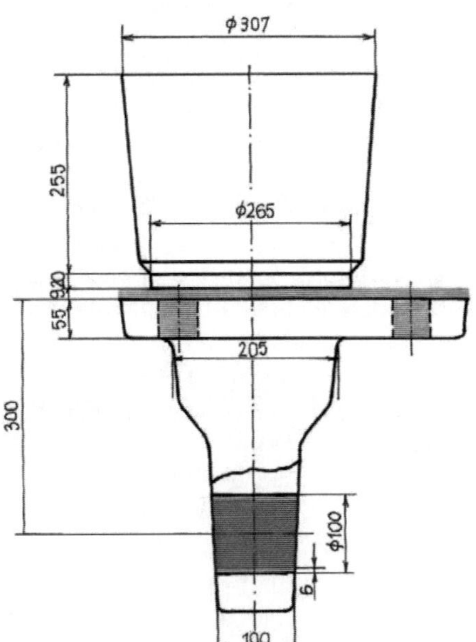

Figure 7. Placing the center riser on the technological design

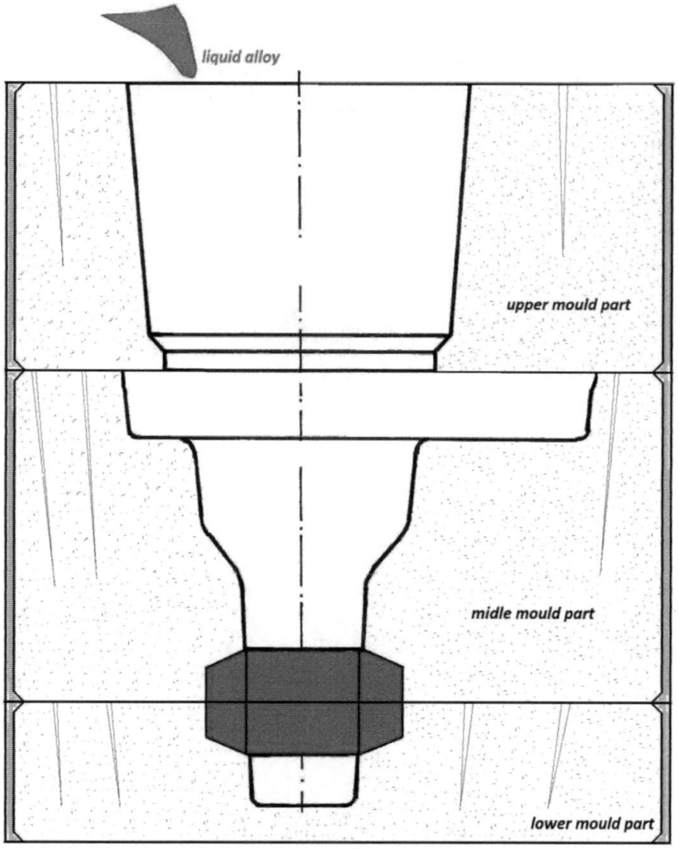

Figure 8. The assembled moulds for casting

3. The modified moulding-casting technology for casting analyzed

Performed the critical analysis of this approach of the possibility of casting the piece is found that the pouring in three mould jacket does not ensure high quality castings so that the percentage of defects (such as burrs and decentering) is high [8-10].

Another disadvantage of this moulding method is very high and unjustified consumption of moulding material.

For optimization of moulding-casting technology of casting studied takes a series of measures regarding [1], [2], [6], [7], [10], [11]:

- Changing current moulding technology (with the application of two plane separation of the pattern and the mould) and the use of moulding technology with a single plane of separation (Figure 9);
- Redimensioning the core and the core box (Figure 10)
- Dimensioning of the gating system (in case of current technology is provided a direct pouring through the open center riser) (Figure 11)
- Changing the casting technology in three moulds and using the casting technology in two moulds (Figure 12).

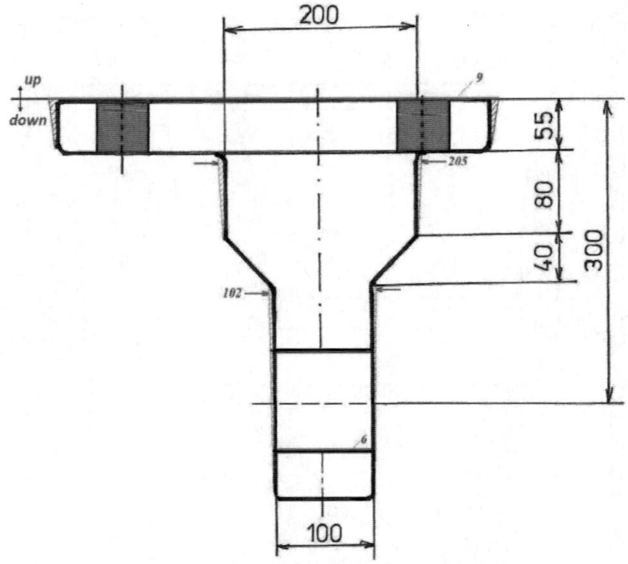

Figure 9. The separation plane

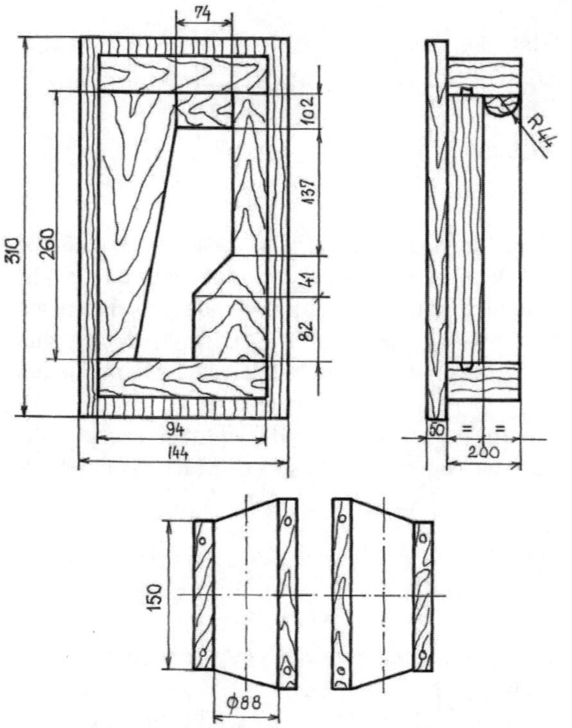

Figure 10. The core box – modified

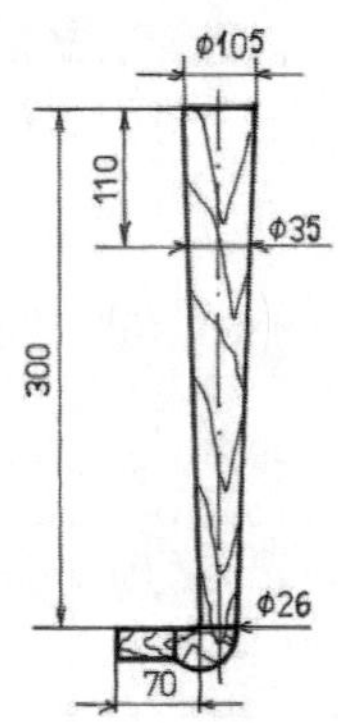

Figure 11. The dimensions of the gating system

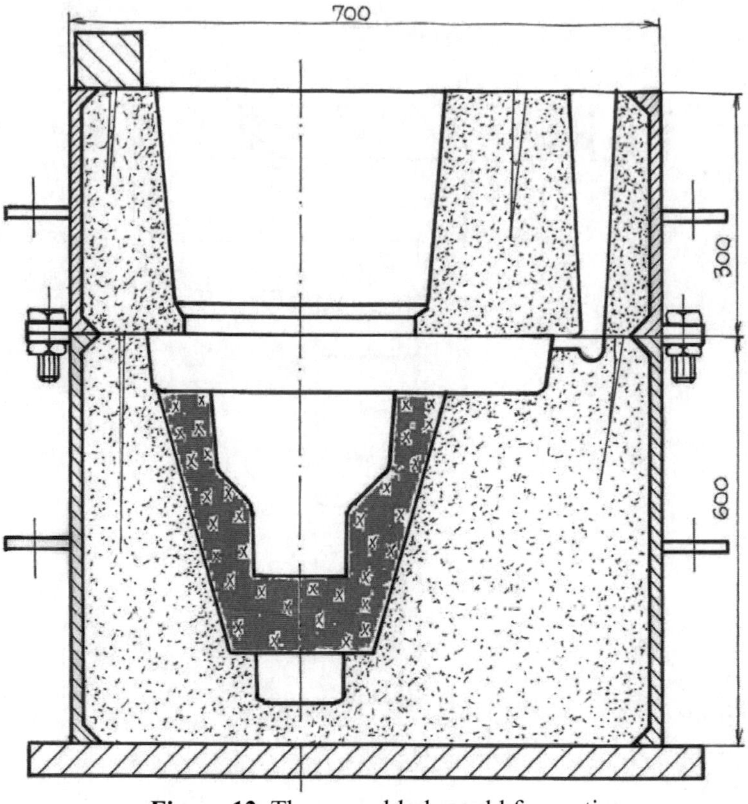

Figure 12. The assembled mould for casting

4. Conclusions

The multitude of problems, large and difficult, facing the manufacturers of castings can not be adequately resolved without a continued increase of the level of theoretical and practical trening. The rejects of castings caused by unremedied defects are too high in many foundries, leading to significant material losses caused by inadequate production.The activity of prevention of castings defects and rejects registered must be intensified with the ultimate aim to increase the castings production effectiveness [2], [8], [9].

Application in industrial practice of all the technological method mentioned lead to the decrease the percentage of rejects registered from 13% to about 4%. This aspect has a positive influence in castings costs respectively in the company costs.

References

[1] Stefanescu C and Cazacu I 1980 *Technologies of realization of the parts by casting*, Technical Publishing, Bucharest
[2] Istrate Gh and Toboc P 1981 *Controlul de calitate și remanierea defectelor pieselor turnate*, Publishing House Didactic and Pedagogical, Bucharest
[3] ***1991 The Standard *SR ISO 3755 - Cast carbon steels for general engineering purposes*
[4] Sofroni L 1975 *The elaboration and pouring of alloys*, Publishing House Didactic and Pedagogical, Bucharest
[5] Socalici A, Ardelean E, Ardelean M, Heput T and Josan A 2007 *The casting and solidification of the steel*, Publishing CERMI, Iasi
[6] Josan A 2012 *Casting and moulding technologies of metal parts – the project guide-book*, Publishing Polytechnic, Timişoara, Romania
[7] Josan A 2002 *The technology of moulding and casting alloys*, Politehnica Publishing, Timisoara

IOP Conf. Series: Materials Science and Engineering **200** (2017) 012010 doi:10.1088/1757-899X/200/1/012010

[8] Josan A 2016 Impoving the quality of castings by optimization of the moulding-casting technlogy, *Acta Technica Corviniensis - Buletin of Engineering* **IX**(4) 75-78

[9] Josan A 2010 *The analysis of casting defects recorded in the metallurgical enterprises*, International Symposium on Advanced Engineering & Applied Management - 40th Anniversary in Higher Education (1970-2010), Hunedoara, Romania, on CD

[10] ***1977 *International Atlas of casting defects*, Technical Publishing, Bucharest

[11] Josan A, Pinca Bretotean C, Ardelean E and Ardelean M 2015 Using lateral cores to casting of carbon steel parts of drive wheel type in a metallurgical enterprise, *IOP Conf. Ser.: Mater. Sci. Eng.* **85** 012016

IOP Conf. Series: Materials Science and Engineering **200** (2017) 012011 doi:10.1088/1757-899X/200/1/012011

Reducing emissions by using special air filters for internal combustion engines

C Birtok-Băneasă, S A Rațiu, V Alexa, A L Crăciun, A Josan, A Budiul-Berghian

University Politehnica Timisoara, Faculty of Engineering Hunedoara, Department of Engineering and Management, 5 Revolution Street, 331128 Hunedoara, Romania

E–mail: sorin.ratiu@fih.upt.ro

Abstract. This paper presents the experimental methodology to carry out functional performance tests for an air filter with a particular design of its housing, generically named *Super absorbing YXV „Air by Corneliu"*, patented and homologated by the Romanian Automotive Registry, to which numerous prizes and medals were awarded at national and international innovations salons. The tests were carried out in the Internal Combustion Engines Laboratory, within the specialization "Road vehicles" belonging to the Faculty of Engineering Hunedoara, component of Politehnica University of Timisoara. The scope of the study is to optimise the air intake into the engine cylinders by reducing the gas-dynamic resistances caused by the air filter and, therefore, to achieve higher energy efficiency, i.e. fuel consumption reduction and engine performance increase. We present some comparative values of various operating parameters of the engine fitted, in the first measuring session, with the original filter, and then with the studied filter. The data collected shows a reduction in fuel consumption by using this type of filter, which leads to lower emissions.

1. Introduction

Motor vehicles are significant sources of pollution that can damage the environment and pose public health issues. Everyone has a stake in limiting pollution. Carbon monoxide, nitrogen oxides, and hydrocarbons are released when fuel is burned in an internal combustion engine and when air/fuel residuals are emitted through the vehicle tailpipe. Gasoline vapors also escape into the atmosphere during refueling and when fuel vaporizes from engines and fuel systems caused by vehicle operation or hot weather. The pollutants in vehicle emissions are known to damage lung tissue, and can lead to and aggravate respiratory diseases, such as asthma. Motor vehicle pollution also contributes to the formation of acid rain and adds to the greenhouse gases that cause climate change [1-3].

Pollutants emitted directly from vehicles are not the only cause for concern. On warm, sunny days, hydrocarbons react with oxides of nitrogen to create a secondary pollutant, ozone. In many urban areas, motor vehicles are the single largest contributor to ground-level ozone which is a common component of smog. Ozone causes coughing, wheezing and shortness of breath, and can bring on permanent lung damage, making it a cause of crucial public health problems.

Proper maintenance of car and truck emission control systems not only limits harmful emissions, but also can improve fuel efficiency and vehicle performance extending the life of the vehicle. Care in storing and handling gasoline and other solvents also reduces evaporative losses to the atmosphere [4].

The fuel mixture is flowing through the intake system under the effect of the depression created by the downward movement of the piston from TDC to BDC, in conjunction with the inertial flow effect and the undulating effect of the pressure waves created by the opening and closing valves. The size of depression in the sucked engine cylinder depends primarily on the gas-dynamic resistances that oppose the fresh load passing through the intake system components. In general, the depression created by the engine depends on many factors, among which the constructive factors are the most important. In practice, it has been found that the depression increases with increasing rotational speed. This depression is directly influencing the quantity of air retained in the engine cylinders during an operation cycle [5].

The air filter is the inlet system component which makes the connection between the gases acting inside the engine and the outside environment [6]. Its main role is to prevent the access of impurities from outside to inside the cylinders. Constructively, it has also the role to reduce the noise produced during the filling process. The air filter should ensure the supply with an air flow rate higher than the maximum one required by the engine. To do not worsen the cylinder filling with fresh charge (by decreasing the air mass actually retained in the engine cylinders), the filter gas-dynamic resistance must be as low as possible. Unfortunately, due to its design, the air filter introduces pressure loss into the system caused by the gas-dynamic resistances generated by its geometric form and the filtration material nature [6].

The air mass found in cylinders after closing the intake valve is the decisive factor in relation to the mechanical work produced by cyclically burning of the fresh load, with direct influence on the engine torque. Consequently, the measures to increase the maximum torque and the maximum power almost always involve creating the conditions for obtaining a maximum possible filling ratio [5].

The scope of the study is to optimise the air intake into the engine cylinders by reducing the gas-dynamic resistances caused by the air filter and, therefore, to achieve higher energy efficiency, i.e. fuel consumption reduction and engine performance increase. We note that this study does not cover the material of the air filter element, which is standardised, but only its housing optimisation for improving the flow conditions.

The novelty of this study is the development of our own air filter concept, as a method of reducing the gas-dynamic losses along the intake route of the internal combustion engines and thereby reducing fuel consumption, with direct implications on emissions. This new concept leads to the multiplication of air filter functions, i.e. the increase of the air capture and intake capacity regardless of the mounting position in the engine compartment [7].

This study is based on our own concept of super absorbing air filter, named *Super absorbing YXV „Air by Corneliu",* patented and homologated by the Romanian Automotive Registry, to which numerous prizes and medals were awarded at national and international innovations salons.

The concept underlying the initiation of this study is subject to the Patent no. 126019 / 28.12.2012 "Super absorbing air filter". This product, along with the related solutions for the optimisation of air intake processes occurring inside the internal combustion engines (the own solutions developed by some of the research team members) were appreciated and honoured with countless awards at the innovations salons at home and abroad (21 gold medals, 4 silver medals, 2 bronze medals and 19 special prizes).

2. Experimental

For performing the experimental measurements, we had an engine test stand, produced by Christiani, consisting of a multi-point fuel injection engine, brand VW, 1.4 MPI, used by the Golf VI 5K1 models, cylinder capacity: 1390 cm3, power: 59 KW/80 CP, 4 cylinders in line, manufacturing years: 2009-2012 (Figure 1).

IOP Conf. Series: Materials Science and Engineering **200** (2017) 012011 doi:10.1088/1757-899X/200/1/012011

Figure 1. Overview of the test stand

The stand offers the possibility to monitor the engine during its operation. It enables the connection to the OBD port of all the testers usually found in garages, and the measurement of signals and electrical quantities, with the possibility of assessing the obtained parameters.

The related software contains a professional interface that allows the PC to be transformed, via an USB port, into a device to be used for diagnostics and visualization of the functional parameters (Figure 2). Also, it enables the visualization and simultaneous storing of three measuring blocks, and can graphically display the essential features of the engine.

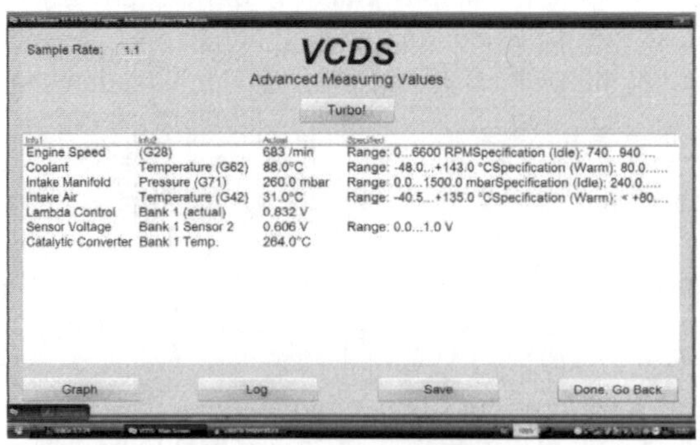

Figure 2. Monitoring the engine parameters

The tested super absorbing YXV filter [8] has the additional function of accelerating the air speed at the output of the filter. Due to the constructive geometry, it provides a significant increase in the air-filling coefficient of the engine cylinders. The filter (Figure 3) is provided with a frontal diffuser, lateral surface with the guiding cells and external diffuser – internal diffuser unit.

a. b. c.

Figure 3. Super absorbing filter YXV „Air by Corneliu": exploded virtual model, realised in AutoDesk Inventor (a); physical models (b and c) [8]

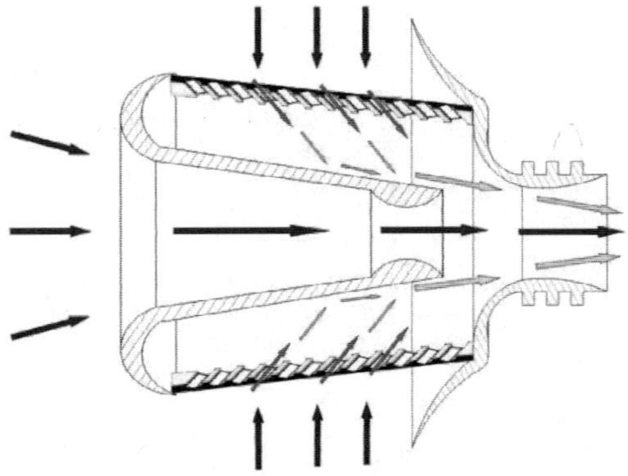

Figure 4. Illustration of the operating principle [8]

The tests consisted in determining the key operational parameters of the engine, at various speeds at idle, the engine being equipped, at a time, with the original filter or with YXV. The measuring sessions were repeated to confirm the results, and the test conditions for the two variants (engine equipped with original filter and with YXV) were maintained strictly constant. The filtering element is the same for both filters.

For the first measuring session, the experimental engine was equipped with the original filter (Figure 5.a). The data acquisition was made at various engine speeds at idle, beginning with the idling speed and continuing with 1500, 2000, 2500, 3000, 3500 and 4000 rpm. We measured: engine speed; throttle position; manifold air pressure; coolant temperature; manifold air temperature; hourly fuel consumption; ambient temperature and atmospheric pressure.

a. b.

Figure 5. Engine equipped with original filter (a) and engine equipped with YXV filter (b)

The next session of measurements was carried out with the engine equipped with YXV filter (Figure 5.b). The same above-specified parameters have been determined, under the same conditions. We mention that the measuring sessions were conducted inside the laboratory, where the temperature and pressure remained strictly constant [9].

3. Results

We present below the comparative values of the above-specified parameters on the performance of the engine equipped with original filter and YXV filter, respectively.

The chart presented in Fig. 6 shows that, at engine speeds below 2250 rpm, when using the YXV filter, the throttle valve opens less, which means less gas-dynamic resistance due to this filter installation, situation beneficial for the process of filling the engine cylinder with fresh load. At higher speed values, it can be seen that the throttle valve position is approximately identical in both cases.

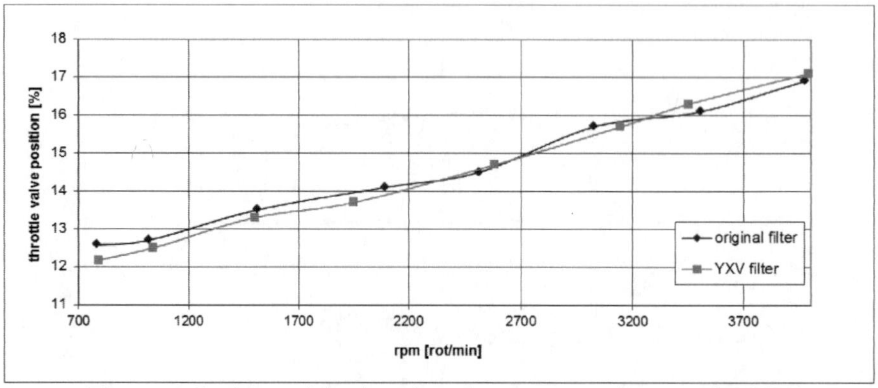

Figure 6. Position of the throttle valve versus engine speed at idle

From Figure 7, we can deduct that, throughout the range of speeds at which the tests have been performed, the pressure in the intake manifold is higher when using YXV than when using the original filter, situation beneficial for the process of filling the engine cylinder with fresh load.

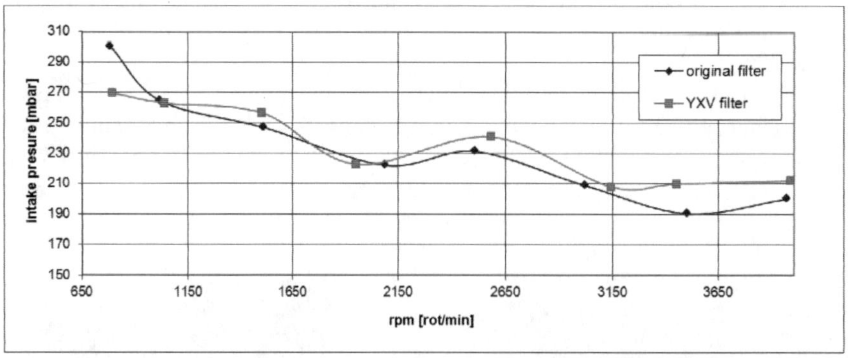

Figure 7. Pressure in the intake manifold versus engine idle speed

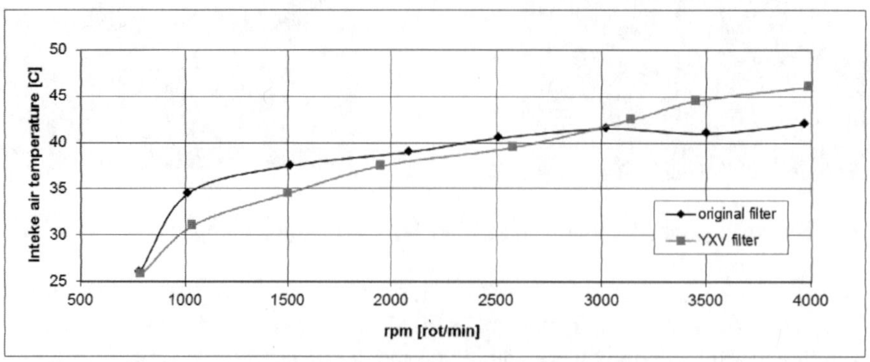

Figure 8. Air temperature in the intake manifold versus engine idle speed

From Figure 8, it can be concluded that, up to a speed of 3100 rpm, the air temperature in the intake manifold, when using the YXV, is lower than that obtained when using the original filter, situation beneficial for the cylinder filling process, because a lower air temperature when entering the engine means a higher density, and therefore a larger amount introduced into the cylinders, with beneficial consequences for the engine power.

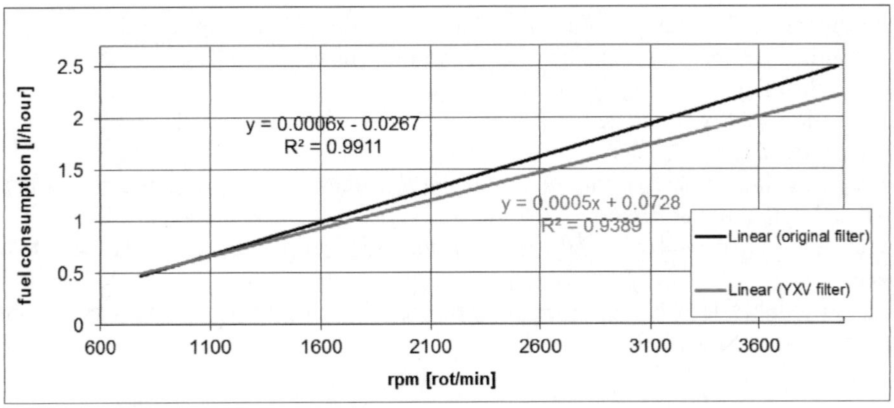

Figure 9. Hourly fuel consumption versus engine idle speed

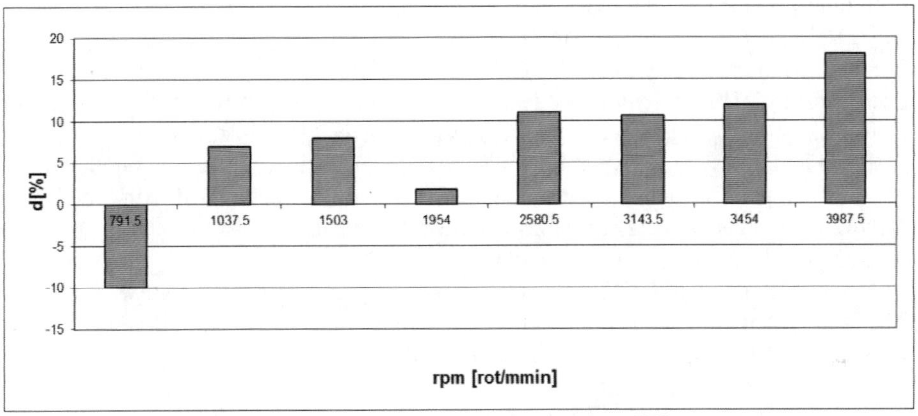

Figure 10. The percentage consumption difference

The most important parameter monitored during the tests was the hourly fuel consumption, in litres/hour. In Figure 9, we can see the variation of the hourly fuel consumption with the engine speed, for the cases when the engine is running equipped with the original filter and with YXV filter, respectively. It appears that, from speed values around 1600 rpm, the fuel consumption begins to decrease considerably when using the YXV filter.

The percentage difference (d) between the hourly consumption with original filter ($Ch_{original.filter}$) and the hourly consumption with YXV ($Ch_{YXVfilter}$) is calculated using the relation (Eq. 1):

$$d = 100 - \frac{Ch_{YXVfilter}}{Ch_{original.filter}} \cdot 100 \text{ [\%]} \tag{1}$$

In Figure 10, we can see that the percentage difference is positive at all the engine speeds, with the exception of idle (more precisely: 791.5 rpm). The positive values show lower fuel consumption than in case of the original filter.
Throughout the speed range used for conducting the tests, the reduction of fuel consumption by using the super absorbing YXV filter is 7.2%.

4. Conclusions

Following the conducted researches and monitoring of the behaviour when installing the filter on various types of engines, we concluded that this one has a number of advantages, such as:

- being in contact with the air, the filter element ensures a minimum gas-dynamic resistance to the absorbed air, increasing thereby the absorption and capture rates;
- the self-cleaning possibility of the filter element;
- visualisation of the filter element, without the prior removal of the filter, for checking its *pollutant loading level*;
- the capacity of air filter housing to significantly increase the speed of absorbed air, both at the input and output of the filter;
- the ability to create a slightly boost, which increases proportionally to the vehicle movement speed.

All these advantages lead to improved performance and thereby reduce fuel consumption, with direct implications in reducing emissions.

References

[1] Stankovic S, Vaskovic V, Petrovic N, Radojicic Z and Ljubojevic M 2015 Urban traffic air pollution: Case study of Banja Luka, *Environmental Engineering and Management Journal* **14**(12) 2783

[2] Hanganu R, Munteanu A, Simion A, Livadaru L, Malanciuc A and Anghel B 2012 Study of a direct current (DC) machine used on automotive low environmental impact applications, *Environmental Engineering and Management Journal* **11**(9) 1725

[3] Li X R, Zhao L M, Li Z J and Liu F S 2015 A simulation study on the diesel engine combustion process based on the field-effect analysis method of combined combustion, *Environmental Engineering and Management Journal* **14**(1) 213

[4] ***http://www.dec.ny.gov/chemical/8394.html

[5] Raţiu S and Alexa V 2016 *Gasodynamic study of the intake route at a spark-ignition engine*, The 9th International Symposium – Machine and Industrial Design in Mechanical Engineering KOD 2016, Balatonfured, Hungary, June 9-12, pp 113-116

[6] Leonte V L 2014 *Contribution regarding the filling process optimization for internal combustion engines*, Technical University „Gheorghe Asachi", Iaşi, Romania, Doctoral Thesis

[7] Raţiu S and Birtok-Băneasă C 2012 *The study of inflow improvement in spark engines by using new concepts of air filters*, In: Internal Combustion Engines - Kazimierz L., Pawet W. (Eds.), InTech, Janeza Trdine 9, 51000 Rijeka, Croatia, 234

[8] Birtok–Băneasă C and Raţiu S 2011 *YXV Super absorbing air filter*, In: Air intake in internal combustion engines – Super absorbing air filters – Dynamical Air Transfer Systems, Politehnica Publisher, Timişoara, Romania, pp 34-37

[9] Raţiu S, Birtok-Băneasă C, Alexa V and Kiss I 2015 Functional performance testing of the universal super absorbing air filters FSU 70 „Air by Corneliu", *IOP Conf. Ser.: Mater. Sci. Eng.* **85** 012028

From design to manufacturing of asymmetric teeth gears using computer application

F Suciu, A Dascalescu and M Ungureanu

Technical University of Cluj Napoca – North University Centre of Baia Mare, Department of Engineering and Technology Management, Dr. Victor Babes str., no. 62A, 430083, Baia Mare, Romania

E-mail: Flavia.Suciu@cunbm.utcluj.ro

Abstract. The asymmetric cylindrical gears, with involutes teeth profiles having different base circle diameters, are nonstandard gears, used with the aim to obtain better function parameters for the active profile. We will expect that the manufacturing of these gears became possible only after the design and realization of some specific tools. The paper present how the computer aided design and applications developed in MATLAB, for obtain the geometrical parameters, in the same time for calculation some functional parameters like stress and displacements, transmission error, efficiency of the gears and the 2D models, generated with AUTOLISP applications, are used for computer aided manufacturing of asymmetric gears with standard tools. So the specific tools considered one of the disadvantages of these gears are not necessary and implicitly the expected supplementary costs are reduced. The calculus algorithm established for the asymmetric gear design application use the „direct design" of the spur gears. This method offers the possibility of determining first the parameters of the gears, followed by the determination of the asymmetric gear rack's parameters, based on those of the gears. Using original design method and computer applications have been determined the geometrical parameters, the 2D and 3D models of the asymmetric gears and on the base of these models have been manufacturing on CNC machine tool asymmetric gears.

1. Introduction

The asymmetric teeth gears are cylindrical gears with different base circle involute teeth profiles.

Designing asymmetric teeth gears contain also the design for symmetric teeth gears because the last one can be considered a particular case of the first one. So the classical gears with symmetric teeth can be considered asymmetric teeth gears, having equal values for the design parameters in relation with the two profiles. In that way it can be considered that it is not necessary to mention like different the asymmetric and symmetric teeth gears. The design algorithm developed is valid and can be used for both.

Designing asymmetric teeth gears offer the possibility to choose the geometrical parameters from many values, not only those corresponding to the standard gears, in order to obtain better or optimal, for one specified objective, functional parameters.

Many results of the paper author's research on this subject have been published as step of the work considering the partial result as objective of the stage, part of the final objective.

The aim of this paper is to present, as a whole considering full research, how the calculus algorithm and the computer applications developed make possible to realize all the step necessary for obtain one optimized asymmetric gears, from the geometrical design, through carrying out the functional parameters, to manufacturing of the gears with standard tools. The calculus algorithm used for obtain the geometrical parameters of the involute asymmetric profiles of the gears have been developed on the base of "direct gear design" [1], [2].

For obtaining the functional parameters have been created calculus algorithms and then for obtain all the results in short time and for comparing different cases have been created and developed a system of computer applications that give for any user many answers to the questions put on the asymmetric gears.

2. The stages of application development

Final application, that offer all the geometrical and functional parameters studied on the steps of the research, uses as subroutines gradually made applications [3], in the same time with establishing the calculation algorithms and the results that can be carry out will be presented using for example one asymmetric gear transmission.

2.1. Setting initial parameters

The initial dates are: the axial distance (a) and the number of teeth (z_1, z_2).

The pressure angles for the direct profile and for the inverted profile $(\alpha_{wd}, \alpha_{wi})$ are considered designing parameters. The pressure angles can be modified and, the computer calculation being very efficiency, it is very easy comparing the functional parameters for asymmetric gear transmission with different values of the asymmetry coefficient.

The geometrical parameters can be also modified by the coefficient of the gear rack angle, noted with "f", and by using one or two gear rack for generating the pinion ad the gear, for emphasize that it has been used the parameter noted with "cr" with two values 1 or 2.

The functional parameters are different if the gear transmission it is used with the pressure angle with greater value on the active profile, called "direct asymmetric gear" or with the pressure angle with smaller value in the active profile, called "inverted asymmetric gear". The variable used for indicate one of the two possibilities noted with "var" can also have values 1, for the first case, or 2 for the second case. So the designing parameters are: $[\alpha_{wd}, \alpha_{wi}, \mathrm{f, cr, var}]$.

It has been chosen for the case studies the following initial dates and design parameters:

$$[\mathrm{a}, z_1, z_2, \alpha_{wd}, \alpha_{wi}, \mathrm{f, cr, var}] = [190, 25, 38, 40, 20, 5, 2, 1]. \quad (1)$$

2.2. Asymmetric gears encoding

For a better systematization of the results have been established that every asymmetric gear transmission will be notated with eight values, all used in design applications. Three of them are the initial date followed of the five designing parameters.

For example if it is used the notation A_as_25_38_190_40_20_0_1_1 that can be read like "Asymmetric gears transmission with the number of teeth 25/48, the axial distance 190, the pressure angles on the involute profile 40/20, with the gear rack angle equal with the pressure angle of the gear, pinion and gear generated with the same gear rack and the ending 1 show that will be used as direct asymmetric gear.

2.3. The geometrical parameters that can be carry out with the computer application

Geometrical parameters it have been considered all the diameters and profiles angles of the two gears of the transmission and all the parameters necessary to represent the single gear rack or two gear racks in relation with the pinion and the gear.

The calculus algorithm that was used for writing the application permit to obtain all the significant diameters like: diameter of the pitch circle (d_{w1}, d_{w2}), diameters of the base circles $(d_{b1d}, d_{b1i}, d_{b2d}, d_{b2i})$, diameters of the outside circles (d_{a1}, d_{a2}), diameters of the dedendum circles (d_{f1}, d_{f2}); all profile angles with significant values, from the profile angles corresponding to the points of beginning of involute profiles (α_{u1d}, α_{u1i}, α_{u2d}, α_{u2i}), profile angles for the first contact point from the bottom of the tooth (α_{p1d}, α_{p1i}, α_{p2d}, α_{p2i}) to the profile angles on the outside circle (α_{a1d}, α_{a1i}, α_{a2d}, α_{a2i}) and the profile angles on the tip circle (α_{v1d}, α_{v1i}, α_{v2d}, α_{v2i}) on the same time for the pinion and for the gear; the contact ratio for the direct and inverted profile ($\varepsilon_{\alpha d}$, $\varepsilon_{\alpha i}$).

Also can be obtained the geometrical parameters of the generating gear racks like: generating rack profile angles (α_{dc}, α_{ic}), generating rack pitch (p_c), diameters of the generating pitch circles (d_{rc1}, d_{rc2}), minimum radial clearance (c_{min}), tip gear racks radius ($R_1 \neq R_2$ or $R_1 = R_2$). The calculation of geometrical parameters can modify the shift rack of the pinion and of the gear (x_1, x_2) for a new set of gears parameters until the achievement of the conditions imposed by the correct meshing. Examples of results are given in table1.

Table 1. Geometrical parameters

Initial dates: $a = 190$, $z_1 = 25$, $z_2 = 38$				
Designing parameters: $\alpha_{wd} = 40^o$, $\alpha_{wi} = 20^o$, f=5, cr=2, var=1				
Parameters	Pinion (1)		Gear(2)	
α_{wd}, α_{wi}	0.6981	0.3491	0.6981	0.3491
d_{a1}, d_{a2}	162.4495		242.7298	
d_{w1}, d_{w2}	150.7937		229.2063	
d_{f1}, d_{f2}	135.0152		215.2955	
d_{b1d}, d_{b1i}, d_{b2d}, d_{b2i}	115.5146	141.6997	175.5823	215.3835
α_{v1d}, α_{v1i}, α_{v2d}, α_{v2i}	0.7888	0.5268	0.7701	0.4937
α_{a1d}, α_{a1i}, α_{a2d}, α_{a2i}	0.7798	0.5110	0.7621	0.4793
α_{p1d}, α_{p1i}, α_{p2d}, α_{p2i}	0.5859	0.1266	0.6375	0.2304
α_{u1d}, α_{u1i}, α_{u2d}, α_{u2i}	0.5633	0.0387	0.6228	0.1722
	Gear rack 1		Gear rack 2	
α_{dc}, α_{ic}	0.6881	0.3252	0.6881	0.3252
p_c	18.7917		18.7917	
d_{rc1}, d_{rc2}	149.5398		227.3005	
R_1, R_2	0.7421		0.5639	
x_1, x_2	0.0802		1.5230	
contact ratio for:	direct profile		inverted profile	
$\varepsilon_{\alpha d}$, $\varepsilon_{\alpha i}$	1.2937		1.7241	

2.4. Generating of the pinion and gear models

All the above mentioned parameters make possible to have the complete profile of the gear and pinion teeth in a graphic representation obtained in MATLAB and also can be used as initial date for write the AUTOLISP application for generating the 2D models of the gears, implicitly the 3D models.

Two applications have been developed: MDA1 – Asymmetric involute teeth modelling for the first gear and MDA2 – Asymmetric involute teeth modelling for the second gear. The representations of the curves that limit the tooth body have been made on the base of the parametrical equations of asymmetric involute profiles [2] and the parametrical equations of the joint profiles [4]. The two gears are in contact on the active profile on the meshing point (Figure 1).

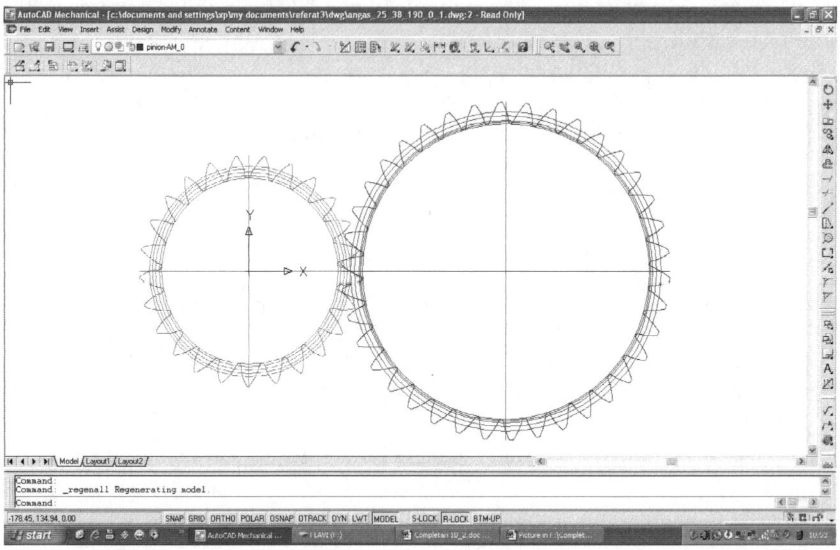

Figure 1. The model of the gear transmission – pinion and gear

The models of the gears offer the possibility of verifying the correct meshing by animation of the models, the study of teeth stress and displacements with the finite element method and also the transfer of the dates on the soft of one cutting machine tool. The geometrical parameters that are given in the first line of the programs are: z_1, z_2, a, α_{wd}, α_{wi}, $\alpha_{dc}, \alpha_{ic}, \alpha_{a1d}$, α_{a2d}, R_1, R_2, x_1, all determined in the previous applications.

2.5. Determination of functional parameters

The aim of designing nonstandard gears, in this case asymmetric involute teeth gears, is to obtain better behavior under the load of the gear transmission.

Having this objective there was first necessary to establish and apply in the computer application the calculus algorithms for some significant features on which one can evaluate the performances and compare different possibilities.

The functional parameters that can be calculated with the developed applications are: the elasticity of the pinion tooth (V1), the elasticity of the gear tooth (V2), and the elasticity of the pair of teeth (V); the normal force on the active profile (Fn); the transmission error due to the displacements of the teeth (Ae); the specific sliding (η'_1, η'_2); the instantaneous efficiency and the medium efficiency (η); the equivalent bending stress (σ_{ech}); the contact stress (σ_H).

The mentioned parameters are calculated in an established number of meshing points, and studying the variation can be carry out the maximum values on the meshing period – Table 2.

Table 2. Functional parameters

Fn [N]	Ae [mm]	η'_1	η'_2	η	σ_{e1} [N/mm^2]	σ_{e2} [N/mm^2]	σ_H [N/mm^2]
2976	0.0024	0.4383	0.3326	0.9873	11.6587	10.1720	272.9418

2.6. Optimization of functional parameters

The advantages of the computer design is the short time for obtain the dates, the comparison between the results obtained using different design parameters, the optimization by comparing the functional parameters and choosing the designing variable in relation with the requirements imposed by the beneficiary.

For studying and comparing a great number of possible asymmetric gears designed with the same initial date, obtained by changing the designing variable and select those that give the best solution for a specified objective function it have been developed an optimization application based on the genetic algorithms. The objective function can be obtained as a weighted average of partial objectives.

2.7. Computer aided machining of the asymmetric gears

In the classical manufacturing, with standard tools, of the involute teeth gears, the curves of tooth profile result as tangent curve to the successive positions of the standard generation gear rack, with symmetric profile materialized by the edges of the cutting tool.

In this case, of the direct design of the asymmetric gears, the design of the gear rack it is made after establishing the useful profiles of the teeth, than it is used for obtain the equations of the joint profiles and it is not a standard gear rack. So for being materialized it is necessary the design of a nonstandard tool in relation with the gears parameters. That it is not possible because it involves extra costs. For this can be avoided it is necessary, for the nonstandard gears, that a standardized tool to follow a trajectory so as to realize the already designed profile of the gear.

The date necessary for programing the movements of the tool are contained in the model of the gear and could be transferred to the soft of the CNC machine tool (Figure 2, Figure 3). Gears manufacturing have been realized (Figure 4) on MICROCUT Challenger model MCV-2416, the material used being PEHD, with a standard tool coded 514L030Z4.0-SIRON-A.

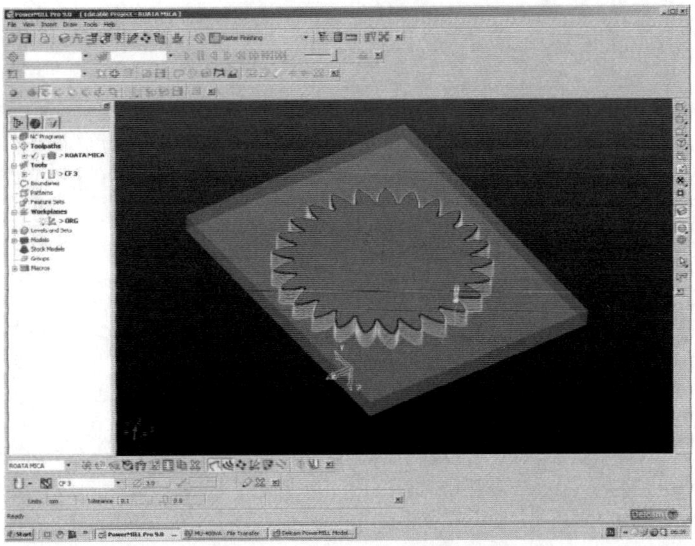

Figure 2. Power Mill Interface containing the pinion

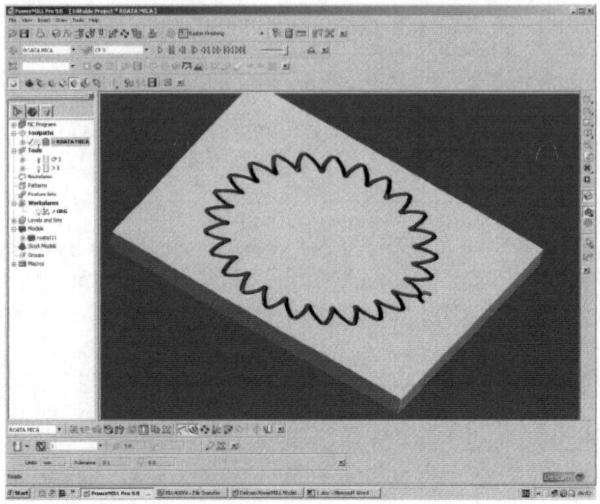

Figure 3. Simulation of the operation

Figure 4. The asymmetric gears designed and manufactured

3. Conclusions

With the original developed computer applications have been realized all the steps necessary for the optimal design of the nonstandard asymmetric gears and those have been obtained on a CNC machine tool with a standard tool. The models of the gears permit to establishing the trajectory of the tool and can be modify, by studying the stress and displacements of the asymmetric teeth, until the best behavior under the load it is attend.

References

[1] Kapelevich A L 2000 Geometry and design of involute spur gears with asymmetric teeth, *Mechanism and Machine Teory* **35** 117-130

[2] Kapelevich A L and Kleiss R E 2002 Direct Gear Design for Spur and Helical Involute Gears , *Gear Technology* september/october

[3] Banica M and Cotetiu R 2006 *Dynamical Optimization of the TipRelief Parameters For an Involute Spurs Gearing with Impose Center Based on Computer Simulation*, 7th International Conference Automation in Production Planing and Manufacturing, Zilina, Slovakia, May, pp11-16

[4] Handra-Luca V and Stoica I A 1982 *Introducere in teoria mecanismelor*, Editura Dacia, Cluj Napoca, Romania

IOP Conf. Series: Materials Science and Engineering **200** (2017) 012013 doi:10.1088/1757-899X/200/1/012013

The impact of overhead lines for employees with stents

P Syrek and M Skowron

Department of Electrical and Power Engineering, AGH University of Science and Technology, Krakow, Poland

E-mail: syrekp@agh.edu.pl

Abstract. The aim of article is to discuss interaction between stents implanted in the body of worker and harmonic magnetic field in the vicinity of electric wires. In last decades, a growing proportion of people has any devices implanted, to list: cardiac pacemakers, cardioverter – defibrillators. Recommendations of International Commision on Non-ionizing Radiation Protection (ICNIRP), and resctrictions imposed in different states, may exlude specific individuals from their duties. The autors focused on the situation, when the employee with stent, works in the immediate vicinity of overhead electric wires, cleaning with dry ice the electric insulators.

1. Introduction

The aim of article is to discuss interaction between stents implanted in the body of worker and harmonic magnetic field in the vicinity of electric wires. In last decades, a growing number of people has any devices implanted, to list: cardiac pacemakers, cardioverter –defibrillators. Together, there is a correlation between an age and the occurrence of cardiovascular diseases' probability. At the same time, to treat the worker as specialist and consider him as worker of extensive experience, many years in the profession are needed. These cause, that there are the situations, excluding part of the personel from their duties. Recommendations of International Commission on Non-ionizing Radiation Protection (ICNIRP), and restrictions imposed in different states, may exclude specific individuals from their duties. The authors focused on the situation, when the employee with stent, works in the immediate vicinity of overhead electric wires, cleaning the electric insulators with dry ice. It is worth to assert, that active screening is complicated to introduce in presented case and in the case of any emergency situation, when the system is invalid may pose the danger for worker. This is the situation where they are exposed to sinusoidal-shape magnetic field. Investigation was conducted with the help of circuit model of stent, instead of field model. And focused on to find the critical location of stent relative to the wires.

2. Low Frequency Magnetic Field

2.1. Parameters of the field generated by three phase line

The band up to hundered kHz is considered to be the limit of low frequency domain [1]. Especially, power lines frequency renders it possible to treat both electric and magnetic components of electromagnetic field (EMF) separately. Human beings are simultaneously exposed to varied frequencies [2], to list: FM radio, mobile phone stations, transformer stations. Focusing on power line frequencies, the band designated as Extremely Low Frequency band, some frequencies may be considered, depending on type of line, number of wires, the power system grounding method and parameters [3], what effects with higher harmonics. However, in the articles the results are for 50 Hz. So the Total Harmonic Distortion (THD) coefficient is set to 0.

2.2. Magnetic field from overhead power lines and its limitation

To obtain the magnetic field distribution of power frequency, three phase systems in vicinity of power lines has to be considered, and each wire has to be considered separately. The instantaneous values of magnetic field distribution to a large extent depend on location point with respect to the power line [4]. Thus, obtained are three distributions of vector field. Instantaneous root mean square values of magnetic component traces an ellipse [5]. If, in any point, the B-vector is in time domain, then all its component may be expressed as a complex number (here, exemplary, *x*-component):

$$B_{\mathrm{X}}^{\mathrm{pn}}(t) = B_{\mathrm{Xmax}}^{\mathrm{pn}} \sin(2\pi f t + \alpha) \rightarrow \underline{B_{\mathrm{X}}^{\mathrm{pn}}} = \frac{B_{\mathrm{Xmax}}^{\mathrm{pn}}}{\sqrt{2}} \mathrm{e}^{\mathrm{j}\alpha} \tag{1}$$

where: B^{pn} – magnetic flux density (the index "pn" – phase number), f – frequency, α - phase angle.

All components of the B-vector, are complex sum of vectors (shifted in time due to phase sequence):

$$\underline{B_{\mathrm{X}}} = \sum_{\mathrm{pn}=1}^{3} \underline{B_{\mathrm{X}}^{\mathrm{pn}}} \tag{2}$$

In fine, root mean square of the magnetic flux density:

$$B_{\mathrm{RMS}} = \sqrt{\mathrm{Re}(\underline{B}_{x})^2 + \mathrm{Im}(\underline{B}_{x})^2 + \mathrm{Re}(\underline{B}_{y})^2 + \mathrm{Im}(\underline{B}_{y})^2 + \mathrm{Re}(\underline{B}_{z})^2 + \mathrm{Im}(\underline{B}_{z})^2} \tag{3}$$

Table 1. Reference levels for exposure to time-varying magnetic field component (rms values) [6]. In the formula the *f* symbol is frequency

Fraquency range (Hz)	Magnetic flux density (mT) for occupational exposure	Magnetic flux density (mT) for public exposure
1–8	200/f^2	40/f^2
8–25	25/f	5/f
25–50	1	0.2
50–400	300/f	0.2

Table 2. Exposure limit values and Actions Levels (ALs) in frequency range [7]

Fraquency range (Hz)	Magnetic flux density (mT), Low ALs	Magnetic flux density (mT), High ALs
1–8	200/f^2	300/f
8–25	25/f	300/f
25–300	1	300/f
300–3000	300/f	300/f

Tables 1-2 give the limits of magnetic component of EMF in low frequency band, due to committees [6], [7].

The below figures present the equipotential lines for the absolute value of magnetic flux density; the levels are adopted for frequency of 50 Hz. The border of zone was added, this is protection zone due to polish regulations [8] – this is the example of very restrictive – its level is equal to 83 μT. In [9], the effect of 100 μT magnetic field at power lines frequency on patients with implanted cardiac pacemaker, was studied.

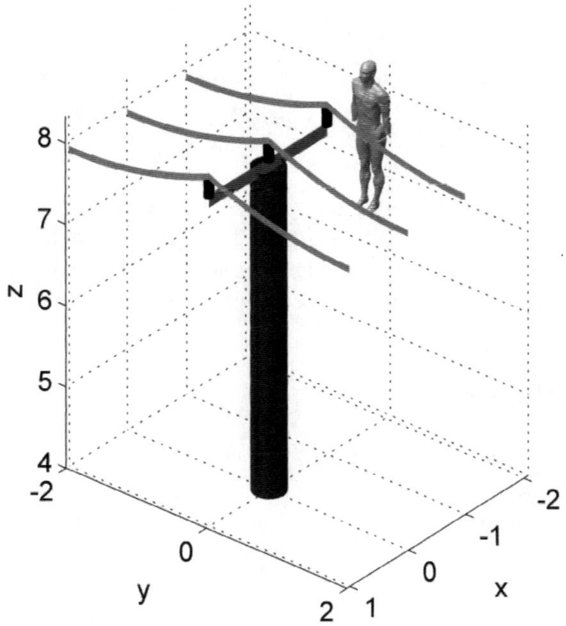

Figure 1. Visualization of human body model, next to medium-voltage line. Exemplary location near the insulators

Figure 1 presents the view of exemplary wires location, and a possible orientation of model representing the worker body. This model comes from Virtual Family [10], and serves to obtain mutual distance between both, stent implanted in chest and wires. The power line is medium voltage power line, but the most important information is its current load. To evaluate the impact of power line on stent, the current root mean square is set to 300 A. However, the power line load, in reality, may differ significantly in time. The tower line is 8 m, and planar orientation of wires is adopted; the distance between wires equals 1 m.

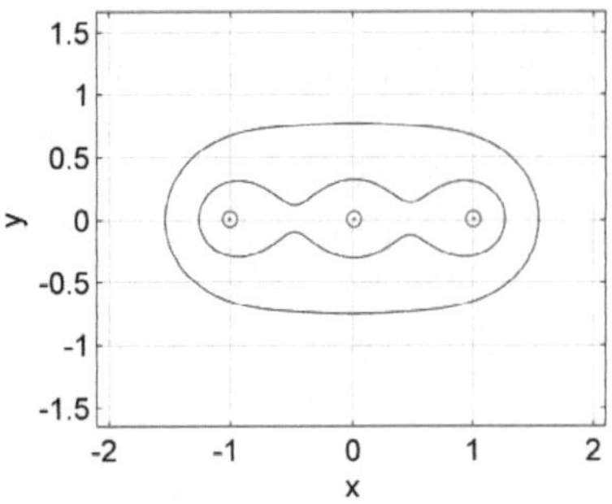

Figure 2. Planar location of wires. Equipotential lines for: 1 mT, 200 μT and 100 μT

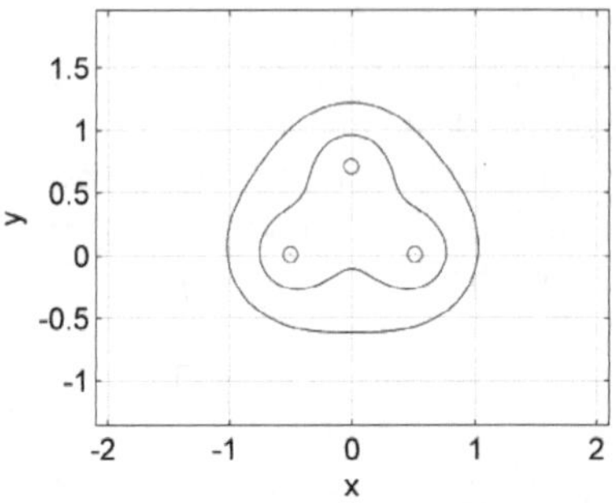

Figure 3. Triangular orientation of wires. The distance between is equal to 1 meter

Figures 2 and 3, present the contour lines, and this scope gives the possibility to evaluate how long any zone reaches. The contours represents the magnetic flux density module of: 1 mT, 200 μT and 100 μT. The level of 100 μT was the border, it was assumption, that the employee is not allowed within the magnetic field higher than such a limit. The range of this zone does not exceed 1 meter from individual wires.

3. Current density in stent branches

3.1. Circuit model for the calculation of stent current

Calculation of the distribution of magnetic component may be performed in various ways, i.e. analytically [11], assuming straight wires or shaping them in any form – numerically, as in many electrical equipment, which can describe the trajectory of the movement of charge (current) by means of parametric curves [12], taking into consideration the conductor sagging between towers.

EMF distribution analysis is carried out using a variety of models. The degree of complexity depends on physical quantity, which is a desirable to determine and thus beginning from the simplified models [13], [14] – reduced in the calculation of the curvilinear integral, and field models, to analyze the area of low electric conductivity [15]. The subareas containing a high electrical conductivity, characterized by such materials as copper, aluminum, titanium, requiring the use of very complex models [16].

However, in the case of a stent, none of these methods do not seem to be adequate, due to the very small volume ratio of the material with high electric conductivity to the surrounding area of low conductivity. Hence the field model should be replaced by circuit model; then the current of each stent branch has to be calculated, to evaluate the response of the object in the context of possible heating up – and to compare the current with relation to acceptable levels for a given section.

The results presented in this paper were determined using circuit model, taking into account the magnetic mutual inductance between the individual wires of both overhead lines and various branches of the stent, and the resistance of the branch stent and stent mutual inductance between branches.

For the calculations the stent is selected, created by branches forming tube (cylinder) of 5 mm radius and 40 mm height. The diameter of the branch stent is 0.8 mm. The stent distinguish 80 branches and 45 nodes. In addition, it is overestimated the electrical conductivity of the material and adopted the value equal to the copper conductivity.

3.2. The results of the different positions of the stent

The results were sought for different stent location, on the assumption, that their location is at the border of a 100 μT area. They were taken into account locations, where the induced currents reach the highest values. Figure 4 shows the results for the stent placed in a sinusoidal field at a frequency of 50

Hz. Results are shown for three different options for the location of – orthogonal to each other. The lowest values were obtained when the axis of the cylinder formed by the stent is parallel to the dominant component of the magnetic field. The other two cases show a stent whose axis is perpendicular to the dominant component.

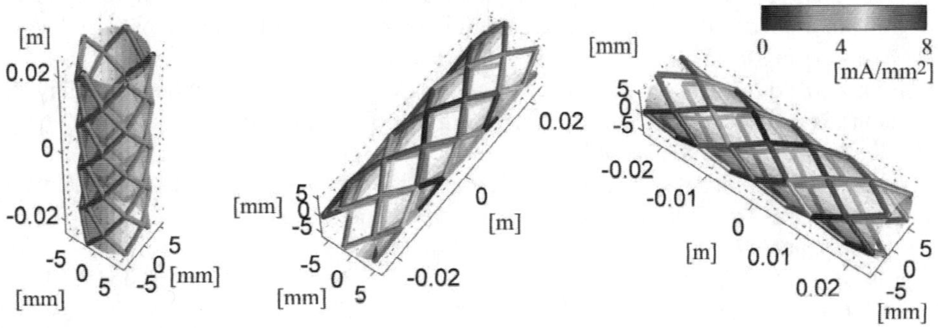

Figure 4. Three mutually perpendicular orientation of the stent relative to the dominant component of magnetic field

The obtained result suggests, that the current density does not pose a danger, for the material forming stent branch, or for the anatomical structures of the worker body. The current density does not exceed 8 mA/mm^2. It is worth to note, that in the simulated examples, the current density does not change under the assumption of a larger stent branch cross–section.

4. Conclusions

The results allow to conclude that if there is a safe distance for employee kept, interaction of line on the stent is negligible.

It is worth of noting that the considerations are valid for a wide range of applied voltage, as the current (line load) is the generative agent impacting the stent implanted in the patient body. Thus, the distance between the employee and the wires determine: the risk of electric shock upon contact with wire.

Thus the stent, unlike other devices implanted in the body, e.g. cardiac pacemakers, does not preclude the continuation of the employment in handling the overhead lines. One should also note that the level adopted for the calculation of a 100 μT, is lower than suggested for public exposure in this frequency range, therefore considerations can be extended to bystanders located in the vicinity of the line.

References

[1] Salceanu A, Paulet M, Ursache S and Poenaru M M 2016 *Evaluating the Cumulative Exposure to LowFrequency Electric Fields,* International Conference and Exposition on Electrical and Power Engineering (EPE 2016), Iasi, Romania, October 20-22, pp 408-412

[2] Nica I, David V, Lazarescu R C and Ciorap M 2012 *Estimation Of Human Exposure In Some Electromagnetic Environments,* International Conference and Exposition on Electrical and Power Engineering (EPE 2012), Iasi, Romania, October 25-27, pp 541-543

[3] Neamt L, Chiver O, Barz C, Costea C and Erdei Z 2014 *Considerations About Power System Grounding for Different Soil Structure,* International Conference and Exposition on Electrical and Power Engineering (EPE 2014), Iasi, Romania, October 16-18, pp 1034-1038

[4] Morega M, Baran I M and Morega A M 2014 *Evaluation of Environmental Low Frequency Magnetic Fields in Occupational Exposure,* International Conference and Exposition on Electrical and Power Engineering (EPE 2014), Iasi, Romania, October 16-18, pp 522-527

[5] Brandao Faria J A and Almeida M E 2007 Accurate Calculation of Magnetic-Field Intensity Due to Overhead Power Lines with or without Mitigation Loops with or without Capacitor Compensation, *IEEE Transactions on Power Delivery* **22**(2) 951-959

[6] ***2010 ICNIRP Guidelines for Limiting Exposure to Time-Varying Electric and Magnetic Fields (1 Hz – 100 kHz), *Health Physics* **99**(6) 818-836

[7] ***2013 Directive 2013/35/EU of the European Parliament and of the Council of 26 June 2013 on the minimum health and safety requirements regarding the exposure of workers to the risks arising from physical agents(electromagnetic fields)

[8] Ciesla A, Kraszewski W, Skowron M and Syrek P 2011 Determination of safety zones in the context of the magnetic field impact on the surrounding during magnetic therapy, *Przeglad Elektrotechniczny* **87**(7) 79-82

[9] Trigano A, Blandeau O, Souques M, Gernez J P and Magne I 2005 Clinical study of interference with cardiac pacemakers by a magnetic field at power frequencies, *Journal of American College of Cardiology* **45**(6) 896-900

[10] Christ A, Kainz W, Hahn E G, Honegger K, Zefferer M, Neufeld E, Rascher W, Janka R, Bautz W, Chen J, Kiefer B, Schmitt P, Hollenbach H P, Shen J X, Oberle M, Szczerba D, Kam A, Guag J W and Kuster N 2010 The Virtual Family – Development of Surface-Based Anatomical Models of Two Adults and Two Children for Dosimetric Simulations, *Physics in Medicine and Biology* **55**(2) 23-38

[11] Lunca E, Istrate M and Salceanu A 2013 Comparative Analysis of the Extremely Low-Frequency Magnetic Field Exposure from Overhead Power Lines, *Environmental Engineering and Management Journal* **12**(6) 1145-1152

[12] Ciesla A and Syrek P 2012 Parametric Curves in the Specification of Windings of Applicators used in Magnetotherapy, *Przeglad Elektrotechniczny* **88**(9a) 213-216

[13] Noetscher G M, Makarov S N, Sciré-Scappuzzo F and Pascual-Leone A 2013 A Simple Absolute Estimate of Peak Eddy Currents Induced by Transcranial Magnetic Stimulation Using the GR Model, *IEEE Transactions on Magnetics* **49**(9) 4999-5003

[14] Syrek P, Skowron M and Ciesla A 2016 *Therapy's Individualization of Bone Injuries with the Magnetic Field Applicators*, International Conference and Exposition on Electrical and Power Engineering (EPE 2016), Iasi, Romania, October 20-22, pp 435-438

[15] Plonecki P, Starzynski J, Wincenciak S and Sawicki B 2009 *Modelling of Eddy-Currents Using Electric Scalar Potential*, XV International Symposium on Theoretical Engineering (ISTET), Lubeck, Germany, June 22-24, pp 294-298

[16] Miaskowski A, Gas P and Krawczyk A 2016 *SAR Calculation for Titanium Bar-Implant Subjected to Microwave Radiation*, 17th International Conference Computational Problems of Electrical Engineering (CPEE), Sandomierz, Poland, September 14-17, pp 1-4

IOP Conf. Series: Materials Science and Engineering **200** (2017) 012014 doi:10.1088/1757-899X/200/1/012014

Study of electromagnetic radiation produced by household equipment

A Iagăr, G N Popa and C M Diniş

Politehnica University of Timisoara, Department of Electrical Engineering and Industrial Informatics, 5 Revolution Street, Hunedoara, 331128, Romania

E-mail: angela.iagar@fih.upt.ro

Abstract. This paper presents the experimental study of electromagnetic radiation produced by different household equipment. Measurements were performed at various distances from the tested devices and in different operating modes of these, using the FA306 electromagnetic field analyzer. Also, it has been analyzed the cumulative effect of multiple sources. The experiments allowed the identification of measures to minimize the effects of exposure to artificial electromagnetic radiation produced by electrical household equipment.

1. Introduction

Cosmic radiation, energetic fields of the earth and man-made technology constantly interfere with the human body, because the human body is a form of electromagnetism.

Constant concern to improve the quality of life and the use of advanced industrial technologies determined an explosive increase in the number of non-ionizing electromagnetic radiation artificial sources, both in household and industry [1-8].

Today, millions of people are exposed to artificial electromagnetic fields of different frequencies and intensities throughout the day, and probably all night.

It is estimated that 63% of the population already use a mobile phone, and this number is growing [1]. In addition, in the last decade occurred the technologies of wireless data connection Wi-Fi, Bluetooth, etc. In fact, Wi-Fi is present almost in every home, school and institution.

Dependency of population (mostly young) from gadgets is growing, and will appear increasingly wide range of devices generating electromagnetic disturbance, with a growing variety of uses.

Electromagnetic pollution is probably the biggest problem of this century, being more dangerous than other forms of pollution [1-5]. In addition, electromagnetic pollution can not be feeling directly, but only through its effects on the health of population.

2. Aspects of electromagnetic biocompatibility

Currently, artificial electromagnetic radiation level has raised exponentially, due to the new disturbance devices. Artificial electromagnetic radiations have an enormous intensity (hundreds of thousands of times) compared to the natural microwave spectrum (Figure 1, [9]).

Although electromagnetic pollution issues have a huge psychological impact on the population, the biological effects of exposure and potential health risks are not sufficiently known.

Since 1960, thousands of studies have demonstrated the harmfulness of long-term exposure to artificial electromagnetic fields.

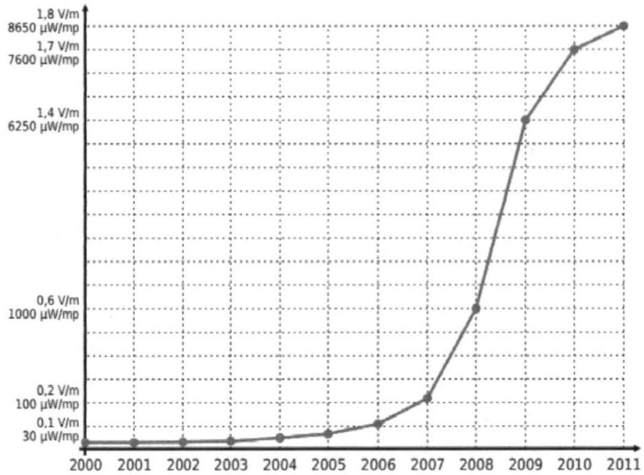

Figure 1. The average increase of artificial electromagnetic radiation in urban areas (900 MHz-2.5 GHz) [9]

The effects of radio frequency fields on human tissue can be thermal and non-thermal. Thermal effects can occur due to electrical conductivity of most biological tissues. The electric fields, induced in the body, generate electrical currents, whose energy causes the increase of local temperature. Because many biochemical reactions are strongly dependent on temperature, the temperature change may cause biological effects. Some studies have shown that hyperthermia causes: burns, behaviour changes, reproduction alteration, blood-brain barrier alteration, cataract, etc. Also, malformations are detected under sustained hyperthermic exposure [3-5].

Non-thermal effects can occur due to many interactions between different components of the electric field and biological tissue.

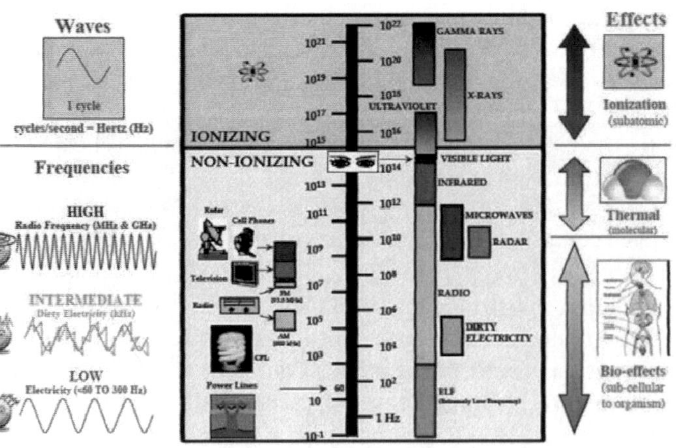

Figure 2. Electromagnetic spectrum and biological effects of artificial electromagnetic radiation [10]

Rules and limitations issued by ICNIRP [11], although obtained through scientific methods, represent some relative reference results, because they took into account only the effects of radiation in the short term.

At national level, was adopted the Government Decision no. 1136, from 30/08/2006 [12] (correspond to European Directive 2004/40/CE and further [13]), on the minimum health and safety requirements regarding the exposure of workers to the risks arising from electromagnetic fields (0-300

GHz). They refers to the safety risks due to recognized adverse effects on the human body at short-term (caused by the circulation of induced currents, by energy absorption and by contact currents), but does not envisage long-term effects.

In the last few decades, many studies established that adverse health effects occur at far lower levels of electromagnetic fields exposure (both extremely low frequency electromagnetic fields, ELF and radiofrequency radiation, RF), where no heating occurs; some effects are shown to occur at several hundred thousand times below the existing public safety limits.

Heart and brain are regulated by internal bioelectrical signals. Environmental exposures to electromagnetic radiation can interact with fundamental biological processes in the human body [1].

Exposure to electromagnetic radiation causes, for many peoples, the electromagnetic hypersensitivity (EHS), manifested by the following symptoms: inability to concentrate, panic state, disorientation, visual disturbances, skin rashes, muscle weakness and even fainting [14-17].

Essentially, artificial electromagnetic fields decrease the hormone melatonin production. Melatonin regulate the sleep, and play an important role in the immune system; therefore appear depression, sleep disturbances, fatigue, increased immune sensitivity.

The principal action of electromagnetic fields on the human body is to accelerate the emergence (or worsening) of cardiac, vascular and neurological disease, and psychiatric disorders. Electromagnetic pollution can cause: headaches, arrhythmia, Parkinson, back pain and even autism [18-28].

It was observed a continuous growth of many types of cancer (glandular and central nervous system), proportionally with increasing levels of electromagnetic fields (particularly in radiofrequency and microwave spectrum) [1], [2], [18]. The risk of acoustic neuroma, a benign tumor of the auditory nerve, was 50% higher in the case of people who used mobile phones for 6 years or more [1-3], [22]. The rate of death from brain cancer among the users of mobile phones without hands-free was higher than the rate of death from brain cancer among those who use hands-free phones, which were kept away from their head [1].

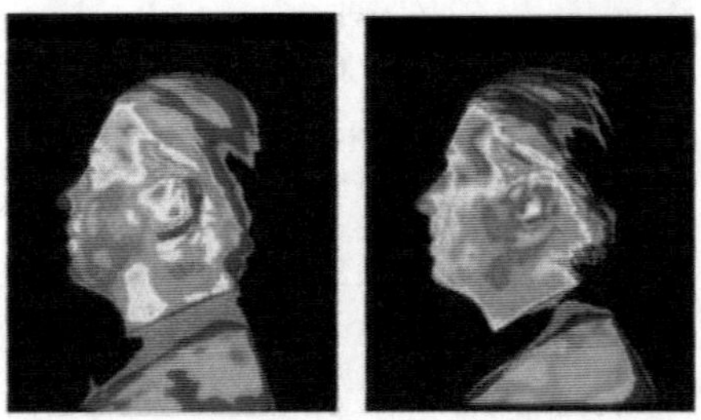

a b

Figure 3. Thermographic image of the head: (a) no expure to mobile phone radiation; (b) after a 15 minute phone call [1]

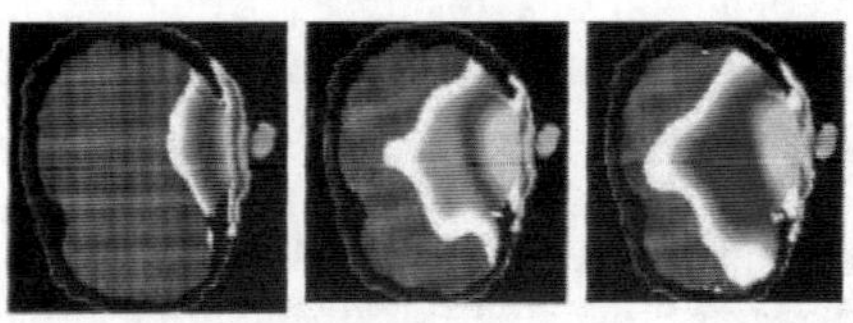

a b c

Figure 4. Penetration of electromagnetic radiation from mobile phone in an adult brain (a), 10 years old child (b) and 5 years old child (c) [1]

The skull of children is deeply penetrated by electromagnetic radiation emitted by a mobile phone (Figure 4). Blood-brain barrier (which protects the brain from toxins blood) can be compromised by this radiation that creates micronuclei in blood cells (characteristic sign of cancer). Thus can be explained the increased number of brain tumors in children [1].

Depending on the value of power density, it was observed the occurrence of following adverse health effects [4], [9], [29]:

- sleep disturbances, weakness, fatigue and pain (if power density is over 0.001 $\mu W/cm^2$);
- cells growth is affected, so epithelial tumors occur (if power density is above 0.1 $\mu W/cm^2$);
- childhood leukemia; motor disturbances, memory and attention disturbances; leukocyte changes in children of school age (if power density is within 1...10 $\mu W/cm^2$);
- hearing and speech disturbances, leukemia, melanoma and other types of cancer (if power density is 10 $\mu W/cm^2$).

3. Experimental measurements for various household equipment

In the experiments we used the FA306 electromagnetic field analyzer. FA306 measures three kinds of fields: high frequency electromagnetic fields (f=800 MHz...2.5 GHz; power density may be in the range of 0.02…39000 $\mu W/m^2$), low frequency electric fields (electric field intensity may be in the range of 0...99 V/m) and low frequency magnetic fields (magnetic flux density may be in the range of 0.01…3.99 μT) [30].

The normal levels and the risk levels of electric and magnetic fields are presented in Table 1. More studies [1-5, 9, 29] revealed that, for intensity of electric field under 6 V/m and magnetic flux density under 0.064 μT, corresponding electromagnetic fields do not cause biological effects, even for a long-term exposure.

Table 1. The risk levels of electric and magnetic fields

Field	*Normal levels*	*Dangerous levels*	*Very dangerous levels*	*Extremely dangerous levels*
Electric, E(V/m)	0...6	6.1...8.9	9...13.9	> 14
Magnetic, B(μT)	0...0.065	0.066...0.069	0.1...0.249	> 0.25
Magnetic, B(mGs)	0...0.65	0.66...0.69	1...2.49	> 2.50
Magnetic, H(A/m)	0...1.625	1.65...2.475	2.5...6.225	> 6.25

In the cases of indoor living or working areas, the safe values are: 1 $\mu W/m^2$ for power density, 10 V/m for electric field intensity, and 0.2 μT for magnetic flux density.

For sleeping areas, the safe values are: 1 $\mu W/m^2$ for power density, 10 V/m for electric field intensity and 0.2 μT for magnetic flux density.

We analyzed following electrical devices: mobile phones, wired landline phone, wireless routers, laptop, microwave, induction hob, TV sets, fluorescent lamps with electronic ballast and electric power cable.

Measurements were performed at various distances from the tested devices and in different operating modes of these. Also, we analyzed the cumulative effect of multiple sources (i.e. wireless router and laptop, wireless router and TV).

3.1. Mobile phone Samsung Galaxy S6 Edge Plus

In stand-by mode, measurement results revealed that the mobile phone is not dangerous in working areas (Table 2). Instead, in stand-by mode, the mobile phone is dangerous in sleeping areas, because the power density exceeds the safety limit.

The experiments showed that dialing is the most dangerous operating mode of mobile phone (Table 3).

In dialing mode, at a distance of 0.5 cm, power density is very close to the limit of measurement range of FA306 analyzer. At a distance of 5 cm, power density decreases 800 times, and at a distance of 10 cm, power density decreases more than 10600 times.

We found that values of power density are lower if touchscreen is turned off. Magnetic flux density and electric field intensity are in the safety domain, in all operating modes of mobile phone.

Table 2. Mobile phone Samsung Galaxy S6 Edge Plus in stand-by

$d(cm)$	$S(\mu W/m^2)$ Peak	$B(\mu T)$	$E(V/m)$
0.5	0.8	0.02	0.2
1	0.63	0.02	0.2
1.5	0.63	0.02	0.2
2	0.63	0.02	0.2
2.5	0.5	0.01	0.2
3	0.4	0.01	0.2
3.5	0.32	0.01	0.2
4	0.32	0.01	0.2

Table 3. Samsung Galaxy S6 Edge Plus mobile phone in other operating modes

$d(cm)$	$S(\mu W/m^2)$ Peak	$S(\mu W/m^2)$ Peak	$S(\mu W/m^2)$ Peak	$B(\mu T)$	$E(V/m)$
	Dialing mode	*Call with touchscreen off*	*Call with touchscreen on*		
0.5	32000	320		0.03	0.2
1	5000	250		0.03	0.2
1.5	4000	200		0.03	0.2
2	3200	160		0.03	0.2
2.5	2000	125		0.03	0.2
3	1500	80	200	0.03	0.2
3.5	1000	63	160	0.03	0.2
4	800	40	125	0.03	0.2
4.5	100	32	50	0.03	0.2
5	40	12.5	40	0.03	0.2
5.5	32	10	25	0.03	0.2
6	16	8	20	0.03	0.2
6.5	10	6.3	12.5	0.03	0.2
7	8	4	8	0.02	0.2
7.5	8	4	6.3	0.02	0.2
8	6.5	3.2	5	0.02	0.2
8.5	5	2	4	0.02	0.2
9	4	1.6	3.2	0.02	0.2
9.5	3.2	1.5	2.5	0.02	0.2
10	3	1.25	2	0.02	0.2

3.2. Digi mobile phone

If Digi mobile phone (2100 MHz and 900 MHz) is in dialing mode, power density values are extremely dangerous, exceeding very much the safety limit (Table 4).

In the first moment of dialing, at shortest distance from the phone, power density is about 13 times smaller than in the case of Samsung Galaxy S6 Edge Plus.

At a distance of 8 cm, power density decreases 50 times compared to the value corresponding to the distance of 0.5 cm, in both operating modes.

Magnetic flux density and electric field intensity are in the safety domain, in all operating modes of Digi mobile phone.

Table 4. Digi mobile phone

d (cm)	*S(μW/m²) Peak Dialing mode*	*S(μW/m²) Peak Call*	*B (μT)*	*E (V/m)*
0.5	2500	2500	0.04	0.3
1	1250	500	0.03	0.3
1.5	1000	400	0.03	0.3
2	800	300	0.03	0.3
2.5	630		0.03	0.3
3	500	250	0.03	0.3
3.5	400		0.03	0.3
4	320	125	0.03	0.3
4.5	250		0.03	0.3
5	200		0.03	0.3
5.5	160		0.03	0.3
6	125	100	0.03	0.3
6.5	100		0.03	0.3
7	80		0.03	0.2
7.5	60		0.03	0.2
8	50	50	0.03	0.2

3.3. Wired landline phone

In stand-by mode of wired landline phone (Table 5, wireless deactivated), electric field intensity has values above 15 V/m (dangerous to health). Magnetic flux density is in the safety domain. Power density exceeds 2.5 times the safety limit at a distance of 5 cm from the phone, and falls below the safety value at distance of 30 cm, in stand-by.

If wireless is activated (Table 6), values of power density increase hundreds of times; electric field intensity is higher about 2.4 times at the distance of 10 cm from the phone, in this case.

Table 5. Wired landline phone in stand-by, wireless deactivated

d(cm)	*S(μW/m²) Peak*	*B(μT)*	*E(V/m)*
5	2.5	0.02	20
10	2	0.02	17
20	1.25	0.02	16
30	0.8	0.02	15

Table 6. Wired landline phone in stand-by, wireless activated

d(cm)	*S(μW/m²) Peak*	*B(μT)*	*E(V/m)*
10	2000	0.02	42
17	800	0.02	23
28	500	0.02	12
48	320	0.02	5.3

3.4. Wireless routers

At a distance of 0.32 m from the first analyzed wireless router (TP-LINK TL-WR941ND, 2.4 GHz-2.4835 GHz, 450 Mbps WAN), power density is extremely high. At a distance of 1.74 m, power density decreases more than 3000 times. Electric field intensity has dangerous levels at distances less than 0.6 m, but at distances greater than 0.8 m, electric field intensity is normal (Table 7).

Table 7. Wireless router TP-LINK TL-WR941ND

$d(m)$	$S(\mu W/m^2)$ Peak	$B(\mu T)$	$E(V/m)$
0.32	39000	0.03	10
0.41	16000	0.02	7.8
0.5	8000	0.02	7
0.6	6300	0.02	6.5
0.8	3200	0.03	5.5
1	200	0.03	4.9
1.2	100	0.03	2.6
1.48	80	0.02	3.36
1.74	12.5	0.02	1.8

Very close to the second analyzed wireless router (Tenda F300, 2.412 GHz-2.472 GHz, 300 Mbps), at 0.1 m, power density is at the limit of measurement range of FA306 analyzer (Table 8). At a distance of 2.37 m, power density decreases 250 times.

Table 8. Wireless router Tenda F300

$d(m)$	$S(\mu W/m^2)$ Peak	$B(\mu T)$	$E(V/m)$
0.10	39000	0.02	20
0.15	25000	0.02	18
0.20	16000	0.02	15
0.40	12500	0.02	10.8
0.52	3200	0.02	9.1
0.97	1000	0.02	6.5
1.69	630	0.02	2.7
2.37	160	0.02	1.2

3.5. Laptop Dell

If wireless is deactivated (Table 9), power density emitted by laptop exceeds 2...4 times the maximum safety value for living and working areas. Magnetic flux density and electric field intensity are low, not dangerous. The negative influence of wireless is extremely high related to power density. At a distance of 5 cm from the laptop, power density is very close to the limit of measurement range of FA306 analyzer (Table 10).

Table 9. Laptop Dell with wireless deactivated

$d(cm)$	$S(\mu W/m^2)$ Peak	$B(\mu T)$	$E(V/m)$
5	4	0.06	0.4
10	4	0.05	0.4
15	3.2	0.04	0.4
20	3	0.04	0.4
25	2.5	0.04	0.4
30	2.3	0.04	0.4
40	2.1	0.03	0.3
60	2	0.03	0.2

Table 10. Laptop Dell with wireless activated

d(cm)	*S(μW/m²) Peak*	*B(μT)*	*E(V/m)*
5	32000	0.06	0.4
10	6300	0.05	0.4
15	1250	0.04	0.4
20	800	0.04	0.4
40	125	0.03	0.3
60	32	0.03	0.2
5	32000	0.06	0.4

3.6. TV sets
If Smart TV is switched on and wireless activated (Table 11), power density increases hundreds of times compared to stand-by mode. At a distance of 0.05 m from the Smart TV, power density exceeds over 800 times the safety value, and decreases about 100 times at a distance of 1.5 m.

At a distance of 0.20 m from TV (Table 12), power density exceeds 280 times the safety level. At a distance of 1.48 m, power density decreases 3.5 times, but remains 80 times higher than the maximum admissible value. Magnetic flux density and electric field intensity are below the safe values.

Table 11. Smart TV Sony

Operation mode	*d(m)*	*S(μW/m²) Peak*	*B(μT)*	*E(V/m)*
Smart TV Sony in stand-by, wireless deactivated	0.05	0.8	0.03	22
	0.1	0.8	0.03	15
	0.5	0.8	0.03	11.5
	1	0.8	0.03	4
	1.5	0.8	0.03	2.8
	2	0.8	0.03	1.1
Smart TV Sony switched on, wireless activated	0.05	800	0.06	20
	0.1	500	0.06	16
	0.5	320	0.06	12.9
	1	20	0.06	5
	1.5	8	0.06	3.5
	2	4	0.05	2

Table 12. Samsung TV switched on, wireless activated

d(m)	*S(μW/m²) Peak*	*B(μT)*	*E(V/m)*
0.20	280	0.02	1.8
0.35	250	0.02	1.2
0.68	200	0.01	0.6
1.10	160	0.02	0.5
1.48	80	0.02	0.2

3.7. Microwave oven
In load operation, electromagnetic pollution produced by microwave oven (with technical data: microwave frequency: 2450 MHz, rated power: 1280 W, output power (microwave): 800 W) is very high (Table 13). At a distance of 2 m from the microwave oven, power density is 32000 μW/m^2.

Electric field has very dangerous values for distances between 5...20 cm from the microwave oven, and drops below 5 V/m at distances greater than 60 cm.

Magnetic flux density has values over 3.99 μT at a distance less than 20 cm, and decreases to 1.17 μT for a distance of 60 cm from the microwave oven.

Table 13. Microwave oven in load operation

d(cm)	*S(μW/m²) Peak*	*B(μT)*	*E(V/m)*
5			23
10			22
20		>3.99	16
30		3.44	4.5
60	>39900	1.17	3.7
200	32000		

3.8. Induction heat plate

In the case of an induction heat plate (with technical data: P=1600 W, f=2500 MHz), in operation mode, magnetic flux density exceeds 3.99 μT at distances less than 0.3 m from the plate, and decreases by almost 6 times at distance of 1 m (Table 14).

At a distance of 0.2 m from the plate, electric field intensity is close to the limit of the measurement range of analyzer, and remains at dangerous levels up to a distance of 1 m from the plate.

Table 14. Induction heat plate

d(m)	*S(μW/m²) Peak*	*B(μT)*	*E(V/m)*
0.2			92
0.3	1.25	>3.99	77
0.5	1	3.48	36
1	0.8	0.6	12.4

3.9. Fluorescent lamps with electronic ballast

We analyzed four fluorescent lamps (FLs) with electronic ballast (4x18 W), connected in parallel (Table 15). At a distance of 0.05 m from FLs, magnetic flux density exceeds 10 times the safety limit, and decreased to a value slightly larger than the safety, at a distance of 1 m.

Electric field intensity has very dangerous values for distances below 0.15 m, but falls below the safe limit at greater distances, over 1 m.

Table 15. Fluorescent lamps with electronic ballast

d(m)	*S(μW/m²) Peak*	*B(μT)*	*E(V/m)*
0.05	0.8	2.03	32
0.1	0.8	0.44	20
0.15	0.8	0.9	11.2
1	0.8	0.03	1
2	0.63	0.03	0.3

3.10. Power cable

At a distance of 2.5 cm from analyzed power cable (Table 16), power density exceeds 125 times the safety level. At a distance of 0.4 m, power density decreases 6.25 times, but remains 20 times higher than the maximum admissible value.

Values of magnetic flux density are below the safe limit. Electric field intensity is extremely high for distances below 0.4 m, the power cable functioning as an antenna.

Table 16. Power cable of a switch

$d(m)$	$S(\mu W/m^2)$ Peak	$B(\mu T)$	$E(V/m)$
0.025	125	0.03	99
0.05	100	0.03	73
0.1	80	0.03	40
0.15	40	0.02	23
0.2	25	0.02	17
0.4	20	0.01	14.8

4. Conclusions

All analyzed household electronic devices loss electromagnetic energy in the environment ("electrosmog"). Level of exposure depends on the proximity to the source that generates electromagnetic radiation.

For all analyzed equipment, operating in the radio frequency band, are available as protective measures, increasing the distance in use and avoid proximity to the source.

The experiments allowed us to identify several specific measures to minimize the effects of exposure to electromagnetic radiation produced by following household equipment:

a). mobile phones:

- when we appeal, we should expect the other person to respond, and then put the phone to the ear. Radiation intensity is highest when dialing (because the phone trying to contact the base), and greatly decreases with increasing the distance from the phone. Switching to speaker and using a hands-free device, increase the distance between the antenna and the head/body;
- call duration should be limited as much as possible;
- SMS communication limited duration of exposure and the proximity of the body;
- when the signal is weak, should not be made calls, because radiation exposure is increased accordingly;
- mobile phones should not be used in enclosed spaces with metal walls (because the metal walls captures the radiation and reflects it back to the occupants of such spaces);
- should not purchase a mobile phone with a high specific absorption rate;

b). wireless router:

- router should be placed as far away from the place where we are staying during the day;
- router must be connected when needed, and disconnected when we sleep;
- it would be much better to use the Internet by optical fiber;

c). laptop:

- if wireless is enabled, keeping the laptop on the legs is very dangerous;
- it should be kept a minimum distance of 40 cm from the laptop;

d). TV/Smart TV:

- watching TV should be at least 2 m away;
- TV should not be kept in the bedroom;
- TV must be unplugged at night;

e). microwave oven:

- it should be used as seldom;
- should not stay near the microwave oven during food heating;

f). induction plate:

- should not stay near the induction plate during food preparation;

g). fluorescent lamps with electronic ballasts and compact fluorescent lamps (CFLs):

- FLs (or CFLs) use should be avoided at distances less than 1 m;
- it would be preferable to use LEDs;

h). power cables of appliances are sources of electromagnetic radiation, functioning as an antenna. Accordingly, should not sit near them, at a distance of less than 1 m.

Romania does not have a well-defined legal framework (as in other European countries, e.g. France), to regulate at least the placement of antennas and wireless use in hospitals and schools. Long-term consequences of this negligence could be serious, because electromagnetic pollution produced by wireless routers is very high.

For the safety of human health, it requires accurate information on exposure levels, and biomedical research concerning possible health effects.

It is necessary to develop an adequate logistics for monitoring electromagnetic fields and their effects, managed by the Public Health Institute, from structure of the Ministry of Health. Besides, national standard guidelines must be reviewed.

References

[1] Drăgulinescu A 2010 *Wireless idols: mobile telephony and electromagnetic pollution (in romanian)*, Christiana, București, Romania
[2] Zamfirescu M, Rusu I, Sajin G, Sajin M and Kovacs E 2000 *Biological effects of radiofrequency electromagnetic radiation and microwave (in romanian),* Editura Medicală, București, Romania
[3] Blake Levitt B 2007 *Electromagnetic Fields. A Consumer's Guide to the Issues and How to Protect Ourselves*, iUniverse, Inc. Bloomington, IN USA
[4] ***Health effects from radiofrequency electromagnetic fields, Report of an Independent Advisory Group on Non-Ionising Radiation, 2013, at https://www.gov.uk/government/uploads/system/uploads/attachment_data/file/333080/RCE-20_Health_Effects_RF_Electromagnetic_fields.pdf
[5] ***International Scientists Appeal to U.N. to Protect Humans and Wildlife from Electromagnetic Fields and Wireless Technology, 11 may 2015, at https://www.emfscientist.org/images/docs/EMF_Scientist_Press_Release.pdf.
[6] Pănoiu M, Pănoiu C and Șora I 2009 Modeling the electromagnetic pollution of the electric arc furnaces, *Revue Roumaine des Sciences Techniques-Serie Electrotechnique et Energetique* **54**(2) 165-174
[7] Pănoiu M, Pănoiu C and Deaconu S 2008 *Study about the possibility of electrodes motion control in the EAF based on adaptive impedance control*, 13th International Power Electronics and Motion Control Conference, Poznan, Poland, September 01-03, pp 1409-1415
[8] Pănoiu M, Pănoiu C and Ghiormez L 2012 *Modeling of the Electric Arc Behavior of the Electric Arc Furnace*, 5th International Workshop Soft Computing Applications (SOFA), Szeged, Hungary, August 22-24, pp 261-271
[9] ***http://www.next-up.org/Newsoftheworld/06Vm.php
[10] ***http://www.apc-romania.ro/ro/i-poluarea-electromagnetica-dauneaza-grav-sanatatii/MzE5LTI.html
[11] ***Guidelines for limiting exposure to time-varying electric, magnetic and electromagnetic fields (up to 300 GHz), International Commission on Non-ionizing Radiation Protection, ICNIRP, 1998.
[12] ***Government Decision no. 1136, from 30/08/2006, on the minimum health and safety requirements regarding the exposure of workers to the risks arising from electromagnetic fields
[13] ***Directive 2013/35/EU of the European parliament and of the council, on the minimum health and safety requirements regarding the exposure of workers to the risks arising from physical agents (electromagnetic fields) (20th individual Directive within the meaning of

Article **16**(1) of Directive 89/391/EEC) and repealing Directive 2004/40/EC
[14] Johansson A, Nordin S, Heiden M and Sandstrom M 2010 Symptoms, personality traits, and stress in people with mobile phone-related symptoms and electromagnetic hypersensitivity, *J. Psychosom Res.* **68**(1) 37-45
[15] Landgrebe M, Hauser S, Langguth B, Frick U, Hajak G and Eichhammer P 2007 Altered cortical excitability in subjectively electrosensitivie patients: results of a pilot study, *J. Psychsom Res* **62**(3) 283-288
[16] Landgrebe M, Frick U, Hauser S, Langguth B, Rosner R, Hajak G and Eichhammer P 2008 Cognitive and neurobiological alterations in electromagnetic hypersensitive patients: results of a case-control study, *Psychol Med.* **38**(12) 1781-1791
[17] McCarty D E, Carrubba S, Chesson AL, Frilot C, Gonzalez-Toledo E and Marino A A 2011 Electromagnetic hypersensitivity: evidence for a novel neurological syndrome, *Int J Neurosci* **121**(12) 670-676
[18] Salama O E, Abou E I and Naga R M 2004 Cellular phones: are they detrimental?, *J Egypt Public Health Assoc.* **79**(3-4) 197-223
[19] Mashevich M, Folkman D, Kesar A, Barbul A, Korenstein R, Jerby E and Avivi L 2003 Exposure of human peripheral blood lymphocytes to electromagnetic fields associated with cellular phones leads to chromosomal instability, *Bioelectromagnetics* **24**(2) 82-90
[20] Hinrikus H, Lass J, Karai D, Pilt K and Bachmann M 2015 Microwave effect on diffusion: a possible mechanism for non-thermal effect, *Biology and Medicine* **34**(4) 327-333
[21] Inomata-Terada S, Okabe S, Arai N, Hanajima R, Terao Y, Frubayashi T and Ugawa Y 2007 Effects of high frequency electromagnetic field (EMF) emitted by mobile phones on the human motor cortex, *Bioelectromagnetics* **28**(7) 553-561
[22] Hyland G J 2000 Physics and biology of mobile telephony, *The Lancet* **356**(9244)1833–1836
[23] Divan H A, Kheifets L, Obel C and Olsen J 2008 Prenatal and postnatal exposure to cell phone use and behavioral problems in children, *Epidemiology* **19**(4) 523-529
[24] Hardell L, Hansson Mild K and Carlberg M 2012 BioInitiative Report Update, Section 11, Use of wireless phones and evidence for increased risk of brain tumors
[25] Hallberg Ö 2016 Cancer incidence versus FM radio transmitter density, *Electromagnetic Biology and Medicine* **35**(4) 343-347
[26] Mahdavi S M, Sahraei H, Rezaei-Tavirani M and Najafi Abedi A 2016 Common behaviors alterations after extremely low-frequency electromagnetic field exposure in rat animal model, *Biology and Medicine* **35**(3) 222-227
[27] Ragy M M 2015 Effect of exposure and withdrawal of 900-MHz-electromagnetic waves on brain, kidney and liver oxidative stress and some biochemical parameters in male rats, *Biology and Medicine* **34**(4) 279-284
[28] Bellieni CV, Pinto I, Bogi A, Zoppetti N, Andreuccetti D and Buonocore G 2012 Exposure to electromagnetic fields from laptop use of "laptop" computers, *Arch Environ Occup Health.* **67**(1) 31-36
[29] ***http://www.bioinitiative.org/report/wp-content/uploads/pdfs/sec24_2012_Key_Scientific_Studies.pdf, Key Scientific Evidence and Public Health Policy Recommendations, Prepared for the BioInitiative Working Group December 2012
[30] ***Users Manual FA306 version 1.0

IOP Conf. Series: Materials Science and Engineering **200** (2017) 012015 doi:10.1088/1757-899X/200/1/012015

The optimization of the position and the magnitude of the clamping forces in machining fixtures

V G Cioată, I Kiss, V Alexa and S A Raţiu

Politehnica University of Timisoara, Engineering and Management Department, 5 Revolution Street, Hunedoara, 331128, Romania

E-mail: cioata.vasile@upt.ro

Abstract. This paper presents a method which allows the optimization of the position and magnitude of the clamping forces for a specific case of processing: milling of a canal in a prismatic workpiece, using a multi-objective method, implemented with ANSYS software. The design variables are location dimensions of clamping elements, whose range of variation is determined by constructive and dimensional considerations of the workpiece-fixture assembly and the magnitude of the clamping forces. The objective functions of optimization are to minimize the maximum total displacement of the selected edge and to minimize the maximum equivalent stress in workpiece. The constraint is the magnitude of the contact forces between locators and the piece's surfaces to be greater than zero. Values of the design variables which meet the objective functions were obtained after the simulation was performed.

1. Introduction

In order to make a workpiece by chip removal processes, it is necessary that the workpiece is installed in the machining fixture or directly on the table of the machine tool in a strictly determined position against the trajectory of the cutting tool. The installation of the workpiece has two functional phases: orientation and clamping.

A strictly determined position of the workpiece against the cutting edges and the trajectory of the cutting tool is determined by orientation, this position being maintained during the chip removal process by clamping. In order to achieve the clamping, it is necessary to apply a system of clamping forces which will make the contact of the workpieces with the locators and maintain this contact during processing, while also ensuring a maximum rigidity to the workpiece-fixture assembly, leading to the reduction or elimination of vibrations.

In the practice of designing machining fixtures, the stage of calculation and choosing the points for applying clamping forces is very important because it determines, largely, the accuracy of the processing of the workpiece. Determination of the clamping forces must be made by taking into account the least favorable situations, even if these situations are very unlikely to occur.

Also, the evaluation of the contact forces between the elements of the workpiece-fixture system (locators, workpiece, clamping elements) is very important because their magnitude is not constant during the processing of the workpiece, depending on the cutting forces and moments, which have position and direction which vary during the processing.

The extreme values of the contact forces are very important. The maximum and minimum values of the contact forces are determined according to [1], [2].

The maximum value of the contact force for locators with a plane contact surface, must not exceed:

$$f_{c\max} = p_{\max} \cdot \pi a^2 = 1{,}5\sigma_y \cdot \pi a^2 \qquad (1)$$

in order to avoid the deterioration of the surface of the workpiece (indentation will appear). In relation (1), $p_{\max}$ is the maximum value of the contact pressure, σ_y is the yield stress of the workpiece material and a is the radius of the contact area.

At the same time, the minimum contact force, $f_{c\min}$, must prevent the loss of contact between the workpiece and the locators, and must also prevent the workpiece to slipping from the locators.

The conditions can be written as follows:

$$f_{c\min} > 0,$$

$$f_{c\min} \geq \frac{\sqrt{q_x^2 + q_y^3}}{\mu} \qquad (2)$$

where q_x and q_y are components on the x and y directions of the tangential forces which act in the contact area and μ is the static friction coefficient between the workpiece and the locators.

2. Literature review

Within the activity of designing the fixtures used at chip removal processes of the workpiece, the optimization of the orientation layout and the optimization of the fixture layout represent a very important phase, as highlighted in various studies and researches.

Therefore, in [3], the authors optimized the clamping forces in order to minimize the elastic deformations in the workpiece-fixture assembly, using an analysis compatible with the ANSYS software, taking into account the material removal effect. They used the balancing force-moment method and Coulomb friction law in order to determine the clamping forces in 5 positions of the machining tool.

In [4], the authors presented a clamping force optimization method, where the locators were assumed as deformable bodies, and the workpiece as a rigid body. They determined the optimal clamping force for which the location error due to the machining forces is minimum.

W. Chen et al., in paper [5], presented an optimization method for minimizing the maximum deformation of the machined surfaces and maximizing the uniformity of the deformation. The optimization method used in this paper is a multi-objective one, implemented using the ANSYS software package.

An optimization method based on genetic algorithm which is used for the optimization of the orientation layout, is presented in paper [6]. The goal of the study is the identification of such orientation layout for which the maximum elastic deformation in various parts of the workpiece is minimum.

In [7], Qin G H et al. lays down the model of the relation between the clamping deformations and the position error. Based on this model, the optimal design of clamping scheme is obtained, which allows the minimization of the position error of the workpiece.

Paper [8] studies the workpiece-fixture interface compliance when the clamping is made using a standard clamping element and respectively, one which is specially designed. The advantages of using a specially designed clamping element, relative to a standard homologue clamping element, are demonstrated by experiments.

In paper [9], authors presents an optimization orientation layout problem of the clamping forces and the number of clamping elements with the goal of minimizing the elastic deformation of the workpiece due to chip removal processes. The workpiece is assumed as elastic body and the fixture elements are assumed as rigid body. The finite element analysis implemented with ANSYS software is

IOP Conf. Series: Materials Science and Engineering **200** (2017) 012015 doi:10.1088/1757-899X/200/1/012015

used in order to determine the elastic deformation of the workpiece, caused by chip removing processes, and the clamping forces.

Paper [10] presents an optimization method of the fixture layout for peripheral milling of the low rigidity workpieces. The study is focused on optimizing the number and the positions of the locators on the secondary locating surface. In the first stage, the initial number and the position of the locators are determined by adding locators in the position with maximum deformation. In the second stage, the number and positions of the locators are optimized, so as the precision of the machining to be maintained.

3. Model development for FEA analysis

A situation which is encountered in practice is assumed, namely the chip removal process (milling is performed using a frontal cylindrical mill with a 25 mm diameter) of a canal in a prismatic workpiece having 225x122x122 mm in dimension (Figure 1).

The workpiece is oriented and fixed in a working fixture of type 3-2-1, which can take over 6 degrees of freedom. The 6 locators – L1, L2,..., L6 – are disposed in three mutual perpendicular planes, which intersect in the point 0, this being the origin of the coordinates system attached to the workpiece – fixture assembly. Their shape is cylindrical, with a 18 mm diameter and a plane contact surface.

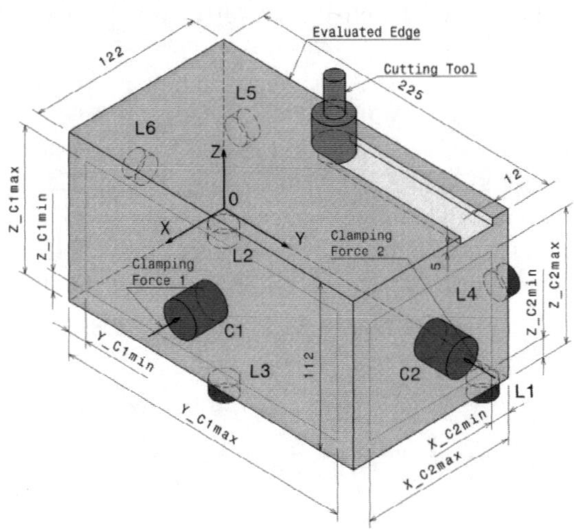

Figure 1. 3D model of the workpiece – fixture system

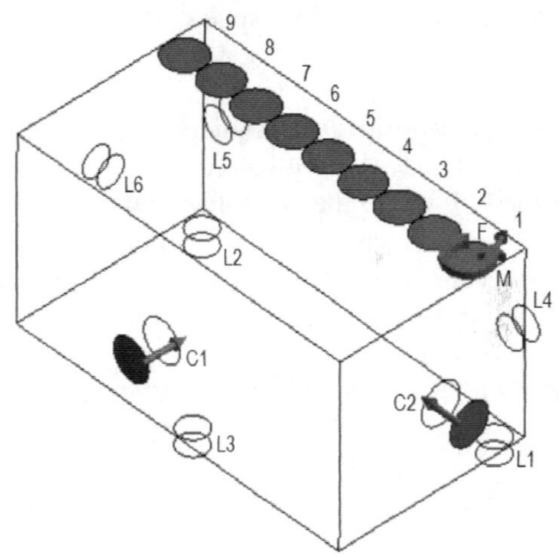

Figure 2. Cutting toll positions considered for analysis

In Figures 1 and 2, is used the next notation: L1,...L6 – locators, C1 and C2 – clamping elements, Y_C1, Z_C1 – position dimensions of the C1 and X_C2, Z_C2 – position dimensions of the C2.

The Clamping Forces 1 and Clamping Force 2 are applied on the side surfaces of the workpiece through the clamping elements C1 and C1, which are cylindrical shape, with a 25 mm diameter. The material of the workpiece is an aluminium alloy and the locators and clamping elements are made of structural steel. The properties of the materials used in the simulation are presented in Table 1.

Table 1. Properties of the materials

Properties	Aluminum Alloy	Structural Steel
Density [kg/m^3]	2770	7850
Young's Modulus [MPa]	71000	200000
Poisson's Ratio [-]	0,33	0,3

Tensile Yield Strength [MPa]	280	250
Tensile Ultimate Strength [MPa]	310	460

The components of the chip removal force and the chip removal moment are calculated according to [11], and their values are: Fx=131 N, Fy=232 N, Fz=55 N and M=2,77 N·m. Total force is:

$$F = \sqrt{Fx^2 + Fy^2 + Fz^2} = 272\ [N].$$

For the simulation of the process, the forces and the moment of the chip removal were applied in 9 equidistant points on the top surface of the workpiece, at a distance of 24,5 mm from the evaluated edge (Figure 2).

The solid model of the system chosen to be analyzed was realized in Ansys Design Modeler and then transferred in Ansys Workbench for structural analysis. All the elements of the system have been modeled as isotropic elastic bodies. The locators and the clamping elements have cylindrical shape, the contact surface is plane. Friction is considered at the contact between the locators and the workpiece. A constant static coefficient of friction (0,1) was used to establish contact properties at the interfaces.

To simulate the locators being rigidly fixed in place, the surface of each locator opposite to the contact was restrained in all three translational degrees of freedom. In order to simulate the clamping force, there have been applied single forces over the surface of both fixture elements, in opposite contact.

4. FEA preliminary analysis

In order to have a term of comparison, a preliminary analysis of the workpiece – fixture system is performed, where the clamping elements are situated as follows: Y_C1 = 112,5 mm, Z_C1 = 56 mm, X_C2 = 61 mm and Z_C2 = 56 mm, and the magnitude of the clamping forces, 640 N and 690 N.

After the analysis, the following results are of interest: the contact forces between locators L1, L2 and L3 and the workpiece, the maximum total deformation of the evaluated edge and the maximum equivalent stress (von Mises stress) in the workpiece. These are presented in the following figures, based on the position of the cutting tool (points 1, …, 9).

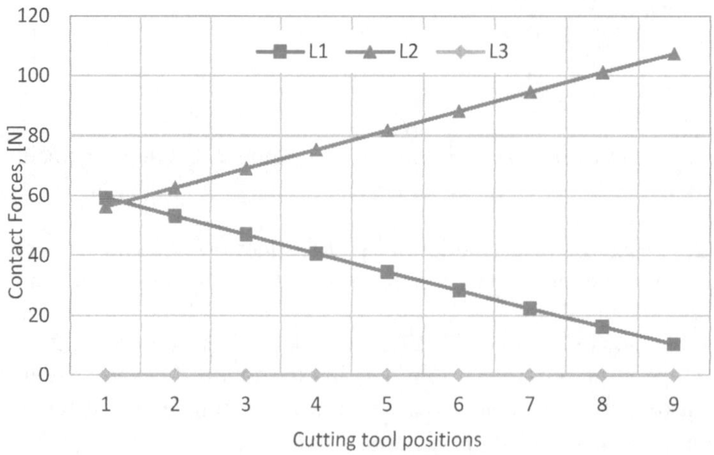

Figure 3. Contact forces between locators L1, L2 and L3 and the workpiece

By analyzing Figure 3, we can observe that the magnitude of the contact forces between locators L1, L2, L3 and the workpiece varies based on the position of the cutting tool. The three forces are larger than zero, and the maximum values do not exceed the critical value which would lead to permanent plastic deformations.

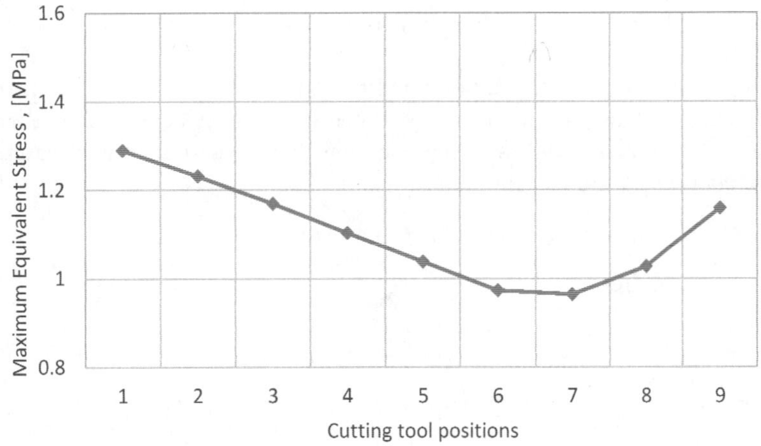

Figure 4. Maximum Equivalent Stress in workpiece

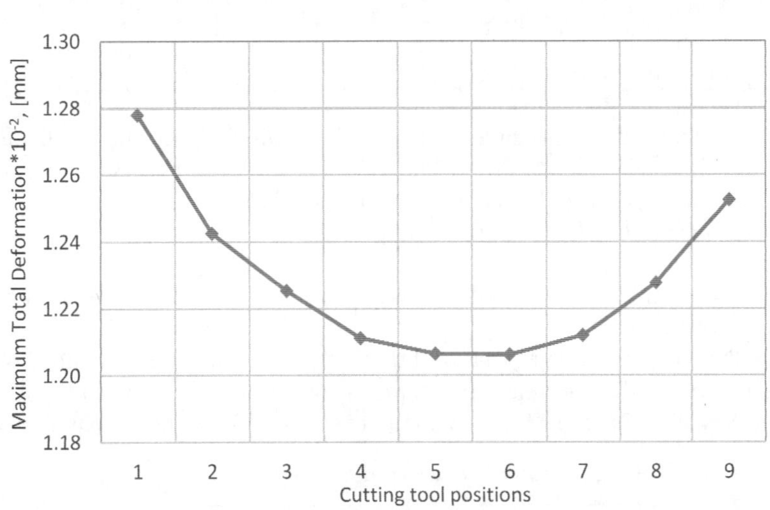

Figure 5. Maximum Total Deformation of the evaluated edge

As regards the maximum equivalent stress which appears in the workpiece, this too varies depending on the position of the cutting tool, a minimum being observed when the cutting tool is between L1 and L2 (Figure 4).

The maximum total deformation of the evaluated edge presents the same bearing of evolution as the equivalent stress (Figure 5).

5. Optimization of the magnitude of the clamping forces and their position

Optimization is made with the goal of determining the optimal magnitude of the clamping forces and their position, so as the stability conditions of the used orientation layout to be accomplished, as well as the precision of the process.

Screening and Response Surface method are used for optimization study. The Screening optimization method uses a simple approach based on sampling and sorting. It supports multiple objectives and constraints as well as all types of input parameters [12].

For this purpose, the following design variables are proposed:

(1) location dimensions of clamping elements and
(2) magnitude of the clamping forces.

The position dimensions of the two clamping elements are specified relative to the origin of the coordinates system: Y_C1, Z_C1 – position dimensions of the C1 and X_C2, Z_C2 – position dimensions of the C2.

The variation ranges of the variables are presented in Table 2. The variation ranges of the position quotas are determined based on constructive and dimensional considerations of the workpiece – fixture assembly, and the variation ranges of the clamping forces are determined based on technical specifications of the clamping system of the fixture.

Table 2. Variation ranges of the design variables

Design variables	Initial value	Min	Max
Y_C1 [mm]	112,5	62,5	162,5
Z_C1 [mm]	56	21	81
X_C2 [mm]	61	31	91
Z_C2 [mm]	56	21	81
Clamping Force 1 Magnitude [N]	640	540	740
Clamping Force 2 Magnitude [N]	690	590	790

The optimization method allows multiple objective functions and constraints. In this case, two objective functions are chosen:

- minimize *Maximum Total Displacement* of the selected edge;
- minimize *Maximum Echivalent Stress* in workpiece;

and three constraints:

- minimum *Values of the Contact Forces* between locators L1, L2, L3 and workpiece must be greater than 0 (zero).

After performing the analysis, optimal values of the design variables were obtained, for which the objective functions and imposed constraints are accomplished. These are presented in Table 3.

The values of the objective functions for the optimal design variables are presented in Table 4.

Table 3. Optimal values of the design variables

Design variables	Initial value	Optimized values
Y_C1 [mm]	112,5	101,26
Z_C1 [mm]	56	50,40
X_C2 [mm]	61	54,90
Z_C2 [mm]	56	50,40
Clamping Force 1 Magnitude [N]	640	576
Clamping Force 2 Magnitude [N]	690	621

The graphical representations from below present a few of the dependences of the objective functions on the design variables, which are useful representations for the stage of orientation and clamping layout, allowing the study of the influence of the variation of the design variables' values on the objective functions.

Table 4. Objective functions for optimal values of design variables

Objective functions	Preliminary values	Optimized values
Total deformation maximum maximum value over time [mm]	0,012	0,0075
Equivalent stress maximum maximum value over time [MPa]	1,275	1,1620
Contact force L1 minimum total [N]	9,997	41,4670
Contact force L2 minimum total [N]	56,678	1,0824
Contact force L3 minimum total [N]	2,395e-09	0,1857

Therefore, the group of figures 6...8 displays the influence of the position of the clamping element C2 on the minimum contact force between the locators (L1, L2, L3) and the workpiece.

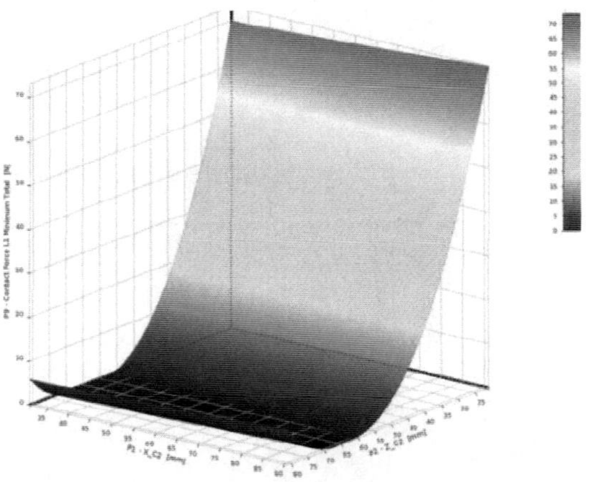

Figure 6. Dependence of the minimum contact force between locator L1 and the workpiece on the variables X_C2 and Z_C2

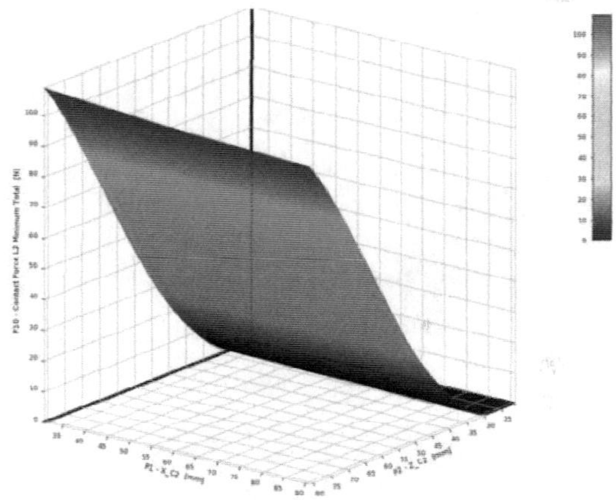

Figure 7. Dependence of the minimum contact force between locator L2 and the workpiece on the variables X_C2 şi Z_C2

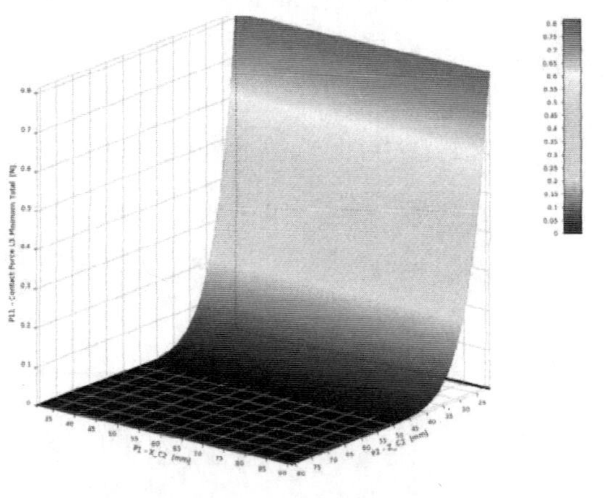

Figure 8. Dependence of the minimum contact force between locator L3 and the workpiece on the variables X_C2 şi Z_C2

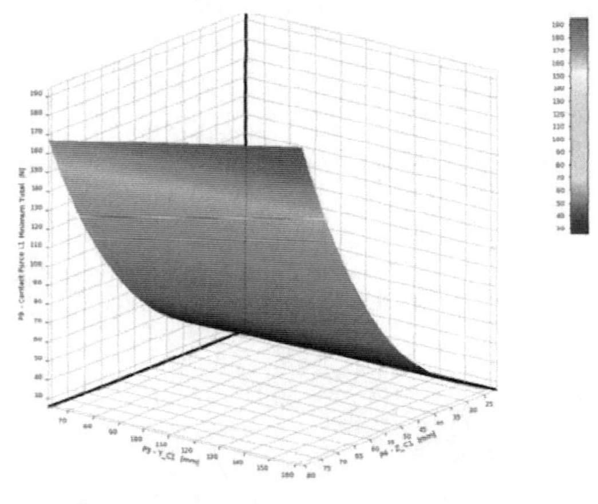

Figure 9. Dependence of the minimum contact force between locator L1 and the workpiece on the variables Y_C1 şi Z_C1

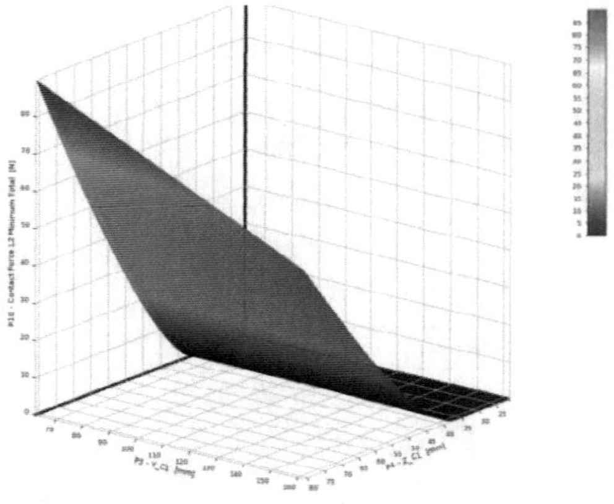

Figure 10. Dependence of the minimum contact force between locator L2 and the workpiece on the variables Y_C1 şi Z_C1

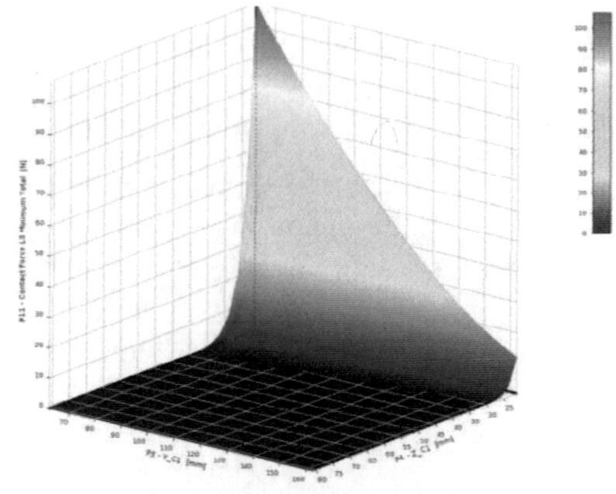

Figure 11. Dependence of the minimum contact force between locator L1 and the workpiece on the variables Y_C1 şi Z_C1

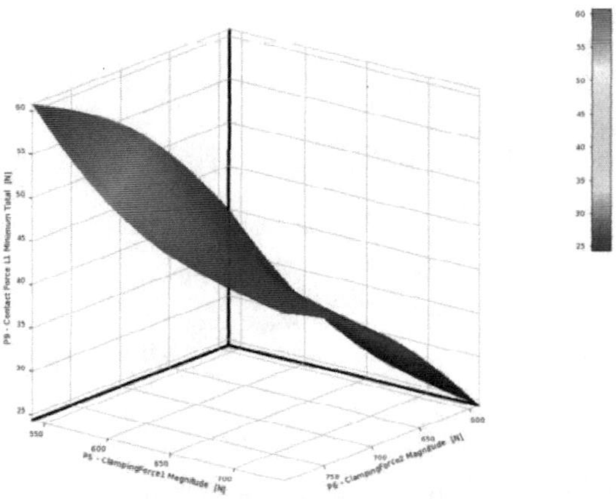

Figure 12. Dependence of the minimum contact force between locator L1 and the workpiece on the magnitude of the clamping forces

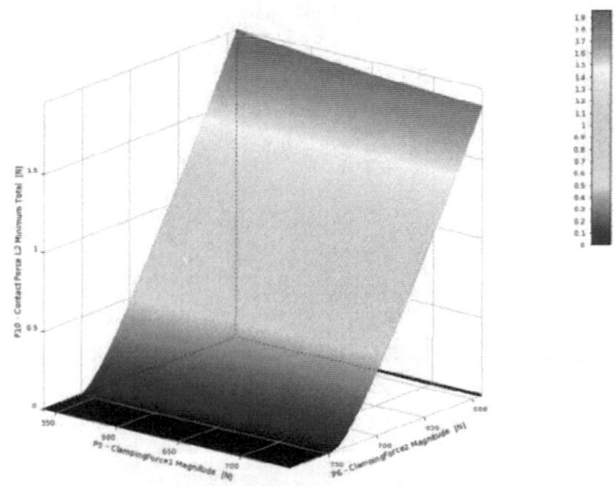

Figure 13. Dependence of the minimum contact force between locator L2 and the workpiece on the magnitude of the clamping forces

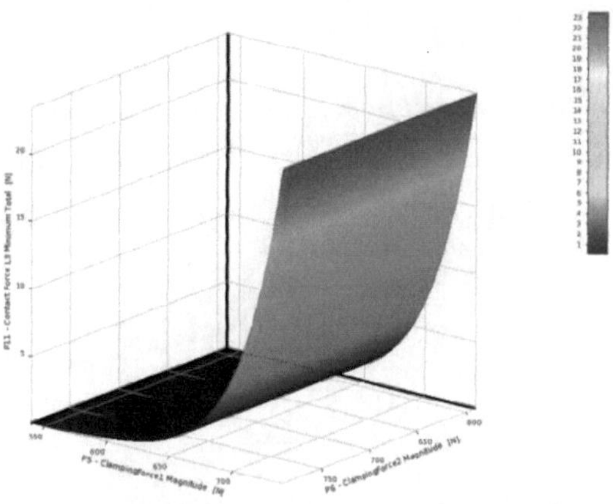

Figure 14. Dependence of the minimum contact force between locator L3 and the workpiece on

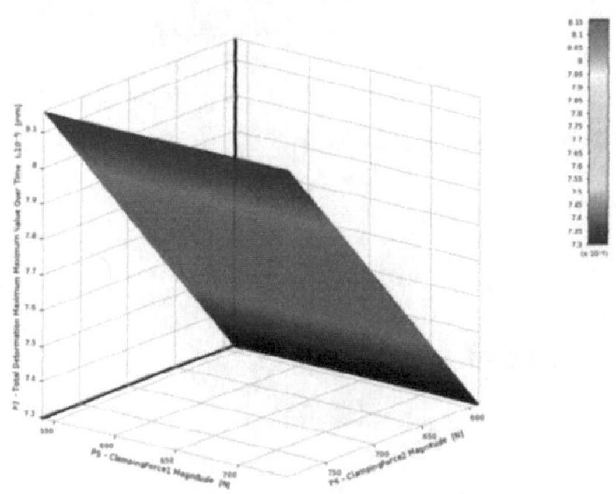

Figure 15. Dependence of the maximum total deformation of the evaluated edge on the

the magnitude of the clamping forces

magnitude of the clamping forces

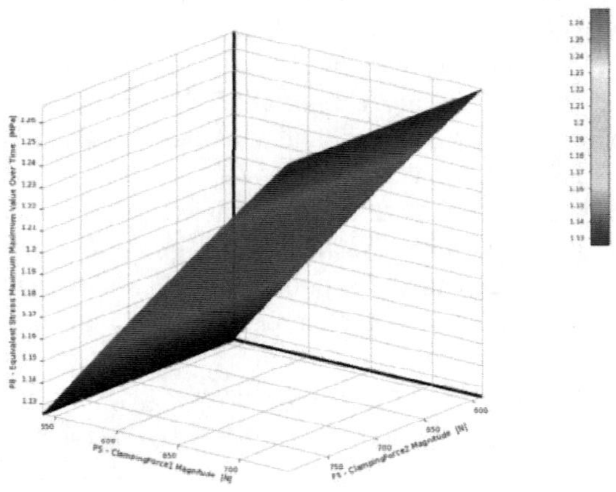

Figure 16. Dependence of the maximum equivalent stress on the magnitude of the clamping forces

The group of figures 9…11 displays the influence of the position of the clamping element C1 on the minimum contact force between the locators (L1, L2, L3) and the workpiece.

The influence that the magnitude of the clamping forces has on the minimum contact force between the locators (L1, L2 and L3) and the workpiece is presented in figures 12…14. Figure 15 displays the influence of the magnitude of the clamping forces on the maximum total deformation of the evaluated edge, and Figure 16 displays the dependence of the maximum equivalent stress – magnitude of the clamping force.

6. Conclusions

It is observed that the values of the objective functions are smaller than those resulted in the preliminary analysis, and the constraints are accomplished, in the sense that the contact forces between the locators and the workpiece are larger than zero, which means that during the processing, the workpiece doesn't lose the contact with the locators, thus the orientation layout is preserved and the process is performed in adequate conditions of precision.

At the same time, comparing the values of the objective functions for the initial values and the optimal ones of the design variables, it is observed that smaller values were obtained for the total maximum deformation of the evaluated edge and the equivalent maximum stress in the workpiece. Also, the optimal values of the clamping forces are smaller, which implies a reduced energy consumption, and the contact forces are bigger than zero, which means that the integrity of the orientation layout is maintained.

By studying the graphical representations in figures 6…16, it can be noticed that for certain values of the position dimensions of the clamping elements, the magnitude of the contact forces is close to zero, which would lead to the compromising of the orientation layout. These representations are useful in order to avoid those areas for applying clamping forces by using clamping elements.

References

[1] Jonhson K L 1987 *Contact mechanics*, Cambridge University Press, Cambridge, USA

[2] Sanchez H T, Estrems M and Faura F 2006 Fixturing analysis methods for calculating the contact load distribution and the valid clamping regions in machining processes, *Int J Adv Manuf Technol* **29**(5) 426–435

[3] Selvakumar S, Arulshri K P, Padmanaban K P and Sasikumar K S K 2010 Clamping force optimization for minimum deformation of workpiece by dynamic analysis of workpiece-

fixture system, *World Applied Sciences Journal* **11**(7) 840-846
[4] Li B and Melkote S N 2001 Fixture clamping force optimisation and its impact on workpiece location accuracy, *Int J Adv Manuf Technol* **17**(2) 104-113
[5] Chen W, Ni L and Xue J 2008 Deformation control through fixture layout design and clamping force optimization, *Int J Adv Manuf Technol* **38**(9) 860–867
[6] Kaya N 2006 Machining fixture locating and clamping position optimization using genetic algorithms, *Computers in Industry* **57**(2) 112-120
[7] Qin G H, Zhang W H and Wan M 2005 Analysis and optimal design of fixture clamping sequence, *Journal of Manufacturing Science and Engineering* **128**(2) 482-493
[8] Vukelic D, Tadic B, Miljanic D, Budak I, Todorovic P M, Randjelovic S and Jeremic B M 2012 Novel workpiece clamping method for increased machining performance, *Tehnički vjesnik* **19**(4) 837-846
[9] Ruksar Begum S D, Sita Ram Babu T and Jagadeeswara Rao M 2015 Finite element analysis and optimization of machine fixture, *International Journal of Current Engineering and Technology* **5**(4) 2842-2848
[10] Liu S G, Zheng L, Zhang Z H, Li Z Z and Liu D C 2007 Optimization of the number and positions of fixture locators in the peripheral milling of a low-rigidity workpiece, *The International Journal of Advanced Manufacturing Technology* **33**(7-8) 668-676
[11] Vlase A, Sturzu A, Mihail A and Bercea I 1983 *Regimuri de aşchiere, adaosuri de prelucrare şi norme tehnice de timp, vol.1*, Ed. Tehnică, Bucureşti, România
[12] *** - Ansys Design Exploration Users Guide

IOP Conf. Series: Materials Science and Engineering **200** (2017) 012016 doi:10.1088/1757-899X/200/1/012016

An analysis of the oscillatory movements of the suspension ropes for an elevator with parabolic speed modeling and continuous deceleration

I Craciun, M Ungureanu, and D Stoicovici

Technical University of Cluj Napoca, North Universitary Center of Baia Mare, Department of Engineering and Management Of Technology, Victor Babes str., no. 62A, 430083 Baia Mare, Romania

E-mail: ioana.craciun@cunbm.utcluj.ro

Abstract. In order to insure passengers' security and smooth functioning, elevators benefit from speed modeling. The paper aims to compare the evolution of the kinematic parameters (acceleration, speed, amplitude) of the oscillatory movements performed by the suspension ropes of the elevator cabin, considering speed variation at start and stop realized using S – curves generated from arches of parabolic curves. Usually, the functioning cycle of an elevator ends with a short period of constant speed traveling, in order to insure levelling of the cabin. We have already made a comparative analysis of the oscillatory movements of the suspension ropes, considering three varieties of speed modeling: with trapezoidal variation of acceleration, with parabolic variation of acceleration and with cosinusoidal variation of acceleration. The paper aims to analyze the possibility to remove the levelling period, and the impact of this upon the kinematic parameters of the oscillatory movements performed by the suspension ropes. We will take into consideration two possibilities of deceleration: with levelling step and with continuous variation of speed. In both cases, for comparison, the same duration of the acceleration period and deceleration period will be considered.

1. Introduction

Speed control, using electronic devices such as frequency convertors, introduced the possibility of accelerating and decelerating the elevators using the so called S curves, which can ensure a smooth variation of speed and acceleration, in order to increase comfort and avoid oscillations of the suspension ropes, generated by shocks. In order to generate S curves, various curves were used, such as polynomial and trigonometric ones [1], [2].

Kinematic parameters of an elevators functioning cycle are speed, acceleration and its derivate, jerk. The functioning cycle of an elevator is divided in three time periods: accelerating time, the period of constant speed travel, and deceleration time. Usually, the last one has three intervals: decelerating, constant speed step, and stop time. We will compare the influence on the kinematic parameters of the oscillatory movements performed by the suspension ropes if deceleration is realized with levelling step and with continuous speed [3], [4].

The time evolution of each parameter was already studied in both cases of deceleration defined above, considering two possibilities of generating S curves by speed control: using arches of parabola and arches of cosinusoidal curve [5].

2. Technical considerations

2.1. The dynamic equation of motion

A theoretical analysis of the cabin's motion during a functioning cycle can be performed starting from the model presented in Figure 1. The cabin has a complex motion, resulted from the superposition of a dampened oscillatory motion over the theoretical linear motion, since suspension cables act like elastic elements [5], [6].

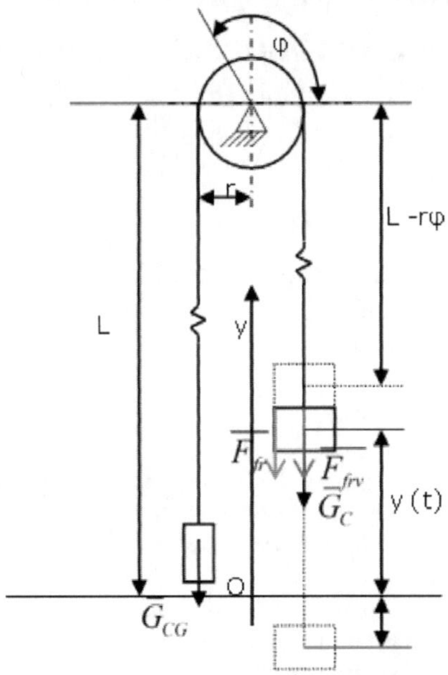

Figure 1. The theoretical model of the elevator

The Dynamic equation of the cabin's motion was elaborated under the following hypothesis: [5-7]

- Only the cabin's movements will be studied, the oscillatory movements of the counterweight and of the friction wheel will be neglected.
- The equation was written only for the vertical motion, defined as axis Oy.
- The weight of the suspension ropes was not taken into consideration.
- The suspension cables were considered elastic elements having a variable stiffness coefficient.
- The damping coefficient of the cables was not taken into consideration.
- The influence of the cabin's suspension system was not considered.

The suspension cable's elasticity coefficient is:

$$k = \frac{E \cdot A}{L - r \cdot \varphi} \quad (1)$$

where:

- E is Young's modulus of the cable,
- L is the initial length of the cable,
- A is the cable's section area
- r is the radius of the friction wheel,
- φ is the rotation angle of the wheel at moment t.

Value r·φ is corresponding to the theoretical displacement of the cabin, considering suspension ropes inextensible.

The dynamic equation of the cabin's motion can be written as follows:

$$M \cdot \ddot{y} = -M \cdot g + k \cdot (r \cdot \varphi - y) - F_{frv} - F_{fr} \tag{2}$$

where M is the mass of the fully loaded cabin, and y(t) is the real (absolute) displacement, measured from ground level, resisting forces are:

- aerodynamic friction force,

$$F_{frv} = 1{,}2 \frac{v_{\max}^3}{27} B \cdot C \tag{3}$$

- dry (coulombian) friction force, due to contact with the slips,

$$F_{fr} = \mu \cdot N \tag{4}$$

The dynamic equation of the cabin becomes: [2], [3]

$$M \cdot \ddot{y} = -M \cdot g + \frac{E \cdot A}{L - r \cdot \varphi}(r \cdot \varphi - y) - 1{,}2 \frac{v_{\max}^3}{27} B \cdot C - \mu \cdot N \cdot \operatorname{sgn}(\dot{y} - r \cdot \dot{\varphi}) \tag{5}$$

2.2. The parameters variation

In order to compare the two types of deceleration, the parameters taken into consideration are the cable elongation and the acceleration. As the fully loaded cabin climbs, the cable length decreases, and its stiffness increases. Hence, the cable's elongation diminishes, and the cabin will accelerate upwards, even if at constant speed. Considering the same time duration for accelerating and decelerating, the time duration of the constant speed period was calculated, for both cases of speed modeling: using arches of parabola and using arches of cosinusoidal curves, and for both cycle types: with and without constant speed step at deceleration.

The study was performed using the parameters of an elevator for personnel, serving 8 levels (22.5 meters), designed for the transportation of 4 people (3000 N weight). Parameters values are: [8]

M = 950 kg,
E = 89 GPa,
L = 27.3 m,
A = 31 mm2,
vmax = 1 m/s,
B·C = 0.882 m2,
μ = 0.15,
N = 428 N.

Considering the equations of variation for the kinematic parameters, at hoisting the fully loaded cabin, the following diagrams were obtained using MatLab ode113 routine (Figures 2 to 9). In order to better understand differences, although diagrams were generated for an entire functioning cycle, figures present only deceleration period.

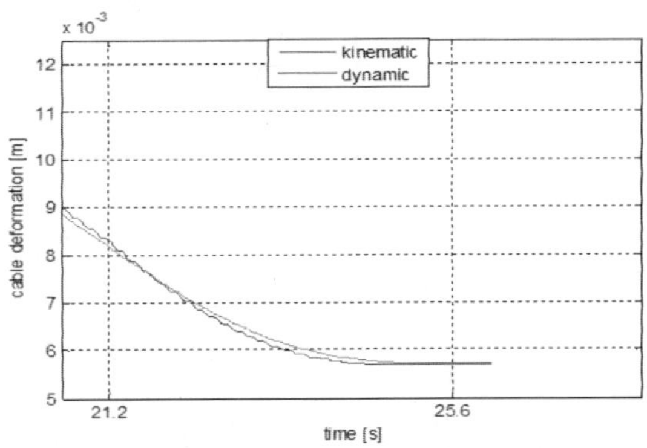

Figure 2. Cable deformation in the case of parabolic modeling of speed, with continuous deceleration

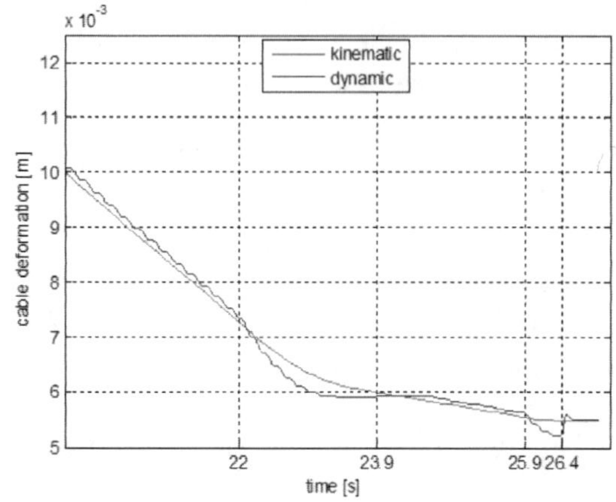

Figure 3. Cable deformation in the case of parabolic modeling of speed, with levelling step at deceleration

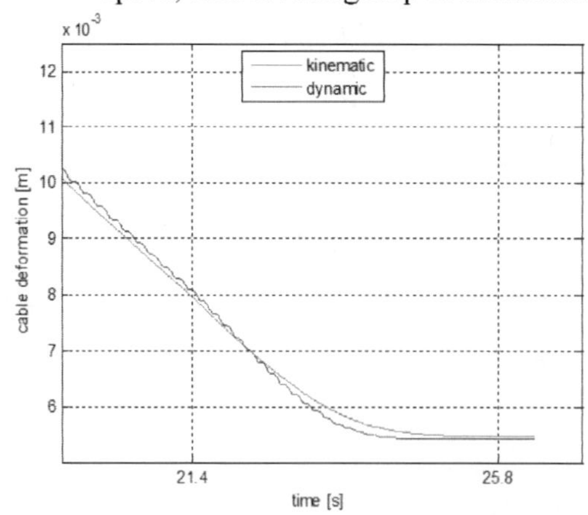

Figure 4. Cable deformation in the case of cosinusoidal modeling of speed, with continuous deceleration

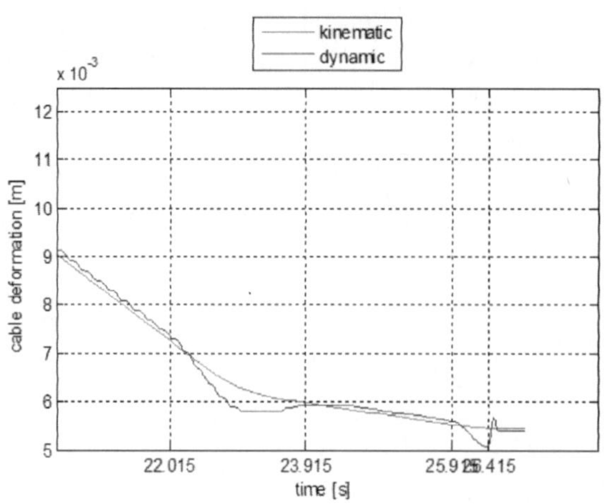

Figure 5. Cable deformation in the case of cosinusoidal modeling of speed, with levelling step at deceleration

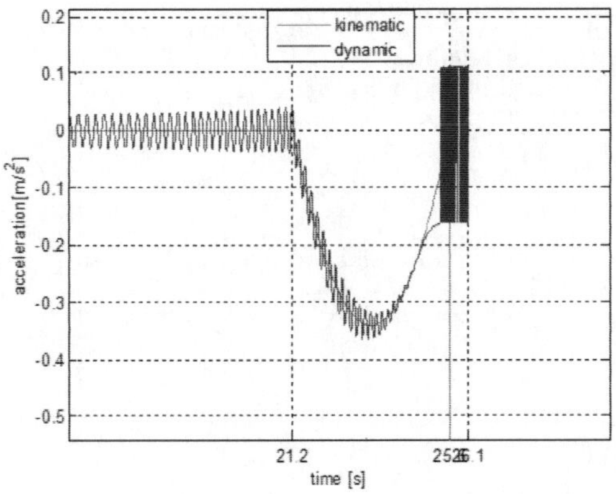

Figure 6. Acceleration in the case of parabolic modeling of speed, with continuous deceleration

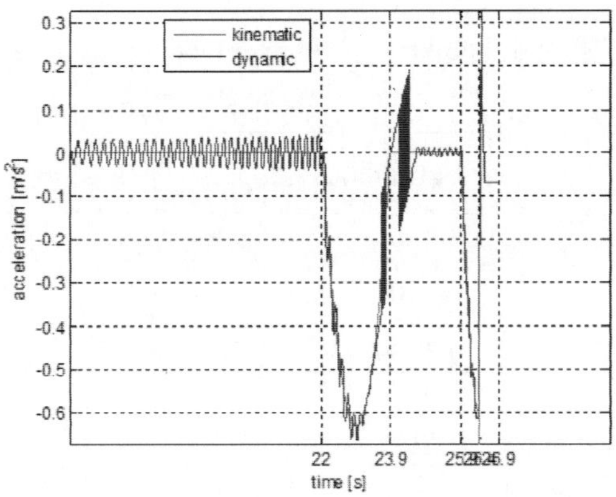

Figure 7. Acceleration in the case of parabolic modeling of speed, with levelling step at deceleration

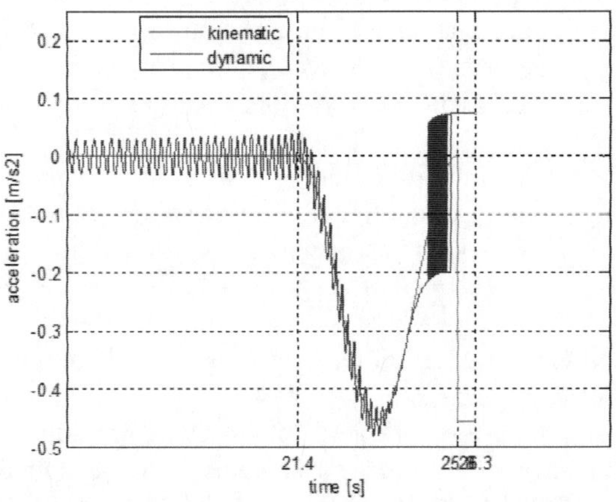

Figure 8. Acceleration in the case of cosinusoidal modeling of speed, with continuous deceleration

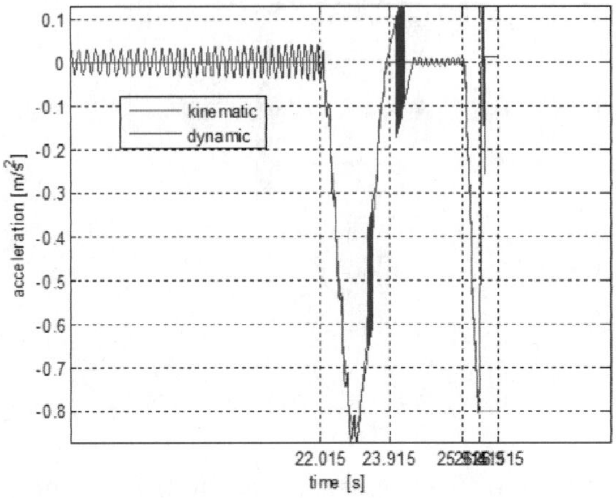

Figure 9. Acceleration in the case of parabolic modeling of speed, with levelling step at deceleration

Table 1. Dynamic parameters values

Parameter	Parabolic		Cosinusoidal	
	With levelling	Continuous	With levelling	Continuous
Time [s]	4.4	4.4	4.4	4.4
Acceleration [m/s^2]	0.625	0.340	0.833	0.454
Jerk [m/s^3]	1.330	0.3099	1.392	0.3245

3. Interpretation and conclusions

Analyzing the diagrams above one can notice that the continuous deceleration significantly reduces cable deformation due to the oscillatory movements of the suspension cables. The variation of acceleration due to the cable elasticity is also reduced in the case of continuous deceleration; hence the dynamical solicitations of the cable are smaller. This is a consequence of the lowest theoretical value of acceleration that can be used when decelerating without levelling step. As shown in Table 1, cosinusoidal modeling of speed conduces, at same time duration of the decelerating period, to higher values of acceleration. Diagrams of deformation and acceleration show that residual oscillations of the cable are diminished in the case of continuous deceleration. Positioning precision, in correlation with the controller, is to be analyzed.

The hypotheses formulated at the beginning of chapter 2 are suitable only for light loads and small travels and speed. The dynamic equation of motion for other values of these parameters should be rewritten, considering the dynamical processes that occur during a functioning cycle [9], [10].

References

[1] Nguyen K D, Ng T C and Chen I M 2008 On Algorithms for Planning S-curve, *International Journal of Advanced Robotic Systems* **5**(1) 99-106

[2] Vladić J, Malešev P, Šostakov R and Brkljač N 2008 Dynamic Analysis of the Load Lifting Mechanisms, *Strojniški vestnik - Journal of Mechanical Engineering* **54**(10) 655-661

[3] Kaczmarczyk S and Picton P 2013 The Prediction of Nonlinear Responses and Active Stiffness Control of Moving Slender Continua Subjected to Dynamic Loadings in a Vertical Host Structure, *International Journal of Acoustics and Vibration* **18**(1) 3944

[4] Arrasate X, Kaczmarczyk S, Almandoz G, Abete J M and Isasa I 2012 *The modelling and experimental testing of the vertical dynamic response of an elevator system with a 2:1 roping*, Proceedings of ISMA2012-USD, pp 2307-23221

[5] Crăciun I 2014 *Contributions to the kinematic and dynamic study of the mechanical systems characterized by fast modifications of motion parameters*, Technical University of Cluj Napoca, Romania, Doctoral Thesis

[6] Ungureanu M, Crăciun I, Bănică M and Dăscălescu A 2016 Study for elevator cage position during the braking period, *IOP Conf. Ser.: Mater. Sci. Eng.* **144** 012010

[7] Ungureanu M, Crăciun I and Stoicovici D 2011 Study for elevator cage position, *Scientific Bulletin Series C: Fascicle Mechanics, Tribology, Machine Manufacturing Technology* **XXV** 73-77

[8] *** MP Ascensores Technical Dossier v3.00, SEP.01 Frequency Changers

[9] Alexandrescu M, Cotetiu R, Cotetiu A and Ungureanu N 2009 Correlations regarding the instantaneous carrying force and pressure distribution to the narrow sliding radial bearing (L/D<0,6) working under hard shocks, *Scientific Buletin, Serie C, Fascicle: Mechanics, Tribology, Machine Manufacturing Tehnology* **XXIII** 9-14

[10] Daraba D, Cosma M and Alexandrescu M 2009 *Setting Up the Maximum Speed at which the Structural Elements of the Fus 25 Milling Machine Resist After Remanufacturing,* The 20th International DAAAM Symposium "Intelligent Manufacturing & Automation: Theory, Practice &Education", Vienna, Austria, November 25-28, pp1063-1064

IOP Conf. Series: Materials Science and Engineering **200** (2017) 012017 doi:10.1088/1757-899X/200/1/012017

The application of the electrodynamic separator in minerals beneficiation

M Skowron[1], P Syrek[1] and A Surowiak[2]

AGH University of Science and Technology, 30 Mickiewicza av., 30-059 Krakow, Poland
[1]Department of Electrical and Power Engineering,
[2]Faculty of Mining and Geoengineering, Department of Environmental Engineering and Mineral Processing,

E-mail: mskowron@agh.edu.pl

Abstract. The aim of presented paper is elaboration of methodology of upgrading natural minerals in example of chalcocite and bornite sample. The results were obtained by means of laboratory drum separator. This device operates in accordance to properties of materials, which in this case was electrical conductivity. The study contains the analysis of the forces occurring inside of electrodynamic separator chamber, that act on the particles of various electrical properties. Both, the potential and electric field strength distributions were calculated, with set of separators setpoints. Theoretical analysis influenced on separator parameters, and hence impacted the empirical results too. Next, the authors conducted empirical research on chalcocite and bornite beneficiation by means of electrodynamic separation. The results of this process were shown graphically in form of upgrading curves of chalcocite considering elementary copper and lead.

1. Introduction

Static electric fields are used in technological processes, among which electric separation can be found. The argument of separation is defined through differences of magnetic properties or/and electrical conductivity. The advantage of electrodynamic separation is the possibility of obtaining efficient separation of individual components in dry conditions. This eliminates many operations such as thickening, dewatering, hydrotransportation, additional drying which cause higher energy consumption and higher processing costs [1-3].

The materials treated by electrodynamic separation require proper preparation, such as crumbing and classification – with purpose of being sure that final products will be properly extracted. Products can be fine-grained, mixed including various ingredients. It often happens that those ingredients are pretty similar to each other in terms of their characteristics, so it is necessary to proceed with further separation processes.

The aim of this paper is to demonstrate possible laboratory applications of electrodynamic drum separator in mineral processing of chalcocite and bornite ores supported with statistical analysis of received results. For the experimental research, the difference in electric susceptibility was chosen as a separation feature.

2. The principle of particles separation in electric field

Electrical separation methods are based on differences between time traces of electrical particles on seeds with different conductance, and, basing on that, different time constants for triboelectric effect. This solution has been applied in electrodynamic drum separator, constructed by Huff in 1905.

The mixture of particles is poured from feeder to coronal zone, where they are electrified with negative charge, and then passed down to the surface of the drum. Particles with low resistivity become electrified positively and are thrown outside by electrical field forces. At the same time, particles of high conductivity keep their charge received in coronal zone and they are transferred by means of electric field and specular reflection forces to another part of the receiver. Then, they fall off or are scraped with a brush. The structure and scheme of the separator are shown on the Figure 1.

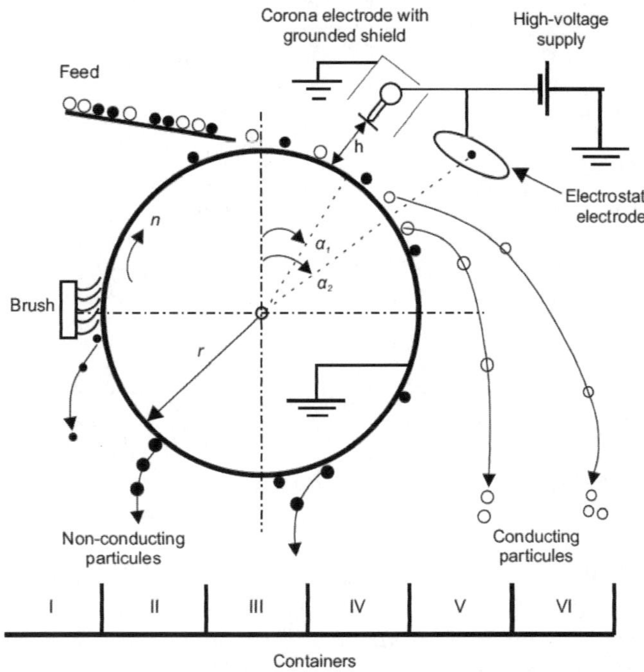

Figure 1. Scheme of drum separator working

2.1. Mineralogical characteristics of copper minerals

The materials used for empirical research are ores of chalcocite and bornite.

Chalcocite is a most common mineral in all three types of ore found on the motherlodes form Lubin to Sieroszowice. The maximum density of chalcocite, up to 90% of total volume, can be found close to nested sandstone structures with anhidritc binders. The average percentage share can vary up to 94% of the total sum of all sulfides.

Bornite (Cu_5FeS_4) – according to quantity, bornite is the second ore mineral in the motherlode after chalcocite. It can be found in all lithological types of ore. In sandstones, the average percentage share of this mineral varies from 0.2-0.7% of volume, in slates – 0.4-1.6%, whereas in dolomites one can notice the variation from 0.1 to 1.0% of total volume. Bornite creates mainly dispersed, nested and veiny structures in all layered rocks. Bornite's chemical composition is very mixed. Bornite taken from Lubin and Polkowice area usually contains more silver.

3. Microscopic model separation with using of drum separator

Figure 2 below shows forces affecting the conductive particles – particle A is located on the drum's surface and particle B has been thrown away from it. As one can see on the forces distribution presented on Figure 2, the main role in terms of conductive particles' separation is played by the sum of electric (F_e) and centrifugal (F_o) force, while looking into the movement of non-conductive particle, the effect of distribution is made by higher electrical (F_e) and specular reflection (F_i) forces [4].

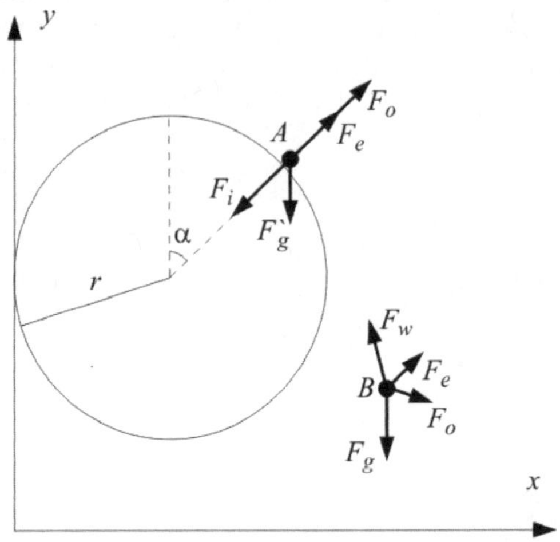

Figure 2. Distribution of forces acting on conductive particles [4]

To determine the dispersion effectiveness of various electrical characteristics, it's necessary to determine the particles' trajectory after leaving the drum's surface, and - as it follows – to create the microscopic separation model. The trajectory of separated particles is affected by many factors, amongst which the most important ones are: the electrodes' arrangement, the high voltage value (both of those factors determine the field distribution between the drums), the drum's angular velocity, the granularity of separated particles and the relative air humidity [4], [5].

The shape of particles' trajectory in drum separator is connected with the affection of forces on the particle while the particle is located on the drum's surface, as well as after being thrown away from it in the area of electric field.

General equation representing acting forces is in form [6], [7]:

$$m\frac{dv}{dt} = \Sigma \boldsymbol{F} \tag{1}$$

where:

m – mass of particle [g], $\frac{dv}{dt}$ – particle's acceleration [ms^{-2}], $\Sigma \boldsymbol{F}$ – sum of forces acting on particle [N].

In time in which a particle is located on drum's surface the forces act on it (their modules are presented).

– of electrical field:

$$F_e = QE \tag{2}$$

– of reflection:

$$F_i = \frac{Q^2}{4\pi\varepsilon_0 \left(d^2\right)} \tag{3}$$

where:

Q – particle's charge [C], E – intensity of electrical field [Vm-1], d – drum's diameter [m].

– centrifugal:

$$F_o = m\frac{v^2}{r} = m\omega^2 r \tag{4}$$

– gravity:

$$F_g = mg \tag{5}$$

where:

ω – drum's axial velocity [s^{-1}], g – gravity [ms^{-2}].

– dynamic medium's drag force:

$$F_w = \Psi S\frac{\rho v^2}{2} \tag{6}$$

where:

Ψ – coefficient of dynamic medium drag [-], S – surface of particle's projection on surface perpendicular to motion [m^2], ρ – density of medium in which particle is moving [kgm^{-3}], v – particle's velocity [ms^{-1}].

4. Numerical analysis of the electric field distribution

The potential distribution and electric field density in drum separator system have been observed using Finite Elements Method using COMSOL Multiphysics environment. This software is a simulation pack, solving sets of partial, non-linear differential equations in one, two or three dimensions [4].

The computed model consist of rotating, grounded cylinder with loader placed above, which aim is to supply the system with particles to separate. The model consist also of two electrodes: the aberrational, motionless, cylinder-shaped one and movable, corona, comb one, which aim is to increase the electric field among its limit.

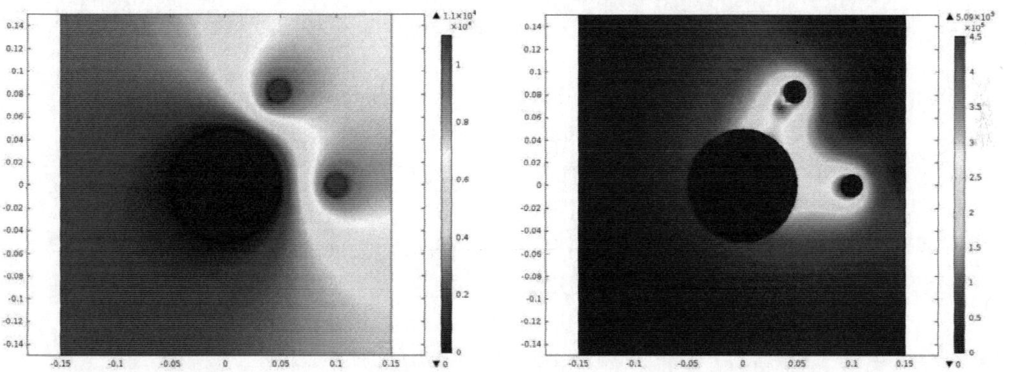

Figure 3. Distribution of the electric potential and electric field for Θ = 60°, φ = 45°, V = 11 kV

Figure 3 shows distribution of electric potential for following angles of corona electrode: Θ = 60°, φ = 45° and voltage V = 11 kV. On the figure 3 the distribution of electric field density for corona electrode described above is presented. The highest density of field can be found around the edge of corona elecrode. However, the surface of drum is also affected by high field density, which means that particles located on the drum, in this area, are under the influence of this field as well.

5. Methodology of research

The investigation of electrostatic separation was performed by means of laboratory drum separator by constant voltage of 11 kV and constant drum rotaries (Figure 4). The samples of chalcocite and bornite were separated which granulation was 0.040-0.063 mm. Chalcocite contained 25.6% of Cu and 16% of Pb and bornite – 20.5% of Cu and 0.9% of Pb. These were minerals of copper ore, selected especially to research and mineralogically qualified as chalcocite and bornite. The materials to investigation were comminuted in laboratory crusher and then the narrow particle fraction 0.063-0.040 mm was sieved with maintained constant conditions of temperature and humidity of area. The separation products were collected in containers and each of them was weighed and marked for contents of individual elements by means of spectrometer XRF.

Figure 4. Photo of electrodynamic drum separator

6. Elaboration of research results

Tables 1 and 2 present shares and distributions of copper and lead contents in sample of chalcocite and bornite of granulation 0.063-0.040 mm after electrostatic separation.

Table 1. Contents of copper and lead in products of chalcocite separation products

No. box	Share [%]	Cu [%]	Pb [%]
1	1.80	50.40	5.30
2	4.90	45.20	4.60
3	10.29	44.50	4.50
4	10.95	44.80	4.90
5	8.66	40.40	6.20
6	7.35	31.30	13.90
7	6.54	32.10	16.10
8	3.10	36.40	10.80
9	0.98	36.50	10.50
10	0.98	30.40	15.40
11	0.98	9.00	22.40
12	43.46	6.30	25.40
13	10.90	31.30	11.50
14	0.10	37.80	6.90

Table 2. Contents of copper and lead in bornite separation products

No. box	Share [%]	Cu [%]	Pb [%]
1	0.00	0.00	0.00
2	1.42	42.60	0.36
3	3.46	33.74	0.63
4	5.51	27.90	0.76
5	4.25	33.50	0.60
6	3.46	31.90	0.65
7	2.05	33.90	0.57
8	1.42	29.90	0.73
9	0.79	26.70	0.81
10	0.31	20.00	1.01
11	4.41	14.30	0.95
12	62.36	17.20	0.99
13	10.55	17.70	0.93

The electrostatic separation is commonly applied in purpose of separating conducting and non-conducting particles [8]. Such separation is usually associated with concentration (beneficiation) of certain elements in certain products (containers) while others will be depleted for certain component. In this paper the attempt of evaluating beneficiation of separated materials divided in drum separator was done on the basis of methods being applied in mineral engineering. To this purpose, coordinates were calculated for Halbich's upgrading curve (called also as Hall's curves) and then the graphs were plotted [9]. It is obtained on the basis of the balance of separation products as dependence of cumulated recovery in function of useful component's contents in concentrate [10]. The cumulated recovery is calculated according to the formula:

$$\sum\varepsilon = \sum\gamma\frac{\beta}{\alpha} \tag{7}$$

where: ε – recovery of analyzed element, γ – yield of product (in his case related to share), β – contents of useful components in concentrate, α – contents of useful component in feed.

On Figure 5 the Halbich's upgrading curves are presented for copper and lead in chalcocite and on Figure 6 in bornite. Halbich's upgrading curves for copper in chalcocite and bornite are similar with growth of copper contents in concentrates. Analogical beneficiation curves for galena in these minerals have course of the same type in chalcocite and bornite, respectively but show growing tendency with growth of lead contents in concentrate. That means that in case of copper in electrostatic beneficiation for containers of lower numbers (close to drum) not much material is transferred but it is rich in copper. For these containers is possible to achieve high quality copper concentrates by small yields. However, in case of lead in tested minerals its cumulation is observed in containers of higher numbers and lead is carried outside the drum. Such distribution of selective cumulation of copper and lead in individual containers is related to many factors, among others to such as: materials granulation, level of components liberation and possibility of absorbing and giving electrical charge on particles' surface in time of acting electric field forces. The electrical properties of elements, such as conductance and specific resistance decide precisely about possibility of absorbing of surface electrical charge by particles which causes that selective transfer of particles to various containers (products) occurs, constructed of elements of characteristic sign being parts of it. For copper the conductance is equal to 58.5 [MS/m] and resistance 0.02 [μΩ·m]. However, for lead the conductance is equal to 4.8 [MS/m] and resistance to 0.02 [μΩ·m]. Various values of these material features for copper and lead decide about various ability to absorb and give electric surface charge and this causes its selective separation.

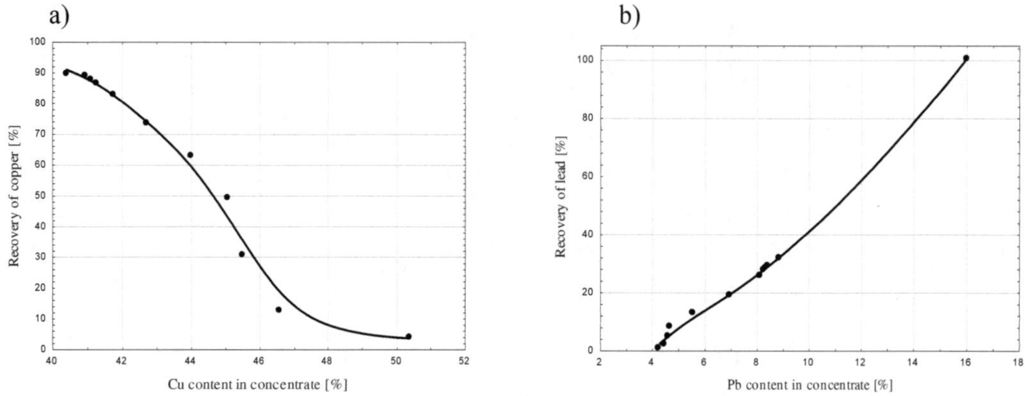

Figure 5. Halbich's upgrading curves: a) for copper – chalcocite, b) for lead – chalcocite

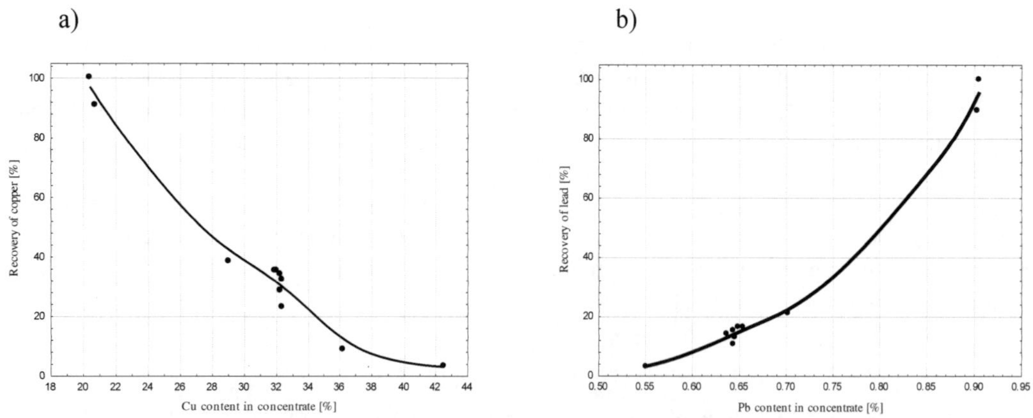

Figure 6. Halbich's upgrading curves: a) for copper – bornite, b) for lead – bornite

It's becoming more and more common that Fuerstenau's curves are used for mineral processing of the material consisting of two different ingredients. To compare the selectivity of the two ingredients' division, the separate balances are made and curves – as the gain relations of one of the ingredients in the concentrate in function of second analyzed ingredient in waste [10] – are drawn. On the Figures 7 and 8 the Fuerstenau's curves – as the relations of gain relations of copper in function of lead loss in waste – are presented for tested minerals. In both cases, the mineral processing curves for chalcocite and bornite are angled from line of coordinates (0,0) and (100,100), which corresponds to situation where there are no reasons to use the particular method of mineral processing due to lack of differences conditioning the separation. The selection indicator F, based on Fuerstenau's curves, has been related to the point of curve's maximum camber point – its value for chalcocite equals to 78/78 and for bornite – to 64/64. This indicator tells about the average lead from copper division in chalcocite and weak lead from copper division in bornite.

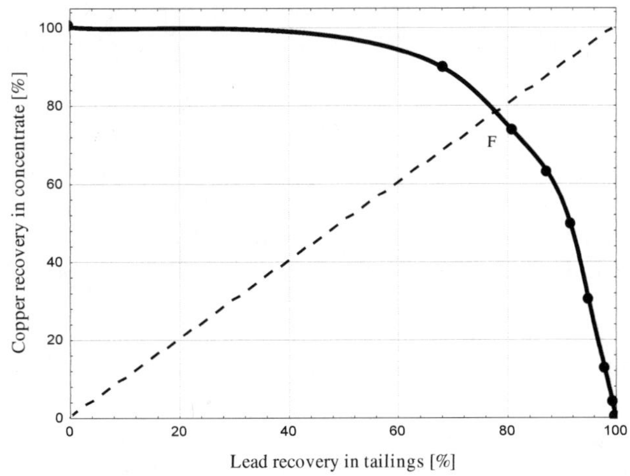

Figure 7. Fuerstenau`s curve for chalcocite

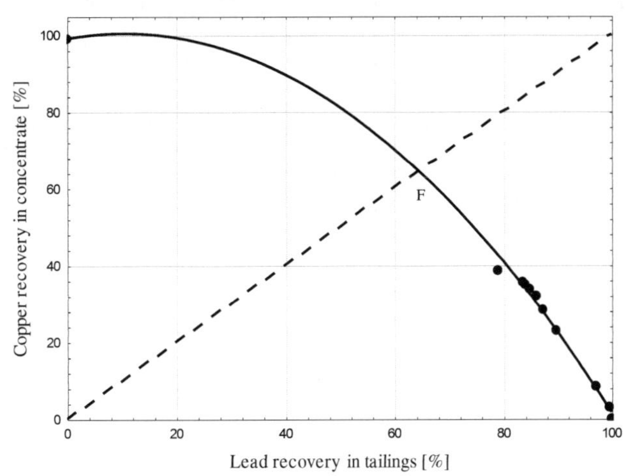

Figure 8. Fuerstenau`s curve for bornite

7. Conclusions

It's possible to use electrostatic separation to divide the conductive from non-conductive elements from the mix of different ingredients or from one material consisting of elements of different electrical features. The experiments done on chalcocite and bornite samples proved that it's possible to use electric separation for dividing copper and lead concentrates with satisfying final effects. However, those conclusions are the result of basic-level experiments and it's necessary to proceed with further works on this subject.

The paper was written as a part of statutory project no. 11.11.100.276.

References

[1] Brozek M and Surowiak A 2004 Distribution of settling velocity of particles in samples of mineral raw materials, *Gospodarka Surowcami Mineralnymi - Mineral Resources Management* **20** 67-84

[2] Brozek M and Surowiak A 2014 Methodology of calculation the terminal settling velocity distribution of spherical particles for high values of the Reynold's number, *Archives of Mining Sciences* **59**(1) 269-282

[3] Surowiak A 2014 Wpływ rozkładu gęstości ziaren na rozkład ich prędkości opadania dla wąskich klas ziarnowych, *Gospodarka Surowcami Mineralnymi - Mineral Resources Management* **30**(1) 105-122

[4] Śliwa U and Skowron M 2016 Analysis of the electric field distribution in the drum separator of different electrode configuration (Analiza rozkładu pola elektrycznego w separatorze bębnowym o różnej konfiguracji elektrod, in polish), *Informatyka Automatyka Pomiary w Gospodarce i Ochronie Środowiska* **2** 79-82

[5] Kurgan E and Gas P 2016 *Mutual Forces Acting on Chains of Particles*, 13th Selected Issues of Electrical Engineering and Electronics (WZEE) , Rzeszow, Poland, May 4-8, pp 1-4

[6] Ciesla A, Kraszewski W, Skowron M, Surowiak A and Syrek P 2016 Wykorzystanie bebnowego separatora elektrodynamicznego do separacji odpadow elektronicznych, *Gospodarka Surowcami Mineralnymi - Mineral Resources Management* **32**(1) 155-173

[7] Ciesla A and Skowron M 2013 *Influence of high voltage on drum electrodynamic separator effectiveness,* ISEF 2013 – XVI International Symposium on Electromagnetic Fields in mechatronics, Electrical and Electronic Engineering, Ohrid, Macedonia

[8] Skowron M, Surowiak A, Syrek P, Ciesla A, Kraszewski W and Lachendrowicz D 2016 *The possibility of separation of electronic waste by means of their electrical properties,* International Conference on the Sustainable Energy and Environment Development (SEED 2016), E3S Web of Conferences **10** 00085

[9] Brozek M 1984 Evaluation of liberation efficiency index of coal mineral substance on the basis of Hall's separation curve, *Przeglad Gorniczy* **11** 384-387

[10] Drzymala J 2001 *Podstawy mineralurgii*, Oficyna Wydawnicza Politechniki Wroclawskiej, Wroclaw, Poland

IOP Conf. Series: Materials Science and Engineering **200** (2017) 012018 doi:10.1088/1757-899X/200/1/012018

Theoretical and methodological foundations of sustainable development of Geosystems

O M Mandryk[1], L M Arkhypova[2], A V Pukish[3], A Zelmanovych[1] and Kh Yakovlyuk[4]

[1]Ivano-Frankivsk National Technical University of Oil and Gas, Institute of Environmental Engineering, Department of Ecology, Karpatska str., 15, 76018, Ivano-Frankivsk, Ukraine
[2]Ivano-Frankivsk National Technical University of Oil and Gas, Institute of Architecture, Construction and Tourism, Department of Tourism, Karpatska str., 15, 76018, Ivano-Frankivsk, Ukraine
[3]Research and Development Institute PJSC "Ukrnafta" North Boulevard., 2, 76019, Ivano-Frankivsk, Ukraine
[4]Ivan Franko National University of Lviv, Faculty of Foreign Languages, Department of English Philology, Universytetska str., 1, 79000, Lviv, Ukraine

E-mail: konsevich@ukr.net

Abstract. The theoretical and methodological foundations of sustainable development of Geosystems were further evolved. It was grounded the new scientific direction "constructive Hydroecology" - the science that studies the Hydrosphere from the standpoint of natural and technogenic safety based on geosystematical approach. A structural separation for constructive Hydroecology based on objective, subjective, and application characteristics was set. The main object of study of the new scientific field is the hydroecological environment under which the part of Hydrosphere should be understood as a part of the multicomponent dynamic system that is influenced by engineering and economical human activities and, in turn, determines to some extent this activity.

1. Introduction

Since the time of the formation of all existing hydrosphere sciences (Hydrology, Hydrochemistry, Hydrobiology, etc.) the principle of systemic approach was set into their theoretical foundation. The doctrine of multifactor formation of quantitative and qualitative composition of water of Hydrosphere is formulated in the works of famous scientists [1-13], who carried out the idea of the unity of physical, chemical and biological processes that occur in all natural waters and about their systematic conditioning by a complex natural and anthropogenic factor. O.M. Adamenko proposed a new constructive trend in the development of ecological science and environmental protection - "constructive Ecology" as the basis of environmental safety [1]. According to the author, constructive Ecology - is the part of the "Great Ecology" by M.F. Reimers, that not only diagnoses the state of the environment and forecast its evolution, but also offers specific ways of its optimization and improvement. It also constructs such natural-technogenic geosystems, that will ensure sustainable balanced harmonious development of Human - Nature - Technosphere. Constructive Ecology - the science and environmental protection sector, which justifies the creation of sustainable natural-technogenic geosystems structures which are the part of the Earth's biosphere and their development should not degrade under the influence of technogenic pressures [2], [3].

We should note that in modern geosciences and practice the term "Geoecology" is widely used, which was introduced to the science by the German geographer (K. Troll, 1970). However, scientist geographers and geologists who use this term at present time, often invest in its content not exactly the same concept [4], [5]. In particular, authors [6], [7] define the term Geoecology as interdisciplinary science of environmental problems of geospheres. According to the famous socioecologists [7], [9], Geoecology is a subdivision of Socioecology and Geography, which studies the spatial variability of the geographical environment for the purpose of sociologically optimal territorial organization of the interaction between society and nature. The authors agree with the broad interpretation of the term Geoecology as a complex science that studies the environmental positions of all covers (areas) of the Earth. The object of the study of the Geoecology is geoecosystems.

2. Presentation of the fundamental material

According to the wording of most environmental scientists, Hydroecology is a new direction in Earth sciences, which studies the hydrosphere as one of the abiotic components of the ecosystems of high level organization [2], [3], [10], [11]. Thus, Hydrosphere is treated and investigated as a multicomponent dynamic system that consists of basin systems and affects the existence and development of biota and the human community.

Please note that all living things supposedly are removed beyond the scope of the system that is studied. The latter is experienced the influence and itself affects living substance, and the only investigation is the interaction of the Hydrosphere and the living beings (always considering the human factor). However, many biologists against such extended interpretation of Ecology and leave it in the theory formation provided by Haeckel.

For any mature science the differentiation and integration of scientific fields are inherent. The appearance of a new object - namely anthropogenically altered water bodies (including artificial water bodies), the existence of such powerful factor as technogenic impact, allows you to believe that there are grounds for isolation of new scientific direction for objective, applicative and even subjective criteria. This direction can be called ***Constructive Hydroecology*** (Figure 1).

The object of study of this direction is hydroecological environment and its components – hydroecosystems that are under the constant influence of human activities - natural technogenic hydroecosystems (NTHES). They also include objects similar to natural, that appeared in result of economic activity. The subject of scientific direction is natural technogenic (environmental) safety of NTHES, balanced water use based on sustainable development.

The authors proposed the definition of *"constructive Hydroecology" as a science that studies hydroecological environment from the standpoint of natural and technogenic safety and balanced sustainable water use.*

The formation of constructive Hydroecology connected both with the development of methodology of different scientific direction in Ecology, Hydrology, Hydrochemistry, Hydroecology, Geology, and the need to ensure natural technogenic safety, balanced sustainable water use, prevention and prediction of adverse risks, which had precedents as a result of efforts to use raw materials, energy, food, recreation, and other resources of the natural ecosystems of the Earth's water shell. At the same time natural hydroecosystems fall under the influence of the technogenic systems (hydraulic, hydropower, transportation, etc.).

The main object of study of new scientific field that is seen in the work is hydroecological environment under which it should be understood *terrestrial or marine part of the Hydrosphere - multicomponent-dynamic system that is influenced by engineering and economic human activities and, in turn, to some extent determines this activity.*

We consider it necessary to research the object of structural Hydroecology of the hydroecological environment with the help of biosphere-ecological concept and the concept of ecosystem-structural approach (see Figure 1). Biosphere-ecological concept and ecosystem-structural approach involves consideration of the biosphere as a global ecosystem rank [5], [6], [7]. That part of the biosphere, where operate modern living organisms, including humans as a species and the humanity as a social structure is defined in the term of "ecosphere" and relevant concept. Such approach allows us to consider the atmosphere, hydrosphere and lithosphere as ecological subsystems of ecosphere.

At the same time the living organisms, or the living matter in the aggregate [6], play the role of the major structural elements of the ecosphere that define the formation of the characteristics of three mentioned areas, subsystems and their relationships.

The need to change the ecological paradigm is emphasized by many outstanding ecologists [7], [8]. The basis of the new paradigm, with no doubt must be the systematic approach that is based on known general scientific position, which goes back to Plato about the irreducibility of properties of the whole to a sum of the properties of its parts. This principle, according to renowned world-class environmentalists, "should serve as the first ecologists working commandment".

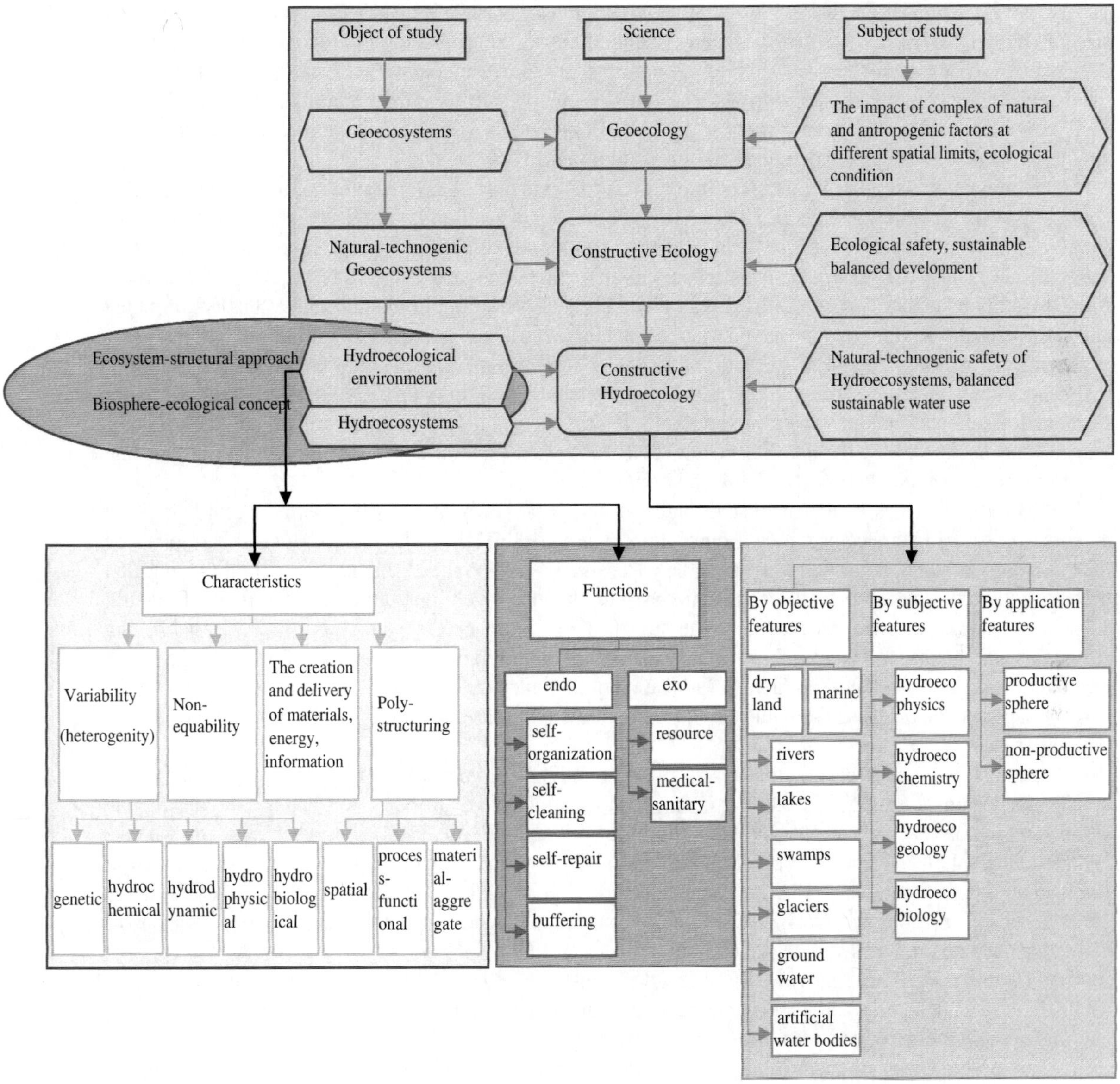

Figure 1. The object-subject structure of the constructive Hydroecology

Some sciences can be distinguished either the object of study (Physics, Chemistry, Biology, Geology, etc) or by unitary methodological approach (Cybernetics, Information theory, etc). Nowadays we distinguish *Systemology* as a special science whose objects of study are complex systems, and which has a specific means of their study [9]. The central concept of Systemology is the concept of *system*. There are more than thousands of its definitions, in the broadest sense the system can be understood as a set of the world's real or imaginary objects that were isolated from the rest in any possible way. During the evolution in time the aggregate is deemed as one system if between its elements at different points of time you can conduct unique correspondence. It is also emphasized that integral properties of the system are the result of the interaction of its parts, and that the relationship between elements of the system must reflect the essential properties of elements and be stronger than the relationship with the elements that are not included in it.

Especially significant the systematic approach to Ecology. So, "the major paradigm of Ecology should be voted as the concept of Ecosystem" [9], here follows and recognition of the need for a systematic approach.

At the early stages of the research of hydroecological environment arise the questions of its structure and size, of hierarchy levels of its territorial dimension and of methodical approaches to solve these procedures.

The analysis of existing approaches to local hydrological and hydro-chemical structures showed that all of them are based only on two basic principles: the zonal (geographically) and the basin (hydrological). The law of geographical zoning, that was introduced by V.V. Dokuchaev in 1898, is also apparent in the territorial structure of hydrological and hydrochemical information fields [12].

Landscape-genetic approach, that takes into account the principle of geographical zoning of the chemical composition of surface waters, was successfully applied for the allocation of elementary hydrochemical structures, such as "hydro-chemical fields" [12]. At the same time, we should take into account that geographical zoning is seen in the spatial change only of those chemicals, the concentrations of which are formed mainly by natural factors. In the spatial distribution of concentrations of substances that are of mixed and anthropogenic origin we not observe any signs of geographical zoning.

Therefore, in researches the application of issues of sustainable water use and environmental safety of hydroecosystem only the scheme of physical-geographic division into districts for the classification of hydroecological environment and its components – hydroecosystem, for which are typical poly-components with increasing in the last decade the predominance of mixed (natural-anthropogenic) and anthropogenic origin substances, is seen to be methodologically groundless.

The basin approach as a method to study integral-functional geosystems, is best suited, according to the authors, to explore hydroecological environment and, within it, NTHES. The methodological base of using the basin approach in organization of sustainable water use is based on the idea that river basin is complicated hierarchically constructed natural-technogenic system, within which the unidirectional flow of natural substances, energy, information facilitates the structuring of natural-technogenic components, establishing strong ties and interaction between them. The systematic relationship of natural and technogenic components of the basin, the exactness and simplicity of allocation of its limits, the possibility to predict the change of the state of natural components and complexes of basin in time and space, form the basis of wide application of basin approach.

The authors proposed structural division in constructive Hydroecology within its objective, subjective, and application features (see Figure 1). In particular, within the objective attributes we distinguish constructive Hydroecology of sea and constructive Hydroecology of land, which can be divided into constructive Hydroecology of rivers, lakes (reservoirs), wetlands, groundwater, glaciers. By subjective characteristics (individual sides of objects) we can identify constructive Hydrogecophysics, constructive Hydroecochemistry, constructive Hydroecobiology, constructive Hydroecogeology. In turn, in these areas we distinguish more narrow and specialized ones.

Due to the applied features we can highlight: constructive Hydroecology of the production sector (industry, agriculture, construction) and constructive Hydroecology of the non-production sphere.

Among the priorities of constructive Hydroecology are:

1. The development of methodology of natural-technogenic (environmental) safety and sustainable balanced water use of Hydroecosystem [10-12];

2. The development of topological classification of NTHES for different regions;

3. The development of the classification of various anthropogenic pressures on the hydroecological environment (the study of natural and with anthropogenic impact modified processes - their direction, intensity, interaction with other processes, the impact on the dynamic stability and changes in NTHES, the creation of complex models of NTHES etc.);

4. The substantiation of quantitative and qualitative indicators of Hydroecosystem potencial and maximum permissible anthropogenic pressures on different types of Hydroecosystem [11], [12];

5. The finding of spatial and temporal regularities of the change of Hydroecosystem potential and mapping of different NTHES types and different anthropogenic pressures on them;

6. The prediction about the changes of NTHES under the influence of various anthropogenic pressures;

7. The development of the recommendations on the conservation the dynamic equilibrium and optimization of NTHES [13].

3. Conclusions

Constructive Hydroecology is an interdisciplinary science that uses achievements and combines the scientific disciplines: Hydrology, Hydrochemistry, Hydrobiology, Hydrogeology, Oceanography, Meteorology and Climatology, general Ecology and others using modern geoinformation technology.

Consequently, constructive Hydroecology - is part of a constructive Geoecology that not only diagnoses the condition of hydroecological environment and forecasts its evolution, but also offers specific models of sustainable water management and environmental safety of hydroecosystems, constructs such NTHES that would ensure "sustainable and harmonious balanced development, the focus of which is a person who is entitled to a healthy and productive life in harmony with nature" (Rio 20+, 2012).

References

[1] Adamenko O M 2007 *Our future home - Ecoeurope. The novel of life, science and love in 4 volumes,* V4, P 428, Symphony forte, Ivano-Frankivsk, Ukraine

[2] Rud'ko G I and Adamenko O M 2009 *Earthlogy: resource and ecology safety of the Earth,* P512, Academpres, Kiev, Ukraine

[3] Rud'ko G I and Adamenko O M 2008 *Constructive geoecology: scientific foundations and practical implementation,* P 320, Maklaut, Kiev, Ukraine

[4] Rosenberg G S and Smelyanskiy I E 1997 Environmental pendulum (paradigm shift in modern Ecology), *Journal general biol.* **58**(4) 5-19

[5] Reimers N F 1990 *Nature usage,* P 634, Nauka, Moscow, USSR

[6] Isakov Y A, Kazanskaya N S and Tishkov A A 1986 *Zonal laws of dynamics of ecosystems,* P 150, Nauka, Moscow, USSR

[7] Gilyarov A M 1998 Ecology in search of a universal paradigm, *Nature* **3** 73-82

[8] Vernadsky V 1944 A few words about the Noosphere, *Successes of modern Biology* **18**(2) 113-120

[9] Fleishman B S, Fedorov V D and Gilmanov T G 1980 *Basics of systemology* P 464, Radio and Communications, Moscow State University, USSR

[10] Korchemlyuk M and Arkhypova L 2016 Environmental audit of Ukrainian basin ecosystem of the Prut river, *Scientific bulletin of National Mining University* **155**(5) 98-106

[11] Arkhypova L M and Pernerovska S V 2015 Forecasting water bodies hydrological parameters using singular spectrum analysis *Scientific bulletin of National Mining University. Scientific and technical journal* **146**(2) 45-50

[12] Adamenko Y O, Arkhypova L M and Mandryk O M 2016 Territorial standard quality hidroekosystem protected areas, *Hydrobiological Journal* **52**(6) 51-59

[13] Mandryk O M, Arkhypova L M, Pobigun O V and Maniuk O R 2016 Renewable energy sources for sustainable tourism in the Carpathian region, *IOP Conf. Ser.: Mater. Sci. Eng.* **144** 012007

IOP Conf. Series: Materials Science and Engineering **200** (2017) 012019 doi:10.1088/1757-899X/200/1/012019

Foreign direct investments and their impact on the economic development of Bosnia and Herzegovina

I Susic[1], M Stojanovic-Trivanovic[2] and M Susic[3]

[1]University of Business Studies, Jovan Ducic Street, No 23A, 78000 Banja Luka, Bosnia and Herzegovina
[2]Independent University Banjaluka, Veljka Mladjenovica Street No 12E, 78000 Banja Luka, Bosnia and Herzegovina
[3]Enterprise Fructa Trade – Kort, Marije Bursac Street No 70, 74400 Derventa, Bosnia and Herzegovina

E-mail: i.susic51@gmail.com

Abstract: From the perspective of macroeconomic indicators, investment is a significant determinant of economic development in general, as well as the development indicator of economic entities in the micro segment. Investments are an essential element of any economic policy, because their implementation provides a platform not only for economic development, but also are prerequisite for the stability of economic and social trends. Foreign direct investment plays an important role in the financing of the global economy, and it represents the most frequent feature in financing the national economies of developing countries and countries in transition. Demand for foreign investment in the global market is large, and thus the governments have been conducting many activities in order to create a more favorable environment to attract investors.

In this paper, special attention was paid to direct investments in financing the economy on a global scale, their importance for the development of the global economy and the impact of foreign direct investment in the economic development of Bosnia and Herzegovina. The major activities, which are necessary to be done to attract investments in the highest possible volume, have been emphasized. With the use of statistical and quantitative analysis, the paper shows that the inflow of foreign capital is one of the basic prerequisite of economic growth acceleration and that the inflow of foreign capital has a positive impact on the economic development of Bosnia and Herzegovina. By monitoring and analyzing the various instruments of foreign capital inflow, with an emphasis on investment in the free zone and a joint venture with foreign investors, it has been clearly pointed out the fact that they have diverse, but proven positive impact on macroeconomic variables in the economy of Bosnia and Herzegovina.

1. Introduction

Foreign direct investments (FDI) represent such a form of investment in which foreign investor keeps the ownership right, provides the control and the management of the firm in which they invested the funds, in order to achieve long-term interests. These investments are the most important instrument of foreign capital inflow because they represent a direct inflow from abroad, i.e. direct inflow of the capital in the economic system of the host country.

Foreign direct investment, as a form of international capital mobility, represents an important contributor to more efficient activities in the economy. They provide faster exit to the international market and as the aftermath are ensuring improved the living standard of the society. Evaluation of investment efficiency is the basis for making investment decisions from one country to another, which will consequently lead to improvement of the economy. Foreign investments are a key development factor in the modern economy, and jointly with the trade, represent the most important leverage of an enterprise, organization of production, supplying goods and services on a global scale.

FDI are supporting the companies in organizing production on a global scale, providing an efficient supply of raw materials, energy, labor as the input, and are facilitating the placement of products and services as the output in the most important markets in a profitable way. On the basis of such activities, the companies can on optimal way use its advantages in technology, expertise, and economies of scale.

Developing countries having high state debt and unfavorable economic situation show huge interest in gaining as higher foreign investments as possible. It has been especially important after bank loans and various financial aid ceased to arrive in some countries. Countries in transition, aiming to integrate into the world economic system, can overcome negative economic tendencies with the help of international capital inflow.

Developed countries, faced with a financial crisis, have been also interested in an increased inflow of foreign capital, since the foreign investments are the most important element of development strategies in general.

With foreign direct investment is not coming just the capital from one country to another, but also the investment package containing new technologies, managerial skills and new markets. In addition, bearing higher risks, FDIs are significantly increasing the opportunities for making profits. Foreign direct investments are autonomous transactions of long-term capital movements, motivated by economic interests, with the profit at the first place.

Observing today's economic systems at the global market, we can notice that the economic systems, in order to accelerate their economic development compete to each other, aiming to attract foreign capital. In fact, the most significant competition is seen in the area of "greenfield" investments. It is important to elaborate, how and why are the countries competing for foreign direct investment, on the one hand, and on the other hand, why the other countries are the main investors of this type of investment?

In order to monitor investments, the following core analytic indicators related to foreign direct investment are examined:

- Flows of foreign direct investment (FDI), which represents new investments during the observed period (usually one year period). They represent the position of the current account. Total flows are divided by instruments used for investment, and are as follows: owners capital (ownership in subsidiaries), and shares in subsidiaries and affiliated companies, re-invested earnings as part of retained earnings of investors that is not distributed; others investment of FDIs, like the borrowing and lending of funds, including various financial instruments and trade credits between investing company and company which has capital inflow.
- Stocks of foreign direct investment represents is the value of investments at the end of monitored period. In Balance sheet, the foreign direct investments are in the assets side of sheet and in the liabilities side if the investment is coming from the host country. Stock of FDIs is divided into: equity capital and reinvested earnings.
- Income of foreign direct investment is income summarized by direct investors during the period. It is divided into three categories: (a) dividends earned in the reporting period and the

profits allocated to direct investor without deduction of income tax; (b) reinvested earnings; (c) interest on loans, i.e. interest earned on loans to subsidiaries, without deduction of income tax.

In addition, an important indicator is intensity of foreign direct investment, measured as a percentage of gross domestic product (the ratio of the average internal and external flows of foreign direct investment and GDP). Higher intensity indicator means more foreign investments in relation to the size of the economy.

One of the goals of this paper is to demonstrate that in developing countries acceleration of the economic development can be achieved with the use of instruments of foreign capital, and with the additional capital inflow from abroad. The purpose of foreign capital should be the achievement of higher investment rates, higher rates of national income, acceleration of economic development in comparison to the development which would be achieved by using only domestic capital.

The second objective of the research is, that different instruments of foreign capital inflow have different impacts on the macro-economic categories in global circumstances. In the paper has been analyzed impact of FDIs on economic development in BH economy. Also, it has been monitored the impact on: employment growth, budget revenues and export volume. Foreign investment can cause negative effects on domestic companies, if foreign investors squeeze domestic producers from the market, and become monopolists. The damage may be made also to the payment balance of the host country due to the high outflow of investors' profits or because of large imports of inputs.

2. Foreign direct investment and effects to economic development

The role and importance of foreign direct investment for the national economy [1] was primarily in improving the key macroeconomic indicators. FDIs are efficient form of usage of private savings in the process of funding economic development and the in reducing the gap between the planned investment and the local savings. Secondly, it is significant contribution of foreign direct investment in overcoming the gap of foreign trade of host country. Thus, the foreign investment is efficiently covering discrepancy between planned revenues and taxes collected, and the expenditures volume in the budget. FDIs are useful because they contribute in transferring managerial and entrepreneurial experiences. Finally, FDI today represent the main channel for the transfer of new technology between countries.

Foreign direct investment can bring many advantages for foreign investors [2] among which, the most important are: savings in transport costs (both, inputs and finished products), lower labor costs, available infrastructure, savings in customs costs and contribution on imported goods, closer position to the customers, the possibility of quick and efficient delivery, and availability of information about preferences and possibility for fast adoption of products in accordance with market requirements.

In particular, it is necessary to analyze the relationship between investment and economic growth. This interdependence can be seen by measurement of macroeconomic aggregates, i.e. in growth rate, movement of investment, foreign exchange level and trends and other. Changes are related to economic developments at the national level, on the basis of which it is possible to assess the success of development policy. When the positive elements have an increase, there are positive structural changes. In periods of crises the negative changes are strengthening (slowing the growth and investment, increasing unemployment rate, increasing in deficits, etc.) [3] investments generate significant positive effects on the economy of the host country. Their impact is recorded in two points: the quantitative growth, measured by the balance and the total inflow measured by gross domestic product, exports and domestic investment: and qualitative through the transfer to the host country the investment, trade, technology and financial flows. To determine the effects of FDI on economic growth of the host country is not an easy task. There are a large number of variables, where the effects are associated with specifics of each country, sectors of the economy and investment. Variables can be viewed from two aspects: first, they are supplementing domestic factors of production and creating the conditions for new employment, and secondly they stimulate the development of the host country through technology transfer, manpower training, liaising with the local economy, and in enabling better presence of local companies at the world market.

The effects of FDI depend on the stage of economic development of the country, and these stages are divided in four phases [4]. In the first phase of development, the most important role is played by natural resources, and at this stage, no significant effects to the host country economy are visible. If the country has an economic development at the second phase, it will be recorded increasing of domestic investment, investing in public goods, communications and transport is present. The state in this phase, shifts its interest from natural resources to the production of labor-intensive goods, and the effects depend on the infrastructure and macroeconomic policy. The third phase covers the period when the development of innovation, knowledge management, organizational benefits, rationalization of production and investments are supported. All these variables have the effects on strengthening the competitiveness of local companies and appearance on new markets. The fourth stage is the highest stage of economic development and in this stage it has been recorded lot of post-industrial service company. The products are direct services and cross-border connections and effects are becoming more intense in this phase. Effects of foreign direct investment on the economic development of the host country depends on strategy which the country has chosen: import substituting strategy or export developing strategy. Studies have shown that investments are extremely important for economic growth of host country, where we want to examine: Does the inflow of foreign direct investment, increases or decreases the total investment volume in the host country?

If foreign direct investments are entering into the sector in which there is competition from domestic companies, there is a competitive struggle whose consequences might be the delaying in investment and exit of certain domestic companies from the sector. This will lead to reduction of the total investment in the sector and thus in the entire economy. If the FDIs are entering the new sector (primary, secondary, tertiary) then the total investment in the country will be increased.

3. Foreign direct investment and growth in the European Union

Total investments in the EU amounted slightly less than one-fifth (below 20%) of brut domestic product. The investment rate (I / Y) is stable in a long-term. The rate of investment in the more developed EU countries is lower and ranges up to one sixth, while in the new Member States and less developed investment rate, reaches up to one quarter of output [5].

The subjects of our interest are flows of foreign direct investment and their impact on the changes of gross domestic product, as an important macroeconomic indicator.

Table 1. Real GDP in the EU for the period in 2003-2015 (Eurostat 2016)

Year	Real GDP (in Billion EUR))	Growth (% GDP)
2003.	9.664,40	1,5
2004.	9.913,20	2,6
2005.	10.127,30	2,2
2006.	10.467,80	3,4
2007.	10.802,70	3,2
2008.	10.842,80	0,4
2009.	10.354,90	-4,5
2010.	10.564,30	2,0
2011.	10.740,30	1,7
2012.	10.697,70	-0,4

2013.	10.696,74	0,0
2014.	10.872,70	1,4
2015.	11.080,30	1,9

Gross domestic product and growth rate for the European Union is presented for the period from 2003 to 2015 (EU expanded from mid-2013). It is obvious that, last few years of the observed period, there is an instability in growth rates as a result of the impact of the global crisis and the debt crisis in Eurozone. Data on GDP and growth rates are presented in Table 1.

Real gross domestic product by 2008 is growing steadily. Following the economic crises in late 2008 and early 2009 there was a fall in the GDP value from 10,842.8 billion Euros, measured in 2008 to 10,354.9 billion at the end of 2009. Thus, in 2009 the economic activities decline due to the recession of the European economy. In 2010 there was a slight increase in GDP, but the rate was below the level that was realized in the economy by the end of 2008.

Since June 1, 2013. European Union expands to 28 members. In this period is observed stagnation of growth of real gross domestic product. According to Eurostat, real GDP in the period from 2013 to 2015 was growing more slowly (in 2014 there was a significant increase compared to 2013, while growth in 2015 points to the long-term stabilization and recovery from the crisis). If we compare only the GDP data from the 2008 with GDP data for 2015, we could conclude that the period of the recession has been behind us.

The change in gross domestic product for the period 2003-2015 in Table 1, are shown in Fifure 1.

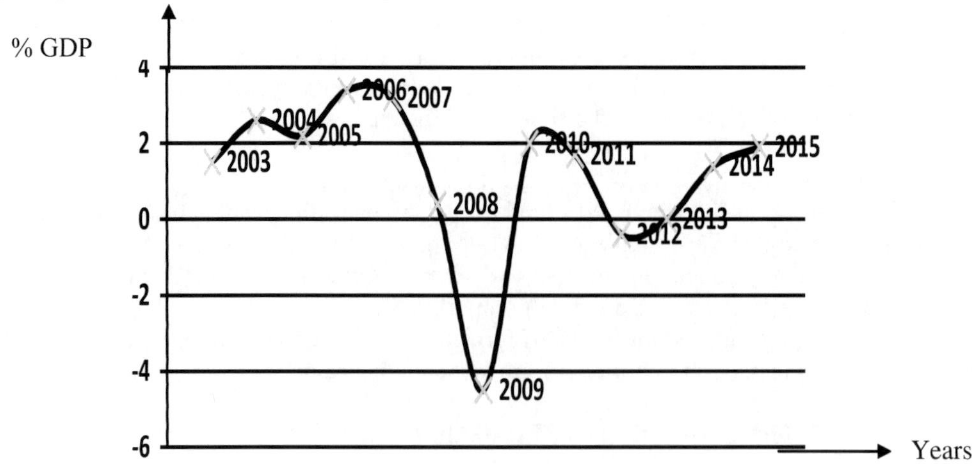

Figure 1. Gross Domestic Product changes (growth in %) (Authors 2016)

The graph shows that from 2003 to 2006, it was recorded a continuous growth and in 2007, before the escalation of the crisis in the US, the growth rate was of over 3%, what is for European circumstances extremely high. In 2008 the growth rate falls sharply to only 0.4%, and it is still positive, although the economy of the European Union, due to the expansion of the American crisis, was already operating in difficult circumstances. As stated, in 2009, it was recorded the decline in GDP, which is an indicator that the economy of the European Union was already in a serious economic crisis.

In 2009 in the European Union had a decrease of outflow, but also to somewhat higher FDI inflow, which, along with other factors, had positive impact on GDP growth in 2010. GDP growth from 2013 to

2015 indicates stabilization in economic developments in the EU, i.e. it was an indicator that the crisis is over.

Table 2 presents the inflows and outflows of foreign direct investment in the period from 2004 to 2015, and data for the first quarter of 2016 for the EU, and the total FDI stock per year of the observed period.

Table 2. FDI flows in millions of euros for EU-27 (Eurostat 2016, European Union, OECD Foreign Direct Investment Statistics, https://knoema.com/OECDFDIS2016/oecd-foreign-direct-investment-statistics [6].

Year	Foreign Direct Investment (FDI) Inflows	Foreign Direct Investment (FDI) Outflows	Stock (total value) FDI	
			Inflows FDI	Outflows FDI
	EUR millions	EUR millions	EUR millions	EUR millions
2004.	244.179	369.134	4.846,365	5.420,856
2005.	591.234	669.041	5.690,696	6.305,140
2006.	726.462	879.818	5.946,830	6.546,792
2007.	1,065.473	1,278.121	7.536,334	8.138,913
2008.	582.517	919.366	6.976,821	8.200,076
2009.	512.626	612.534	7.644,430	9.589,730
2010.	480.892	585.530	7.596,391	9.111,480
2011.	652.062	725.924	7.716,407	9.370,997
2012.	345.096	392.379	8.223,772	9.716,007
2013.	315.559	333,559	8.789,142	10.397,848
2014.	264.794	226.858		
2015.	494.272	508.522		
2016 Q1..	235.085	204.060		

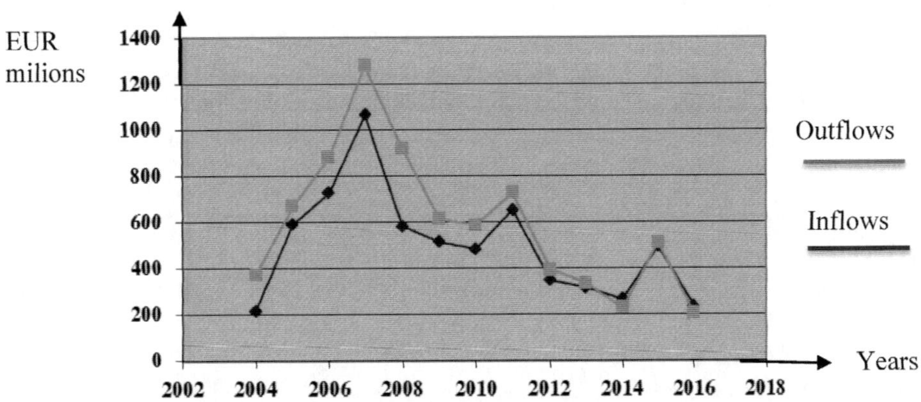

Figure 2. FDI inflow and outflow in EU (Author 2016)

From Table 2 and Figure 2, it can be seen, that inflows and outflows of foreign direct investment in the initial years of the period from 2004 to 2007, recorded steadily increase. If, we take year 2004 as the base year, then the outflow of foreign direct investment in 2005 grew enormously, from 369,134 million to 669,041 million euros, while the inflow increased from 244,179 million to 591,234 euros, i.e. over 100%. Upward trend continued until the end of 2007. During 2008, a decline in flows, in the inflow and outflow, of foreign direct investment was recorded due to the global crisis.

Therefore, the crisis which was present at global financial markets has the reflection on the economy of the Union. Being caused by mismatch of financial and real sectors of the economy, its effects were manifested in the decline in aggregate demand, slowing growth, deflation and rising unemployment. Regardless of the contagion of the crisis from the US financial markets, the European Union still has a close relationship with the US as the largest foreign trade partner, with which realized huge traffic of goods, services and capital, which is on a daily basis, significantly higher than one billion euros. Likewise, the EU continuously making efforts to: increase participation in world trade, achieve the growth of all forms of investment and innovation, and to make upgrade in entrepreneurship skills and to improve corporate social responsibility level.

It is indisputable that the EU countries, same as the most of the world economy, reduced investment activity during the financial crisis, especially outside its borders. Among others, by these measures they try to reduce the negative effects of the global economic crisis. Negative trend and big drop in foreign direct investment continues in 2009, in which the crisis escalated. Many of the world economies were temporarily closed because of concerns for internal economic situation. Similarly, in 2010 it was recorded negative inflows and outflows of foreign direct investment.

Since 2007, for the first time in 2011, the EU 27 noticed the recovery and growth of foreign direct investment. The growth was negligible and did not recover the overall level of investment, bearing in mind that foreign direct investments in 2007 were significantly higher, on the both sides, in inflows and outflows. In 2012, again the decrease of FDI flows was recorded in comparison to the previous year, due to the second wave of the global crisis and the specific financial and fiscal crisis, which includes the individual EU member states, especially Greece.

Flows of foreign direct investment vary each year and it has an increase in periods of growth, and decrease in period of recession. In the European Union, after a drop of 60% in 2008, foreign direct investment is recovering in 2009 to 28%, which is largely the result of growth of own capital and reinvestment of earnings (Eurostat, 2015).

There is an issue in displaying the stock of foreign direct investment, due to the lack of official data of FDI stock in 2012 for territory of the European Union. Since there are databases from previous years, the value of the stock in 2012 has been calculated as the average growth rate for all years back to 2004. Unlike foreign direct investment flows, which fluctuate over the year, the total stock of FDI per year is not decreased, but is constantly growing, and that fact shows that EU countries are attractive for foreign investment.

4. Impact of foreign direct investment on the economic development of Bosnia and Herzegovina

The main objective of the host country is to achieve long-term growth based on increased investments and on adoption new technologies with increased competitiveness of the products on the world market. Research has shown that economic growth can be most efficiently through the „greenfield" investments, on which is mostly oriented the host country while attracting the foreign investment.

Investments for the renewal and increase of domestic capital, for keeping up with new technologies, are necessary prerequisite for making a long-term growth. Foreign direct investments are considered to be the best source of capital, and have an advantage over other sources of capital, such as debt, portfolio investment and foreign aid. The advantages of foreign direct investment, compared to credits are, that they do not represent a significant capital outflows in the future, which can disrupt the current balance and economic growth in the future. Also, the portfolio investments are seen as significantly instable. Foreign direct investment is related to investments in fixed assets and represent a long-term investment. Analysis of financial effects show that foreign investments represent the inflow of foreign resources which are contributing to increase of total investment in the host country. Their share in total domestic investment is increasing, especially in countries in transition. In addition to the basic (initial) capital inflows, foreign direct investment is presenting also all reinvested profits of branches of foreign companies in a host country.

The effects of the total investment in the country depend on forms of investment. Greenfield [7] investments have a greater effect on the increase in the total investment than the mergers and acquisitions. Mergers and acquisition (M & A) are, in the essence of the change of ownership of existing resources, rather than creating new ones. Mergers and acquisitions have an impact on the increase of the total investment if the investors (foreign investors) are investing in new facilities to its subsidiaries. These forms of investment have an effect on raising the total investment stock, in the event that the result of basic investments by foreign companies, there is also an increase in investment from other foreign companies, or if there is an increase in domestic investment in the following way: First, foreign direct investment can increase the entry of other transnational companies (TNCs) that are in business connections with an initial investor, and who follow him in all other markets. Second, the investment of a company gives to other TNK signals that there is a favorable investment climate in the host country, at the same time showing that the investment risk is reduced. Third, foreign direct investment stimulates domestic companies to invest, if there is complementarity in the production of domestic companies and branches in the country. In such an environment, foreign affiliates and domestic companies do establish a stable connection, and from those conditions the domestic companies create additional motivation to implement new projects and invest additional funds. Fourth, the experience of the host country show that branches of foreign companies have a propensity towards investing more than domestic, and thus increase the investment in the host country.

4.1. Inflow of foreign capital into business and free zone in Bosnia and Herzegovina

Business zones (BZ) represent a space for wide variety of business activities. They operate and are managed by specialized team of professionals or a specific organization, and are formed as a result of organized development concept of the state, region or municipality. To the entities that operate within the business zone, are typically offered certain benefits, such as cheaper rent or annuity, various tax benefits, deferred payment options, assistance with exports of goods and services, faster and easier access to capital and so on.

Business zone can be organized and managed by the local government or regional organization, which is proved to be the most effective approach to managing and developing the business areas. Efficiency of the business area depends on the characteristics and imposed operating policies in the zone. So, business zones are a tool for the development of entrepreneurship and attracting investment, both domestic and foreign.

The free zone is part of the customs territory of a particular country, which is specially separated by fence and marked, in which the economic activities are performed under special conditions. Usually, these conditions are defined in detail by national laws on free zones. The users of these zones do not pay import duties and value added tax - VAT. Customs duties and tariffs are not charged on imports, and import of equipment that will be used for production and operations within a free zone, is exempted from customs and other duties. It is important to point out that investments in the free zone, transfer of profit and transfer of investment are not charged at all. Free Zone, first of all, implies more liberal regime of business. The need for regulatory standardization requirements for the formation, function and defining benefits for work in free zones is stated in the regulation, which was initially defined by the entities' Laws on Free Zones, and finally established by the Law on Free Zones BH (Off. Gazette of BH 37/02 and 137 / 03).

The relevant legislation that touches the issues of operations in free zones, refers to the Law on Customs Policy of BH, the Law on Free Zones in BH and the Law on foreign direct investment in BH.

Using the survey method, it was obtained the information on the operations of the free zones which are shown in Table 3 and refer to the basic elements of operations of free zones in Bosnia and Herzegovina.

Table 3. Foreign Trade Chamber Bosnia and Herzegovina 2016 The basic elements of operations of free zones in Bosnia and Herzegovina, www.komorabi [8]

Elements	YEAR (in 000 BAM)				Index 2006/03	Average Index
	2003	2004	2005	2006*		
1. The value of income	130.000	290.000	370.000	520.000	4,00	2,05
2. The value of exports	96.506	252.572	332.827	487.754	5,05	2,35
3. Share of export in total income of Zone in % (2/1)	74,2	87,0	90,0	93,7	1,26	0,8578
4. Number of employed	2.140	2.310	2.339	3.311	1,55	1,3475
5.The value of imported equipment	1.741	2.464	28.495	4.971	2,85	1,5130
6. Customs duty on equipment	192	271	1.709	298	1,55	0,9563
7. Paid customs duty on equipment	30	30	150	30	1,00	0,80
8. Export from BH (mil. BAM)	2.330	2.994	3.826	5.271	2,26	1,60
9. Share of export from Free Zones in total export of BH (%)	**4,14**	**8,43**	**11,6**	**9,25**		

*estimation

As it is evident from the presented table, the activity in the free zones in BH recorded high dynamics in the period 2003-2006. i.e. after 2002, when new regulation on free zones (FZ) was adopted, known under name "Bulldozer initiative", in 2004 was recorded reducing in percentage of exports from free zones, from 75% to 50%. During this period, the value of production has increased 4 times, and the value of exports was increased 5.05 times. Number of employees has increased by 55%, while the share of exports in total exports of FZ in BH increased from 4.14% to 9.25%. At the same time, it is evident that the issue of quantification of duty-free imports of equipment have a relatively low effect of below 2,5 million of BAM.

4.2. An analysis of positive and negative impacts (loss-profits) for each of the analyzed entities (segments)

Impact of two regulatory measures was analyzed:

a. harmonization of tariffs for industrial products and equipment, to the provisions of the Stabilization and Association Agreement (SAP)
b. implementation of the provisions of the duty-free import of equipment to operate in the free zones (according to Article 176 of the BH Law on Customs Operations-BH LCO)

Table 4. Assessment and effects of regulations relating to free zones in Bosnia and Herzegovina regulation by stakeholders [9], (Estimation of author dr. A. Domazet 2007)

Stakeholders	Reduction of customs duties in accordance to SAP		Adjustments with Article 176 of SAP	
	Positive effects	Negative effects	Positive effects	Negative effects
1.Business community	Lower costs of R & D of about 0.75 million. BAM		Lower costs of R & D of around 0.825 mi-	

IOP Conf. Series: Materials Science and Engineering **200** (2017) 012019 doi:10.1088/1757-899X/200/1/012019

within Free Zone	annually The attractiveness for domestic and foreign investors, and higher export competitiveness		llion. BAM, the attractiveness for domestic and foreign investors, and higher export competitiveness	
2.Business community out the Free Zone	Lower costs of R & D of about 111.75 million. KM annually The attractiveness for domestic and foreign investors The higher export competitiveness			
3. BH Government	Gains in VAT revenues of 85.5 million BAM annually and gains from contributions and labor taxes 21.4 million. BAM	The loss of revenue from customs duties around 112.5 million. BAM	Gains in VAT revenues of 0.5 million and gains from contributions and labor taxes 1.26 million. BAM	The loss of revenue from customs duties around 0,83 million BAM
4. Employee	Gains in wages to 4,250 workers a total of 30.6 million. KM plus 15.3 million. KM of pension and health fund		Gains in wages to 250 workers a total of 1.8 million BAM plus 0.63 million BAM of pension and health funds	
5. Public sector	The inflow of funds for social funds of 2.2 million. BAM, solidarity for pension and health funds and social cohesion		The inflow of funds for social funds 90,000 BAM plus solidarity for pension and health funds	
Total	250,0 million BAM	112,5 million BAM	4,395 million BAM	0,83 million BAM
Net effect		137,5 million BAM		3,565 million BAM

It can be concluded that the business community engaged in free zones have no special effects, from implementation of Article 176 of the BH LCO, in the sense of benefit in the monetary aggregates (about 830,000 BAM) due to the low level of investment in equipment. If the level of investment changes, the benefits could be greater. Undoubtedly, the greatest impact would have the business community generally speaking, a measure of the horizontal reduction of customs duties on equipment originating from the EU, with effect from around 111.75 million BAM. The greatest loss is to the government with about 112.5 million BAM, i.e. in reduced revenues from customs duties, but it is mostly compensated through increased value and increased VAT, with the total effects of the 106.9 million BAM. Winners are employed with the effects of the new salary and contributions for pensions and health care of 45.9 million BAM.

IOP Conf. Series: Materials Science and Engineering **200** (2017) 012019 doi:10.1088/1757-899X/200/1/012019

5. Investigation of the influence of foreign direct investment on economic development Bosnia and Herzegovina

In this section will be shown the correlation between FDI inflows and GDP trends, as main determinants of economic growth. There is no doubt that foreign direct investments have a positive impact on the growth of gross domestic product, which is main hypothesis of this paper. In order to better examine the intensity and quality of the effects of FDI inflow to movements in gross domestic product, and other features of this relationship, it will be used two methods, regression and correlation analysis.

5.1. Regression method- theoretical approach

Regression methods are an integral part of any analysis of the data, which is concerned with describing the links between dependent and independent variables.

The objective of the analysis that uses this method is to find the model that is best suited to data, and at the most economical, but still acceptable way describes the relationship between a dependent variable and a set of independent variables that influence the first variable. In a great number of studies and experiments has been observed a connection between two or more variables. In such a case it is expected from the researchers to determine whether there is connection, and if yes, what are direct functional dependences between these sizes. The term "regression" and methods for finding connections between two variables is in use more than 100 years. The first usage was by Francis Galton, the famous British biologist, in 1908, when he studied inheritance. One of its conclusions was that the children of parents that are higher than the average are not as high as their parents. Thus, the „regression toward mediocrity" given this statistical method name. The term regression and its evolution mainly describes the statistical relationship between the variables. Special, free regression is a regression method that examines the connection between a dependent variable, Y, and one independent variable, X. The simple linear regression model is usually given in the form:

$$Y = \alpha + \beta X + \varepsilon \,, \tag{1}$$

where Y is the dependent variable, α is the intercept on the y-axis, tilt β is a real simple linear regression, X is the independent variable, and ε is a random annoyances. Depending variable is also called variable of response and independent variable is called explanatory variable, or predictor. The explanatory variable explains the causal variable changes in response. More generally presenting the regression model can be written as

$$Y = EY + \varepsilon \,, \tag{2}$$

where EY mathematical expectation of variable of response. When the EY is linear combination of the predictor X1, X2,. . . , Xk, then the regression is linear. If k = 1, the regression is simple linear. If the EY is non-linear function of the predictor X1, X2,. . . , Xk, the regression is non-linear.

It is necessary to thoroughly examine the evaluation of the linear connection between the two variables, and Yi Xi, which has the form:

$$Y_i = \alpha + \beta X_i + \varepsilon_i \,,\ i = 1,2,3, \dots ,n \tag{3}$$

where Yi is highlighted as *i*-observing in the line of observing the dependent variable Y, which may represent GDP, with Xi we marked i-observing of the independent variable X, which can be FDI. These data can be collected from the Central Bank of Bosnia and Herzegovina or the Agency for Statistics BH, in which case such data can be collected over time for the state and then this data is called time series. In such case "n" is the number of data, where it may be the number of years if the data is collected over time.

α and β are, as it was said earlier, slice and slope of regression line that describes a simple linear relationship between Y and X. These are the unknown parameters which are to be assessed on the basis of the data. Graphic data (Xi, Yi), i = 1,. . . , n, can on very illustrative way show the type of connection that empirically exists between these two variables. For example, if Y represents GDP and X represents available FDIs, then it could be expected a positive relationship between these two variables. If α and β were known, we could draw a straight line α + βXi. It is clear that to all the data (Xi, Yi) do not belong to straight line, α + βXi. In fact, equation (1) shows the difference between each value for Yi and the corresponding values of α + βXi that occurs due to disturbances of εi.

In the analysis of the representativeness of the regression line it is used the coefficient of determination. The coefficient of determination is a relative measure of the regression line adaptation to empirical data. Total deviation of empirical data (variable y) from the average value of the variable y is disassembled to part of deviation interpreted by regression model (the difference between the regression values and average values) and the part of the not interpreted in the model (the difference between the measured values and regression values). The coefficient of determination takes values in the interval 0 and 1.

$$R^2 = \frac{a\sum y + b\sum xy - n\bar{y}^2}{\sum y^2 - n\bar{y}^2}, for\ the\ law\ y = a + bx + \varepsilon_i \tag{4}$$

With the usage of regression analysis, it has been confirmed a relation that explains the nature of dependence between two variables, but if we want to determine the degree of this dependence it is necessary to use the correlation analysis. The correlation coefficients represents the measure of connection between two variables. Correlation studies the connection and mutual relation between phenomena. There are several correlation coefficients, which are used in different cases. In practice, working with linear models it is commonly used Person's correlation coefficient. Person's correlation coefficient is used in cases where the observed variables are linear and are normally distributed. Values of Person's correlation coefficient ranges from +1 (a perfect positive correlation) to -1 (perfect negative correlation) and is marked with r.

The simplest case of describing the analytical connection between the two variables is if the diagram wastage determine that the relationship is linear. In such cases, it is calculated so-called Pearson's coefficient of simple linear correlation (code: r). Assuming that we have a total of n ordered pairs (xi, yi), the values of the variables X and Y, the coefficient r is calculated by the formula:

$$r = \frac{\sum_{i=1}^{n} x_i y_i - \frac{1}{n}\sum_{i=1}^{n} x_i * \sum_{i=1}^{n} y_i}{\sqrt{\sum_{i=1}^{n} x_i^2 - \frac{1}{n}\left(\sum_{i=1}^{n} x_i\right)^2} * \sqrt{\sum_{i=1}^{n} y_i^2 - \frac{1}{n}\left(\sum_{i=1}^{n} y_i\right)^2}} \tag{5}$$

It can be shown that the coefficient r can take on all the values from the segment [-1, 1], and therefore, its minimum value is equal to -1, and the largest is +1. Correlation analysis interprets a special sign of Pearson's coefficient of simple linear correlation, especially its absolute value. The absolute value of the correlation coefficient (code: r) indicates the strength of linear relationships among variables. What is r closer to zero, the connection is weaker, and what is r closer to the number one, the connection is stronger. Person's correlation coefficient is based on a comparison of the actual impact of observed variables to one another in relation to the maximum possible impact of two variables for the samples with a maximum of 30 elements.

To calculate the correlation coefficient is required to have three different sums of squares: the sum of squares of variable X, the sum of the squares of variable Y and the sum of the products of the variables X and Y, that is:

$$r = \frac{\sum_{i=1}^{n}(x_i - \bar{x})(y_i - \bar{y})}{\sqrt{\sum_{i=1}^{n}(x_i - \bar{x})^2 * \sum_{i=1}^{n}(y_i - \bar{y})^2}} \tag{6}$$

Standardized measure of the strength of the statistical link between the phenomenon presented in two quantitative variables is the correlation coefficient [10]. Since it is an arranged phenomena, the data can be displayed in the coordinate system. A set of these points is called a scatter diagram [11] from which one can see the dependence between variables. The correlation between phenomena can be positive and negative [12]. In case of a positive linear correlation, if one variable is increasing corresponding linear increase in other variables will appear as well. If negative correlation is complete, sign r takes the value of -1.

5.2. Research using single linear regression model

In this part of the work it is carried out statistical and quantitative analysis of the impact of foreign capital in the economic development of Bosnia and Herzegovina. This model applies a single linear regression [13] In these models, the dependent variable is GDP, which is the function of foreign direct investment (FDI).

The model is expressed as follows:

$$Y = a + bX, \tag{7}$$

where is:

Y - dependent variable,
X - independent variable,
a and b - the parameters.

Based on the above, the first application of the single linear regression model in this study can be expressed as follows:

$$GDP = f(FDI), \tag{8}$$

GDP - dependent variable,

FDI - an independent variable.

Data of the Central Bank of Bosnia and Herzegovina is used in the application of the single linear regression, i.e.: GDP = f (FDI).

Table 5. The data in a single linear regression model (in millions of euros) (Central Bank of BH, July 2016 Author 2016)

YEAR	Xi (FDI)	GROWTH (% FDI)	Yi (GDP)	GROWTH (% GDP)	XY	X^2	Y^2
2009	180	-73,68	12.700	-2,90	2.286.000	32.400	161.290.000
2010	307	70,56	13.000	2,36	3.991.000	94.249	169.000.000
2011	357	16,29	13.400	3,07	4.783.800	127.449	179.560.000
2012	307	-14,00	13.400	0,00	4.113.800	94.249	179.560.000
2013	208	-32,20	13.700	2,24	2.849.600	43.264	187.690.000

IOP Conf. Series: Materials Science and Engineering **200** (2017) 012019 doi:10.1088/1757-899X/200/1/012019

2014	399	91,80	14.000	2,19	5.586.000	159.201	196.000.000
2015	244	-38,90	14.600	4,28	3.562.400	59.536	213.160.000
Σ	**2.002**		**94.800**		**27.172.600**	**610.348**	**1.286.260.000**

Figure 3 shows a diagram of FDI and GDP growth (in %) per year for the period from 2009 to 2015.

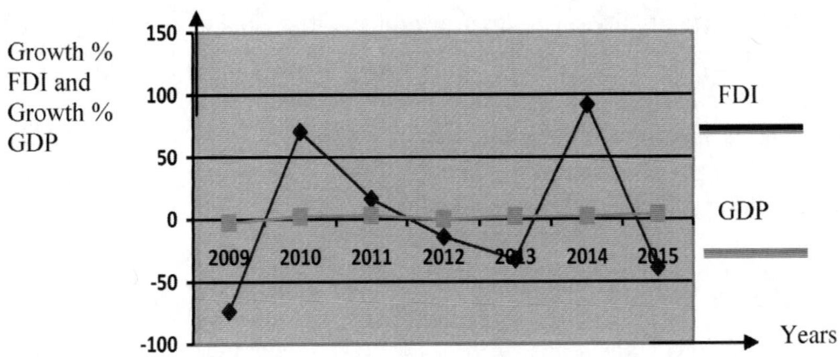

Figure 3. FDI and GDP growth (in %) per year for the period from 2009 to 2015 (Author 2016)

The diagram shows that, in Bosnia and Herzegovina is measured a permanent decline in foreign direct investment from 2010 to the 2013. Small change for the better was in 2014, but after that year, the trend of further decline in foreign capital inflow continued.

Analyzing the regression line for the dependent variable (GDP), which is shown as follows:

$$Y = a + bX,$$

and follows:

$$a = \bar{Y} - b\bar{X} \tag{9}$$

By inserting the data into forms, it follows:

$$\bar{X} = \frac{1}{n}\sum_{i=1}^{n} X_i = \frac{2002}{7} = 286$$

$$\bar{Y} = \frac{1}{n}\sum_{i=1}^{n} Y_i = \frac{94800}{7} = 13.543$$

Using the following formula it is calculated b:

$$b = \frac{\sum_{i=1}^{n} X_i Y_i - \bar{X}\sum_{i=1}^{n} Y_i}{\sum_{i=1}^{n} X_i^2 - \bar{X}\sum_{i=1}^{n} X_i} = \frac{27.172.600 - 286\times 94.800}{610.348 - 286\times 2.002} = \frac{59.800}{37.776} = 1{,}583$$

Using formula (9) it is calculated the a:

$$a = \bar{Y} - b\bar{X} = 13.543 - 1{,}583 * 286 = 13.090$$

Based on the above, linear regression for the dependent variable is:

$$Y = a + bX = 13.090 + 1{,}583 * X$$

For this research function of GDP is:

$$GDP = a + b * FDI = 13.090 + 1{,}583 * FDI$$

The coefficient of determination is calculated using the formula (4):

$$R^2 = \frac{a\sum y + b\sum xy - n\bar{y}^2}{\sum y^2 - n\bar{y}^2} = \frac{13.090\times 94.800 + 1{,}583\times 27.172.600 - 7\times 13.543^2}{1.286.260.000 - 7\times 13.543^2} = 0{,}024$$

Since the coefficient of determination is less than 1, we can conclude that the points are scattered around the line (y). Which means that the higher the linear dependency is between X and Y, if the coefficient of determination closer to 1, and vice versa.

The coefficient of correlation is calculated using the formula

$$r = \sqrt{R^2} = \sqrt{0{,}024} = 0{,}155$$

The absolute value of the correlation coefficient (r) indicates the strength of linear relationship between the variables. Value of r closer to zero shows that the connection is weaker, and opposite, if closer to one, the connection is stronger. The correlation in this research is positive, indicating that the linear increase in one variable corresponding to a linear increase in other variables.

Since the calculated correlation coefficient is 0.155, we can conclude that there is a slight influence of independent variable (FID) on the dependent variable (GDP). Linear regression shows that the increase in foreign direct investment in the fixed assets of 1 million euros, leads to an increase in GDP in value of 1,583 million euros.

Conclusion

The foreign capital may affect economic development in three ways. First, it enlarge the domestic investment rates. To the extent that a country has a mismatch between savings and investment as a result of balance of payments deficit, the inflow of foreign capital can help boost investments in the country. Second, it increases business efficiency. The increase in productivity is resulting from the increase in efficiency, based on the transfer of experiences, new knowledge and new technologies that are brought in with foreign capital. Third, it stimulates the economic system and the business. The existence of efficient companies in the market can stimulate local competitors to innovate business in order to survive in the market. In this way, it increases the quality and diversity of local producers in a country that is the recipient of capital.

Acceleration of the economic development and the rapid pace of the development in Bosnia and Herzegovina is linked to the rate of accumulation. Accumulation is one of the main factors of development. The main problem of modern economic systems of developing countries is to ensure the mass and the rate of accumulation that are sufficient to finance investments leading to economic development. Second goal is maximizing the efficiency of accumulation. The main issue of our country, which is preventing the development of the economy, is not that the economy is not able to produce, but that the fact that capital ends up in the hands of those social classes who use the capital on unproductive manner. In addition to the investments, that are very important factor that determines the speed of economic growth and development, there are other factors of the development such as: better and more efficient organization, better and more efficient use of production capacity, the higher level and quality of workforce education, on time application of new technologies and technical progress and so on. The basic criterion for determining the speed of development of the economic system is, apart from the increase in per capita income, also the amount of investments together with other factors which are available means of achieving economic development for society as a whole.

Based on the survey results, the positive impact of foreign capital on economic development is a proven by statistical-quantitative analysis, where it is recorded a positive correlation between FDI and GDP. Calculated the correlation coefficient is 0.155, so we can conclude that there is a slight influence of independent variable (FDI) on the dependent variable (GDP). However, a little correlation coefficient shows that foreign capital has an insignificant impact on the enabling and acceleration of the economic development in Bosnia and Herzegovina, but also that it is not the basic premise of the generation and acceleration of the economic development of Bosnia and Herzegovina. In addition, the economy of BH has no investment capacity, and must properly provide environment for the inflow of capital, which will be directed into those branches that will achieve the greatest benefit for the progress of the society as whole.

Different instruments of foreign capital inflows have different impacts on macroeconomic ratios of Bosnia and Herzegovina in global circumstances. This is a proven by usage of statistical-quantitative analysis, and it has been proven that FDIs have different effects on economic development in different periods. Positive correlation has confirmed the positive impact of FDI on economic development of BH. Survey method has proven that the instruments of capital inflows through the free zones, in the reporting period, showed outstanding results, i.e. the value of production increased 4 times, and the export value was higher 5.05 times, the number of employees has increased by 55%, while the share of exports from the free zones, in total exports of Bosnia and Herzegovina, increased from 4.14% to 9.25%.

References

[1] Todaro M P and Smit S C 2006 *Economic development*, TKD Sahinpasic, Sarajevo
[2] Jovanovic and Gavrilovic P 2006 *International business finance*, Faculty of Economics, Belgrade
[3] Samuelson P and Nordhaus W 1992 *Economic growth in the long run the most important factor in the economic success of nations*, McGraw-Hill
[4] Dunning J 1982 *A Note on the Intra-industry Foreign Direct Investmen,* Banca Nazionale del Lavoro (Rome), March, pp 265-284
[5] Popović G 2009 *Economics of the European Union, the macroeconomic aspects and common policies*, Faculty of Economics, Banja Luka
[6] ***OECD 2016 *Foreign Direct Investment for Development*, https://knoema.com/OECDFDIS2016/oecd-foreign-direct-investment-statistics
[7] ***UNCTAD 2002 *World Investment Report*, pp 319-327
[8] ***Foreign Trade Chamber Bosnia and Herzegovina 2016 *The basic elements of operations of free zones in Bosnia and Herzegovina*, www.komorabi
[9] ***The Ministerial Council of Bosnia and Herzegovina 2007 *Report of assessment and efects of regulations relating to free zones in Bosnia and Herzegovina regulation by stakeholders*
[10] Sosic I 1988 *Problems in statistics,* University of Zagreb, Faculty of Economics, Zagreb
[11] Yan X and Su X G 2009 *Linear Regression Analysis: Theory and Computing*, World Scientific Publishing, New Jersey
[12] Fox J 2008 *Applied Regression Analysis and Generalized Linear Models*, Sage Publications, New Jersey
[13] Montgomery D C, Peck E A and Vining G G 2012 *Introduction to Linear Regression Analysis*, 5 th Edition, John Wiley & Sons, New Jersey

IOP Conf. Series: Materials Science and Engineering **200** (2017) 012020 doi:10.1088/1757-899X/200/1/012020

The application of CAD / CAM technology in Dentistry

I Susic[1], M Travar[2] and M Susic[3]

[1]University of Business Studies, Jovan Ducic Street, No 23A, 7800 Banja Luka, Bosnia and Herzegovina
[2]University of Business Studies, Jovan Ducic Street, No 23A, 78000 Banja Luka, Bosnia and Herzegovina
[3] Enterprise Fructa Trade – KORT, Marije Bursac Street, No 70., 74400 Derventa, Bosnia and Herzegovina

E-mail: i.susic51@gmail.com

Abstract: Information and communication technologies have found their application in the healthcare sector, including the frameworks of modern dentistry. CAD / CAM application in dentistry is the process by which is attained finished dental restoration through fine milling process of ready ceramic blocks. CAD / CAM is an acronym of english words Computer-Aided-Design (CAD) / Computer-Aided-Manufacture (CAM), respectively dental computer aided design and computer aided manufacture of inlays, onlays, crowns and bridges. CAD / CAM technology essentially allows you to create a two-dimensional and three-dimensional models and their materialization by numerical controlled machines. In order to operate more efficiently, reduce costs, increase user/patient satisfaction and ultimately achieve profits, many dental offices in the world have their attention focused on implementation of modern IT solutions in everyday practice. In addition to the specialized clinic management software, inventory control, etc., or hardware such as the use of lasers in cosmetic dentistry or intraoral scanning, recently the importance is given to the application of CAD / CAM technology in the field of prosthetic. After the removal of pathologically altered tooth structure, it is necessary to achieve restoration that will be most similar to the anatomy of a natural tooth. Applying CAD / CAM technology on applicable ceramic blocks it can be obtained very quick, but also very accurate restoration, in the forms of inlays, onlays, bridges and crowns.

The paper presents the advantages of using this technology as well as satisfaction of the patients and dentists by using systems as: Cercon, Celay, Cerec, Lava, Everest, which represent imperative of modern dentistry in creating fixed dental restorations.

1. Introduction

Modern dental practice implies an increased application of information and communication technologies. There are numerous advantages to facilitate the work of the dentist, but also users of dental services that are becoming more demanding in terms of aesthetics, with the clearly expressed desire for the minimum of staying and delaying in the dental office. The computer, as a means of interactive communication, have a greater role in prosthodontics in terms of practice in dental office, but also in dental technical laboratories. Namely, when it is necessary to substitute the removed pathologically-altered tissue, and producing a fixed prosthetic inlays, onlays, crowns and veneers are indicated, or when it is necessary to make up missing teeth, and therefore bridges are prodcued, it comes to the fore the application of CAD / CAM technology.

In the early 90s, over 70% of private dental practices in the United States used PCs [1]. Undoubtedly the advantage of such a work organization is to increase the speed of work, communication with patients and reducing the space for data storage [2]. An important role is also to reduce the possibilities of entering wrong, illogical or incomplete data [3]. This computer application today represents far the most common form of using in our profession. The use of computers in therapy is a challenge for enthusiast and visionaries who developed a whole new field: computerized dentistry. CAD / CAM systems represent the pinnacle of computer technology with lots of realized and potential applications in dentistry. CAD / CAM systems in dentistry consist, basically, of three components [4]:

• The first component is a device that reflects the preparation of teeth and other supporting tissues and is responsible for spatial data digitalization (CAI - Computer Aided Inspection);

• The second component consists of computer which plans and calculate body form of restoration, equivalent to the area of CAD-s;

• The third component represents a numerically controlled milling machine which from the basic shape produces dental restoration, corresponding CAM area. As a rule, there are recommended additional processing such as polishing or individual preference by a dental technician or doctor [5].

2. CAD/CAM tehnology in dentistry

Year 1985 is the key to the introduction of CAD / CAM technology in dentistry. In fact, this year, with the help of triangular cameras it is executed multidimensional measurement, enabling the transfer of information on the measurement to the computer screen. With the help of a PC, software for image processing and connections with CNC milling machine, it is obtained the first silicate inlay restoration at the University of Zurich. Then it was almost unimaginable technology and practical creation of a new concept in dentistry. In Germany, the CAD / CAM technology is introduced in the dental practice in 1988 [6]. Modern software provides such a possibility where the minimum thickness of the restoration is less than the recommended, it alerts the dentist to the existing problem. Also, on the virtual model is marked and clearly identifiable critical areas that can be corrected with the offered tools. (Www.sirona.com)

The development of technology went from the machine copy milling through to fully computer-controlled system, with a large base form of the tooth, which allowed the automatic production of crowns and bridges. Today, there is several of these systems (Cerec, Cercon Celaya, Lava, Everest) and they are considered to be the ones in the future to have a much greater use in producing fixed restoration. Figure 1 shows the fixed restoration produced by means of computer-controlled system.

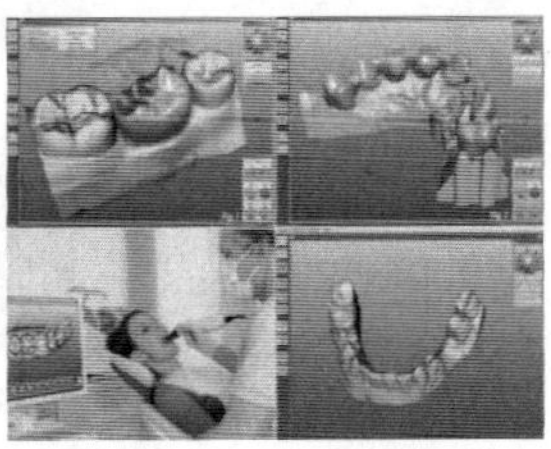

Figure 1. Producing fixed restorations with the aid of computer-controlled system

Thanks to the CAD-CAM technology and numerous studies, it resulted in formula for making extremely faithfull restoration, which not only possess outstanding aesthetic characteristics, but also extremely biocompatible. It is a non-metal ceramics. Depending on the defect in the teeth, these materials may be used for making crowns and bridges, dental veneers, but also for special fillings.

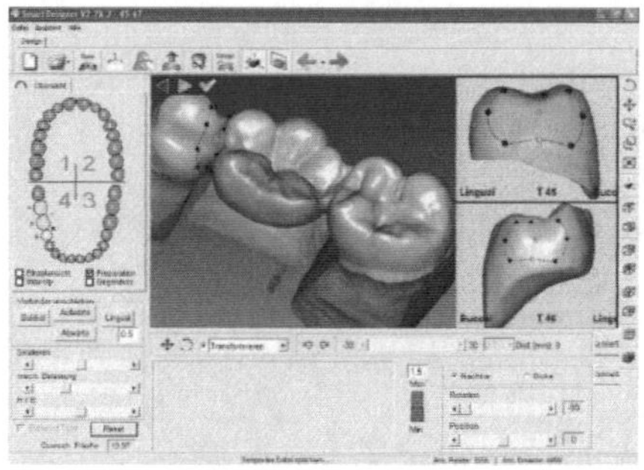

Figure 2. Innovative CAD/CAM technology

All of these restorations are produced in laboratories for dental technology, equipped with CAD-CAM technology (computer), what guarantee outstanding precision and aesthetics, as shown in Figure 2.

By forming a 3D image of teeth and gums on the screen, it allows the dentist that by cursor form a very precise and appropriate anatomical design of missing tooth substance. The resulting 3D models provide an ideal basis for restoration design. When designing it is in a very simple manner taken into account the relationship with adjacent teeth, the teeth in the opposite jaw that establish appropriate contacts, but also the relationship between restorations and the soft tissue and gums.

CAD/CAM machine through further milling process of finished ceramic blocks produces restoration of teeth which is the exact replica of the 3D drawings, ie. design of restoration, which is done by dentist via the CAD / CAM software, as shown in Figure 3. Factory ceramic blocks which are processed by milling process, are made in a number of different shades, so that the color corresponding to the requirements of patients, as well to the parameters that determine the high level of aesthetics.

By applying modern machines type MC XL for grinding of finished ceramic and zirconia blocks, fixed prosthetic restoration can be today made for only 2-3 hours in the office. The advantages of this machine are multiple, as shown in Figure 4. Precision of milling is moving in the range of +/- 25 microns, while the time required for grinding circular bridge does not exceed 6 minutes. Thanks to a grinding resolution of 7.5 microns, prosthetic restoration surfaces that are produced this way are certainly adhere better, compared to conventionally made works (www.kalmar.hr/usluge/cadcam).

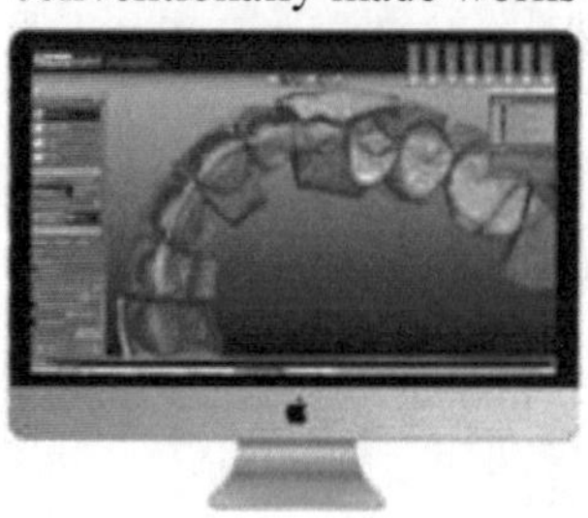

Figure 3. CAD/CAM technology for manufacturing dental crowns and bridges

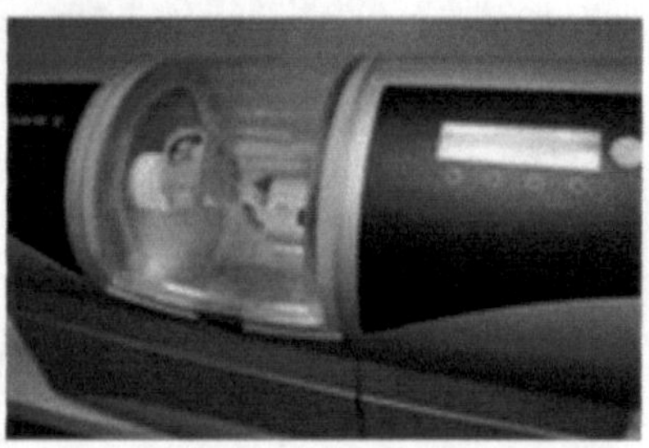

Figure 4. CAD/CAM allows us to quickly restore damaged teeth with natural-colored ceramic illings

3. Proces od producing metal-free restorations by CAD/CAM tehnology

The process of producing ceramic restorations by CAD-CAM technology is more precise than the conventional process of producing metal-ceramic crowns and bridges. Figure 5 provides an overview of the CAD / CAM system in the process of producing crown bridges.

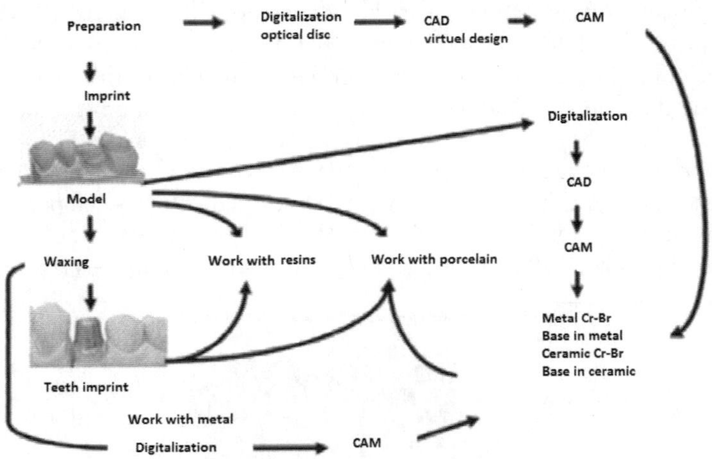

Figure 5. Display of dental CAD / CAM system in the process of producing crown-bridges [7]

Prosthetic restoration are made in several phases, which are allocated by the following order [8]:

1. *Overview and history-* Based on the indications and the status of the tooth, as shown in Figure 6, the dentist diagnoses and recommends several options, explaining the pros and cons, depending on the indication.

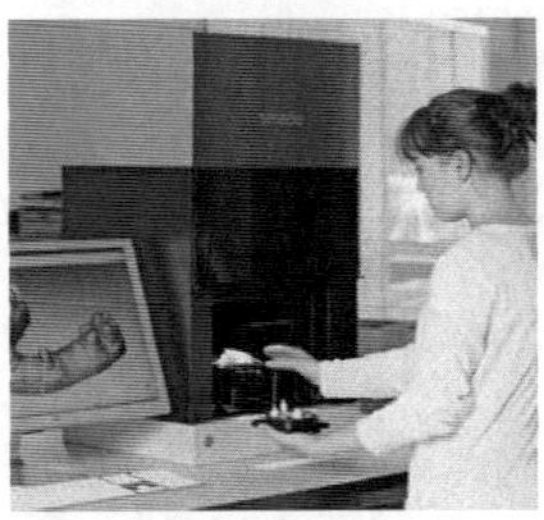

Figure 6. Overview and history

2. *Preparation of teeth for placing prosthetic restorations.* Process begins by grinding of teeth and its suppression, which is carried out by dentist depending on the type of the ceramics to be used for the certain clinical case, i.e. to create prosthetic restoration.

3. *Taking the tooth imprint.* The dentist performs the tooth imprint (one or more, depending on which prosthetic restorations works), on which it will carry out further construction and casting of prosthetic restoration.

4. *Casting of the model.* Based on the tooth imprint plaster model is casted, on which is carried out further work, on the basis of tooth imprint.

5. *3D scan of the model.* 3D oral camera captures teeth, after which the image is transferred to a computer and processed using the software. These cameras allow a high degree of accuracy and efficiency, and are particularly suitable for the restoration of individual crown.

6. *Modeling.* CAD / CAM software modeling the teeth based on the entered requirements.

7. *3D teeth printing.* Before you start teeth printing, you need to install ceramic blocks in the milling. The ceramic block is fixed on the wheel that allows block to be inserted. Bridge is produced by milling process on the basis of the 3D model from the block set in the CAD-CAM device. Milling machine develops the desired shape in accordance with the instructions of a computer. The ceramic block is processed by turning on its axis, a diamond disk rotates, moves up and down around the ceramic block and processes it. The movement of the diamond disc is enabled via electric rail.

8. *Cementation.* Prosthetic restorations are cemented with special aesthetic cement for metal-free ceramics, as presented in Figure 7. There are two types of cementing - temporarily and definitively. Temporary cementing of restoration is done in the period of adaptation of prosthetic restoration to the jaw, while definitive cementing is done after ensuring that the prosthetic restorations is accepted.

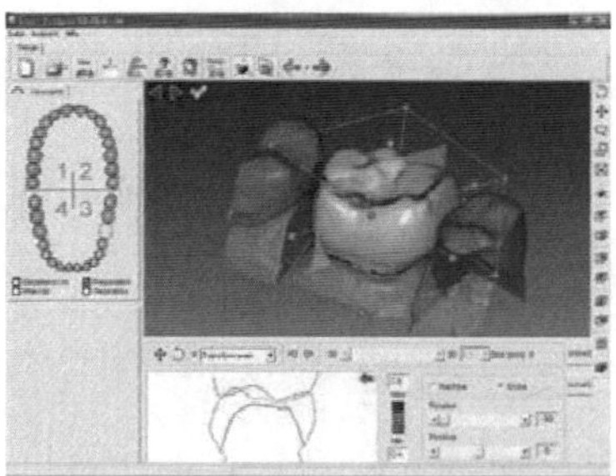

Figure 7. Display of the metal-free restoration by CAD / CAM technique

The advantages of metal-free ceramic compared to metal-ceramic works:

- Complete biocompatibility of materials,
- The absence of allergy to this material (a large number of patients with metal-bridges suffer from allergic reactions because of the large amount of nickel in the metal alloy),
- Absence of bimetallism at metal-free works (creating low-voltage levels between the two metals, eg. between metal-ceramic crowns),
- Firmness of works is 4 times higher than the metal used for metal-ceramic works,
- Persistence and not changing its physical and chemical properties even after long years spent in the mouth,
- The aesthetic superiority compared to metal-ceramic works,
- Beneficial effects on the gums, ie. „gingiva“ with which it comes into contact,
- The absence of dark discoloration of the "gingiva" at the junction of crown and gums.

The disadvantages of metal-free ceramic compared to metal-ceramic works:

- Price. Due to the expensive and long-term development of this technology, expensive CAD-CAM machines and expensive process for manufacturing, metal free crowns are more expensive than metal-ceramic works. However, taking into account the relationship between price and quality, it can be said that the ratio is on the side of metal-free ceramics.

IOP Conf. Series: Materials Science and Engineering **200** (2017) 012020 doi:10.1088/1757-899X/200/1/012020

4. Types of metal-free works

4.1. Metal
Metal-free crowns are aesthetic restorations that are made in the dental laboratory of special blocks by using CAD-CAM techniques. Blocks have characteristics very similar to natural teeth, depth and transparency, so that the final product represents a faithful copy of natural teeth. Cemented by a special cement, which further contribute to the aesthetic characteristics of the crown.

4.2. Inlay – Onlay
These are dental restorations which represent transition between crown and filling. They are used when there is no much remaining tooth structure, and avoiding producing crowns. In the cases when dental caries too much devastated the tooth structure, and when after the removal of caries, the resulting cavity can't be adequately compensated by classical fillings (either of amalgam or composite), then it is produced inlays.

Inlays are usually made of ceramic or metal (although now days rarely made of metal). The main difference between the inlay and the fillings is, in addition to the material from which it is made, that the inlay is made outside of the mouth. Therefore, for the manufacturing of inlays it is needed at least two visits to the dentist. In the first visit to remove the carious mass and preparing the tooth for an inlay. Then making tooth imprints. Based on tooth imprints technician in the lab creating an inlay, which is then in the second visit cemented in the mouth of patient.

Depending on the size of the inlay, ie. the extent of the cavity that is formed after the removal of caries, we distinguish two forms of this restoration, as presented in Figure 8, including:

- inlays (inlay) - which affects maximum of two surfaces of the tooth (for example making inlays can be seen in the Dental SPA center).
- onlays (onlay) - that affect three or more surfaces of the tooth.

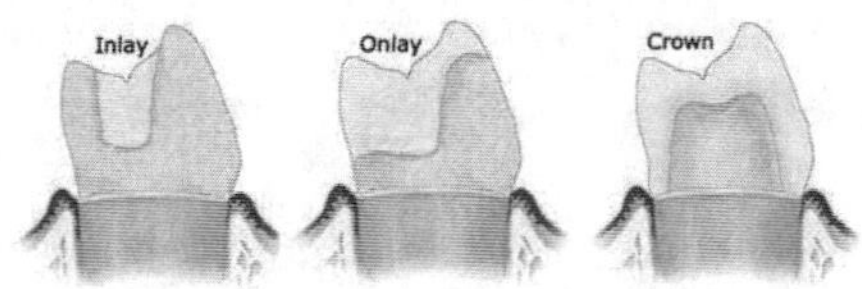

Figure 8. Inlay-Onlay bridges

4.3. Inlay Bridges
Inlay bridge is minimal invasive method for metal-free dental restorations, excluding implants. In this type of bridges adjacent teeth are grounded to a minimum in the form of fillings, as presented in Figures 9a and 9b. For all other types of bridges adjacent tooth is grinded and onto it is set crown in order to carry the missing tooth.

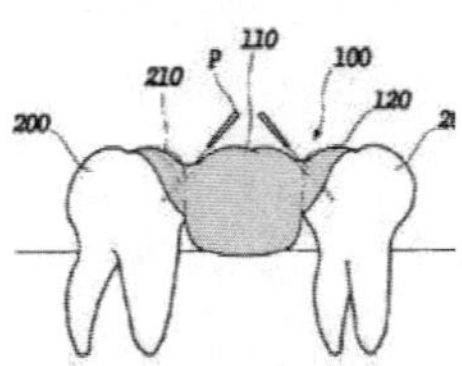

a)

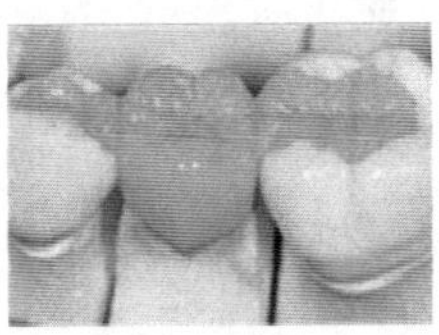

b)

Figure 9a and **9b** : Inlay bridges

IOP Conf. Series: Materials Science and Engineering **200** (2017) 012020 doi:10.1088/1757-899X/200/1/012020

5. Optical methods of spatial digitalisation

Optical methods of spatial digitalisation, like mechanical, based on the criteria of space where the scan is performed, are divided as:

- intraoral and,
- extraoral methods.

In relation to the size of the scanning area, they are classified on the dotted and striped (surfaced). *Intraoral scanning* means work in a dental office, while the *extraoral methods*, mainly related to laboratory work. Both methods have been developed side by side, but today in the practical application is present only a single intraoral (two are in announcement) and the great number of extraoral systems. The requirements set for them are different. For ergonomic reasons, intraoral scanner should not be fixed to the remaining teeth. This affects the request of its shape, size, weight and ability to maintain hygiene, but above all the speed of scanning. Empirically it is proven that trained user can keep the scanner head immovable and still versus the scanned tooth, mostly for 0.5 seconds. The data on the speed of data measurements acquisition, in addition to the resolution, is one of the most important in the choice of the system and its broad applicability. Size of the scanning field is minimally 14x14mm, and optimally 25x14mm. The range of scanning depth should be at least 10 mm, but should not be greater than 14 mm. Scanner resolution should be at least ± 25µm [9]. The most famous representative of intraoral optical method is Cerec system, as shown in Figure 10 (Sirona Dental Systems GmbH, www. Sirona.de) [10].

Figure 10. Sirona, 3M Dental CAD/CAM System

This technique is using more light rays, in the form of lines, projected on the preparation (line hatched area). The rays in rapid oscillations move across the object, so that in a short period of time is obtained three-dimensional shape of preparation. Similarly to conventional photography, the camera at the time of recording should be kept as still as possible. Fixing the camera opposite to the object at this system is not necessary because the time required for data processing from all the 340,000 pixels is less than 0.5 sec. – as illustrated in Figure 11. During the year 2005 are shown two more intraoral scanners Evolution 4D and HintEls.

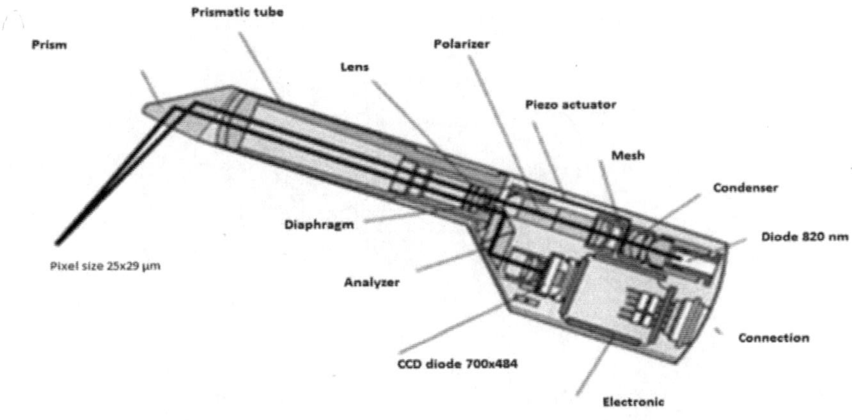

Figure 11. Scheme of Cerec 2 scanner head section

Extraoral systems scan is carried out on the model, and for this reason there is a need for dental technical laboratory. In these systems it is not critical high speed data collection, because the head of the scanner and the object that is scanned are immovable, but the width of the scanning and precision measurements.

A different solution, to achieve the third dimension by using the CCD chips, gives the *laser triangu procedure,* after Lelandais and Clainchard-in (1984). If you focus laser dot air with oscillating mirror for CCD camera there will be a clear limited laser line. The great advantage of this system is the possibility of scaning undermined surfaces. This mode is for now only possible as extraoral methods.

Representative of extraoral dot scanner is Cerec Scan, illustrated in Figure 12, and Cerec inLab, illustrated in Figure 13. The scanner is fixed to one of the motors of the milling machine and object of scanning is movable. Scanner resolution is similar to the intraoraoral scanner, but the scan time of one tooth is much longer. For the "four-member" bridge it takes 2-3 minutes. This year, the factory has developed a new high-resolution scanner, in which this time is reduced to about 40 seconds (Cerec INEOS) [11].

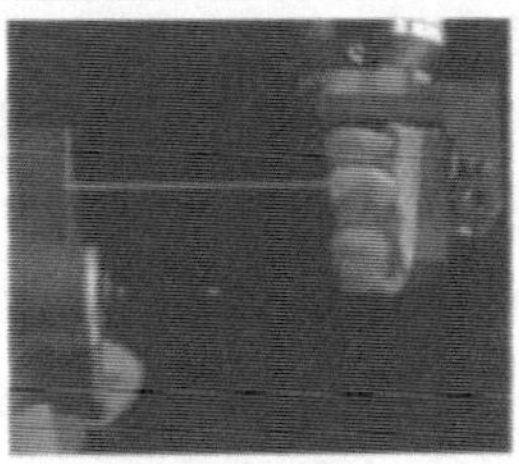

Figure 12. Cerec Scan – *integrated laser scannerr, dot scanner in the left engine*

Fig. 13. Cerec inLab system – *from left to right: inEos ekstraoral Cerec in Lab with inbuilt dot scanner, PC computer with 3D softver*

6. Advantages of using CAD / CAM technology in dentistry

In dental practice, CAD / CAM technology is primarily used to provide patients with one or multiply-restorations in the mouth or on the teeth themselves. Application of this technology provides a simplified procedure compare to the current way of producing. Restorations made this way are long-term, as the first studies saying. Figure 14 shows restorative dentistry with CEREC® 3D computerized. Sirona study shows that in five years 95-97% of CEREC's crowns are still intact, while the degree of preserved inlays and onlays is 90-95% after ten years.

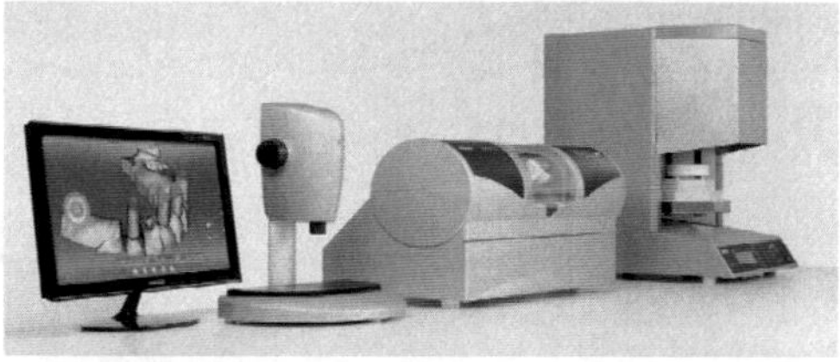

Figure 14. Restorative dentistry with CEREC® 3D computerized

The patient is provided with comfort and complete treatment is performed in one session in the context of dental practice, of course with condition that a dentist has so called *Chairsaide* CAD / CAM. This actually means that the entire equipment can be placed in the area of dental office. This is a remarkable advantage, because it eliminates the need to continue work in dental technical laboratories, as well eliminates additional costs. However, if the CAD / CAM technology is located in the dental laboratory and is applied by dental technician, it requires two visits to the dentist. By using this technology very precise restorations are produced, which so far is certainly not the case, whether it is about complicated multi-member bridges, crowns, inlays, onlays or veneers.

Advantages of using CAD / CAM technology for dentists are:

- The patient spends less time in the office;
- A simplified procedure;
- Significantly reduced costs for dental technical laboratories;
- Reduced consumption of materials;
- Increased productivity;
- Easier way of producing;
- Precisely produced restorations;
- Increased productivity.

Advantages of using CAD / CAM technology in dental-technical laboratory:

- Easier way of producing;
- More precisely made restorations;
- Lower consumption of materials;
- Higher productivity.

Advantages of using CAD / CAM technology to produce onlays:

- Very often saves the tooth structure compared to traditional crowns.

Advantages of using CAD / CAM technology to produce inlays:

- Much better restoration than traditional fillings.

7. Survey of patient satisfaction using CAD / CAM technology in Bosnia and Herzegovina

In general, bearing in mind that for providing the services patients are essential, because who would provide services if there would be no demand, and crucial is continuously conducting research and analysis of feedback from patients. When analyzing opinions, feelings and experiences of patients, it is necessary to take all factors into account that in any way touch the patient. These are above all: speed of service delivery, the pain that is (not) felt when providing services, aesthetic effects, and the price of services as one of the most important elements, kepping in mind the economic crisis and the turbulent market developments in our and as well regional areas.

Accordingly, it is formulated ten questions, formed into the questionnaire for assessing the level of satisfaction of patients after making dental implants using CAD / CAM technology. After treating the patient, while still in the dentist's chair, ask the patient to set aside a few minutes to fill out the questionnaire, expressing their opinions and feelings about the experience with dental restorations. Also, in addition to the above questions, to which the patient responds, indicating the level of agreement with the above findings, the patient has the ability to verbally and in writing to state its suggestions that would contribute to the further improvement and advancement of providing these services.

Table 1. Structure of questions for patients with five suggested answers on the issue (Author 2016)

Questions	Disagree	Partially agree	Agree	Mostly agree	Strongly agree	Total:
1. The process of preparations for the prosthetic restoration is pleasant and does not last long.	0	10	13	28	49	100
2. It is not necessary repeatedly to come to the dentist, in order to complete the preparation process.	0	0	38	42	20	100
3. Producing restorations takes surprisingly short.	0	0	4	22	74	100
4. During producing restoration , it is involved surprisingly professional staff	10	16	12	18	44	100
5. Installing restoration is completely painless.	0	11	32	24	33	100
6. Restorations do not differ entiate from natural teeth, look completely natural.	0	0	10	19	71	100
7. Once the restoration is set up, it is needed a certain period of adaptation.	0	2	30	22	46	100
8. Dental restorations give completely natural feelings, no need for restrictions of certain foods.	0	0	0	0	100	100
9. Recommend the proces tom its frineds.	0	1	50	25	24	100
10. Price is afrodable.	0	0	9	12	79	100
Total:	**10**	**40**	**198**	**212**	**540**	**1000**
Participations in %.	1	4	19,80	21,20	54	-

In order to investigate the satisfaction level of patients, 100 patients are examined and asked 10 questions per patient, in the past 6 months. The results showed that patients exceptionally satisfied with making dentures by CAD / CAM technology. Since the questions are formulated mainly that the first column (I disagree) reflects the lowest level of satisfaction of the patient (1%), while the last two columns reflect the highest level of patient satisfaction (75.20%).

The research results and responses participation depending on the customer satisfaction, is shown in the following Figure 15.

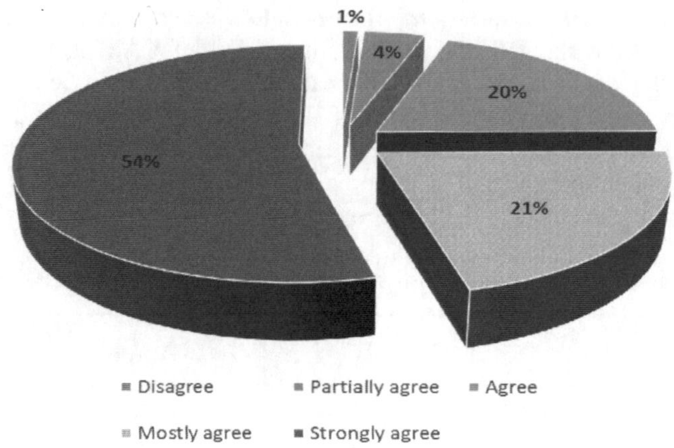

Figure 15. Structure of the responses to the questionnaire on patient satisfaction (Author 2016)

From Figure 15 can be noted that the most significant participation in the structure of responses take final answers to the set of 10 questions: most agree and strongly agree, that is the highest level of patient satisfaction in the amount of 75.20%. A minimum participation to the set of 10 questions take answers: do not agree, that is the lowest level of patient satisfaction of 1%. **The introduction of new technologies in Bosnia and Herzegovina leads to the improvement of services in the field of dental medicine, as confirmed by patient satisfaction (95%).** Accessibility prices to patients is very satisfactory (91%). On the issue of "producing restorations is short", patient satisfaction was 96%. Asked whether dental restorations giving a natural feeling?, patient satisfaction is 100%.

Broken down by one question from the questionnaire, structure response is as follows - Figure 16:

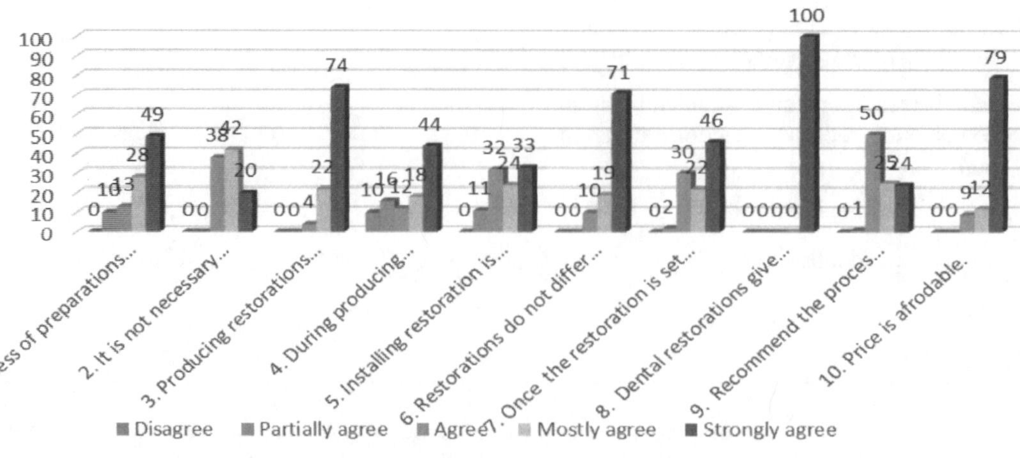

Figure 16. The structure of the questionnaire responses on the issues of patient satisfaction (Author 2016)

8. Conclusions

If the dentist claims to be a leader in the development of inlays, onlays, crowns and bridges, it is necessary to practice the application of CAD / CAM technology in aesthetic fixed restorations. The use of this technology provides a high quality, professionalism, profit, but also a steady increase of a "new" and satisfied patients.

Any progress in the field of information and communication technologies at the same time finds its application in various fields of medicine, including in dentistry. This work is further evidence of the necessity of applying IT in dentistry. The expansion of information and communication technologies will in the future contribute to even stronger impact on the timely diagnosis, timely and adequate treatment, monitoring of therapeutic effect, but also to achieve a high level of aesthetics in all branches of medicine and dentistry where would be necessary.

"The ideal CAD / CAM system" for many years is a dream of many researchers. Since the precision of restorations made by CAD / CAM technology in the function of all the individual errors of procedures and equipment, and that scanning is the initial source of possible inaccuracies, the higher resolution scanner will most significantly contribut to the quality of the entire system.

The introduction of new technologies leads to the improvement of services in the field of dental medicine, as confirmed by patient satisfaction (95%). Accessibility prices to patients is very satisfactory (91%). On the issue that producing restoration is short, patient satisfaction was 96%. Asked whether dental restorations are giving a natural feeling?, patient satisfaction is 100%. The new technologies development of services in the dental medicine acquire aesthetic character. The use of CAD / CAM technology significantly shortens the time of creating prosthetic work, and CAD / CAM systems are easy to use.

References

[1] Casanova A W and Marshall W 1986 Computer applications in large group practices, *Den Clinic of North America* **30** 673-681

[2] Gilboe D B and Scott D A 1991 Computer system for dental practice management, *J Can Dent Assoc* **57** 782-786

[3] Chasteen J A 1992 Computer databa se approach for dental practice, *J Am Dent Assoc* **123** 26-33

[4] Rekow D 1987 Computer aided de signand manufacturing indenti stry: A review of the state of art, *J Prosthet Dent* **58** 512-516

[5] Todorović A 2005 *The application pf CAD/CAM technology in dental prosthetics*, Belgrade: Copyright edition

[6] Ritter A V 2002 Chairside CAD/CAM in Dentistry, *JERD* **15**(2)

[7] Miyazaki T, Hotta Y, Kunii J, Kuriyama S and Tamaki Y 2009 A review of dental CAD/CAM: current status and future perspectives from 20 years of experience, *Dental Materials Journal* **28** 44-56

[8] Joda T and Buser D 2016 *Digital implant dentistry - a workflow in five steps,* URL: http://www.dental-tribune.com/htdocs/uploads/printarchive/editions pp 16-20 pdf

[9] Pfeiffer J Dental 1998 CAD/CAM technologies: the optical impression (I), *Int J Comp Dent* 29-33

[10] Schneider W 2000 Cerec 3, *Int J Comp Dent* 33-46

[11] Monkmeyer U R et al. 2005 The prefabricated anatomical polychrome CAD/CAM crown for the inLab system, *Int J Comp Dent* 16-78

Statistical analysis using the multiple regression research in areas of the indefinite chilled cast–iron rolls manufacturing

I Kiss , V Alexa and V G Cioată

University Politehnica Timisoara, Faculty of Engineering Hunedoara, Department of Engineering and Management, Hunedoara, Romania

E–mail: imre.kiss@fih.upt.ro

Abstract. To analyze the metallurgical processes is used, mainly, the statistical fundamental methods that permit to draw conclusions, from the observed values, about the repartition of the frequencies of various parameters, about their interaction, about verification validity of certain premises, and about the research of the dependencies among different parameters. In this sense, the realization of optimum chemical compositions of the cast–iron can constitute a technical efficient way to assure the exploitation properties, the material from which the rolling mills rolls are manufactured having an important role in this sense. This paper reviews key aspects of roll material properties and presents an analysis of the influences of chemical composition upon the mechanical properties of the indefinite cast iron rolls. Now, using the multivariate research, we present some mathematical correlations and graphical interpretations between the hardness and the chemical composition. Using the double and triple correlations variation boundaries for the chemical composition, in view the obtaining the optimal values of the hardness of indefinite cast iron rolls, are obtained. The partial results and evidence obtained by actual experiments are presented. For the multiple regression equations, correlation coefficients and graphical representations the software MATLAB was used.

1. Introductory notes

All of the rolls must have high exploiting qualities, which are determined by hardness, wear resistance and stability at high temperatures [1–3]. When the roll hardness is higher, the wear resistance is greater. Within the framework of the rolling operations predominate the wear, it is necessary that rolls to have good hardness [1–3]. Since the properties of any cast iron parts are determined by microstructure which are formed during solidification in the casting mould and under the influence of cooling rate, the criterion that determines the basic physic–mechanical properties of the rolls is the structure. Therefore, the composition of the alloy is an important parameter [1–3]. Hardness on the crust depends on the degree of alloying elements and the mode of distribution of the graphite quantity does the metal mass.

Carrying out the hardness of the cast iron rolls, prescribed by the technology, with the perlite–cementite–graphite type structure, shall be ensured by strict compliance with the chemical composition, conditions of the iron melting and casting, and the thermal cooling applied regime. The hardness variation in the entire section of the roll, it is obtained through the establishment of a precise correlation between the speed of cooling and the chemical composition, after cooling, the hardness on the crust must be greater, due the cementite in the near of the metal chill, i.e. in the area of the roll body [1–3]. In the middle as well as on the necks of roll, the separation shall be in the form of carbon graphite, in increased quantities, and avoid the formation of the cementite in this area, in order to ensure a high resistance needed into the rolling process. Increasing the graphite, starting with peripheral zone and working toward to core, takes place gradually, through a scheme related the heat technology, which make the hardness to vary gradually, through passing zone, in the case of roll with indefinite crust and by a sudden transition, in the case of the defined crust. In this context, the indefinite chilled cast iron rolls, in which missing a clear and distinct boundary between crust, passing zone and the core area, occupy an important place, and specialized studies in this field are missing almost completely [1–3].

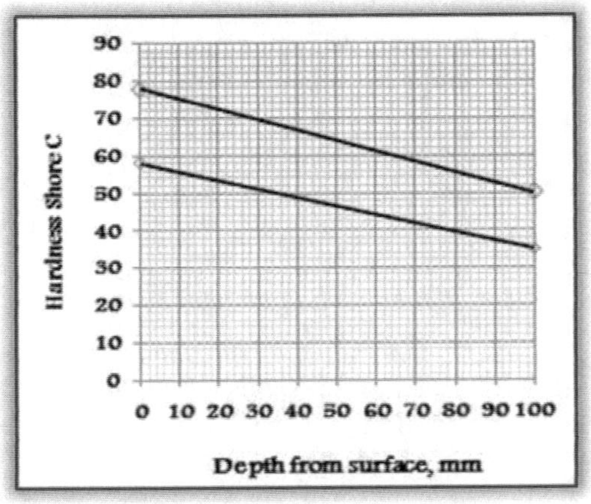

Figure 1. Indefinite chill rolls

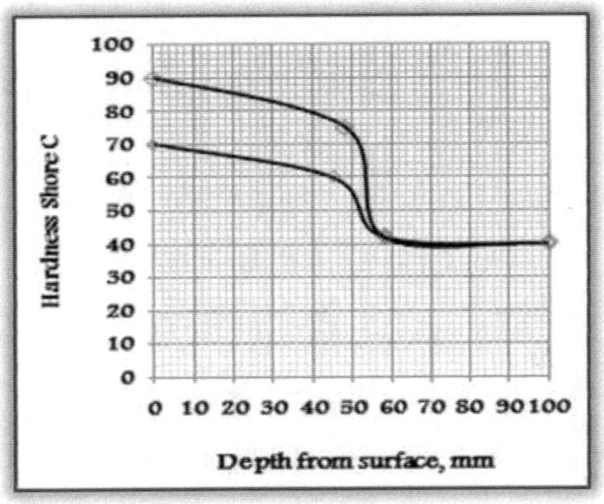

Figure 2. Clear chill rolls

The main structural constituent of cast–iron is the graphite, on the quantity, shape, dimensions and the manner of the distribution of mass basis, depending on the physico–mechanical and wears resistance of the entire rolls [1–3]. Part of the carbon forms flake graphite and the remainder forms carbides. The amount of graphite is less prominent in higher hardness grades and increases towards the centre of the roll. (Figure 1 and Figure 2) Thus, a hardness gradient is developed between the rolls core and the body. The depth of the chilled layer is difficult to define and the hardness decreases gradually towards the centre of the roll. (Figure 1 and Figure 2) The chilled outer surface is very hard and has a white fracture face for a distance beneath the surface (known as the chill zone), signifying that the formation of free graphite in that area had been suppressed by the rapid cooling. The surface layers contain very small particles of graphite and the structure changes smoothly into the grey core. The hardness decreases slowly at first from the surface and more quickly towards the soft core. (Figure 3)

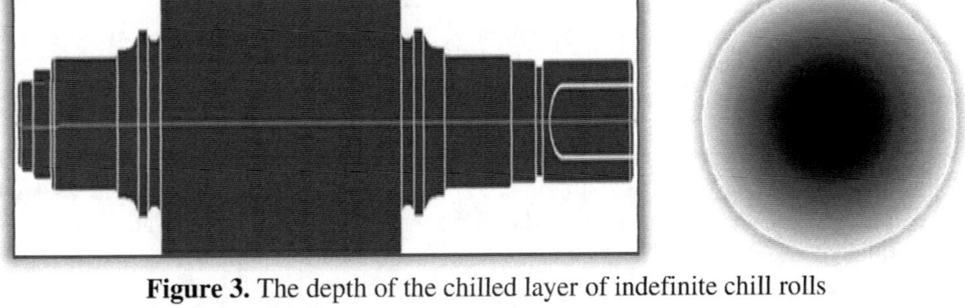

Figure 3. The depth of the chilled layer of indefinite chill rolls

In the cast iron rolls, are to be found all the structural constituents, each of them having its own hardness, well determined. One of the basic factors that determine the structure is the chemical composition of the cast–iron, the basically (C, Si, Mn, P, S) and the alloying (Cr, Ni, Mo) too [1-3]. The non–compliance of the each sort of rolls chemical composition, will lead to rejection of them. If on the crust the hardness is determined by the quantity of cementite, the rolls core must contain graphite, in order to ensure a high resistance to flexing, which they are subject into the industrial exploitation. Graphitization of cast irons designated for the working layer of rolls occurs under isothermal conditions at high temperatures, which vary over the cross section and resulted from casting the roll core [4].

The rolls must present high hardness at the surface and lower hardness in the core, adequate with the high working regimes. If in the crust the hardness is assured by the quantities of cementite, the core must contain graphite to assure that. While indefinite chill rolls can be produced within large ranges, the composition and resulting properties of the chill roll can be more easily controlled and are more desirable if the compositional ranges are limited to the requirement's values [1], [3]. The roll casters need to appreciate that minor changes to the chemical ranges and also substitution of comparably active elements can be made to the indefinite chill roll composition, while maintaining the desired exploitation properties. Silicon contributes to the formation of graphite and to maintaining the character of the cast iron. Nickel is added to the indefinite chill roll composition to promote the formation of free graphite in the alloy, having in view that an excess of Nickel will tend to destabilize the iron's structure. Therefore, the composition of the melt is set using alloying methods by fixing the concentrations of Carbon and Silicon in the presence of Nickel and the effective sum of the carbide forming elements in such a manner that, at its solidification, a microstructure is formed [3], [5].

Control of microstructure and properties of cast iron prediction is in perfect control of the melting process and the metallurgical melt processing. Required metallurgical quality of a particular type of iron and its properties are determined by chemical composition and a proper melting processing. Prediction of the properties of the resulting cast iron is a prerequisite for the management of metallurgy, especially in material destined to the cast rolls.

2. Methodology

The analytical approach aims original line through interdisciplinary aspects, but also through the mathematical analysis approach, the method resulting several equations and graphics. Planning and research on the industrial data, obtained from the rolls manufacturing area, are impressed by creation of a database necessary to the statistical processing, mathematical interpretations, modelling and optimization alike. In this sense, we implemented into the series research by lots of industrial data analyses, by the way of the statistical experiment, in separate series. This analysis gathers within the framework of research, the cast roll material and aims to characteristics of the material. From this point of view a mathematical modelling is made, which is effectuated on the basis of differentiation on parts of the rolls (body and necks), taking account the industrial data [3], [6–11].

As results of the applied mathematical modelling, we described a number of multi–component regression equations, determined to the 3rd and 4th dimensions spaces. Also, we generated several regression surfaces and matrices of the level curves, which can be represented and interpreted by the roll makers and may be considered as correlation charts between the analyzed variables. For the multiple regression equations, correlation coefficients and graphical representations the software Matlab was used.

3. Multicomponent relations determination

The variables and the hardness variation limits are presented in Table 1. The average values for the three variables ([C], [Si], [Ni]) and the hardness (HS), necessary for the calculation of the optimal form of modelling are presented in Table 2 [6].

Table 1. The variation domains of variables

Carbon, [C]		Silicon, [Si]		Nickel, [Ni]		Hardness, $HS_{(body)}$		Hardness, $HS_{(necks)}$	
min	max	min	max	min	max	min	max	min	max
3.17	3.32	0.77	1.42	2.06	3.89	56	78	57	76

Table 2. The average values of variables

Carbon, [C]	Silicon, [Si]	Nickel, [Ni]	Hardness, $HS_{(body)}$	Hardness, $HS_{(necks)}$
med	med	med	med	med
3.2261	1.0344	3.0235	70.39	67.88

The optimal form of modeling in the case of $HS_{(body)} = f([C], [Si], [Ni])$ is given by the equation (1), where the correlation coefficient is $rf_{HS(body) = f([C], [Si], [Ni])} = 0.6462$ and the average square aberration from the regression surface is $sf_{HS(body) = f([C], [Si], [Ni])} = 3.0771$.

The optimal form of modeling in the case of $HS_{(necks)} = f([C], [Si], [Ni])$ is given by the equation (2), where the correlation coefficient is $rf_{HS(necks) = f([C], [Si], [Ni])} = 0.6499$ and the average square aberration from the regression surface is $sf_{HS(necks) = f([C], [Si], [Ni])} = 2.9500$.

$$HS_{(body)} = 120.0595\,[C]^2 - 29.9622\,[Si]^2 - 0.9482\,[Ni]^2 - 18.6234\,[C]\,[Si] - 1.9188\,[Si]\,[Ni] + 55.4181\,[Ni]\,[C] - 918.9241\,[C] + 127.6439\,[Si] - 166.3666\,[Ni] + 1725.0823 \quad (1)$$

$$HS_{(necks)} = -2.9071\,[C]^2 - 34.1476\,[Si]^2 - 0.7563\,[Ni]^2 - 63.8114\,[C]\,[Si] - 3.1912\,[Si]\,[Ni] + 43.3294\,[Ni]\,[C] - 40.6041\,[C] + 286.6411\,[Si] - 128.0087\,[Ni] + 163.9398 \quad (2)$$

In the technological field, the behavior of these hyper surfaces in the vicinity of the saddle point, or of the point where three independent variables take their average value, can be studied only tabular, which means that the independent variables are attributed values on spheres concentric to the studied point. Because these surfaces, described by the equation (1) and (2), cannot be represented in the three–dimensional space, the independent variables ([C], [Si], [Ni]) were successively replaced with their average values ($[C]_{med}$, $[Si]_{med}$, $[Ni]_{med}$).

Therefore, in the case of $HS_{(body)} = f([C], [Si], [Ni])$, the equations (3)–(5) were obtained. In the case of $HS_{(necks)} = f([C], [Si], [Ni])$, the equations (6)–(8) were obtained.

$$HS_{(body)}[C]_{med} = -29.9622\,[Si]^2 - 0.9482\,[Ni]^2 - 1.9188\,[Si]\,[Ni] + 67.5632\,[Si] + 12.4177\,[Ni] + 10.0869 \quad (3)$$

$$HS_{(body)}[Si]_{med} = -0.9482\,[Ni]^2 + 120.0595\,[C]^2 + 55.4181\,[Ni]\,[C] - 168.3514\,[Ni] - 938.1881\,[C] + 1825.0583 \quad (4)$$

$$HS_{(body)}[Ni]_{med} = 120.0595\,[C]^2 - 29.9622\,[Si]^2 - 18.6234\,[C]\,[Si] - 751.3676\,[C] + 121.8423\,[Si] + 1213.4061 \quad (5)$$

$$HS_{(necks)}[C]_{med} = -34.1476\,[Si]^2 - 0.7563\,[Ni]^2 - 3.1912\,[Si]\,[Ni] + 80.7791\,[Si] + 11.7762\,[Ni] + 2.6921 \quad (6)$$

$$HS_{(necks)}[Si]_{med} = -0.7563\,[Ni]^2 - 2.9071\,[C]^2 + 43.3294\,[Ni]\,[C] - 131.3097\,[Ni] - 106.6105\,[C] + 423.9038 \quad (7)$$

$$HS_{(necks)}[Ni]_{med} = -2.9071\,[C]^2 - 34.1476\,[Si]^2 - 63.8114\,[C]\,[Si] + 90.4024\,[C] + 276.9922\,[Si] - 230.0087 \quad (8)$$

4. Generating the regression surfaces and the correlation charts

These regression surfaces, belonging to the three–dimensional space, can be represented graphically, resulting matrices of level curves, which can be interpreted by roll makers as correlation charts. In this way, the correlation of the values of the twos independent variables can be made, so that the dependent variables (i.e. $HS_{(body)}$ and $HS_{(necks)}$) can be obtained in between the requested limits, for both tehnological cases (the roll's body surface and the roll's necks and core). The regression surfaces and the correlation charts (matrices of the level curves) are presented in the Figures 4–15.

IOP Conf. Series: Materials Science and Engineering **200** (2017) 012021 doi:10.1088/1757-899X/200/1/012021

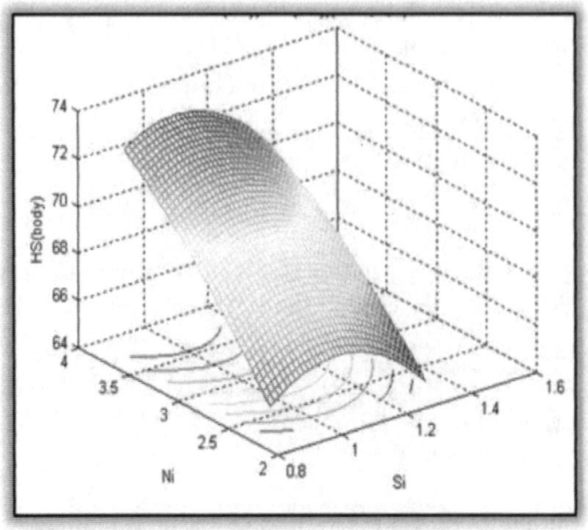

Figure 4. The regression surface, case of $HS_{(body)} = HS_{(body)}([C]_{med}, [Si], [Ni])$

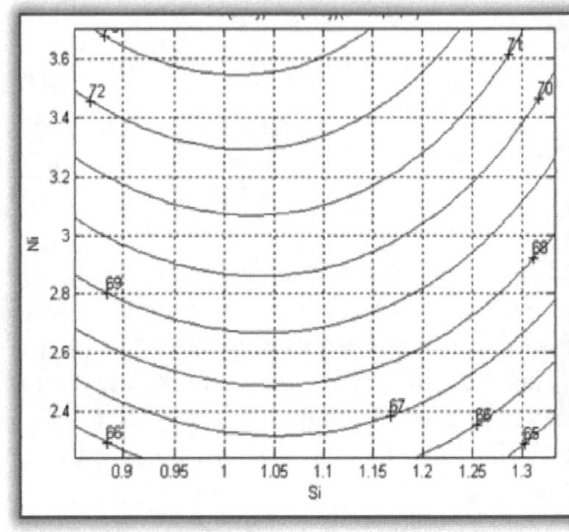

Figure 5. The matrix of the level curves, case of $HS_{(body)} = HS_{(body)}([C]_{med}, [Si], [Ni])$

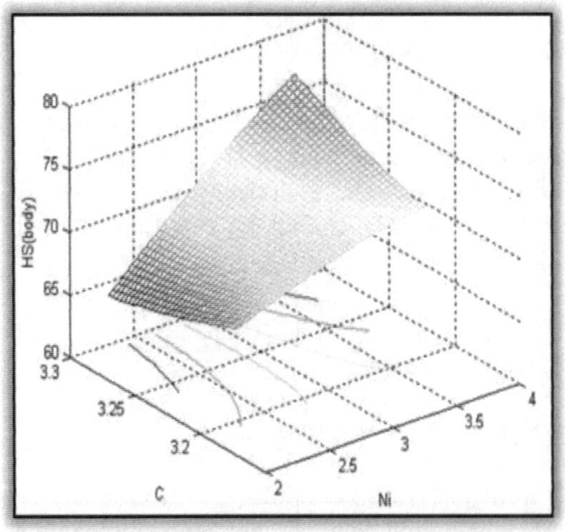

Figure 6. The regression surface, case of $HS_{(body)} = HS_{(body)}([C], [Si]_{med}, [Ni])$

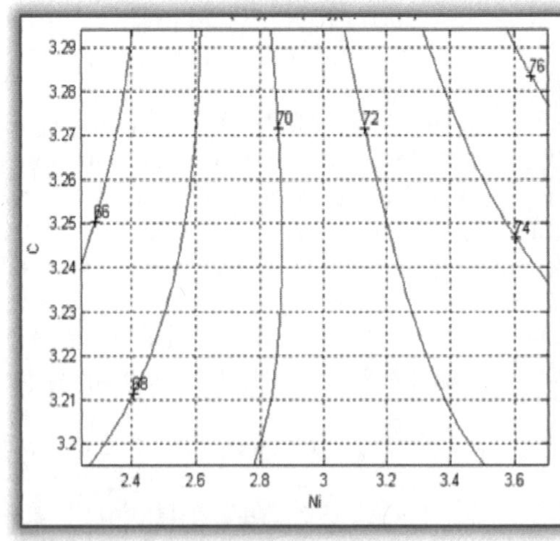

Figure 7. The matrix of the level curves, case of $HS_{(body)} = HS_{(body)}([C], [Si]_{med}, [Ni])$

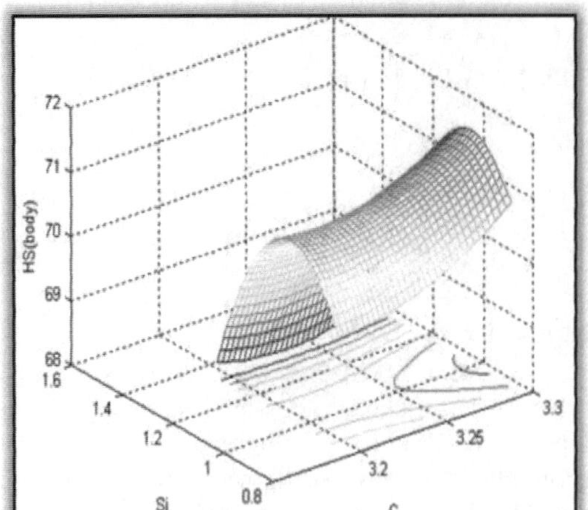

Figure 8. The regression surface, case of $HS_{(body)} = HS_{(body)}([C], [Si], [Ni]_{med})$

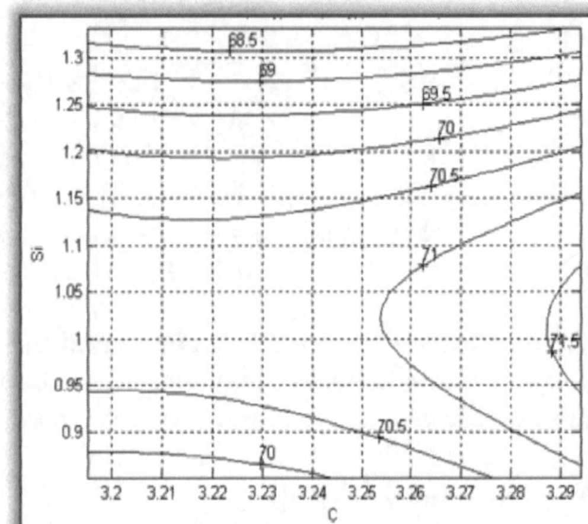

Figure 9. The matrix of the level curves, case of $HS_{(body)} = HS_{(body)}([C], [Si], [Ni]_{med})$

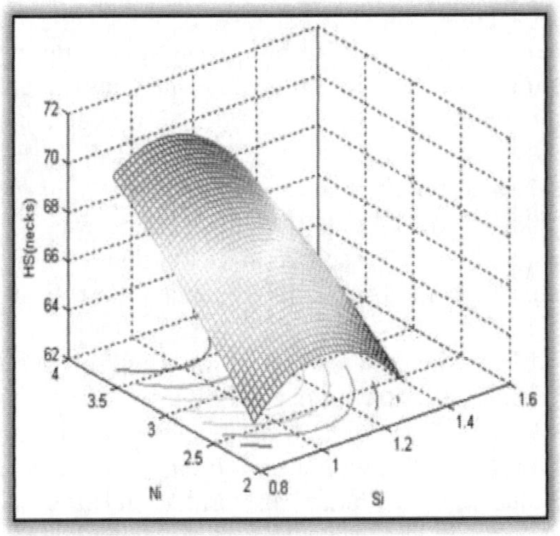

Figure 10. The regression surface, case of $HS_{(necks)} = HS_{(necks)}([C]_{med}, [Si], [Ni])$

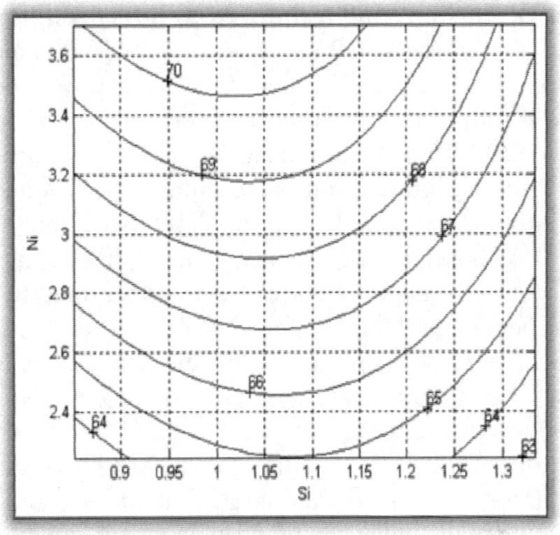

Figure 11. The matrix of the level curves, case of $HS_{(necks)} = HS_{(necks)}([C]_{med}, [Si], [Ni])$

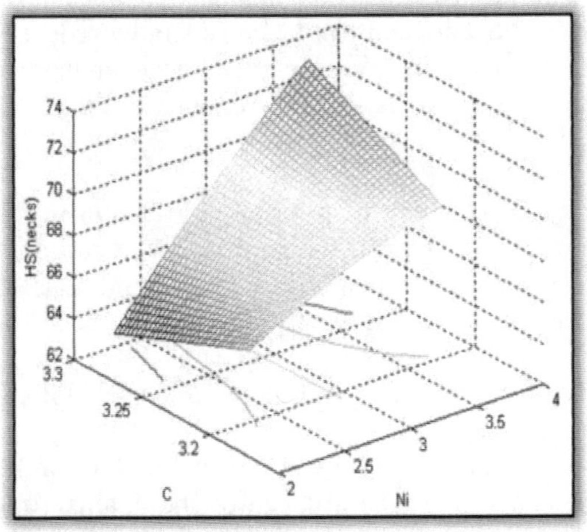

Figure 12. The regression surface, case of $HS_{(necks)} = HS_{(necks)}([C], [Si]_{med}, [Ni])$

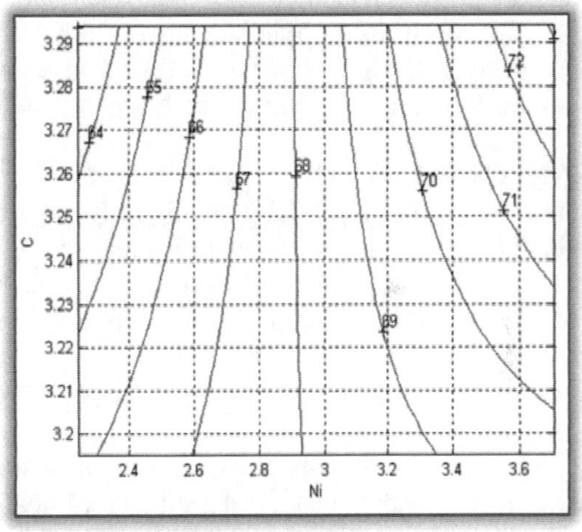

Figure 13. The matrix of the level curves, case of $HS_{(necks)} = HS_{(necks)}([C], [Si]_{med}, [Ni])$

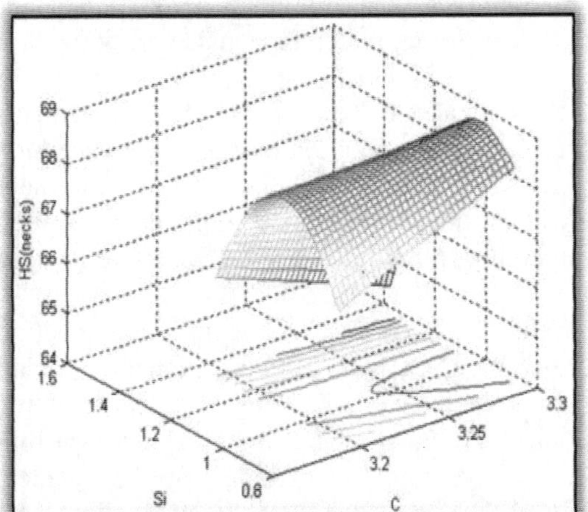

Figure 14. The regression surface, case of $HS_{(necks)} = HS_{(necks)}([C], [Si], [Ni]_{med})$

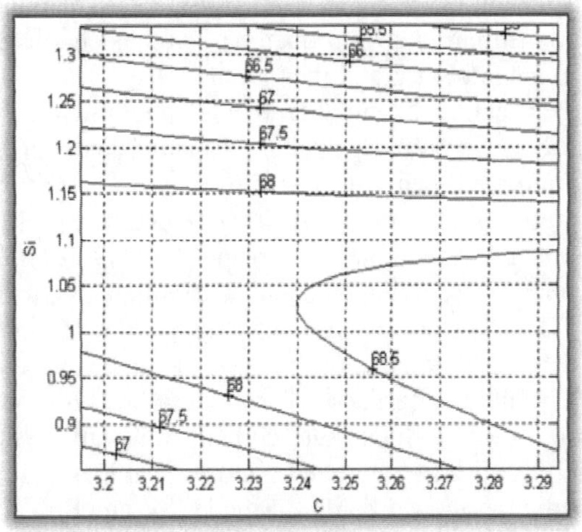

Figure 15. The matrix of the level curves, case of $HS_{(necks)} = HS_{(necks)}([C], [Si], [Ni]_{med})$

5. Discussions and technological interpretations

In the interests of an especially good balance of the graphite and carbide distribution during solidification, and in order to further improve the operating characteristics of the indefinite chilled cast iron roll, it has been proven favorable for a final Silicon content of more than 1.0 to 1.2%, preferably 1.05 to 1.15, to be provided.

In the optimization of material properties and material quality, although it is advantageous for the Carbon content of the melt to be set to a value of more than 3.2 to 3.24%, preferably 3.2 to 3.22%.

With regard to a controlled graphite content and an intended assumption of hardness of the material, it is preferred for the irons destined to the indefinite chilled cast rolls to contain 2.6 to 3.4% of Nickel, preferably 3.0 to 3.4%.

When, according to the current approaches described here, the composition of the iron melt is set such that, in the presence of 2.8 to 3.0% Nickel, the concentration ratio of Carbon to Silicon is preferably less than or equal to 2.6, the graphite precipitation and/or the graphite fraction in the material can be held within the desired range with high precision and within tight limits. At a ratio of Carbon content to Silicon content that exceeds this value (2.6), carbides are formed and graphite formation is adversely affected as well.

The Nickel addition leads to the improvement of the mechanical properties (including the of the cast rolls). If we do not allow this element to increase the graphitisation degrees and the white solidification in the peripherical area of the rolling surface, this content can be considerably reduced. Accordingly, the Silicon content of the irons is modified, as this element replaces the Nickel content.

Thus the optimal additions can be determined in these elements to assure the proper hardness's.

6. Concluding remarks

The rolls requirements may not be fully met, forcing to the granting of priorities in relation to the specific nature of the rolled products, so to compromise. In general, the problem is reduced to choosing the correct roll's material, relieved the lush experience available in current conditions of production, but complicated by wide diversity of the used materials. By identifying specific needs of manufacture of the indefinite chilled cast iron rolls, the current approaches responds to the identified the rolls casting sector's needs by increasing the quality of the rolls and contribute to their exploiting satisfaction.

Although the rolls manufacturing is in continuous updating, however requirements to claim to be high–quality rolls are not yet satisfied completely, in many cases, the absence of the quality rolls preventing the optimum quality laminates or to achieve productivities which are capable the rolling mills. The choice of material for the rolling mill rolls is the operation which take into account both their own usage of the rolling process corresponding with the rolled type of products (half–finished or finished laminated products), as well as characteristics of the different materials considered optimal in the manufacturing of the various rolls.

In fact, an essential feature of indefinite chill rolls is the critical balance between alloying elements such as Carbon [C], Nickel [Ni] and Silicon [Si], which promote the formation of graphite and carbide forming elements such as Chromium. The proper balance of graphite and carbides requires extremely careful selection of melting metallic stock, closely controlled melting conditions (charging, temperatures regimes, time) and rigid control of chemical composition (alloying stages, corrections) to obtain the required type and distribution of graphite.

The performed research had in view to obtain correlations between the chemical composition (defined by the one of the basic element – Silicon [Si], and one of the main alloying element – Nickel [Ni]) and the hardness of the indefinite chilled cast iron rolls (on the roll's working surface and on the both necks). Indefinite chilled rolls are basically controlled under actual molding and casting conditions, therefore the desired hardness is obtained by changing the carbide/graphite balance, including the Carbon [C]/Silicon [Si] and Silicon [Si]/Nickel [Ni] ratios or Carbon [C] – Silicon [Si] – Nickel [Ni] relationship, which is achieved by a close control of chemistry and process parameters.

References

[1] Schröder K H 2002 *Questions, answers, more questions, Twenty – five years of experience in discussing rolls and rolling technology*, ESW Handbook

[2] Schröder K H 1988 Rolling conditions in hot strip mills and their influence on the performance of work rolls, *Metallurgical Plant and Technology* **4** 44-56

[3] Kiss I 2008 *Rolling rolls – Approaches of quality in the multidisciplinary research*, Mirton Publishing House, Timisoara

[4] Vdovin K N, Zavalishchin A N, Gorlenko D A and Feoktistov N A 2016 Structure and properties of cast iron designated for working layer of rolls, *Journal of Materials Science Research* **5**(1) 77-88

[5] Krawczyk J 2011 Microstructure and tribological properties of mottled cast iron with different chemical composition, *Archives of Materials Science and Engineering* **51**(1) 5-15

[6] Kiss I 2012 Investigations upon the indefinite rolls quality assurance in multiple regression analysis, *Revista de Metalurgia, CENIM* **48**(2) 85-96

[7] Kiss I, Cioata V G and Alexa V 2010 Increasing the rolling–mill rolls quality in some multidisciplinary research, *Acta Technica Corviniensis – Bulletin of Engineering* **III**(2) 31-36

[8] Kiss I 2008 Research upon the quality assurance of the rolling–mill rolls and the variation boundaries of the chemical composition, *Revista de Metalurgia, CENIM* **44**(4) 335-342

[9] Kiss I 2013 Investigations on materials used for manufacturing the rolling rolls in few durability experiments, *International Journal of Surface Engineering and Interdisciplinary Materials Science* **1**(1) 46-56

[10] Kiss I, Alexa V, Ratiu S A and Cioata V G 2014 The quality assurance in the rolling industry: Methods, approaches and tendencies, *Machine Design* **6**(4) 131-136

[11] Kiss I, Cioata V G, Alexa V and Ratiu S A 2011 Technological behaviour and interpretations in some multidisciplinary approaches, *Annals of F.E.H. – International Journal of Engineering*, **IX**(4) 203-206

IOP Conf. Series: Materials Science and Engineering **200** (2017) 012022 doi:10.1088/1757-899X/200/1/012022

Multivariate research in areas of phosphorus cast–iron brake shoes manufacturing using the statistical analysis and the multiple regression equations

I Kiss, V G Cioată, V Alexa and S A Rațiu

University Politehnica Timisoara, Faculty of Engineering Hunedoara, Department of Engineering and Management, Hunedoara, Romania

E–mail: imre.kiss@fih.upt.ro

Abstract. The braking system is one of the most important and complex subsystems of railway vehicles, especially when it comes for safety. Therefore, installing efficient safe brakes on the modern railway vehicles is essential. Nowadays is devoted attention to solving problems connected with using high performance brake materials and its impact on thermal and mechanical loading of railway wheels. The main factor that influences the selection of a friction material for railway applications is the performance criterion, due to the interaction between the brake block and the wheel produce complex thermos–mechanical phenomena. In this work, the investigated subjects are the cast–iron brake shoes, which are still widely used on freight wagons. Therefore, the cast–iron brake shoes – with lamellar graphite and with a high content of phosphorus (0.8–1.1%) – need a special investigation. In order to establish the optimal condition for the cast–iron brake shoes we proposed a mathematical modelling study by using the statistical analysis and multiple regression equations. Multivariate research is important in areas of cast–iron brake shoes manufacturing, because many variables interact with each other simultaneously. Multivariate visualization comes to the fore when researchers have difficulties in comprehending many dimensions at one time. Technological data (hardness and chemical composition) obtained from cast–iron brake shoes were used for this purpose. In order to settle the multiple correlation between the hardness of the cast–iron brake shoes, and the chemical compositions elements several model of regression equation types has been proposed. Because a three–dimensional surface with variables on three axes is a common way to illustrate multivariate data, in which the maximum and minimum values are easily highlighted, we plotted graphical representation of the regression equations in order to explain interaction of the variables and locate the optimal level of each variable for maximal response. For the calculation of the regression coefficients, dispersion and correlation coefficients, the software Matlab was used.

1. Theoretical aspects

Brake shoes are made in a variety of materials (cast iron, composite, sintered), all of which have their own drawbacks (hard on the wheels, a low friction coefficient, cost etc.) [1–4]. Traditionally, freight wagons were fitted with brake systems using cast iron blocks. Passenger wagons still resist and are nowadays equipped with the original solution of cast iron brake blocks [5], [6].

Careful investigations on the parameters given by brake users, combined with the vast experience of the brake makers in the friction materials area, often lead to improved solutions for finding a shoes which gives the best possible technically compromise. In this way, the brake failure problems can be solved by good co–operation [4–6]. Beside the mechanical properties of the brake shoes materials, the process technology and the parameters in the exploitation are decisive for the brake's performance [5], [6]. Discussions are sometimes useful for and may help to create new ideas in area of the friction materials expertise and especially, in the brake shoes manufacturing (Figure 1 and Figure 2). However facts about brakes performance are most important for braking development [5–7].

In fact, the best way for brake makers to achieve better shoes is to ensure that better materials and improved manufacturing processes are used and that shoes users take account of friction conditions and improved braking processes. The main problem in shoe's manufacturing processes is the optimal chemical composition assurance [5], [6], [8].

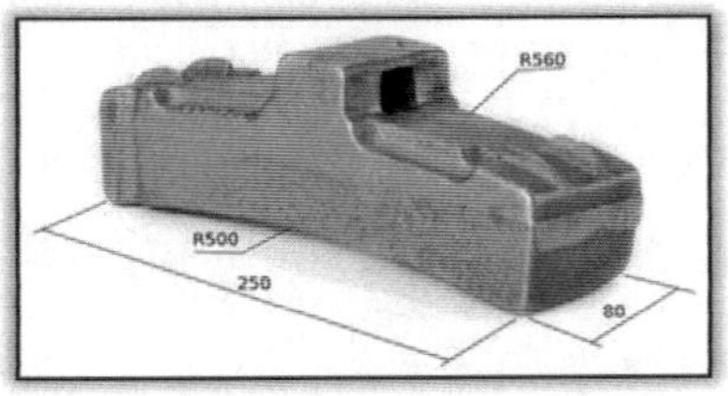

Figure 1. Grey cast iron brake shoes for passenger trains [5], [6]

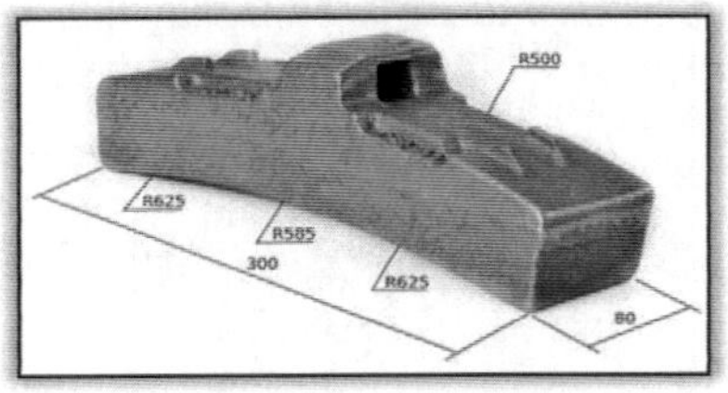

Figure 2. Grey cast iron brake shoes for freight wagons [5], [6]

Therefore, the statistical analysis is useful for controlling the industrial production of braking shoes, due to the shoe's chemical composition is not specified by railway customers. Due to industrial data analysis, the improving of the braking shoes technological process of manufacturing, according to the specifications, is needed. Statistical evaluations are helpful for brake shoe–makers in determining their actions to improve the shoe's quality [5], [6]. Our studies on the multivariate analysis combine the statistical analysis techniques and the multivariate mapping for easier data interpretation.

Several interpretations of the correlations between the cast irons chemical components – Carbon (C), Silicon (Si), Manganese (Mn), Phosphorus (P) and Sulphur (S) – and the obtained brake shoes hardness (HB) was enounced in the recent literature related with the grey phosphorus cast–iron destined to the brake shoes, which proposed three kinds of correlations, using the Matlab area: [5], [6]

» statistical analysis related the general behavior of Carbon content in relation with the rest of chemical elements, which have influence on the hardness of brake shoes. As result, regression surfaces and correlation charts are revealed; [5], [6]
» statistical analyze related the combined behavior of all chemical elements of grey phosphorus cast iron on the hardness of brake shoes, in several correlations. As result, regression surfaces and correlation charts are revealed; [5], [6]
» statistical analyze related the equivalent carbon content value behavior on the hardness of cast iron brake shoes. As result, several regression surfaces and correlation charts are revealed [5], [6].

The current performed research upon a number of 100 charges of phosphorous cast iron brake shoes had in view to obtain multivariate three–dimensional correlations between the cast–iron brake shoes' hardness and its chemical composition, defined by several representative elements – Manganese (Mn), Phosphorus (P) and Sulphur (S) –, which determine the microstructure.

2. The phosphorus cast iron

The properties of grey iron are primarily dependent on its composition, being influenced by the elements present in irons. The lower strength grades can be consistently produced by simply selecting the proper melting stock, but the higher strength grades require close control of their processing as well as their composition. Within a class of grey iron, hardness is a good indicator of its engineering properties and many of these properties are directly related to its hardness. The hardness is affected by the processing of the grey iron as well as the composition because these factors influence the microstructure [5], [6], [9–11].

Along with Carbon (C), Silicon (Si), Manganese (Mn) and Sulphur (S), Phosphorus (P) is traditionally regarded as one of the five elements normally present in grey cast iron. In the absence of carbide–forming components and with low cooling rates (cast in sand mould), Phosphorus form a low melting point phase in grey iron that is commonly referred as steadite (Fe_3P) [12].

The minor elements in grey iron are Phosphorus and the two interrelated elements: Manganese and Sulphur [12]. Sulphur is a very significant element because it exerts marked effects on the solidification behavior of iron. The influence of Sulphur needs to be considered relative to its reaction with Manganese in iron. For this reason, the Sulphur content in iron is usually controlled within limits and with a selected ratio of the Manganese content. Sulphur combines chemically with Manganese to form manganese sulphide (MnS), and with Iron (Fe) will form iron sulphide (FeS) and segregates on grain boundaries during freezing and precipitates during final stage of freezing [12].

Table 1. Chemical compositions of cast iron [5], [6]

Main components	Proportion, [%]
Carbon, [C]	3.0–3.5
Phosphorus, [P]	1.3–1.5
Silicon, [Si]	1.5–2.0
Sulphur, [S]	0.1–0.15
Manganese, [Mn]	0.5–0.8

Typically the alloy has a composition corresponding to conventional grey cast iron, except for the high phosphorus content. Phosphorus (P) is used as an alloying element in irons for special applications, the classical example being the brake shoes destined to trains. Due to its high wear resistance, high phosphorus grey cast iron is the best material selection for brake shoes [5], [6], [8–11]. Preferably, the Phosphorus is added to cast iron in the form of Ferro–phosphorus, which may be incorporated into cast iron in the proportions necessary to provide an alloy with the desired Phosphorus content [8]. Increasing phosphorus content, the hardness will increase. In fact, increasing steadite increases the hardness of cast iron overall [8–11].

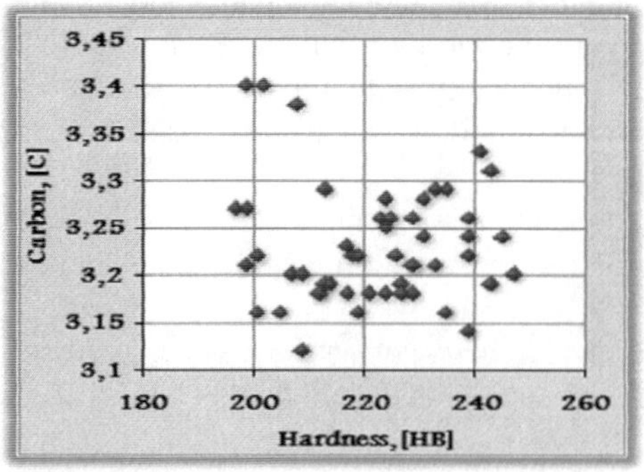

Figure 3. Effect of Carbon on Hardness

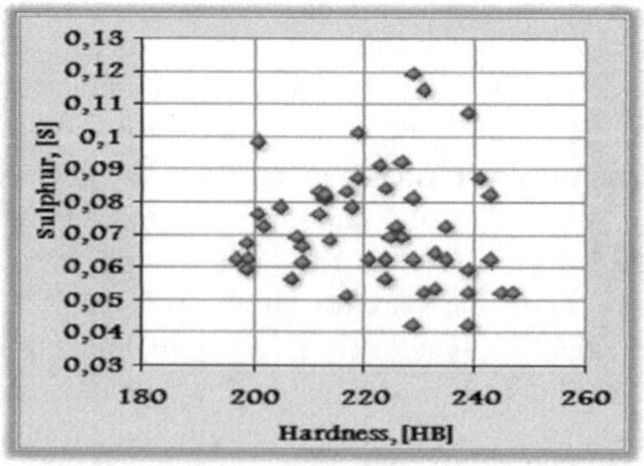

Figure 4. Effect of Sulphur on Hardness

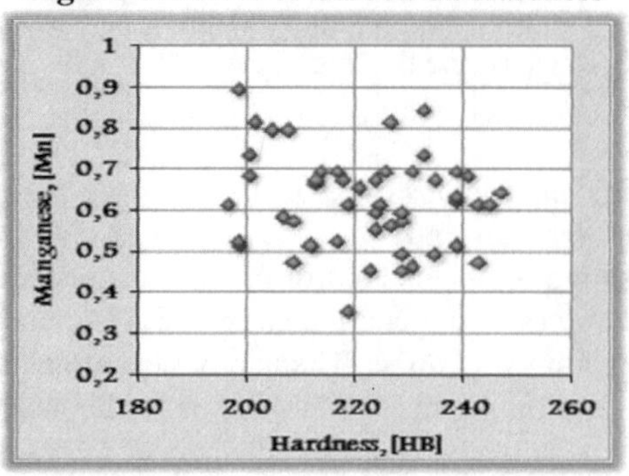

Figure 5. Effect of Manganese on Hardness

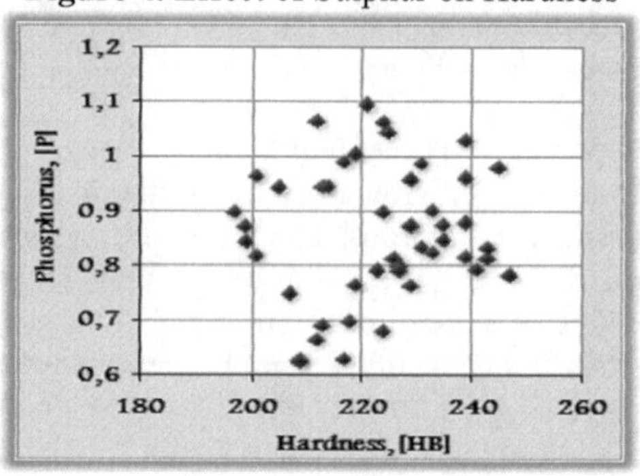

Figure 6. Effect of Phosphorus on Hardness

The single effects of the main elements on the hardness of phosphorous grey cast iron are shown in Figures 3–6. Two–dimensional rendering represents only the tendency of an influence of the chemical composition, by decreasing or increasing the analyzed parameter (hardness), according to one of component or the other. The polynomial functions appreciate only on the whole influence of the chemical elements upon the hardness, at different points of the shoes, indicating only the limits of variation. For this reason, it requires an analysis of the influence of a number of cumulative items on the shoe's hardness. In this sense, the chemical composition is an important cumulative parameter.

3. Methodology

The mathematical modeling establishes a methodology for the determination of the technological parameters values, for which a mechanical characteristic (the hardness) has the desirable values. Because is disposed of real data, the optimization model is based on industrial data, obtained from cast–iron brake shoes [5], [6].

Multiple linear regression is one of the fundamental models in statistics used to represent the relationship between a dependent variable and several independent variables and, finally, for predict (or estimate) the value of a variable by knowing the value of other variables. This article focuses on expressing the multiple linear regression model related to the hardness assurance by the chemical composition of the phosphorous cast irons destined to the brake shoes, having in view that the regression coefficients will illustrate the unrelated contributions of each independent variable towards predicting the dependent variable. The main goal of a multiple regression analysis is to see whether two or more variables co vary, and to quantify the strength of the relationship between these, whereas regression expresses the relationship in the form of a regression equation [5], [6].

In fact, multiple regression allows us to ask (and hopefully answer) the general question: What is the best predictor from the chemical composition of the phosphorous cast irons for a proper hardness of the brake shoes? In the current statistical experiments we use the correlation and regression for the following scopes:

» the interest related the cause–and–effect relationships, commonly used for investigating the relationship between two or more quantitative variables. In this case, we try to determine if the variation of the chemical composition of the phosphorous cast irons causes variation in hardness value, using the regression equations.

» the interest related the associated variables (if exist), without necessarily inferring a cause–and–effect relationship, knowing that a strong association between two or more components results in an increase in the accuracy of the prediction of proper hardness.

4. Results of modelling

Starting from the above–mentioned affirmations related to the values of Manganese (Mn), Phosphorus (P) and Sulphur (S) in the cast–irons, two new statistical experiments are effectuated. The equations of regression hiperplanes, which describe the mathematical dependency between the above–mentioned elements and the hardness, are determined. Therefore, is searched a mathematical solution which can determine the optimum chemical composition for the hardness desirable values. Thus the optimal additions can be determined in these elements to assure the proper hardness [5], [6].

In the first statistical experiment, the cumulative effect of Carbon (C), Manganese (Mn) and Sulphur (S) is analyzed. The optimal form of modeling in the case of HB = f([C], [Mn], [S]) is given by the regression equation (1), where the correlation coefficient is $rf_{(1)} = 0.5773$.

In the second statistical experiment, the cumulative effect of Carbon (C), Phosphorous (P) and Sulphur (S) is analyzed. The optimal form of modeling in the case of HB = f([C], [S], [P]) is given by the regression equation (2), where the correlation coefficient is $rf_{(2)} = 0.6301$.

$$HB = a_1[C]^2 + a_2[Mn]^2 + a_3[S]^2 + a_4[C][Mn] + a_5[Mn][S] + a_6[S][C] + a_7[C] + a_8[Mn] + a_9[S] + a_{10} \quad (1)$$

where: $a_1 = -553.9699$; $a_2 = -114.4673$; $a_3 = 2598.4966$; $a_4 = -73.6059$; $a_5 = 4435.1292$; $a_6 = -2868.1394$; $a_7 = 3325.7384$; $a_8 = 592.9469$; $a_9 = -13164.8786$; $a_{10} = -4884.8396$

$$HB = a_1[C]^2 + a_2[S]^2 + a_3[P]^2 + a_4[C][S] + a_5[S][P] + a_6[P][C] + a_7[C] + a_8[S] + a_9[P] + a_{10} \quad (2)$$

where: $a_1 = 580.2331$; $a_2 = 11760.6092$; $a_3 = -65.3734$; $a_4 = 3184.6546$; $a_5 = -38.2403$; $a_6 = -1017.4488$; $a_7 = -3894.4233$; $a_8 = -11324.9699$; $a_9 = 317.2420$; $a_{10} = 6703.7864$

Because these surfaces, described by the equation (1) and (2), cannot be represented in the three–dimensional space, the independent variables ([C], [Mn], [S] and [P]) were successively replaced with their average values ($[C]_{med}$, $[Mn]_{med}$, $[S]_{med}$ and $[P]_{med}$). Therefore, in the case of correlation HB = f([C], [Mn], [S]), the equations (3)–(5) were obtained, in which the correlation coefficient are $rf_{(3)} = 0.4046$, $rf_{(4)} = 0.4572$ and $rf_{(5)} = 0.4218$.

$$HB\ [C]_{med} = b_1[Mn]^2 + b_2[S]^2 + b_3[Mn][S] + b_4[Mn] + b_5[S] + b_6 \quad (3)$$

where: $b_1 = -114.4673$; $b_2 = 2598.4966$; $b_3 = -2868.1394$; $b_4 = 355.2145$; $b_5 = 1159.7018$; $b_6 = 77.8333$

$$HB\ [Mn]_{med} = c_1[C]^2 + c_2[S]^2 + c_3[C][S] + c_4[C] + c_5[S] + c_6 \quad (4)$$

where: $c_1 = -553.9699$; $c_2 = 2598.4966$; $c_3 = 4435.1292$; $c_4 = 3280.3235$; $c_5 = -14934.5206$; $c_6 = -4562.5678$

$$HB\ [S]_{med} = d_1[C]^2 + d_2[Mn]^2 + d_3[C][Mn] + d_4[C] + d_5[Mn] + d_6 \quad (5)$$

where: $d_1 = -553.9699$; $d_2 = -114.4673$; $d_3 = -73.6059$; $d_4 = 3644.0920$; $d_5 = 387.0719$; $d_6 = -5816.4262$

In the case of correlation HB = f([C], [S], [P]), the equations (6)–(8) were obtained, in which the correlation coefficient are $rf_{(6)} = 0.4902$, $rf_{(7)} = 0.4473$ and $rf_{(8)} = 0.4618$.

$$HB\ [C]_{med} = b_1[S]^2 + b_2[P]^2 + b_3[S][P] + b_4[S] + b_5[P] + b_6 \quad (6)$$

where: $b_1 = 11760.6092$; $b_2 = -65.3734$; $b_3 = -1017.4488$; $b_4 = -1039.1723$; $b_5 = 193.7335$; $b_6 = 178.3426$

$$HB\ [S]_{med} = c_1[P]^2 + c_2[C]^2 + c_3[P][C] + c_4[P] + c_5[C] + c_6 \quad (7)$$

where: $c_1 = 580.2331$; $c_2 = -65.3734$; $c_3 = -38.2403$; $c_4 = -3665.8287$; $c_5 = 244.2096$; $c_6 = 5951.4751$

$$HB\ [P]_{med} = d_1[C]^2 + d_2[S]^2 + d_3[C][S] + d_4[C] + d_5[S] + d_6 \quad (8)$$

where: $d_1 = 580.2331$; $d_2 = 11760.6092$; $d_3 = 3184.6546$; $d_4 = -3925.4866$; $d_5 = -12151.4641$; $d_6 = 6918.3509$

5. Generating the regression surfaces and the correlation charts

These regression surfaces, belonging to the three–dimensional space, can be represented graphically, resulting several variation domains, which can be interpreted as correlation charts. In this way, the correlation of the values of the twos independent variables can be made, so that the dependent variables (HB) can be obtained between the requested limits.

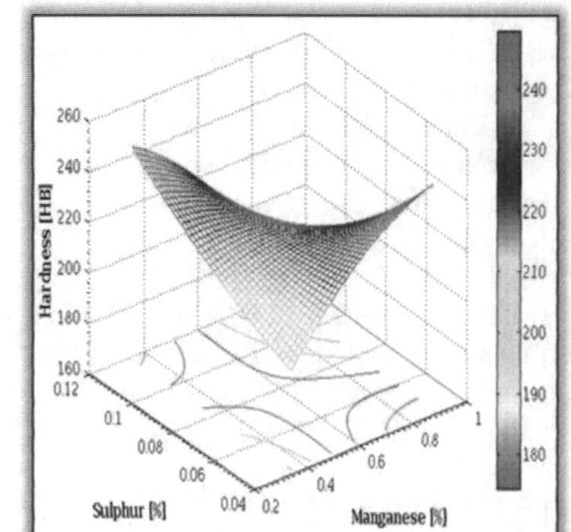

Figure 7. The three–dimensional regression area HB = f([S], [Mn], [C]), when $[C] = [C]_{med}$

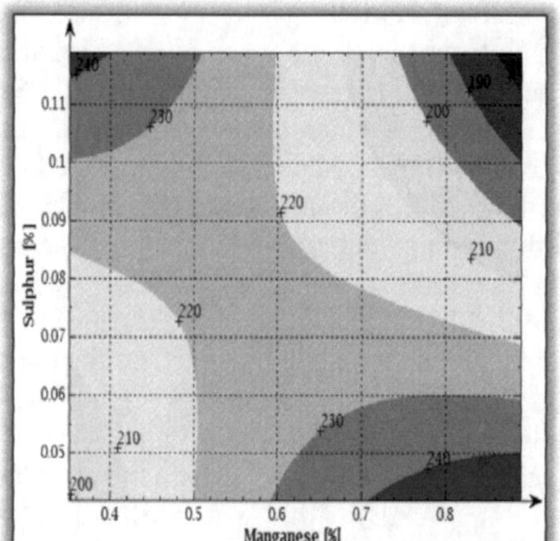

Figure 8. The correlation chart HB = f([S], [Mn], [C]), when $[C] = [C]_{med}$

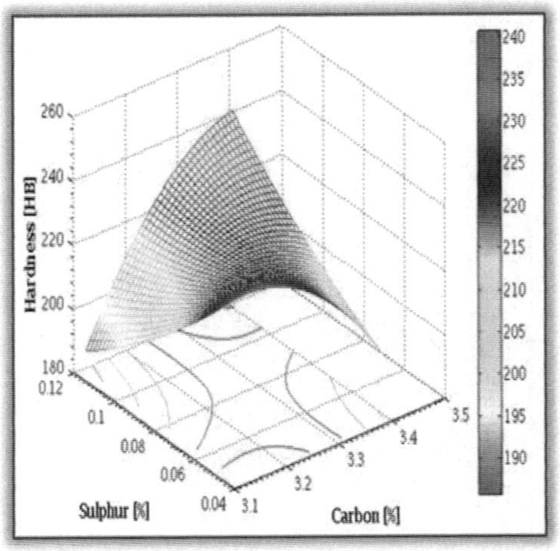

Figure 9. The three–dimensional regression area HB = f([S], [Mn], [C]), when $[Mn] = [Mn]_{med}$

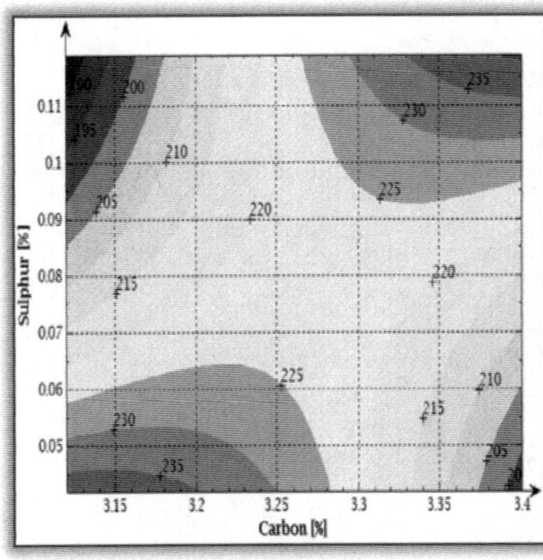

Figure 10. The correlation chart HB = f([S], [Mn], [C]), when $[Mn] = [Mn]_{med}$

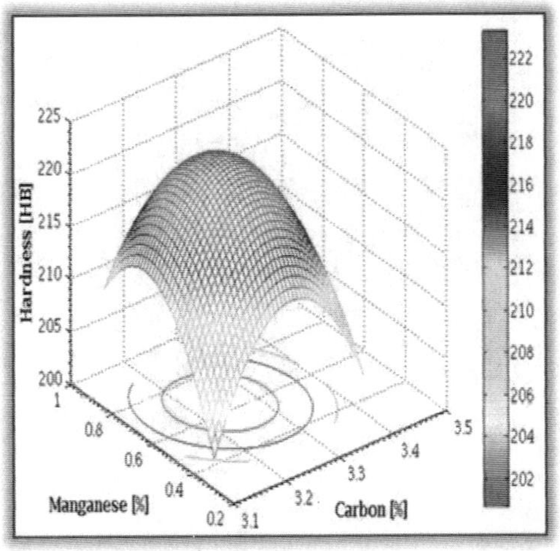

Figure 11. The three–dimensional regression area HB = f([S], [Mn], [C]), when $[S] = [S]_{med}$

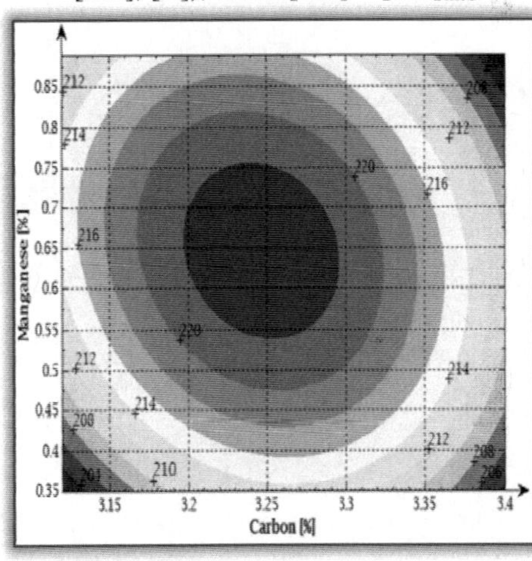

Figure 12. The correlation chart HB = f([S], [Mn], [C]), when $[S] = [S]_{med}$

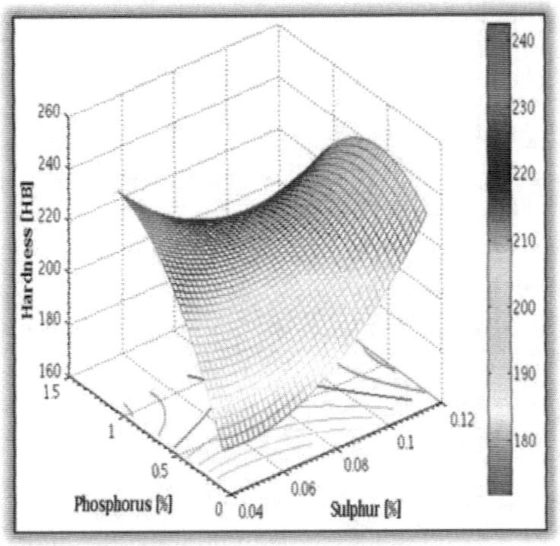

Figure 13. The three–dimensional regression area HB = f([C], [S], [P]), when $[C] = [C]_{med}$

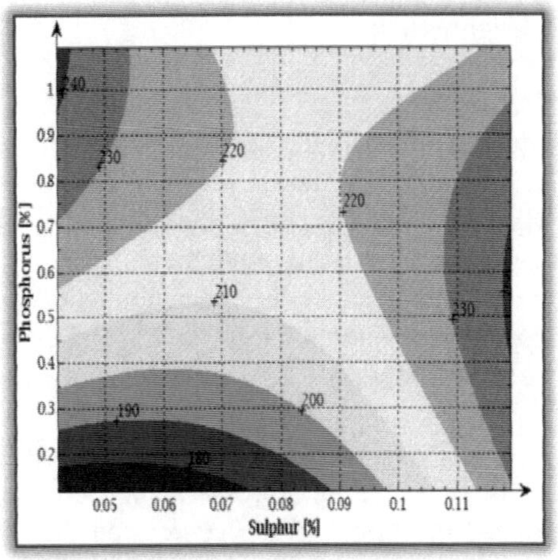

Figure 14. The correlation chart HB = f([C], [S], [P]), when $[C] = [C]_{med}$

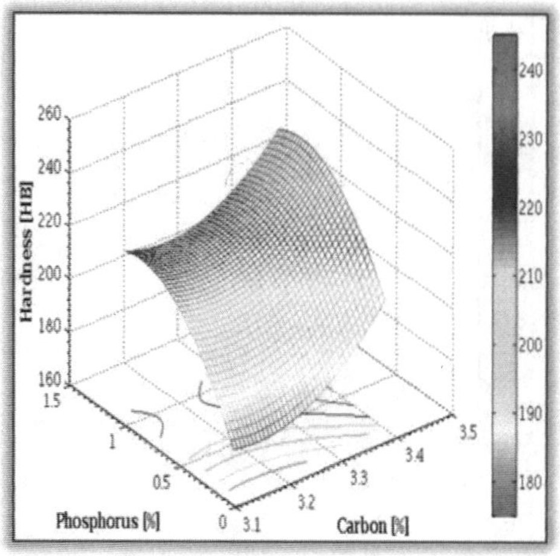

Figure 15. The three–dimensional regression area HB = f([C], [S], [P]), when $[S] = [S]_{med}$

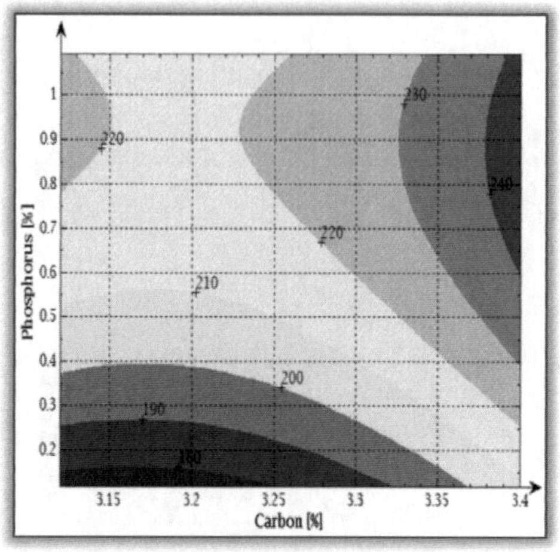

Figure 16. The correlation chart HB = f([C], [S], [P]), when $[S] = [S]_{med}$

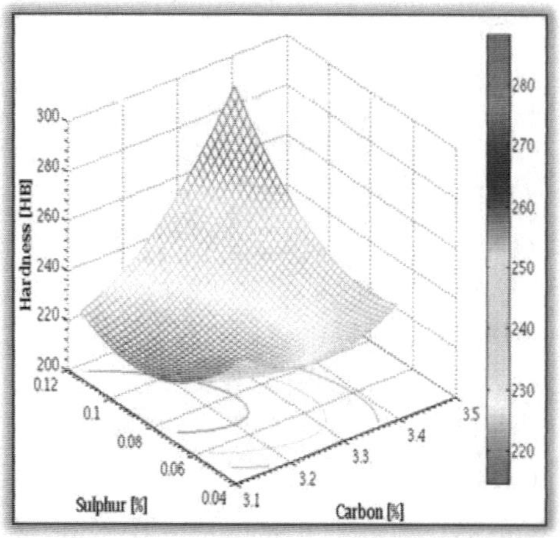

Figure 17. The three–dimensional regression area HB = f([C], [S], [P]), when $[P] = [P]_{med}$

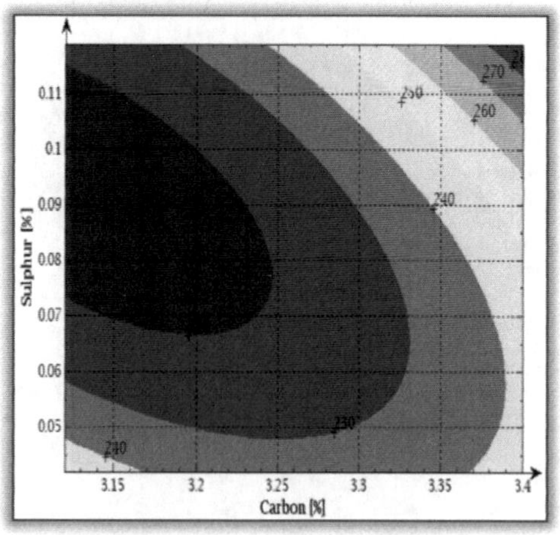

Figure 18. The correlation chart HB = f([C], [S], [P]), when $[P] = [P]_{med}$

The regression surfaces and the correlation charts, in both statistical experimental cases, are presented in the Figures 7–18.

6. Discussions

The mathematical model quantifies the influences of the independent variables ([C], [Mn] and [S], respectively [C], [S] and [P]) on the dependent variable ([HB]), and it was established by a statistical analysis. The real values of the process variables were chosen statistically, their limits of variation being given by the industrial data from the phosphorous cast–iron brake shoes manufacturing.

In order to settle the correlation between the [HB], and the above–mentioned parameters (independent variables, i.e. [C], [Mn], [S] and [P]) the following model of polynomial equation type has been proposed:

$$Y = a_1[X_1]^2 + a_2[X_2]^2 + a_3[X_3]^2 + a_4[X_1]\,[X_2] + a_5[X_1]\,[X_3] + a_6[X_2]\,[X_3] + a_7[X_1] + a_8[X_2] + a_9[X_3] + a_{10} \quad (9)$$

where $a_1,..., a_{10}$ are the regression coefficients.

For visualization of the regression surfaces and for the calculation correlation coefficients the software Matlab was used. The spatial representations of the response surfaces [HB] for constant values of variables reveal that all data are located inside the considered surface.

In order to estimate the independent variables overall influence on hardness [HB], the multiple correlation coefficients were calculated. The obtained values are: $rf_1 = 0.5773$ and $rf_2 = 0.6301$, respectively. Analysing the values of regression coefficients one can conclude that the process performance it is noticeably influenced by all three parameters, in both statistical experiments. From the value of the determination coefficient resulted that the proposed mathematical model is adequate for a confidence interval of 57–63%. In other words, in the chosen intervals the independent variables influence the hardness in an extent of approximately 60%. The rest of approximately 40% can be attributed to the effect of other chemical elements or manufacturing factors ignored during these statistical experiments.

The three–dimensional regression surface is a graphical representation of the regression equation. It is plotted to explain interaction of the variables and locate the optimal level of each variable for maximal response. Each regression surface represents the different combinations of two variables at one time while maintaining the other variable at the average value. These regression surface (Figures 7, 9, 11, 13, 15 and 17) and its respective correlation charts Figures 8, 10, 12, 14, 16 and 18) provide a visual interpretation of the interaction between two factors (when the third factor is constant) and facilitate the determination of optimum experimental conditions. The point for which the response is optimized is the point called the stationary point. The stationary point may be a point of maximum response, minimum response or a saddle point (point of inflexion).

7. Conclusions

The regression analysis provides an opportunity to specify hypotheses concerning the quality assurance of brake shoes, as well as explanatory factors in the manufacturing practice. When the analysis is correctly executed, having on base the statistically valid adjustment, this can produce quantitative effects on the quality assurance process. In the practice of the phosphorous cast–iron brake shoes manufacturing many variables interact with each other simultaneously. Multivariate research and visualization comes to the fore when the engineers have difficulties in comprehending many dimensions at one time. In order to establish the optimal condition for hardness assurance in the phosphorous cast–iron brake shoes manufacturing we proposed a mathematical modelling study by using the statistical analysis and regression equation.

Based on the experiments, on the results obtained from data processing and on the technical analysis of these data, we concluded that realization of the optimum chemical compositions of the phosphorus cast–iron can constitute a technical efficient way to assure the exploitation properties (like hardness), the material from which the brake shoes are manufactured having an important role in this sense. We believe that a chemical composition within acceptable limits, correlated with the general requirements on brake shoes manufacturing process, can give the required or proper hardness. A three–dimensional regression surfaces with variables on three axes is a common way to illustrate multivariate data, which can be applied to the brake shoes manufacturing. The advantage of this approach is that the maximum and minimum values are easily highlighted.

References

[1] Boghani A B 1981 *Characterizing the performance of cast–iron, high–phosphorus cast–iron, and composition freight car brake shoes under simulated drag braking conditions*, Federal Railroad Administration, pp 54–61

[2] Akkus A and Yeğin M 2014 Research on wear rate and mechanical properties of brake sabots (shoes) used in railway rolling stocks, *International Journal of Applied Science and Technology* **4**(7) 76–84

[3] Dungan L I and Dungan M A 2010 Materials used in couplings mechanical friction braking system of high speed railway vehicles, *Buletinul AGIR* **2–3** 121–126

[4] Cruceanu C 2012 *Train Braking, Reliability and Safety in Railway*, Intech

[5] Kiss I 2016 The chemical composition of phosphorous cast irons behavior in the manufacturing of brake shoes meant for the rolling stock, *Acta Technica Corviniensis – Bulletin of Engineering* **IX**(3) 77-84

[6] Kiss I, Cioata V, Alexa V and Ratiu S 2016 Investigations on the selection of friction materials destined to railway vehicles applications, *Annals of Faculty Engineering Hunedoara – International Journal of Engineering* **XIV**(4) 231-240

[7] Pascu L 2015 *Research regarding increasing quality of brake shoes destined to railway vehicles applications (Cercetări privind îmbunătățirea calității saboților de frână destinați materialului rulant)*, University Politehnica Timisoara, Doctoral Thesis

[8] Socalici A, Pascu L, Ardelean E and Heput T 2016 The influence of Carbon, Manganese and Silicon content upon the hardness of phosphorus cast iron used in manufacturing rolling stock braking shoes, *Tehnički Vjesnik* **23**(6) 1643-1648

[9] Zhang Y, Chen Y, He R and Shen B 1993 Investigation of tribological properties of brake shoe materials – phosphorous cast irons with different graphite morphologies, *Wear* **166**(2) 179-186

[10] Tsujimura T 1994 Improved performances of cast iron brake shoe, Railway Technical Research Institute, *Quarterly Reports* **35**(4) 215-217

[11] Tsujimura T, Arai H and Fujiwara N 1990 Development of high performance special cast iron brake shoes, Railway Technical Research Institute, *Quarterly Reports* **31**(4) 218-224

[12] Abbasi H R, Bazdar M and Halvaee A 2007 Effect of phosphorus as an alloying element on microstructure and mechanical properties of pearlitic grey cast iron, *Materials Science and Engineering: A* **444**(1) 314-317

IOP Conf. Series: Materials Science and Engineering **200** (2017) 012023 doi:10.1088/1757-899X/200/1/012023

Innovative methods of popularizing technical education

L Y Shkitsa[1], V G Panchuk[2] and V A Kornuta[1]

[1]Ivano-Frankivsk National Technical University Of Oil And Gas, Department Of Engineer And Computer Graphics, Karpatska str., no. 15, 76019, Ivano-Frankivsk, Ukraine

[2]Ivano-Frankivsk National Technical University Of Oil And Gas, Department of computer engineering production, Karpatska str., no. 15, 76019, Ivano-Frankivsk, Ukraine

E-mail: shkitsa@gmail.com

Abstract. There have been analyzed reasons of the loss of technical education's popularity. Also, the analysis of known educational and production methods, oriented at the innovative model of development of society, was performed. It is stated that the acquisition of 21st century's skills as a result of competition of technical education are natural for the DIY ideology, which was realized in the institutions like Fab Lab. The new educational strategy, based on project-based learning, is proposed to be implemented as a special laboratory with equipment, which would be a center of innovative development for students at the Technical University. Moreover, the list of projects planned for implementation, that includes not only projects, specific to a particular university, but also projects, demanded by society as a whole, is specified. It is worth to implement trendy projects in the laboratory, such as toy-like, ecological projects; projects of the energy dependence decrease or the energy efficiency increase, modern digital or innovative projects etc. The student should gain knowledge, skills and, possibly, equipment that are available for immediate usage on the labor market or for the realization of his own projects or the community's projects in everyday life after the realization of the particular project at the laboratory

1. Introduction

The majority of world's countries have faced the problem of lack of technical education's popularity, except the IT sphere. Popularity of technological technical education diminished in Ukraine although recently such tendencies contradict the labor market [1-4].

Stereotypes play an important role in choosing a future occupation. Among many stereotypes the following can be highlighted: it is difficult (to be competent in technology, to develop something new); hard (it is necessary to learn new abstract subjects with broad usage of means and methods of calculus); long (there is relatively long period of adaptation to specific production conditions, respectively, a badly-paid job); marginally profitable (the absence of "success's flashes" like quick enrichment against the background of low starting capital without taking into consideration expenditure on education, compared with designers, programmers, economists and other creative jobs).

The production equipment, which has been developing for a long time, is a product of many technological processes. Products [received with such equipment] exceed dramatically the similar ones, obtained by simpler means (for instance, the drilling machine provides the opportunity to make holes with such well-regulated accuracy that is impossible for a hand drill). Results are replicated easily by the worker without university education, then labor costs can't be expensive. The equipment is highly productive that is why a relatively small number of professionals can satisfy the production's demand of the market.

The high technical level of equipment in the traditional educational system defines the necessity of learning of considerable amount of theoretical and methodical information for understanding the equipment's work. The necessity in development of new equipment appears relatively seldom in steady markets, so a demand for specialists/professionals of the level that allows to complete such a job is small compared with the volume of labor market. The collateral consequence/stereotype – technical specialties are considered uncreative.

There is a similar situation concerning social sciences but a relatively low cost for entering the labor market and promoted success stories conduce predominance of the inflow of social specialties' neophytes compared with the technical ones.

Thus, in terms of traditional educational system [of Ukraine], stereotypes, that lead to the lack of popularity of technical education, have certain basis. The aforementioned reasons lead to attenuation of historical impulse of desire to be educated in general and to technical education in particular. However, the sustainable development, which is based on high technologies and innovations, is impossible without people, who have been educated in the technical sciences.

All successful corporations demonstrate a need for innovations [5], the innovation policy is declared by the main paradigm of all successful countries and blocs, particularly the European Union [6], because the industrial development is unsustainable under conditions of limited resources [7].

2. Search

A number of approaches, concerning on overcome of the aforementioned or similar stereotypes, are known. This approaches have been started with pragmatic education and project's method, that are described in works of J. Dewey [8], W. H. Kilpatrick [9].

The most famous modern educational models and methods have a similar direction. They are suggested by economists and representatives of a number of other social sciences, for example Thom Markham [10], [11], the similar works in Ukraine are also known [12], [13].

An application of the similar methods in technical education for gaining skills of the 21st century [14] is known on example of the concept of fab lab [15], [16], proposed by MIT's professor Neil Gershenfeld. The fab lab concept also grew out of a popular class at MIT (MAS.863) named "How To Make (Almost) Anything". The fab lab also is known as a modern, well equipped, workspace for DIY (Do It Yourself). The concept of reforming of engineering education CDIO is proposed in MIT [17].

Providing of access to modern means of production, mainly on the CNC basis, and a kit of electronic components for providing a possibility to create one's own product of CNC is the main offer of Fab Lab.

Up to date the fab lab community is worldwide (Figure 1). Three fab labs are in Ukraine, all of them are in Kyiv.

The following projects of creation the manufacturing equipment can be taken as an example of DIY: a 3D printer or a milling. Numerous descriptions of realization individually, as a rule, of such projects in the DIY format are known (there are many examples on YouTube). Both people with special education and initiative autodidacts are authors of these descriptions. The possibility of realization of such projects appeared as a result of the progress in manufacturing of engines, systems of numerical data processing and CNC in particular. The globalization of markets, the possibility of direct cooperation between manufacturers and potential buyers, the development of common education and the broad usage of numerical devices in everyday life played an important role in providing of such a possibility.

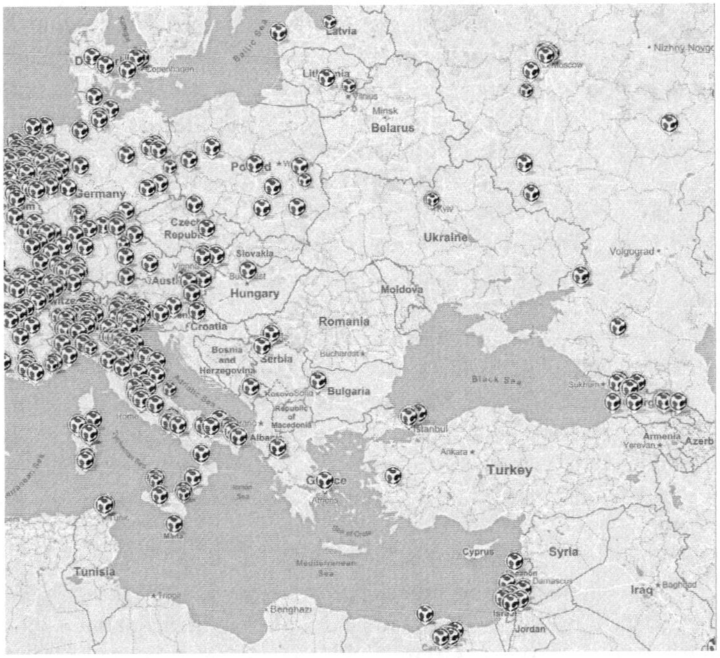

Figure 1. Up to day map of distribution of fab labs in Europe [17]

The framework CDIO provides students with an education stressing engineering fundamentals set in the context of Conceiving — Designing — Implementing — Operating (CDIO) real-world systems and products [17]. CDIO is mostly based on PBL [10]. 77 world's universities have joined the CDIO initiative as of today (no universities in Ukraine, Figure 2).

Figure 2. Part of the map of the current CDIO member schools [16]

The core of CDIO is CDIO Syllabus and CDIO Standards. CDIO Syllabus is a detailed list of competences, needed for today's engineer, opened to discussion and changes. These guiding principles (CDIO Standards) were developed in response to program leaders, alumni, and industry partners who wanted to know how they would recognize CDIO programs and their graduates.

The review and change of educational programs and work plans according to the changes of market's requests is basic in CDIO Standards. The realization of such a plan of the educational process can be imagined only in terms proposed by O. M. Telizhenko and others [14] in according to the Order of the Ministry of education of Ukraine "...about academic mobility of students..." [18], in terms of the existing [in Ukraine] educational system with a plan for 4+1(2) years.

3. Reflections and Suggestions

Based on the aforementioned, a hypothesis can be expressed: it is necessary to withdraw from attempts to cover the major part of the history of machinery's development and applied theoretical disciplines for popularization of the technical education, taking into consideration the vast amount of data about technical systems of the applied level and the high speed of their accumulation in the modern world.

The hypothesis about the new task of the technical education can be also articulated: the alumnus of the technical education's establishment should have [as an addition to the resume] a portfolio of executed projects. The educational establishment should provide a verification of such a portfolio (an access of interested people/employers to technical descriptions of resolutions and the corresponding theoretical part, necessary while clarifying of accepted technical solutions; a video depicting the process of devices' creation etc.).

Traditional university can be characterized as a set of the following functions: a knowledge base (a library, own collections, the knowledge and experience of teachers, a set of machinery and/or samples, details (a museum) etc.); a repeater, that is broadband and individual, of knowledge according to a schedule (classroom work of the teacher during lectures and other pursuits); a rating of a qualifying level (based on results of supervisory measures regarding the audit of learning); less often – an agent of licensing, that provides licenses for completing a particular kind of works. New knowledge, which can replenish the knowledge base, can also be created on the basis of the knowledge base in university's laboratories.

Conditions of the knowledge dissemination have completely changed during the last decades. The appearance of the broadband access to the Internet allows getting the access to knowledge from many sources in a different media appearance. There is a large number of courses in the free access developed in leading universities of the world. Search engines and specialized knowledge bases provide an opportunity to perform a research of the factual material easily. A certain obstacle is the modern model of copyright protection that is aimed at assistance of authors in gaining income in the result of spreading/realization of their works. Nevertheless, the value of traditional functions of the university as the base and repeater of knowledge has decreased.

In other words, the majority of people, acknowledged with the work on the Internet (this indicator is near 100% among young people nowadays), gain necessary information on request online free or for money. In our opinion, the key assertion is "on request". It is quite natural in terms of humans' fast accumulation of data, as the majority of other ways can provide irrelevant/outdated data.

In terms of fast exchange of data control means of qualifying level's rating reduced to struggle against the access to the Internet. Although in real conditions during the implementation of professional tasks such a challenge does not appear as a rule. The university creates the challenge for the preservation its traditional model of work. If student's profession does not foresee that he will work in a closed information space (without an access to knowledge) in the future, such a politics will be, at least, ineffective, at most – harmful. The harm is shown up in the creation of psychological block and/or lack of ability to find the relevant information quickly in a stressful situation.

Of course, there are professions that demand the availability of knowledge and skills "here and now" because it influences the reaction rate critically, and, consequently, the result. We assume that the majority of engineering professions are not referred to these ones.

Industry data creates more and more powerful tools for the extracting of knowledge, including unstructured data (knowledge bases, expert systems). Therefore, the functions of the fact search and generalization for the production of new knowledge, finally, answers to the questions, are now widely automated. That is why, in our opinion, universities should change approaches to learning in accordance with the hypothesis 1.

Qualification level is assigned, based on points for the subject. The list of subjects and marks are given in the Appendix to the diploma. However, it is not always possible to restore the contents of acquired knowledge and their relevance to the requirements of /workplace/office by the name of the subject. The content of the discipline is not often shown by the university because it is intellectual property. At the same time, the content of the project performed by the student is the intellectual property of the author. Thus, access to the results of the project will not infringe the rights of ownership of the university, but allows to estimate easily his (student's) knowledge and skills. Evaluation of the worker/contractor for the portfolio is quite common among the employers/customers.

It is necessary to set up a task of completing the concrete project for a student for the increase of motivation of gaining knowledge that defines the modern technical education. The student should gain knowledge, skills and, possibly, equipment that are available for immediate usage on the labor market or for the realization of his own projects or the community's projects in everyday life after the realization of the particular project at the laboratory. The possibility of obtaining better results of working (the higher quality of results of the designed equipment's usage; development of equipment by correct design is cheaper etc.) can become a motivation to working in such groups, contrary to the unorganized self-realization of DIY projects.

Apart from the readiness to use modern methods of education, it is necessary to create the institutional system of innovative education's logistical support, for example, a laboratory for the popularization of the technical education. The inventory set of Fab Lab is a good model for such a providing. However, Fab Lab is oriented at the attraction of the broadest circle of citizens without any relation to the educational direction of the university or priorities of the development of the society. Providing of these specifics will not necessarily lead to lower prices of equipment or the ability of the laboratory. But not taking the specifics into consideration is able to lead to not providing the ability of realization of needed projects.

The list of projects, planned for realization, should be defined for identifying the list of equipment. Such a list ought to comprise not only projects, specific for the concrete university, but also innovative projects, which are in a high demand of the community or the society in general, taking the aim of increasing the popularity of the technical education into consideration. It is worth to implement trendy projects in the laboratory, such as toy-like, ecological projects; projects of the energy dependence decrease or the energy efficiency increase, modern digital or innovative projects etc. For example, in the IFSTUOG laboratory the following list of projects can be introduced: design CAD model of logo of team and print it on a 3D printer; assemblage of 3D printer from a DIY kit; operate with 3D scanner; storage with automated access control; modular-vendor machine; 3D scanner; 3D CNC; machine, adjusted to CNC; "Heat of solar" (collector, heat pump, heat accumulator, heated water/coolant with automated control); robot cleaner; automated pumping station; automated wellhead equipment; automated washing rig system with remote control. The list of equipment for realization of proposed projects should include (but not bounded): 3D printer, CNC milling machine; solder station; solder for PVC pipes; electrically operated components of pipelines (shut-off and control valves); electrical and electronic components, controllers, stepper and servo motors.

The proposal for the implementation of projects, proposed by potential employers or through own challenges/problems of potential students, the availability of infrastructure for the projects, the availability of supplies, spare parts, contacts with suppliers, the executed projects (portfolio) is the basis for the interest in technical education. The modern university has to perform a sort of business incubator and "first job" for its graduates. This exactly is able to return the popularity of high technical education.

Thus, ways of increasing the popularity of technical education as a result of the innovative teaching methods' review and the analysis of causes why technical education lost its popularity have been proposed.

IOP Conf. Series: Materials Science and Engineering **200** (2017) 012023 doi:10.1088/1757-899X/200/1/012023

References

[1] ***Bilyns'ka O *Most popular professions in the world: current state and development trends* Ukr http://znoclub.com/dovidnik-zno/211-najpopulyarnishi-profesiji-svitu-suchasnij-stan-ta-tendentsiji-rozvitku.html
[2] ***Which professions in Ukraine are the most paid and popular 2016 Ukr http://24tv.ua/yaki_profesiyi_v_ukrayini_ye_naybilsh_oplachuvanimi_ta_populyarnimi_n725521
[3] ***Popular professions who learn to become successful 2016 Ukr http://nv.ua/ukr/style/beauty/populjarni-profesiji-na-kogo-vchitisja-shchob-stati-uspishnim-198832.html
[4] ***The most popular profession in 2017 in Europe. What specialty to choose in 2017 2016 Ukr http://2017pik.pp.ua/novini-2017/507-sam-zatrebuvan-profesyi-2017-v-yevrop.-yaku-specalnst-obrati-v-2017-roc.html
[5] Dovhopolyy D 2016 Why do corporations need startups Ukr http://www.epravda.com.ua/columns/2016/11/4/610088
[6] ***What is Horizon 2020? https:/ec.europa.eu/programmes/horizon2020/en/what-horizon-2020
[7] ***World Charter for Nature 1982 *United Nations, General Assembly, 48th Plenary Meeting, October 28* http://www.un.org/documents/ga/res/37/a37r007.htm
[8] Dewey J 1938 *Experience and education* New York: Macmillan 116
[9] Kilpatrick W H 1929 *The Project Method: The Use of the Purposeful Act in the Educative Process* New York: Teachers college, Columbia university 18
[10] Markham T 2012 *Project Based Learning* HeartIQ Press, San Rafael, California 152
[11] MindShift 2013 *10 Ways to Teach Innovation* https://ww2.kqed.org/mindshift/2013/04/01/10-ways-to-teach-innovation/?_ga=1.115482049.615115458.1478342938#content
[12] Telizhenko O M, Luk'yanykhin V O and Baystryuchenko N O 2016 Project-oriented approach to the educational process *Visnyk NTU "KhPI"* **1** (1173) 30-35 http://repository.kpi.kharkov.ua/handle/KhPI-Press/21627 DOI: 10.20998/2413-3000.2016.1173.62016
[13] Lutsenko H V 2015 Organization Of Future Engineers' Project-Based Learning When Studying The Project Management Methodology *Information Technologies and Learning Tools* **45** 1(2015) 123-133 http://journal.iitta.gov.ua/index.php/itlt/article/view/1152
[14] ***Stratford High Shool 21st Century Learning Expectations http://stratfordhigh.stratfordk12.org/Content/21st_Century_Skills_1.asp
[15] ***What Is A Fab Lab? *FabFoundation* http://fabfoundation.org/index.php/what-is-a-fab-lab/index.html
[16] ***The Fab Charter 2012 http://fab.cba.mit.edu/about/charter/
[17] ***The CDIO™ Initiative is an innovative educational framework for producing the next generation of engineers http://www.cdio.org/about
[18] ***About the statement of Approximate position about the academic mobility of students of higher educational institutions of Ukraine 2013 Ukr *The Order of Ministry of Education of Ukraine* 635 *29.05.2013*. http://www.mon.gov.ua/img/zstored/files/635.docx

Safety of the shaft-wheel assembly of electric locomotives

M Todic and V Golubovic-Bugarski

University of Banjaluka, Faculty of Mechanical Enginering, Stepe Stepanovica 71, 78000 Banjaluka, Bosnia and Herzegovina

E-mail: mladentodic@unibl.rs

Abstract. Shafts of electric locomotives are exposed to complex normal and tangential stresses during its exploitation. These stresses could have extremely high level producing the breakage of the shaft. It is well known that shafts have much longer service life than wheels. However, since the stresses in shaft's material are high, it is possible to micro-cracks appear and propagate until the shaft's breakage. The breakage of the shaft may cause the great human and material losses. Because of that, during manufacturing these assemblies it must be taken into account all parameters which can initiate shaft crack. Geometric measure of seating and shaft are recommended by UIC regulations having great influence to quality and safety of realized assembly. The influence of contact surfaces and their lubrication during manufacturing the shaft - wheel assembly is shown in this paper.

1. Introduction

The safety of railway trains is largely dependent on the shaft-wheel assembly. For passenger's and freight's railway cars having not its own power, the bending stress is dominated in axle's material under load. For traction rail vehicles (e.g. electric locomotive), shafts are loaded with bending and twisting stresses due to the static and dynamic loads [1]. Due to the complex stresses, the special attention must be paid in manufacturing of individual elements of these assemblies and their assembling [2]. At the initial moment of motion and stopping of the train there is a significant impact load. Due to this complex load, the UIC regulations require the regular control of these assemblies and the written evidence of it is obligatory. Due to the irregular control and improper detection of changes of the shaft a serious railway accident can be happen, with human and material losses as the consequences. By the statistical monitoring it has been found that the fractures often occur at the beginning of motion of the locomotive, due to extremely high tangential stresses in the shaft, especially when the composition is heavy loaded. An example of dynamic fracture by which some material losses were made, but fortunately there were no human losses, is shown in Figures 1 and 2, [5], [7]. The fracture usually occurs between the wheel hub and the driver gear. In addition to shaft fracture, some disturbance of the solid assembly between wheels and shaft can happen, that is slipping between the inner contact surface of the hub and the fitting surface on the shaft. To avoid this, the technological process of mounting of that assembly is very important. Technical service life of the shaft can be up to 45 years and more, while the service life of the wheel is significantly lower, just between 5-10 years, depending on the kilometers traveled [8].

Figure 1. Place of the accident - fracture of the shaft

Figure 2. The fractured surface of the shaft

2. The shaft-wheel asembly

In order to ensure a safeness of the shaft-wheel assembly, technological preparation of geometrical and other parameters before mounting is very important. It is important to establish that there are no micro-cracks on the shaft, as well as geometric dimensions and parameters are technological achieved at the press fit of shaft and wheel. Also, prescribed geometric measures and technological parameters must be checked at the inner surface of the wheel hub during machining of these surfaces. UIC regulations define indicative parameters to be followed in the preparation and realization of press fit of wheel-shaft assembly. However, micro-technological parameters for the realization of press fit are manufacturer's secret knowledge that are difficult to be found because they significantly affect to safety and reliability of the assembly. The newly formed shaft-wheels assembly is shown Figure 3.

Figure 3. The shaft-wheels assembly

The quality check of the resulting press fit is performed by measuring the electrical resistance and force-displacement dependency. Electrical resistance for the new assembly (new shaft and new wheels) must be less than 0.01 Ω, while for the assembly composed by an old shaft and new wheels electrical resistance must be less than 0.1 Ω. In the force-displacement diagram, curve gradient must always be rising except at the position of oil supply channel for dismantling of shaft-wheels assembly and at the end of mounting [9].

2.1. Realisation of the press fit

General dependence between force and displacement during realisation of the press fit is shown by curves in Figure 4.

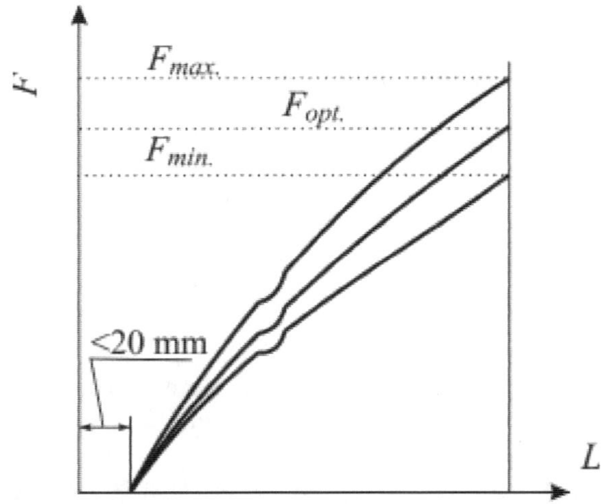

Figure 4. The force - displacement dependence of the shaft-wheels press fit [5]

The shape of the curve is affected by many parameters such as: the parallel alignment between contact surfaces, roundness of wheel hub and shaft, roughness of contact surfaces, direction in which the surfaces is realized in machining process, the quantity of lubricants and its distribution and the thickness during assembling. These parameters have a special significance for the realization of a safe assembly regarding the prescribed force-displacement curve shape, because they considerable impact on the rate of change of curve gradient. Deviation of curve gradient is permitted only at the position of oil supply channel [3]. The outer surface of wheel is coated with heat-activated color (used for daily visual inspection) that changes its shade when thermal-stresses appear in the wheel-shaft press fit that is slipping between contact surfaces. Thus, in the daily visual inspection of these assemblies, if change of the shade of the heat-activated color is detected on the surface of wheels, the electric locomotives should be excluded from traffic. After that, it is necessary to send it to the nearest authorized service workshop for detection and removal of defects.

3. Pressing force

Limit values of the pressing force are defined from empirical equation (1)

$$F = D \cdot K \tag{1}$$

where:

D - a shaft diameter at the hub seat in [mm],

K - intensity of 3,5 to 6 kN/mm.

Minimum value of the pressing force is $F = 3.5 \cdot D$ kN, while maximum value is $F = 6 \cdot D$kN. If the force value is out these limits, than assembly must be dismantled, potential faults removed and assembly once again pressed fit [6]. Beside this, if residual stresses in assembly are out of permited values, than shaft-wheel assembly sholud not be built in locomotive. Becouse of this, it is necessary to start with investigation and definition of lubricant's parameters influencing to press fit process.

Theoretical dependence of the pressing force is given by equation (2)

$$F = \pi \cdot s \cdot E \cdot l \cdot tg\alpha \cdot \left(\mu_p + tg\alpha\right)\left(1 - \psi_e^e\right) \tag{2}$$

where:
E - elastic modulus
α - the half cone angle
l - length of the active assembly block
μ_p - coefficient of stickness
ψ_e - the ratio of diameter d / D.
Since the coefficient of friction is dependent on a few parameters, it is necessary to investigate these parameters that is its influence to press fit process.

4. Experimental research

The aim of experimental research is to investigate the impact of lubricants on the pressing force curve gradient. Lubrication is carried out with a lubricant MoS_2. Speed of pressing is 50 [mm/min]. Other geometric and technological parameters are within the limits allowed by the regulations. Press fit is performed on the press intended exclusively for pressing of shaft assemblies for railway vehicles. Press is controlled by software which records the force-displacement curve in the real time. The recorded data is stored in a database. Force-displacement diagram can be printed in the form as shown in Figures 5 and 6. The curve flow in the force-displacement diagrams shown in Figures 5 and 6 satisfies. The lubricant MoS_2 is used in the anticipated amount, measured electrical resistance of the assembly meets (it is far less of permitted value). Maximal generated forces are equal. Jump of the force at the end of displacement of the first wheel-shaft assembly on the side of the driving gear, Figure 5, is caused by the contact between wheel hub and gear hub. However, changing in the amount of lubricant and its distribution on the contact surface leads to the pressing force below permissible limits, Figure 7.

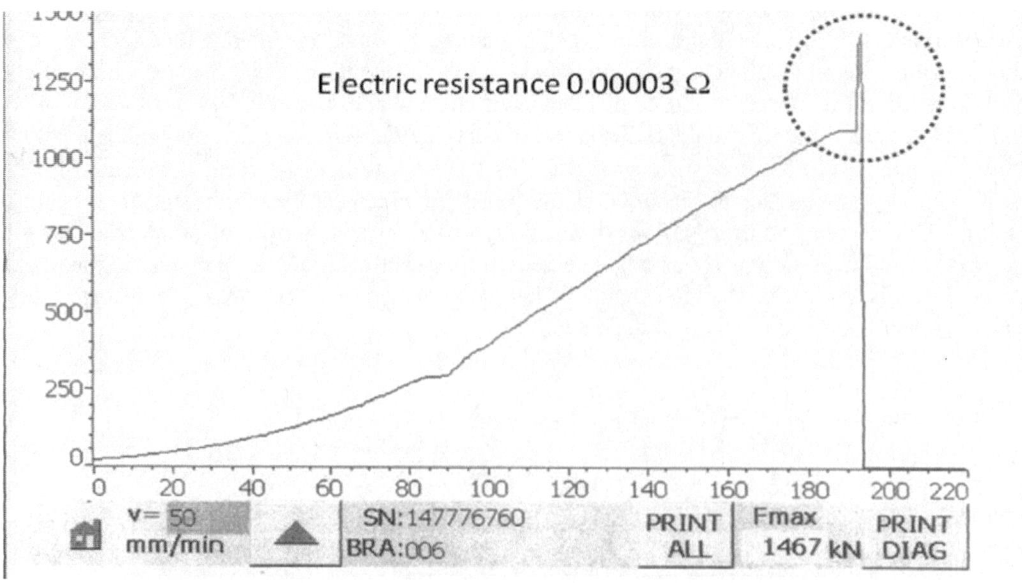

Figure 5. Diagram of the press fit of the shaft and wheel A

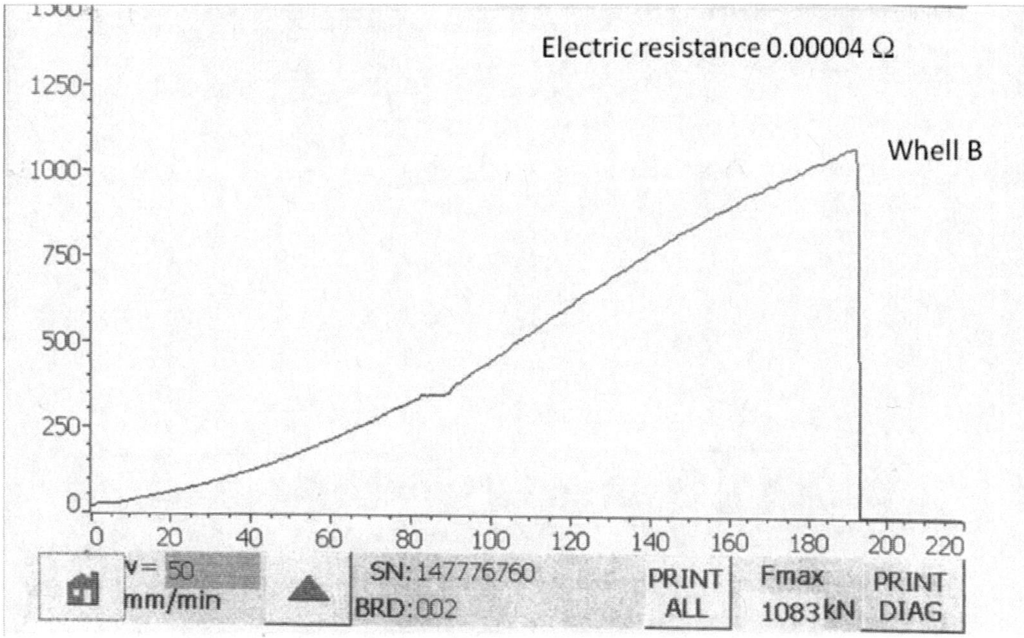

Figure 6. Diagram of the press fit of the shaft and wheel B

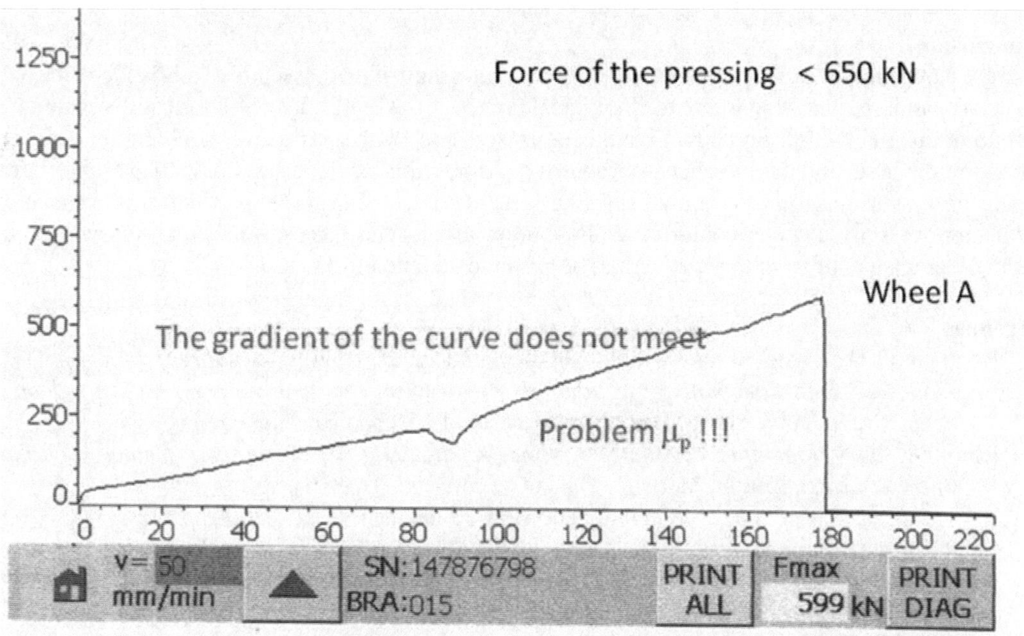

Figure 7. Force intensity below permissible value (wheel No. 015; shaft 147876798)

Assembly is dismantled and visual inspection showed there is no mechanical faults. Pressing process with defined amount of lubricant distributed properly on the contact surface is done again, Figure 8. Pressing force is about 900 kN.

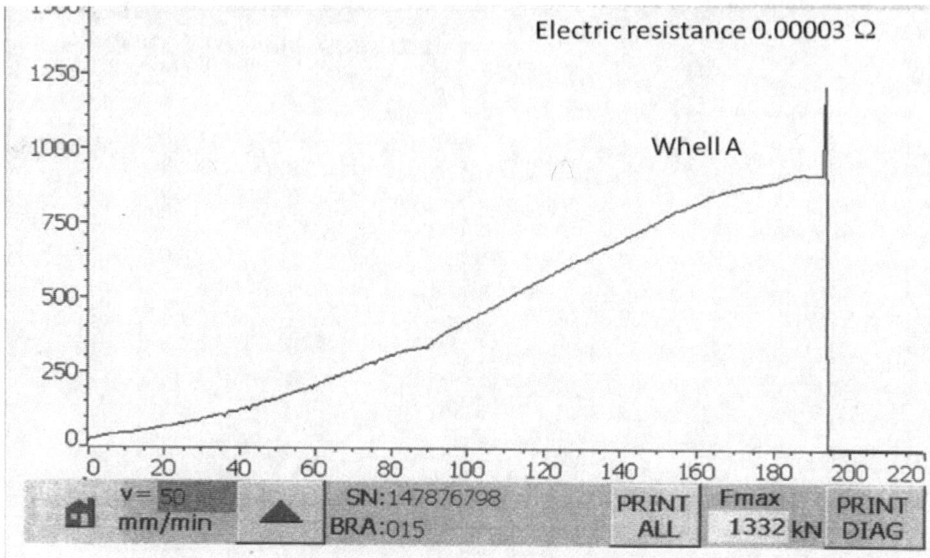

Figure 8. Re-pressed assembly (wheel No. 015, shaft 147876798)

One can conclude from force-displacement diagram, Figures 7 and 8, that the amount and distribution of lubricant have a significant influence on a quality of shaft-wheel assembly.

5. Conclusion

Research shows that in order to achieve quality and safe shaft-wheel assembly, lubrication of contact surfaces has an important role in the technological process of assembling. The lubricant's quantity and distribution must be within limits that have been established by theoretical research and experimental verification of these studies. In order to properly distribute the lubricant it is essential to monitor the pressing process in real-time by monitoring the force-displacement curve gradient on the machining system monitor. Any change in lubricant distribution affects the force-displacement curve gradient. The other parameters of assembly parts must be within calculated limits too.

References

[1] BlagojeviC D, Todic M and Golubovic-Bugarski V 2015 *Studija napresavanja osovinskih sklopova elektrolokomotive (Study of shaft-wheel assembly pressing of electrical locomotives)*, University of Banjaluka, Faculty of Mechanical Engeenering

[2] Rinjiku T 2008 *Committee of Studies on Wheelsets for High Speed Railway Vehicles*, (*Railway Wheelset*), First Edition, Maruzen Planet Corporation, Tokyo, pp 53

[3] ***European Committee for Standardization *EN 13262: 2004+A2: 2011*, pp 12

[4] ***Railway Group Standard GM/RT2466 2010 *Railway wheelsets*, Issue Three

[5] ***UIC 510-2 2004 *Trailing stock, Wheels and wheelsets, Conditions concerning the use of wheels of various diameters*, 4th edition

[6] ***UIC 813 2003 Specification for the supply of wheelsets for tractive and trailing stock-tolerances and assembly

[7] Iwnicki S 2006 *Handbook of Railway Vehicle Dynamics*, Taylor and Francis Group

[8] Chaar N and Berg M 2005 Vehicle-track dynamic simulations of a locomotive considering wheelset tructural flexibility and comparison with measurements, *Journal of Rail and Rapid Transit* **219** 225-238

[9] Kappes M, Kröning B, Rockstroh H, Salzburger J and Walte F: Non-Destructive testing of wheel- sets as a contribution to safety of rail traffic, Fraunhofer Institute Non-destructive Testing (IzfP) – Universität, Gebäude 37 - (66123) Saarbrücken - Germany

IOP Conf. Series: Materials Science and Engineering **200** (2017) 012025 doi:10.1088/1757-899X/200/1/012025

The heavy metal ions (Cu^{2+}, Zn^{2+}, Cd^{+}) toxic compounds influence on triticale plants growth

V M Brezoczki and G M Filip

Technical University of Cluj Napoca, The North University Centre of Baia Mare, Department of Mineral Resources Engineering, Materials and Environment, Dr. Victor Babes 62 A Street, 430083, Maramures, Romania

E-mail: brezoczkivaleria@gmail.com

Abstract. The presence of the heavy metals toxic compounds ($CuSO_4 \cdot 5H_2O$, $ZnSO_4 \cdot 7H_2O$ and $3CdSO_4 \cdot 8H_2O$) in water and soil can be observed by their negative effects on the germination and growth process for different vegetable (barley, oat, maize) who are used for human and animal consumption. This paper it aims the determination of germination and growth inhibition negative effects for triticale plants in the heavy metals ions presence by ecotoxicological laboratory tests. The triticale plants was chosen for their different characteristics to the other grasses respectively: a very good resistance for a wide range of diseases, an accelerated growth and a very good tolerance for aluminum ions presents in acid soils. The determinations were conducted step by step, first, we put the triticale grains in contact with the heavy metal solutions with different concentration then for 3 days we noticed the triticale germination inhibition effects and finally we noticed the growth inhibition process for triticale plants respectively in 7th and 9th day from the start of the experiment. At the end of the tests we can conclude that the triticale roots have a very great sensibility to a $CuSO_4$ solutions compared to the effects for their stalks. A positive effect for triticale stalks we can see for low $CuSO_4$ solution concentrations thus for 5 mg Cu/l the growth is 19,44%. A positive effect for triticale roots it can see for low $ZnSO_4$ solution concentrations so for 5 – 15 mg Zn/l the growth is 24,4%. In the presence of the $CdSO_4$ solution all the processes are inhibited (germination and growth for triticale plants) even for a low concentrations for this toxic.

1. Introduction

The concept of ecotoxicity refers to the processes that disrupt the ecosystems's structure and functions, as a consequence of direct or indirect action of some agents of pollution, manifested through reversible and irreversible ecological imbalances [1].

In the current context, regarding the residual industrial pollution with heavy metal ions of soils in mine sites, particularly mining waste dumps contaminated with heavy metals, for their bioremediation have been carried out a series of tests on the subject of a variety of graminaceous mixtures for grassing (wheat, oat, rye) in order to lead to the installation of vegetation on the polluted fields as well as to extract the heavy metal ions that are in the soil.

The tests proposed in this paper have been focused on using as a working material the triticale seeds obtained artificially by hybridizing wheat and rye.

The resulted hybrid presents a series of biological and economical characteristics that recommend it: greater resistance to low temperatures; rapid growth rhythm, vigorous plants and a high berry and mass production; genetic resistance to a large range of diseases, tolerance for aluminum ions toxicity that determines the cultivation on acid soils without the necessity of liming; triticales spring more rapidly as a result of a greater water absorption capacity and of a more intense activity of the alpha-amylase hydrolyser enzyme; the triticale cultures do not require the use of herbicides.

2. Ecotoxicity of the heavy metal ions

Prolonged exposure of certain plants to high concentrations of heavy metals originating from the atmosphere or directly form the soil constitutes the premise of their accumulation in quantities which can affect plant metabolism and can lead to visible changes of their structure.

Through transfer via the food-chain, these plants which bio-accumulate heavy metal ions represent a potential hazard for the consumers' health.

2.1. Copper toxicity

Copper is an essential element for the growth and development of plants, and its absorption in superior plants is achieved mainly as Cu^{2+}. Copper has redox properties and is a structural and catalytic component for the protein and enzyme molecules involved in the metabolic processes.

The toxicity of copper resides in its ability to form free radicals which cause an oxidative stress. Copper is assimilated by plants as divalent ions or as chelates, and can penetrate the plant both through its roots and its leaves. The visible effects of the toxicity of copper upon terrestrial plants are chlorosis, necrosis, plant growth diminution, or the drying of plants.

The soil PH directly influences the organic compounds formed by copper in the soil, as well as the plant absorption process, i.e. soils with an alkaline PH bond copper more easily than acidic soils. Copper deficiency in crops appears when the soil is unable to supply an adequate amount of the element in a form that is absorbable by plants. Copper deficiency causes similar symptoms in barley, oats, rye and wheat [2].

In case of excess, copper can become extremely toxic, which manifests itself as chloroses, necroses, leaf discolorations. At cellular level, toxicity can be the consequence of copper bonding to the sulfides groups of proteins, thus preventing the activity of enzymes and proteins.

Plants that are sensitive to copper excess are the clover, the alfalfa, the poppy, the potato, and the strawberry [3].

During tests with triticale one could notice the positive influence of copper ions in small quantities, i.e. 5 mg Cu/L, upon the germination and growth of the triticale plants. As one uses experimental variants with higher concentrations of copper ions, the germination of the triticale seeds is slowed down or even inhibited completely [4].

Copper in a concentration of 6.25 mg/L has a negative effect upon the growth of the roots (12% growth diminution) and the stem (16% growth diminution) of the alfalfa seedlings after 14 days. As regards the Lolium perenne plants, copper in concentrations lower than 6.25 mg/L does not inhibit the growth of either roots or stems. At concentrations of 100 mg Cu/L the inhibition of root growth is almost complete (84 %) [5].

For duckweed, several studies indicated that copper excess interferes with respiration, photosynthesis, synthesis of pigments and enzymatic activities. Copper toxicity for duckweed has the value of CL_{50} equal to 1.51 mg/L [6].

2.2. Zinc toxicity

One noticed that for the alfalfa, zinc, even in low and moderate concentrations (between 6.25 and 50 mg/L) reduces the growth of the plant, in comparison to the control sample, by about 20%; only in high concentrations (100 mg/L) does zinc strongly inhibit growth (i.e. 40% diminution of the root growth and 30% of the stem growth) [5].

Copper and zinc in low concentrations have positive effect upon the germination and growth of plants.

2.3. Cadmium toxicity

Cadmium pollutes the soils through atmospheric precipitation and deposits in areas in the neighbourhood of metallurgical plants, as well as through employment in agriculture as chemical fertilizers in the group of super-phosphates. Cadmium in the soil is considered to be bio-available; it can be easily absorbed by plants through radicular absorption.

Cadmium negatively influences the growth of alfalfa even at 0.9 mg/L, but in Lolium there is a slightly positive effect. The toxicity tests for cadmium in grains highlighted the fact that at a concentration of 62.5 mg Cd/L the growth of roots is practically stopped (a decrease by 89%), and the growth of seedlings is also drastically affected (a decrease by 60%) [7].

The accumulation of cadmium in plants follows predominantly this pattern: roots> stems> leaves > seeds. Cadmium is an inhibitor of photosynthesis, and the effects which occur are mainly linked to the diminution of chlorophyll, stomatal closure (in plants such as the clover, beans, soy, alfalfa); it induces the activity of peroxidation in roots and stems (fact proven in rice); in is involved in the generation of oxidative stress in adult peas plants; it accumulates in the mitochondria, and inhibits certain enzymes in the respiratory system of the plants [5].

The toxicity of cadmium for plants is most often clearly identified, yet it can also be the result of the interaction of the main toxic ions with other ions, as well as other environmental factors. The various degrees of cadmium tolerance in various plants definitely involve differences in the structure and functioning of the cellular membrane [8].

3. Materials and manner of work

3.1. Required materials

Ecotoxicological laboratory tests for triticale were carried out in various types of laboratory dishes and glasswear, which had been cleaned and sterilized accordingly.

The laboratory devices and materials used for the tests are:

✓ Solutions for toxic medium:
- copper sulfate solution ($CuSO_4 \cdot 5H_2O$) at 5% concentration, reactive p.a.;
- zinc sulfate solution ($ZnSO_4 \cdot 7H_2O$) at 5% concentration, p.a;
- cadmium sulfate solution ($3CdSO_4 \cdot 8H_2O$) at 1% concentration;

✓ Triticale seeds;
✓ Petri dishes;
✓ Laboratory glasswear (graded cylinders, glass flat-bottom flasks, and Erlenmeyer glasses);
✓ Clean plastic cups and support sieves for germinated triticale seeds;
✓ Automate droppers;
✓ Tap water;
✓ Filter paper;
✓ Analytical balances.

3.2. *Manner of work*

The experimental tests were carried out in stages, by exposing the triticale seeds to toxic solutions (copper sulfate, zinc sulfate and cadmium sulfate) at various concentrations and observing the germination process and subsequently the plant growth over a period of 7 to 9 days.

For the control sample only tap water was used. For each variant, three repetitions were carried out (marked A, B and C).

The preparation of the experimental toxic concentrations started with the following stock solutions:

✓ S_1 - solution in concentration 5000 mg Cu/L using chemically pure reactive copper sulfate ($CuSO_4 \cdot 5H_2O$);

✓ S_2 - solution in concentration 5000 mg Zn/L using chemically pure reactive zinc sulfate ($ZnSO_4 \cdot 7H_2O$);

✓ S_3 - solution in concentration 1000 mg Cd/L using chemically pure reactive cadmium sulfate ($3CdSO_4 \cdot 8H_2O$).

The test solutions were prepared by diluting the stock solutions with tap water. Thus, the following toxic concentrations were used in the experiments, as illustrated in Table 1.

Table 1. Toxic concentrations of test solutions

No. crt.	**Toxic ($CuSO_4 \cdot 5H_2O$)** S_1 = 5000 mg Cu/L	**Toxic ($ZnSO_4 \cdot 7H_2O$)** S_2 = 5000 mg Zn/L	**Toxic ($3CdSO_4 \cdot 8H_2O$)** S_3 = 1000 mg Cd/L
1.	C_0(M) = 0 mg Cu/L	C_0(M) = 0 mg Zn/L	C_0(M) = 0 mg Cd/L
2.	C_1=5 mg Cu/L	C_1=5 mg Zn/L	C_1=1 mg Cd/L
3.	C_2=15 mg Cu/L	C_2=15 mg Zn/L	C_2=2 mg Cd/L
4.	C_3=45 mg Cu/L	C_3=45 mg Zn/L	C_3=4 mg Cd/L
5.	C_4=135 mg Cu/L	C_4=135 mg Zn/L	C_4= 8 mg Cd/L
6.	C_5= 405 mg Cu/L	C_5= 405 mg Zn/L	C_5= 16 mg Cd/L

Tests were carried out in two stages: the germination stage of triticale seeds in toxic medium and the triticale seedling growth stage after germination.

Germination tests – the tests were carried out in Petri dishes, properly sterilized, 12 cm in diameter, in which two layers of regular filter paper were placed, which were then properly soaked in the toxic solution, according to the work variant. Over the filter paper, 5 seeds of triticale were placed in each Petri dish and were left to germinate for 3 days. The Petri dishes were kept closed under a lid at room temperature (18 ÷ 20^{0}C) (Figure 1).

After 3 days, the number of germinated triticale seeds was recorded and measurements were carried out in order to gauge the average length of the roots, as well as the average length of the stems. Moreover, particular observations were recorded for each sample (root ramifications, occurrence of fungi, stem coloration, plant vigor etc).

Tests for growth inhibition – after the 3 days germination period, the triticale seeds were moved from the Petri dishes into 150 ml plastic cups on the support sieves. The triticale roots were positioned in such a way that they hung through the net of the sieves directly into the culture medium (Figure 2). The samples were kept in the laboratory at constant temperature, and fresh toxic solution was added to them when necessary.

The plants were kept for 9 days in these solutions, in the laboratory in natural light at room temperature (Figure 3). After 7 days of exposure and 9 days after the beginning of the tests, new measurements were carried out to gauge the length of the roots and the stems, and observations regarding the vigor of plants, the root ramification etc. were also recorded.

Figure 1. Sample preparation for germination of triticale [9]

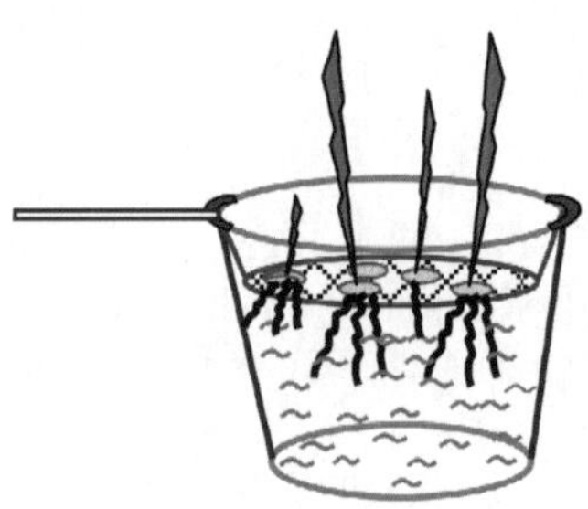

Figure 2. Sketch positioning triticale roots, growth stage

Figure 3. Triticale samples to 9 days [9]

4. Test results and discussions

4.1. Test results in the presence of the toxic ($CuSO_4{\cdot}5H_2O$)

The data of the research regarding the germination and development of triticale plants in the presence of the toxic $CuSO_4{\cdot}5H_2O$ at various concentrations are centralized in Table 2.

Table 2. Data on germination and development of triticale in the presence of toxic $CuSO_4{\cdot}5H_2O$

No days	Parameters	U.M.	Control sample	Concentration of toxic ($CuSO_4{\cdot}5H_2O$)				
				$C_1 = 5$ mg Cu/L	$C_2 = 15$ mg Cu/L	$C_3 = 45$ mg Cu/L	$C_4 = 135$ mg Cu/L	$C_5 = 405$ mg Cu/L
3 days	The average length of roots, L_{MR}	(mm)	18.8	17	11,2	7,93	5,06	1,5
		(%)	100	90,42	59,57	42,18	26,91	7,96
	The average length stems, L_{MT}	(mm)	7,2	8,6	7,13	6,6	6,11	2,72
		(%)	100	119,41	99,03	91,66	84,86	37,77
7 days	The average length of roots , L_{MR}	(mm)	21,18	18,53	11,73	8,14	5,25	0,33
		(%)	100	87,48	55,38	38,43	24,78	1,55
	The average length stems, L_{MT}	(mm)	36,53	36,45	25,4	20,75	16,22	3,83
		(%)	100	99,78	69,53	56,8	44,4	10,48
9 days	The average length of roots , L_{MR}	(mm)	23,08	15,13	12,0	7,73	4,91	0,33
		(%)	100	65,55	51,99	33,49	21,27	1,43
	The average length stems, L_{MT}	(mm)	63,56	46,45	41,86	31,33	25,73	2,97
		(%)	100	73,08	65,85	49,29	40,48	4,67

Laboratory tests highlighted interdependence between the concentration of the toxic and the accentuation of the inhibition process both during the germination and the development stage of triticale plants, and the results are presented in Figures 4 and 5.

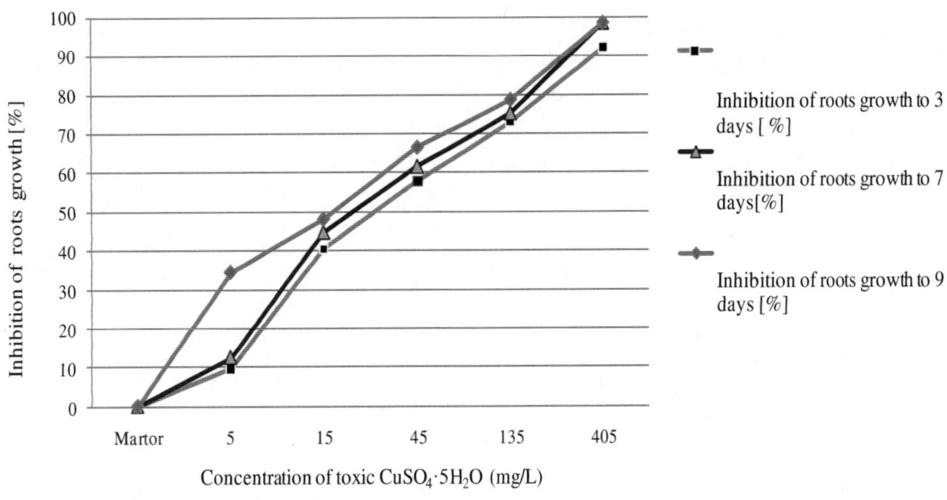

Figure 4. Variation of inhibition of the triticale roots growth in the presence of toxic $CuSO_4{\cdot}5H_2O$

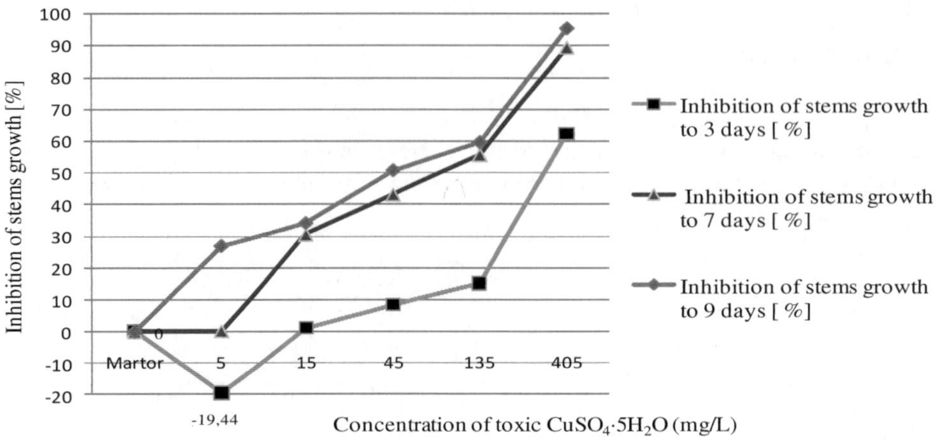

Figure 5. Variation of inhibition of the triticale stems growth in the presence of toxic $CuSO_4 \cdot 5H_2O$

After analyzing the data from the germination and development stages of the triticale exposed to various concentrations of copper sulfate for 3, 7 and 9 days respectively, the following observations were made:

✓ for the experimental variants C_1=5 mg Cu/L and C_2=15 mg Cu/L, all triticale seeds germinated, all having 2-3 well-developed root ramifications;

✓ after 3 days of exposure at a concentration of 5 mg Cu/L a positive evolution of the stem growth by 19.44 % was recorded as compared to the control sample;

✓ for the experimental variants C_3=45 mg Cu/L and C_4=135 mg Cu/L germination is slowed down, roots and stems have limited sizes, and the inhibition of the growth and development of the plants is between 66.51 and 78.72 % for the roots and 50.71-59.51% for the stems.

✓ for the experimental variant C_5= 405 mg Cu/L the germination process (after 3 days) is inhibited by 92.03 %, and after 9 days in contact with the toxic medium, the growth of the roots is inhibited by 98.57% and that of the stems by 95.33%.

4.2. Test results in the presence of the toxic ($ZnSO_4 \cdot 7H_2O$)

The data of the research regarding the germination and development of triticale plants in the presence of the toxic $ZnSO_4 \cdot 7H_2O$ at different concentrations are centralized in table no. 3, and the inhibition of the growth and development of the triticale plants in Figures 6 and 7.

Table 3. Data on germination and development of triticale in the presence of toxic $ZnSO_4 \cdot 7H_2O$

No days	Parameters	U.M.	Concentration of toxic ($ZnSO_4 \cdot 7H_2O$)					
			Control sample	$C_1 = 5$ mgZn/L	$C_2 = 15$ mg Zn/L	$C_3 = 45$ mg Zn/L	$C_4 = 135$ mg Zn/L	$C_5 = 405$ mg Zn/L
3	The average length	(mm)	10,53	13,1	12,81	8,93	5,1	3,3
days	of roots, L_{MR}	(%)	100	124,4	121,65	84,8	48,43	31,33
	The average length	(mm)	8,53	9,06	7,45	8,73	6,26	4,7
	stems, L_{MT}	(%)	100	106,21	102,34	87,33	73,38	55,09
7	The average length	(mm)	30,47	14,16	13,3	10,9	5	4,09
days	of roots, L_{MR}	(%)	100	46,47	43,64	35,77	16,4	13,32
	The average length	(mm)	59,3	47,68	37,86	35,86	27,2	23,2
	stems,L_{MT}	(%)	100	80,4	63,84	60,47	45,86	39,12
9	The average length	(mm)	35,46	13,86	13,8	11,86	4,45	3,7
days	of roots, L_{MR}	(%)	100	39,08	37,16	33,44	12,54	10,43
	The average length	(mm)	86	70,39	63,26	32,45	28,8	23,4
	stems , L_{MT}	(%)	100	81,84	73,55	37,73	33,48	27,2

The data analysis highlights the following aspects:

➢ During the first stage of the process, the germination stage of the triticale seeds, the presence of the toxic $ZnSO_4 \cdot 7H_2O$ for the experimental variants C_1=5 mg Zn/L and C_2=15 mg Zn/L, show a positive influence upon the growth of the triticale roots (24.4% and 21.65%) as compared to the control sample;

➢ Also during the germination stage of the triticale, the presence of the toxic in the concentrations C_1=5 mg Zn/L and C_2=15 mg Zn/L leads to a growth of the triticale stems by 6.21% and 2.34% as compared to the control sample.

The increase in the concentration of the toxic and in the exposure time from 7 to 9 days shows an inhibition of the growth of the stems with values between 18.16% at the concentration C_1=5 mg Zn/L and 72.8% at a concentration of the toxic C_5=405 mg Zn/L, as well as the inhibition of the growth of the roots with values between 60.82% at the concentration C_1=5 mg Zn/L and 89.57% at a concentration of the toxic C_5=405 mg Zn/L.

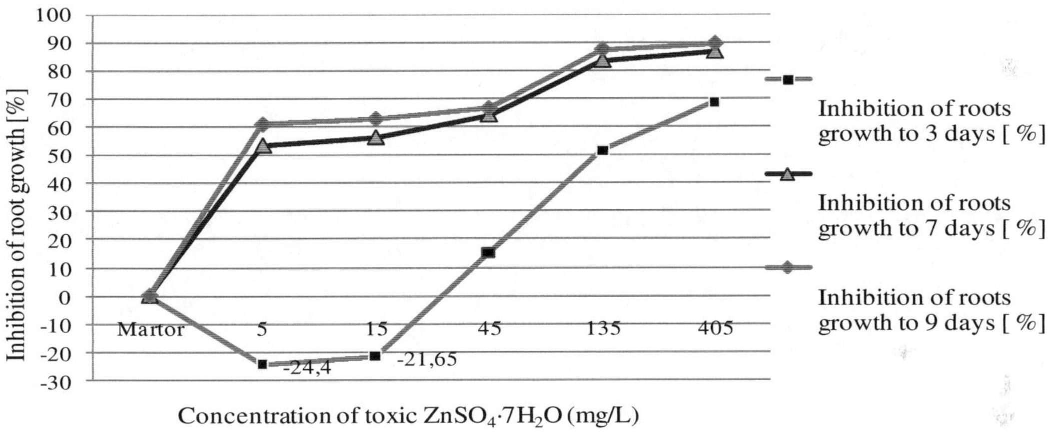

Figure 6. Variation of inhibition of the triticale roots growth in the presence of toxic $ZnSO_4 \cdot 7\ H_2O$

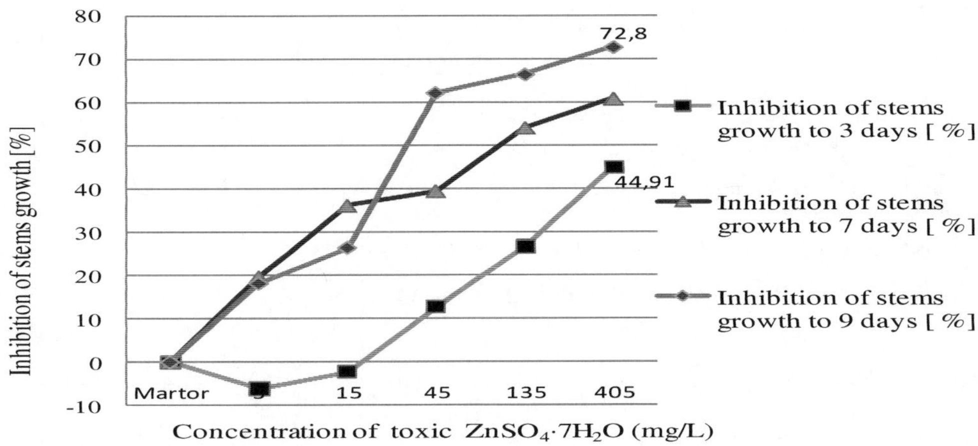

Figure 7. Variation of inhibition of the triticale stems growth in the presence of toxic $ZnSO_4 \cdot 7H_2O$

4.3 Test results in the presence of toxic $3CdSO_4{\cdot}8H_2O$

Inhibition of the growth and development of triticale plants in the presence of toxic $3CdSO_4{\cdot}8H_2O$ at various concentrations can be observe in Figures 8 and 9.

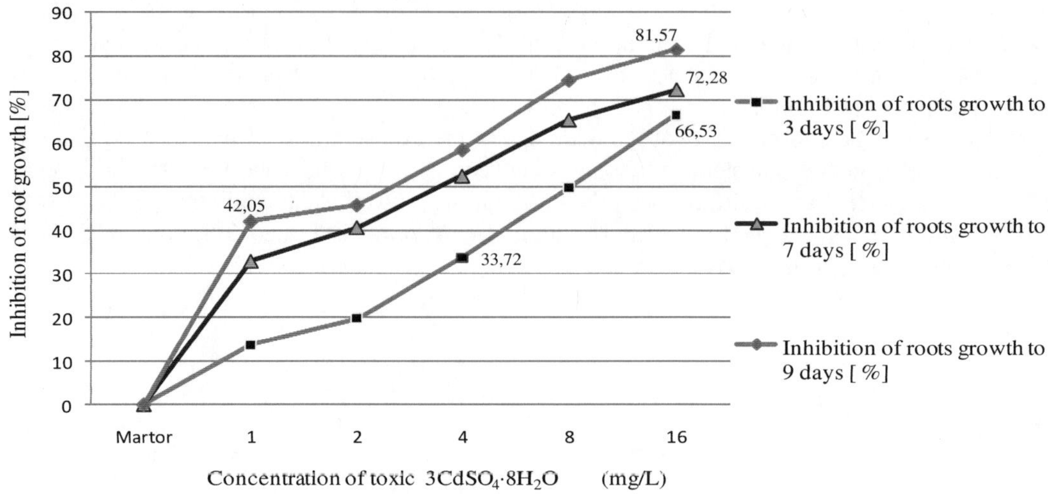

Figure 8. Variation of inhibition of the triticale roots growth in the presence of toxic $3CdSO_4{\cdot}8H_2O$

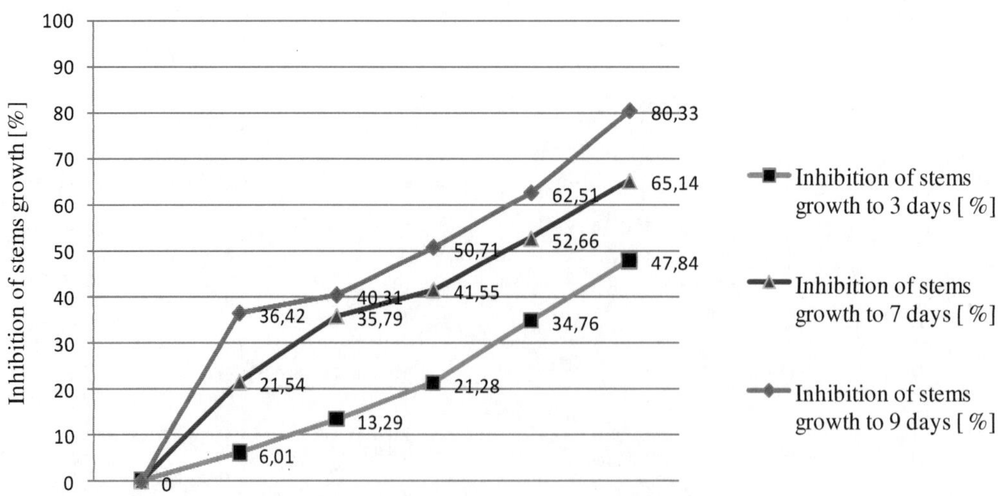

Figure 9. Variation of the inhibition of the triticale stems growth in the presence of toxic $3CdSO_4{\cdot}8H_2O$

After analyzing the test results, one can state the following:

✓ The triticale roots are more sensitive to the effects of cadmium than the stems, one noticing an accentuated inhibition of root growth since the first days of tests (at concentrations of 4 mg Cd/L root inhibition is 33,72%);

✓ Concentrations of 16 mg Cd/L lead to a major inhibition of roots growth, starting from the germination stage by 66.53% and reaching 81.57% inhibition after 9 days of tests;

✓ The triticale stems are less sensitive in comparison with the roots, tests showing the inhibition of their growth by 47.84% after 3 days of tests, at a concentration of 16 mg Cd/L.

✓ The occurrence of toxic cadmium in the environment negatively influences the germination and growth of triticale seedlings even at low concentrations and at the increase of the time spent in contact with the toxic element there appear necroses on the brims of the leaves and the triticale plants dry wither.

5. Conclusions

Another important aspect to be observed is the fact that, as compared to their stalks, the triticale roots are more sensitive to the presence of copper ions as toxic compound. Basically, this is the confirmation of the absorption process of ionic copper in soil and water, especially in the plant roots through radicular absorption.

The reduced copper concentrations (5 mg Cu/L) had a positive influence on triticale stems growth (by 19,44% in relation to the control sample) on a short duration exposure (3 days), while the 5mg Zn/L and 15 mg Zn/L concentrations highlighted a positive influence on roots as well as stalks.

A long exposure to any of the three types of used toxic compounds influences the triticale plants in a negative way, so that after 9 days of exposure the tests show a major inhibition of the growth and evolution of new plants.

References

[1] Marian P and Teodorescu A 2009 *Biotechnology Environmental Protection*, Second Edition, added and revised CD PRESS Publishing, pp 113-116

[2] Shorrocks V M and Alloway B J 1988 *Copper in plant, animal and human nutrition,* Copper Development Association, CDA Publication TN35

[3] Lixandru Gh et ad 1990 *Agrochemistry*, Bucharest, Publisher Didactic and Pedagogic

[4] Oros V and Brezoczki V M 2013 Preliminary ecotoxicological test on heavy metals effects on triticale seedling growing, *Scientific Bulletin North University of Baia Mare* **27**(2)

[5] Oros V 2011 *Elements of ecotoxicology and ecotoxicological tests,* Publisher Risoprint Cluj-Napoca, pp 254-300

[6] Oros V 2013 Aquatic phytotoxicity of heavy metals: Cu, Cd and Zn ecotoxicological tests with duckweed plants (Lemna minor), *Environmental Engineering and Management Journal* **12**(2) 1535-1542

[7] Oros V 2013 Effects of cadmium on the growth of the wheat seedlings, *Scientific Bulletin North University of Baia Mare* **27**(1) 33-48

[8] Das P, Samantaray S and Rout G R 1997 Studies on cadmium toxicity in plants: a review, *Environmental Pollutions* **98**(1) 29-36

[9] ***2016 Personal album

IOP Conf. Series: Materials Science and Engineering **200** (2017) 012026 doi:10.1088/1757-899X/200/1/012026

Audit in public administration's information systems – External or internal?

D Drljača[1] and B Latinović[2]

[1]Europrojekt centar, Banja Luka, Republic of Srpska, Bosnia and Herzegovina
[2]Pan European University APEIRON Banja Luka, Faculty of Information Technologies, Republic of Srpska, Bosnia and Herzegovina

E-mail: drljacad@gmail.com

Abstract. Audit of the information system, thanks to the increased use of ICT and related cyber-crime, becomes a very important process in modern companies and institutions. It is usual to engage or outsource a third party for independent financial audit. But what about auditing of the information system of public administration institutions? This paper gives an introduction to possible aspects of information system's audit with the aim to discuss possible answer on the question in the title.

1. Introduction

The computer systems and machines are modernising and automate the business process in private (industry) but also in public sector (public administration and institutions). Modern business is not any longer possible without computer systems and Internet-based communication. Traditional (paper-based) information systems become a part of history while computer-based (supported) information systems are more and more in use.

Even the business processes in public administration and institutions introduced this concept and heading fast towards paperless offices. This is important in order to meet expectations from the citizens and business entities since all aspects of living are influenced with the digitalisation.

All of these changes have proven to be positive and usefulness for the overall progress of society, but they are followed by misuse causing fraud and significant problems. Therefore, institutions - especially those dealing with personal and other sensitive data – have to invest additional efforts in protecting such data. Public administration has even more responsibility to do this since their information systems and repositories are overloaded with such kind of information, starting from financial data, personal data up to the data on overall safety and security (such as military data).

Audit of such information systems (IS) can be a really difficult task, from its beginning. But this task is important to be done if we want to secure proper functioning of an IS and protection of data.

This paper will present some aspects of IS audit and will provide insight into the necessity to make it internally.

2. Purpose of IS audit

The audit of information systems (wider than an audit of IT) appeared with the introduction of the information technologies in the accounting systems. During 1960-ies this was known as Electronic Data Processing (EDP) Audit. The increased the use of computers in businesses (as a trigger) resulted with the need for auditors to become more familiar with EDP concepts in business. Business dictionary defines EDP as the “use of computers in recording, classifying, manipulating and summarising data” [1].

Since mid-1950-ies, many books have been written about EDP, but as the most significant was the works of Richard G. Canning from 1956, titled “Electronic Data Processing for Business and Industry” and Felix Kaufman from 1961 titled “Electronic Data Processing and Audit” [2]. To simplify and to connect it with the public administration we can say that EDP is primarily concerned with the use of computers and IT in basic administrative and accounting functions.

According to Dube and Gulati [3], one of the most important reasons for the establishment of information systems auditing separately from financial is insufficient knowledge of computer that negatively and adversely affects the ability of auditors to carry out the attestation functions. With other words, auditors do not possess sufficient IT skills for adequate attestation of claims.

One of the first definitions of information systems’ audit was given in 1998 by Ron Weber [4] that is in use even today. It seems that other scholars favour this definition, so Cangemi sets almost identical one, defining it as the process of collecting and evaluating evidence on the basis of which it can be determined whether the information system keeps the property of the company adequately, maintain the integrity of data, allows you to effectively achieve the set goals, and whether they use available resources efficiently [5].

So, the key phrases are: collecting and evaluating evidence, keeping the property of the entity, maintain data integrity, effective and efficient use of resources. These are the main guidelines for a successful audit. This means that IS audit is becoming more important considering spectrum of controls it has to check and validate against best practices, audit frameworks, and standards.

There were some efforts to improve above mentioned definition of which is good to mention the effort of Panian from Croatia. Namely, Panian intention is to deepen the definition by linking it more closely to the definition of the financial one, adding part of sentence saying “and determining whether the organisation meets the appropriate regulations, rules and conditions” [6].

Taking in the consideration the complexity of IS audit, Piattini considers that IS audit, once being a complement to the financial one, is having own existence and can be considered as a strictly professional discipline [7].

Therefore, it is possible to conclude that IS audit is wider than the financial one and requires a set of specific skills and knowledge of auditor, from both economy and ICT. As such complex process, it also requires specific agreement between the auditor and the client about the confidentiality of audited data and processes.

Nevertheless, it is also obvious that IS audit is becoming more and more important part of usual business practices, especially for entities dealing with e-commerce, or providing e-services such as public administration. Therefore, IS auditor should not be seen as a person of whom organisation should be afraid of, but as a business partner willing to help and to assist in overcoming problems. Therefore, the role of IS auditor crucially differs from the role of the traditional financial auditor (see below Table 1). The overall holistic approach in IS audit requires new skills and more important – new way of thinking and approaching clients as business partners.

Table 1. The role of IS auditor against financial auditor

IS auditor	Financial auditor
To prevent misuse or fraud	To detect misuse or fraud
To act as a partner in the team	To punish as authority
To act as risk preventing manager	To act as evaluator and auditor
To focus on business process	To focus on audit findings

3. Types and aspects of audit

In order to have better insight into the issue of the audit, it is necessary to see different types and aspects of the audit process. In a literature, it is possible to find different definitions and aspects of audit in the company. This often depends on the aspect and point of view. But, summarising experiences from the different literature sources, a possible classification of audit types is given in Figure 1 below.

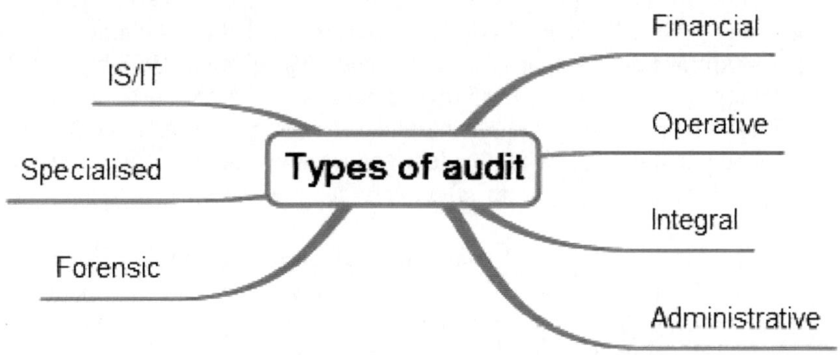

Figure 1. Types of audit in the institution (author)

The most common type of audit is **the financial audit**. The purpose of this audit is to evaluate the validity of financial reports. It relates to the integrity and reliability of financial information. This audit in public administration institutions is obligatory by law and is performed usually by contracted auditing company such as PricewaterhouseCoopers, Deloitte etc. Also, it can be done by authorised independent and licensed auditor under the condition that there is no conflict of interest and auditor is not an employee of the institution auditing.

Operating audit has a purpose evaluating the structure of internal controls of given process or work area. An example of this type of audit is the audit of application controls and logical security systems. This is a specific and targeted audit.

Integral audit combines both – financial and operative audits. The integral audit, in essence, is implemented to evaluate organisational goals related to the financial information, preserving of property, efficiency and harmonisation with overall goals of the audited institution.

Administrative audit aims to evaluate issues related to the efficiency of operative productivity within the organisation or institution. An administrative audit can be agreed even as the part of more complex reviews and audit.

IS audit aims to collect and to evaluate evidences to determine if the information system and related resources adequately preserves the organisation's assets, maintain integrity and availability of the data and the system itself, provide relevant and reliable information, achieve the business aims of the organisation with effective and efficient use of resources and have internal controls. This should assure achievement of business, organisational and control aims and that the unwanted events will be discovered, prevented and/or corrected.

Specialised audit presents a specific type of targeted audit. It is about the larger number of specialised audits delivered or undertaken by the third parties. Given business often and in large depends on services from the external service provider, the demand for this kind of audit is growing.

Forensic audit is a special case of the IS audit. Its task is to detect, publish and follow-up of threats and fraud. The primary purpose of this audit is to collect forensic evidence for eventual court processes and procedures.

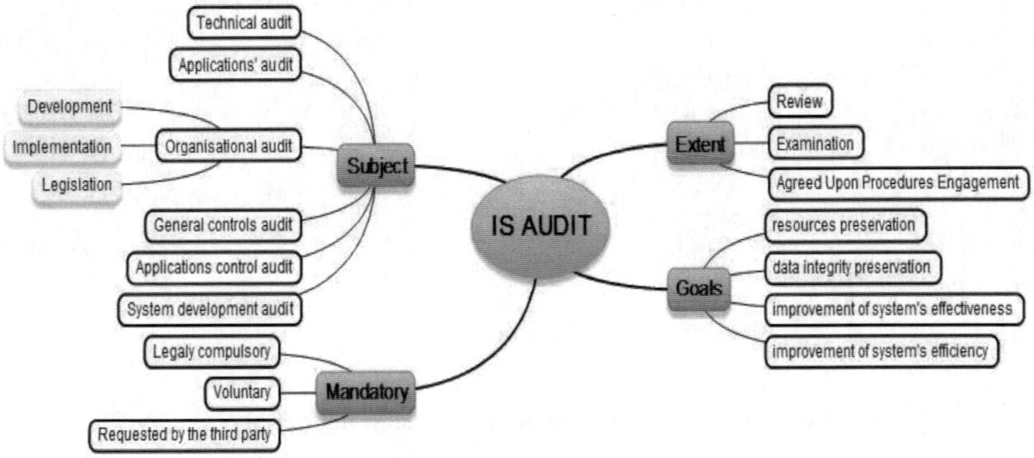

Figure 2. Aspects of IS audit in the institution (author)

4. External or internal?

Having in mind variety of aspects and points of interests for audit, it is very interesting to decide whether this audit should be outsourced and contracted by external audit company? This is a very important issue that requires determination and reasoning of the CEO or Management Board, which are in charge to make a decision on the audit. This is a very important issue especially for public administration organisations and their decision-making bodies having in mind that sensitivity of data being analysed.

The audit process can be organised internally and externally depending primary on the decision from top management and on the volume of the audit.

The audit of an information system can be a part of the internal audit. It can also function as a separate or integrated audit together with the financial audit. Integration with the financial audit should provide more assurances related to the IT linked controls for financial auditors. It is up to the management of the institution to decide on the modality for implementation of the audit before making Audit Charter. But it has to be strictly defined in the purpose of the audit – either the audit of IS functionalities, or audit of IS in the function of support for financial audit, or audit in the function of business operations improvement.

The Audit Charter (the programme of the audit) has to include the decision on which type of IS audit shall be implemented. It also must have a clear statement from the management of objectives and aim of the audit as well as delegated authorities for implementation of the audit. Besides these more general, the management should make a decision on volume of the audit – which can be defined together with the assistance of auditing authority or audit implementing company.

As results of the audit will be used for more tactical and strategic decision-making process and at the operative level used for implementation of corrective measures, these Terms of References for Audit should be approved by the senior manager or owners of the company or in some cases an audit board (if present).

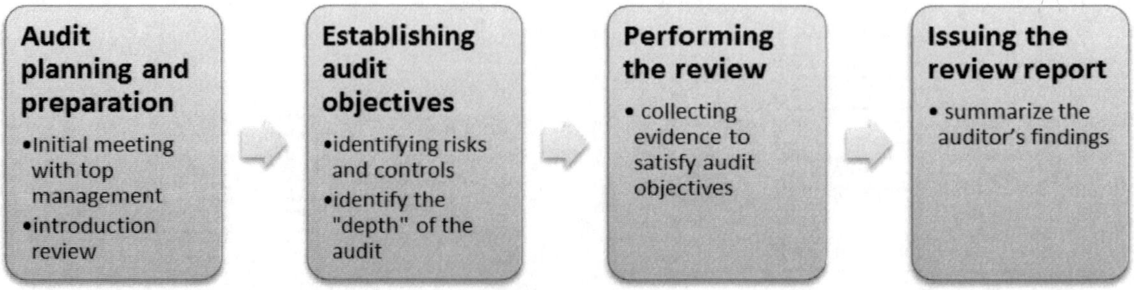

Figure 3. Usual steps in IS audit in the institution (author)

The IS audit is very complex and wider than a financial one. Therefore, IS auditor should be a technically complete person, with sufficient skills and knowledge needed for implementation of IS audit. IS auditors have to update and review their competencies and knowledge due to the large and fast changes in the field of information technologies, communication technologies, but also in the overall area of information system components. For IS auditor, it is important to understand the processes and if they are harmonised with best worldwide practices. However, it is necessary also to know the legal framework for business operations in the company or the institution. There are two main lines related to familiarity with the legal framework: one is to know legal aspects related solely to the audit of information systems and the other one is to know legal aspects of the topic of the audit, which is more specific and more precise.

In the planning phase, the auditor must be aware of the complete area that is under examination. This may include (but also not limit to) understanding and knowing of different business practices and functions that are in direct relation with the topic of the audit. However, for IS auditor is necessary to learn about the type of information system, supporting technologies and many other things.

If implemented as a part of the internal audit, the management should make all necessary efforts to make this audit as independent as possible. However, a hint of subjectivity will always be presented in such audit. Internal auditors are responsible to the audit board or to the top management and owners. The aims of internal audit should be aligned with the mission of the organisation or institution and should be based on the short- and long-term plans for the audit process. The aim of short-term audit plans is to plan a specific arrangement with known initial point and end with concrete goals. While on the other hand, long-term plans for audit deal with guidelines and definition of resources for efficient audit implementation. As in the financial audit, in the core of IS audit is also evaluation of related risks. Figure 4 shows usual steps in IS audit in the institutions or organisations.

It is obvious that even the most objective persons from the institution or organisation can be subjective in evaluating the IS and its functionalities. The internal auditor will be very skillful to evaluate the status of the technology and processes due to working experience and affiliation, while external auditor may blindly follow the instructions from frameworks and standards and not specifically experienced in the field.

The internal auditor will pay less attention to the achievement of customer satisfaction, while the external will look exactly to this and the two opinions may be very different. The internal auditor

IOP Conf. Series: Materials Science and Engineering **200** (2017) 012026 doi:10.1088/1757-899X/200/1/012026

might focus more on security aspects of IS, while external can focus easily on overall functioning as independent one, especially when dealing with communication and information flow.

From the safety aspect, the internal auditor is more desirable one since all the knowledge and data are staying "in the house" unlike the external auditor. The organisation may perceive external auditor as a potential point for "leaking of information" and as security problem despite signed confidentiality agreements. From the other hand, if not providing quality access to data and information to the external auditor may cause false interpretation of the collected data and also mistrust from the auditor on its skills and independent view.

However, the quality of data collected and opinion presented is mostly in the interest of the organisation or the institution. Therefore, this is always an open and ongoing issue in the planning phase of the IS audit process.

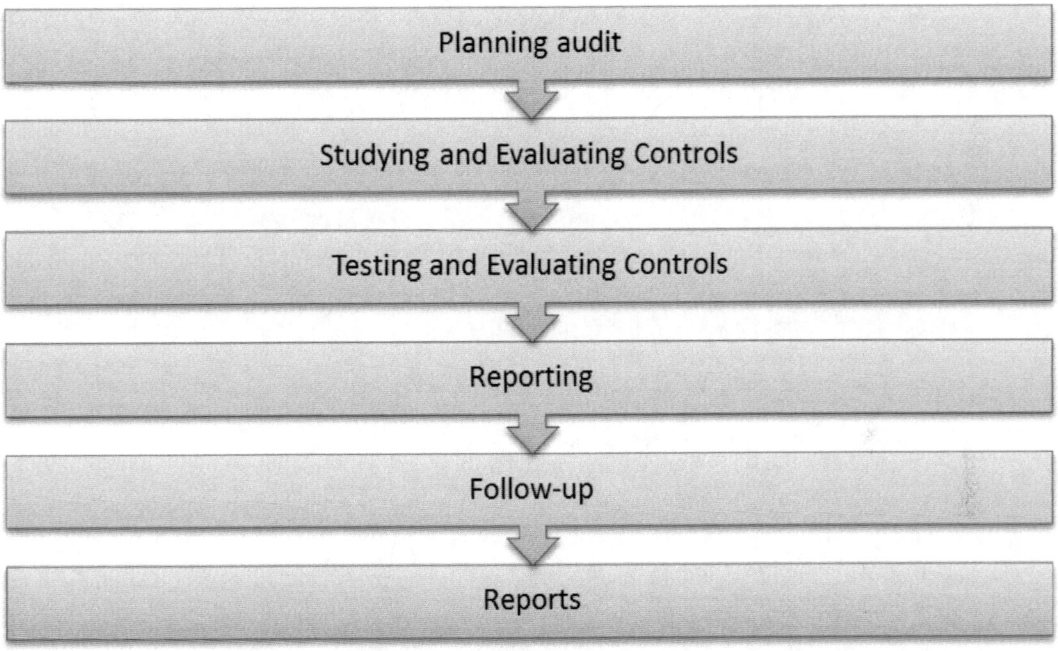

Figure 4. Usual steps in IS audit in the institution (author)

5. Conclusion

The audit of IS becomes more important for business operations regardless the size and the functionalities of the IS, especially in those dealing with sensitive data and with a large number of clients. The primary interest of the audit is security and safety of the IS, but there should be considered many other controls influencing IS role in the business processes.

Public administration IS are very sensitive to any kind of audit due to the importance of data. Although there are no legal requirements for IS audit in public administration in Bosnia and Herzegovina, this should be enforced in legislation such as it is a case in the banking sector.

Also, there is an ongoing dilemma - who should perform this audit? Should it be internal auditor or external? What do we get by engaging and paying external one? Do we really need these services and will this external auditor possess sufficient knowledge and skills to do it better than our internal auditor? Do we have a skilful internal auditor, since this audit is completely different from the financial one?

There are just some of questions that should be clear in minds of those making a final decision on implementation of the IS audit in the organisation or institution. One should consider all benefits and all potential risks in the implementation of the IS audit. IS including related IT, is not only a tool for

IOP Conf. Series: Materials Science and Engineering **200** (2017) 012026 doi:10.1088/1757-899X/200/1/012026

business operations but is a crucial element of each business and should be kept and maintained with more care and with more attention.

References

[1] ***Business dictionary 2016 *Electronic Data Processing*, available at http://www.businessdictionary.com/definition/electronic-data-processing-EDP.html, accessed on 21.10.2016

[2] Wayne S B 1965 *Auditing with the Computer*, University of California Press

[3] Dube D P and Gulati V P 2005 *Information System Audit and Assurance*, Tata McGraw-Hill, New Delhi

[4] Weber R 1988 *EDP Auditing--Conceptual Foundations and Practice*, McGraw-Hill, 1988

[5] Cangemi M P 2000 *Managing the Audit Function: A Corporate Audit Department Procedures Guide 3rd ed.*, John Wiley & Sons, New York, USA

[6] Panian Ž 2001 *Kontrola i revizija informacijskih sustava*, Sinergija, Zagreb, Croatia

[7] Piattini M 2000 *Auditing Information Systems*, Idea Group Publishing, USA/UK

IOP Conf. Series: Materials Science and Engineering **200** (2017) 012027 doi:10.1088/1757-899X/200/1/012027

Particulate matter urban air pollution from traffic car

G M Filip and V M Brezoczki

Technical University Cluj-Napoca, Engineering Faculty, Mineral Resource, Material, Environment Engineering Department, Baia Mare, Romania

E-mail: gabrielamfilip@gmail.com

Abstract. The particulate matters (PM) are very important compounds of urban air pollution. There are a lot of air pollution sources who can generate PM and one of the most important of them it is urban traffic car. Air particulate matters have a major influence on human health so everywhere are looking for PM reducing solutions. It is knows that one of the solution for reduce the PM content from car traffic on ambient urban air is the fluidity of urban traffic car by introduction the roundabout intersections. This paper want to present some particulate matter determinations for PM10 and PM2.5 conducted on the two types of urban intersection respectively traffic light and roundabout intersections in Baia Mare town in the approximate the same work conditions. The determinations were carried out using a portable particulate matter monitor Haz – Dust model EPAM – 5000, who can provide a real time data for PM10, PM 2.5.Determinations put out that there are differences between the two locations regarding the PM content on ambient air. On roundabout intersection the PM content is less than traffic light intersection for both PM10 and PM 2.5 with more than 30%.

1. Introduction

Particulate matters are one of the most important compounds of air pollution. The risks of air pollution with PM are extremely serious for human health and for environment.

The effects of particulates on the environment depend on the particle size and their chemical composition and may occur as follows:

- short and medium influencing environmental factors (air, water, soil) and the condition of fauna and flora at local and regional level;
- long-term effects by climate change globally

Particulate matters negatively influence the development of flora and fauna. Decrease the brightness, the dissolving oxygen in water, so it is a negative influence for plants photosynthesis, disturbing animals breath can be harmful animals' health.

Coarse (bigger) particles, called PM10, can irritate eyes, nose, and throat. Fine (smaller) particles, called PM2.5, are more dangerous they can play a role in causing serious illnesses and death because they are small enough to be inhaled deep into the lungs or even into blood. Once fine particles are in the lungs, they can affect the heart, blood vessels, and lungs, Figure 1 [1].

People exposed to fine particles over a long period of time have more heart and lung problems

Estimates of the health impacts attributable to exposure to air pollution indicate that PM2.5 concentrations in 2012 were responsible for about 432 000 premature deaths originating from long-term exposure in Europe [2].

So it is very important for our health and environment health to maintain a proper concentration for PM on ambient air. The first step for achieving this goal it's to know the origin sources for particles emission and then to find the best method/technologies for reduce the PM concentration to a limit value.

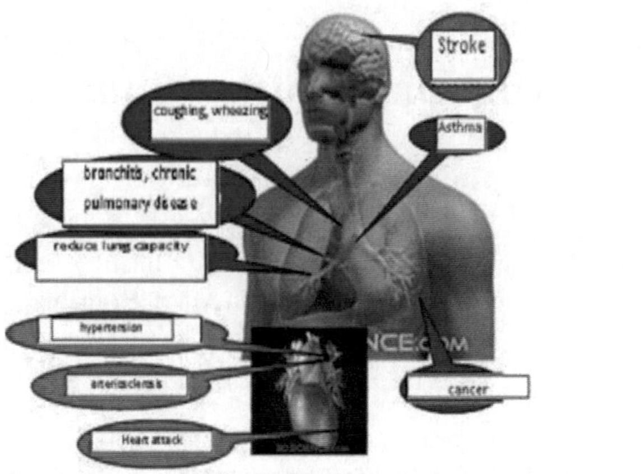

Figure 1. Diseases caused by PM pollution

It is very well known that there are two important sources that can provide PM pollution respectively:

- natural sources such as volcanoes eruptions, forest fires, sandstorms;
- anthropic sources, power plants, factories, cars and trucks, construction sites.

The pollutant emission from both mentioned sources can contain all type of particulate matter, PM10, PM2.5 and PM1.

In the last years the accelerated development of transport and the increasing of car numbers it have change the major balance sources of air pollution and the car traffic became one of the most important source for PM emissions especially in ambient urban air.

The environment scientists have found that one of the solutions for reduce the PM content from car traffic on ambient urban air is the fluidity of urban traffic car by introduction the roundabout intersections instead of traffic light. Modern roundabouts can improve traffic flow as well as cut down vehicular emissions and fuel consumption by reducing the vehicle idle time at intersections and thereby creating a positive impact on the environment [3].

This paper presents some particulate matter determinations for PM10 and PM2.5 conducted on the two types of urban intersection respectively traffic light and roundabout intersections in Baia Mare town, Romania. Through these determinations we wanted to see if and how much PM concentration on ambient air decreases.

2. Method

2.1. The status of fleet car from Baia Mare town. Consideration about the air pollution sources on Baia Mare

Baia Mare is a municipality the capital of Maramures County in NV Romania with a population about 150000. The city is located along the Săsar River and it is encircled by hills. The economic activity has been based on the mining and metallurgical activity but for about 8 years ago almost all industrial activity has ceased but behind this activity it remained a big tailings pond at a 15 km distance from Baia Mare [4].

So the most important air pollution sources are Bozânta tailing pond (PM), the heating fuel systems (PM and combustion gas) and car traffic.

IOP Conf. Series: Materials Science and Engineering **200** (2017) 012027 doi:10.1088/1757-899X/200/1/012027

The reports from Environmental Protection Agency Maramures emerges that one of the most important source for PM10 pollution in Baia Mare for the last two years is road traffic. For the last year there were a total of 32 exceedances of the daily PM10 limit value.

Whereas that one of the causes of increase the amount of air pollutants generated by motor vehicles is car fleet age and the type of use fuel so we have identified the following characteristics of the car fleet of Baia Mare:

- for the moment of our experiment the vehicles number it was more than 80000;
- average cars age has dropped from 9 years in 2007 to 5 years in 2014;
- out of total vehicles number, 75% they are cars and 60% of which operate on gasoline.

2.2. Experiment method

The determinations show in this paper were conducted for the two type of urban intersection respectively traffic light and roundabout intersections in Baia Mare.

Determinations were conducted identically for both traffic intersections. During the experiment we determined the content of PM10 and PM2.5 in ambient air in this area. They were made in the early summer during 5 days in a row (from Monday to Friday).

The used monitoring system it was the ***EPAM-5000*** portable environmental PM monitor a highly sensitive and accurate portable monitor for ambient and environmental monitoring with interchangeable size-selective sampling heads for PM10, PM2.5.

Working conditions were as follows:

1. determinations were carried out at rush hour car traffic respectively in the afternoon between 3 pm and 5 pm;
2. monitoring time it was 30 minutes for each traffic intersection;
3. weather conditions were almost the same throughout the experiment period respectively sunny day with little wind.

We must mention that there are certain important factors that we could not control them, which are the numbers of cars, the cars model, the used fuel, the technical condition of cars. Considering this we can say that the results presented in the paper are not very accurate.

3. Results

3.1. PM determinations for light traffic car intersection

For our determinations we choose one of the biggest light traffic cars intersection in Baia Mare city between two large avenues and we placed the monitoring system according to Figure 2 at about 6 m from the center of the intersection.

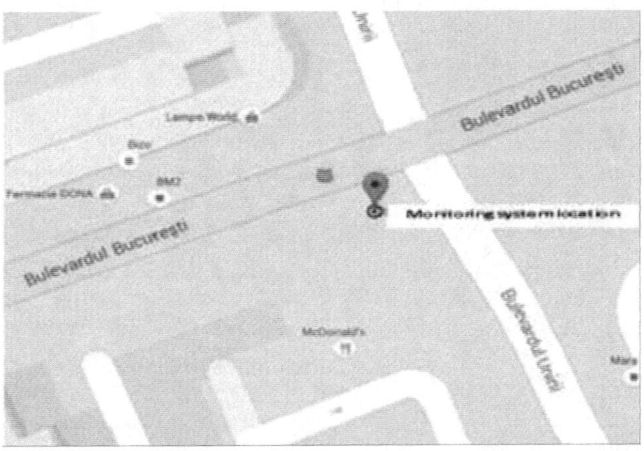

Figure 2. Monitoring system location on light traffic car intersection

The content for PM10 and PM2.5 average values on ambient air are presented in Table 1.
In the air quality directive (2008/EC/50) [5], the EU has these limit values for particulate matter:

- the PM10 daily mean value may not exceed 50μg/m^3 more than 35 times in a year;
- the PM2.5 may not exceed 25μg/m^3.

Table 1 PM10 and PM2.5 average values - for light traffic car intersection

Day	PM10 average value, μg/m^3	EU reference value	PM2.5 average value, μg/m^3	EU reference value
1	55	50 μg/m^3	32	25μg/m^3
2	47		39	
3	61		42	
4	53		37	
5	69		43	

It can see that for all 5 days the average value for PM10 and PM2.5 exceed EU reference value.

3.2. PM determinations for roundabout traffic car intersection
The location for monitoring system EPAM5000 on roundabout intersection is showed in Figure 3.

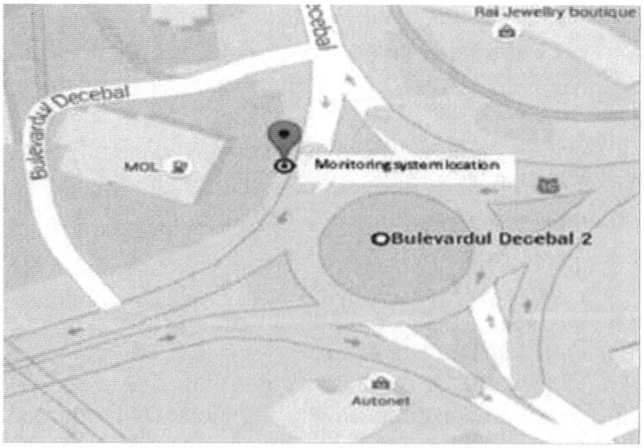

Figure 3. Monitoring system location on roundabout traffic car intersection

The contents for PM10 and PM2.5 average values on ambient air on roundabout traffic car intersection are presented in Table 2.

Table 2 PM10 and PM2.5 average values - for roundabout traffic car intersection

Day	PM10 average value, μg/m^3	EU reference value	PM2.5 average value, μg/m^3	EU reference value
1	35	50 μg/m^3	15	25μg/m^3
2	41		24	
3	33		17	
4	43		23	
5	37		18	

The determined values for PM10 and PM2.5 on roundabout traffic car intersection are below the allowed EU reference value.

3.3. Comparative results

The average PM values for the experiment period are given on Table 3.

Table 3 Comparative PM emissions results

Location /Emission	Light intersection	Roundabout intersection	Difference
PM10 average value,µg/m^3	57	37	20
PM2.5 average value,µg/m^3	38,6	19,4	17,6

The comparative results for PM contents determinations in those two intersections types are shown in the graphic form in Figure 4 for PM10 contents determinations and Figure 5 for PM2.5 contents determinations, It can see that the PM contents on air decrese for both PM10 and PM2.5 for the roundabout intersection traffic.

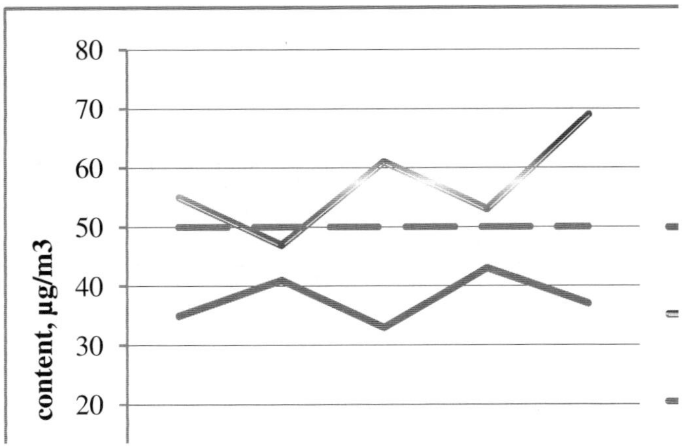

Figure 4. PM10 content on traffic intersections

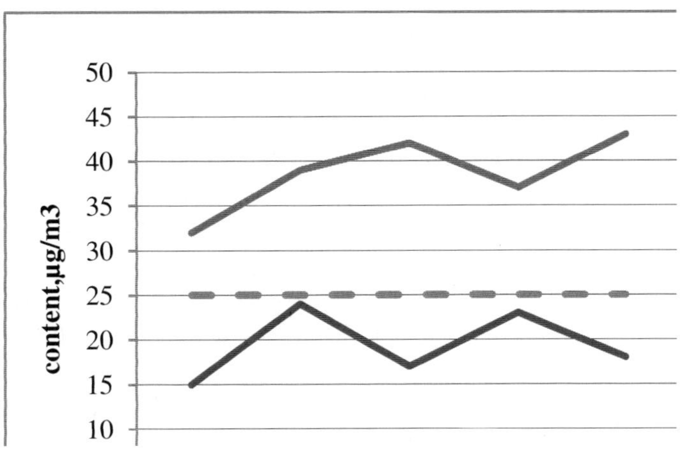

Figure 5. PM2.5 content on traffic intersections

4. Conclusion

Considering the above summary the roundabout intersection is very effectively in reducing PM cars emissions.

Thus it is noted that PM10 content on ambient air decreases with 33,68% and PM2.5 content decreases with 49,74% compared with light traffic intersection.

All the values for PM10 and PM2.5 on roundabout intersection for all 5 days experiment are below the allowed EU reference value.

As mentioned before the results shown are not very accurate but they may give a signal as regards the importance of the traffic cars organization in the cities in such a way as to reduce the ambient air pollution

References

[1] ***http://www.systemlife.ro/un_risc_pentru_sanatate.php
[2] ***European Environment Agency, Air quality in Europe — 2015 report Luxembourg: Publications Office of the European Union, 2015 ISBN 978-92-9213-702-1 ISSN 1977-8449
[3] Mandavilli S, Russell E and Rys M 2003 *Impact of Modern Roundabouts on Vehicular Emissions*, Proceedings of the 2003 Mid-Continent Transportation Research Symposium, Ames, Iowa, August 2003, pp 1-10
[4] ***http://www.baiamare.ro/ro/
[5] ***http://ec.europa.eu/environment/air/quality/legislation/time_extensions.htm

IOP Conf. Series: Materials Science and Engineering **200** (2017) 012028 doi:10.1088/1757-899X/200/1/012028

Migration of optical core network to next generation networks – Carrier Grade Ethernet Optical Transport Network

D Glamočanin

Mtel a.d. Banja Luka, Bosnia and Hercegovina

E-mail: dusanka.glamocanin@mtel.ba

Abstract. In order to maintain the continuity of the telecom operators' network construction, while monitoring development needs, increasing customers' demands and application of technological improvements, it is necessary to migrate optical transport core network to the next generation networks - Carrier Grade Ethernet Optical Transport Network (OTN CE). The primary objective of OTN CE is to realize an environment that is based solely on the switching in the optical domain, i.e. the realization of transparent optical networks and optical switching to the second layer of ISO / OSI model. The realization of such a network provides opportunities for further development of existing, but also technologically more demanding, new services. It is also a prerequisite to provide higher scalability, reliability, security and quality of QoS service, as well as prerequisites for the establishment of SLA (Service Level Agreement) for existing services, especially traffic in real time. This study aims to clarify the proposed model, which has the potential to be eventually adjusted in accordance with new scientific knowledge in this field as well as market requirements.

1. OTN/DWDM network

The current topology of main cable OTN/DWDM transport network of one operator in Bosnia and Herzegovina is shown in the Figure 1.

As it is shown, all equipment is from one manufacturer (Alcatel-Lucent), and on most locations is mounted equipment 1830PSS-32. This equipment offers a significant capacity and performance to transfer large amounts of data over long distances in order to fulfill needs of a growing number of services and capacity. They should be used to add equipment that prepare transport network for further development of existing, but also technologically more demanding, new services. Also, it is necessary to realize the preconditions for the support real-time services.

At four locations, beside DWDM equipment, IP / MPLS core routers are installed.

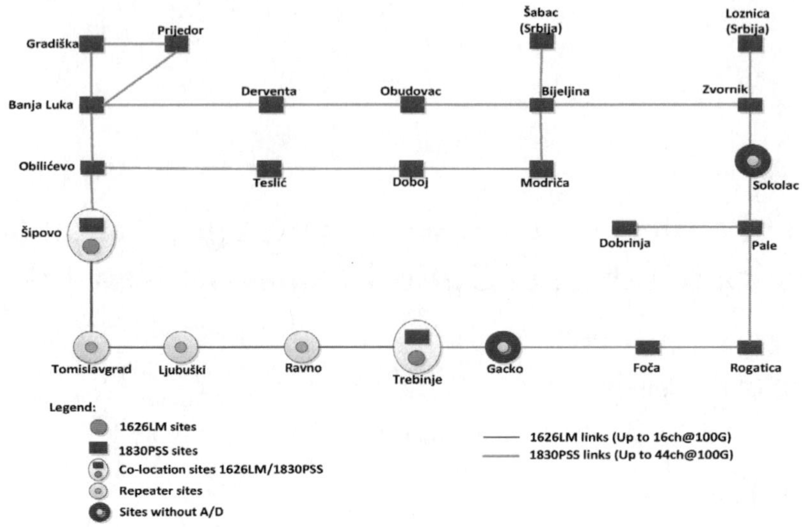

Figure 1. Topology of OTN/DWDM network

1.1. Existing IP/MPLS of transport network

Architecture is typical two-layer MPLS network with four routers in core and eighteen edge routers, as shown in Figure 2.

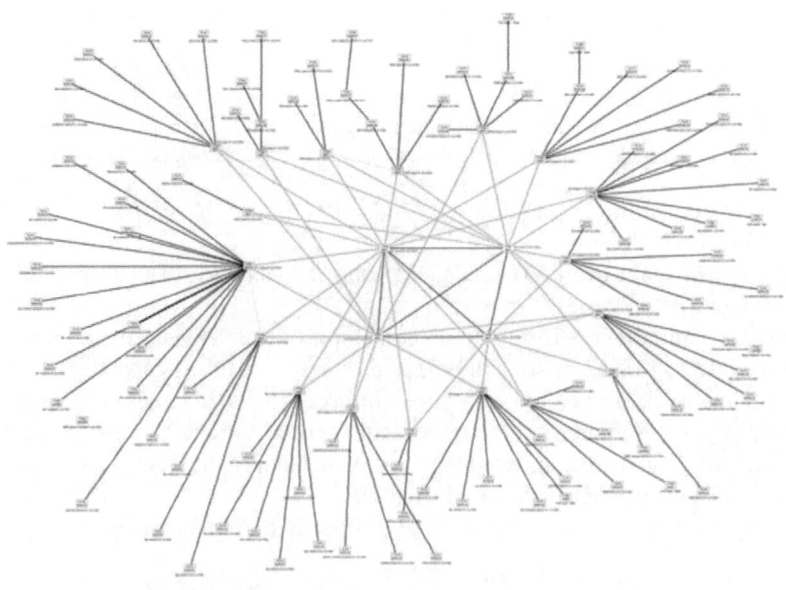

Figure 2. Topology of transport MPLS network

It can be seen that topology of grid type is from core to edge routers. Each edge router has an interface to two core routers, and there is a double connection as the basis of physical reliability.
Because tandem network elements increase network price, operators provide a different level tolerance cancellation for edge routers and network core. For example, core network is protected from router failure with grid topology, as shown in Figure 2. In this way, alternative routes are quickly establish and prevent losses. Network core use additional routers and links in order to ensure fault tolerance.

IOP Conf. Series: Materials Science and Engineering **200** (2017) 012028 doi:10.1088/1757-899X/200/1/012028

In contrast, thousands of users often have connection via one end router which is the most sensitive part of network. In that case, instead of additional routers and links, pairs of panel control processors are used, duplicate liner board are used or Tandem links are used (such as SDH APS - Synchronous Digital Hierarchy Automatic Protection Switch) to provide tolerance [1], [2].

In addition, edge routers could be directly linked to each other, and that should be aim, where it is economically feasible.

According to one-year study IP router core network availability, conducted by University of Michigan in the study of regional IP services provider, it is found that average cancellation time ruter interface is 955 minutes per year, which present interface availability of just 0,998 [3].

As reference value is taken cancellation time of 5.2 minutes per year, because it gives the availability 99.999% of time [4]. This is standard value of availability for telecommunication systems.

In listed study reason of network cancel is reached, and in 23% because router cancellation (software or hardware, "denial of service attack"), 32% due to a link termination (termination of optical cables, network congestion), 36% because router maintenance (error while configuring, upgrading software or hardware), 9% from other reasons [5].

From these data it can be observed that most networks cancellation hapened as a result of the router cancellation. According to another study, failures of routers software are the biggest reason (35%) of router cancellation. More precisely, the control plane failures represent 60% of software cancellation [6]. Calculations of routers reliability and availability and their control levels, indicate the necessity of software and hardware control level redundancy, in order to obtain a high level of services availability [7], [8].

According to information obtained from operators sources, over 90% mobile and fixed networks traffic is done via the MPLS network, mainly using VPLS (Virtual Private LAN Service) mechanism, typical for traffic management on line-level in Ethernet type networks.

A key part of physical layer IP operator's network is fiber network, presented in Figure 1.

Integrated Packet Transport is based on the integration of the Alcatel-Lucent Service Router Operating System (SR OS) into multiservice WDM Alcatel-Lucent 1830 Photonic Services Switch (PSS). Integrated Packet Transport (IPT) is leaning on the end-to-end transport layer 2 architecture, using a common management model and operational guidance from core to user equipment.

2. Technical demands

Taking into account current state of the operator’s transport network, and the need to maintain continuity of further construction, with monitoring of development needs and the application of technological improvements, idea is to do a project of migration existing main optical networks to next generation networks - Carrier Grade Ethernet Optical Transport Network (OTN CE), which implementation would be a prerequisite for the development of many other operator’s services.

The primary objective of OTN CE is to realize an environment that is based solely on the switching in the optical domain, i.e. the realization of transparent optical networks and optical switching to the second layer of ISO / OSI model.

First, it is necessary to make a detailed analysis of equipment condition in the operator’s transport network and choose one of the possible approaches to migration: upgrading existing transport networks or building parallel overlapping networks. For the purposes of the above it is necessary to do the following activities:

- Basic authentication, ie. measurement of all throughput parameters, availability and QoS (Baseline) which has to be done before the implementation of new applications, as well after implementation, because of measurement of expectations for new solution. Result of these activities should be information about performance, particularly critical infrastructure segments that should be urgently addressed;
- Setting up technical model OTN. Due to the prevailing representation of equipment from one manufacturer's in transport operator’s network, it is assumed equipment expansion from same

manufacturer. Architectural basis for migration approach is existing IP/MPLS/SDH/DWDM network.
- Setting up financial model OTN. Cost model include investment costs (CAPEX) for the purchase of basic equipment, implementation costs of the installed devices (IMPEX) and operating expenses (OPEX).

The new technical solution should indicate operators ability to provide greater scalability, reliability, security and quality of service QoS, as prerequisites for the establishment of SLA (Service Level Agreement) for existing services, especially traffic in real time.

It is necessary to provide the following functionality [9]:
- Resistance to certain errors of network sections - Possibility of network dynamic reconfiguration, without flow limitation for end-users on primary routes, backup routes and ability to overcome multiple sequential or parallel faults on connections in the network topology.
- Flexibility - Ability to dynamically or programmed network configuration or changing topology depending on current needs.
- Priorities - Possibility of higher priority channels transmission, instead of the channel with lower priority, in case of network failure.
- Project should be done in accordance with applicable technical regulations, complying with the current plans for cable and wireless access networks construction and operators service platforms.

3. Set up a technical model for introduction Carrier Ethernet OTN transport network

The operator should have strategic vision to realize the main network center in one central location and another redundand center.

In these centers all service platforms equipment hould be concentrate, OSS and BSS. Also, these centers would contain strong OTN/CE nodes from which real-time traffic from regional hubs should be direct to service platforms (IPTV hedendom, Softswitch, SGSN, GGSN, etc.). These hubs will be based on the 1830 PSS switches.

This assumes and aimes for a solution where each switch PSS 1830 is going to be connected with two pairs of optical fibers (by separate routes) to each of the central 1830 switches.

The basic idea of real time services migration project is in first phase, gather real-time traffic at the edge PSS 1830, instead of the edge 7750 SAR, and direct traffic with special fiber on two central PSS 1830 in a central location and the second redundant center . The same traffic would be directed to the edge MPLS routers 7750 SAR with one 10 Gbps port, instead of many GE and FE ports that are used in the present state on some locations. MPLS shuold networks serve as a backup in case of failure in the OTN/CE network.

Control plane should be based on ASON/GMPLS management, from the physical to the transmission network layer. In accordance with the declared capabilities of 1830 PSS switch, Ethernet frames switching can be based on PBB (Provider Backbone Bridging) combined with ELS (Ethernet Label Switching) and GMPLS techniques.

In the second phase, OTN/CE network should be spread to DSLAM/MSAN nodes and base stations. There, existing hubs (TSS 7210 and 7705 SAR and other) would be replaced or upgrade to give OTN/CE outputs to edge nodes 1830 PSS. Existing cascade DSLAM configurations would be replaced EPON structures, where from EPON nodes shuold come out OTN/CE. For better utilization of the existing optical fiber in the access network can be used ROADM 1626LM, which would be dismantled from the network after the introduction of 1830 PSS switches.

Existing TDM transmission systems (SDH, PDH) would be transformed into EoSDH and EoPDH systems. On existing SDH and PDH multiplex equipment would add appropriate adapters which preserve existing investments in TDM. This would enable CE diagnostics and synchronization to endpoints, unified implementation and management services on packet technology.

IOP Conf. Series: Materials Science and Engineering **200** (2017) 012028 doi:10.1088/1757-899X/200/1/012028

In the third phase, in addition to existing DSLAM, based on the ATM over ADSL to ADSL modems or concentrators, would set CE-based DSLAM, from which CE over ADSL go to CE-based user-Gateway, which beside telecommunications terminals (STB IPTV and VoIP phone), should support telemetry devices and smart meters (water, electric and gas). This will provide the MAC operational management, synchronization and diagnostics to the end devices.

3.1. Technical model of CE OTN network

As noted above, future transport network topology model will be redundant PetaWeb Model [10]. This means that should form a star redundant structure (double star with interrelated major nodes connected with all boundary nodes). Transport network architecture practicaly adjusts for future Software Defined Network (SDN) network [11].

This structure will be based on 100 Gbps links which connect two central nodes 1830 PSS-32 with eighteen regional nodes 1830 (PSS 32, 16, 4 and 1). It is similar with typical backhauling architectures, as shown in Figure 3.

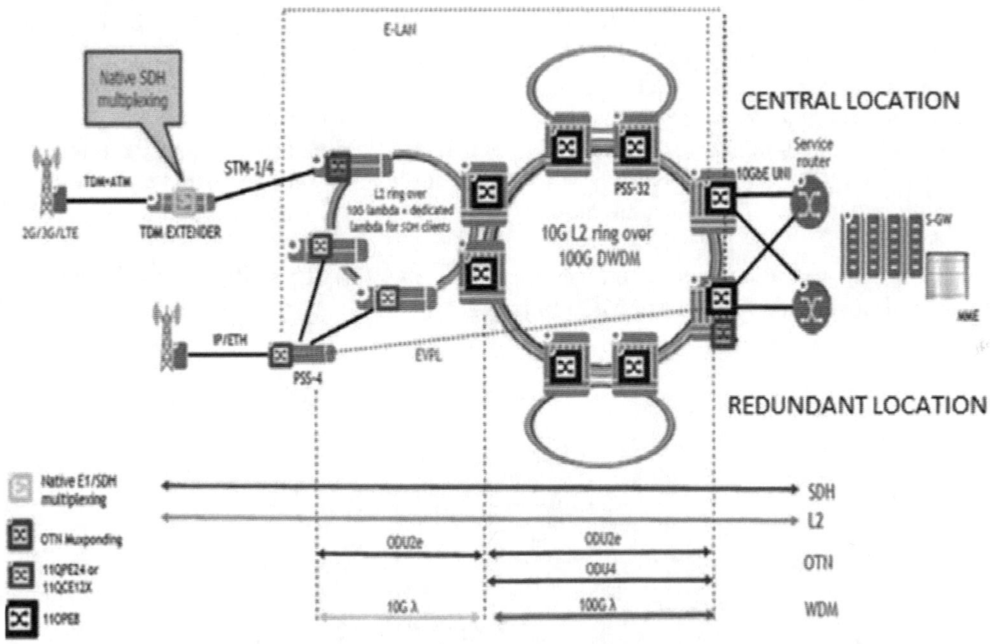

Figure 3. Topology of transport MPLS network Model of future CE-OTN networks as backhauling network for RT services

On the right side is, with core elements of the mobile network, represented total service platform which will in future be concentrated core elements of all network operators, including OSS/BSS. Elements of service platform will be with Ethernet ports connected to two main 1830 PSS-32 switches. On the two main 1830 PSS-32 switches (central location and the redundant center) are connected two main service routers 7950 XRS, which will have all functions of BNG's (Border Network Gateway) for real time services. They are MEF certified, so they are going to be connected on 1830 PSS with two OTU-4, 100 Gbps.

On the left side of larger ring in Figure 3 are shown nodes 1830 PSS in regional centers that will be located along existing MPLS edge routers. All access lines on edge MPLS lines, coming from the DSLAM/MSAN or the base stations (Figure 2), directly or agregate (via access routers or 7705 SAR

7210 SAS switches), were 1Gb Ethernet. They should be transferred to Carrier Ethernet. This will enable Carrier Ethernet OAM and synchronization possibilities to end devices.

MPLS transport network transmit only non real-time Internet traffic. In new architecture, every edge router would through its edge 1830 switch connect to every other router in the network. This would virtually establish connection between edge routers each with each, which would achieve far greater reliability than existing. In case of failure of any MPLS routers, his role can take over any other regional router.

With releasing MPLS transport network from real time traffic, it could be much more favorable to offer Internet access, hosting, virtualization, data centers and so on. It will free a large number of GbE ports on the edge MPLS routers for this purpose. In the next 5 years is expected migration of the majority of internet access on mobile devices, through micro, pico and femto cells. This new blow broadband Internet traffic will be able to accept MPLS transport network. MPLS transport network, also can serve as a backup in case of failure of the Carrier Ethernet OTN network.

3.2. 1830 PSS switch as a key node in the OTN-CE network

Integrated Packet Transport is based on the integration of Alcatel-Lucent Service Router Operating System (SR OS) into multiservice WDM Alcatel-Lucent 1830 Photonic Services Switch (PSS). Integrated Packet Transport (IPT) is leaning on the end-to-end transport layer 2 architecture, using a common management model and operational guidance from core to user equipment.

Need for sophisticated point-to-point and multipoint services is realized through the integration of Ethernet (L2) into WDM. Applying statistical traffic multiplex and services from multiple locations provides better utilization of bandwidth/wavelength and supports delivery of E-LAN and E-Tree multipoint services.

1830 PSS switch has a key role in integrated packet transport IPT, offering by Alcatel-Lucent, which is base of migration proposal, shown in Figure 4.

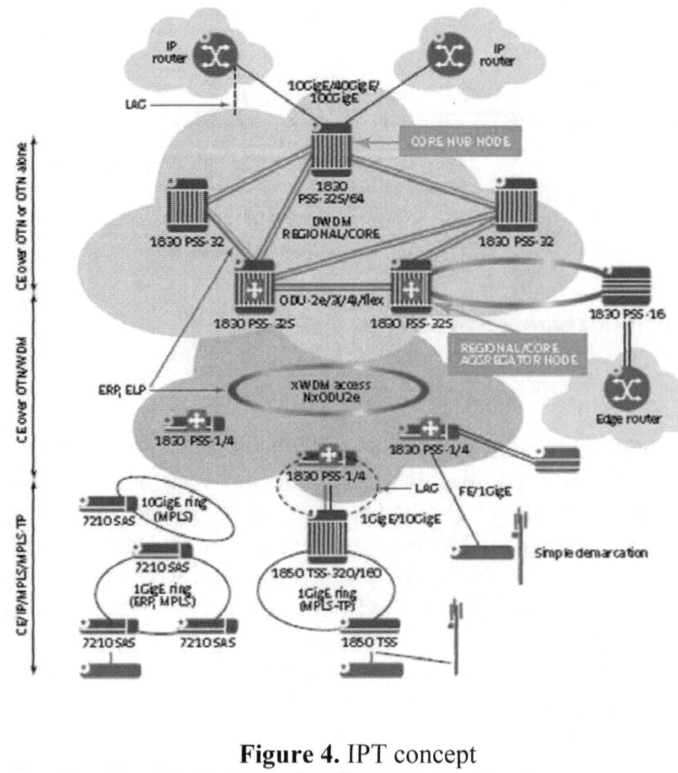

Figure 4. IPT concept

Service providers must develop its network to fulfill demands for bandwidth due to increased deployment of FTTx networks. Growth of bandwidth, as a result of the transition from Digital Subscriber Line (DSL) to Gigabit - capable Passive Optical Network (EPON, GPON), requires implementation WDM aggregation rings, as shown in Figure 4.

By introducing 1830 PSS in regional hubs, operator will be able to dismantle a significant amount of existing ROADM 1626LM because 1830 PSS has already builted Add-drop multiplexers. Dismantled equipment will be used to resolve aggregation rings in the access network, where necessary.

The key electro-optical elements of 1830 PSS switch are shown in Figure 5:

- Point-to-point Ethernet service muxponders,
- Ethernet switching muxponders,
- Ethernet switching OTN client cards,
- Network interface demarcation devices, miniaturized in a Small Form Factor Pluggables (SFPs).

Point-to-point (P2P) muxponders perform L2 statistical multiplex in order to aggregatie Gigabit Ethernet (GigE) traffic in the access network. Improved P2P muxponders offer greater scalability of services and a higher level of Maintenance and Operational Management (OAM), protection and Synchronous Ethernet possibility (SyncE).
Options: EPL, EVPL services, flexible classification, MEF input range profile, CE OAM (IEEE 802.1ag and ITU-T Y.1731), Performance monitoring (PM), Link aggregation (LAG), SyncE, 1588v2 P2P phase timing.

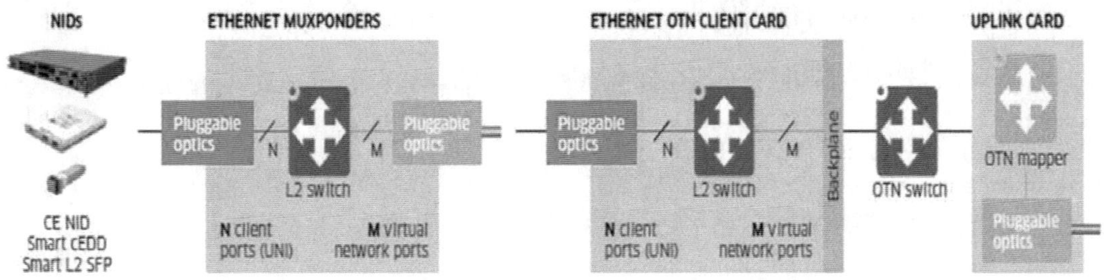

Figure 5. Key E/O elements of 1830 PSS switch

Ethernet switching muxponders and Ethernet switching OTN client cards are guided by SR OS. They perform L2 switching in order to aggregate traffic from the access network and the optional higher-capacity line, for providing optimal scalability aggregation and optional commutation of clients and lines for optimization of OTN applications, for example between the router and 100G WDM lines. Beside these features P2P muxponders have: full opportunities of networking with other node elements that support Carrier Ethernet (Alcatel-Lucent 7750 Service Router, Alcatel-Lucent 1850 Ethernet Service Switch, Alcatel-Lucent 7210 Service Access Switch), E-Line, E-LAN and E-Tree services, SR OS service model (service access point [SAP], virtual private LAN services [VPLS]), MEF CE 2.0, QoS classification at the entrance, measuring and policing per service class (class 8, 8 rows per port with deployment and preventing congestion).

3.3. Subnet models for real-time services

3.3.1. Mobilne Backhaul. MPLS network should be unloaded from mobile traffic or it should be serve as a backup in case of failure DWDM Ethernet mobile backhaul network. SDH lines can be directly implement into the 1830 PSS and native transmitt through fiber network to controller or service platform. In an effort to support migration from TDM to packet solution for all mobile network

IOP Conf. Series: Materials Science and Engineering **200** (2017) 012028 doi:10.1088/1757-899X/200/1/012028

traffic, these backhaul architecture resolves challenges such as: high resistance/availability, highly precise synchronization of clocks and management from end to end. This is achieved by supporting earlier and packet synchronization approaches, such as Synchronous Ethernet SyncE (frequency) and IEEE 1588v2 (phase/time) synchronization. Beside this, solution provides advanced troubleshooting and SLA (Service Level Agreement) monitoring.

3.3.2. Broadband (BB) backhaul and Triple Play distribution. Convergences of packet and WDM in a single platform improves economics from CAPEX perspective - based on reducing number of ports and network elements, as well as due to the increased utilization of fiber. Operational efficiency is also increased because of a large number of services on the same transmission basis.

1830 PSS is essential for this purpose because they are more efficient than purely electronic CE switches with multiple 10Gbps capacity with external filters, colored optic, and multiple stacked 10G WDM rings.

PSS 1830 features for the purpose of broadband aggregation:

- Termination: IP DSLAM, GPON OLT, CMTS / CCAP cable backhaul
- E-LAN and E-Tree perfectly suitable for multicast and dual of homming router
- A wide range of options for protecting interconnection
- IPTV (streaming and on-demand video distribution) in combination with internet access for residential users, and LAN interconnect for enterprise
- Better utilization of wavelength and bandwidth using L2 multicast and IGMP snooping, the IGMP Proxy and Fast Leave for complete optimization and QoE.

4. OTN CE network investment

4.1. Current status

Operator has about 20 locations (nodes) at which aggregation of traffic is carried. Since the aforementioned locations at the same time are part of the MPLS and DWDM networks as their hub, they should be used as a hubs of ONT network.

Currently, corporate services are provisioning differently on various network levels. The largest number of network services are L2VPN (usually VPLS) type, shown in Figure 6.

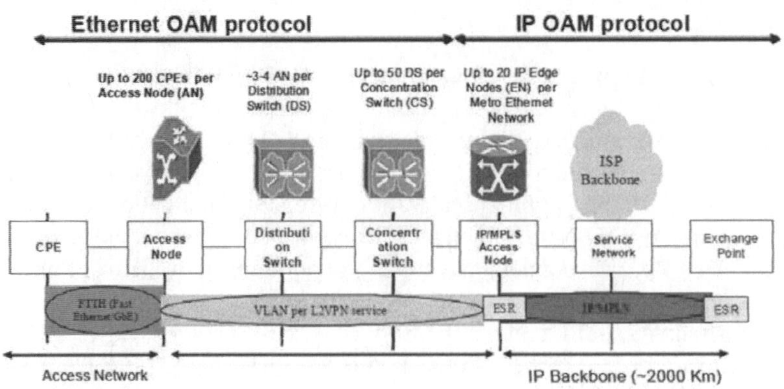

Figure 6. Referent network for L2VPN services

As can be seen, each P2P connection between two access nodes use L2 VLAN in access network to the access node (DSLAM or OLT), L2-MPLS (VPLS) from the access node to the edge router in the MPLS network, and tunnel in the IP/MPLS network.

This approach requires a lot of manual configuration.

On the other hand, it is used many different protocols for maintenance and operational guidance in any part of the network. The number of network elements (potential point of failure) is also great. Therefore, the complexity of operational management and maintenance (OAM) L2VPN service is currently unresolved issue.

4.2. *Future status*

As shown in Figure 7, with introduction of L2 switching technology, as it was proposed, it should be get unique end-to-end OAM system, as well as multi-domain control plane mechanisms, which enable the dynamic establishment, recovery or realocation GMPLS tunnels between different access domains.

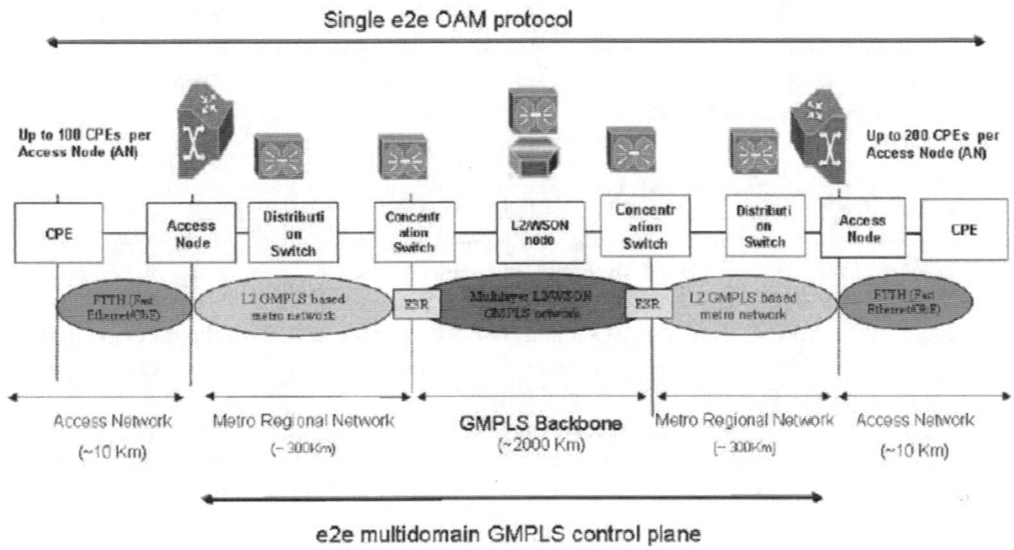

Figure 7. Unique L2 switched network with GMPLS

Beside this, PSS 1830 node combines switchings on the first (optical) and on the second electrical level (OTN-CE) with one ASON/GMPLS control plane. With this it is achieved significantly reducing management complexity, maintenance and operational management, on a much smaller number of network elements.

For ease of handling of L2 switches, operator's expert teams can concentrate in two operational centers, one center and one redundant center, where daily maintenance in rest of the network can be left to the technician teams. This will reduce OPEX by 50%.

5. Conclusions

The first paragraph after a heading is not indented (Bodytext style). At the present time, the development of telecom operators is often dictated by the requirements of the user. It is necessary to maintain the continuity of the telecom operators' network construction and application of the latest technological solutions to meet the changing requirements of users. In this connection, it is necessary to migrate optical transport core networks to the next generation networks - Carrier Grade Ethernet Optical Transport Network (OTN CE). These are transparent optical network-switched exclusively to the optical domain - the second layer ISO / OSI model. Such networks enables higher scalability, reliability, security and quality of service QoS, as well as prerequisites for the establishment of SLA (Service Level Agreement) for existing services, especially traffic in real time.

The idea is, in the first phase, gather real-time traffic at the edge OTN/CE nodes and and direct traffic with special fiber on two central OTN/CE nodes in a central location and the second redundant center. In the second phase, the OTN/CE network should be spread to the DSLAM / MSAN nodes and base stations. Existing nodes (TSS, 7210 and 7705 SAR and other) would be replaced or upgrade to give OTN/CE outputs to edge OTN/CE nodes. In the third phase would be set CE-based DSLAM's, which bind to CE-based Gateway user. This will provide the MAC operational management, synchronization and diagnostics to the end devices.

References

[1] Zhou D and Subramanian S 2000 *Survivability in optical networks,* In IEEE Network, pp 16–23

[2] Anelli P and Soto M *Evaluation of the APS protocol for SDH rings reconfiguration*, http://ieeexplore.ieee.org/abstract/document/789674

[3] Craig Labovitz, Abha Ahuja and Farnam Jahanian *Experimental Study of Internet Stability and Backbone Failures,* University of Michigan, Department of Electrical Engineering and Computer Science, 1301 Beal Ave, Ann Arbor, Michigan 48109-2122

[4] Glamočanin D 2015 *Levels of application and measurements to provide quality of services*, International Conference on Social and Tehnological Development STED 2015

[5] ****Survivable Network Design*, http://www.pitt.edu/~dtipper/2110/Slides12.pdf, https://www.coursehero.com/file/5963244/Slides12

[6] ***http://www.ciscopress.com/articles/article.asp?p=361409&seqNum=4

[7] Yi C, Abraham J, Afanasyev A, Wang L, Zhang B and Zhang L 2013 *On the Role of Routing in Named Data Networking,* University of Arizona, UCLA, University of Memphis, NDN, Technical Report NDN-0016

[8] Capone A, Cascone C, Nguyen A Q T and Sanso B 2015 *Detour Planning for Fast and Reliable Failure Recovery in SDN with OpenState,* Dipartimento di Elettronica, Informazione e Bioingegneria, Politecnico di Milano, Italy

[9] Glamočanin D 2016 *Service Level Agreement - SLA, Service reliability and quality of system in telecomunication*, The 39th International Convention on Information and Communication Technology, Electronics and Microelectronics, MIPRO 2016, Opatija, Croatia, May 30-June 3, pp 755-760

[10] Huang A, Kabranov O and Makrakis D *Performance analysis of the PetaWeb optical network architecture,* http://ieeexplore.ieee.org/document/1012971

[11] King D, Rotsos C, Aguado A, Gaorgalas N and Lopez V 2002 *The Software Defined Transport Network: Fundamentals, findings and futures,* http://ieeexplore.ieee.org/document/7550669

IOP Conf. Series: Materials Science and Engineering **200** (2017) 012029 doi:10.1088/1757-899X/200/1/012029

Grain refinement of 7075Al alloy microstructures by inoculation with Al-Ti-B master alloy

V Hotea[1], J Juhasz[1] and F Cadar[2]

[1]Technical University of Cluj-Napoca, North University Center of Baia Mare, Faculty of Engineering, Department of Mineral Resources Engineering, Materials and Environment, dr.V. Babes str., no. 62A, 430083, Baia Mare, Romania
[2]Universal Alloy Corporation Europe SRL, Dumbravita, Romania

E-mail: vasilehotea50@yahoo.com

Abstract. This paper aims to bring some clarification on grain refinement and modification of high strength alloys used in aerospace technique. In this work it was taken into account 7075 Al alloy, and the melt treatment was carried out by placing in the form of master alloy wire ternary AlTiB the casting trough at 730°C. The morphology of the resulting microstructures was characterized by optical microscopy. Micrographs unfinished and finished with pre-alloy containing ternary Al5Ti1B evidence fine crystals, crystal containing no columnar structure and highlights the size of the dendrites, and intermetallic phases occurring at grain boundaries in Al-Zn-Mg -Cu alloy. It has been found that these intermetallic compounds are $MgZn_2$ type. AlTiB master alloys finishing ensures a fine eutectic structure, which determines the properties of hardware and improving the mechanical properties of aluminum alloys used in aeronautical engineering.

1. Introduction

One of the most common aluminum alloys used for structural applications is alloy 7075 T6 from Al-Zn-Mg alloys series due to its attractive properties, such as low density, high resistance ductility, toughness and fatigue resistance [1-4]. This alloy is extensively used in aircraft structural parts and other structural applications highly demanded from mechanical point of view and corrosion.

The structure of the grain is an important feature and easily observable in casting high-strength aluminum alloys. Refining is one of the prevalent techniques in strength and hardness control of processing by hot plastic deformation. Refining and finishing grains by adding ternary master alloys such as Al-Ti-B, mentioned as inoculants is the best way due to its simplicity [5], [6].

The specialized literature indicates a number of potential finishers grain for aluminum alloys, but some of them have a limited industrial application [7], [8].

Spittle [9] and others have studied a number of finishers and they have shown that at Ti /B ratios close to the stoichiometric ratio, a lower finishing is achieved than when using an excess of Ti or B. The industrial practice suggests the addition of 0,1% excess Ti to obtain good performance for stoichiometric finishers [10], [11].

Easton and John have tested the effect of increasing titanium content in the maintenance of the content of TiB2, finding a negligible change of the grain size [12].

In recent years, the finisher products market has increased by the special use of TiB Alloy and Strobloy. Using Alloy TiB for hypoeutectic alloys used in foundries allowed carrying out studies referring to porosity and lack sedimentation of the borides and they are available as coiled wire, rod shaped cut and tablets.

Grain finishers, added to liquid alloys of aluminum may provide nucleation and favor the production of fine grain and uniform material [13]. Research on new generations of master alloys finishers / modifiers are still young and require a great deal of experimentation to establish the optimal pre-alloy complex and for studying the kinetics and the mechanism of the amendment process and finishing, practically running concurrently. There is also needed clarification on the role of alloying elements and impurities on the effect of titanium and boron introduced to melt together in the form of pre-alloy complex. In this context, the present work aims to bring some clarification on finishing/changing of medium and high strength alloys used in aeronautical engineering.

2. Experimental procedure

2.1. Materials

In this work it was considered aluminum alloy 7075 and the melt treatment was carried out by placing in the form of ternary alloy wire AlTiB directly to the pouring spout at 730 ^{0}C. Microstructure morphology results were characterized by optical microscopy. In this work it was considered 7075 aluminum alloy according to the American Association of Al classification, the samples being taken from UAC Europe SRL (Dumbravita). This alloy is commonly used in the aeronautical industry for the high strength structural components (fuselage, wings). The alloy composition is shown in Table 1 [14].

Table 1. The chemical composition of the 7075 alloy (% wt.)

Element	Al	Mg	Si	Mn	Cu	Zn	Fe	Cr	Ti	Others
% wt.	difference	2.1-2.9	0.40	0.30	1.2-2	5.1-6.1	0.50	0.18	0.20	0.5

The treatment of Al 7075 melt alloy was achieved by introducing AlTi5B1 ternary master alloy wire on the pouring spout at 730 °C. The parameters of the grain finishing process in Al-Zn-Mg alloys system are shown in Table 2.

Table 2. Process parameters for finishing Al 7075 alloy with pre-alloy AlTiB [13]

Phase	Process parameters	
	Temperature, ^{0}C	Time, min.
Alloying with pre-alloy AlTiB	730	10
The degassing with argon (55 l/min.)	730	12
Ternary pre-alloy $AlTi_5B1$ (2,8 kg/t)	5% Ti	1% B

By treating the metal melt with AlTi5B1 pre-alloy occurs:

- Improvement of the mechanical properties and refractoriness;
- Improvement of the casting properties (high fluidity).
- Shrinkage porosity reduction and increased density;
- Easier achievement of the uniformity response to heat treatment;
- Improved extrudability and processability by chip cutting, surface finishing due to the fine structure.

In principle, the master alloy is (Figure 1) as a wire passing through a feeder or drive system, a guide tube for casting gutter where it is injected.

IOP Conf. Series: Materials Science and Engineering **200** (2017) 012029 doi:10.1088/1757-899X/200/1/012029

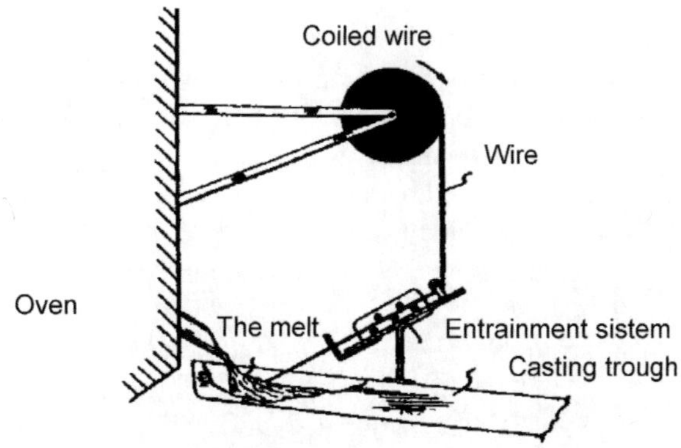

Figure 1. The scheme of finishing with titanium as Ti-B master alloys of Al-Zn-Mg alloys

In Figure 2 it is sequentially shown the practice of supply and injecting process on the pouring spout at the UAC Europe SRL Company.

Figure 2. Supply system of the Al-Ti-B master alloy (Source: UAC Europe SRL)

$AlTi_5B1$ finisher as wire is more efficient than the tablets as they have higher dissolution speed in the melt, leading to an effective grain finish at solidification, within 30 s after the addition. It is available as coiled wire of 9.5 mm (3/8 in) diameter in standard coils of 180 kg per pallet packaged in three coils protected from moisture. For the addition of $AlTi_5B1$ pre-alloy finisher, the titanium content recommended and the added amount are listed in Table 3.

Table 3. The content of titanium and the added amount recommended [12]

Zn	% mass	5	5.2	5.4	5.6	6
Ti	% mass	0,05	0,04	0,03	0,02	0,01
$AlTi_5B1$	g/100 kg	1000	800	600	400	200

If cast at high speed cooling (with thin sections or pressure casting), a smaller amount of finisher is added. In some cases, it is only necessary to prevent the formation of large grains while in other cases it is necessary to form a fine structure to optimize the properties of molded products. In some alloys there are residual titanium content from recycled materials. If the percentage of titanium exceeds 0.1%, benefits could be easily achieved by adding pre-alloy TiBAl.

It is necessary to avoid excessive addition to prevent the formation of agglomerates borides and / or aluminides which may lead to their sedimentation in long maintenance issues and problems in chip cutting processing. For titanium addition to the content above 0.1% is recommended inoculation of Al-Ti master alloys and products 100% ALTABTM.

2.2. Optical metallography

The castings samples for chemical and microscopic analysis, shown in Figure 3 were made and taken by the metallographic process: the selection of samples before and after finishing with Al-Ti-B, grinding, polishing and etching. Microscopy samples were cleaned with grit sandpaper (400-1200) and polished with paste (10, 7, 5, 3.5, 2.5) followed by etching with a Keller freshly prepared solution (1.0 ml HF, 1.5 ml HCl, 2.5 ml HNO_3 and 95.0 ml H_2O).

Figure 3. The collected samples for chemical and microscopic analysis

The morphology of the microstructures before and after finishing for two magnification shown in Figures 4-5 were characterized by optical microscopy.

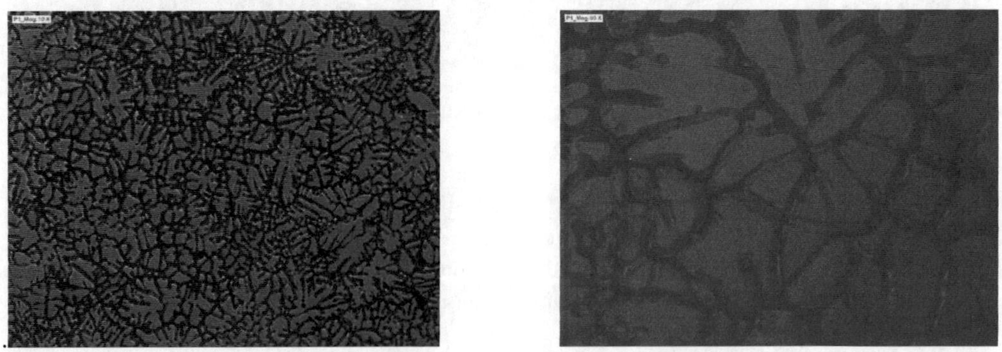

(a) **(b)**

Figure 4. Optical micrographs: a) before treatment, $AlTi_5B1$ finisher; b) after treatment (100x)

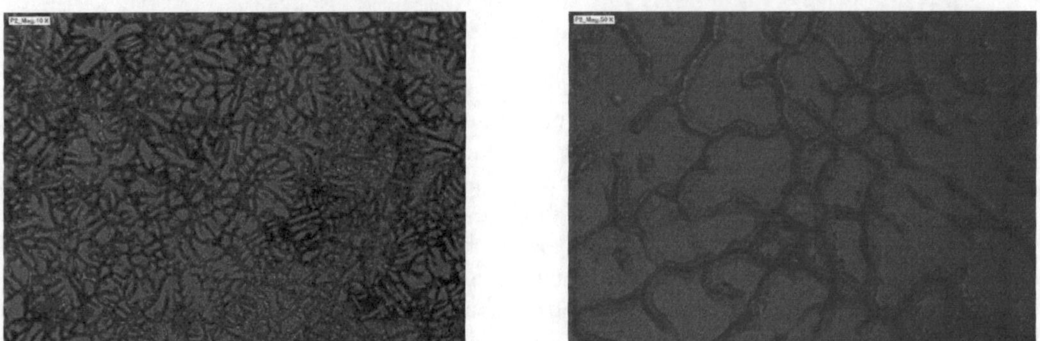

(a) **(b)**

Figure 5. Optical micrographs: a) before treatment, $AlTi_5B1$ finisher; b) after treatment (500x)

The Micrographs of the samples obtained by optical microscopy (Leica DM2500 microscope) unfinished and finished with AlTi5B1 ternary pre-alloy contain fine crystals, the structure containing no columnar crystals and highlights the size of the dendrites and intermetallic phases occurring at grain boundaries in the alloy Al-Zn-Mg-Cu. It has been found that these intermetallic compounds are MgZn2 type. These compounds may form during solidification below the solidus as cast aluminum alloy 7xxx series due to the metals redistributing (Mg, Zn) in solidification [15].

In the case of cast molding, at high cooling speed (with thin sections or die casting), a small amount of the finisher is added. In some cases, it is only necessary to prevent the formation of large grains while in other cases it is necessary to form a fine structure to optimize the properties of molded products.

In the practice of aluminum alloys grain finishing there are a series of issues such as: decreasing degree of the finishing time due to the particles agglomeration and sedimentation, the "poisoning" phenomenon (eg. due to the zirconium presence, dissolved in the melt) and the performance of the operation [16]. The borides sedimentation appears in unstirred melts due to the high density of TiB2 (4.5 g / cm^3) compared to the density of the liquid aluminum alloy (~ 2.3 g / cm^3).

The borides agglomeration reduces the performance of the grain finishing by agglomerates fast decanting. Partial reactivation can be achieved by rapid stirring of the melt, which, however, it is sometimes is difficult to achieve. Also, the borides agglomeration can block the ceramic filters.

The "poisoning" of the grain finisher is a very serious problem because it can be produce very quick and generally it is irreversible. "Poisoning" most often caused by Zr is considered as a result of the Ti substitution from $TiAl_3$ or TiB_2 structure with the formation phase $(Zr,Ti)Al_3$ or $(Zr,Ti)B_2$ having unfavorable network parameters for nucleating.

3. Conclusions

- ✓ Finishing is a treatment method of the melts before casting, which ensures a fine eutectic structure, which leads to an improvement of molding properties and the physical-mechanical properties of aluminum alloys used in aeronautical engineering.

- ✓ In recent years there has been a tendency of modifying and finishing the high strength aluminum alloys, in a single step, by treating the melt with complex master alloys containing particles of grain finishers ($TiAl_3$ and TiB_2).

- ✓ In this paper, a type of pre-alloy ternary system Al-Ti-B has been characterized, with reduced Ti and B, as well as treating Al-Zn-Mg liquid alloys (alloy high strength Al 7075) in bringing together the grain finishing and modifying operations to improve the mechanical characteristics and the cost of the molded products.

- ✓ The treatment in a liquid state of 7075 alloy with $AlTi_5B1$pre-alloy and its characterization with optical microscopy, using Leica DM2500 optical microscope, revealed the unfinished samples microstructure and the one finished with pre-alloy ternary $AlTi_5B1$, that contains fine crystals, the structure containing no columnar grain and highlighting the size of the dendrites and intermetallic phases that occur in grain boundaries in the of Al-Zn-Mg-Cu alloy. It has been found that these intermetallic compounds are $MgZn_2$ type. These compounds may form during solidification below the solidus as cast aluminum alloy 7xxx series, due to metals (Mg, Zn) redistributing in solidification.

References

[1] Heinz A and Haszler A 2000 Recent development in aluminum alloys for aerospace applications, *Materials Science and Engineering A* **280**(1) 102-7
[2] Williams J C and Starke Junior E A 2003 Progress in structural materials for aerospace systems, *Acta Materialia* **51**(19) 5775-5799

[3] Clark R, Coughran B, Traina I, Hernandez A, Scheck T, Etuk C et al. 2005 On the correlation of mechanical and physical properties of 7075-T6 Al alloy, *Engineering Failure Analysis* **12** 520-526
[4] Adeyemi Dayo Isadare, Bolaji Aremob, Mosobalaje Oyebamiji Adeoye, Oluyemi John Olawale and Moshood Dehinde Shittu 2013 Effect of Heat Treatment on Some Mechanical Properties of 7075 Aluminium Alloy, *Materials Research* **16**(1) 190-194
[5] Crossley F A and Mondolfo L F 1951 Mechanism of grain refinement in aluminum alloys, *Trans AIME* **191** 1143-1148
[6] Mondal D P, Nidhi J H A, Badkul A and Das S 2012 Effect of Al_TiB master alloy addition on microstructure, wear and compressive deformation behaviour of aluminum alloys, *Trans. Nonferrous Met. Soc. China* **22** 1001-1011
[7] Kashyap K T and Chandrashekar T 2001 Effects and mechanisms of grain refinement in aluminium alloys, *Bull. Mater. Sci.* **24**(4) 345-353
[8] Tondel P A 1994 *Grain refinement of hypoeutectic Al-Si foundry alloys*, University of Trondheim, Norway, PhD Thesis
[9] Spittle J A and Keeble J M 1999 The grain refinement of Al7Si alloys with boron-containing refiners, *Light Metals 1999*, Edited by C.E. Eckert (Warrendale, PA: The Minerals, Metals & Materials Society), pp 673-677
[10] Cook R and Spooner S J 1998 Grain refiners for aluminum-silicon casting alloys, Castcon 98 (The Institute of British Foundrymen), pp 61-72
[11] Bondhus E and Sagstad T, *Light Metals 1999*, pp 693-698
[12] Easton M and StJohn D 1999 Grain refinement of aluminium alloys: Part I the nucleant and solute paradigms – a review of the literature, *Metall. and Mater. Trans. A* **30** 1613-1623
[13] Moldovan P 2003 *Alliages non ferreux*, Editura Matrix Rom, Bucuresti, Romania
[14] Mondolfo L F 1976 *Aluminum Alloys: Structure and Properties*, Butterworts, London-Boston
[15] Isadare D A, Aremo B, Adeoye M O, Olawale J O and Shittu M D 2013 Effect of Heat Treatment on Some Mechanical Properties of 7075 Aluminium Alloy, *Mat. Research* **16**(1) 190-194
[16] Zamkotovicz Z, Stuczynski T, Augustyn B, Lech-Grega M and Nezyk W 2004 *Light Metals 2004*, pp 807-816

IOP Conf. Series: Materials Science and Engineering **200** (2017) 012030 doi:10.1088/1757-899X/200/1/012030

Challenges of human resource management in the institutions for care of elderly people

Lj Lepir, D Šćepović and A R Radonjić

University of Banjaluka, Faculty of political science, Department for Social work, Duke Petar Bojović boulevard 1a, Banjaluka 78 000, Bosnia and Herzegovina

E-mail: ljubolepir@gmail.com

Abstract. Human resources are the most important resources of any work organization or institution. They are the bearers of the work process and are key in determining the quality of a product or service. The staff employed in an institution contributes with their work to its business success which is why the institution management should pay close attention to human resources management. Functionally, the successful management of social protection institutions implies the attainment of balance in the process of satisfying interests of beneficiaries, interests of the institution management (founder-owner), and interests of the employees (service providers to elderly people). Interests of beneficiaries (elderly people) who are placed in residential care are reflected in the need for high quality and accessible services. Interests of the institution management are recognized as the need to achieve economically sustainable and profitable institution. An interest of direct service providers (employed caregivers, social workers, etc.) is in the safe premises and good working conditions.

The term "human resources" in institutions of social protection implies overall knowledge, skills, abilities, creative capabilities, motivation, loyalty and personal characteristics owned by employees in the institution. It is the overall intellectual and emotional energy of employees that the management can count on and that can be engaged to achieving the objectives of the institution.

The objectives of human resource management in social protection institutions are related to providing capability with job demands, fulfillment of professional and optimal number of competent workers, improving socio-economic status of employees, ensuring full time engagement of workers, improving the quality of work conditions and work environment, creating and maintaining a flexible and adaptive potential of employees, reducing resistance and increasing the sensitivity of employees to changes in the institution.

The activities carried out in pursuit of the objectives of human resource management in institutions for the elderly are analysis and planning of working positions, recruitment and selection of candidates, socialization, training and professional development of workers, performance assessment and evaluation, reward and motivation, the determination of salaries, information, development of interpersonal relations, supervision and support to employees (prevention of "professional burnout"), ensuring the realization of the legal rights (regularity of salaries and contributions, affairs and records related to employment, health and safety at work etc.)

The absence of these activities or lack of attention to achieving the objectives of human resources management calls in question the basic sense of existence of institutions for care of the elderly, and this is to provide a decent life to an elderly person.

1. Introduction

Demographic trends in contemporary society indicate sudden increase in the proportion of elderly people in the total population. The increased proportion of elderly in the total population imposes the need for well-organized means of social protection. Challenges faced by social protection as a key area of support for the elderly, point to the need to create new management models based on the application of modern methods in planning, organization and management as well as the use of control mechanisms. Today, social protection, especially protection of the elderly, is faced with the need to find optimal ways of management that will accept the fact that the needs for social services are increasing and those system resources, human and material, are insufficient. In such adverse conditions, social protection system tries to look for possible way out through introduction of elements of management, where the control mechanisms are placed in the service of achieving efficiency of overall social actions aimed at the senior citizens.

Placement in social care institutions (homes for the elderly) is even today the most recognized and most common way of systematic care for the elderly. Institutional (residential) placement has become a synonym for systematic care for the elderly. It represents the element of tertiary level of social protection and is usually the last option in the care for the elderly. Residential care has become an indispensable way of caring for the elderly, especially those who are no longer able to care for the most basic life necessities, and who have no help from their relatives. Regardless of more frequent criticism of residential placements, this model of care provides for the most demanding needs of the elderly, especially needs related to long term care and continuous medical supervision. In some situations, residential care is the only possible option of achieving high-quality protection of the elderly.

The institutional form of care for the elderly implies integral protection that is achieved in particularly isolated areas, in conditions of collective placement, under the supervision of professional services and taking into account the specific needs of the elderly caused by the aging process and age status. Institutional form of care for older people is realized in social protection institutions (homes for the elderly) that provide housing, food, clothing, personal hygiene, care and rehabilitation, medical surveillance/supervision, psychological and social support, cultural, entertaining, recreational and occupational activities. In addition to the standard services provision, homes for the elderly may provide temporary shelter and assistance to the homeless, provide assistance and care in their homes, organization of day care centers (clubs) for elderly, organization of cultural, entertaining, educational and informative activities, humanitarian activities, etc.

In order for the placement in homes for the elderly to achieve a satisfactory quality of services, it is necessary to achieve efficient management process. The management of social protection institutions is the process of planning, organizing, managing and routing the resources in order to achieve high quality of social protection of the elderly. Functionally, the successful management of social protection institutions implies the attainment of balance in the process of satisfying interests of beneficiaries, interests of the institution management (founder-owner), and interests of the employees (service providers to elderly people). Interests of beneficiaries (elderly people) who are placed in residential care are reflected in the need for high quality and accessible services. Interests of the institution management are recognized as the need to achieve economically sustainable and profitable institution. Interests of direct service providers (employed caregivers, social workers, etc.) are in the safe premises and good working conditions.

By resources of social protection institutions are implied organizational, material and human capacities that institution has. Human resources are the most important resources of any work organization or institution. They are the bearers of the work process and are key in determining the quality of a product or a service. The staff employed in the institution with their work contributes to its business success which is why the management of an institution should pay close attention to human resources management. This is emphasized in the most commonly used definitions of management in the literature which suggest that the field of management is focused on the art of doing things (work) by people [1] or, that it is coordination of activities related to the job, so that they are performed

IOP Conf. Series: Materials Science and Engineering **200** (2017) 012030 doi:10.1088/1757-899X/200/1/012030

efficiently and effectively - with the people and with the help of other people [2]. The term "human resources" in institutions of social protection implies overall knowledge, skills, abilities, creative capabilities, motivation, loyalty and personal characteristics owned by employees in the institution. It is the overall intellectual and emotional energy of employees that the management can count on and that can be engaged to achieving the objectives of the institution. Professional and business results of social care institutions depend on the state of "human resources".

2. Human resources management in social protection institutions

Human Resource Management (hereinafter HRM) is one of the strategic functions in the management process of social protection institutions. It is an unavoidable topic of modern management in social protection. This component of management is very important for the establishment of an adequate method of organization of the services provision process in homes for the elderly. In order for the organization to respond to the requirements set before it, managers must establish an adequate human resource management or human potentials management [3]. HRM presents a basic management tool through which the manager affects the employees, because the focus of his/her interest is placed on human resources, their potentials and their motivation. Some authors believe that HRM is special managerial function because it includes "all management decisions and activities that affect the nature of the relationship between the company and employees - its human resources" [4]. However, HRM can be seen as an integral part of the function of organization, because the professional structure, potential and motivation of employees are precondition for effective and advanced business organization. "The quality of an organization is mainly only a result of the quality of people employed and retained at work" [5]. Therefore, human resources are an indispensable item in the performance analysis of social protection institutions. Recruiting and hiring a competent professional staff is an essential condition for sustainability of the quality of work of the institution and its performance/success.

The activities of an organization, which are focused on HRM, are related to three general processes: the process of identifying and selecting competent staff, the development and adoption of new knowledge by the employees and the process of retaining skilled and productive employees. Implementation of activities in these processes can guarantee good results in human resource management. In the institutions of social protection, these processes involve the engagement of management in assessing personal competencies of employees and in terms of possession of empathy for the beneficiaries. The selection process, the development process and the process of retaining workers must be justified in assessment of the degree of their ability to understand the needs of elderly people and willingness to provide more than just technical performance of the assignment.

In homes for elderly employed are workers of different profiles, from professional to support staff. Among the workers who work directly with beneficiaries are mostly represented caregiver and medical technicians. In addition, in working with beneficiaries engaged are social workers, occupational therapists, psychologists, physiotherapists and physicians. Technical and administrative staff is responsible to provide material and technical conditions of placement and accommodation of beneficiaries in the resident care, such as hygiene, nutrition, purchasing, financing, etc. The task of management is to achieve efficiency in the work of institutions which includes, in addition to economic sustainability and customer satisfaction, achieving the optimum balance between the interests of employees and the results of their work. In particular, this applies to workers who have direct contact with beneficiaries and who provide different types of protection services to the elderly. Human resource management in homes for the elderly implies precisely achievement of this primary objective/goal.

The specific objectives of HRM in social protection institutions are relate to providing competences in accordance with the requirements of the position/job, professional and optimal number of competent workers, improving the socio-economic status of employees, ensuring full-time employment of workers, raising the quality of working conditions and working environment, creating and maintaining flexible and adaptive potential of employees, reducing resistance and increasing the sensitivity of employees to changes in the institution. The first two objectives are aimed at satisfaction of interests of the institution, while others are aimed at achieving employee satisfaction, which is a prerequisite for business success of the institution.

Provision of abilities in accordance with the requirements of the position/job and meeting professional and optimal number of competent workers are related to the field of human resource planning in the institution. Human resource planning is one of the most important areas of HRM. Planning is a process in which managers perceive the optimal need for number of employees, their qualification structure and functional deployment. In doing so, it is essential to have overview of the existing human resources in order to adequately analyze the educational qualification, experience and knowledge of employees and the nature of their work, description of activities and specification of knowledge that are needed to perform a specific job [4]. Assessment of needs for human resources, in fact, reflects the objectives and strategy of development of the company or organization. The relationship between the analysis of the current situation and development needs of the organization is the basis on which the manager builds estimation of deficit of human resources.

According to the research conducted [6] in homes for the elderly in Republic of Srpska, the chronic problem is the lack or shortage of workers and inadequate deployment of workers in positions for direct care for beneficiaries. The public homes owned by the state, which often have excess of workers, there is a mismatch between the professional and administrative workers at the expense of the first, and in private homes there is mainly a small number of employees who often do different kinds of work in order for the management of institution to pay less amount for salaries of workers.

Human resource planning plays an important role in ensuring the economic sustainability of an institution. In the market conditions of business, when institutions for placement of elderly must offer economically acceptable price, while at the same maintaining the quality of services, the number and professional competence of employees are becoming a significant factor in its success. The goal of the management of an institution is to offer high quality and as attractive as possible service for fewer costs. Excess of workers and the low level of their expertise is a major threat to the future business of that institution. Therefore, the institution's management should approach to assessment of the need for workers very systematically which includes conducting analysis of the situation and optimization of number of users, accommodation capacities and coordination of work processes. The potential of an institution to cope with business risks in the market of social services for the elderly will depend on how coordinated these elements are.

The planning of human resources is also affected by processes that are not directly related to the needs of the institution itself. Education is very important area, which indirectly affects the offer of qualified specialists. Lack of full time education programs for caregivers has opened possibilities for additional training and requalification. Increased interest in the opening of private homes for the elderly in BiH has increased the demand for educated profiles of caregivers which opened space to conduct additional training programs [7]. Lack of qualified staff on the labor market makes it more difficult to plan human resources in this area. In countries where the care and nurturing of elderly people is highly developed, there is constant lack of such profile of the workforce. So, for example, in Germany, the demand for caregivers in recent years has doubled. It is anticipated that by 2020 Germany will have around three million people that should be provided with long-term care services. According to the analysis, in 2009 there was shortage of around 110 thousand of new caregivers and total of 4,300 new homes for care of people [8]. More recent data from the Federal Labor Office (*Arbeitsamt*) say that this demand has increased, particularly for the purpose of care of the elderly people. According to their forecasts, by 2030 Germany will have shortage of about 200 thousand caregivers. In the end of 2013, Germany had shortage of about 30 thousand caregivers for the elderly [9]. The situation is similar in other Western European countries where they annually employ tens of thousands of nurses and caregivers from other countries, specifically on positions of caregivers of the elderly.

Very poor offer of caregivers on the labor market and the increased need for caregivers refer management of institutions to create and have a human resource plan. Human resource planning in the field of social protection of the elderly includes the creation of employment dynamics of trained and experienced competent experts. The results of the research conducted in social protection institutions in Republic of Srpska [10] show that there is very poor practice of making plans of human resources. According to the survey, only 17% of managers of these institutions said that they have a plan of human resources. Most managers do not use human resource planning as a relevant mechanism for new employment. The need for new workers in social protection institutions of Republic of Srpska is usually determined *ad hoc* and is often not caused by reasons that have essential connection with the developments in the field of social protection. The results showed that when hiring new employees in the institutions of social protection, much less attention is paid to the needs of the work process of organizations. In most cases, people who are employed do not have any experience in working with the elderly people. The results of researches carried out in the social protection system of Republic of Srpska confirmed that the management of human resources is not developed and thereby makes the work of institutions for the care of the elderly difficult as well as their functionality and their sustainability.

In order to achieve good business results, institution's management should implement activities in the field of human resource planning which includes, in addition to recruiting new employees, the investment in the existing professional staff. Namely, it is difficult to achieve good operating results if employees are not satisfied with the work that they do, and this is achieved through their motivation. Work motivation is an essential factor of HRM and it is directly aimed at employees whose results affect the business success of the institution. Older research in psychology and sociology of organizations indicate the relationship between work motivation, job satisfaction and the possibility of worker participation in the process of planning the workflows [11]. Recent motivation strategies indicate that unmotivated employees have lower productivity, they are not interested in quality of products, do not identify with the organization and are not interested in organizational problems, development and success [3]. For the management of social protection institutions of special importance is the identification of employees with the institution. The mechanisms that the institution's management will use to achieve this goal may be different, from wage/salary increases to the increased participation in the process of services planning. Which mechanism will the management apply depends on the structure of employees and established interpersonal relations, both among workers and in the relation workers - management.

3. Working with employees

A special part of HRM in institutions for placement of the elderly is work with employees. The aim of that work is to achieve the highest possible commitment to the work tasks and thereby achieve the higher operating results. In addition to individual characteristics that the workers bring to the institution by their employment, the results of each employee are affected by a range of interventions, measures and activities undertaken by the management of the institution towards them. These activities are the backbone of the work with staff employed in institutions for placement of the elderly.

Working with employees can be seen as: recruitment and candidate selection, socialization, training and development of workers, performance assessment and evaluation, reward and motivation, the determination of wages/salaries, information, development of interpersonal relations, supervision and support to employees (prevention of "professional burnout") ensuring the fulfillment of legal rights (regularity of salaries and wages, affairs and records related to engagement, occupational safety, etc.).

Recruitment and selection of personnel is done through the implementation of the procedure which is generally defined by law. It is very important that the conditions for the selection of candidates complies with the necessary professional skills needed for the job which involves having a formal qualification, required experience and social competencies. However, direct work with the beneficiaries requires developed skills of emotional intelligence among workers [12]. Empathy, communication and other social skills, self-awareness and self-control as components of emotional intelligence, represent the major skills that should be owned by caregivers, social workers, medical technicians and other workers who directly provide services to the elderly.

Socialization, training and development involve direct investment in workers which raises the level of trust of the workers to the institution's management and thus creates a sense of community. These activities are dedicated, first and foremost, to the new workers who just started working in the institution. To the newly recruited, managers must provide mechanisms for easy integration into the organization, so that they areas soon as possible ready to work, but they also have to offer support mechanisms to the employed. Their goal is to provide newly employed as easily and as quickly as possible to "fit" into the existing organizational structure and work processes. However, those activities can also be provided for the other employees as an investment in development. Any investment in the development of employees is at the same time allowing the transfer of new knowledge and skills necessary to work with the elderly. Besides that, this way of working with the workers has an impact on creating a good working atmosphere which is a prerequisite for creating a desirable living environment for the beneficiaries.

Performance assessment and evaluation, reward and motivation are important activities in HRM. These activities bring together the work of employees and their expectations with the results achieved, thus it rounds up the process of management in the institution. These aspects are mutually connected and are part of the control function in the management of the institution, so it is very important that they are performed in accordance with clearly defined criteria, transparently, and with the participation of employees. Expectations of employees are related to the improvement of their socio-economic status (higher income), ensuring permanent employment and raising the quality of working conditions and working environment. The combination of activities arising from the achievement of these objectives and activities aimed at the development of professional competencies and psycho-social potential of employees is the basis of work with the employees in the institutions of social protection.

Not taking into account only individual reasons (values and character traits), employee satisfaction is affected by how much attention and time the manager devoted to their needs. Employees have their expectations and achieved level of employee satisfaction is an indicator of the success of HRM.

Determining the wages/salaries and ensuring the realization of legal rights of workers means fulfillment of those obligations by the institutions management which are expected by the employees. Through the implementation of these activities, employees have access to the legality of operations of the employer. Any non-fulfillment of obligations towards the worker provided by law undermines business and interpersonal relationships in the institution, reduces trust, and ultimately affects the quality of services provided. For all these reasons, it is important that at the beginning of engagement worker is introduced to legal rights and obligations, and to have a clear insight into how his/her performance is evaluated and how the wage/salary is determined. Fulfillment of these goals is supported by the practice of regular, accurate and timely information of employees that can be implemented in different ways: meetings, talks "face to face", notices, newsletters, etc.

Supervision and support to employees is a key intervention of HRM in social protection institutions. It is focused on prevention of burnout of employees, especially those who work directly with beneficiaries. Professional burnout in social protection is very common and frequent phenomenon which has a significant impact on the effectiveness and performance of employees.

Occurrence of burnout syndrome, in addition to individual characteristics of service providers, is affected by factors arising from the specificity and nature of their work [13]. The characteristics of the workplace, professional characteristics related to the job of nurturing, the specifics related to customers, work that involves dealing with death and organizational characteristics of the work process are factors that play a significant role in the high-intensity of professional burnout in services providers to elderly people. This requires continuous provision of professional support to employed staff. Supervision presents the effort to strengthen professional competence in the work process with professionals through learning from their own experiences, to seek their own solutions to problems encountered during the work, including effective coping with stress, and it provides the overview of situations and resources of beneficiaries, their thoughts, emotions and resources and the client relationship from different perspectives. By integrating these different aspects of professional situations, a professional creates the preconditions to act as a professionally competent person [14].

4. Conclusion

Human resources are an important segment of management in social protection institutions. All specifics of work in social protection institutions indicate the importance and the significance of having high-quality professional staff. At a time of market-oriented system of social protection and increased demand for services of institutional care of the elderly, human resource management involves harmonization of actions of an institution with the services quality requirements and economic sustainability of the institution. This is one of the strategic management process that focuses on human resource development in order to ensure the competitiveness of the institution in the services market. A wide range of activities that make the human resources management provide for social protection institutions for the care of older people to maintain their competitiveness in the market. The main problem for the provision of human resources in homes for elderly people can be found in the absence of formally trained and competent workers in the labor market. Most employees in homes for the elderly for the first time face with the business of nurturing, thus the investment their professional development is the task that management of an institution should not forget. Employees must be provided with constant expert and professional development, and to have opportunity of enriching their knowledge and acquiring new skills. For them it will mean certainty in achieving economic security for them and their family, for an institution it will provide secure and efficient staff, and for customers it will mean quality and satisfactory social services.

References

[1] Šehić Dz and Rahimić Z 2006 *Management*, Faculty of Economics, Sarajevo, Bosnia and Herzegovina
[2] Robbins S and Coulter A 2005 *Management*, Data status, Belgrade, Serbia
[3] Bahtijarević-Sunroof F 1999 *Human Resources Management*, Golden, Zagreb, Croatia
[4] Šunje A 2002 *Top manager*, Tirade, Sarajevo, Bosnia and Herzegovina
[5] Robbins S and Coulter A 2005 *Management*, Data status, Belgrade
[6] Lepir Lj 2009 *The role of management in the social welfare of the elderly in the Republic of Srpska*, Faculty of Political Sciences, Sarajevo, Master thesis
[7] ***SEC 2009 Available at: http://sec.ba/formalne-obuke/
[8] Knüppel J 2015 *Zahlen - Daten – Fakten,* Pflege. DBfK, Berlin
[9] Leubecher M 2013 *Deutschland-sucht-die-Welt-nach-Pflegekraeften* Available at: http://www.welt.de/politik/deutschland/article123381446/
[10] Lepir Lj 2013 *The social welfare of the elderly in the Republic of Srpska*, Faculty of Political Sciences, Banja Luka, Doctors thesis
[11] Sekulić D 1982 *The motivation and social action*, in: Obradović J ed. 1982 *The psyhology and sociology of organizations*, Školska knjiga, Zagreb
[12] Šćepović D 2009 *Emotional intelligence of managers in social work*, Social Thought, No. 3/09th, pp 101-122
[13] Dedić G 2005 Burnout syndrome at work, *Vojnosanit Pregl* **62**(11) 851-855
[14] Ajduković M and Cajvert L 2004 *In community*, Society for Psychological Assistance, Zagreb

IOP Conf. Series: Materials Science and Engineering **200** (2017) 012031 doi:10.1088/1757-899X/200/1/012031

Agile methodology selection criteria: IT start-up case study

Lj Micic

TA Business Informatics, University of Banja Luka, Faculty of Economics, Majke Jugovia 4, 78 000 Banja Luka, Bosnia and Herzegovina

E-mail: ljubisa.micic@efbl.org

Abstract: Project management in modern IT companies is often based on agile methodologies which have several advantages compared to traditional methodologies such is waterfall. Having in mind that clients sometimes change project during development it is crucial for an IT company to choose carefully which methodology is going to implement and is it going to be mostly based on one or is it going got be combination of several.
There are several modern and often used methodologies but among those Scrum, Kanban and XP programming are usually the most common. Sometimes companies use mostly tools and procedures from one but quite often they use some of the combination of those methodologies. Having in mind that those methodologies are just a framework they allow companies to adapt it for their specific projects as well as for other limitations.
These methodologies are in limited usage Bosnia but more and more IT companies are starting to use agile methodologies because it is practice and common not just for their clients abroad but also starting to be the only option in order to deliver quality product on time. However it is always challenging which methodology or combination of several companies should implement and how to connect it to its own project, organizational framework and HR management. This paper presents one case study based on local IT start up and delivers solution based on theoretical framework and practical limitations that case company has.

1. Introduction

Modern software development is, to a greater extent, based on applying some of the agile methodologies, among which particularly highlights the Scrum. Agile methodologies are accepted in the IT companies in a large scale and usually result in better delivery of products and greater efficiency and effectiveness in the developing world.

However, there are no "complete solutions" in selecting the methodology. Defining the appropriate methodology and its adaptation to the particular company is, on many occasions, really serious process and requires an analysis of the company itself, employees, the product concerned but also the plans for the future. It is the basic motive for this paper – an attempt to define criteria in the selection of the appropriate methodology in software development in the IT companies as well as the factors for assessing the success of the application of that methodology.

However, the introduction of some of the agile methodologies, to a lesser or greater extent, depends on certain preconditions that must be met. Theory knows the common prerequisites but they may vary somewhat from region to region, company to company, and of course of the types of project which is expected to implement specific methodology. Also, the success in the implementation of the methodology depends on a number of factors, from organizational to physical. It is also part of the issue concerned in this paper. In addition, the performance of applications is greatly determined by the profiles and characteristics of individuals who perform key roles in the team. For example, in scrum, as the most spreadable methodology, these roles are Scrum Master, Product Owner and team members. Each member of the team should possess certain traits that make application to be successful. Identification of these characteristics, based on the opinion of experts with experience in dealing with the scrum, is one of the goals of this paper.

The introductory part of the paper provides a basic clarification of concepts of agile software development, with an emphasis on scrum as the most common methodology that has shown the best results. The second part of the paper gives a preview of study case with local IT start-up with provided solutions to specific issues that this start up is dealing with.

The methodology used in the preparation of this paper includes:

A review of literature in scientific areas as well as the use of technical literature available on the Internet (web articles, newspaper articles, tutorials, etc.),

Business case study on the selected issues and provided solutions,

Materials from specialized scrum master training held in Banja Luka in February 2015.

2. Agile methodology

Agility, based on the Cambridge and Oxford dictionaries, is defined as the ability of fast and easy movement, but also of thinking, acting and problem solving.

For these reasons it is not hard to conclude why exactly that word was part of the name of specific group of methodologies, which are characteristic for the modern software development. Agile methodologies include whole set of methodologies and they are often intertwined or used combined with ones which are traditional in the IT companies that are focused on efficiency and effectiveness in the software development. As a priority and crucial issue by the agile methodologies, but also by traditional ones imposes collecting and defining requirements, that are often immeasurable, imprecisely defined and without limitation by time or space.

The application of the agile methodologies in companies, as a PriceWatersCoopers states [1], is work with high risk in deadlines exceeding and delaying the launching of the products in companies with little or without experience in the a application of agile methodologies. Selecting an appropriate framework for the software development can be of crucial importance in efficiency, as well as the effectiveness of business, but also in achievement in defined deadline of the taken commitments. The appropriate framework and approach can enable the success through collaboration and work with stakeholders. Modern and global business, considering the fact it changes fast, requires high level of flexibility and acceptance of modern tools and business modes. Regarding to these reasons, it is not strange at all that agile methodologies are highlighted as a most accepted methodologies in software development, because they enable high level of flexibility considering the collaboration with all stakeholders. This is really important in IT companies, which are focused on software development, because their business requires high level of compatibility between technology and operating, which is not quite often in the other areas. As PwC states in its research [1], the importance of the agile methodologies we can easily notice through statistics, but also through Gartner's and Forrester's prediction: in 2000 less than 1% IT department knew what the agile methodologies are, nowadays between 60 and 80% of developing teams use this type of methodologies as the primary in the software production. Also, they predict the continuation of the tendency of popularity of the agile methodologies, and increment of use this type of methodologies in software development. However, same study cites the fact that despite using the agile methodologies in high percentage, same can't afford objective results without quality basic management by planning and communication.

The initial focus in the agile methodologies is on what should be the result or outcome of software production; also they are less focused on the way how to get it. Regarding to this is the fact that, in the prior methodologies, the requirement and the characteristics of the expected product were defined often incorrectly or imprecisely, which eventually led to wasting time on something which wasn't client's requirement. The agile methodologies focus on the precision definition of the requirements, so the time and energy of team would be invested in the creation of the appropriate product and/or service that is also expected.

Through iterations, team improves the way how they get the desired product, but also team increases own efficiency in the precisely definition of the activities, stakes, deadlines etc. But also, it is really important how the expected product is defined and precision, which certainly work on definition of the methodology (specific way of software development) makes easier. Also, involvement of team in the process is really important, because that helps them to figure out what type of product is about, or what characteristics should it own.

The main objective of the agile methodologies in the process of collecting and defining the requirements and distribution through iterations, is improvement of the communication, work compatibility and but also involvement of clients from the beginning, so the objectives and needs of the organization could be achieved.

Ola Sundin in his brochure "Scrum in 5 minutes" beside Scrum as the most popular agile methodology cites also Kanban, Extreme Programming (XP) and Lean Software Development noting that all of them can be combined and applied together: Kanban for the effectiveness issues, Lean for defining of principles for development of the organization, Scrum for organization and planning of the projects, and XP as an operating methodology that is based on 12 principles [2].

2.1. Most popular agile methodologies

The most wide-spread agile methodologies or by variation of access, management style and methodology, are, by group, author of the book "Learning agile" [3]:

- Scrum
- XP — extreme programming
- Lean
- Kanban

Ultimately, each of these four methods are possible to implement on its own but the agile approach to software development in modern IT companies includes the variation of some or all of the four approaches. Therefore, it is frequently encountered that IT companies use Kanban and work on a self-organized teams of scrum. Also, XP is often utilized with Scrum and often, especially in societies such as Japan, the Lean philosophy is applied with any of the other agile methodologies. According to The VersionOne State of Agile Development in 2013 the most popular agile methodologies is the scrum. It is followed by a hybrid of Scrum and XP, then customized individual hybrids and the fourth place is certainly a variation of Scrum and Kanban, the so-called Scrumban [4]. Stellman defines agile in the following way: "Agile method is you and methodologies that help team to think more effectively, work more efficiently and make better decisions" [3]. The same author further states that these are practices that are directed and optimized to maximize flexibility [3]. Also, agile is a way of thinking and acting team, particularly in scrum, where self-organization is one of the priorities of the functionality and operation. Most of the values of the agile software development are contained in the Agile manifesto [5] although different environments and conditions require different practices and preparations from different approaches.

3. The requirements collection: traditional vs agile methodologies

Considering the speed at which technology is being changed, but also the effect of same on business, especially business of IT companies, it's not hard to conclude the importance of the organizational adjustments and continually improvement of work. Modern business requires improvement of connectivity and collaboration with clients or users, often through the use of these modern technological tools.

The use of technology becomes part of modern market competition, often a crucial part of business strategies of the companies. Continual improvement became the requirement for the companies when keeping their market position, or eventually improving it. The application of the agile methodologies, beside their popularity, can be harmful for organization (too complicated or too confusing). This is partly justified, because of the process of transition from the traditional methodology, which is concentrated around the documents, and is quite define in advance, to methodology, which is focused on the requirements of work and the effectiveness of it as well.

However, the traditional methodologies, such as, for example Waterfall are characterized by "separation" of phases, and lack of communication from one phase to other, often lead to development of software, which doesn't match the need of client. On the other hand, possible re-start of the process can be too long and too complicated, which may cause dissatisfaction of clients, but also dissatisfaction within organization. In the Waterfall in the initial phase, we have an intensive and often long process of collecting of requirements, however at the of this phase, communication with client and business analyst almost stops, which disables changes in the requirements, that are often appear in the process of development.

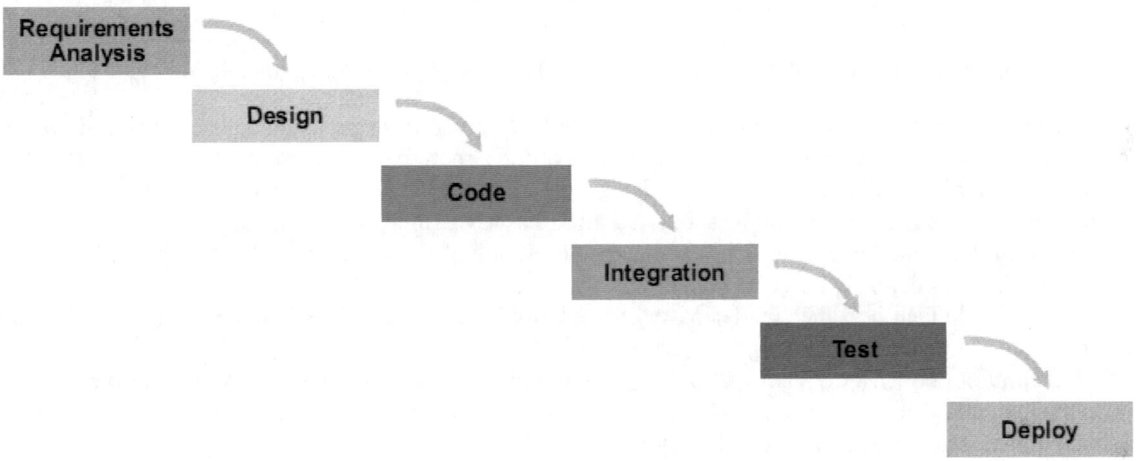

Figure 1. Display of Waterfall model [6]

Abundant documentation, as a result of the phase of defining the requirements is forwarded to the development team, but also is being forwarded through the phases of development, also leaving the possibility of different defining of initial need of the client.

Within the team there are usually different definitions of what is actually the requirement of the work, which leaves possibility of different interpretation, and inharmonious work. That brings as to crucial flaw of the Waterfall methodology, but also of the other traditional methodologies: lack of communication and precision in the work, which ultimately can lead to delay of the product launches but also much lower quality product that does not meet the requirements of the client.

Involvement of the team in development from initial to end phase improves their understanding of the expected product, which affects, not just the efficiency in the work, but also it increases the quality of outcome of the work.

On the other hand, the goal of the agile methodologies is the more quality launch of the product through the promotion of collaboration, and teamwork through the incremental process and optimizing of resources and the budget.

The agile methodologies have main requirement to improve the speed of reaction, also to increase precision in the definition of the requirements, which at the end should give a result through distribution of the products, which are optimally meet the client's requirements.

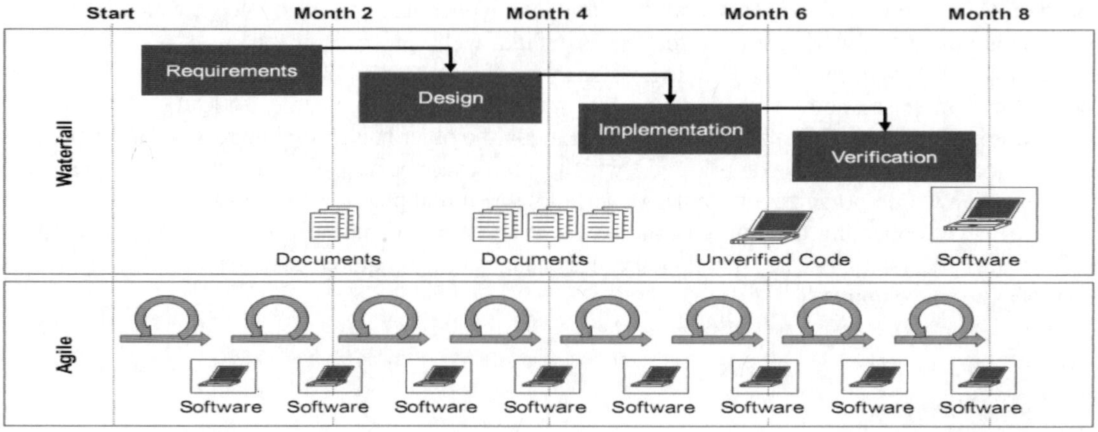

Figure 2. Display of the difference between the Waterfall and the agile methodologies [1]

However, although the agile methodologies initially arose from the need for greater flexibility, it doesn't minimize the importance of the existence of the structure and the project planning. In fact, the project planning and the definition of stakes, the scope of work, and time limits are even more important in the agile methodologies, but in a different way than in the traditional methodologies.

The agile methodologies divide whole expected scope of the work into the small parts, known as "EPICS". Epics define the crucial segments, which are necessary to develop through the project. Epics are further divided into the smaller parts of the work, usually called "STORIES", within which are defined special tasks – TASKS.

Definition of the individual and mutual conditionality, which are for sure very important in the implementation of the project, is crucial in the agile project planning. Also, it is very important to determine correctly and precisely the priorities in the work, for what are especially interested in team members. Determination of the priorities is mostly caused by the requirements of the work, but also by the continual communication with clients, as a characteristic of the agile methodologies.

At the end, we have SPRINTS, which represents time sections through which we implement epics and stories. Sprints are usual in the Scrum agile methodology, although they are different named and appear in the different methodologies.

Members of the project team in the agile methodologies must have common and precisely understanding of the project which they work on, that is partly achieved through their daily involvement in the project planning, determination of complexity of the tasks etc.

4. Criteria for selection of the appropriate methodology

4.1. The relationship of the methodology, organization and selection criteria

Each methodology carries certain qualities but also threats. The important thing that matters for the organization is to predict that there is no single solution to a potential issue and that each situation requires a special approach and further consideration. Sometimes, outdated methodologies, which are not adjusted for modern way of doing business, can be appropriate in some situations, whether combined or used independently. The choice of methodology, among other things, greatly depends on the size of the organization, the type of technology used, the style of management, and the structure of employees, locations, companies, clients/users and a number of other factors that need to be taken into account. Additionally, the choice of methodology is followed by the application that can identify whether the methodology is relevant or the modification is necessary. The application can also show that the management team or the organization as a whole is too rigid and inapplicable and that the changes cannot be implemented.

Overall, methodologies can be grouped into four larger groups [7]:

- Linear (example: waterfall)

- Archetypical – iterative
- Incremental (a combination of linear and iterative)
- Other methodologies (spiral, PAPER and similar).

The above mentioned work has advantages and disadvantages of each of them. Among other things, linear is recommended for teams with less experience as well as an organization in which management has no experience of running projects. Also, it is recommended when the system can easily measure progress. However, taking into account its rigidities, heaviness and high control it is not recommended for projects that require quick response and those exposed to frequent changes of priorities and tasks.

What the study States [7] should not be used in real-time systems, conditioned systems and leading-edge applications. Prototype development is specifically recommended for the development of a system for unknown clients as well as with systems where the team members are experienced in preparation and launching the products. However, this methodology is not recommended when resources are limited, and when they are demanding strict control of deadlines of execution i.e. final product launch. Incremental development is recommended when there is a need for reducing risks through the "breaking" of the overall job into smaller parts where eventual failure of one does not reflect as a failure of the whole project. Some consider this methodology as a combination of two preliminary because, for each process, partially or fully controlled and planned process is implemented as it is in linear methodologies.

4.2. A review of the criteria for selection of the appropriate methodology

Caper Jones, after their research, states that "... the choice of methodology development has more in common in association with the cult of technical decision making. ..." [8] which stresses that the process of selecting the methodology, as part of a managerial decision, is classified in the area of work of social sciences that by themselves are more subjective and less precise than technical. Daniel Krige, site leader in AWS support in Amazon Web services, lists several criteria that got the survey [9]:

- Clearly defined risks
- Defined the scope and complexity of the work
- "Maturity" (if the client recognizes his own needs)
- Development costs
- Relations on the team
- The flexibility of the interested parties
- Experience team
- Necessity of documentation
- The complexity of the applications/systems
- The size of the application/system
- Security environment
- Team size
- Knowledge of technology
- Programming languages used
- Contracting costing methodology and
- The impact of the unit (if the project develops within the larger enterprises in the IT Department).

5. IT start-up business case study: Logica Technology, Banja Luka

In order to get valuable information regarding my research I tested above mention in case study of an IT start-up, Banja Luka's company Logica Technology.

5.1. Basic information

Name of the company Logica Technology Ltd
HQ: Banja Luka

Year when it was founded: 2014
Number of partners: 2
Number of employees (in time of interview): 18 and two interns

5.2. History
Logica Technology is an IT start-up which is founded by extra capitalization of already existing start-up Code Blue Studio (CBS) which has already existed in Banja Luka in period 2012-2014. Central focus of CBS was WordPress (WP further more) platform and similar web services with some work not connected with it.

New investor (after investment one of two partners) is adding one new extra focus: creation of web platform for online gambling and casinos (CS furthermore) because he has already has some experience in this industry abroad. Investors main focus of the work is on-line media, marketing, SEO, specialization in one type of portal, content management, organization of port of the company, HRM and general management. Another partner, the one who has CBS previously is expert in WP technologies with very good knowledge of other technologies (HTML, CSS, PHP etc.) and his focus is continuation of the WP theme work and cooperation with technical and design part of the company.

5.3. Condition of the company in the time of interview
Company has two main streams:

- Word press themes (eventually other themes such are HTML/CSS) and
- Development of the new online service for advising on gambling, casinos and CS.

First stream is consisted of several smaller scale projects with 24/7 need for client support. Development of CV platform is a long term internal project and there are no specific clients for this one – it is completely rounded by second partner in company. There are not specific deadlines for CS stream. In time of the interview only WP stream was profitable and although it produced small income it was still the only profitable stream so far.

5.4. Resource and condition of the work
Company has excellent working conditions. Employees are using adequate space in center of the city, 115m^2, and they can use similar underground space almost same size.

Figure 3. Working space of the company

Space has a small kitchen, two toilets and several balconies. Lighting is good and it secures successful work with no obstacles. Space is organized in four offices, different size and Management of the company has praxis to have joint activities and improve motivation and cooperation during and after working hours.

5.5. Employees

Having in mind the company's streams, there is a need for employees with certain profile and working experience: front and back-end development, content management/content creation, web editing, and design. Formal education of the employees is variable and does not fully correspond to the scope of the company however formal education often has small importance for the success of the company itself. Profiles of employees, by education, include: electrical engineers (self-taught in IT) graduated journalists, foreign language bachelors and economists and lawyers (graduates). The company has hired a modest team of developers and so far one senior designer.

Positions inside company are: front/back-end development & programming, content management and consulting, web editing and editing content for the portal in different languages and design. In addition, there are also regular general functions such as management and general administration.

The organization of the working and space in time of the interview was following:

- One co-owner (co-owner A onwards) is focused on the development and maintenance of WP (HTML/CSS) stream. He has relatively regular meetings with another co-owner. Most of the time he is working with developers and designer. He is in the office 3 with the designer and one front-end developer and works closely with developers and translator in the next office 4.
- Co-owner A is primarily engaged with WP theme: communication with existing clients as well as developing new ones, having in mined the potential for high profitability of the same. A smaller portion of time he is devoted to strategic planning and the organization of the rest of the company which is largely left to the other co-owner.
- The other co-owner (co-owner B onwards) is in the office 1 with company's lawyer, and he is in charge of general affairs and administration. There are also two experienced journalists, responsible for the organization and management "journalistic team," which is consisted of eight foreign language teachers, translators and "content creators" (who are in office 2, which is also the largest in the company).
- Co-owner of B is devoted to ideological and conceptual development of CS web services doing most of the time things connected to SEO and creation of the content. He left direct communication and organization of the journalistic team to two experienced journalists that are with him in the office.

In the time of interview, most employees were working on the development of CS web services and few have been working on the development and maintenance of WP themes.

5.6. Proposed solution

Given the fact that the company has two base streams the first suggestion is certainly the introduction of scrum in both development teams. Of course, before that it is necessary to precisely define who constitutes the teams and who performs the general function of project management as well as management in general.

The process of change would consist of the following steps:

- Precisely define the management team
- Establishment of two development teams: WP team and CR team
- Define the roles in both (scrum) teams
- If necessary: additional training for those who take roles in scrum teams
- Define the period for the introduction of the new methodology, after which a comparison will be made (the need for the continuous measurement of effectiveness, efficiency and employee satisfaction)

- Start with the full implementation of the scrum
- Define the period of performance evaluation and the need for additional changes.

Management team

The management team in this solution would consist of two co-owners, lawyer and senior journalists (one for every of two teams).

Co-owner A would be part of the WP team while the other should stay outside the CR team because it would be desirable CR team sees him as a "Client". This would leave him more time for general management issues, HR issues, marketing, SEO and similar. It should be consider hiring an additional employee (psychologist, or economist) for Human Resources management.

WP team

WP team would be consisted of: co-owner A, 3 developers and 1content manager/interpreter. Roles in the WP team: co-owner A would be the team leader who would carry out fully-functions of the Product Owner, 1 front-end developer (N., a girl, with a good reputation and communication skills) would be a Scrum Master and ensure that the scrum process function smoothly and that the satisfaction of the employees in this team is satisfactory. It should be consider employing a QA (which could be QA for the other team too) and additional developers, given the potential revenues that this team can accomplish. It should be consider employing intern developers.

CR team

CR would be consisted of: 2 journalists, 8 associates for content. Roles in a team: a journalist, which is part of the management team of the company, would in this team be a Product Owner-a (conclusion based on the basis of good cooperation with the B co-owner), and second journalist would be a Scrum Master.

Company organization and functions

Strategic meetings should be organized at least once in two months in order to define further steps. Additionally, at the end of each month there should be an evaluation meeting in teams and then in the overall company in order to improve business processes and trigger further change.

6. Conclusion

Agile methodologies especially scrum as the most widespread and the one that gives the greatest effect, are an integral part of contemporary IT companies. Bearing in mind that it is largely confirmed that scrum or other agile methodologies ensure a well-functioning systems and processes, with a focus on people and the outcomes. It is reasonable that more and more companies recognize their quality and usefulness in the application.

However, the introduction of agile methodologies in itself does not guarantee the success of the process and work of the organization itself. Classic elements of management, organization, management of human resources and technical quality of work are certainly a prerequisite for the success of each company including IT start-up focused on the development of software and information systems.

Additionally, one of the main advantages of these methodologies is to continuously work improvement which is provided through meetings of classic scrum. These recommendations are actually taken into account and usually become an integral part of the process and the quality policy.

Through this paper there is a review of literature on this subject and the scientific field and preview of the basic characteristics of the agile based on the scrum as the examples of agile methodology. Through a case study I wanted to make an analysis of the work and the organization of smaller IT company in Banja Luka and on the basis of the data obtained make proposals variants and ways of introducing scrum into their daily work.

References

[1] ***PricewaterhouseCoopers LL 2014 *Adopting an Agile methodology,* PricewaterhouseCoopers LL, USA
[2] Sundin O 2015 *Scrum Master Training Banja Luka*, Softhouse, Sweden

[3] Stellman J A 2015 *Learning agile*, Sebastopol, CA O'Reilly Media Inc
[4] ***VersionOne 2013 *7th Annual State of Agile Development Survey*, VersionOne.
[5] Beedle M, Bennekum A V, Cockburn A, Cunningham W, Fowler M, Highsmith J, Hunt A, Jeffries R, Kern J, Marick B, Martin R C, Sutherland J and Thomas D 2001 *Agile Manifesto*, http://www.agilemanifesto.org/iso/sr/
[6] James M and Walter L 2012 *Scrum Reference Card,* http://www.collab.net/sites/default/files/uploads/CollabNet_scrumreferencecard.pdf
[7] ***CMS - Center for a Medicare and Medicaid Services 2008 *Selecting Development Approach,* http://www.cms.gov. http://www.cms.gov/Research-Statistics-Data-and-Systems/CMS-Information-echnology/XLC/Downloads/SelectingDevelopmentApproach.pdf
[8] Jones C 2013 *Evaluating Agile and Scrum with Other Software Methodologies*, InfoQ
[9] Krige D 2015 *Selection criteria for a development methodology*

IOP Conf. Series: Materials Science and Engineering **200** (2017) 012032 doi:10.1088/1757-899X/200/1/012032

Empirical evidences of owners' managerial behaviour - the case of small companies

G Lobontiu, M Banica and S Ravai-Nagy

Technical University of Cluj-Napoca, North University Centre of Baia Mare, Dr.V.Babes 62A, 430083, Baia Mare, Romania

E-mail: Mihai.Banica@cunbm.utcluj.ro

Abstract. In a small firm, the founder or the owner-manager often leaves his or her own personal "stamp" on the way things are done, finding solutions for the multitude of problems the firm faces, and maintaining control over the firm's operations. The paper aims to investigate the degree to which the owner-managers are controlling the operations of their firm on a day-to-day basis or even getting involved into the management of the functional areas. Our empirical research, conducted on a sample of 200 small and medium-sized enterprises (SME) from the North-Western Romania, Maramures (NUTS3 level - RO114), shows that owner-managers tend to be all-powerful, making decisions based on their experience. Furthermore, the survey highlights the focus of owner-managers on two functional areas, namely the production, and sales and marketing. Finally, the correlation analysis states that in the case of small firms, the owner-manager is more involved in managing the functional areas of the firm, as compared to the medium-ones.

1. Introduction

Over the years, the importance the small business sector is playing for the health of a nation's economy was widely recognized, and so the interest in small business management flourished.

Despite the relevance of the topic and the big number of papers analysing the SMEs sector, there are still some areas insufficiently covered, and not being matched by adequate theoretical concepts and models. One such insufficiently covered area is the managerial behaviour of the small business owner, and his/her locus of control, with the associated impact of the latest upon the way the small firm is organized and managed.

It is therefore worthwhile to carry out a series of studies designed for constructing conceptual categories and a theory that could help identifying and discussing managerial behaviour of the owner-manager and the way this influences the organizational processes in small firms from a management perspective.

Our paper introduces the results of such a study, which intended to reveal the extent to which the owner-manager is centralizing the decision in his/her hands, and sometimes, even the execution of some operational activities. We tried to add new contributions to the field, by addressing the case of small business owner (managers) from a CEE country (particularly, from North-West Romania).

The paper is structured in four major sections. Following Introduction, the next section reviews the dedicated literature on the definition of SMEs, their owners' managerial behaviour, and the way they manage the firm, and its eventual growth. Section 3 presents the empirical findings, while the final section concludes.

2. Theoretical development

2.1. Defining a small firm

Rigorously defining the small business has always been difficult even controversial. Mainly, small businesses have been distinguished from larger companies by such criteria as financial turnover, assets, market share, numbers employed and ownership [1].

For many years, since the Bolton Report, in 1971 [2], the most used conceptual definition of a small business in Europe was that of a "legally independent organization; with a small market share; and owned and managed by the same individual or group of individuals" [3].

More recently, the European Commission tried to unify the member states approaches on defining a SME, and adopted a new recommendation (Commission Recommendation 2003/361/EC 2003): "the category of micro, small and medium-sized enterprises (SMEs) is made up of enterprises which employ fewer than 250 persons and which have an annual turnover not exceeding 50 million euro, and/or an annual balance sheet total not exceeding 43 million euro" [4].

When compared with their bigger counterparts, small firms are not just scaled down versions of large ones [5]. The SME presents several dimensional, structural and procedural characteristics and they go about their business in several fundamentally different ways [6], being marked by [7]:

- ✓ Personal entrepreneurship and direct personal work relations;
- ✓ Less division of work;
- ✓ Stronger dependency on the environment.

2.2. Entrepreneur or owner-manager?

As the owner-managed small-businesses depend on the owner-manager for their management [8], his/her analysis is germane to understanding the managerial processes within such firm.

Although there is a considerable overlap between small business and entrepreneurship, the concepts are not the same [9].

Because the definition of entrepreneurship denotes the creation of some combination that did not previously exist, entrepreneurship is often equated with small business ownership and management. But even though the terms were considered interchangeable, more and more researchers of the field start to separate the two.

Storey and Sykes distinguish between owner-managers and entrepreneurs starting from the type of firms they are running [10]: a) the lifestyle firms, which are intended to grow, the management is mostly tactical, and they are set up primarily to undertake an activity that the owner-manager enjoys or gets some comfort from whilst also providing an adequate income, and b) the growth firms, which are set up with the intention of growth, usually by entrepreneurs.

Entrepreneurs are said to have an internal locus of control, a high sense of control over the events in their environment [11]. Moreover, the study performed by Stewart Jr. et al. [12] concluded that entrepreneurs were higher in achievement motivation, risk-taking propensity, and preference for innovation than were both the corporate managers and the small business owners.

2.3. A functional approach to managerial behaviour of the small business owner

Researchers following functional approaches have regarded managers as performers of a set of functions who apply specific skills [13]. They suggested that the process of managing a business consists of different functions, including planning, organising, leading, and controlling [14]. The functional approach focuses on individual performance in the operational process.

In many small businesses, the owner or a member of the owner family chooses to run the firm him/herself [15]. On one hand, owner-management provides a solution to the inherent agency problem involved in operating the business. On the other, it has been claimed that this combination of the roles of owner and top manager may have unfavourable consequences for the efficient operation of a firm, say Barth et al. [15].

In an owner-managed family business, the top manager is taken from a much more restricted pool of talent than when the manager is recruited from the general market for managers [15]. This situation generally leads to a lower quality among owner-managers than among professional managers and may be detrimental to the productivity of the firms [16], [17].

In the early years, some functions of the firm are performed by the founders (or owner-managers), such as marketing activities, human resource management, product development, research and development, information systems and financial control activities, state Wilson and Bates [18]. Small firms will in majority have no specialized staff for finance, personnel or marketing, and certainly not for legal affairs [19].

3. An empirical study on small companies' functional management

3.1. Methodology

The method used for collecting inputs from small business owners from the area of Maramures, Northern Romania (NUTS3 level - RO114) was the survey method. The sample of small businesses contained 200 items, selected from the Maramures County Council's database using two criteria:

(1) Firm size: businesses with fewer than 249 employees.

(2) Industry affiliations: seeking for a more complex organizational structure of the investigated small firms, the incidence of the manufacturing industry was prevailing.

For the statistical analysis of the collected data, we used the correlation coefficient method, trying to find out if a mathematical correlation between the average number of employees and the way the functional areas are managed appears. In this regard, we chose to calculate the Pearson product-moment correlation coefficient, being one of the most used tests in verifying correlations, and then to secondly confirm the existence of such correlation by calculating the Kendall tau rank correlation coefficient, which is a nonparametric test that works very well for any kind of data distribution.

3.2. Survey instrument

Aiming for a transversal study, a self-administered questionnaire, with multiple-choice answers, consisting of two sections and a scale of 8 items, was designed. Section I began with general items querying the firm's characteristics: its legal form of existence, the industry to which it belongs, the number of employees. In Section II, the items elicited information regarding the incidence of functional specialization within firm, and the level of personal involvement of the owner-manager in managing each functional area.

The average firm of the investigated sample is a manufacturing small firm (10÷49 employees), where the ownership and the management lay in the hand of the same person. The distribution of firms based on the industry affiliation and size are as follows - see Figures 1 (a) and (b), where the division into the three categories was made in accordance with the Commission Recommendation [4].

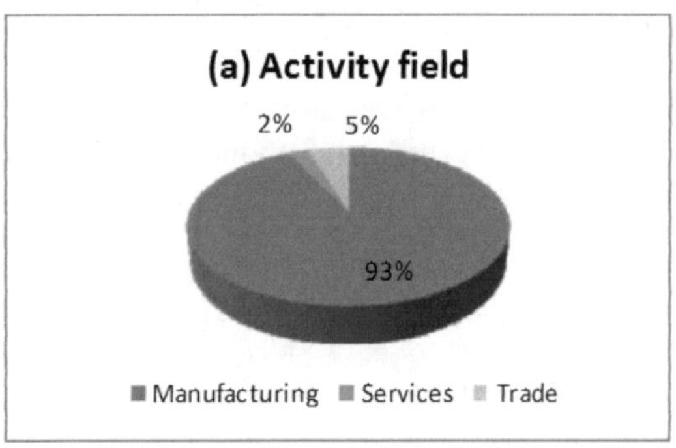

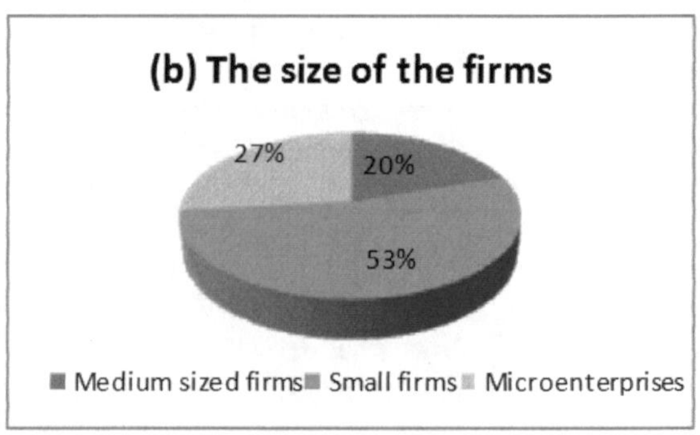

Figure 1. The investigated sample structure:
(a) per the activity field; (b) according with size (by no. of employees)

As observed, the main analysed industry was manufacturing (93% of the sampled firms belonging to this sector), while service industry and trade industry accounted only 5%, respectively 2% of the total number of firms. Regarding firms' size, 27% of the sample was constituted by microenterprises, with 53% small firms, and 20% medium sized.

3.3. *Practical Results*

Trying to indirectly count for the incidence of non-owning managers appointed to coordinate the functional areas of the firm's, the investigated owner-managers were firstly asked to mention which are the functional areas that are directly coordinated and controlled by them (if such functional areas are defined and structured). The answers are centralized in Table 1.

Table 1. Functional areas personally coordinated by the owner-manager

The Functional Area	**Personally coordinated**		**Coordinated by an appointed manager**		**Undefined Functional Area**	
	No.	Percentage	No.	Percentage	No.	Percentage
Research & Development	100	50%	60	30%	40	20%
Production	166	83%	34	17%	0	-
Sales & Marketing	166	83%	34	17%	0	-
Finance & Accounting	60	30%	140	70%	0	-
Human Resources	114	57%	86	43%	0	-

Looking at the results, we can't back-up Bridge et al. [6] when stating that functional managers are absent in a SME. What we can notice is that Finance & Accounting, Human Resources, and Research & Development are the functional areas which are more preferred for delegation, with percentages that exceed the share of the medium sized enterprises amongst the sample (precisely 20% - see Figure 3b). We made this remark, since it's mostly assumed that enterprises that exceed one hundred employees are expected to have a cadre of non-owning managers. Nevertheless, even more, the results synthesized in the above table confirm the assumption that owner-managers (at least those from the investigated sample) do appoint others to be, even if partially, responsible for the management of others not only in medium sized firms.

It's also interesting to notice that the sampled owner-managers are quite unwilling to share the management of Production, and Sales & Marketing, the latest being mentioned by Nooteboom [19], as well as Wilson and Bates [18] as one function to be performed by the owner-manager. Nonetheless, another study, performed by Voicu and Sava [20], proved that managers of the Romanian industrial companies generally tend to be aware of the widespread impact the marketing has on the overall company's performance.

This approach leads to a kind of integrated management, with an overall view of the firm [6]. A simultaneous coordination allows the owner-manager to link the customers, and their specific demands, with manufacturing, such ensuring a greater flexibility in production and an increased possibility to fulfil and overcome customer expectations. By keeping a total control over these two areas, the owner-manager makes sure that customers' satisfaction, on-time delivery and other related aspects are guaranteed. When new products are being developed, the requirements, needs, desires, and the expectations of consumers are translated into technical, i.e. product specifications [21].

These outcomes are also consequent with the results of Stanworth and Gray [3], ranking production and marketing as areas with high level of expertise, and consequently importance.

Moreover, Draghici and Draghici [22] underlined that competitive advantage is often related to the core competencies of the organization, and such competencies are primarily found in Production and Marketing. Core competencies are frequently based on implicit expertise or tacit knowledge, which is an intangible, unstructured asset. Additionally, since the owner is the main source of such knowledge and expertise, the best way to fully exploit them is by assuming the main decisions concerning the above mentioned functional areas.

Human Resources scores 57% as directly coordinated by the owner-manager and this confirms the characteristic of personal entrepreneurship and direct personal work relations [7], [18], [23].

Summarizing, it appears that most of the investigated owners still prefer to manage their company by direct physical intervention, rather than to control the business by monitoring information, as recommended by Burns [23]. If we take into consideration the high level of personal implication in the functional management of the owner-managers, we can state that Burns spider's web [23] occurrence amongst the investigated firms is quite widespread, regardless their development stage.

3.4. Determining the correlations between the average number of employees and the organizational management of the investigated sample of firms

Looking at the general trend of the previous results, we expect that the smaller the firm, the greater the involvement of the owner-manager in personally managing the functional areas of the firm will be.

Table 2. Distribution of answers amongst categories and classes

Official categories[1]	Micro (1÷9 employees)			Small (10÷49 employees)						Medium-sized (50÷249 employees)								
Total	**54**			**106**						**40**								
Classes	1÷9			10÷24			25÷49			50÷99			100÷174			175÷249		
Class total	**54**			**72**			**34**			**22**			**12**			**6**		
	Type of coordination																	
Functional area	Personally coordinated	Appointed manager	Non-existent	Personally coordinated	Appointed manager	Non-existent	Personally coordinated	Appointed manager	Non-existent	Personally coordinated	Appointed manager	Non-existent	Personally coordinated	Appointed manager	Non-existent	Personally coordinated	Appointed manager	Non-existent
Research & Development	26	0	28	58	6	8	4	30	0	8	12	2	2	8	2	2	4	0
Production	54	0	0	72	0	0	26	8	0	12	10	0	2	10	0	0	6	0
Sales & Marketing	54	0	0	70	2	0	26	8	0	10	12	0	4	8	0	2	4	0
Finance & Accounting	24	30	0	28	44	0	8	26	0	0	22	0	0	12	0	0	6	0
Human Resources	40	14	0	42	30	0	16	18	0	12	10	0	4	8	0	0	6	0

[1] http://ec.europa.eu/enterprise/policies/sme/files/sme_definition/sme_user_guide_en.pdf

Consequently, we initiated a statistical analysis to check if a mathematical correlation between the average number of employees and the way the functional areas are managed can be established. For this purpose, the sample of firms was further divided, by the number of employees, into six classes - where dedicated literature recommends at least five [24]. The questionnaire's results, grouped in classes, are introduced by table 2.

The existence of such correlations was graphically verified, by the means of two statistic methods, a nonparametric one (Kendall tau rank correlation coefficient), and a parametric one (Pearson's product-moment coefficient), as presented below, for all five functional areas.

Based on data introduced by Table 3, we calculated the regression lines (Figure 2) and then the Kendall and Pearson coefficients for Research & Development (Table 4). The point clouds in Figure 2 are quite diffuse, which means weak correlations, and the Kendall and Pearson coefficients have small values, while the levels of significance for the one-tailed tests reaches high values. These results indicate that, in the case of Research & Development, no correlation can be established between the average number of employees and the way the function is managed.

Table 3. Distribution of answers amongst classes for Research & Development

RESEARCH & DEVELOPMENT			
Average number of employees	Personally coordinated [%]	Appointed manager [%]	Non-existent [%]
5	48.15	0.00	51.85
17	80.56	8.33	11.11
37	11.76	88.24	0.00
74,5	36.36	54.55	9.09
137	16.67	66.67	16.67
212	33.33	66.67	0.00

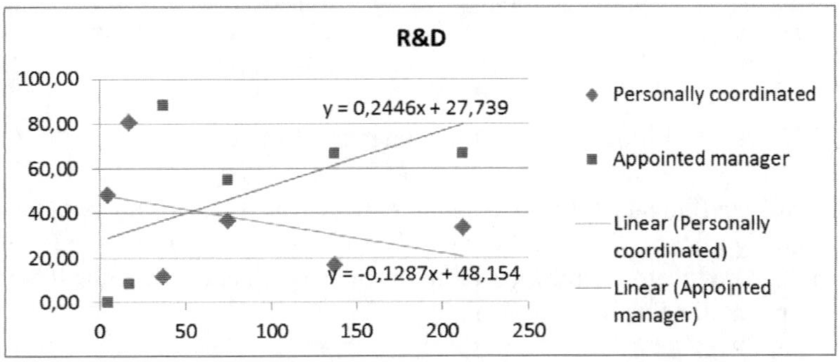

Figure 2. Regression lines for Research & Development (R&D)

Table 4. The Kendall and Pearson coefficients for Research & Development

Research & Development	**Average number of employees** Kendall Test		**Average number of employees** Pearson Test	
	Calculated coeff., **τ**	Calculated risk, **α**	Calculated coeff., **r**	Calculated risk, **α**
Personally coordinated	-0.333	0.174	-0.416	0.206
Appointed manager	0.552	0.063	0.555	0.126

One possible reason for such a deviation resides in the fact that the Research & Development covers activities that are not always present within a firm (like Technical conception, for example, which is more specific to complex manufacturing activities, from bigger firms). Also, the other Research & Development activities tend to be underdeveloped or unstructured, so the owner-manager may want to maintain a direct coordination and control of those, even when the firm gets bigger.

Table 5. Distribution of answers amongst classes for Production

PRODUCTION			
Average number of employees	Personally coordinated [%]	Appointed manager [%]	Non-existent [%]
5	100.00	0.00	0.00
17	100.00	0.00	0.00
37	76.47	23.53	0.00
74.5	54.55	45.45	0.00
137	16.67	83.33	0.00
212	0.00	100.00	0.00

From these data, we obtained the regression lines (Figure 3) and the Kendall and Pearson coefficients for Production (Table 6). The point clouds from Figure 3 suggest strong connections (correlations) between the average number of employees and the way the Production is managed.

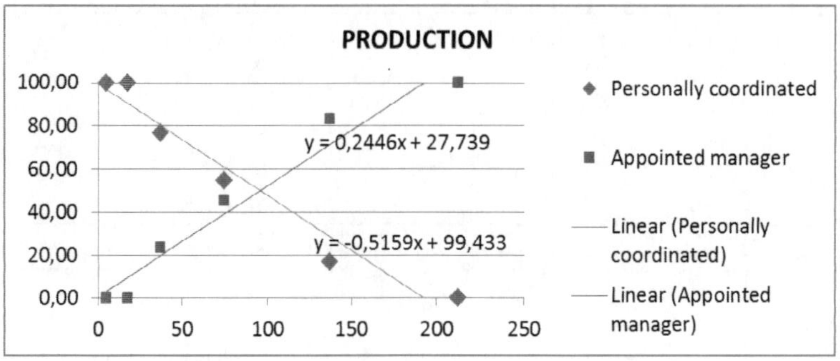

Figure 3. Regression lines for Production

Also, the Kendall coefficients reach high values, which prove a strong correlation between the parameters, while the significance level for a one-tailed test is having a very small value (0.4%). The Pearson coefficients also indicate a strong correlation, the significance level for the one tailed test in this case having a very small value too (0.04%).

Table 6. The Kendall and Pearson tests coefficients for Production

Production	**Average number of employees** Kendall Test		**Average number of employees** Pearson Test	
	Calculated coeff., **τ**	Calculated risk, **α**	Calculated coeff., **r**	Calculated risk, **α**
Personally coordinated	-0.966	0.004	-0.977	0.0004
Appointed manager	0.966	0.004	0.977	0.0004

Table 7 introduces the data for Sales & Marketing, while Figure 4 is presenting the regression lines, and Table 8 the Kendal and Pearson test results for the same functional area. Looking at the point clouds from Figure 4, we can again observe strong correlations that appear, in the case of Sales & Marketing, between the average number of employees and the way this functional area is managed.

Table 7. Distribution of answers amongst classes for Sales & Marketing

	SALES & MARKETING		
Average number of employees	Personally coordinated [%]	Appointed manager [%]	Non-existent [%]
5	100.00	0.00	0.00
17	97.22	2.778	0.00
37	76.47	23.53	0.00
74.5	45.45	54.55	0.00
137	33.33	66.67	0.00
212	33.33	66.67	0.00

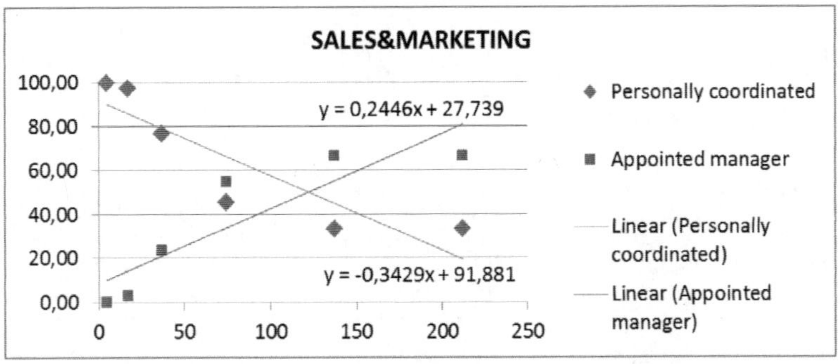

Figure 4. Regression lines for Sales & Marketing

Table 8. The Kendall and Pearson tests coefficients for Sales & Marketing

Sales & Marketing	**Average number of employees** Kendall Test		**Average number of employees** Pearson Test	
	Calculated coeff., **τ**	Calculated risk, **α**	Calculated coeff., **r**	Calculated risk, **α**
Personally coordinated	-0.966	0.004	-0.889	0.009
Appointed manager	0.966	0.004	0.889	0.009

In addition, the Kendall coefficients reach again high values, which make us conclude that there is a strong correlation between the average number of employees and the way the Sales & Marketing are coordinated, the significance level for a one-tailed test having again a very small value (0.4%). The Pearson coefficients also indicate a strong linear correlation, the significance level for the one-tailed test in this case being of a small value (0.9%).

Table 9 data allowed us to establish the regression lines (Figure 5) and then the Kendall and Pearson coefficients for Finance & Accounting (Table 10). The regression lines for Finance & Accounting and the associated point clouds indicate once more the presence of strong correlations in this case too. There is an obvious dependency here: as the number of employees gets higher, the occurrence of an appointed manager for Finance & Accounting is taking place.

On the other hand, the Kendall and the Pearson coefficients are reaching high values in this case too, and the significance level for one-tailed tests is very low for the Kendall test (0.8%), respectively low for the Pearson test (2.2%). Both these results are indicating a strong correlation between the average number of employees and the management of Finance & Accounting.

Table 9. Distribution of answers amongst classes for Finance & Accounting

	FINANCE & ACCOUNTING		
Average number of employees	Personally coordinated [%]	Appointed manager [%]	Non-existent [%]
5	44.44	55.56	0.00
17	38.89	61.11	0.00
37	23.53	76.47	0.00
74,5	0.00	100.00	0.00
137	0.00	100.00	0.00
212	0.00	100.00	0.00

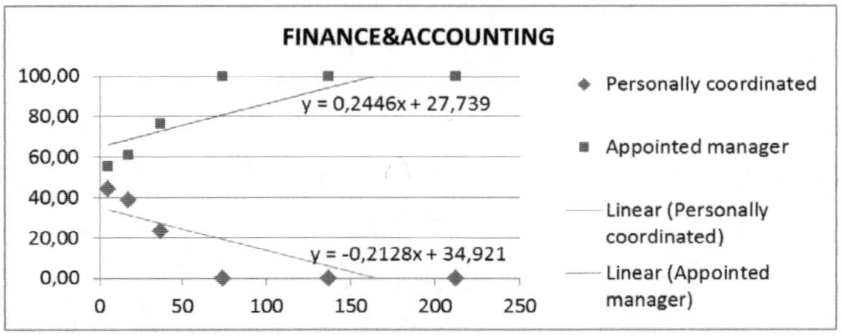

Figure 5. Regression lines for Finance & Accounting

Table 10. The Kendall and Pearson tests coefficients for Finance & Accounting

Finance & Accounting	**Average number of employees** Kendall Test		**Average number of employees** Pearson Test	
	Calculated coeff., **τ**	Calculated risk, **α**	Calculated coeff., **r**	Calculated risk, **α**
Personally coordinated	-0.894	0.008	-0.825	0.022
Appointed manager	0.894	0.008	0.825	0.022

From the data introduced by Table 11, we obtained the regression lines (Figure 6) and the Kendall and Pearson coefficients for Human Resources (Table 12). Due to the very low degree of diffusion in the point clouds, Figure 8 suggests that in case of this last functional area, Human Resources, we can identify powerful correlations in the way the function is managed and the average number of employees.

Table 11. Distribution of answers amongst classes for Human Resources

HUMAN RESOURCES			
Average number of employees	Personally coordinated [%]	Appointed manager [%]	Non-existent [%]
5	44.44	55.56	0.00
17	38.89	61.11	0.00
37	23.53	76.47	0.00
74.5	0.00	100.00	0.00
137	0.00	100.00	0.00
212	0.00	100.00	0.00

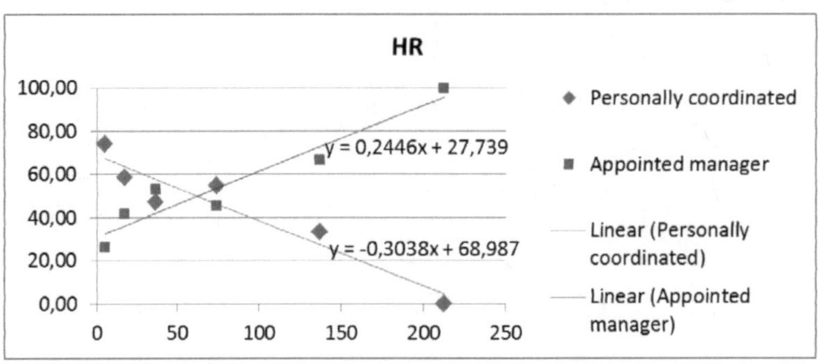

Figure 6. Regression lines for Human Resources (HR)

Table 12. The Kendall and Pearson tests coefficients for Human Resources

HR	**Average number of employees** Kendall Test		**Average number of employees** Pearson Test	
	Calculated coeff., **τ**	Calculated risk, **α**	Calculated coeff., **r**	Calculated risk, **α**
Personally coordinated	-0.867	0.007	-0.951	0.002
Appointed manager	0.867	0.007	0.951	0.002

Performing the Kendall tests, we got high value coefficients, with a very low significance level (0.7%). The Pearson test also indicates a powerful linear correlation for the Human Resource management, due to the very low significance level that was obtained (0.2%).

Looking at the results introduced so far, we can observe clear evidences of strong correlations between the average number of employees and the direct implication of the owner-manager in the management of each functional area of a firm in four out of five cases (the exception was Research and Development). Therefore, we decided to perform an overall analysis, applying the same tests for the aggregated data. The results are presented in the Tables 13, 14, and Figure 7.

The point clouds from Figure 7 indicate that overall correlations also appear: precisely, as the number of employees grows, there is a clear tendency to have an appointed manager designated to coordinate the functional areas of the firm.

Table 13. Distribution of answers amongst classes for the aggregated data

	AGGREGATED DATA		
Average number of employees	Personally coordinated [%]	Appointed manager [%]	Non-existent [%]
5	73.33	16.30	10.37
17	75.00	22.78	2.22
37	47.06	52.94	0.00
74.5	38.18	60.00	1.82
137	20.00	76.67	3.33
212	13.33	86.67	0.00

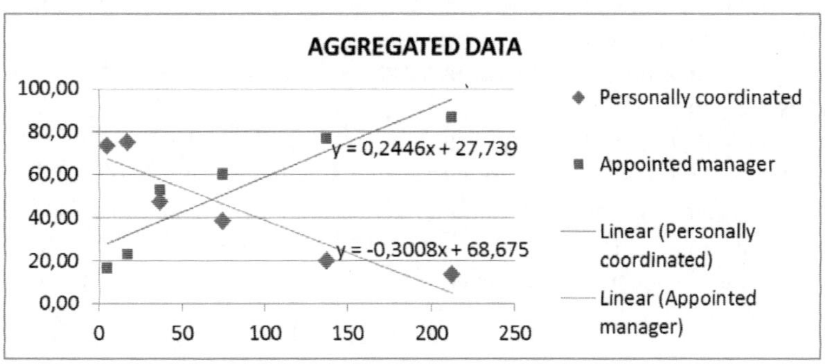

Figure 7. Regression lines for the aggregated data

Table 14. The Kendall and Pearson tests coefficients for the aggregated package of data

Aggregated data	**Average number of employees** Kendall Test		**Average number of employees** Pearson Test	
	Calculated coeff., **τ**	Calculated risk, **α**	Calculated coeff., **r**	Calculated risk, **α**
Personally coordinated	-0.867	0.007	-0.928	0.004
Appointed manager	1	0.000	0.918	0.005

The Kendall coefficients are significantly high, and this suggests an obvious linear correlation between the above mentioned two aspects (the personal involvement of the owner-manager and the average number of employees) in the case of the overall management of the firm's functional areas; the significance level for the one-tailed test achieves very small values (0.7%, and 0.0% respectively).

The Pearson coefficients also denote the same powerful correlations as the Kendal coefficients did. The one-tailed test' significance level registers low values, of 0.4%, and 0.5% respectively.

4. Conclusions

Concluding on the performed investigation and the dedicated literature review, we can iterate that the small business owner-manager tends to be all-powerful, more inclined to listen to his or her own inclinations, and to rely on his/her personal experience.

Small business managers are responsible for a wide variety of tasks in the company (from operational to strategic), same as Verhees and Meulenberg [25] pointed out.

The owner-managers from the investigated sample put a great emphasis on two functional areas: Production and Sales & Marketing, these two functional areas being personally managed by a large amount of the questioned entrepreneurs. Such a behaviour may be the source of some important features a small business exhibits and which ground their main competitive advantages, such as customized products, high degree of customer satisfaction, good connection between business and market, flexibility, autonomy and successful niche-market strategies. By controlling the two most important functions of the enterprise (Sales & Marketing and Production), the owner-managers make sure that all the specific demands of the customers are acknowledged, understood (by quickly disseminating the information amongst the staff, to whom the owner-manager maintain a close, personal relationship), implemented into products, and then manufactured accordingly. So, the investigated small firms are substituting the functional specialization for combined and integrated management, with a strong touch of informality. The founders understand the fragility of competitiveness and are prepared to bend over backwards to provide high and rising levels of service, as Wilson and Bates [18] outlined too.

Verifying if a mathematical correlation between the average number of employees and the personal implication of the owner-manager in the functional management of the firm exists, we can confirm such correlations for four out of five functional areas (the exception was the Research & Development). Thus, the smaller the firm, the more involved in managing the functional activities of the firm the owner-manager is, aspect that was confirmed when the aggregated data were tested too.

References

[1] Curran J and Blackburn R A 2001 *Researching the Small Enterprise*, London, Sage

[2] ***Bolton Report 1971 *Report on the Commission of Enquiry on Small Firms*, London, Cmnd 4811, HMSO

[3] Stanworth J and Gray C (Editors) 1991 *Bolton 20 Years On: The Small Firm in the 1990s*, published on behalf of Small Business Research Trust and commissioned by the National Westminster Bank, London, Paul Chapman Publishing Ltd

[4] ***European Union 2003 Commission Recommendation 2003/361/EC, *Official Journal of the European Union* **L124**(36)

[5] Welsh J A and White J F 1981 A Small Business Is Not a Little Big Business, *Harvard Business Review* **59**(4) 18-32
[6] Bridge S, O'Neill K and Cromie S 1998 *Understanding Enterprise, Entrepreneurship and Small Business*, London, Macmillan Business
[7] Driel H J van and Poutsma E 1989 SME and Control: Autonomy for Flexibility, Poutsma E and Walravens A (Editors) *Technology and Small Enterprises. Technology, Autonomy and Industrial Organisation*, Delft University Press, pp 105-123
[8] Lepoutre J and Heene A 2006 Investigating the Impact of Firm Size on Small Business Social Responsibility: A Critical Review, *Journal of Business Ethics* **67** 257-273
[9] Carland J W, Hoy F, Boulton W R and Carland J A 1984 Differentiating Entrepreneurs from Small Business Owners: A Conceptualization, *The Academy of Management Review* **9**(2) 354-359
[10] Storey D J and Sykes N 1996 Uncertainty, innovation and management, Burns P and Dewhurst J (Editors) *Small Business and Entrepreneurship*, Second Edition, Basingstoke, Palgrave Macmillan Business, pp 73-93
[11] Shaver K G and Scott L R 1991 Person, Process, Choice: The Psychology of New Venture Creation, *Entrepreneurship Theory & Practice* **16**(2) 23-45
[12] Stewart Jr. W H, Watson W E, Carland J C and Carland J W 1999 A proclivity for entrepreneurship: A comparison of entrepreneurs, small business owners, and corporate managers, *Journal of Business Venturing* **14**(2) 189-214
[13] Ateljevic J 2007 Small tourism firms and management practices in New Zealand: The Centre Stage Macro Region, *Tourism Management* **28** 307-316
[14] Miller N J, Mcleod H and Oh K Y 2001 Managing family businesses in small communities, *Journal of Small Business Management* **39**(1) 73-87
[15] Barth E, Gulbrandsen T and Schøne P 2005 Family ownership and productivity: the role of owner-management, *Journal of Corporate Finance* **11** 107-127
[16] Coleman J S 1990 *Foundations of Social Theory*, Cambridge, MA, The Belknap Press of Harvard Univ. Press
[17] Burkart M, Panunzi F and Shleifer A 2002 *Family firms*, Working paper 8776. Cambridge, MA: National Bureau of Economic Research
[18] Wilson P and Bates S 2003 *The Essential Guide to Managing Small Business Growth*, Chichester, John Wiley & Sons
[19] Nooteboom B 1994 Innovation and Diffusion in Small Firms: Theory and Evidence, *Small Business Economics* **6**(5) 327-347
[20] Voicu I V and Sava A 2015 The Attitude of Romanian Industrial Companies towards the Market during the Transition, *Acta Polytechnica Hungarica* **12**(15) 81-99
[21] Dejanović A M, Nikolić S T and Stanković J 2015 Integral Model of Strategic Management: Identification of Potential Synergies, *Acta Polytechnica Hungarica* **12**(8) 115-133
[22] Draghici A and Draghici G 2007 New Business Requirements in the Knowledge-Based Society, Cunha M M, Cortes B C and Putnik G D (Editors) *Adaptive Technologies and Business Integration: Social, Managerial and Organizational Dimensions*, Idea Group Publishing, Information Science Publishing, IRM Press, CyberTech Publishing and Idea Group Reference, pp 211-243
[23] Burns P 2002 *Entrepreneurship and Small Business*, Hampshire, Palgrave
[24] Mendenhall W, Beaver R J and Beaver B M 2009 *Introduction to Probability and Statistics*, 13th Edition, Belamont CA, USA, Brooks/Cole
[25] Verhees F J H M and Meulenberg M T G 2004 Market Orientation, Innovativeness, Product Innovation, and Performance in Small Firms, *Journal of Small Business Management* **42**(2) 134-154

IOP Conf. Series: Materials Science and Engineering **200** (2017) 012033 doi:10.1088/1757-899X/200/1/012033

Software architecture for a multi-purpose real-time control unit for research purposes

S Epple[1], R Jung[1], K Jalba[2] and V Nasui[3]

[1]University of Applied Sciences Kempten, Research Centre, Innovapark 20, 87600 Kaufbeuren, Germany
[2]University of Applied Sciences Ulm, Production Engineering and Production Economics, Prittwitzstraße 10, 89075 Ulm, Germany
[3]Technical University Cluj Napoca, Northern University Centre of Baia Mare, Faculty of Engineering, Strada Doctor Victor Babeş 62A, Baia Mare 430083, Romania

E-mail: steffen.epple@hs-kempten.de

Abstract. A new, freely programmable, scalable control system for academic research purposes was developed. The intention was, to have a control unit capable of handling multiple PT1000 temperature sensors at reasonable accuracy and temperature range, as well as digital input signals and providing powerful output signals. To take full advantage of the system, control-loops are run in real time. The whole eight bit system with very limited memory runs independently of a personal computer. The two on board RS232 connectors allow to connect further units or to connect other equipment, as required in real time. This paper describes the software architecture for the third prototype that now provides stable measurements and an improvement in accuracy compared to the previous designs. As test case a thermal solar system to produce hot tap water and assist heating in a single-family house was implemented. The solar fluid pump was power-controlled and several temperatures at different points in the hydraulic system were measured and used in the control algorithms. The software architecture proved suitable to test several different control strategies and their corresponding algorithms for the thermal solar system.

1. Introduction

The needs and approaches for a control unit suitable in research environments was analyzed in [1]. This paper gives an overview of the software structure for this control unit.

The hardware architecture is shown in Figure 1. The hardware consists of a heterogeneous internal communication. The digital multiplexers are connected via SPI to the Microcontroller. The analog multiplexer is connected to one ADC input pin of the Microcontroller and switching is performed with an external integrated circuit. This is controlled by five port pins of the microcontroller (four pins addressing and one pin for inhibit). The real time clock, not shown in Figure 1, is connected via Two Wire Interface (TWI) protocol with chip select signals to normal port pins. LCD display is connected in four bit mode to one of the four 8-bit digital multiplexing output ports. One free pin of this port is used to control the backlight. The push buttons and indication LEDs are each connected directly to an output or respectively input pin of the digital multiplexer.

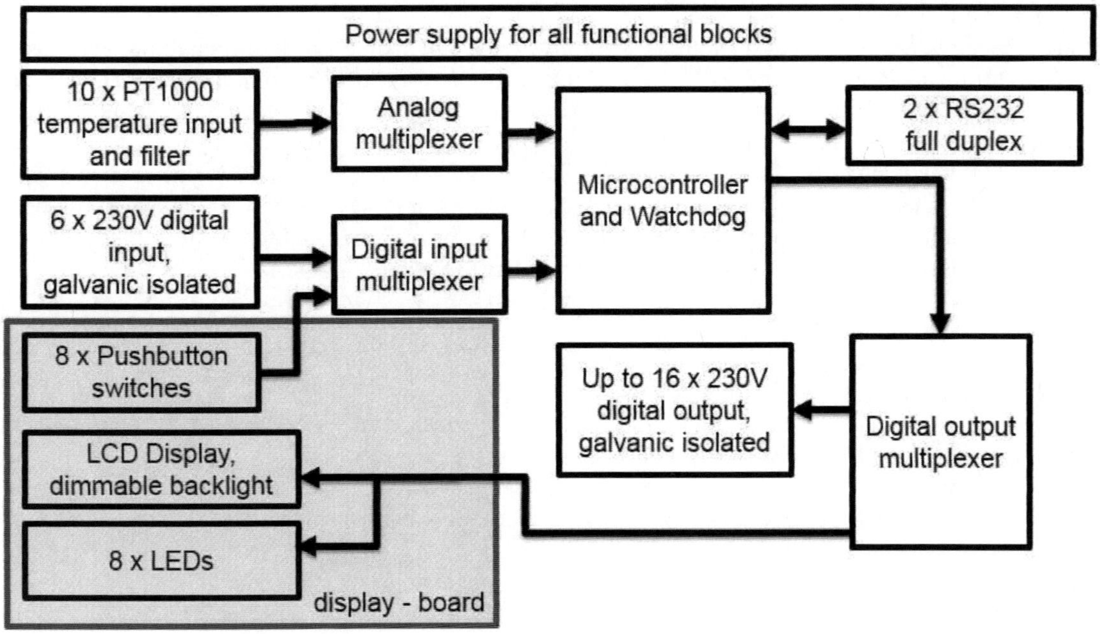

Figure 1. Simplified block Diagram of the intended functions [1]

2. Why another free programmable control unit?

Conventional programmable logic controllers (PLC) are usually built up by blocks of predefined software, like PID block, Logic blocks and others. A great advantage is the reduction of development time by using these predefined blocks of software. On the other hand a great disadvantage in the case of different control strategies and algorithms to be tested, conventional PLCs usually cannot be implemented appropriately.

Another great advantage of free programmability is timing control. Whenever faster processing is needed, it can be implemented. For instance in a detected dangerous state of the system under control, the free programmable PLC can be programmed to act upon detection of the situation within a few clock cycles of the Microcontroller. On standard PLCs hazard detection is part of the control blocks within the normal control cycle. No background test routine can be run at higher speed for instance.

3. Goals of the software design for the new control

In the hardware design the number of input and output signals needed for the experiments and the type of these signals were defined.

In the software design the routines to control the hardware functions were implemented.

-We implemented some functions in different ways and compared their performance or behavior in the test case application.

- As we used free software for the hardware design, we continued this approach with the software and used the open source WinAVR [2], a windows distribution of the AVR Libc. It comes with gcc compiler and other preconfigured tools [3].

- For static code analysis the open source tool Splint was used [4].

- To gain a deeper understanding of the Atmel mega microcontroller series and its peripherals implementation and to give our students in future a possibility to easily "dive into" the matter of microcontrollers, we kept the implementations close to the datasheet of the controller [5].

- Another goal was to separate the logic and control-logic functions from the implementation of the timing and IO-management. The idea was to change only one .c – file when the application changes.

4. Safety and reliability considerations

Unlike the hardware design considerations, the software design should fulfill one small but no easy to reach point: It should not "get stuck". This is important, as already mentioned in the hardware part [1] in our test case with the solar collector: The sun does not care about our control. So the collector will heat up as long as the sun shines on it. The software must not switch the solar fluid pump on, when the collector is filled with steam. This might cause a steam explosion.

On the other hand, if the controller gets stuck and temperature rises above steam temperature, then the controller must wait until it is safe again to switch on. So the measured temperature should be as exact as possible. Otherwise "sunshine time" might get lost for energy harvest.

5. Conceptual approach

5.1. Microcontroller system

The applied Microcontroller natively offers 128kBytes of Flash, which can be used for program and constants storage, and four kilobytes of RAM.

These memory resources of the microcontroller are pushed with external RAM circuitry to 64 Kbytes of hardware addressable RAM. With manual memory management, 512Kbytes can be used.

64kbytes of RAM turned out to be sufficient for control purposes with the specified control concept. However, we used the rest of the free 64kbytes memory for data logging.

5.2. Operating system

To realize many fast control functions with very little resources, we thought that it is necessary to reduce the operating system functions to an absolute minimum. So we had the idea to leave away any operating system completely and replace it with an architecture that enables real time capability on one hand and allows events on the other hand.

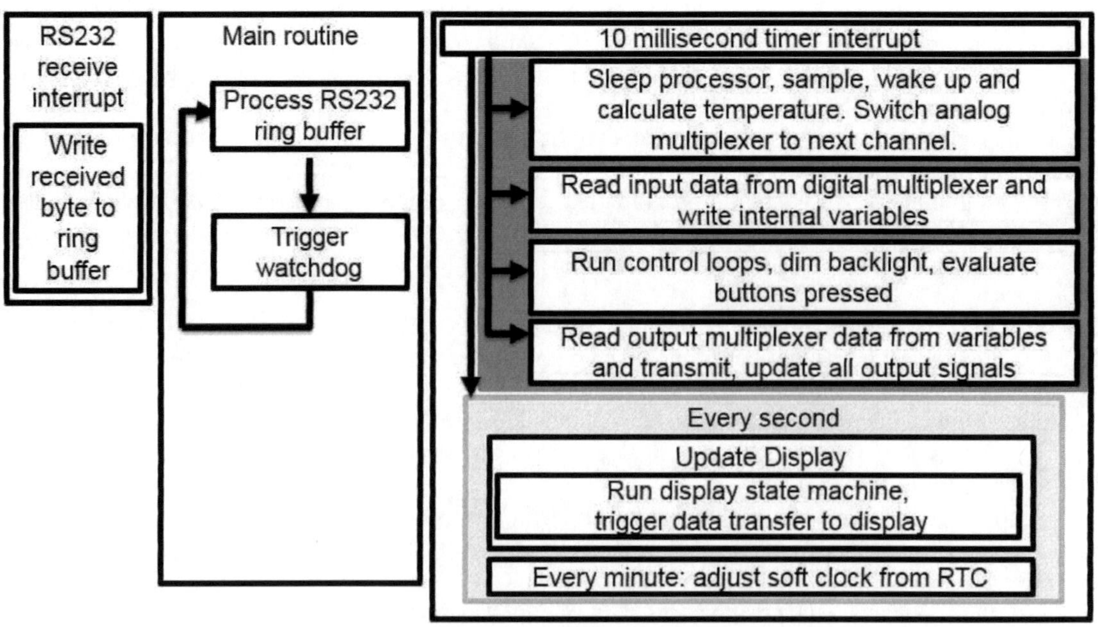

Figure 2. Conceptual structure

Legend: green: comfort functions, blue: real time or time critical functions

The approach is, to use a single large timer interrupt, as shown in Figure 2, which triggers all the time sensitive actions and calls their respective functions. The main timer interrupt was described as suitable for very small systems only in [6]. The timer interrupt was split up to trigger the sampling of only one channel every 10 milliseconds. This behavior was described in [7, chapter 5] and named "to be in time".

5.3. Communication

To allow event based communication, a standard interrupt to receive RS232 bytes was implemented.

The RS232 Data is processed within the main routine. This way, the processing of serial data can be interrupted by the timer and does not disturb the time critical functions. This construct allows real time communication with other controls or RS232-sensors as well as PC communication. Under RS232-sensors we understand laboratory measurement equipment like mass spectrometers, optical spectrometers and so on.

Although we tested only PC communication yet, the second serial port can also be used for daisy chaining. Care must be taken that the controller is not fed with too much serial data within short time. This would lead to data loss of some extent as soon as the receiving ring buffer overflows. This effect could be prevented by a hardware-handshake. Unfortunately the hardware-implementation of the mega128 does not offer hardware-handshake (RTS/CTS) for the serial lines. An implementation in software is a future option. To be able to use the four line LCD display, the display communication algorithms were completely rewritten, since the used display behaved slightly different to several two line variants used before.

5.4. Timing and performance

Much care had to be taken to estimate the runtime of each function to avoid that the last function within the timer interrupt has finished when it is triggered again.

Such overrun situations were prevented by implementing a detection and clearing mechanism. For debugging, occurrences of overrun situations are reported to RS232 and can be logged by a PC. When properly designed, overruns should not occur.

All programming was done with "programmers' notepad" which came along with the "WinAVR" package.

Due to low bit rates in the communication protocols of display and real time clock, communication between the microcontroller and its external peripherals is very time consuming. Many small delays must be implemented in the protocols. Without a dispatcher, this time can hardly be used for other calculations. A not negligible part of processing time is wasted for very small "waiting" purposes. This is particularly the case when protocols are implemented in software and run over normal IO-pins, - like the two wire interface (TWI) communication with the real time clock (RTC).

5.5. Display state machine

To be able to use different manipulation options in each menu, we did not use a list-based implementation for the menu system. Instead we implemented a rather large state machine. Each "menu" (=state) with user input options contains another state machine to process the input actions. In this way we realize modification of values with up/down and +/- buttons operating different in various menus.

6. Conclusion

The chosen software design proved very well in the tested configurations. We could implement and test several different control strategies for a solar system and could see great differences in solar performance, behavior and electric power consumption of the solar system.

Power control in the millisecond timing range proved to work well, with little restrictions due to jitter issues caused by the necessary slow communication with the LCD controller.

The tested power control strategies worked very well with the used solar fluid pumps. One pump was used for desktop-testing, a different brand and power was used in the test system.

Splint proved well to find systematic programming failures that gcc warnings did not complain. At least with the compiler options we used [4].

Whenever we had obscure behavior of the software during the implementation phase, a run through the static code analyzer Splint gave a clear answer where to take a closer look.

The control unit we tested controlled the solar system for several months without a watchdog-reset and without malfunction, proving its stability.

7. Outlook

Further improvements of the software may include hardware handshake on RS232 lines (RTS/CTS) as well as improvements of execution speed, primarily the display update-rate. This may depend on major changes in hardware design. Eventually the PC-RS232 will be replaced by an Ethernet connection with some web content capabilities. The free ram already allowed data logging for a period of about one week at one complete and uncompressed dataset per minute.

References

[1] Epple S, Jung R, Jalba K and Nasui V 2016 *Real time capable control design with increased life expectancy for research purposes*, Innovative Ideas in Science IIS2016, Baia Mare, Romania, November 11-12
[2] ***WinAVR: https://sourceforge.net/projects/winavr/files/
[3] ***http://www.nongnu.org/avr-libc/
[4] Evans D and Larochelle D 2002 *IEEE Software*, Jan/Feb http://www.splint.org/ and http://www.cs.virginia.edu/~evans/pubs/ieeesoftware-abstract.html
[5] ***Atmel Corporation, 2011 8-bit Atmel Microcontroller with 128Kbytes, In System Programmable Flash ATmega128 ATmega128L
[6] Rob W 2006 *Real-Time Systems Development*, Oxford, Butterworth-Heinemann
[7] Heinz W and Uwe B 2005 *Echtzeitsysteme, Funktionsweisen, Anwendungen*, Springer

IOP Conf. Series: Materials Science and Engineering **200** (2017) 012034 doi:10.1088/1757-899X/200/1/012034

Real time capable control design with increased life expectancy for research purposes

S. Epple[1], R. Jung[1], K. Jalba[2] and V. Nasui[3]

[1]University of Applied Sciences Kempten, Research Centre, Innovapark 20, 87600 Kaufbeuren, Germany
[2]Production Engineering and Production Economics, University of Applied Sciences Ulm, Prittwitzstraße 10, 89075 Ulm, Germany
[3]Faculty of Engineering, Technical University Cluj, Northern University Centre of Baia Mare, Strada Doctor Victor Babeş 62A, Baia Mare 430083, Romania

E-mail: steffen.epple@hs-kempten.de

Abstract. A new, freely programmable, scalable control system for academic research purposes was developed. The intention was, to have a control unit capable of handling multiple PT1000 temperature sensors at reasonable accuracy and temperature range, as well as digital input signals and providing more powerful output signals at 230V AC than conventional control units. To take full advantage of the system, control-loops are run in real time. The whole system runs independently of a personal computer. The two on-board RS232 connectors allow to connect further units to use more sensors or actuators or to connect other laboratory equipment, as required. To allow usage for long-time experiments, systematically electronic components with low failure-in-time (FIT) rate have been chosen in order to achieve high life expectancy. This paper describes the third prototype, which now provides stable measurements, and an improvement in accuracy compared to the previous designs. A rough estimation about the expected mean time between failures is given. As test case, a thermal solar system to produce hot tap water and assist heating in a single-family house was implemented. The solar fluid pump was power-controlled and several temperatures at different points in the hydraulic system were measured and used in the control algorithms. The hardware design proved suitable to test several different control strategies and their corresponding algorithms for the thermal solar system.

1. Introduction

When preparing experiments in the test and education laboratories, again and again we encountered similar situations: Deciding which control is the most suitable for the intended specific application. When setting up test benches in the field of air conditioning, more temperature sensors, than standard control systems usually offer, were needed. When setting up the thermal solar energy field test system later described, more sensors were needed, more input channels were needed. For planned future tests more powerful outputs to drive electric valves and mixer motors at 230V will be needed than the systems we looked at could offer at a price that is suitable for unsponsored, independent research. Many of the conventional controls offer relay outputs. These are great for high currents, but not suitable for power control in a mix of pulse width modulated- and leading – edge manner we intended.

The commercially available smaller control units were designed with few input channels and few output channels on board. Additional input/output modules are expensive. The larger commercial control units had enough input/output channels but were financially out of reach and did not offer the capability to be freely programmable.

2. Yet another free programmable control unit?

In order to prepare several experiments for student's lab courses, the same above described difficulties arose.

Additionally none of the available systems was completely free programmable. Real time response times in the millisecond range we could not find. But for fast process control the short response times of control loops are essential. To realize short control loops free programmability is necessary.

The open source control boards like the Arduino and raspberry families are built-up on a modular basis which is great to test ideas and strategies on the desk. But for test benches intended to run for several years, we had objections concerning the lifetime. And the fine granular modularity of Arduino and raspberry input/output modules doesn't make mounting "on-site" an easy task.

So we decided to see if we can build a system that serves our needs and gives our students a good chance to learn how control systems can be built with little resources.

3. Goals of the new design

As goals of the design we defined to have, in contrast to the first version with sixteen temperature inputs, only ten PT1000 temperature inputs to save space on the board. Additionally we wanted to have six inputs at mains voltage of 230V, up to sixteen semiconductor outputs at 230 Volts to power single phase motors.

To be able to modify parameters on-site without the need of a PC, an additional, but optional board was designed to have a small text display (four lines at sixteen characters), eight input buttons and eight LEDs for indication purposes.

A second filter stage at the analog input ports should be added to further reduce the influences of electromagnetic interference (EMI) and hum when long sensor lines are connected. The length of the sensor lines could exceed twenty meters.

To allow timed actions a hardware real time clock (RTC) was added.

4. Safety and reliability considerations

In the laboratory test bench controls are switched off most of the time, so reliability issues that come along with "runtime-hours" are usually insignificant. Functional safety has been taken into account. A risk analysis has been performed. In case of air conditioning the controlled equipment doesn't cause harm in case of a malfunction. No safety measures have to be included.

In the considered case of the air condition test bench the electronic equipment was assumed to be mounted in a place inside the housing where temperatures are relatively high during use. Mounting in a cooler area inside the housing was avoided due to expected condensation problems.

The control application for the solar field test has need for safety, since the solar fluid pump must not run when the collector is filled with steam. If fluid is pumped into the hot collector in this state, a steam explosion is likely to occur.

Hot temperatures although, are known to reduce lifetime of electronic components and likely cause earlier malfunctioning behavior of a device.

To keep the risks low, a big design goal was to minimize malfunction behavior caused by component wear.

To allow easy repair the PCB layout was decided to be fitted on two layers of copper.

5. Conceptual approach

The block diagram in Figure 1 is derived according to sections 3 and 4.

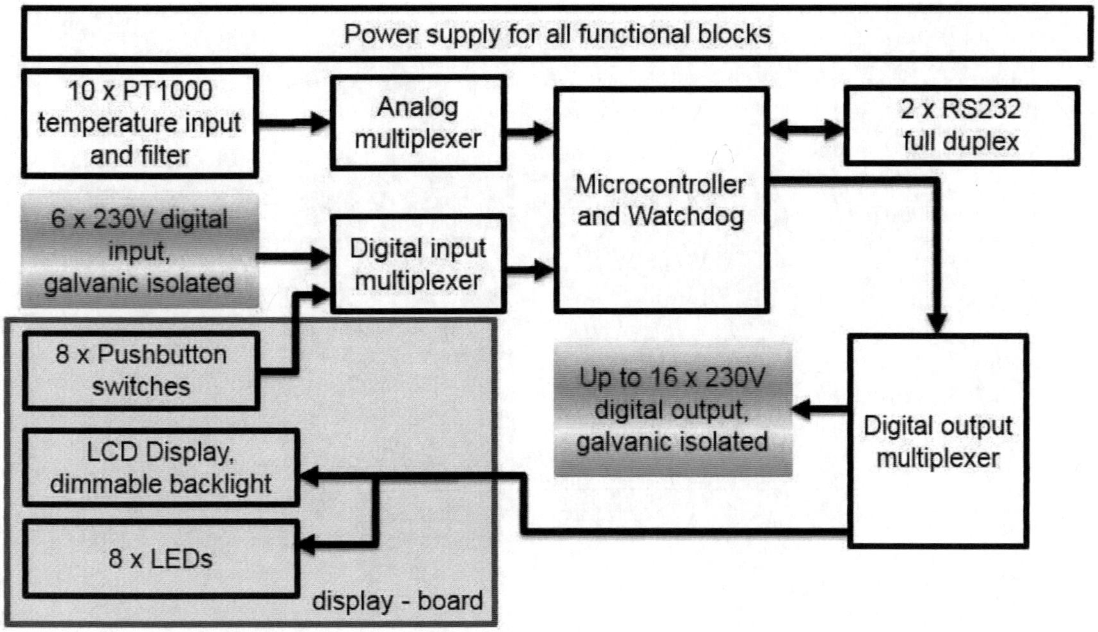

Figure 1. Simplified block diagram of the intended functions

5.1. Explanations of intended functions

In Figure 1 the main functional units are shown as defined in previous sections.

In addition the 230V galvanic isolated parts are marked yellow-green to symbolize that care must be taken to ensure the clearance and creepage distances in the layout according to [7]. The violet block indicates the optional display- and button unit (man-machine-interface, MMI).

Not shown are additional RAM and real time clock.

5.2. Details of selected hardware functions

The digital multiplexers are connected with SPI lines to the Microcontroller.

The multiplexers are structured in a way that allows the reading of the input signals at the same time the outputs are written. Transmitting and receiving of one byte on the SPI bus is done by the built-in USART of the microcontroller. This allows fast Input and Output transitions of the multiplexed input and output signals at relatively low microprocessor operation time.

The real time clock is connected to normal input/output pins of the Microcontroller and operated by SPI protocol implemented in software.

As Controller system we decided to use the Atmel MEGA128 "AU" version. With its 5Volt technology and integrated clamping diodes it is electrically very robust and relatively tolerant to electromagnetic interference (EMI) [1].

To reduce power consumption and self-heating, the Microcontroller runs at only twelve Megahertz.

This frequency is sufficient for all of the tested functionality, including the dimming of the LCD-backlight by pulse width modulation (PWM). To enhance reliability, a ceramic resonator is used instead of a quartz. The reduced accuracy in resonator frequency, compared to a quartz, limits communication speed on the RS232 interfaces to approximately 56k baud.

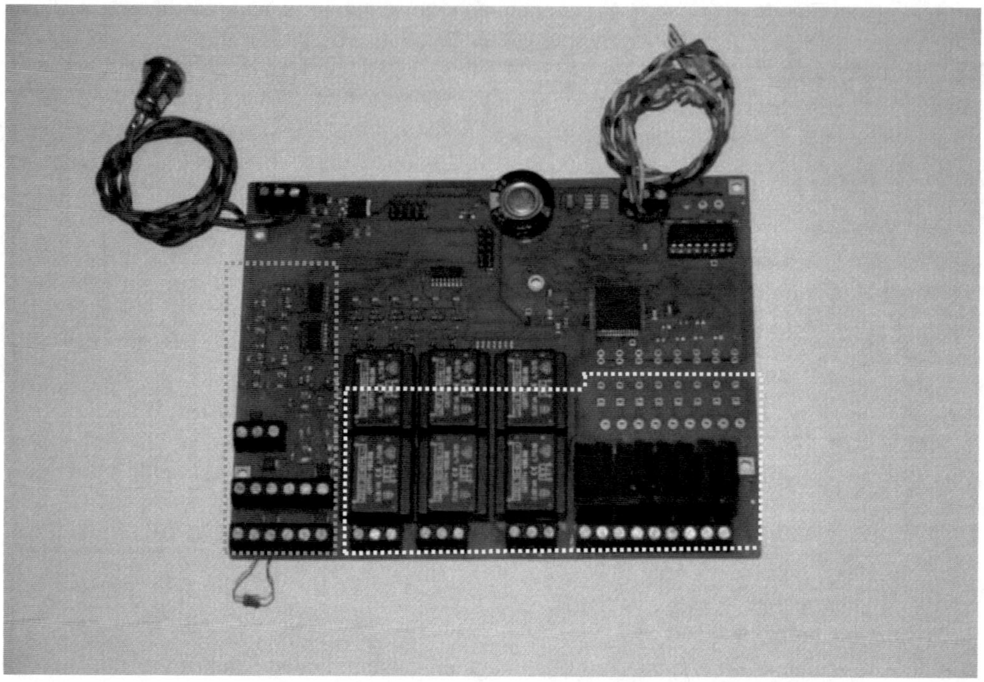

Figure 2. Top view of the main control board.

Legend: yellow: 230Volts domain, blue: analogue domain

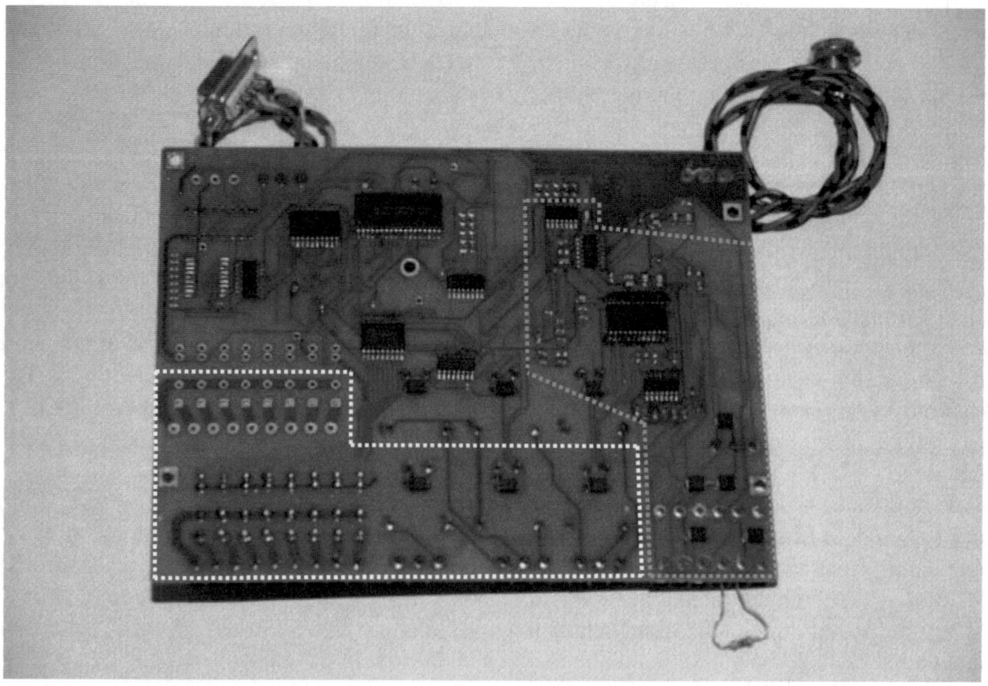

Figure 3. Bottom view of the main control board, colours see Figure 2

5.3. Power supply

To supply the whole system with energy a switching wall plug is used at 7Volts output. To bring these 7 Volts down to the 5V system voltage, a linear voltage regulator is used. This simplifies the handling of EMC filtering and at the very low currents used enhances reliability compared to switching regulators.

To give a rough estimation of power consumption, Table 1 shows the main energy consumers.

Table 1. Main power consumers in design

Microcontroller AT Mega128 AU	About 3mA with output pins low
230V switches (solid state relays)	Each about 8 mA on DC-side in activated state
LCD backlight LEDs	About 160 mA at full power, dimmed about 100mA
Overall current consumption	The measured values were below 200mA with eight power outputs in "on"-state and display backlight dimmed.

5.4. The analog PT1000 temperature measurement circuit

The experiments carried out showed, that the temperature stability of the measurement electronics is far better than expected.

However, the absolute accuracy had to be adjusted manually for each temperature sensor.

The test scenario with the air condition showed that at and below zero degrees Celsius accuracy is not as good as the simulations resulted in. We suppose, that the multiplexer circuits introduce injection currents that influence the measurements in a negative way. Future redesigns will take into account these effects by simulations [5].

The specified temperature range is -30°C to approximately +170°C. Using analog signal conditioning circuits similar to the ones described in [3,4] turned out in our constellation to give relatively large errors and offsets. Experiments have shown that more accurate measurements can be obtained when using oversampling techniques [5] and omitting large signal conditioning circuits. For all measurements the built-in ADC of the microcontroller was used. Massive filtering is needed to handle the noise picked up by long measuring lines. Shielded lines did not significantly reduce the effect.

Wiring was decided to be carried out as two lines per sensor. The change in resistance is much greater than the resistance induced by the wiring as well as the change of the wiring's resistance with temperature. Four line measuring was not implemented due to limited space on the PCB.

5.5. The display unit

The display unit consists of a two-layer PCB with large ground plane (Figure 4) and eight mechanical pushbutton switches. (Figure 5)

These "old fashioned" mechanical switches give an immediate "click"-sound feedback when pressed. This is important in combination with the slow display update rate of one per second.

Measured at the number of activation events during the lifetime of the control, these SMD switches are not a lifetime limiting factor. In figure 4 close to the lower left corner of the PCB an unused double row header can be seen. This is an intended design alternative to have the choice of which HD44780 compatible display module to use.

The non-graphical character display was chosen due to several reasons:

- If graphics are needed, a PC can be connected by RS232.
- The capabilities of the eight bit controller were not intended to be used to the outmost extent.
- The limited RAM of the controller is used to store variables and measurement results for later download by a PC.

IOP Conf. Series: Materials Science and Engineering **200** (2017) 012034 doi:10.1088/1757-899X/200/1/012034

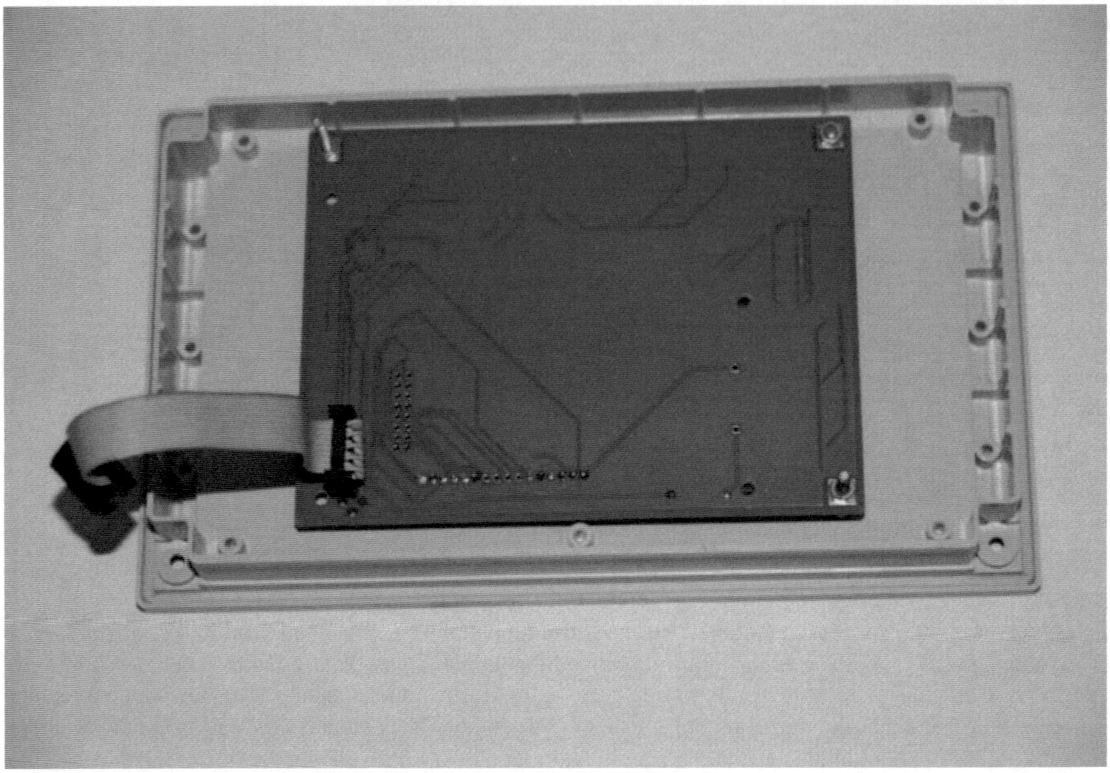

Figure 4. Bottom view of the display board, mounted to the backside of a housing lid

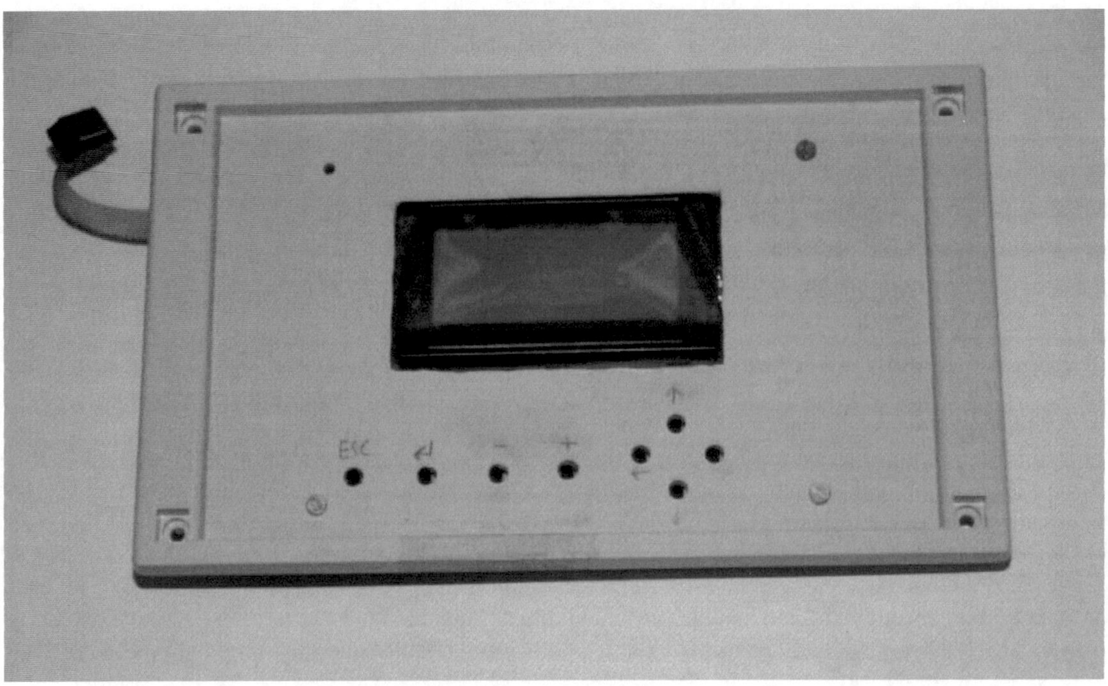

Figure 5. Front view of the display board. LEDs are hidden under the top cover

IOP Conf. Series: Materials Science and Engineering **200** (2017) 012034 doi:10.1088/1757-899X/200/1/012034

6. Reliability estimation according to SN29500

According to ISO 61709 a reliability calculation using the FIT rates of Siemens standard SN29500 was carried out. [6,7,8]

The result is an estimated rate of 1313 FIT, which equals an MTBF value of approximately 86years. Calculation was done for the main board including 8 solid state relays for the digital output. Not included in these calculations is the display-board.

According to the fact, that the display board does neither have an own voltage regulator nor a microcontroller, the only part that may contribute to a non-negligible fit-rate is the four-lines of sixteen characters HD 44780 compatible display which is replaceable by almost any similar type.

7. Conclusion

In our Test scenario with the solar collector it could be shown, that the reliability and functionality is well suited to operate equipment for research purposes. The reset function by watchdog usually should not have to do anything. It occurred, that systematic programming failures sometimes lead to a watchdog reset. As simple the function of the on-chip watchdog is, it turned out to help a lot to ensure the intended and safe functioning of the overall control behavior, resolving “dead-locks” by restart.

As a side effect of the design process we have seen that complex projects like this can be well done with free software. In this case we used TinyCad for the schematics and FreePCB for the board layout. [9, 10]

8. Outlook

Further improvements will be to increase the absolute accuracy of temperature measurements. The practical experiments have shown that exact temperature measurement is a very important point in maximizing solar energy gain.

Especially at low temperature differences e.g. at cloudy weather solar energy gains can be achieved whereas with less precise measurement the system must be stopped much earlier. Depending on the weather conditions, on many days no gain can be realized at all with unprecise measurement – especially when heating support is an intended function during the winter months.

For future improvements the focus will be to enhance absolute accuracy and to make the temperature inputs more flexible, for example by supporting standard 4-20mA industry transmitters.

References

[1] Atmel Corporation 2011 8-bit Atmel Microcontroller with 128Kbytes In-System Programmable Flash ATmega128 ATmega128L

[2] http://www.linear.com/designtools/software/#LTspice

[3] Tietze Ulrich, Schenk Christoph 2002 Halbleiter-Schaltungs-technik, Springer 2002 ISBN 3-540-42849-6

[4] Pfeiffer Wolfgang 1999 Elektrische Messtechnik, VDE 1999, ISBN 3-8007-2316-6

[5] Atmel Corporation, Application Note AVR121: Enhancing ADC resolution by oversampling, Atmel 2005

[6] IEC 61709: 2011 Electric Components – Reliability – Reference conditions for failure rates and stress models for conversion

[7] DIN EN 60664-1 VDE 0110-1:2008-01 Isolationskoordination für elektrische Betriebsmittel in Niederspannungsanlagen

[8] Siemens Standard SN 29500, Ausfallraten Bauelemente – Erwartungswerte, 2004-2010 Siemens

[9] Pyne Matt, Beischer Magnus, Lucas Don: TinyCAD user manual, 2009, https://sourceforge.net/projects/tinycad/

[10] Wright Allan, FreePCB User Guide Version 1.4, 2007, freepcb.com

IOP Conf. Series: Materials Science and Engineering **200** (2017) 012035 doi:10.1088/1757-899X/200/1/012035

Retrofit concept for small safety related stationary machines

S Epple[1], C K Jalba[2], A Muminovic[2] and R Jung[1]

[1]University of Applied Sciences Kempten, Research Centre, Innovapark 20, 87600 Kaufbeuren, Germany
[2]University of Applied Sciences Ulm, Production Engineering and Production Economics, Prittwitzstraße 10, 89075 Ulm, Germany

E-mail: steffen.epple@hs-kempten.de

Abstract. More and more old machines have the problem that their control electronics' lifecycle comes to its intended end of life, whilst the mechanics itself and process capability is still in very good condition. This article shows an example of a reactive ion etcher originally built in 1988, which was refitted with a new control concept. The original control unit was repaired several times based on manufacturer's obsolescence management. At start of the retrofit project the integrated circuits were no longer available for further repair of the original control unit. Safety, repeatability and stability of the process were greatly improved.

1. Introduction

Retrofitting is especially interesting for companies who want to keep their old machines for backup purposes or for production of small quantities at today's quality standards.

This raises interest equipping old machines with new controllers, refitting to up-to-date safety regulations and, regarding modern industry concepts, to add connectivity and remote supervision functions. The considered equipment was a reactive ion etcher model 1540-S RIE/PE from VacuTec AB. At the time of the retrofit the machine played a central role in the production process working in three shifts at sixteen hours machine operation time per day. Manual operation was necessary, since the old control did not support automated processing, what was required in this special case. Small interferences on almost any electric line caused an immediate shutdown without logging process parameters and product data of processed material. Finding the failure without knowing the cause of the shutdown was a difficult task. Documentation of the processes was done manually and the set points for the processing media (different gas flows) and vacuum pressure were done by ten-turn adjustable potentiometers. Today's safety needs were missed (e.g. starting pump down at open process chamber might damage the pump). As the machine was operated manually [1] an operator had to survey the machine at all times.

2. Goals of the retrofit

Three top priority goals of the retrofit were identified: First to meet the up-to-date needs of safety regulations.

Second to improve reproducibility and accuracy of the processes.

And third to save operator time by automating the processes as far as possible. Another intended benefit was to stay independent of certain component manufacturers and allow the exchange of components independent of a specific supplier.

3. Working principle of the machine

Reactive ion etching (RIE) is used as a method to chemical and mechanical etch the surface of a masked material. Adequate gases like SF_6, CHF_3 and O_2 are set in a state of high frequency plasma at low pressure.

A DC-offset voltage accelerates free radicals out of the plasma. Now two different processes happen in parallel:

Firstly the chemical etching process: accelerated radicals impinging the surface of the substrate intended to be etched, react chemically with surface parts and the gaseous reaction products (usually fluorine compounds, sulfuric compounds and oxidized photoresist molecules) are pumped out of the process chamber.

Secondly the mechanical etching process: some of the accelerated ions are less reactive, e.g. carbon or argon. These ions crash into the surface and sputter out small parts of the surface.

Due to incomplete reactions the gases pumped out are noxious.

4. Analysing the machine structure and primary conclusions

The lack of accurate documentation and the fact that many undocumented changes to the machine were done by the owner made it necessary to analyze the complete technological structure and measure the connections of each single wire. As a result an accurate wiring diagram and a set of structural representations of the machine were determined:

- The vacuum system consisting of the process chamber, Valves and two pumps in line, a pressure controller and control valve
- The high frequency system consisting of Electrodes in the process chamber, wiring, a matching network and a Radio Frequency (RF) generator,
- The gas inlet control system consisting of several valves and mass flow controllers,
- A set of sensors consisting of Vacuum gauges, control component feedback signals, set point generators and actual flow feedback from the mass flow controllers.
- Some signal changing components, e.g. magnetic valves to control the pressurized air operated gas flow on-off valve and larger vacuum valves as well as voltage to frequency converters to read back the actual value input.
- The safety circuits, including a main relay, cooling water flow-switches, emergency-power-off button, relay logic.
- The (original) CPU-500 control unit.
- Power supplies

The analysis showed that several components are likely often to be replaced by spare parts of different manufacturers, like the turbo pump and its controller, or the pressure controller. Other components are, in this case, technologically obsolete. E.g. the voltage-frequency converters. Some components have already been replaced by newer and more accurate models, like the mass-flow controllers.

5. Safety analysis

Regarding the components of the machine, the process itself, and the handling of the machine by the operator, the main risks were identified and the according measures were derived.

The main risks are:

- Contact of the operator with reaction products
- Contact with high frequency/voltage at the electrodes
- Rotating masses getting out of control (turbo pump)
- Getting body parts crushed in the lid-mechanism of the chamber

An analysis according to IEC 61508 [2], [3] came to the result that Safety Integrity Level (SIL) 2 is required to keep the rotating masses under control and SIL1 is required for the other main hazards.

6. Things we considered about standard PLCs in Retrofit constellation

Standard programmable logic controllers used in retrofits, have certain intrinsic risks:

Adding input/output modules can rapidly become very expensive and bears the same risk of "getting outdated" as the original controller. When the version of a PLC changes, software can be hard to be "re-used". When machine components wear out and are replaced, often new communication standards are required and PLCs need new or additional communication modules.

Procurement is often time-consuming and when PLCs get older, these modules are no longer available which may end up in an unexpected early next retrofit.

The Cycle-time of standard PLC does not necessarily meet the reaction times required for certain processes.

7. Conceptual approach

The concept is based on five main ideas:

- Stacking of three medium sized mobile electronic control units (ECUs) with safety capabilities interconnected safely by CAN-Bus and signaling lines.
- Parameterization by a standard-PC via CAN-Bus.
- To ensure that the PC does not lock up the processes the control routines run entirely on the control units.
- Running the safety-critical considered software-functions completely on the individual controls, independent of the other control units and independent of the PC ensures that at least two of the three controllers shut down safely if one controller fails. In case of a more important failure, the hardware-logic will shut down the failed controlling unit safely.
- Programming the Control units in "C", using the provided multitasking environment of the manufacturer allows short reaction times in the range of 1-6 milliseconds [4].

This concept is shown in Figure 1. The black arrows indicate the CAN-Bus.

To allow PC-independent process step control, a direct communication between the control units via CAN bus, indicated by the green arrows is realized.

Black and green colored lines show functionality. Both are realized on the same single physical CAN bus.

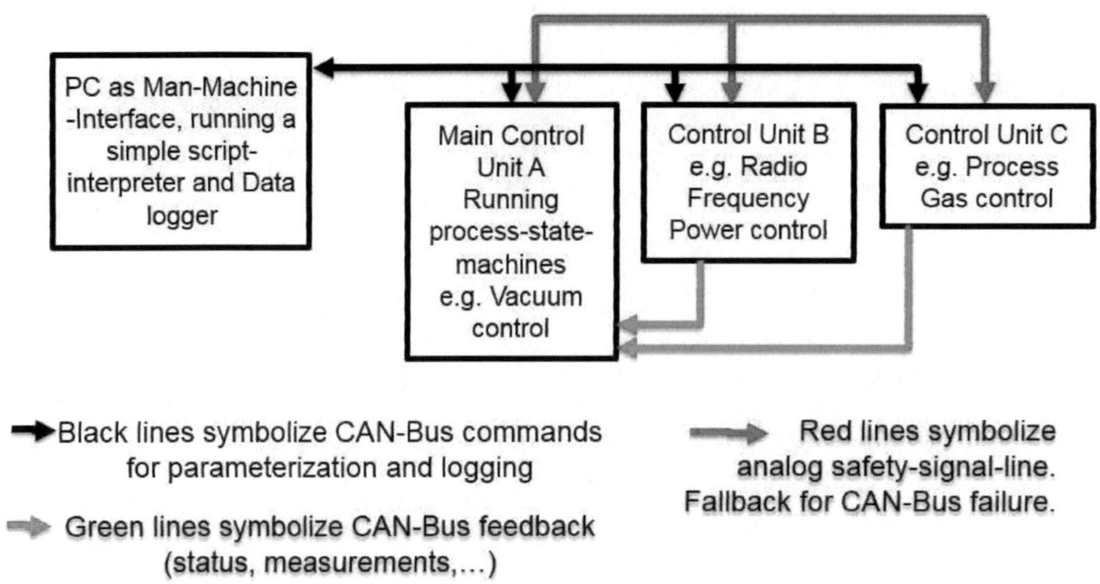

Figure 1. Communication architecture of the new control design

According to ISO 13849 [5] a second path as a very simple fall back communication was implemented, indicated by the red lines in Figure 1. Any of the three control units can pull this signal low to indicate a detected fault state. This results in a controlled shutdown of the other two controllers. Non-answering to a CAN-Bus command within a timeout is also detected as malfunctioning and triggers this mechanism.

While the CAN-Bus is operating in normal mode, all expected and unexpected events will be logged. In case of a failure the process engineers will have a chance to find out the reasons that led to a shutdown or process interruption.

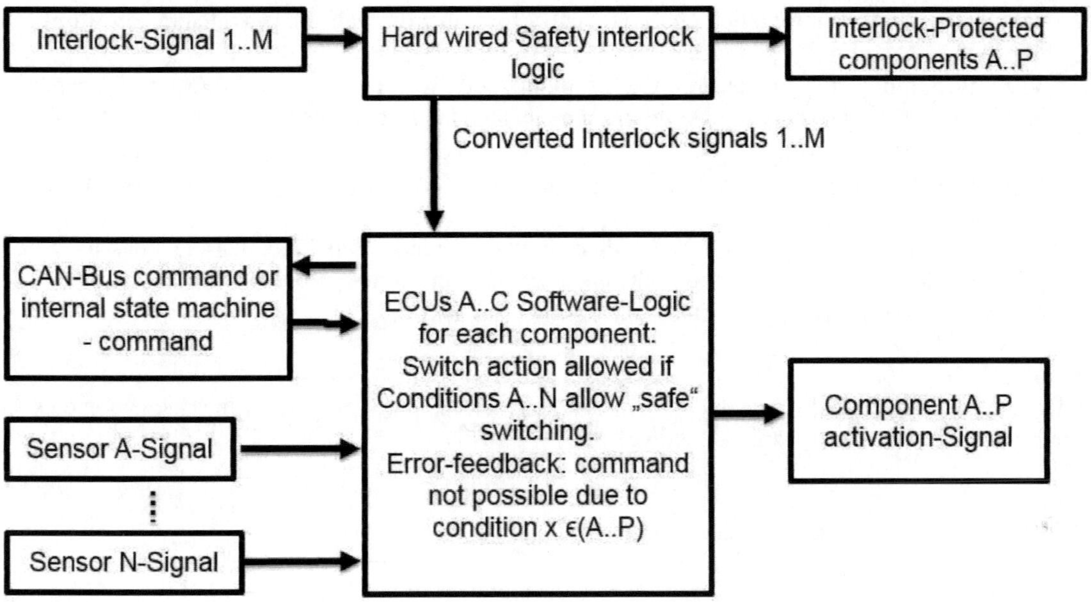

Figure 2. Safety mechanisms by design

To avoid the machine getting into hazardous states, before each switching action certain checks are performed as shown in the lower part of Figure 2.

Figure 3. Machine before retrofit

Figure 4. Machine after retrofit

IOP Conf. Series: Materials Science and Engineering **200** (2017) 012035 doi:10.1088/1757-899X/200/1/012035

8. Conclusions

- It is possible to successfully use Electronic Control Units of the type "ESX 2, by STW" that were built and intended for use in 16-32V mobile machine environments, as well as in stationary machines. All three controllers and their power supply fit into a standard nineteen inch rack case, as shown in the upper left corner of the machine in Figure 4. For comparison the machine before the retrofit is shown in Figure 3.
- It is possible to implement safety relevant control functions with Electronic Control Units independently of their intended usage.
- In this case it was possible to achieve the required levels of safety without the need of safety-software or safety control units. The safety paths for critical signals are provided in hardware logic.
- The much more rugged construction of mobile controls performs a lot better in harsh electromagnetic interference (EMI) environments than the original industrial standard control.
- The "worst case" considered crash happened unexpectedly after the retrofit. Nobody was injured, the machine had no further damage. The safety concept has been proven to be successful.
- Reproducibility and accuracy of the processes were greatly improved. Failure rates of the processes are almost zero, as long as the operator starts the correct process.
- Operator time is greatly saved. Once started, the machine runs a process fully automated.

9. Outlook

The machine is in operation since its retrofit without serious failures. No further developments concerning the retrofit or safety, are planned yet.

References

[1] ***VacuTec AB Sweden, 1988, 1540-S RIE/PE user manual
[2] ***IEC 61508
[3] Börcsök J 2015 *Funktionale Sicherheit*, VDE-Verlag
[4] **Sensor Technik Wiedemann GmbH, ESX-Hilfe V1.03R0, 19.06.2007
[5] ***ISO 13849

Research on the injectors remanufacturing

D Daraba[1], I M Alexandrescu[1] and C Daraba[2]

[1]Technical University of Cluj-Napoca, North University Center of Baia Mare, Engineering and Technology Management Department, Dr. Victor Babes Street, no. 62A, 430083 Baia Mare, Romania
[2]Darmold SRL, Vasile Lucaciu Street, no. 162, 430402, Baia Mare, Romania

E-mail: dinu.daraba@cunbm.utcluj.ro

Abstract. During the remanufacturing process, the injector body - after disassembling and cleaning process - should be subjected to some strict control processes, both visually and by an electronic microscope, for evidencing any defects that may occur on the sealing surface of the injector body and the atomizer. In this paper we present the path followed by an injector body in the process of remanufacturing, exemplifying the verification method of roughness and hardness of the sealing surfaces, as well as the microscopic analysis of the sealing surface areas around the inlet. These checks can indicate which path the injector body has to follow during the remanufacturing. The control methodology of the injector body, that is established on the basis of this research, helps preventing some defective injector bodies to enter into the remanufacturing process, thus reducing to a minimum the number of remanufactured injectors to be declared non-conforming after final verification process.

1. Introduction

Remanufacturing is an industrial activity, where a product, which is in some state of physical or moral wear or complete disrepair, it is transformed into a product having at least the performance of a similar new one, whether or not having the same use, reusing as many components and giving it a new period of operation, identical to a newly manufactured product.

Remanufacturing eliminates, to a large extent, the possibility of bringing a product to its primary value, given by the value of raw materials, a situation encountered in case of the primary and secondary recycling processes [1].

Remanufacturing is often mixed up with refurbishing or even repair of a product, in the technological equipment manufacturing industries. Remanufacturing, refurbishing and repair of a product are distinguished by the amount of the required work, but especially by the quality of the resulted products. It can be said that remanufacturing, which is using the greatest amount and the most qualified work, develops the highest quality, followed by refurbishing and repair [1], [2].

In the process of remanufacturing injector that are fitted in the internal combustion engines, particular attention is paid to the injector body, as it is the component that meets all the requirements of the remanufacturing process.

2. Work assumptions

Once in the workshop, the injector has to be cleaned, checked on the test bench and completely dismantled. The internal components are analyzed to determine their wear degree. The remanufacturing process seeks complete removal of wear, compensation of the allowances to the original tolerances as well as restoring the free movements of the moving parts. The worn-out parts and those that could not be remanufactured have to be replaced with new ones.

The studied subject of this paper resulted from the analysis of the problems arisen during the process of injectors refurbishing, from various specialized units.

Thus, some injectors did not pass the final check on the test bench after refurbishing, later finding that the malfunctioning according to parameters was caused by loss of leak tightness in connection of the injector body and atomizer, resulting in too much fuel on return, in the case of most injectors [3]. In the process of the injector remanufacturing, the atomizer and the atomizer needle are to be replaced with new ones. Between the injector body and atomizer there is a metal-to-metal seal on a flat surface, into which more holes are made [3], [4].

Since the injector body is subjected to outstanding stress, operating under high pressure and temperatures, a series of shape defects and surface quality (lack of flatness, high roughness caused by the emergence of erosion, camber holes etc.) on its sealing surface with the atomizer. In this paper, the authors' aim is to study the root causes of these sealing problems, and they propose a technology to eliminate them.

3. Research stages

One batch of four injector bodies that are shown in Figure 1 was used for the research purposes. These are the stages proposed for the research: checking the roughness of the sealing surface of the injector body and atomizer; checking the hardness of the injector body in the sealing area and the microscopic analysis of the sealing surface.

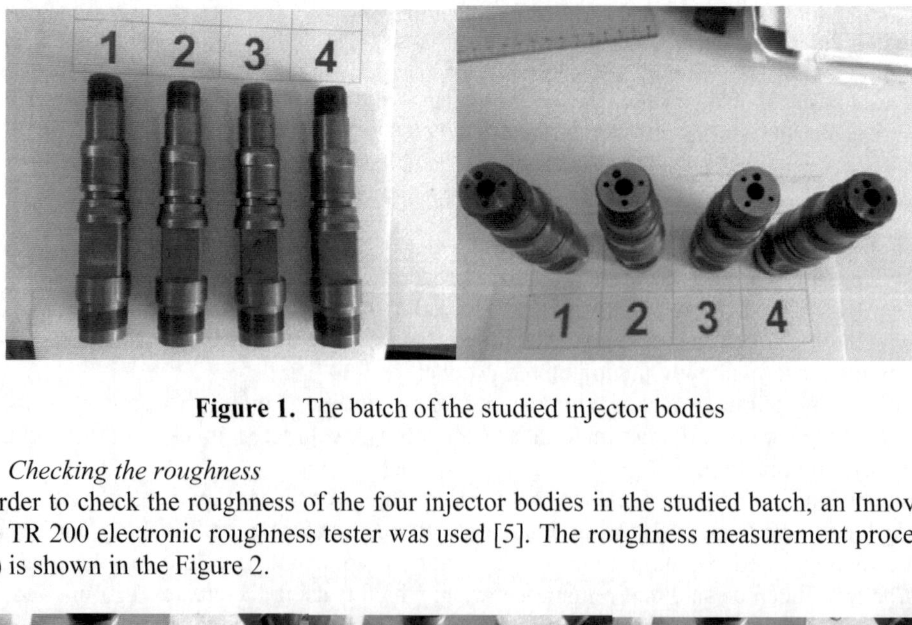

Figure 1. The batch of the studied injector bodies

3.1. Checking the roughness

In order to check the roughness of the four injector bodies in the studied batch, an Innovatest type TR 200 electronic roughness tester was used [5]. The roughness measurement process of (Ra) is shown in the Figure 2.

Figure 2. Measuring the roughness of the injector body

The roughness checking areas are shown in the Figure 3, in which (1) and (4) are the holes for the positioning pins, (2) is the return hole, and (3) is the intake hole.

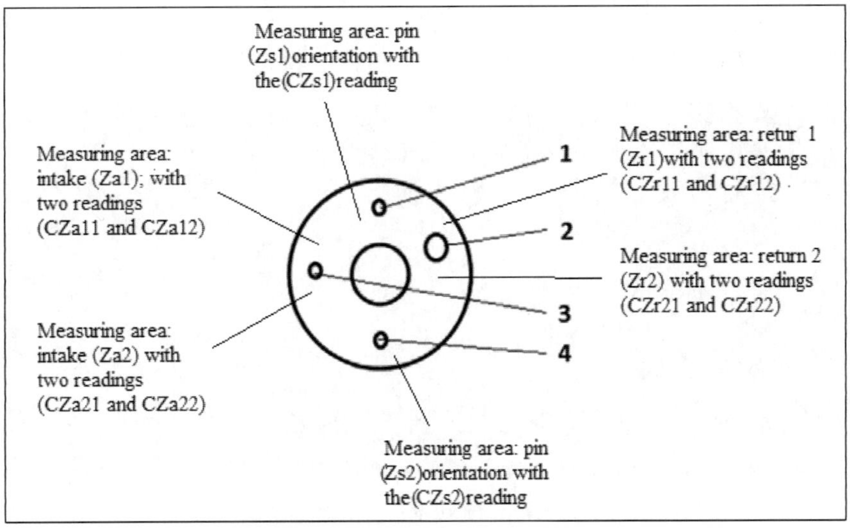

Figure 3. The areas on the sealing surface in which the roughness is measured

The results of the measuring are shown in the Table 1.

Table 1. The results of the roughness measuring

	Measuring area									
	Pin				Return				Intake	
	Za1		Za2		Zr1		Zr2		Zs1	Zs2
Injector	CZa11	CZa12	CZa21	CZa22	CZr11	CZr12	CZr21	CZr22	CZs1	CZs2
Injector 1	0.015	0.013	0.030	0.022	0.016	0.027	0.014	0.040	0.014	0.020
Injector 2	0.048	0.052	0.032	0.054	0.032	0.086	0.028	0.043	0.033	0.028
Injector 3	0.021	0.058	0.019	0.022	0.025	0.106	0.086	0.044	0.056	0.036
Injector 4	0.066	0.029	0.010	0.013	0.018	0.022	0.019	0.055	0.027	0.020

The roughness variation is shown in the Figure 4.

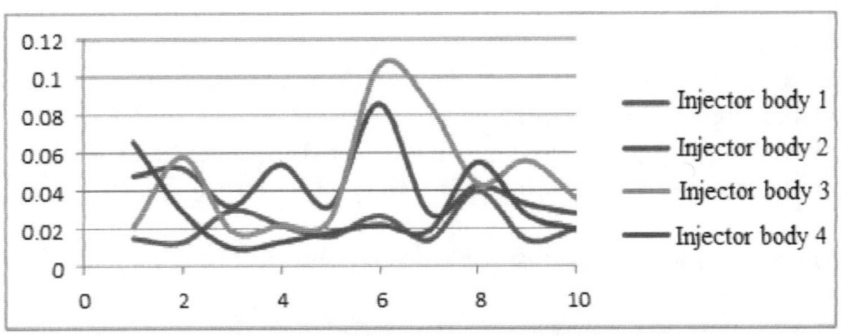

Figure 4. Roughness variation

3.2. Checking the roughness

Checking the injector body hardness aims to identify possible changes in hardness in certain areas of the body, a phenomenon which requires a different approach to remanufacturing.

The hardness was measured using a CV-450 AAT type Vikers Hardness Tester, manufactured by CV-Instruments [6]. The hardness measuring is shown in the Figure 5.

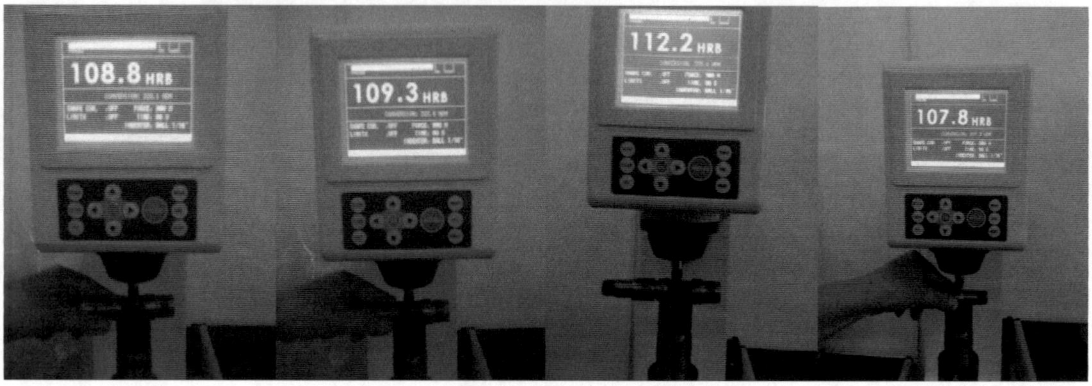

Figure 5. Hardness measuring

The results are shown in the Table 2.

Table 2. The results of the roughness measuring

Injector 1	Injector 2	Injector 3	Injector 4
112.2 HRB	107.8 HRB	109.3 HRB	108.8 HRB

Analyzing the results, it follows that no change of hardness of the studied injectors bodies appeared. The body hardness was obtained by a thermochemical nitriding treatment.

3.3. Microscopic analysis of the sealing surface

The purpose of the analysis is to identify certain physical defects appeared on the studied sealing surface and/or the total or possibly partial structural modifications of the material structure, that were caused by the thermal cycles.

The appearance of possible structural changes that would remove the injector body from the cycle remanufacture is checked.

The emergence of possible structural changes of the material of the injector body can be caused by heating and keeping at a high temperature in conjunction with possible hidden material defects [7].

The heating of the extremity of the injector is made by heat exchange between the space in the combustion chamber and the extremity of the atomizer, where there are exhaust gases containing oxygen, carbon dioxide, nitrogen oxides, water vapors and other corrosive gases favoring the phenomena of oxidation and decarburization.

Carbon oxides can detach from the contact surface between the injector and atomizer, especially from the area of the intake hole and entering into the injection circuit, causing the destruction of the piston.

The metallographic analysis was performed with a Neophot 21 type metallographic microscope shown in the Figure 6.

The surface preparation of the injector body that is to be studied is performed using a laboratory grinding equipment shown in Figure 7.

Figure 6. Metallografic microscope

Figure 7. Equipment for preparing metallographic samples

A microscopic analysis of the sealing surface, at a magnification of 1575x, without attacking the surface, in order to identify any cracks, tear or other defects that can not be seen with the naked eye was conducted in the first phase. The results of this analysis are shown in Figure 8. This stage of the study could not reveal any defects in any of the four injector bodies.

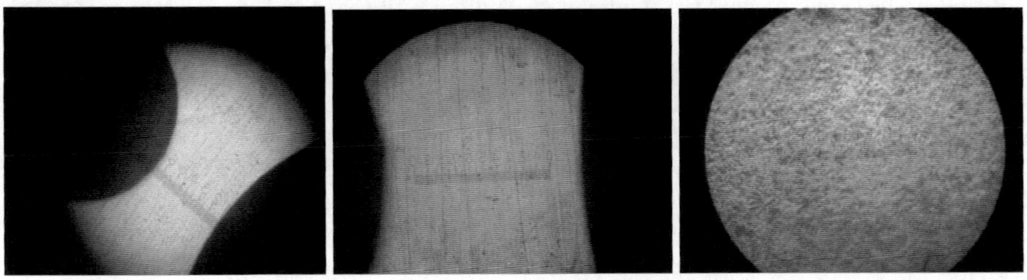

Figure 8. The structure of the material without attacking the analyzed surface

The microscopic structure was evidenced by attacking the analyzed surface with Nital reagent in the area of the intake holes, in the next stage of the research. We chose the intake holes whereas the highest stress appears there during the operation of the injector, due to the high pressure and the cavitation phenomenon which may occur. The results of this phase of the research are presented in the images in Figure 9.

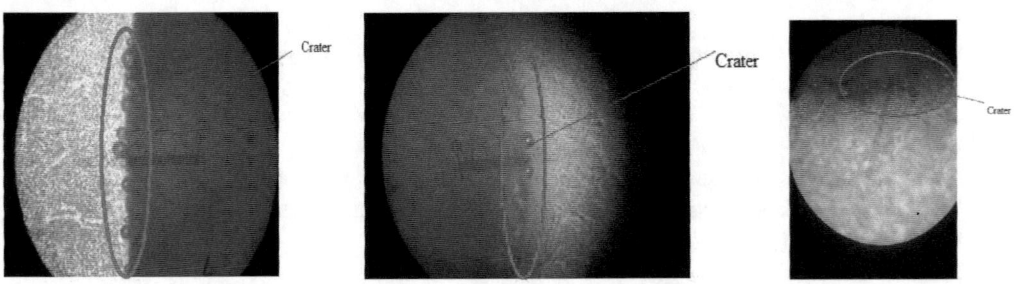

Figure 9. Highlighting the craters appeared around the intake holes

The presence of craters in the area of the intake hole was revealed by metallographic analysis. Therefore, a simple microscopic analysis which is usually done in the event of repair / refurbishing of the injectors is not sufficient.

4. Interpretation of the results

It was found that, the roughness obtained in the manufacturing process was maintained after one operation cycle (injector body 1 and 4). The restoration of the flatness of the sealing surface was attempted using sandpaper placed on a flat surface, in this way affecting the roughness of that surface, at the injection bodies 2 and 3. The hardness of the sealing surface between the injector body and the atomizer does not change after one operation cycle; therefore, based on this criterion, the body of the injector can be used in the remanufacturing process of the injector. The appearance of some craters in the area of the intake holes leads to loss of tightness between the body of the injector and the atomizer.

The sealing surface of the injector body should be brought back to the initial parameters, for the remanufacturing purposes. In this regard, it is recommended that the remanufacturing technology of the injector to provide the lapping operation of the sealing surface of the body. Lapping is a metal cutting operation for obtaining a flat, low roughness surface, using a process of removal of metal particles from a surface by mechanical or mechanical-chemical methods, using free, abrasive, very fine grains mixed into a viscous mass.

5. Conclusions

An appropriate technology should be outlined, depending on the type of the injector and the type of defects that occur during the exploitation process, before starting the process of the injector remanufacturing.

The stage of checking the injector parts under the remanufacturing process is essential, and the metallographic checking of the sealing surface should not miss

In this research we did not measure the dimensions of the craters highlighted by metallographic analysis.

Corroborating the research results with data from the companies which are carrying out injectors repair or refurbishing activities, we can say that during exploitation, due to the aggressive operating conditions (high temperature and pressure) and the cavitation phenomenon, a erosion process takes place at the intake hole area, which leads to the appearance of those craters that affect the sealing between the injector body and atomizer.

Based on the presented research results, we may conclude that the processing of the front surface of the injector body is required, so that to obtain an injector having the same operation characteristics of an injector manufactured as new, through the remanufacturing process.

The mentioned surface processing would be performed on a depth that would lead to the elimination of any craters. This requirement imposes a metallographic study by the means of an electronic microscope that allows measuring the depth of the appeared craters.

References

[1] Daraba D 2007 Remanufacturing – Opportunities and Barriers, *Scientific Bulletin, Serie C, Fascicle: Mechanics,Tribology, Machine Manufacturing Technology,* **XXI** 165-170
[2] ***https://apra.org/?page=Remanufacturing/
[3] ***https://www.boschautoparts.com/en/auto/diesel-parts/remanufactured-injectors
[4] ***www.e-automobile.ro/categorie-motor/19-diesel/25-motor-diesel-injectie-directa.html
[5] ***www.innovatest-europe.com
[6] ***http://www.purduemetrology.co.uk/testing_products
[7] Moldovan St A, Cosma M and Nasui V 2016 *Experimental research about positioning correction of CNC machine with X-axis up to 12000 millimeters using laser measurement equipment,* 4th International Conference on Computing and Solutions in Manufacturing Engineering CoSME'16, Brasov, Romania, November 3-4

IOP Conf. Series: Materials Science and Engineering **200** (2017) 012037 doi:10.1088/1757-899X/200/1/012037

System monitoring feedback in cinemas and harvesting energy of the air conditioning condenser

P P Pop[1], A Pop-Vadean[1], C Barz[2], T Latinovic[3] and O Chiver[4]

[1]Technical University of Cluj Napoca, Faculty of Mechanical Engineering
[2]Technical University of Cluj Napoca, Center from Baia-Mare, Romania
[3]University of Banja Luka, Faculty of Mechanical Engineering, Bosnia and Herzegovina
[4]Technical University of Cluj Napoca, Center from Baia-Mare, Romania

E-mail: poppaulpetrica@yahoo.com

Abstract. Our article monitors the degree of emotional involvement of the audience in the action film in theaters by measuring the concentration of CO2. The software performs data processing obtained dispersion sensors and displays data during the film. The software will also trigger the start of the air conditioning condenser where we can get harvesting energy by installing a piezoelectric device. Useful energy can be recovered from various waste produced in cinema. The time lag between actions and changes in environmental systems determines that decisions made now will affect subsequent generations and the future of our environment.

1. The principles of our research

We have made this research guided by the need of getting a real feedback from the spectator's emotional involvement in the action of the movie they are watching. The course of the research is linked to the breathing process. The emotional involvement of the human body physiologically translates into certain hormones being released into the bloodstream depending on each state of mind - uncertainty, immediate danger, dread, fear, joy, love, happiness, etc. These substances modify the heart rate, namely increase the pressure of the blood flow and thus the individual's oxygen intake and CO2 release varies depending on the stimulus. If we measure the amount of CO2 produced over certain periods of time we could obtain data that reflects the spectator's degree of emotional involvement, either at a specific moment or for the duration of an entire movie.

This study can be applied to any cinema room with an audience. The direction of developing this research aims at realizing an automated measuring system that can support itself mostly by harvesting energy from the analyzed environment, but without having any effect on the measured values.

2. The objectives of the research

Our first objective is to make the measurements without affecting their quality and precision.

The second is to integrate the energy harvesting equipment into the data collection environment without modifying the precision or the quality of the results. They will be used for produce more electrical power needed to power the equipment.

The third objective represents our wish to constantly improve this method of harvesting energy in order to make it more and more independent from the electrical network of the original monitoring system.

The forth objective is to store the obtained measurements for a longer period of time in order to be interpreted and to get the expected results. To do this we will use a database management system stored on a server.

The last and probably the most important objective is to make an integrated software that monitors can interprets the results we obtain. This program will be mostly developed by using high level coding that will provide a user friendly interface. The interface will have intuitive menus and will include commands for real time access to the information stored in the database.

3. Measuring methods. The best alternative

There is a number of measurement methods that we have taken into account in the course of our research. This allows us to improve new measuring techniques and to develop the research for several fields of interest. In the future we estimate the possibility of extending these methods in order to diversify the value and the quantity of the data acquired from the measurement area.

From all the methods used, we deem worth mentioning only those we considered appropriate for the objectives of our research. We believe that our research opens new directions of development using the above mentioned methods. This way we will bring our contribution to ensuring a better academical cooperation in this particular field and in the adjacent fields as well. The ideas integrated into the development of this system will not only guarantee a scientifical sustainability of our research, but will also be a component of development in the technical field.

The first method is to use the methematical integral applied on the function of the space determined by the room. Namely Cauchy's integral formula of obtaining values from points that are inaccesible from the inside of the measured area, knowing the values of the functions on the border [1].

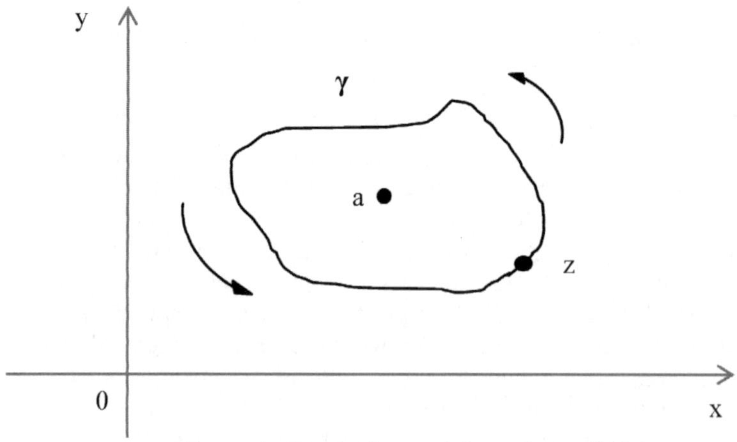

Figure 1. Cauchy integral formula applied

Let G be open subset, non-void, conex in $\mathbb{C}$. The function f is completely derivable in a point $z_0 \in G$ if there is the limit:

$$f'(z_0) = \lim_{z \to z_0} \frac{f(z) - f(z_0)}{z - z_0}.$$

In the case where the function *f* is completely derivable in each point in the vicinity of z_0, this is called holomorphic function in the point z_0.

The notion of holomorphy extends the notions of derivability and continuity from real analysis to the complex one.

If we refer to the values of CO2 amount we can rearch the research level where we precisely establish the concentration in any inaccesible point within the measuring area. We have to underline the fact that the audience cannot be disturbed to make measuements during the show, therefore each spectator represents an inaccesible point in the cinema. This method represents a direction for our future research due to its complexity, which involves high costs and a large amount of resources, but offers in return integrating developments in the fields of mechatronics and technical systems.

The second measuring method consists of using sensors that are installed at a certain level in the room. The sensors are attached to the electronic equipment used to transfer the data to the database server. The method consists of making a complex network of sensors that need a high amount of resources and a high amount of power source. The results of such a mesurement can be influenced by the existence of some currents inside the room or by the audience moving across the room during the performance. There is definitely some lack of precision involved. This is due to the amount time when the CO2 is produced and the moment of the measurement. We cannot precisely control the lapses of time between which we perform the measurements. The gas spreads into the room unevenly, making it very difficult to establish correlations between certain scenes in the show and how the spectator emotionally connects to those scenes.

Figure 2. CO2 gas sensor

According to our research, the most efficient method of achieving the objectives we have set is to measure the CO2 concentration in the tubes of the air conditioning system. This particular method has the advantage of not being influenced by external factors. We can always measure new quantities of CO2. We precisely establish the time periods over which we make the measurements in order to link them to the sequences in the movie. And the most environmentally friendly advantage is the fact that the system harvests energy to power the sensors. The costs of implementing such a system are reduced because is no need to have a vast network of sensors (Figure 2), like in the case of the previous measuring method.

4. Harvesting energy for sensors

In the last few years, radio, microprocessor and sensor technologies have made considerable advancements enabling the use of wireless sensor nodes (WSN) for the monitoring of several surveillance and control systems. Battery technology evolution not have the same rhythm with these enabling technologies for wireless sensor networks, so battery powered nodes must be replaced frequently.

An ingenious solution is the use of energy harvesting devices will be created so that will feed from ambient sources such as: electromagnetic, light, heat or vibration. An energy harvester could extend the lifespan of network node without having to change batteries.

Harvesting localized flow power from HVAC systems then offers a method to power a sensor and low power radio that can be run on low duty cycle with as little as 100 μW **[1]**. HVAC systems represent a convenient point to monitor air quality and safety for public buildings.

These examples will begin to reach a large potential for energy harvesting and HVAC systems will provide power for many sensors that could be used around air ducts or inside them.

There are plenty of strategies harvesting fluid flow, some specific to HVAC systems and some with broader applications.

Even though many studies have been conducted to harvest energy from the fluid flow, there is still a gap in the ability to harvest.

Many systems are capable of providing 100 μW need to power as a wireless sensor node, but few can do this across the range of flow rates specific to HVAC ducts, especially at lower speeds of 3 m/s.

Myers et al have added turbines piezoelectric and managed to obtain a reduction in the rate of about 2.5 m / s [2], but the equipment rotation is not desired performance.

The aim of the experiment is to achieve 100 μW in HVAC.

When the piezoelectric material is deformed, it converts mechanical energy into electrical energy through the crystal's structure.

This version offers high performance at relatively low flow speed, and ideally will provide the required performance for a wider range of flow rates than other variants of Harvester.

Small cross-sectional area exposed to the air flow and the size of this device make it an ideal energy harvester for HVAC systems. Figure 1 shows a line drawing to familiarize with the basic principle that operates harvested.

This design is suitable for environments where the flow direction is unknown.

For HVAC unidirectional environment, this issue is not important.

For these reasons, this device offers competitive performance in its intended application of harvesting from HVAC flow.

Heating, ventilation and air conditioning (HVAC) is technology for indoor and vehicular environmental comfort.

Its aim is to provide thermal comfort and acceptable indoor air quality.

For larger buildings, Variable Air Volume (VAV) HVAC systems have proven to be much more efficient which significantly improves efficiency and lowers cost.

Power generation between 100 - 3000 μW for flow speeds in the range of 2 - 5 m/s are sufficient for powering a sensor and wireless radio of a node.

The simple design of this energy harvester, if produced in bulk, offers an inexpensive solution for powering wireless sensor nodes in and around VAV HVAC systems.

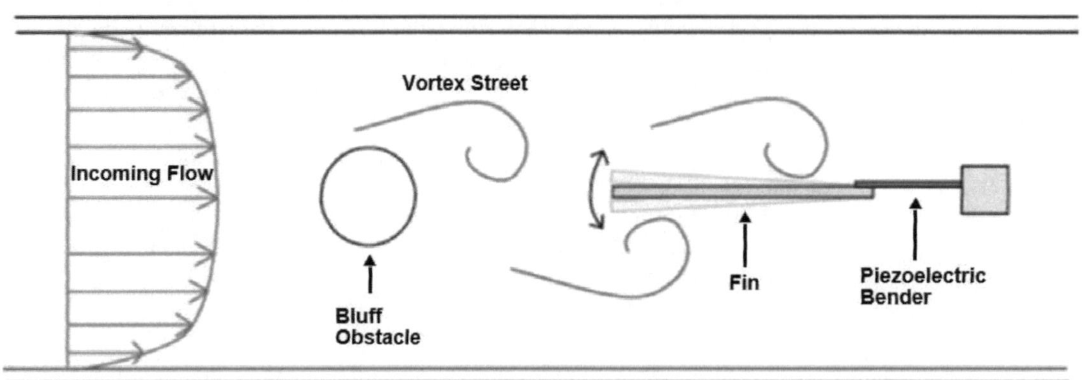

Figure 3. Diagram illustrating idea behind the operation of the harvester

IOP Conf. Series: Materials Science and Engineering **200** (2017) 012037 doi:10.1088/1757-899X/200/1/012037

5. Storing the measured data

In order to have a rigorous research one needs to store the data over a medium-to-long period of time. The information has to be available for processing and interpreting for a statistical, or any other kind of nature. The data obtained by the sensors is stored into the database installed on a SQL server. It is designed on the criteria of integrity and easy access. For the chart sequence we use the hierarchic model in designing and operating the database with primary and secondary key. The software automatically authorises CRUD operations (Create, Read, Update, Delete) according to the level of conectivity to the database. The access levels are activated by using credentials. The system administrator has full rights and different users can have, according to the access data, certain rights over the database operations [4].

6. System interface

For easy access to the system we came up with a software that integrates all the technology used. This is designed by using Microsoft Visual Studio, more specifically the C # (Sharp) language [5]. We realised a software product to access and process the obtained data in real time. The program is available for Windows platforms and for Windows compatible ones, thanks to the Windows Forms system we have integrated (Figure 4) [6].

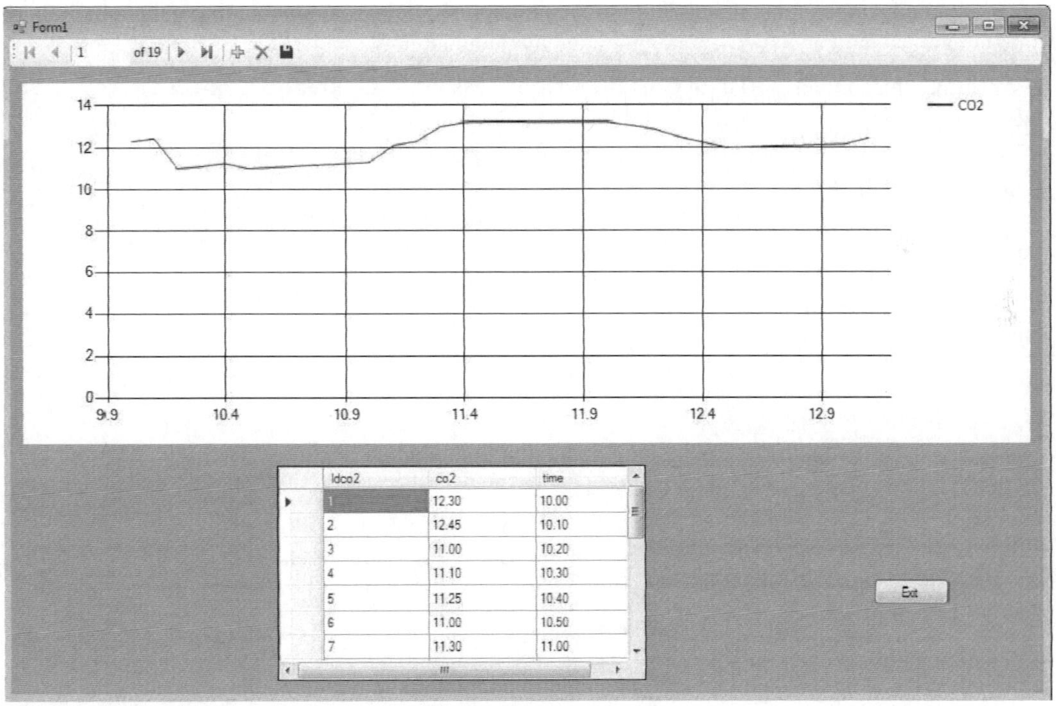

Figure 4. Software interface

As shown in (Figure 4) the data is taken from he database tables and is automatically processed. The program creates a real time chart through which we can observe the measurements taken over periods of time (5 minutes in our case). The interpretation can be thus correlated with the key moments of the movie. We can therefore know which part of the movie the audience was most involved in and to what extent [7], [8] .

7. Conclusions

Our system has attained the objectives we set at the beginning of our research. We have managed to establish a connection between the moments of the artistic act and the way in which the audience gets emotionally involved during these moments. This mechatronic system can be installed into any cinema hall with an audience.

Our research has opened new horizons for the artistic production, especially from a marketing perspective. The ideas that we presented also launched new methods of research. We hope to develop them into new technical approaches. We have therefore offered new ways of research for the centres that can afford to invest serious resources in order to get results that are vital for this field.

References

[1] Stănăşilă O 2014 *Resurse matematice şi informatice,* Editura Niculescu, Bucureşti, Romania
[2] Rabaey J, Ammer J, Otis B, Burghardt F, Chee Y H, Pletcher N, Sheets M and Qin H 2006 *Ultra-Low-Power Design*, IEEE Circuits Devices Mag., July, pp 23-9
[3] Eastham J F, Profumo F, Tenconi A, Hill-Cottingham R J, Coles P C and Gianolio G 2002 Novel Axial flux Machine for aircraft drive: design and modeling, *IEEE Transactions on Magnetics* **38**(5) 3003-3005
[4] Priya S, Kim H, Vickers M and Myers R 2007 Small Scale Windmill, *Appl. Phys. Lett.* **90** 05410
[5] Donis M 2006 *Programming Microsoft Visual C# 2005: The Language*, Microsoft Press, Seatle, USA
[6] Jhon S 2010 *Microsoft Visual C Sharp 2010. Step by Step*, Microsoft Press, Washington, USA
[7] Paul K 2002 *Advanced C# Programming*, McGraw Hill, Osborne, Australia
[8] ***https://msdn.microsoft.com/enus/library/dd831853 (v=vs.100).aspx.

IOP Conf. Series: Materials Science and Engineering **200** (2017) 012038 doi:10.1088/1757-899X/200/1/012038

Modelling pressure rolling of asymmetric rolling process

V Alexa, S A Ratiu, I Kiss and V G Cioata

University Politehnica Timisoara, Faculty of Engineering Hunedoara, Department of Engineering and Management, Hunedoara, Romania

E–mail: vasile.alexa@fih.upt.ro

Abstract. The paper presents a comparative analysis between experimental results and modelling in order to interpret the value of the contact pressure on the asymmetric longitudinal rolling. It is also intended action and the different behaviour of upper cylinder compared to the lower cylinder action in situations when both are driven, or only one operates. In the modelling will be presented on the basis of boundary conditions imposed rolling pressure variation in the degree of reduction and also re size arc length of contact. Determining a curve is also important to determine the locus of points which characterize symmetry conditions partial rolling process between unequal diameters cylinders.

1. Establishing engineering mathematical models for the rolling process

This paper is part of the modern trends of gradual replacement of the physical experiments with numerical ones in the design and optimisation of the technological processes of plastic deformation by rolling. By its nature, this paper requires a multidisciplinary approach, starting with the deep understanding of the theory of complex phenomena of metallic materials plastic deformation by rolling, and continuing with the modern techniques of processing and interpretation of the experimentally obtained data.

1.1. The model used for the contact arc length [1]

The length of the contact arc between the metallic material and the rolls is one of the decisive parameters influencing the strength, temperature and kinematic conditions of the rolling process. To determine the length of the contact arc for symmetrical and asymmetrical cold rolling, in addition to the roll reduction and radius, we must also know the elastic deformations of the rolls and the strip undergoing the rolling process, values that depend on the strength of the rolling process.

This problem can be solved by using the rolling theory equations, as well as the elasticity theory equations, based on the well-known models and methods for calculating the length of the contact arc for cold rolling.

Using the Hertz's concept (relationship 1), a number of models have been obtained, as following:

$$l_c = \sqrt{\frac{2 \cdot R_1 \cdot R_2}{R_1 + R_2} \Delta h} \tag{1}$$

a) The *Hitchcock's* model, which takes into account the elastic compression of the rolls:

$$l_c = x_{lc} + \sqrt{R \cdot \Delta h + x_{lc}^2} \tag{2}$$

$$x_{lc} = 8 \cdot p_{mc} \cdot R \cdot \theta_c \tag{3}$$

$$\theta_c = \frac{1-\gamma_c^2}{\pi \cdot E_c} \tag{4}$$

R – the radius of the working rolls;
Δh – the absolute reduction;
p_{mc} – the mean normal contact stress;
γ_c – the Poisson's ratio;
E_c – the elastic modulus of the material which the rolls are made of
Relationship 2 does not take into account the elastic deformations of the strip undergoing the rolling process, and the normal stress diagram is adopted in the shape of semi ellipse.
b) *The Roberts* model takes into account the contact between the rolls and an ideal rigid plan:

$$l_c = \sqrt{R \cdot \Delta h} + 1{,}08\sqrt{\frac{2 \cdot R \cdot p_b}{E_c}} \tag{5}$$

p_b – The normal force acting per unit length of the strip;
c) *Țelicov and Grişcov* proposed the following relationship to calculate the increase of the contact arc length by taking into account the asymmetry of the normal stresses diagram:

$$x_{lc} = 8 \cdot p_{mc} \cdot R \cdot \theta_c \tag{6}$$

This relationship (6) helps reducing the errors between the calculated values and the experimental ones.
d) The *Țelicov's* model introduces also the influence of elastic deformations of the strip:

$$l_c = x_l + \sqrt{R \cdot \Delta h + x_l^2} \tag{7}$$

$$x_{lc} = 8 \cdot p_{mc} \cdot R \cdot (\theta_c + \theta_b) \tag{8}$$

$$\theta_b = \frac{1-\gamma_b^2}{\pi \cdot E_b} \tag{9}$$

The mean normal contact stress, the contact arc length and its increase along the plane passing through the centres of the rolls are calculated by taking into account the elastic deformations of the rolls and the strip;
e) *Dinnik* shows in his model that the determination of the contact arc length increase x_l without taking into account the length of the strip and the diagram symmetry for normal stresses is irrational, proposing x_l to be determined using the relationship:

$$x_l = 0{,}5 \cdot a \cdot x_{lc}\left(1 + \sqrt{1 + 1{,}74\frac{h_l}{x_{lc}}}\right) \tag{10}$$

a – Coefficient considering the asymmetry of the normal stresses diagrams:
h_l – Thickness of the strip after rolling
For determining the value of x_{lc}, Dinnik recommends the Relationship 3.
f) The relationship proposed by *Cepurchin* determines the contact arc length taking into account only the elastic compression of the rolls, as follows:

$$l_c = \sqrt{R \cdot \Delta h + c \cdot p_b \cdot \theta_c \cdot R} \tag{11}$$

c – Shape coefficient of the normal stresses diagram
g) Based on the previous model, *Poluhin* obtained the dependence for determining the length of the contact arc by taking into account the elastic deformations of the rollers and strip. Therefore:

$$l_c = \sqrt{R \cdot \Delta h + c_1 \cdot p_b \cdot (\theta_c + \theta_b) \cdot R} \tag{12}$$

Where:

$$c_1 = 8 + 2\sqrt{\frac{4 \cdot \Delta h}{p_b(\theta_c + \theta_b)} + 16} \tag{13}$$

1.2. The model used for the mean normal contact stresses [1]

Since the elastic deformations of the strips are neglected in all the proposed models, we are going to analyse the plastic contact portion between the strip and the rolls to determine the mean contact pressure.

Thus, there are a number of models proposed in the literature, which are distinguished by simplicity and compactness, and are obtained considering a constant value of the resistance to deformation for the mean values of the respective parameters. The models proposed by these authors, without taking into account the influence of elastic deformations of the rolls, are shown in the following table:

Table 1. The variation domains of variables

Stone	$p_m = \beta \cdot \sigma_{cm} \cdot \frac{E^{f\frac{1}{h_m}} - 1}{f\frac{1}{h_m}}$	(14)
Bland and Ford	$p_m = \beta \cdot \sigma_{cm}\left(1{,}08 + 1{,}79 \cdot \varepsilon \cdot f \cdot \sqrt{\frac{R}{\Delta h}} - 1{,}02 \cdot \varepsilon\right)$	(15)
Roberts	$p_m = \beta \cdot \sigma_{cm}\left(1 - 1{,}25 \cdot \varepsilon + \frac{f \cdot l}{2 \cdot h_m}\right)\frac{1}{1 - \varepsilon}$	(16)
Lugovschi	$p_m = \beta \cdot \sigma_{cm}\left(1 + \frac{f \cdot l}{2 \cdot h_m}\right)$	(17)
Telikov	$p_m = \frac{\beta \cdot \sigma_{cm}}{\Delta h}\left\{\frac{h_0}{\delta - 2}\left[\left(\frac{h_0}{h_n}\right)^{\delta-2} - 1\right] + \frac{h_1}{\delta + 2}\left[\left(\frac{h_n}{h_{n1}}\right)^{\delta+2} - 1\right]\right\}$	(18)

where:

$$\delta = \frac{2 \cdot f \cdot l}{\Delta h} \quad (19)$$

$$l = \sqrt{R \cdot \Delta h} \quad (20)$$

$$h_n = \sqrt[2\delta]{h_0^{\delta-1} \cdot h_1^{\delta+1}} \quad (21)$$

$$\Delta h = h_0 - h_1 \quad (22)$$

$$\varepsilon = \frac{\Delta h}{h_0} \quad (23)$$

For the convenience of further analysis of the model, we replace the "p_m" value with the dimensionless parameter "n", which is the stress state coefficient:

$$n_\sigma = \frac{p_m}{\beta \cdot \sigma_c} \quad (24)$$

And dependencies: for determining: $l, \delta, \frac{h_1}{\Delta h}, h_n, \frac{h_0}{h_n}, \frac{h_n}{h_1}$ and h_m:

$$l = \sqrt{R\Delta h} = h_0\sqrt{\frac{R}{h_0} \cdot \varepsilon} \quad (25)$$

$$\delta = \frac{2 \cdot f \cdot l}{\Delta h} = 2 \cdot f \cdot \sqrt{\frac{R}{h_0} \cdot \frac{1}{\varepsilon}} \quad (26)$$

$$\frac{h_1}{\Delta h} = \frac{1 - \varepsilon}{\varepsilon} \quad (27)$$

$$h_n = \sqrt[2\delta]{h_0^{\delta-1} \cdot h_1^{\delta+1}} = h_0\sqrt[2\delta]{(1 - \varepsilon)^{\delta+1}} \quad (28)$$

$$\frac{h_0}{h_1} = \frac{h_0}{h_0 \sqrt[2\delta]{(1-\varepsilon)^{\delta+1}}} = \frac{1}{\sqrt[2\delta]{(1-\varepsilon)^{\delta+1}}} \quad (29)$$

$$\frac{h_n}{h_1} = \frac{h_0 \sqrt[2\delta]{(1-\varepsilon)^{\delta+1}}}{h_0 - \Delta h} = \frac{1}{1-\varepsilon} \sqrt[2\delta]{(1-\varepsilon)^{\delta+1}} \quad (30)$$

$$h_m = 0{,}5(h_0 + h_1) = 0{,}5(2-\varepsilon)h_0 \quad (31)$$

With consideration of relations, models of the Table 1, take the form

Table 2. The variation domains of variables

Stone	$n_\sigma = \frac{1}{\frac{2\cdot f\cdot\sqrt{\frac{R}{h_0}}\cdot\varepsilon}{2-\varepsilon}}\left(E^{\frac{2\cdot f\cdot\sqrt{\frac{R}{h_0}}\cdot\varepsilon}{2-\varepsilon}} - 1\right)$	(32)
Bland and Ford	$n_\sigma = 1{,}08 + 1{,}79\cdot\varepsilon\cdot f\cdot\sqrt{\frac{R}{h_0}\cdot\frac{1}{1-\varepsilon}} - 1{,}02\cdot\varepsilon$	(33)
Roberts	$n_\sigma = \left(1 - 1{,}25\cdot\varepsilon + \frac{f}{2}\sqrt{\frac{R}{h_0}}\cdot\varepsilon\right)\frac{1}{1-\varepsilon}$	(34)
Lugovschi	$n_\sigma = \left(1 + \frac{f}{2-\varepsilon}\sqrt{\frac{R}{h_0}}\cdot\varepsilon\right)$	(35)
Telikov	$n_\sigma = \frac{1}{\varepsilon(\delta-2)}\left\{\left[\frac{1}{\sqrt[2\delta]{(1-\varepsilon)^{\delta+1}}}\right]^{\delta-2} - 1\right\} + \frac{1-\varepsilon}{\varepsilon(\delta+2)}\left\{\left[\frac{1}{1-\varepsilon}\sqrt[2\delta]{(1-\varepsilon)^{\delta+1}}\right]^{\delta+2} - 1\right\}$	(36)

Dependencies can be simplified to some extent, if adopted:

$$A = f\sqrt{\frac{R}{h_0}} \quad (37)$$

$$\delta = 2A\sqrt{\frac{1}{\varepsilon}} \quad (38)$$

After entering in we obtain:

Table 3. The variation domains of variables

Stone	$n_\sigma = \frac{1}{\frac{2A\sqrt{\varepsilon}}{2-\varepsilon}}\left(E^{\frac{2A\sqrt{\varepsilon}}{2-\varepsilon}} - 1\right)$	(39)
Bland and Ford	$n_\sigma = 1{,}08 + 1{,}79\cdot\varepsilon\cdot A\cdot\sqrt{\frac{1}{1-\varepsilon}} - 1{,}02\cdot\varepsilon$	(40)
Roberts	$n_\sigma = \left(1 - 1{,}25\cdot\varepsilon + \frac{A}{2}\sqrt{\varepsilon}\right)\frac{1}{1-\varepsilon}$	(41)
Lugovschi	$n_\sigma = \left(1 + \frac{A}{2-\varepsilon}\sqrt{\varepsilon}\right)$	(42)
Telikov	$n_\sigma = \frac{1}{\varepsilon(\delta-2)}\left\{\left[\frac{1}{\sqrt[2\delta]{(1-\varepsilon)^{\delta+1}}}\right]^{\delta-2} - 1\right\} + \frac{1-\varepsilon}{\varepsilon(\delta+2)}\left\{\left[\frac{1}{1-\varepsilon}\sqrt[2\delta]{(1-\varepsilon)^{\delta+1}}\right]^{\delta+2} - 1\right\}$	(43)

The relationships confirm the fact that the stress state coefficient "*n*", found in all the analysed models, finally represents a function of three dimensionless parameters:

1 – the coefficient of friction;

2 – the ratio between the radius of the working rolls and the thickness of the bar at the entrance into the deformation zone;

3 – the relative reduction, which is also the criterion of similarity;

2. The experimental plant and testing method

The research for this theme purpose have been made on a 170 mm reversing two-high rolling mill, created and installed in the no conventional technologies and plastic deformation laboratory of the Engineering Faculty from Hunedoara [1-3].

Figure 1, [1] shows a general view of the installation, which has been built by the author of this paper and which, compared to all the other mills, allows a simultaneous registering of the symmetric and asymmetric rolling process parameters, which are the rolling force (F_d and F_a), the side efforts (X_d and X_a), the pressure on the surface of contact with the superior roll (p_{ms}) and the with inferior one (p_{mi}), the real lengths of the springs of contact with the superior roll (l_s) and with the inferior one (l_i).

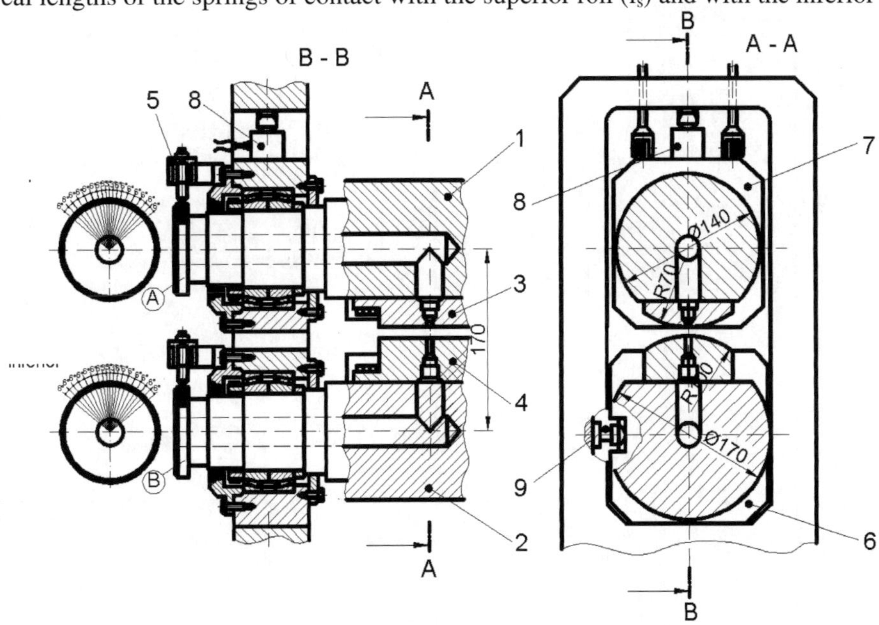

Figure 1. The installation for the research on the force parameters of the symmetric and asymmetric longitudinal rolling process

1 – 140 [mm]- "A" divided head superior roll; **2** – 170 [mm]-"B" divided head inferior roll;
3 – R = 70 [mm] -segment for the superior roll; **4** – R =100 [mm] – segment for the inferior roll;
5 – device for registering the real length of the contact springs; **6 -** bearing
7 – superior roll bearing; **8**– rolling force (F) detector; **9** – side force (X) detector

Since the intended purpose was to study the qualitative aspect of phenomena related to the asymmetric longitudinal rolling, for eliminating the inevitable influence of the iron oxides (scale) on the process, the tests were carried out on samples of aluminum and copper, with the following dimensions [4]:

$$h_0 = 1, 2, 6, 12 \ mm$$
$$b_0 = 40 \ mm \ and \ l_0 = 150 \ mm$$

In this paper, by analyses and calculations, we are going to use the experimentally obtained values for pressure, and the actual contact lengths of the arcs with the upper and lower roll, for equal diameter and different diameter.

3. Discussions and technological interpretations

Figure 2 shows the dependence of the real length of the contact arc – with the upper roll (smaller) and the lower roll (bigger) – versus the reduction, when rolling the samples whose thicknesses are h_0=12; 6; 2 and 1 mm [4].

As can be seen in the figure, the curves are increasing differently, their points of intersection being found at different degrees of reduction, depending on the thickness of the rolled samples.

Up to the points of intersection (ε=30-35% for h_0 =12 and 6 mm), the length of the arcs afferent to the smaller-diameter roll (upper) is greater, and by further increasing the reduction, the actual lengths of the contact arcs afferent to the smaller-diameter roll become smaller.

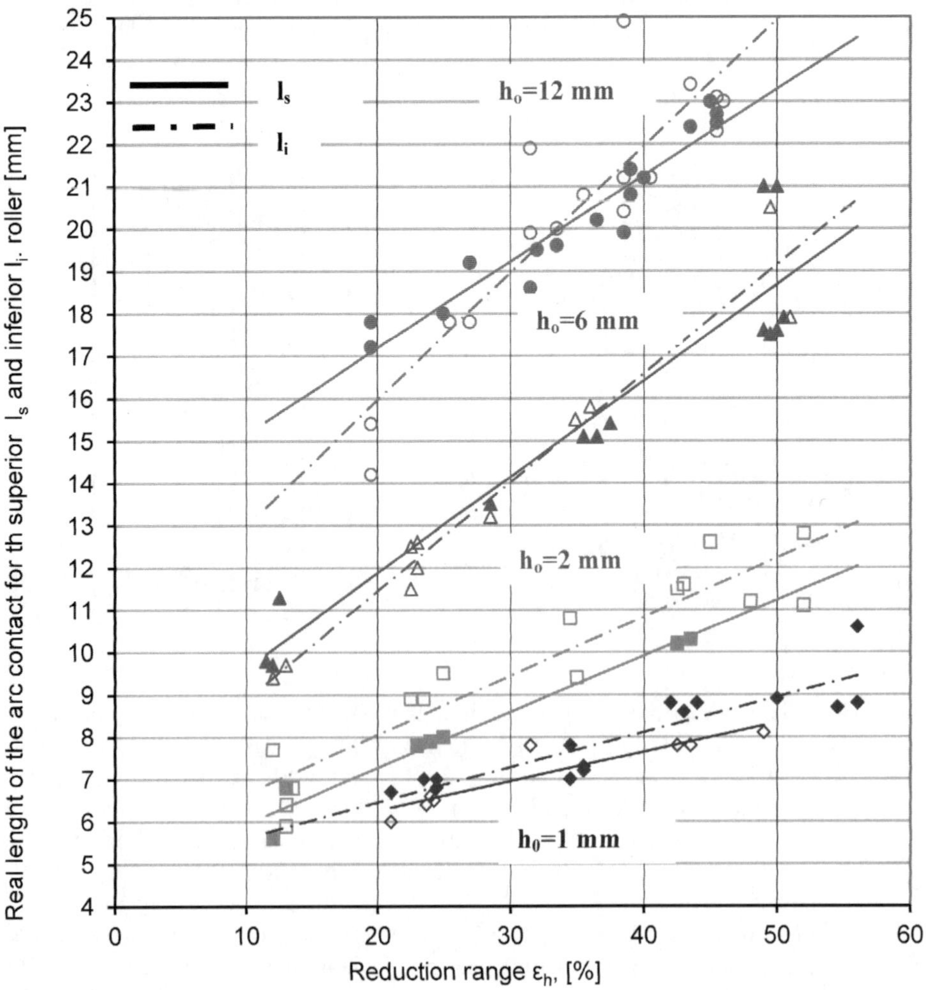

Figure 2. The dependence of the real length of the contact arc – with the upper roll (smaller) and the lower roll (bigger) – versus the reduction, when rolling the samples whose thicknesses are h_0 = 12; 6; 2 and 1 mm

By overlapping the areas where the mean pressure is exerted by the upper roll (smaller) and the lower roll (bigger) for rolling the samples whose dimensions are h_0 =12 mm (Figure 3 and 4), we discover a similar picture, with the only difference that the chart dependencies for the upper and lower rolls are changing their places, i.e. to the highest values of the length arcs correspond to the lowest pressure values and vice versa.

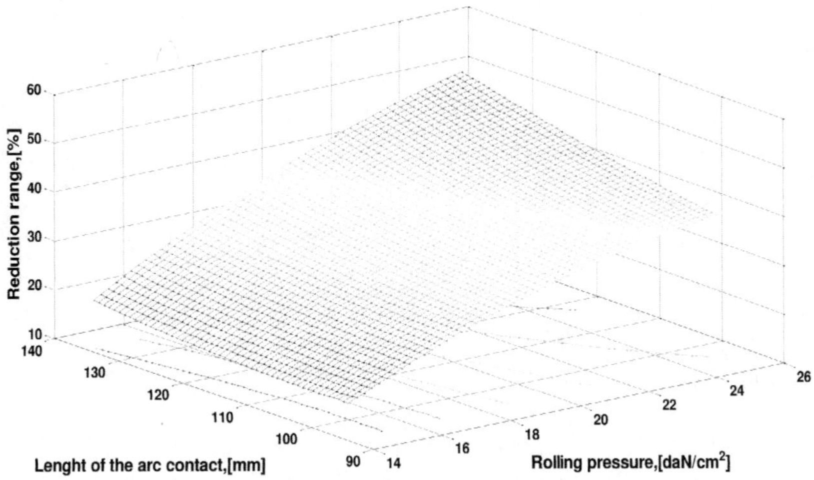

Figure 3. Variation of the rolling pressure and the contact arc length, h_0=12 mm samples between different diameter cylinders $\frac{D_s}{D_i} = \frac{140}{170}$ mm, applying various reductions, lower cylinder

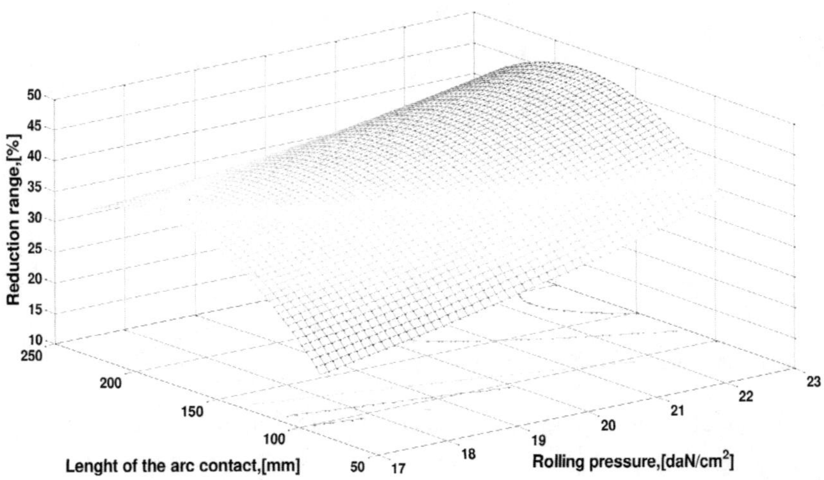

Figure 4. Variation of the rolling pressure and the contact arc length, h_0=12 mm samples between different diameter cylinders $\frac{D_s}{D_i} = \frac{140}{170}$ mm, applying various reductions, upper cylinder

For each thickness, the intersection points of the curves representing the mean pressure and the actual length of the contact arc are placed at the same value of the reduction, confirming the correctness of the experimentally obtained data [5].

For the samples whose dimensions are h_0 = 2 mm (Figure 5 and 6), the intersection points of the curves $l_{rs}=f(\varepsilon, p_{ms})$ and $l_{ri}=f(\varepsilon, p_{mi})$ are moving considerably and correspond to the reduction ε=15-25%, and the length of the contact arcs afferent to the larger-diameter roll (lower) is greater, $l_i > l_s$. It is obvious that, for reductions smaller than the mentioned range, $l_{ri} < l_{rs}$.

IOP Conf. Series: Materials Science and Engineering **200** (2017) 012038 doi:10.1088/1757-899X/200/1/012038

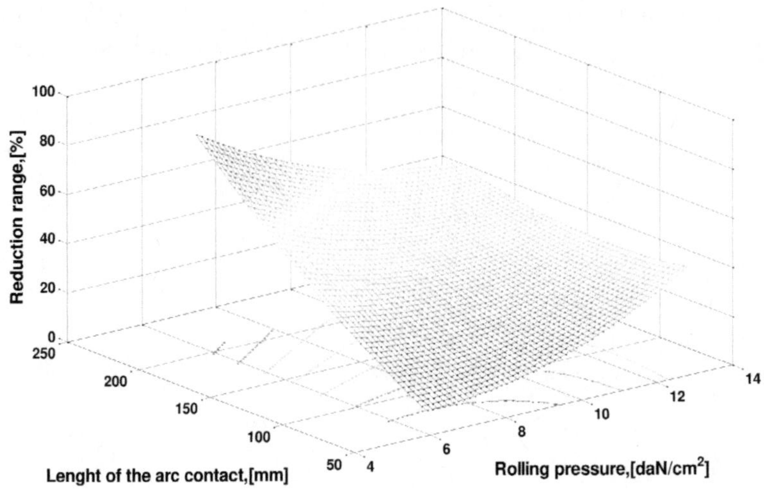

Figure 5. Variation of the rolling pressure and the contact arc length, h_0=2 mm samples between different diameter cylinders $\frac{D_s}{D_i} = \frac{140}{170}$ mm, applying various reductions, lower cylinder

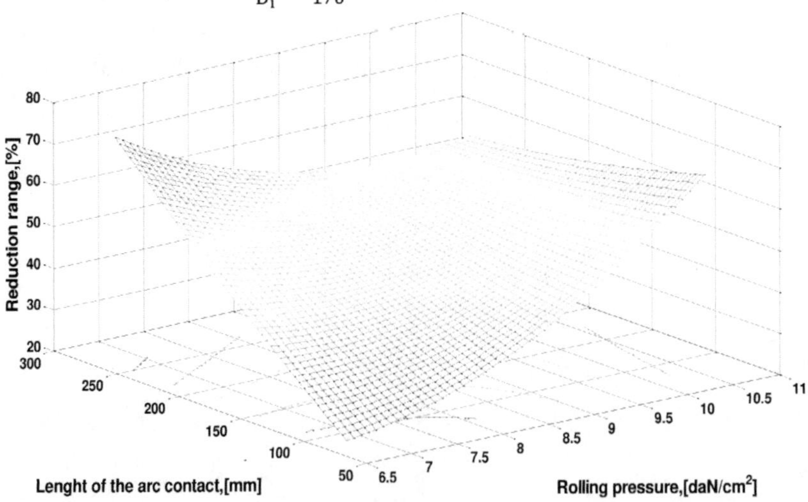

Figure 6. Variation of the rolling pressure and the contact arc length, h_0=2 mm samples between different diameter cylinders $\frac{D_s}{D_i} = \frac{140}{170}$ mm, applying various reductions, upper cylinder

The distribution of the pressure exerted by each roll, for the samples whose dimension is h_0 =2 mm, is shown in Figures 5 and 6. In this case, there is a total concordance with the results presented so far. Therefore, within the reduction range ε =10-25% and $p_{ms} = p_{mi}$.

With increasing degree of reduction, the mean pressure on the smaller-diameter roll (p_{ms}) has a growing value compared to the mean pressure exerted by the larger-diameter roll (p_{mi}).

At reductions smaller than 10-25%, similar to the lengths of the contact arcs, the inequality $p_{mi} > p_{ms}$ is the one which probably applies. The results obtained in the experimental research carried out when using rolls with unequal diameters are fully consistent with the laws of mechanics.

Indeed, for any difference between the diameters of the working rolls and any degree of reduction (ε) applied, $F_s = F_i$, or: $\frac{p_{ms}}{p_{mi}} = \frac{l_i}{l_s}$

Joining the intersection points l_s and l_i shown in Figure 2, for various thicknesses of the rolled samples, as well as the intersection points of the curves describing the mean pressure on the smaller roll (p_{ms}) and the bigger one (p_{mi}) in Figures 3, 4, 5 and 6, we obtain a parabola passing through the origin of the coordinate axes (Figure 7), which in general can be expressed by the equation:

$$y = aX^n \tag{44}$$

where: y – represents the mean pressure (or the lengths of the contact arcs, respectively);
a – coefficient taking into account the rolling conditions;
X – critical degree of deformation;
n – index depending on the degree of asymmetry of the process.

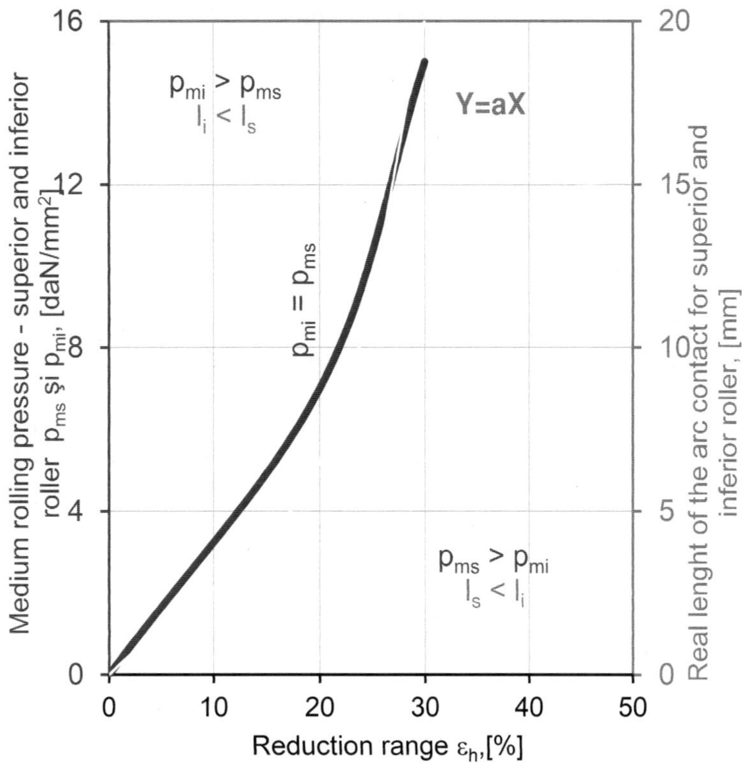

Figure 7. Variation of the rolling pressure and the contact arc length, between different diameter cylinders $\frac{D_s}{D_i} = \frac{140}{170}$ mm, applying various reductions

The obtained curve is in itself the geometrical locus of the points which characterise the partial symmetry conditions of the rolling process between unequal-diameter rolls when the following equalities occur, determining this symmetry:

$$\begin{aligned} p_{ms} &= p_{mi} \\ l_s &= l_i \end{aligned} \tag{45}$$

In this case, regardless of the ratio between the working diameters of the rolls, the metallic material comes out in a straight line.

The field located above this curve corresponds to the situation when, on the larger-diameter roll (lower), the pressure is higher and the contact arc length is lower, i.e.:

$$\begin{aligned} p_{ms} &> p_{mi} \\ l_s &< l_i \end{aligned} \tag{46}$$

In this case, all the rolled samples are bending upwards when coming out from the rolls (on the smaller-diameter roll).

The respective field is located under this curve, where the pressure exerted by the smaller-diameter roll (upper) is higher, and the contact arc is shorter:

$$\begin{gathered} p_{ms} < p_{mi} \\ l_s > l_i \end{gathered} \tag{47}$$

In this case, the rolled samples are bending downwards when coming out from the rolls (on the larger-diameter roll). Therefore, in the field located above the curve plotted in Figures 3, 4, 5 and 6, the pressures exerted by the larger-diameter roll are high, and in the field located under the curve, the pressures exerted by the smaller-diameter roll are much higher.

Obviously, when $p_{mi} > p_{ms}$, the bigger is the difference between the working diameters of rolls and the bigger is the initial thickness of the sample, the larger is the field. Related thereto, we should mention that the mean pressure is decreasing with increasing the deformation unevenness.

4. Concluding remarks

The research results confirmed that the current views on the deformation theory, according to which, at asymmetrical rolling, the pressures exerted by the smaller-diameter roll are always higher, are flawed. To characterise the rolling process based on the results obtained from this research, we consider useful to introduce a coefficient able to characterise the degree of asymmetry of the process:

$$K_a = \frac{p_{ms}}{p_{mi}} = \frac{l_i}{l_s} \tag{48}$$

Interpreting the dependencies presented in Figure 7, it results:

If $K_a=1$, this situation corresponds to the case when $y=aX^n$, and characterises the adequate conditions required by a symmetrical process to occur between the unequal-diameter rolls.

The more differs the K_a coefficient from the $K_a=1$ value, the greater is the difference between the contact arc lengths and the mean pressures exerted by the unequal-diameter cylinders, and the more is increasing the degree of unevenness of the deformation.

At the asymmetrical process, when the rolls have unequal diameters, the p_m value is much lower than in case of a symmetrical process under identical conditions of reduction. This phenomenon is fuelled by the denatured shape of the deformation zone, which creates favourable conditions for developing longitudinal tensile stresses, whose action over the bar during the rolling process can match the previous and subsequent stress applied, which considerably reduces the mean pressure exerted on the contact surfaces of the rolls.

References

[1] Alexa V 2002 *Contributions to research asymmetric rolling process*, Politehnica University of Timişoara, Romania, Doctoral Thesis

[2] Alexa V, Ratiu S and Kiss I 2005 Theoretical and experimental analysis of the rolling strain in the deformation area, *Metalurgia International* **5** 20-28

[3] Alexa V, Ratiu S and Kiss I 2016 Metal rolling – Asymmetrical rolling process, *IOP Conf. Ser.: Mater. Sci. Eng.* **106** 012019

[4] Alexa V, Ratiu S and Popa G 2010 *Method and program for automatic calculation forces and momentum to asymmetric longitudinal rolling*, MACMESE'10 Proceedings of the 12th WSEAS international conference on Mathematical and computational methods in science and engineering, Faro, Portugal, November 3-5, pp 303-306

[5] Alexa V 2004 Analysis of the lateral efforts in asymmetrical longitudinal rolling, *Annals of the Faculty of Engineering Hunedoara* **2** 45-50

IOP Conf. Series: Materials Science and Engineering **200** (2017) 012039 doi:10.1088/1757-899X/200/1/012039

Allelopathy relationship between plants and their use in organic farming

M Marian, Z Voşgan, O Mare Roşca and L Mihalescu

Technical University of Cluj-Napoca, North University Center of Baia Mare, no.76 Victoriei Street, Romania

E-mail: monica.marian@cunbm.utcluj.ro

Abstract. Allelopathy is a process still little studied in the plant world, if we refer to the diversity of biochemical compounds, through which plants can interact with each other, with fungi or bacteria. Biochemical "dialogue" between organisms may have stimulatory or inhibitory effects, contributing to numerical setting of the populations, the assertion of some species over others, the establishment and strengthen of plant communities. Practically, the allelopathy can be exploited in organic farming in understanding and identifying compatibility between species, to finding natural substances with herbicide potential. In experiments conducted, diluted extracts of *Taraxacum officinale* and *Cirsium vulgare* strongly inhibited the germination and growth of corn and beans, while *Hedera helix* produced the same effect, but at higher concentrations of the extract. *Humulus lupulus* extract has a stimulating effect on the two species and it is possible to use as natural fertilizer. Extracts of *Chenopodium album* works as stimulator for *Beta vulgaris*, and potent inhibitor for *Triticum aestivum*. *Agropyron repens* is an aggressive competitor for *Lycopersicon esculentum* and *Capsicum annuum*, producing a strong inhibition. *Juglans regia* extracts and especially those of *Satureja hortensis*, can be used as natural herbicides for up delay germination and growth suppression for the species: *Echinochloa crus-galli* and *Setaria glauca*.

1. Introduction

Allelopathy is a process of biochemical interaction between plant species, and between plants and microorganisms and plants and fungi. In nature, plants and animals tend to wear out strategies by which to compete for food resources and space. Allelopathy part of these strategies [1], [2].

Competition natural biochemical mechanisms can be exploited in artificial ecosystems, such crops in weed control as an alternative to synthetic herbicides [3].

Experimental studies have proceeded to the extraction of various plant tissues of phytotoxic substances that inhibit the germination and growth of other species [4]. Such substances, known generically as allelochemicals, are derived from secondary metabolism, especially in the way acetate and shikimic acid pathway [5-8].

Drugs considered allelochemicals are present in almost all plant species in various tissues of roots, stems, leaves, flowers, fruits and seeds. Are generally water-soluble substances, released into the environment through root exudation, volatilization or decomposition of plant that are found [9].

Many factors influence the production and toxicity of allelochemicals. Among them there are: the availability of nutrients in soil texture and soil moisture, the amount of solar radiation, temperature, and internal factors related to the biology of the species involved. Production of allelochemicals compounds is conditioned genetically essential being, donor plant genotype [10-12]. Plant age is an important factor in the amount of allelochemicals, it diminishes with age [13], [14].The interaction between plant species and pathogens, parasites and herbivores can stimulate production of allelochemicals [15], [16]. Allelopathy is a phenomenon closely linked to the interaction between the donor species and the environment. For example in ecosystems facing repeated fires, there is a high production of allelochemicals [17], [18]. Production may be a consequence of physical damage, or a strategy to reduce flammable species in the vicinity of producing and donating allelochemicals [19]. Some studies have shown an increased yield of allelochemicals and their's concentration in soil, both in the stage climax ecosystems, and in those that are in the final stages of succession [20]. Allelochemicals belonging to different categories of compounds, determine their solubility in water, volatility in the air, their affinity to soil particles and degradation over time [21-23].

Allelochemicals effects on the ecosystem level, including of agro ecosystems, generate changes in the rate of germination, seedling growth inhibition, functioning micorizelor, insects and bacteria, inhibiting nitrification or decomposition of litter and deadwood [24]. Also allelopathy regulate the density of target species [25]. Most studies of allelopathy, were based on interactions between cultivated plants and weeds, and more recently on identifying those chemicals involved, and how they act [24], [26], [27].

Allelopathy properties of plants can be successfully used as a tool to reduce pathogens and weeds. The current trend is to identify biological solutions that reduce the impact of herbicides and insecticides in crops [28]. Testing combinations of species, aimed at identifying those species with the strongest allelopathy properties against pathogens and weeds. Testing must follow some certain criteria: a. Allelopathy process monitoring, which have no interference with competition for resources, b. Testing a large number of plants; c. to use simple, unsophisticated reagents; d. the use of inexpensive experiments [28].

Also, testing existing allelopathy interference between weeds and crop plants, it is a step forward to understanding the mechanisms of competition that occurs between these categories, in order to sustain crops, with a minimum intake of synthetic chemicals.

2. Matherials and methods

They were made of plant tissue extracts of eight species of plants from spontaneous flora, commonly present in the environment. The extracts were tested on germination and growth of other eight species, grown for food. Combinations of species used:

Pigweed (*Chenopodium album)* tested on the species weat (*Triticum aestivum*) and *Beta vulgaris.*

Quackgrass (*Agropyron repens)* tested on the species *Capsicum annuum* and tomato (*Lycopersicon esculentum*).

Taraxacum officinalis tested on the species *Zea mays* and *Phaseolus vulgaris.*

Cirsium arvense tested on the species *Zea mays* and *Phaseolus vulgaris.*

Hedera helix tested on the species *Zea mays* and *Phaseolus vulgaris.*

Humulus lupulus tested on the species *Zea mays* and *Phaseolus vulgaris.*

2.1. Obtaining of the extracts from vegetal matherials

It was performed one aqueous extract from each of the eight plant species analyzed. Each aqueous extract was made from: 500 g each of young leaves and branches of the plant species; they were washed five times with tap water and finally rinsed with distilled water. The plant material was dried at 60° for 5 days, and subsequently was pulverized. 40 grams of powder was mixed with 1 liter of distilled water. The mixture was macerated for 24 hours at 23° C to carry out extraction. Subsequently, it was filtered under the pump vacuum, and the filtrate was regarded as a stock solution with a concentration of 100%. From this solution, dilution was made 1: 1, 1: 2. Final solutions were tested separately on some seed lots, then the young plants.

2.2. Processing of test plants seeds

The seeds of the species tested, were used for germination assays. The seeds were sterilized with 5% sodium hypochlorite for 10 min and five times rinsed with distilled water. In each experiment 100 seedlings were used and the experiments were repeated three times.

2.3. Processing of test plants seeds

To evaluate the allelopathic effect to water soluble extracts, 50 seeds of each tested species, placed in a plate, containing two layers of filter paper moistened with 10 ml of aqueous extract, respectively, in different dilutions, as Figure 1 suggests. The plates were maintained, for 7 days, in a Sanyo environmental test chamber (temperature 23^{0}C, illumination 10.000 lx and humidity 82%). The experiments were replicated three times.

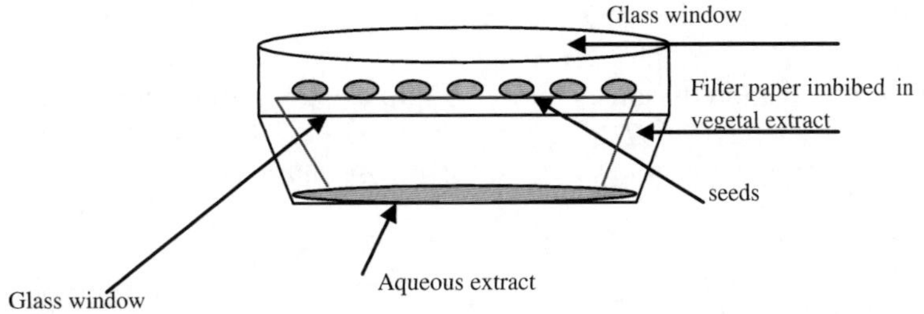

Figure 1. Testing of plant extracts on germination and growth of seedlings

For each combination of plants, experimental variations were made:
V0 - control group, germinated seed Placed on the substrate material wetted with distilled water;
V1 - Placed germinated seed material on the substrate wetted with diluted extract in a 1: 1
V2 - Placed germinated seed material on the substrate wetted with the extract 1: 2;
V3 - germinated seeds Placed on the substrate material wetted extract with 1: 3.

3. Results and discussions

Plant species of the extracts were carried out, are the components of the spontaneous flora. Some, such as *Cirsium arvense, Chenopodium album, Taraxacum officinale* and *Agropyron repens*, being segetal and ruderal plants occupy the same habitat with species cultivated. The experiment aims to highlight possible biochemical influences between weeds and cultivated plants. Other species of those tested: *Humulus lupulus, Hedera helix*, recognized that they have a rich biochemical composition, and also have a diversified secondary metabolism. We are looking to identify reactions allelopathy on those species with the potential to become either fertilizers or biocides to organic crops.

Two of the most cultivated species: corn (*Zea mays*) and beans (*Phaseolus vulgaris*) were subjected to the influence of the four species with potential allelopathy assumed, heightened dandelion (*Taraxacum officinale*), thistles (*Cirsium vulgare*), hops (*Humulus lupulus*) and ivy (*Hedera helix*). Both, germination and seedlings growth are heavily influenced by the treatment with plant extracts. Three of the species tested: dandelion, ivy and thistles and exert powerful inhibitory influences both the germination and the growth on crops. Analysis of extract of *Taraxacum officinale* on the species *Zea mays* and *Phaseolus vulgaris*.

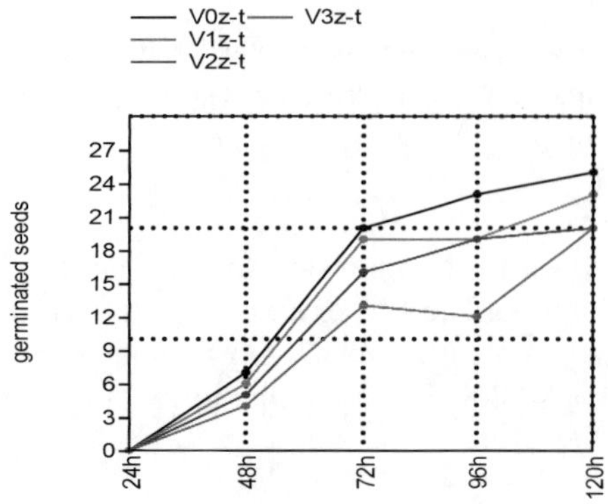

Figure 2. *Zea mays* germination dynamics under the influence of the extracts of *Taraxacum officinale*

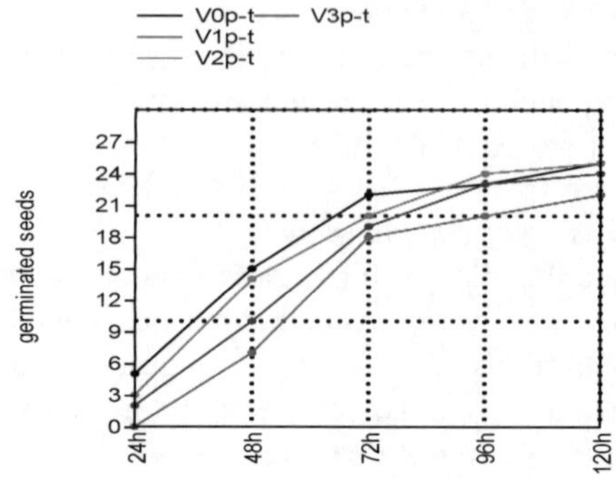

Figure 3. Germination dynamics under the influence of the *Phaseolus vulgaris* extracts of *Taraxacum officinale*

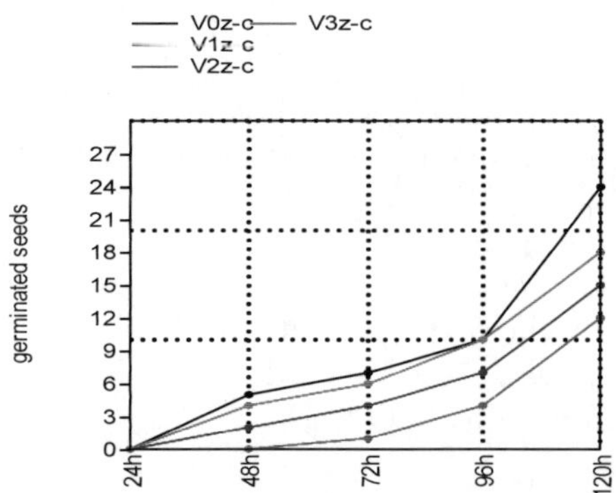

Figure 4. *Zea mays* germination dynamics under the influence of the extracts *Cirsium vulgare*

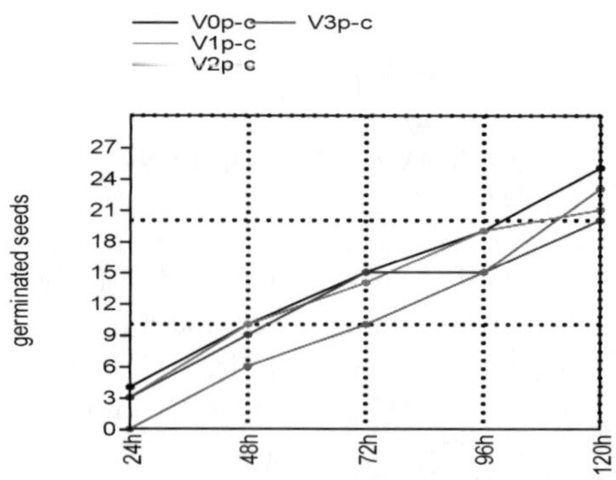

Figure 5. Germination dynamics under the influence of the *Phaseolus vulgaris* extract *Cirsium vulgare*

Dandelion (*Taraxacum officinale)* and thistles (*Cirsium vulgare)*, are two species often present in crops. Among competitive mechanisms for space and food resources, enroll and allelopathy competition. Although corn is more sensitive to the influence of both weed species it proved that thistles have more aggressive in generating inhibitory effect. In all four combinations analyzed variant V3, consisting of dilution largest, it has demonstrated a maximal effect in reducing agricultural germination in both species (Figure 2, 3). Variant V1 containing extract the most concentrated, has an influence very similar to the control V0 (Figure 6). V3 is closest to the natural concentration of metabolites, dosed so, to achieve maximal effect in the competitive mechanism between species (Figure 4, 5). The samples with the highest concentrations, V1 and V2, exceeding much value metabolites, naturally occurring in plants. Beans responded differently to the action of dandelion and thistles. *Phaseolus vulgaris* germination was stimulated by the concentrated extract (V1) of *Taraxacum officinale. Cirsium vulgare* extract was shown to inhibit germination *Phaseolus vulgaris*, independent of concentration, taking effect only as their metabolites and less quantity.

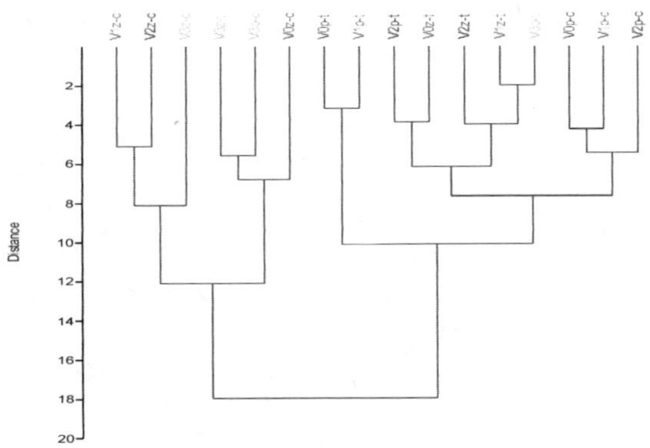

coph. Corr= 0,705

Figure 6. Cluster analysis based on Euclidean equation, on the influence of different concentrations of extracts of the tested species

Influence of *Hedera helix* extract on species *Zea mays* and *Phaseolus vulgaris.*

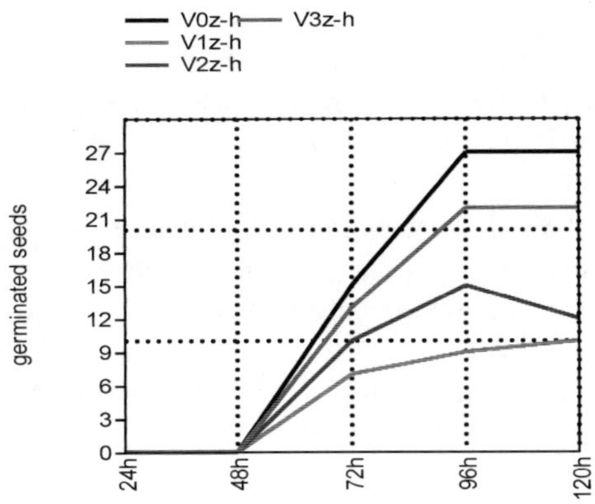

Figure 7. Influence of the extract of *Hedera helix* on the germination of *Zea mays*

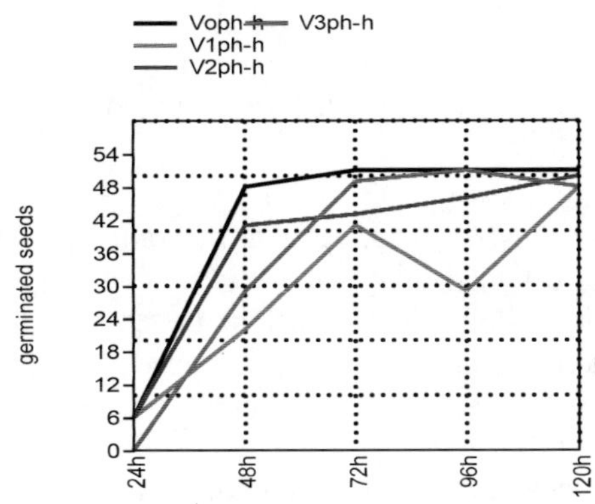

Figure 8. Influence of the extract of *Hedera helix* on the germination *Phaseolus vulgaris*

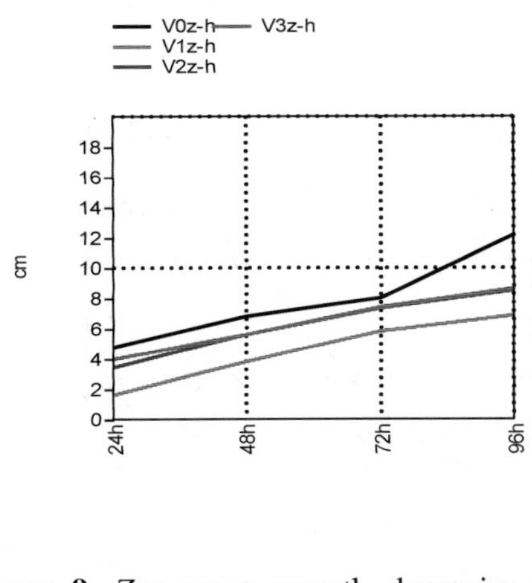

Figure 9. *Zea mays* growth dynamics under the influence of the extract of *Hedera helix*

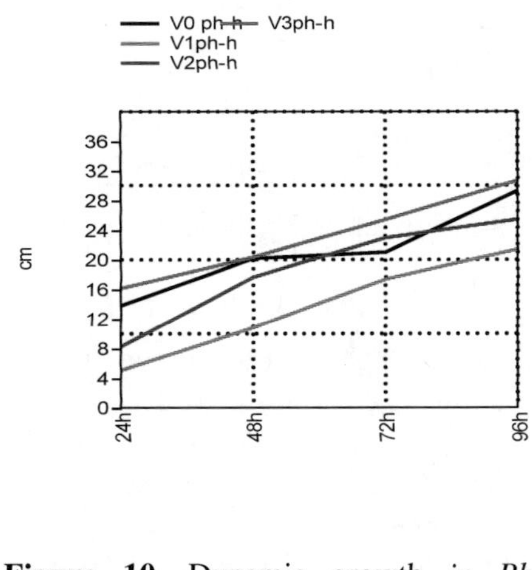

Figure 10. Dynamic growth *in Phaseolus vulgaris* under the influence of extract of *Hedera helix*

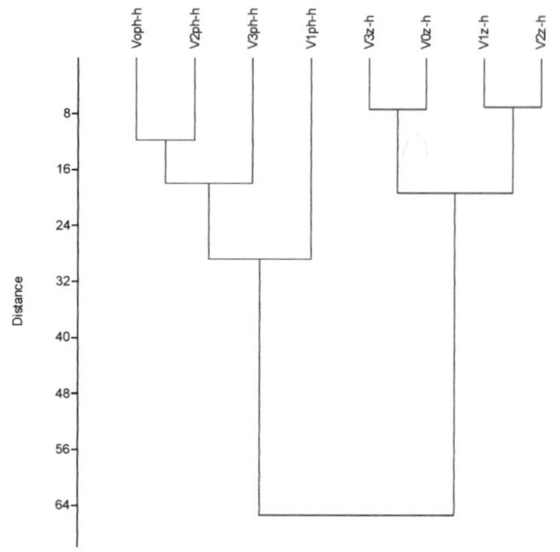

Coph.corr. 0,92

Figure 11. Cluster analysis based on Euclidean equation, on seed germination under the influence of extracts of *Hedera helix*

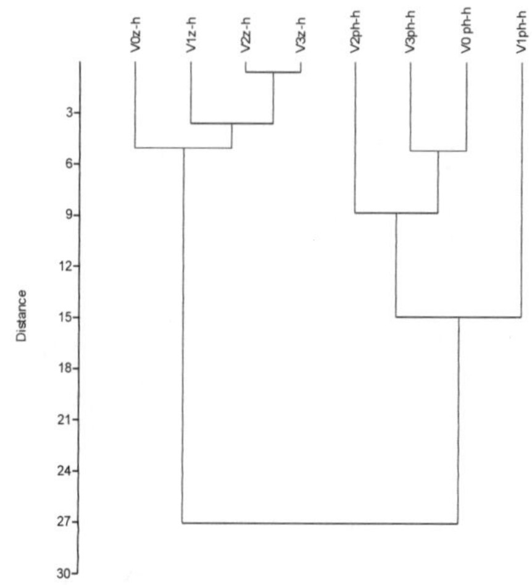

Coph.corr. = 0,8834

Figure 12. Cluster analysis based on Euclidean equation, on seedling growth under the influence of extracts of *Hedera helix*

Hedera helix is a species that naturally has different habitats of the tested crops. In nature the two species are found rarely and only accidentally together. However, ivy, generates an inhibitory effect on species tested, which means its metabolites shows the spectrum without specificity is, or targeting other target species. Corn is noticeably inhibited much stronger than beans (Figure 7, 8). Compared with typical crop weed species, ivy, generates a stronger allelopathyc response to higher concentrations of the extract (Figure 9, 10). Therefore, metabolites showing a small specificity, require high concentrations, for generating allelopathyc effect. In terms of growth dynamics, extract the diluted (V3), it proved valence stimulating to *Phaseolus vulgaris* (Figure 11, 12). This concentration can be used as stimulant / fertilizer for raising seedlings.

The influence of *Humulus lupulus* extract of the species *Zea mays* and *Phaseolus vulgaris*

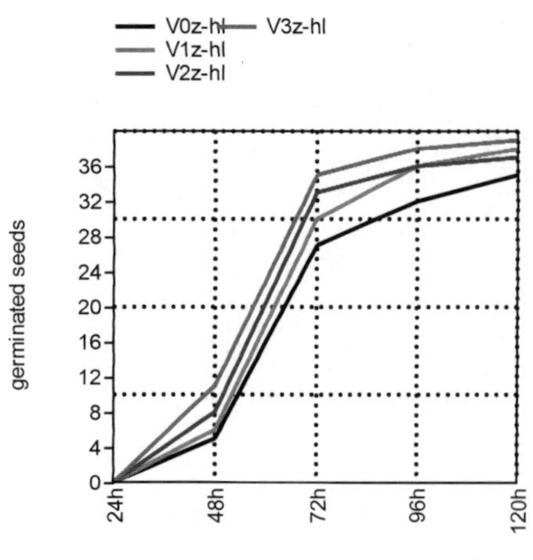

Figure 13. *Humulus lupulus* extract influence on the germination of *Zea mays*

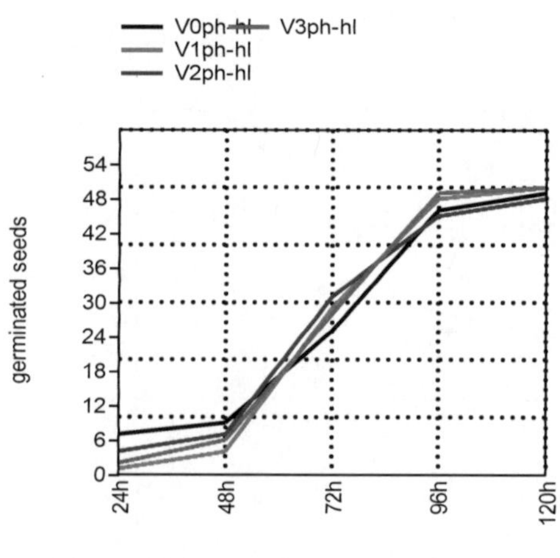

Figure 14. *Humulus lupulus* extract influence on the germination of *Phaseolus vulgaris*

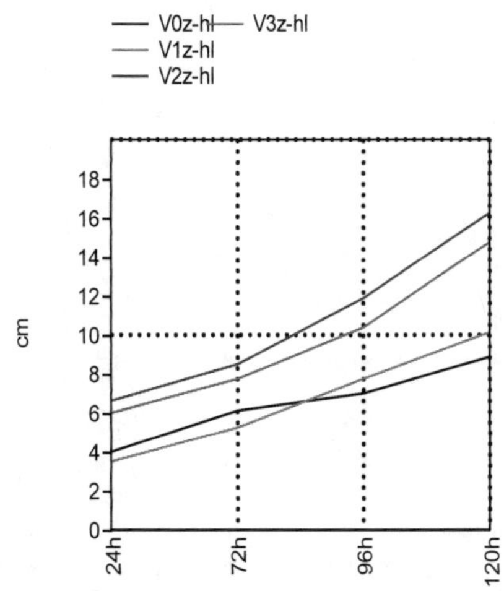

Figure 15. *Zea mays* plant growth dynamics under the influence of *Humulus lupulus* extract

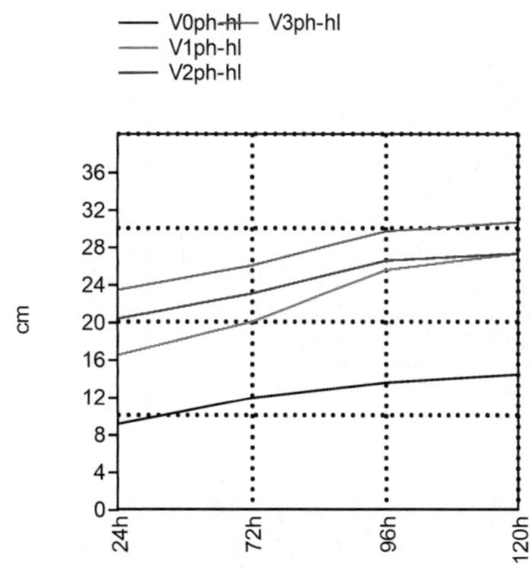

Figure 16. *Phaseolus vulgaris* plant growth dynamics under the influence of *Humulus lupulus* extract

Hops is a species in nature do not take similar biotypes, with corn and beans. Hop extracts in all tested concentrations, they exerted stimulatory effects on the two cultivated species (Figure 13, 14). The less concentrated extracts proved the strongest effect (Figure 15, 16). In accord with experimental observations, hops can be exploited not only for beer production, but also to obtain potential natural fertilizers for organic agriculture.

Chenopodium album influence on the species *Triticum aestivum* and *Beta vulgaris.* Pigweed (*Chenopodium album*) is a ruderal species, often present in crops that generate very different responses, ranging both the target species but with the stage of development thereof.

Allelochimicals soluble in *Chenopodium* species (*C. murale*) proved strong inhibitory to a number of species tested: *Trifolium alexandrimum, Triticum aestivum, Melilotus indicus, Lycopersicon esculentum, Cucumis sativus* [29].

Similarly, wheat germination is strongly inhibited by extract pigweed, the stronger, the more it is more concentrated (Figure 17). The experiment highlights the necessity of preparing land for sowing by removing debris weeds that by escaping metabolites, reducing the rate and pace of germination of wheat. Instead, beets metabolic benefits from the support of the *Chenopodium album*. Germination is more stimulated, the more extract concentration is higher (Figure 18).

This distinction suggests that the inhibitory effect is manifested more strongly between phylogenetically more distant species, pigweed and *Beta vulgaris* were from related families. The results are consistent with other experimental observations, that the same concentration of a plant extract can stimulate the growth and development of a species (Figure 9), but may inhibit the growth of another species [30].

It is therefore essential to identify the concentration at which each species give an answer, and that interaction occurs allelopathy useful in weed management [7], [8], [31], [32].

IOP Conf. Series: Materials Science and Engineering **200** (2017) 012039 doi:10.1088/1757-899X/200/1/012039

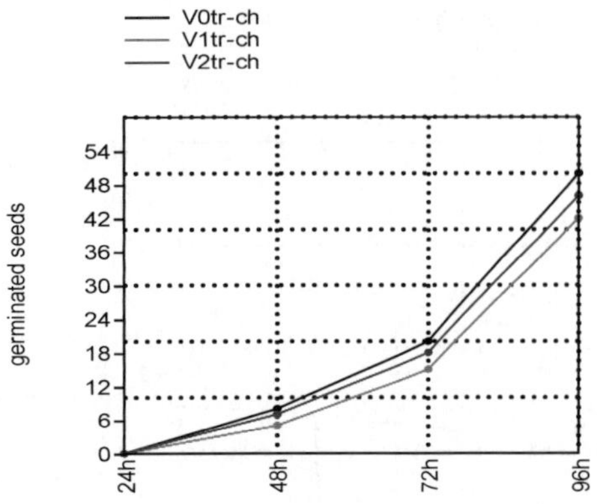

Figure 17. The dynamics of germination from *Triticum aestivum* influenced by *Chenopodium album* (pigweed)

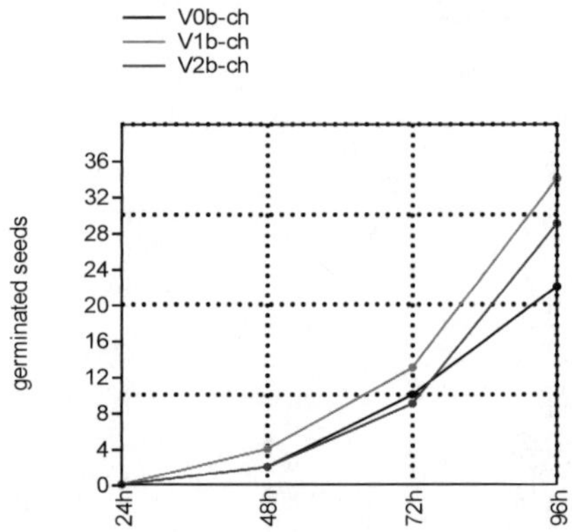

Figure 18. The dynamics of germination from *Beta vulgaris* influenced by *Chenopodium album*

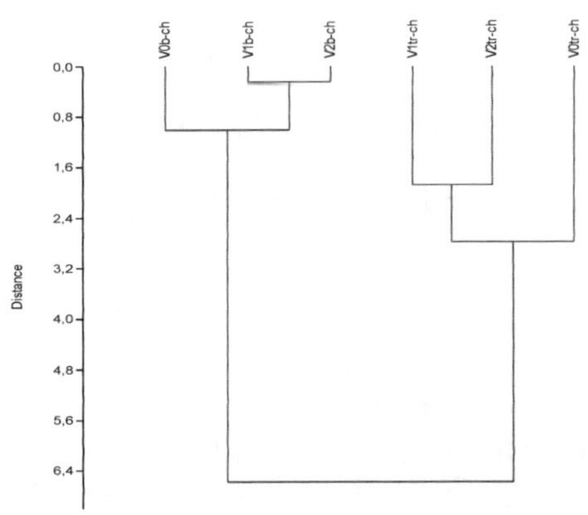

Figure 19. Cluster analysis based on Euclidean equation, on seed germination under the influence of *Chenopodium album* extracts

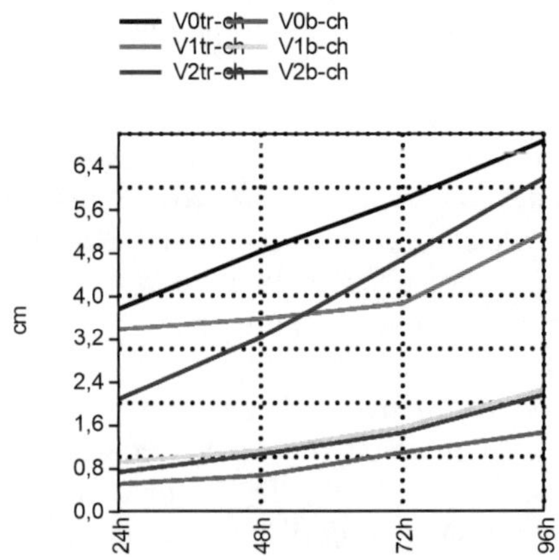

Figure 20. Growth dynamics of the seedlings

Regarding the growth rate of seedlings, the situation is similar, wheat is strongly inhibited, the stronger the higher the concentration is higher (Figure 20). Instead highest concentration of extract is obvious incentive for beet, the slowest growing plants in the control group V0 (Figure 19).

Quackgrass, *Agropyron repens* influence on species *Lycopersicon esculentum* and *Capsicum annuum*

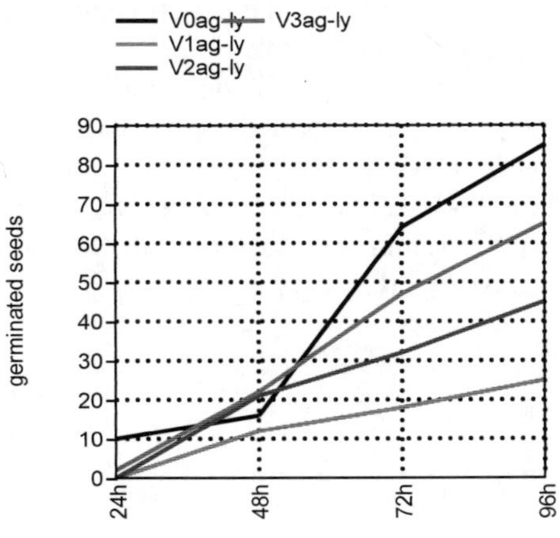

Figure 21. *Lycopersicon esculentum* plant growth dynamics under the influence of *Agropyron repens* extract

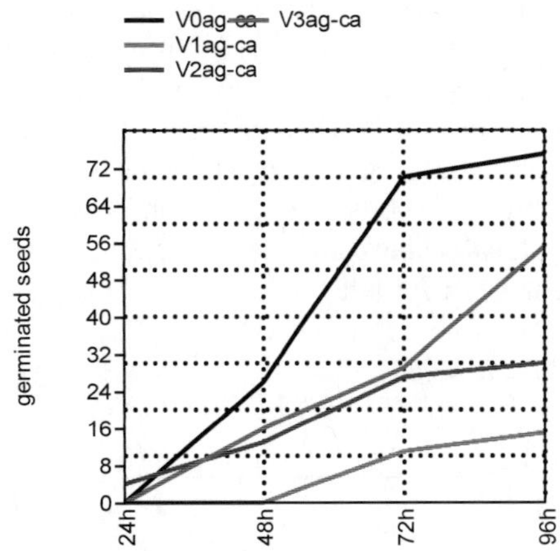

Figure 22. *Capsicum annuum* plant growth dynamics under the influence of *Agropyron repens* extract

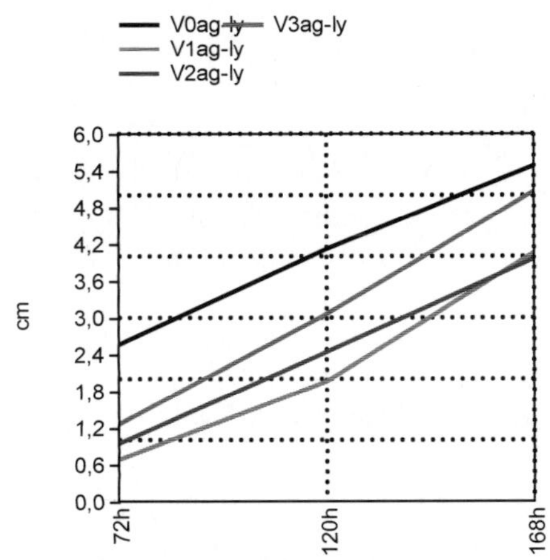

Figure 23. *Lycopersicon esculentum* seedlings growth dynamics under the influence of extract of quackgrass

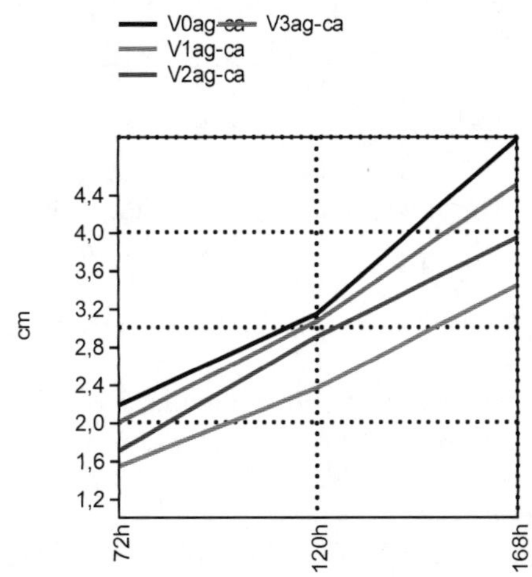

Figure 24. *Capsicum annuum* growth dynamics under the influence of extract of quackgrass

Couch grass, *Agropyron repens*, is a widespread weed and "aggressive" in vegetable gardens. Testing the influence of this species on vegetable species, aimed at identifying competitive mechanisms like "weed - cultivated species' behavior to adopt suitable for growing vegetables and avoid the use of herbicides.

Tests performed with extracts of couch grass on tomato and pepper plants, highlights a strong allelopathy mechanism inhibitor of couch grass (Figures 21, 22). It manifests the stronger, the more concentration is higher in both the germination stage, and in the growth of seedlings (Figures 23, 24). Therefore, it is beneficial that vegetable planting material, to be free from inhibiting competition generated by quackgrass. It is preferred that the couch grass be removed by mechanical actions.

Influence of extracts of *Juglans regia* and *Satureja hortensis* on the species: *Setaria glauca* and *Echinochloa crus-galli*

Increased interest on allelopathy studies, it shows the possible inhibition of weeds in crops by using herbal extracts that are clean. To this end, they were tested and walnut extracts of thyme, segetal on some species, spread in gardens: *Setaria glauca* and *Echinochloa crus-galli*. Similar experiments were conducted to test the properties of maize and sorghum herbicides on wild barley [32].

Both species have been shown to have potent inhibitory for both germination and for the development of weeds. Development of both weed species was strongly inhibited by the extract of *Satureja hortensis* (Figure 25). The experimental results are preliminary for further testing to identify natural herbicide consisting of plant extracts with certain allelopathyc behavior (Figure 26). Bibliographical studies show similar results. Extracts of corn and sorghum significantly reduces both, the percentage of sprouted grains of wild barley, and root and hypocotyl length.

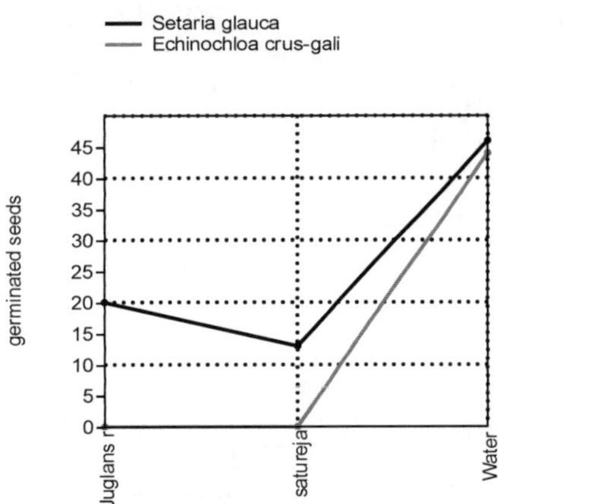

Figure 25. Influence of extracts of *Juglans regia* and *Satureja hortensis* on the species: *Setaria glauca* and *Echinochloa crus-galli*

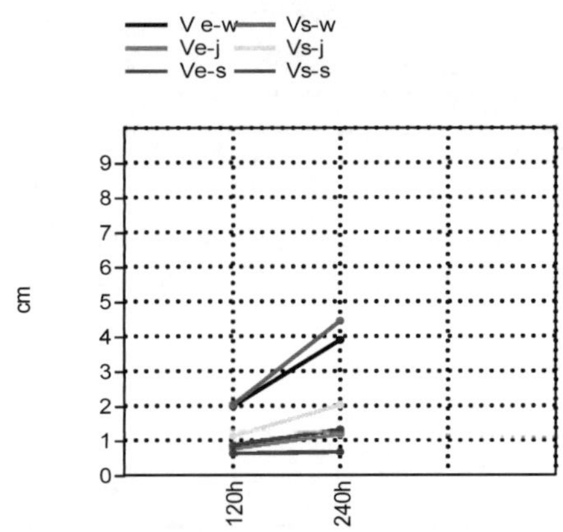

Figure 26. Influence of extracts of *Juglans regia* and *Satureja hortensis* on the species: *Setaria glauca* and *Echinochloa crus-galli*

Considering that in general, weeds germinate much earlier and faster than cultivated species, can consider, that the land be treated with thyme extract to suppression of unwanted species. Future studies should identify and isolate the most effective herbal alelochimicale studied, and use them in the development of natural herbicides for weed control.

References

[1] Grab S 1961 Differential growth inhibitors produced by plants, *Bot. Rev.* **27**(3) 422-443

[2] Dubey B, Shukla J B, Sharma S, Agarwal A K and Sinha P 2010 A mathematical model for chemical defense mechanism of two competing species, *Nonlinear Analysis: Real World Applications* **11**(2) 1143-1158

[3] Bhowmik P and Inderjit C 2003 Challenges and opportunities in implementing allelopathy for natural weed management, *Crop Protect.* **22**(4) 661-671

[4] Turk M A and Tawaha A M 2003 Allelopathic effect of black mustard (*Brassica nigra* L.) on germination and growth of wild oat (*Avena fatua* L.), *Crop Protection* **22**(4) 673-677

[5] Rice E L 1984 *Allelopathy Second Edition*, Science Academic Press Inc, Orlando

[6] Hall M H and Henderlong P R 1989 Alfalfa autotoxic fraction characterization and initial separation, *Crop. Sci.* **29** 425-428

[7] Chon S U and Kim J D 2002 Biological activity and quantification ofsuspect ed allelochemicals from alfalfa plant parts, *J. Agron. Crop Sci.* **188**(4) 281-285

[8] Tawaha A M and Turk M A 2003 Allelopathic Effects of Black Mustard (*Brassica nigra*) on Germination and Growth of Wild Barley (*Hordeum spontaneum*), *Journal of Agronomy and Crop Science* **189**(5) 298-303

[9] Aminidehaghi M, Rezaeinodehi A and Khangholi S 2006 Allelopathic potential of *Alliaria petiolata* and *Lepidium perfoliatum*, two weeds of the Cruciferae family, *Journal of Plant*

Diseases and Protection **20** 455-462
[10] Bell E A and Charlwood B V 1980 *Secondary Plant Products*, Springer Heidelberg
[11] Miller D A 1996 Allelopathy in Forage Crop Systems, *Agronomy Journal* **88**(6) 854-859
[12] Quader S T, Bello-DeOcampo D, Williams D E, Kleinman H K and Webber M M 2001 Evaluation of the chemopreventive potential of retinoids using a novel in vitro human prostate carcinogenesis model, *Mutat Res.* **496**(1-2) 153-161
[13] Weston L A, Harmon R and Mueller S 1989 Allelopathic potential of sorghum-sudangrass hybrid (sudex), *Journal of Chemical Ecology* **15** 1855-1865
[14] Wolfson J L and Murdock L L 1990 Growth of Manduca sexta on wounded tomato plants: role of induced proteinase inhibitors, *Entomologia Experiment. et Applicata* **54**(3) 257–264
[15] Farkas G L and Kiraly Z 1962 Role of phenolic compounds in the physiology of plant diseases and disease resistance, *Phytopathol. Z.* **44** 105–150
[16] Baldwin I T 1989 Mechanism of damage-induced alkaloid production in wild tobacco, *J Chem Ecol.* **15**(5) 1661-1680
[17] Williamson G B and Black E M 1981 High temperature of forest fires under pine as a selective advantage over oaks, *Nature* **293**
[18] Coley P, Bryant J, Chapin F 1985 Resource availability and plant antiherbivore defense, *Science* **230** 895–899
[19] Williamson G B, Richardson D R and Fischer N H 1992 Allelopathic mechanism in fire-prone communities. In S. J. Rizvi and V. Rizvi (Editors), *Allelopathy: Basic and applied aspects*, Chapman & Hall, London, pp 59-75
[20] Rice E L and Pancholy S K 1973 Inhibition of nitrification by climax ecosystem. II. Additional evidence and possible role of tannins, *American Journal of Botany* **60** 691-702
[21] Cheng H H 1992 A conceptual framework for assessing allelochemicals in the soil environment. In: Rizvi S J H and Rizvi V (Editors), *Allelopathy: Basic and applied aspects*, Chapman & Hall, London, pp 21-29
[22] Peñuelas J 2005 *Biogeoquímica y termodinámica margalefianas de los cambios globales en los ecosistemas terrestres,* in Proceedings of Unity in Diversity, a conference on ecology afte the legacy of Ramón Margalef, ComoCaixa Barcelona, Spain, November 16-18, pp 29
[23] Birkett M A, Campbell C A M, Chamberlain K, Guerrieri E, Hick A J, Martin J L, Matthes M, Napier J A, Pettersson J, Pickett J A, Poppy G M, Pow E M, Pye B J, Smart L E, Wadhams G H, Wadhams L J and Woodcock C M 2000 New roles for cis-jasmone as an insect semiochemical and in plant defense, *Proceedings of the National Academy of Sciences* **97**(16) 9329-9334
[24] Blanco J A 2007 The representation of allelopathy in ecosystem-level forest models Review, *Ecological modelling* **20**(9) 65-77
[25] Weidenhamer J, Hartnett D C and Romeo J T 1989 Density-Dependent Phytotoxicity: Distinguishing Resource Competition and Allelopathic Interference in Plants, *Journal of Applied Ecology* **26**(2) 613
[26] Pellissier F and Souto X C 1999 Allelopathy in nordhern temperate and boreal semi-natural woodland, *Crit. Rev. Plant Sci.* **18** 837-652
[27] Einhellig F A 2004 Mode of allelochemical action of phenolic compounds, in: Macias F A et al. (Eds.), *Allelopathy: chemistry and mode of action of allelochemicals*, London, CRC Press, LLC, pp 217-238
[28] Xuan T D, Shinkichi T, Khanh T D and Min C I 2005 Biological control of weeds and plant pathogens in paddy rice by exploiting plant allelopathy: an overview, *Crop Protection* **24** 197–206
[29] El-Khatib A A, Hegazy A K and Galal H K 2004 Does allelopathy have a role in the ecology of Chenopodium murale? *Annales Botanici Fennici* **41**(1) 37–45
[30] Narwal S S 1994 *Allelopathy in crop production*, Scientific Publishers, Jodhpur
[31] Economou G O, Gani T A, Yannitsaro A and Bilalis D 2002 Allelopathic effect of Conyza

IOP Conf. Series: Materials Science and Engineering **200** (2017) 012039 doi:10.1088/1757-899X/200/1/012039

albida on Avena sativa and Spirodela polyrhiza, *J. Agron. Crop Sci.* **188**(4) 248-253

[32] Al Tawaha A R M and Odat N 2010 Use of Sorghum and Maize allelopathic properties to inhibit germination and growth of wild barley (*Hordeum spontaneum), Not. Bot. Hort Agrobot Cluj* **38**(3) 124-127

IOP Conf. Series: Materials Science and Engineering **200** (2017) 012040 doi:10.1088/1757-899X/200/1/012040

The possibility of developing hybrid PV/T solar system

M Dobrnjac[1], P Zivkovic[2] and V Babic[1]

[1]University of Banja Luka, Faculty of Mechanical Engineering, Stepe Stepanovica no 71, 78000 Banja Luka, Bosnia and Herzegovina
[2]University of Nis, Faculty of Mechanical Engineering, Aleksandra Medvedeva no 14, 18000 Nis, Republic of Serbia

E-mail: dobrnjac.mirko@gmail.com

Abstract. An alternative and cost-effective solution to developing integrated PV system is to use hybrid photovoltaic/thermal (PV/T) solar system. The temperature of PV modules increases due to the absorbed solar radiation that is not converted into electricity, causing a decrease in their efficiency. In hybrid PV/T solar systems the reduction of PV module temperature can be combined with a useful fluid heating. In this paper we present the possibility of developing a new hybrid PV/T solar system. Hybrid PV/T system can provide electrical and thermal energy, thus achieving a higher energy conversion rate of the absorbed solar radiation. We developed PV/T prototype consisted of commercial PV module and thermal panel with our original solution of aluminium absorber with special geometric shapes. The main advantages of our combined PV/T system are: removing of heat from the PV panel; extending the lifetime of photovoltaic cells; excess of the removing heat from PV part is used to heat the fluid in the thermal part of the panel; the possibility of using on the roof and facade constructions because less weight.

1. Introduction

Increased energy consumption and environmental degradation to climate change is one of the biggest global problems in the 21st century. Solution to both problems is in more intensive use of renewable energy sources, with an important role of solar energy [1]. The aim of this paper is to explore the possibilities for the development of combined PV/T solar panel that simultaneously converts solar radiation into electricity and heat. This panel is composed of a photovoltaic (PV) and thermal (T) modules assembled into a single unit.

Popularity of PV/T solar systems is increasing as they are reliable, flexible and environmentally friendly. Well-designed PV/T system, with appropriate efficiency, solves the problem of electricity and thermal energy supply of various residential buildings. The system is fully automatically controlled and maintains heating and cooling of thermally well insulated building, CO_2 emission-free.

2. Production and technical and geometric characteristics of new PV/T solar panel

For development of efficient PV/T system the structural characteristics and efficiency of PV/T panels that can be found on the market were analyzed. Obtained results showed that critical part of such hybrid system is mainly the thermal side. Thermal panels are usually made of copper pipes with flat absorber and the different quality of surface absorption. Considering the advantages and disadvantages of existing solutions, to increase efficiency and improve the design, the new solution for thermal panel is designed [2].

It takes into account the weight of the panel, primarily due to the load on the roof or facade construction, which can be significant, especially if the larger number of panels are installed [3]. Aluminium has good technical characteristics in terms of heat transfer and the possibility of adequate long-term surface protection. New thermal panel represents the integration of functional design and appropriate materials with special geometric shapes. The panel consists of a series of specially designed aluminium fins, placed in a aluminium thermo-insulated casing.

Figure 1. Absorber surface of solar panel with a schematic connection of absorber lamellae (cross-section)

The plate thermal panel is geometrically designed so that it can be combined with a commercial photovoltaic panel. All the necessary connections are adjusted for the system to function as a single unit – PV/T panel.

The PV panel characteristics:

Dimensions	1654x989x40 mm
Dimensions of cell	156x156 mm
Number of cells	60
Type of cell	Polycrystalline cells with 3 busbar technology
Maximum permissible load	6000 Pa
Front cover type	Micro structured solar glass in 3.2 mm thick
Frame material	Anodized aluminum
Panel mass	18,2 kg
Water content	5 kg
Nominal electric output	250 W

Figure 2. PV panel

When installing the PV/T system it was noticed that the most delicate is the problem of merging PV and T panels, where the most important is to achieve good surface contact for better transfer of heat from the PV to the T panel. This contact is improved by applying thermal paste [4].

Figure 3. PV and thermal part of solar panel before finishing

The PV/T panel characteristics:

Dimensions	1654x989x40 mm
Effective area	1,62 m^2
Panel mass	28 kg
Water content	5 kg
Nominal electric output	250 W

Figure 4. Finished PV/T solar panel

3. Conclusion

- The dissipation of heat from PV to the of thermal panel (water-cooling), results in longer service life of PV cells and the whole PV/T panel due to the better working conditions, while producing more electricity than a conventional PV panel;
- The excess heat which is removed from the PV part is used to heat the fluid in the thermal part of the panel;
- Less surface area is required for the installation of PV/T system compared to conventional PV and thermal panels which results in lower load to the bearing structures in the period of exploitation;
- More precise effects and results of this model PV/T panel can be expected after the completion of thermal testing, and through monitoring in operation for a longer period of time.

References

[1] Tripanagnostopoulos Y, Souliotis M, Makris Th, Georgostathis P and Sarris M 2010 *Design and Performance of a Hybrid PV/T Solar Water Heater,* CP1203, 7th International

Conference of the Balkan Physical Union, pp 1019-1024
[2] Dobrnjac M, Latinović T, Dobrnjac S and Živković P 2016 Increasing the Efficiency of Solar Thermal Panel, *IOP Conf. Ser.: Mater. Sci. Eng.* **144** 012011
[3] ***Deutsche Gesellschaft für Sonnenenergie 2005 *Planning and Installing Solar Thermal Systems: A Guide for Installers, Architects and Engineers*, Earthscan
[4] Kozić Đ and Dobrnjac M 2013 *Development of solar collectors regarding efficiency increasing and the price decreasing*, International Conference on Sustainable Energy & Environmental Protection SEEP 2013, Maribor, Slovenia

IOP Conf. Series: Materials Science and Engineering **200** (2017) 012041 doi:10.1088/1757-899X/200/1/012041

A combined approach of AHP and TOPSIS methods applied in the field of integrated software systems

A D Berdie[1], M Osaci[1], I Muscalagiu[1] and C Barz[2]

[1]Politehnica University of Timisoara, Department of Electrical Engineering and Industrial Informatics, Engineering Faculty of Hunedoara, 5 Revolution Street, 331128, Hunedoara, Romania
[2]Technical University of Cluj-Napoca, Baia Mare Northern University Center, Department of Electrical Engineering, Electronics and Computers, Engineering Faculty CUNBM, Str. Dr.V. Babeş, 62A, 430083, Baia Mare, Maramureş, Romania

E-mail: adela.berdie@fih.upt.ro

Abstract. Adopting the most appropriate technology for developing applications on an integrated software system for enterprises, may result in great savings both in cost and hours of work. This paper proposes a research study for the determination of a hierarchy between three SAP (System Applications and Products in Data Processing) technologies. The technologies Web Dynpro -WD, Floorplan Manager - FPM and CRM WebClient UI - CRM WCUI are multi-criteria evaluated in terms of the obtained performances through the implementation of the same web business application. To establish the hierarchy a multi-criteria analysis model that combines the AHP (Analytic Hierarchy Process) and the TOPSIS (Technique for Order Preference by Similarity to Ideal Solution) methods was proposed. This model was built with the help of the SuperDecision software. This software is based on the AHP method and determines the weights for the selected sets of criteria. The TOPSIS method was used to obtain the final ranking and the technologies hierarchy.

1. Introduction

The implementation of a business project on the SAP NetWeaver platform requires a proper study regarding the appropriate technology selection. A flexible technology is always responsive to the customer requirements. Selecting an inappropriate technology for the implementation of a project, leads not only to a high execution time and memory consumption, but also to a high maintenance. Therefore, the performance of the project is significantly affected [1].

This paper, introduces a hybrid model for selecting between three SAP ABAP UI technologies, based on the concepts of multi-criteria analysis and the methods AHP and TOPSIS. The main particularity of the multi-criteria analysis is based on the judgment strengths of the decision makers to determine the objectives and the criteria, to estimate the relative weights of criteria and partly to evaluate the contribution of each alternative in order to achieve each criterion [2-4].

This proposed model combines the AHP and the TOPSIS methods. The AHP is used to score and set the share of importance of the relative weights for the decision criteria-sets and the TOPSIS method is used to obtain the final ranking.

2. Objectives of the study

The aim of this study is to develop a hybrid multi-criteria model for selecting the most efficient ABAP UI technology used in the development of SAP business applications. The proposed research methodology combines the TOPSIS and AHP methods, usually used in the decision-making process. The model requires the prioritization of the technologies using the TOPSIS method, in which the relative weights of the criteria result from the combination, between the weights of importance defined and the values determined with AHP method. According to the concepts of multi-criteria analysis related in [5] and [6], to determine a hierarchy for the SAP UI technologies, the objectives of this work are outlined as follows:

- Determining the alternatives.
- Finding the most relevant criteria sets that normally results from implementing and testing the same business application with Web Dynpro ABAP, Floorplan Manager and CRM WebClient UI technologies.
- Building the performance matrix that describes the consequences or the performance of alternatives regarding the fulfilment of the decision criteria.
- Scoring and setting the share of importance of the relative weights for the decision criteria-sets through the AHP method.
- Selecting a normalization process for the criteria values expressed by different units, to be transformed into a common scale.
- Implementing the TOPSIS algorithm in order to obtain the technologies prioritization.

3. Results and interpretation

In the context of achieving the objectives set, the alternatives in the hierarchy are the three ABAP UI technologies: Web Dynpro, Floorplan Manager and CRM WebClient UI.

The process of identifying the most relevant and variable criteria that are normally used in the selection process is facilitated using the SAP Business Transaction Analysis and HTTP Watch for Internet Explorer tools. For each business application are measured a set of parameters: response time (RT), database (DBRT) request time, CPU time consumed at the end of a transaction and the time obtained in front-end (HTTP Watch time). Another set of parameters considered to be relevant and frequently mentioned in the literature [7-10] are the effort for: development (DE), maintenance (ME) and enhancement (ME) a business application implemented with a certain ABAP UI technology. These two sets of parameters will represent the criteria for our selection process in the model.

The elements of the Table 1 represent the values of the criteria regarding the alternatives. For the first set of criteria (CPUT, DBRT, HTTPWT, RT), the values derived from the measurements performed by implementing the business application through the three UI technologies. For the second set of criteria (DE, ME, EE), the values are selected according to the conversion scale from the Table 2, which represents the Saaty's intensity scale [11]. In the proposed hybrid multi-criteria model the composition of the Table 1 corresponds to the decisional matrix.

Table 1. The decisional matrix

ABAP UI Technology/ Parameters	Criteria Set1(c1,c2,c3,c4) Set2(c5,c6,c7)						
	C1	C2	C3	C4	C5	C6	C7
	CPUT (ms)	DBRT (ms)	HTTPWT (ms)	RT (ms)	DE	ME	EE
WD	315.00	187.24	1830.2	476.53	7	3	3
FPM	384.57	300.43	2254.33	628.76	5	5	5
CRM WCUI	2249.25	805.43	5824.00	2722.24	3	7	7

IOP Conf. Series: Materials Science and Engineering **200** (2017) 012041 doi:10.1088/1757-899X/200/1/012041

Table 2. The Saaty rating scale [11]

Intensity of importance	Definition
1	Equal importance
3	Somewhat more important
5	Much more important
7	Very much more important
9	Absolutely more important
2,4,6,8	Intermediate values

In order to obtain the relative weights for the decision criteria, a decision model is achieved with the Super Decision software. The share of importance considered from the decision makers in providing weights for the two sets of criteria is different: the first criteria set will be considered with a share of 65% importance and the second criteria set with a share of 35% importance.

The Super Decision tool implements the AHP and the Analytic Network Process (ANP) methods for decision making. Decisional models using Super Decision uses the weighting ranking approach in evaluation and selection mode. The basic principle of the AHP method consists of splitting a problem in sub-problems that can be analysed independently of each other. The elements of the hierarchy are evaluated by comparing them sequentially two by two and orderings are converted into numerical values, which allow the aggregation of results in order to obtain the solution [12] and [13].

The main steps underlying the AHP algorithm are:

- Determining the relative weights of the decision criteria;
- Determining the relative rankings of the alternatives.

The decision model from Figure 1 built as hierarchy of objectives, criteria and alternatives makes the determining of relative weights for each decision criteria possible. Through the model, intangible information from our experience and intuition and tangible information provided as data are combined. The obtained weights for selection criteria are used in the hybrid model.

The three levels of the model contain nodes that define the elements of the problem: goal, criteria and alternatives. These are connected to each other in top-down direction. The criteria level is evaluated according to their importance, compared to the goal and the alternatives according to the preference regarding the criteria [14-16].

The comparisons between each criterion were performed using the Saaty's scale from Table 2, where the values between 1 and 9 depend on the relative importance of the criterion. The calculated inconsistency value 0.01976, from the Figure 2, shows that the values of subjective judgment are considered acceptable.

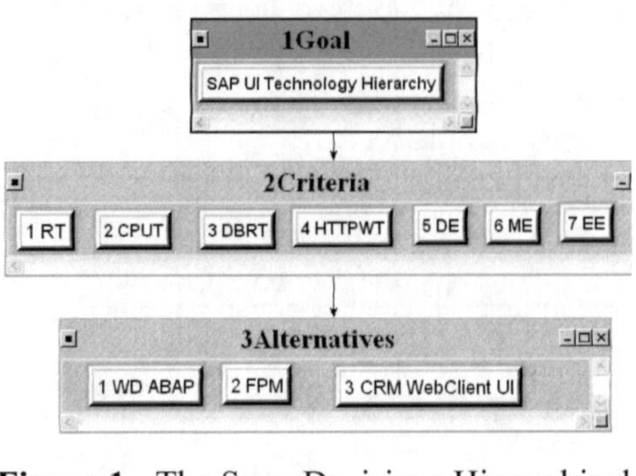

Figure 1. The SuperDecisions Hierarchical Model

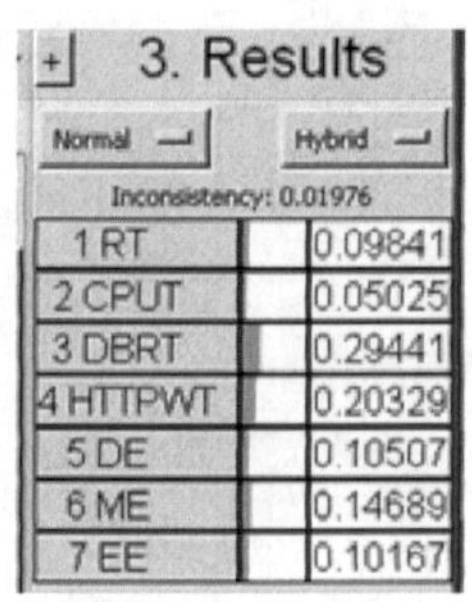

Figure 2. The criteria relative weights

The final stage in order to achieve the proposed objectives is to apply the TOPSIS method. The basic principle of this method is that the optimal decision shall be as close as the most advantageous solution and as far away as the worst solution. The ideal solution should have a rank of one and the worst alternative should have a rank approaching 0 [17] and [18].
Through the proposed model, the TOPSIS method involves the following steps:

- Construction of the decision matrix. This step leads to the Table 1 whose configuration is the decisional matrix.
- Calculation of the normalized decision matrix. The goal of normalization is to transform different dimensional attributes into dimensionless attributes.

The literature review [19-22] describes several methods of normalization. One of these methods is the vector normalization who is applied to the performance matrix.

The corresponding equation is:

$$CN_{ij} = \frac{C_{ij}}{\sqrt{\sum_{i=1}^{m} C_{ij}^2}} \quad (1)$$

where i is the criterion index (i= 1… n), j is the alternative index (j= 1… m) and the C_{ij} elements refer to the criteria i with respect to alternative j.

The normalized values CN_{ij} from the Table 3 denote the normalized decision matrix which represents the relative performance of the fixed alternatives. Table 3 contains also the relative weights provided by the Super Decision hierarchical model and the share of importance from the decision maker for the two sets of criteria.

Table 3. The normalized values related to the criteria

Technologies	V_j	The normalized values related to the criteria CN_{ij}						
		C_1	C_2	C_3	C_4	C_5	C_6	C_7
		CPUT (ms)	DBRT (ms)	HTTPWT (ms)	RT (ms)	DE	ME	EE
WD	V_1	0.1367	0.2128	0.2812	0.1681	0.7683	0.3293	0.3293
FPM	V_2	0.1669	0.3415	0.3464	0.2218	0.5488	0.5488	0.5488
CRM WCUI	V_3	0.9764	0.9155	0.8949	0.9605	0.3293	0.7683	0.7683
Relative weights		0.05	0.3	0.2	0.1	0.01	0.31	0.03
P_j		0.65				0.35		

Table 4. The weighted normalized matrix

Technologies	V_j	Weighted normalized values of the criteria CNP_{ij}						
		C_1	C_2	C_3	C_4	C_5	C_6	C_7
		CPUT	DBRT	HTTPWT	RT	DE	ME	EE
WD	V_1	0.0068	0.0638	0.0562	0.0168	0.0085	0.1021	0.0089
FPM	V_2	0.0083	0.1024	0.0693	0.0222	0.0060	0.1701	0.0148
CRM WCUI	V_3	0.0488	0.2746	0.1790	0.0960	0.0036	0.2382	0.0207

- Determination of the weighted normalized matrix. Because the selection criteria are of different importance weight, the weighting decision matrix shown in Table 4 is constructed by multiplying each element of each column of the normalized decision matrix by the relative defined weights P_j from Table 3 in according to the equation

$$CNP_{ij} = p_j * CN_{ij} \quad (2)$$

- Identification of the positive (C_j^+) and negative (C_j^-) ideal solution are calculated according to the weighted decisional matrix. The equations below take into account that the type of criteria are minimized.

$$C_j^+ = \min_{1 \le i \le m} \{CNP_{ij}\}$$
$$C_j^- = \max_{1 \le i \le m} \{CNP_{ij}\} \quad (3)$$

The calculated values are presented in the Table 5.

Table 5. The positive (C_j^+) and negative (C_j^-) ideal solution

	C_1	C_2	C_3	C_4	C_5	C_6	C_7
	CPUT	DBRT	HTTPWT	RT	DE	ME	EE
C_j^+	0.0068	0.0638	0.0562	0.0168	0.0036	0.1021	0.0089
C_j^-	0.0488	0.2746	0.1790	0.0960	0.0085	0.2382	0.0207

- The separation distance of each competitive alternative from the ideal and non-ideal solution is calculated with the equations

$$S_i^+ = \sqrt{\sum_{j=1}^{n} \left(CNP_{ij} - C_j^+\right)^2} \quad (4)$$

$$S_i^- = \sqrt{\sum_{j=1}^{n} \left(CNP_{ij} - C_j^-\right)^2} \quad (5)$$

The relative closeness of each location to the ideal solution S^* is measured for each alternative through the equation

$$S_i^* = \frac{S_i^-}{S_i^- + S_i^+} \quad (6)$$

Based on the obtained values presented in Table 6, the competitive alternatives are ranked in ascendent order towards the ideal solution.

Table 6. Ranking of alternatives

The ABAP UI Technologies	S^+	S^-	S_j*	The preference order
WD	0.0048	0.2936	0.9838	I
FPM	0.0798	0.2312	0.7435	II
CRM WCUI	0.2936	0.0048	0.0162	III

After applying the TOPSIS algorithm the preferred technology in therms of performance is the Web Dynpro ABAP technology, because she results with minimum score. This hierarchy is considered valid for the defined selection criteria and their corresponding weights.

4. Conclusion

This work proposed a new methodology of selecting between three ABAP UI technologies through a hybrid multi-criteria selection model. The model underlies the process of multi-criteria analysis and is based on a combination between the AHP and TOPSIS methods. The relative weights for the seven selected criteria determined from the Super Decision model with AHP, were combined with a share of importance defined by the decision makers. The TOPSIS method was applied to achieve the final ranking for the selected technologies. Web Dynpro ABAP was identified as the best technology that can be used to develop a SAP UI project. If there are taken into account other selection criteria or other methods to determinate the relative weights for criteria, it is possible to have a modified order of preference.

References

[1] Keller H and Thümmel W H 2009 *Official ABAP Programming Guidelines*, SAP Press
[2] Filip F G 2012 A Decision-Making Perspective for Designig and Building Information Systems, *Int J Comput Commun* **7**(2) 264-272
[3] Dobre I, Badescu A V, Pauna L, 2007, *Teoria deciziei*, Ed. ASE, Bucureşti
[4] Filip F G 2007 *Sisteme suport pentru decizii*, Ed. Tehnică, ed II completată şi revizuită, Bucureşti
[5] Velasquez M and Hester P T 2013 An Analysis of Multi-Criteria Decision Making Methods, *International Journal of Operations Research* **10**(2)56-66
[6] Onder E and Dag S 2013 Combining Analytical Hierarchy Process and Topsis Approaches for Supplier Selection in a Cable Company, *Journal of Business Economics and Finance* **2**(2) 56-74
[7] Tzanko St, Sezikeye A and Mal S 2011 *SAP WebClient, A Comprehensive Guide for Developers*, SAP Press
[8] Schneider T 2010 *SAP Performance Optimization Guide*, SAP Press, pp 123-156
[9] Chaitanya M 2012 *The Complete Guide to SAP NetWeaver Portal*, Galileo Press
[10] Wiegenstein A, Schumacher M, Schinzel S and Weidemann F 2009 *Sichere ABAP-Programmierung*, SAP Press, pp 54-62
[11] Saaty T L 2008 Decision making with the analytic hierarchy process, *Int. J. Services Sciences* **1**(1) 83-98
[12] Filip F G 2008 Decision support and control for large-scale complex systems, *Annual Reviews in Control* **32** 61-70
[13] Choi H 1994 Analytic hierarchy process - It can work for group decision support systems, *Computers & Industrial Engineering* **27**(1-4) 167-171
[14] ***Creative Decision Foundation 2012 *Super Decision Software for decision making* [Online]. Available: http://www.superdecisions.com
[15] Saaty T L 1994 *Fundamentals of Decision Making and Priority Theory with the Analytic Hierarchy Process*, Pittsburgh, RWS Publications
[16] Saaty T L 1990 *Decision Making for Leaders: The Analytic Hierarchy Process for Decisions in a Complex World*, Pittsburgh, PA RWS Publications
[17] Huang I B, Keisler J and Linkov I 2011 Multi-criteria decision analysis in environmental science: ten years of applications and trends, *Science of the Total Environment* 3578-3594
[18] Jadidi O, Firouzi F and Bagliery E 2010 TOPSIS Method for Supplier Selection Problem, *World Academy of Science, Engineering and Technology* **47** 956-958
[19] Yoon K 1980 *Systems selection by multiple attributes decision making,* Kansas State University, Manhattan, Kansas, PhD thesis
[20] Rao R V and Davim J P 2008 A decision-making framework model for material selection using a combined multiple attribute decision-making method, *International Journal of Advanced Manufacturing Technology* **35** 751-760
[21] Milani A S, Shanian A, Madoliat R and Nemes J A 2005 The effect of normalization norms in multiple attribute decision making models: a case study in gear material selection, *Industrial Applications* **29** 312
[22] Shih H S, Shyur H J and Lee E S 2007 An extension of TOPSIS for group decision making, *Mathematical and Computer Modelling* **45**(7-8) 801-813

Critical study of current situation of Vrănicioara tailing pond on Cavnicului Valley, risks and consequences

I Bud, S Duma, D Gusat, I Pasca and A Bud

Technical University of Cluj Napoca, North University Centre of Baia Mare, Faculty of Engineering, Romania

E-mail: greenminingteam@gmail.com

Abstract. In northern Romania, there are numerous tailing ponds, resulting from mining activities that present significant environmental risks. Some of them, including Vrănicioara tailing pond, were the subject of technical projects for ecological rehabilitation. Vrănicioara pond is located on the right side of Cavnic Valley, downstream Cavnic town, about 4 kilometers far. It has about 500 m length and is located parallel to the road linking Baia Sprie and Cavnic localities. Chemical and physical stability of the tailing pond before rehabilitation interest the research, analysis and conclusions were published in several scientific meetings. In addition, close to the pond at less than 100 m, an open pit has developed, exploiting andesite by mining blast, increasing the risk of physical stability by continuous exposure to vibration. This activity currently continues, advancing towards the tailing pond body. The critical study addresses the current state of Vrănicioara Tailing Pond, analysis of some rehabilitation works done incorrectly, analysis of chemical stability that was not a priority during rehabilitation. Research intention is heading to water analysis confirming the existence of acid drainage that was not stopped or at least reduced. The scientific approach is based on the Technical Standards for Waste Deposits, in force in Romania, providing the rules to ensure physical and chemical stability.

1. Introduction

Mining closure in Romania has had and continues to have a strong social, economic and environmental impact, due to the fact that it has not been carried out on the basis of skilled and responsible specialists analysis. These aspects have been studied and presented in numerous works and attitude by authors, in scientific events, specialized journals and media with the intention to draw attention upon these serious environmental problems [1-3].

By analyzing, in the last period of time, a part of rehabilitation works in Baia Mare mining area we continuously find disparities between the rules of good practice for waste deposits and the execution. Such a case is presented in the current work by analyzing the principles of minimum requirements imposed by Technical Normative and theoretical and practical foundations which have led, along the time, to these rules and the manner of Vrănicioara tailing pond rehabilitation which presents a multitude of vulnerabilities, meaning risks with major consequences.

2. Situation of Vrănicioara tailing pond before rehabilitation

2.1. Location and construction of Vrănicioara tailing pond

Vrănicioara tailing pond is located on the left bank of the Cavnicului Valley, 3,5 kilometres downstream Cavnic city. It has been in activity in the period 1957...1977. As a result of stability problems of Plopiş tailing pond, located in the vicinity, in 1985-1986 and 1988, Vrănicioara has been used as reserve deposit.

Figure 1. Vrănicioara tailing pond location, open pit placed at less than 100 m distance

The construction of the pond was accomplished by simple settling with an average slope of 16° on the bottom and a slope much more accentuated 25 ... 26° toward the top. The tailing pond location land has a slope of 8 ... 9°, with an inclination from N to S. The tailing pond capacity was exceeded by the additional storage, representing more than 2,5 million m^3, with a height of 39 m and extending on 15,5 ha surface. In the upper part the deposit was streaked by numerous ravines generated by precipitation waters and the complete absence of vegetation. In the lower area, a part of the material entrained by water has covered existing vegetation.

Figure 2. The effects of surface water

2.2. Physical and chemical stability before rehabilitation

External erosion could be seen all over the pond embankment being more relevant on the upper part of the slope because of material deposition emphasized in last years of the deposit activity. In the case of tailings ponds, surface erosion is caused by the presence of rainfall waters and exfiltrations. Rainfalls produce the phenomenon of gully erosion, meaning dragging different-sized particles in the direction of water flow. Exfiltrations produce numerous erosions on the embankment surface, some with important dimensions.

Laboratory tests have shown that the material particle size distribution is below 50 microns in variable proportions, 3 ... 24%, which leads to difficulties in draining the water from the pond body with consequences on stability.

The calculations of physical stability achieved by analytical and numerical methods have shown that for static loads the stability coefficient was around 1,1 (close to the limit stability and lower than the value of the safety coefficient) and for dynamic loads less than 1 (stability range). For dynamic loads (seismicity) or special weather events appears the risk of tailing pond rupture [4].

Close to the tailing pond, at less than 100 m, it has been opened an andesite mining exploitation which use perforation and mining blasting. The seismicity calculations have taken account the vibrations generated by explosions showing that in the situation of a classic blasting the stability risks of the pond are imminent. The dynamic loads increase the risk of liquefaction or thixotropic phenomena appearance [4].

Tailings ponds formed by the deposit of mining wastes resulted from complex ore processing have the particularity of generating acid drainage due to the presence, in significant amounts, of sulphates. Vrănicioara tailing pond is a regular deposit of polymetallic sulphides, chemically active: the water collected at the base of the pond is acid and loaded with heavy metals, visible phenomenon due to the formation of a precipitate [4].

Figure 3. a) Vrănicioara Tailing Pond before rehabilitation

Figure 3. b) Vrănicioara Tailing Pond after rehabilitation

These aspects were included in numerous research papers, along the time, by the authors, underlining the gravity of physical stability (for static and dynamic loads – seism and blasting) and chemical stability problems generated by Vrănicioara tailing pond.

3. Actual situation of the tailing pond after rehabilitation and discrepancies with Technical Normative rules concerning waste deposits

Vrănicioara tailing pond was the subject of a project funded by the European Union through the European Regional Development Fund in the total amount of 6 432 311 lei without VAT. Contract denomination: "Greening Works and Mining Closure of Upper Bolduţ-perimeter - Vrănicioara Tailing Pond, Maramureş County".

IOP Conf. Series: Materials Science and Engineering **200** (2017) 012042 doi:10.1088/1757-899X/200/1/012042

3.1. Technical Normative

Implementation of solutions for the tailing pond rehabilitation is subject of regulations in the construction, management and rehabilitation of mining wastes deposits: Judgment No. 856/2008 on the management of waste from extractive industries that provides a general legal framework and the provisions of Technical Normative concerning the waste storage from Romania which transposes the European Directive in this field [Order No. 757 of 26/11/2004, published in Official Monitor No. 86/26 I 2005] [5]. This regulatory framework provides some very strict rules which involve a large volume of cover material and hence, significant costs.

According to the Official Monitor of Romania, part I, no. 86bis/26.01.2005 the Methodology Rules detail the stratigraphy for closing cover material of hazardous waste. There are presented and accepted two variants, differentiated from each other by waterproofing layer nature: clay in first version (Figure 4a) and geocomposites in second version (Figure 4b).

For hazardous waste deposits, the coverage options provide minimum 2,3 m for successive layers - from the carrier layer up to soil in case of clay weatherproofing (Figure 4a) and minimum 1,8 m in the case of weatherproofing with geocomposites (Figure 4b). These rules applied to areas exceeding tens of hectares lead to the necessity of finding viable technical solutions that optimize the realization of rehabilitation.

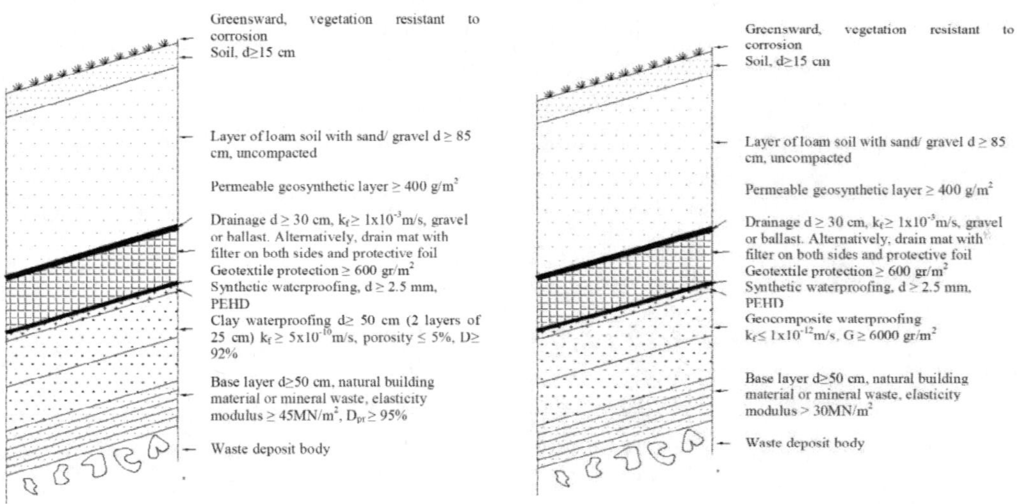

(a) Clay waterproofing *(b) Geocomposite waterproofing*

Figure 4. Stratigraphy of cover material for hazardous waste deposits

The most important issues that arise are: identification of sources of material, meaning carrier layer, waterproofing clay layer, drain layer, loamy soil layer with sand/gravel, soil; the distance of materials transportation; the implementation, etc.

3.2. The slope and the final geometry of the deposit

Using a GPS Garmin Montana 650t type with topographical map it was established measure points as in Figure 5. In each point altitude was measured as in and the values were used to calculate inclinations in different zone of the slope.

IOP Conf. Series: Materials Science and Engineering **200** (2017) 012042 doi:10.1088/1757-899X/200/1/012042

Figure 5. Measure points to identify slope inclination (system WGS 84 – EPSG:4326)

Table 1. Elevation values in measurement points

No.	Name	Elevation
1	Iaz00	491.599121
2	Iaz16	528.486816
3	Iaz15	532.483276
4	Iaz14	527.540588
5	Iaz13	516.193359
6	P12	511.677063
7	Iaz11	499.605347
8	Iaz01	508.326782
9	Iaz0	497.110168
10	Iaz10	500.052826
11	Iaz9	531.369446
12	Iaz8	530.373657
13	Iaz7	529.049072
14	Iaz6	528.88031
15	Iaz5	532.326843
16	Iaz4	532.27002
17	Iaz3	534.102112
18	Iaz2	531.896851
19	Iaz1	517.903137
20	Car	511.080078

It was found that:

- on the top of the slope there are areas where the slope exceeds the Technical Normative, requirements i.e. under max 33% (1: 3) in absolute 18,26°;
- the final slope after stopping settlements, according to Technical Normative, must be at least 5%, but the tailing pond geometry remained the same with canopy, beach (slope reverse) and lake/sink (between points Iaz3 Iaz 4 and 5 in Figure 5) .

3.3. Surface waterproofing system

Being given the fact that the tailing pond is composed of reactive material - sulphides are oxidized in the presence of rainwater and oxygen - it is necessary, according to Technical Normative [5], a waterproofing meeting the requirements for closing dangerous waste deposits (Figure 4a and b).

The reality on the field proves the absence of any cover layer (of support, of mineral or artificial waterproofing, textile as a protective layer for the membrane, layer for rainfall water drainage, textile as separating layer (Figure 6).

Figure 6. "Rehabilitation works"

The ultimate goal of waterproofing layers imposed by Technical Normative is to confine the tailing pond over a long period of time and ensure vegetation restoration, so that the reactive material has no interaction with the environment and the deposit reintegrates in landscape.

3.4. Recultivation layer

Technical Normative [5] provides that the re-cultivation layer must have a minimum thickness of 1 m and bushes and trees shall be planted. On the slope surface the existence of re-cultivation is not relevant and an acacia plantation was made (Figure 6 and Figure 7).

Figure 7. Acacia plantation on the tailing pond slope

IOP Conf. Series: Materials Science and Engineering **200** (2017) 012042 doi:10.1088/1757-899X/200/1/012042

Trees do not contribute to stability because landslides occur at greater depths than fixing root area; instead they contribute to raising hydrostatic level, having as consequence the reduction of stability coefficient. If the waterproofing - drainage layers would be properly placed (Figure 7), the tree roots would affect their structure.

In the case of granular materials deposits (as mining waste deposits) with thicknesses of tens of meters where the sliding bed forms at greater depths than the roots length, soil erosion may occur at important depths, dragging granular material.

The apparent rehabilitation of slope surface with trees, in the cases of inclination angles close to the limit stability is proven as an improper solution (Figure 8).

Figure 8. The case of Berbești tailing pond, Vâlcea county– instability phenomenon in area cultivated with acacia

Figure 8 highlights phenomena of sliding and erosion in a mining waste deposit (tailing pond) rehabilitated by acacia plantation, without correcting the slope embankment in the area of safety.

3.5. The hydrostatic level in the tailing pond body

Piezometers were placed in the deposit to indicate the water level in the tailing pond body, although it consists of reactive material, showing the nonsense of this way of rehabilitation. The measurements have shown a high hydrostatic level on an alignment of 1,5 – 2,5 m from the slope surface, affecting both physical stability to static and dynamic loads and chemical stability by reactivating sulphide oxidation, causing acid drainage.

3.6. The ditch guard location

The ditch guard is designed to collect surface water from the slopes upstream the deposit and is to be placed around the deposit body, outside its perimeter. In the case of Vrănicioara tailing pond, the ditch guard is positioned in the body (length 500 m) in the area where the material has minimum particle size (Figure 9).

Before rehabilitation there was a ditch guard securely positioned upstream of the dam body with a length of over 700 m, that in time was clogged and partially destroyed. If rehabilitation would be made under the Technical Normative rules, after a new geometrization respecting minimum and maximum slope, the new ditch guard position could be changed by location upstream of the dam body plus the overlays. The construction of a ditch guard in the tailing pond body just increases the contact surface between the reactive material, water and oxygen.

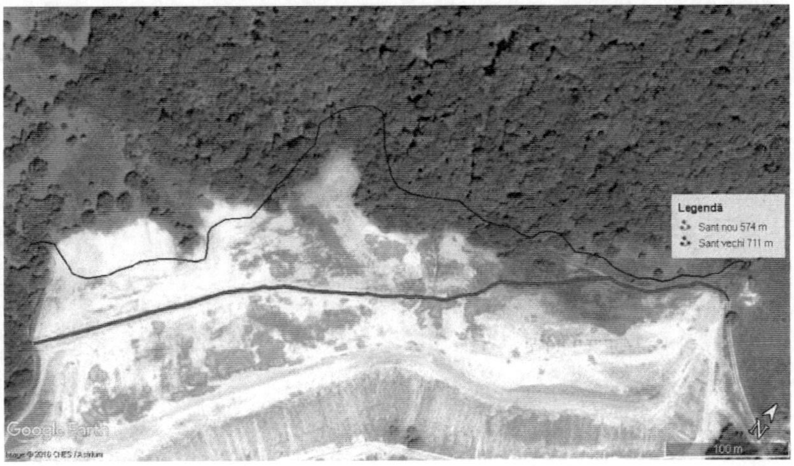

Figure 9. Old (711m) and new (574m) ditch guard

4. Conclusions

Vrănicioara tailing pond is composed of a reactive material with content of sulphides, located on a hill side where downstream exist strategic objectives, a county road heavily traveled, which links to the tourist resort of Cavnic and sights of historical Maramures, the gas pipeline, River Cavnic and downstream localities ([6], [7]).

Ensuring stability of this deposit is of utmost importance and imposes the consideration of the most demanding requirements to ensure physical and chemical stability.

This tailing pond was the subject of a rehabilitation project with a considerable financial value, but numerous deficiencies in design and execution. The project should be made public, but we did not have access to it, and therefore we analyzed the execution works which were compared with the requirements of the technical standards [5] on waste disposal.

In this way, many inaccuracies were identified on geometrization and waterproofing of this deposit: geometrization does not comply with the minimum and maximum slopes; lack of cover layers (waterproofing and re-cultivation); lack of construction of a perimeter ditch guard; lack of berms gutters every 10 m high; inadequate acacia plantation; construction of a ditch guard in tailing pond body, in fine material area; high hydrostatic level at depths of 1,5 – 2,5 m from the slope surface, etc.

In the same time, it was found not satisfied physical stability but only reducing of surface erosion and to ensure chemical stability it has not been taken any measure, pointing out the seepage phenomena at the base of the pond through the retaining wall and the embankment, flushing directly into the Cavnic River.

The existence of an open pit in the proximity of the tailing pond endanger its stability, particularly when mining drilling and blasting are happening, because hydrostatic level is high and increase the liquefaction and thixotropic risk.

References

[1] Bud I and Duma S 2010 Closing mines in Romania. Tailing dumps rehabilitation, *Revista Minelor* **5** 23-28

[2] Duma S, Bud I, Gusat D, Pasca I and Bud A 2015 *Analyses of "Aurul" tailing pond surface rehabilitation – by using regional resources*", 15th International Multidisciplinary Scientific GeoConference SGEM 2015, Vol **3**, pp 33-39

[3] Bud I, Duma S, Gusat D, Pasca I and Bud A 2015 *Environmental risks of abandoning a mining project already started: Romaltyn Mining Baia Mare, IOP Conf. Ser.: Mater. Sci. Eng.* **144** 012004

[4] Duma S and Bud I 2010 Environmental risk of an open pit placed close to a tailing pond,

Mining Revue **6** 34-38
[5] ***Ordin nr. **757** din 26/11/2004 published in M.O. Nr. **86** bis/26.I.2005 – *Normativ tehnic privind depozitarea deşeurilor*
[6] Bud I, Duma S, Denuţ I and Taşcu I 2007 *The environmental impact of Cavnicului Valley Tailing Dams*, BHM - Berg-und Huttenmannische Monatshefte, Springer Wien NewYork
[7] Bud I, Denuţ I and Duma S 2006 Physical stability analyse concerning Cavnicului Valley tailing dams, *Journal of Mining and Metallurgy* **42A** 29-35

IOP Conf. Series: Materials Science and Engineering **200** (2017) 012043 doi:10.1088/1757-899X/200/1/012043

Harvesting energy an sustainable power source, replace batteries for powering WSN and devices on the IoT

A Pop-Vadean[1], P P Pop[1], T Latinovic[2], C Barz[3] and C Lung[3]

[1]Technical University of Cluj Napoca, Department of Mechatronics and Machine Dynamics, Muncii str., no. 103-105, 400641 Cluj Napoca, Romania
[2]University of Banja Luka, Faculty of Mechanical Engineering, Republic of Srpska, Bosnia and Herzegovina
[3]Technical University of Cluj Napoca, Center University of North from Baia-Mare, Department of Electrical Engineering Electronics and Computers, dr. Victor Babes str., no.62A, 430083 Baia Mare, Romania

E-mail: adinaluciapopvadean@yahoo.com

Abstract. Harvesting energy from nonconventional sources in the environment has received increased attention over the past decade from researchers who study these alternative energy sources for low power applications. Although that energy harvested is small and in the order of milliwatt, it can provide enough power for wireless sensors and other low-power applications.

In the environment there is a lot of wasted energy that can be converted into electricity to power the various circuits and represents a potentially cheap source of power.

Energy harvesting is important because it offers an alternative power supply for electronic devices where is does not exist conventional energy sources. This technology applied in a wireless sensor network (WSN) and devices on the IoT, will eliminate the need for network-based energy and conventional batteries, will minimize maintenance costs, eliminate cables and batteries and is ecological. It has the same advantage in applications from remote locations, underwater, and other hard to reach places where conventional batteries and energy are not suitable.

Energy harvesting will promote environmentally friendly technologies that will save energy, will reduce CO2 emissions, which makes this technology indispensable for achieving next-generation smart cities and sustainable society.

In response to the challenges of energy, in this article we remind the basics of harvesting energy and we discuss the various applications of this technology where traditional batteries cannot be used.

1. Introduction

In the last period,it was a focus of attention on the Internet of Things (IoT). The rapid development of the Internet of Things (IoT) ,will mean that almost everything we can imagine can be connected to the Internet (Figure1).This technology refer to the wireless interconnection of the millions of “things” and devices through the Internet or local area networks, to increase efficient utilization.

Peter Middleton, research director at Gartner said that the development of IoT will far exceed the development of other connected devices.

By 2020, the number of smartphones and tablet PCs in use will reach about 7.3 billion units [1].

IOP Conf. Series: Materials Science and Engineering **200** (2017) 012043 doi:10.1088/1757-899X/200/1/012043

Gartner says that 6.4 billion connected devices will be used in 2016 to 30 percent more than in 2015 [2].

Figure 1. Internet of Things (IoT) (Source Blue Sky Center)

2. IoT trends

Internet of Things (IoT) is one of the new technology in IT as noted in Gartner's IT Hype Cycle (Figure 2).

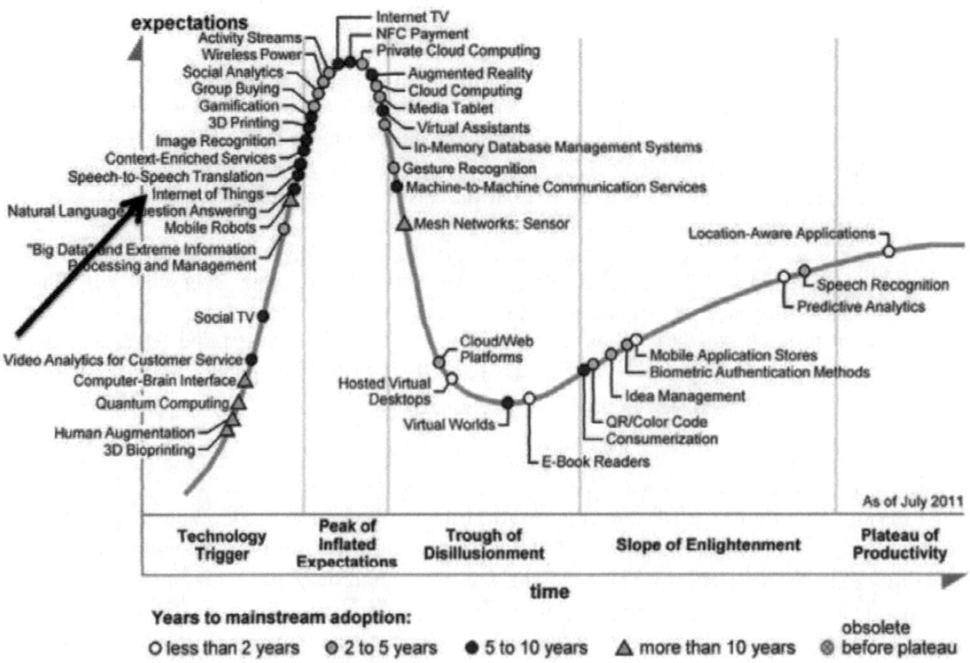

Figure 2. Gartner 2011 Hype Cycle of Emerging Technologies (Source: Gartner Inc. [3])

Hype cycle [3] is a method to represent the occurrence, maturity and impact of specific technology applications. IoT will take 9-10 years for adoption by the market. Web search popularity, as measured by Google search trends for terms in the past 10 years the IoT, computing and WSN are presented in Figure 3 [4], where:

A. Algorithms and Protocols for WSN Provides You with A Comprehensive Resource Market Watch - Nov 18 2008
B. IoT -- From Vision to Reality Market Watch - Apr 14 2010
C. CCID Consulting: China‘s IoT Industry Sees a Landscape Characterized by Clustering in Four Regions Market Watch - Oct 4 2011
D. China Hi-Tech Fair Highlights IoT Market Watch - Nov 21 2011
E. ARM unveils low-power chip for the (IoT)Reuters UK - Mar 13 2012 F. Web connected objects get a „voice" on the (IoT) Winnipeg Free Press - Apr 25 2012[4]

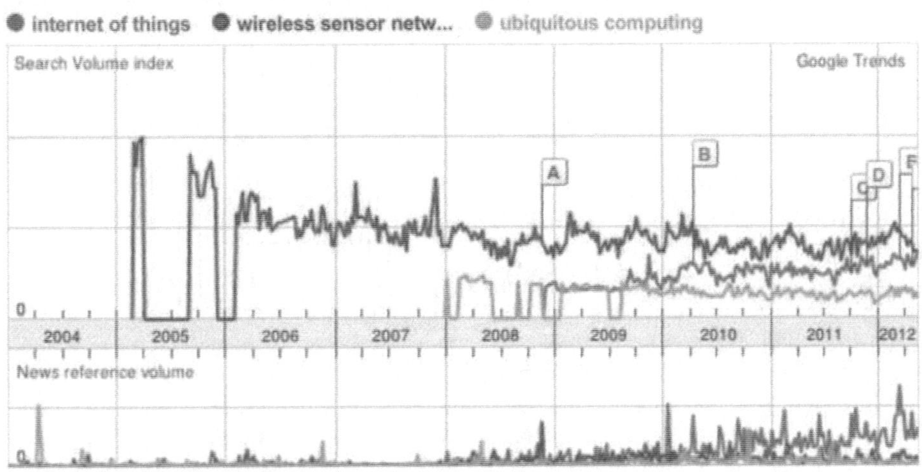

Figure 3. Google search trends since 2004: Internet of Things, Wireless Sensor Networks, Ubiquitous Computing. SPOT points are listed

We can seen, since IoT has come into existence, search volume is growing with the falling trend for Wireless Sensor Networks. This trend is to continue in the next decade as well as other generic technologies that converge to form a true Internet of Things.
From the Internet be observed that Internet of Things has started obtain popularity. This thing reflects the social acceptability of the technology.

3. Energy Harvesting or how can power IoT

With those millions of things come millions of batteries that must be purchased, maintained and disposed of. Unfortunately, battery innovation not increased proportionally with digital innovation. Battery disadvantages, threaten to limit growth across the entire Internet of Things (IoT) landscape.
Here comes the Energy Harvesting (EH) with technology innovations around ultra low power wireless sensor networks, network protocols and standards. EH soon becomes an important part of IoT ecosystem for many areas of industry but not only, it all comes down to manufacturing.

Energy harvesting (EH) represents a simple solution for easily powering those remote devices using clean energy. Energy-harvesting technologies it removes the need for batteries, and so eliminates an obstacle to the success of the Internet of Things (IoT).

Through things and devices on the IoT are included wireless terminals equipped with sensors which are connected to a network and will collect information about the environment around the terminal sensor.

Miscellaneous sensor types are used by the wireless sensor terminals, such as the temperature, humidity, light, motion, pressure, stress, distortion, position, velocity, and gas flows.
Placement of many terminals will determine the variety and accuracy of data collected. This information, will help achieve control devices, monitoring and prediction that were previously impossible.

IoT development will have a big impact on our world, thanks to the evolution of semiconductor devices and wireless technology.

What is necessary for the Internet of Things, even mandatory, is the ability to place the wireless sensor terminals in all sorts of locations to collect the most important data. All are nice and good, but a problem arises when we installed power-distribution wires, or, when using the battery: life of the battery or battery replacement time.

Nobody would find this a problem with 10 or 20 batteries, but when there are 10,000 or a million or a hundred million, there are concerns not only for battery costs, but also for the enormous scale of maintenance expenses. This is one reason for debate the topics about wireless sensor terminals and is necessary to find urgent solutions.

Energy harvesting may provide this urgent solution. Energy harvesting isn't a new idea.
We have data on harvest light energy (photovoltaic) since the 19th century, when Alecsandre - Edmond Becquerel observed photoelectric effect in 1839 [5].

The techniques for harvesting energy use elements generators convert light (sun), heat (thermoelectric), vibration (piezoelectric), or RF energy (such as those emitted from cell phone towers) into electrical energy in a stably and without a great loss.

Energy harvesting allows designing systems capable of operating for years being fed to these energy sources in the environment, removing the battery change problem.
This does not mean that there is no use for a battery in these performant systems. After it gathered ambient energy , must then be stored to provide the current required at a time in which:

a. necessary (nodes have low duty cycles, meaning that an air temperature sensor can be active only a few milliseconds per hour and can be rest time in sleep mode)
b. when AC power is not available (sunlight, for example, are not present at night). Conventional rechargeable cells can be used in this way (semiconductor or supercapacitors).

4. Method that balances the power generation with the power consumption

When necessary energy harvesting, we must consider the necessary balance between power generation and power consumption. This is because the device will not work if it will generate less energy than can be consumed by the device. Although all of the power generating elements are continuously improved, it is difficult to produce enough power for a device that must operate continuously.

One way to solve this problem is to collect power generated in a fuel sensor capacitor and at times, resulting in a method that balances energy production with energy consumption. To achieve this, the designer must understand the perfect environment for generating power generator element, the power generated, the time needed, energy consumption and consumption device.
Figure 4 illustrates the important points using power generating time, power collecting time, and power consumption time to solve the balance of power generation, power collection, and consumption.

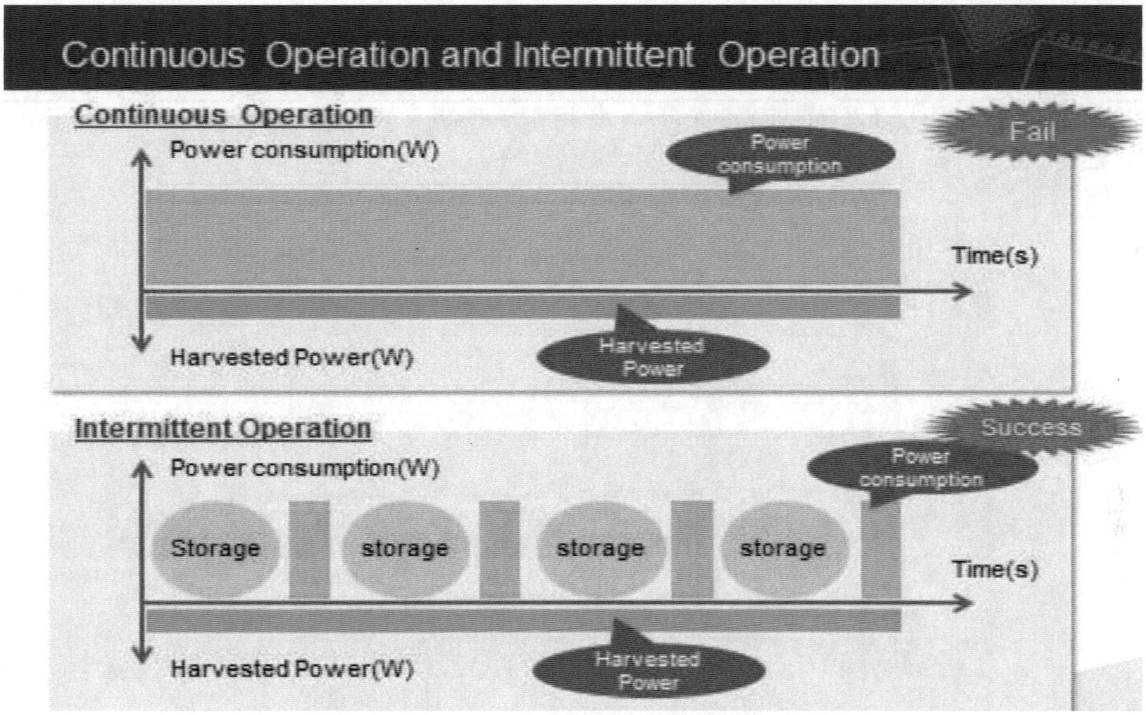

Figure 4. A way of balance the energy harvested power required by the device is to collect energy continuously, but wireless sensor terminal operation to be done at intervals

If we can harvest efficiently this energy we transform these relatively low levels of energy in an amount which can provide enough power for IoT node.

5. Elements of a wireless IoT node and categories of ambient power

In Figure 5 shows the main components of a system of wireless sensors including EH encoder processor energy (including energy conversion and storage) sensor, microcontroller and wireless radio.

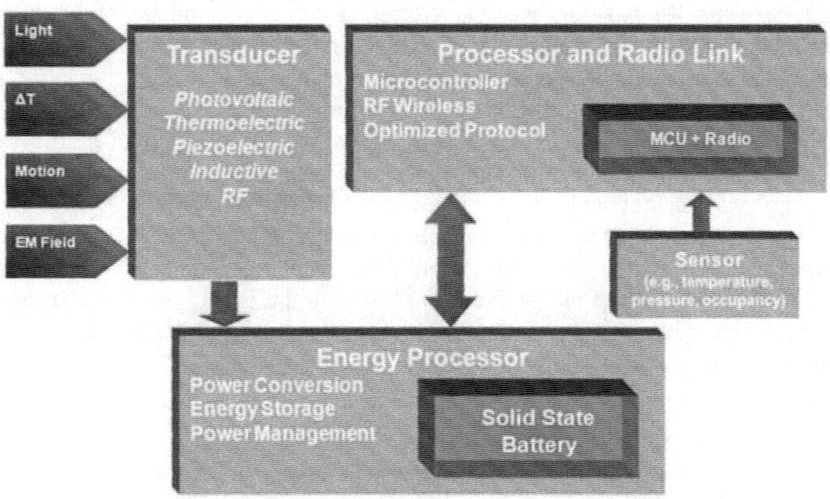

Figure 5. Elements of a wireless IoT node (Source: Cymbet)

IOP Conf. Series: Materials Science and Engineering **200** (2017) 012043 doi:10.1088/1757-899X/200/1/012043

We will talk a little about the four major categories of ambient power: light, temperature difference,vibration and RF.

Figure 6 shows the amount of harvested power available from these energy sources with solar (ambient outdoor light) and vibration (such as from industrial machinery) offering a considerable power advantage over the others; keep in mind, however, that not all applications will require that much energy.

Energy Source	Harvested Power
Vibration/Motion	
Human	4 μW/cm^2
Industry	100 μW/cm^2
Temperature Difference	
Human	25 μW/cm^2
Industry	1–10 mW/cm^2
Light	
Indoor	10 μW/cm^2
Outdoor	10 mW/cm^2
RF	
GSM	0.1 μW/cm^2
WiFi	0.001 mW/cm^2

Figure 6. Energy-harvesting sources.and harvested power. (Source: Texas Instruments)

5.1. The sun shine power

One of the newest technologies in permanent development and operation of energy is solar energy, which has been used for a long time to power small devices and, more recently, to provide back-up emergency power for mobile phones . Solar power collects and converts sunlight into electricity, but conversion efficiency is not very high. Regular solar panels are rated at 15% or 20% efficiency in optimal conditions.

Solar panels work in rain or cloudy (reducing production to 30% of the electricity generated in bright sunlight), in any season (solar intensity is reduced in winter), but changing the number of hours available light impact negative on the energy. To solve the problem semiconductor companies have developed controllers to optimize energy harvesting from solar panels.

These controllers use Maximum Power Point Tracking (MPPT). This technique optimizes the tranfer between the solar array (PV panels), and the battery bank or utility grid. The function of MPPT has described as analogous to the transmission in a car. When the transmission is in the wrong gear, the wheels do not receive the maximum possible power (Figure 8).

An MPPT algorithm monitors the input current and voltage and controls the duty cycle to maintain the MPP set point needed to maximize energy output from the photovoltaic module.

A controller IC verify the output of the panels, compares it to the battery voltage and then determines which is the best power, then the panel can put out to get the maximum Amps into the battery. Most MPPT's are around 95-97% efficient.

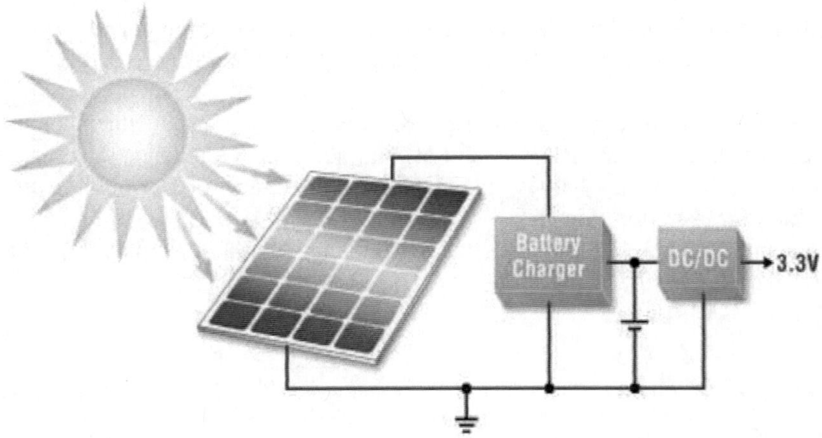

Figure 7. Solar Harvesting energy system (Source: Linear Technology)

5.2. *Thermoelectric energy*

Thermal energy can be used for the IoT by taking advantage of the available transducers and converter ICs.Using energy harvesting from thermal gradients, low-power circuits can operate many years without the need for battery replacement.

Two scientific "effects" are present here and we discuss for this. The "Peltier effect" is used to create a temperature difference by applying a tension between two electrodes connected to a sample of semiconductor material, usually to provide solid-state cooling. The "Seebeck effect" is a phenomenon in which a temperature difference between two dissimilar electrical conductors or semiconductors is converted into an electrical potential, or voltage. The simplest example, an energy autonomous sensor for measuring airflow temperature use copper to conduct warm air inside the tube where air is flowing, while a heat sink on the outside conducts the cooler ambient air, creating a difference in temperature between the sides of a Peltier element, and generating power for the sensor.

Residual heat from renewable plant equipment such as pumps and motors for industrial applications energy is converted into Internet objects (IIoT).

In the future, useable waste heat could be trapped and the human body to feed some of wearable sensors for medical implants developing (a harvester attached to a person's skin will be able to use a ΔT of up to 10°C).

5.3. *Shake it for energy*

The kinetic energy can be converted into electricity by piezoelectric effect; mechanically deforming a piezoelectric crystal (Micro Electro Mechanical System piezo MEMS) voltage or pressure to generate electrical charge, which can be measured as a voltage across the electrodes of the piezo element.

Piezoelectric vibration energy harvesters convert this mechanical vibrational energy into alternating electrical energy (AC).

This AC is then converted into DC electronic format that can be used to drive wireless IoT applications or recharge a battery.

Piezoelectric energy transducers supplies energy when operating at maximum resonance frequency of the vibration source, and when it works in a task designed to match the output impedance piezoelectric.

The secret for harvesting the vibrational energy with any piezoelectric material is to understand the vibration environment, and the best method to do this is to measure the vibration using an accelerometer.

IOP Conf. Series: Materials Science and Engineering **200** (2017) 012043 doi:10.1088/1757-899X/200/1/012043

5.4. RF harvesting energy

Due to the widespread growth of wireless communications and a concomitant increase in the number of radio transmitters, in particular mobile base stations, mobile phones, there is a lot of energy "free" RF around us.

Besides Wi-Fi sources, researchers can find sources of RF energy from short-range wireless technologies such as Bluetooth and ZigBee, and cellular services, long-range. A reception coil serves as an energy source, RF energy harvesting, generating a voltage in response to electromagnetic coupling with the RF transmitter.

The challenge in RF energy harvesting lies in maximizing the output from the transducer at a given ambient energy level (Figure 9).

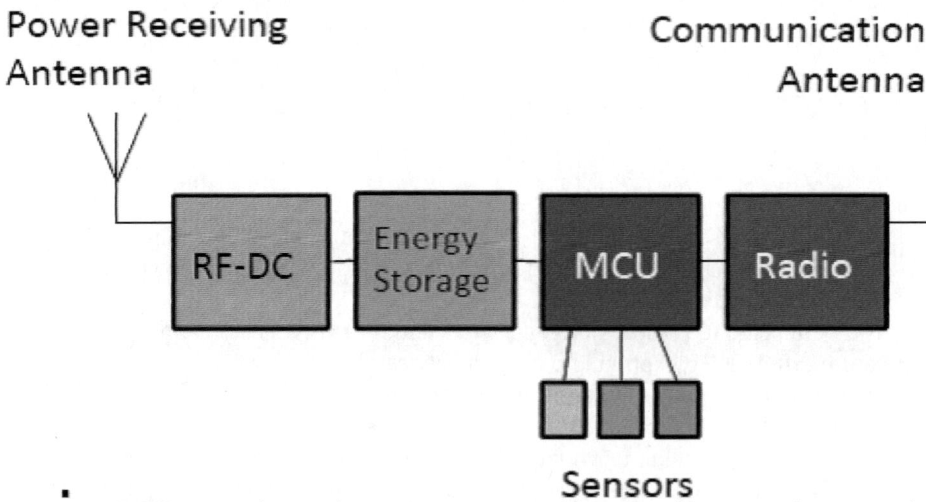

Figure 8. RF energy harvesting systems can provide renewable energy by converting radio waves to DC power for IoT applications (Source: Powercast)

6. IoT in Europe

Internet of Things activities is present, with numerous projects underway in industry, academic research and various levels of government.They will work together to achieve technological evolution in this area.

In Europe, is deposited a substantial effort to strengthen inter-domain activities of the research groups and organizations in a unified IoT framework.

Supported by the European Commission 7thFramework program (EU-FP7) this includes the Internet of Things European Research Cluster (IERC) and its objectives are: to establish a cooperation platform for IoT activities in Europe and become a contact point for IoT research in the world.

It includes projects such as CASAGRAS2, a consortium of international partners from Europe, the USA, China, Japan and Korea exploring issues surrounding RFID and its role in realizing the IoT
IERC includes the Internet of Things Architecture (IoT-A) project established to determine an architectural reference model for the interoperability of Internet-of-Things systems .

IoT Initiative (IoT-i) is coordinated action formed to support the development of IoT European community. IO project bring together a consortium of partners to create a common strategic vision and technical IO Europe.

In the same time, the project SmartSantander is under way to develop a large city testbed IoT for providing research and implemented in the city of Santander, Spain, and at locations in the UK, Germany, Serbia and Australia [6].

7. Conclusion

Development of actual devices that use energy harvesting power ICs is moving forward in many locations and applications. Sometimes energy harvesting feeds battery free wireless sensors terminals. In other cases, use of batteries and energy harvesting, together extend battery life.

In this way, there is an acceleration of the development of wireless terminals which incorporates sensors for energy harvesting.

In the next period, we should see the location of wireless sensor terminals with this technology in all manner of locations. Power management ICs designed for energy harvesting, as well as low-power MCUs, will help advance the growth of the Internet of Things.

References

[1] ***http://www.gartner.com/document/2625419?ref=QuickSearch&sthkw=G00259115
[2] ***http://www.gartner.com/newsroom/id/3165317
[3] ***Gartner's Hype Cycle Special Report for 2011, Gartner Inc http://www.gartner.com/technology/research/hype-cycles/ (2012)
[4] ***Google Trends, Google. http://www.google.com/trends (n.d.)
[5] Becquerel A E 1839 Recherches sur les effets de la radiation chimique de la lumiere solaire au moyen des courants electriques, *Comptes Rendus de L'Academie des Sciences* **9** 145-149
[6] ***https://arxiv.org/ftp/arxiv/papers/1207/1207.0203.pdf

IOP Conf. Series: Materials Science and Engineering **200** (2017) 012044 doi:10.1088/1757-899X/200/1/012044

Software for improving the quality of project management, a case study: international manufacture of electrical equipment

D M Preradović [1], Lj S Mićić[1] and C Barz[2]

[1]University of Banja Luka, Bosnia and Hercegovina, Faculty of Economics, Ul. Majke Jugovića, no. 4, 78000 Banja Luka, Bosnia and Hercegovina
[2]Technical University of Cluj-Napoca, North University Centre of Baia Mare, Victor Babes str., no. 62A, 430083 Baia Mare, Romania

E-mail: dragana.preradovic@unibl.rs

Abstract. Production conditions in today`s world require software support at every stage of production and development of new products, for quality assurance and compliance with ISO standards. In addition to ISO standards such as usual metrics of quality, companies today are focused on other optional standards, such as CMMI (Capability Maturity Model Integrated) or prescribing they own standards. However, while there is intensive progress being made in the PM (project management), there is still a significant number of projects, at the global level, that are failures. These have failed to achieve their goals, within budget or timeframe. This paper focuses on checking the role of software tools through the rate of success in projects implemented in the case of internationally manufactured electrical equipment. The results of this research show the level of contribution of the project management software used to manage and develop new products to improve PM processes and PM functions, and how selection of the software tools affects the quality of PM processes and successfully completed projects.

1. Introduction

Now days in the era of information technology (IT) it's never be easier to find information, knowledge and example of good practices in area of project management. Volume of useful and available information is enormous. In spite of all this, there is still a worrying amount of failed projects on a global level. Companies which efforts focus on project management (PM) spending 13 times less money for reaching the business goals, because developing of different PM structure is result of better selection and alignment of projects with long period strategy [1]. The practice recognizes relevance of PM in business but this fact is still not fully adopted.

Research problem in this paper is about large number of failed projects that are not implemented on time, in defining budget and not meeting foreseen purposes. Reports on global level show increase of numbers of unsuccessful projects and costs that are connected with projects. These have failed to achieve their goals, within budget or timeframe. This paper focuses on checking the role of software tools through the rate of success in projects implemented in the case of internationally manufactured electrical equipment.

First secondary hypothesis is that properly selected projects that are in harmony with strategic goals of company contribute decrease of costs and growth of performance (meet goals, finished on time and in budget frame).

Second secondary hypothesis is that use of advanced software has direct impact on increase the number of successfully finished projects.

Main hypothesis is that advanced software tools have impact on increase of maturity level of PM process.

First part of the paper is about global trends in PM and situation about status of realized project for the last 4-5 years on global level during the few most important parameters. The rest of research is case study in manufacture of international company from electrical area about maturity level of PM process for developing and introduction of a new products in the production line. At the end are conclusions and recommendation for future researching.

2. Global trends

This part of paper shows situation on global level about projects success status, impact factors on success, risks etc. Also, here is part of report about CMMI level of assessment on global level. It shows how much more must be done in area of project management and moving up maturity level of organizations in today's world.

2.1. Projects success status

Balance sheet of status of realization project on global level is not promising for last few years. According to the Standish Group Report (figure 1) status of realization 50.000 IT project shows that only 27-31% of all projects finishes successfully [2]. And in average, more then 50% of all started projects finished, but with transgress of time, budget and purpose.

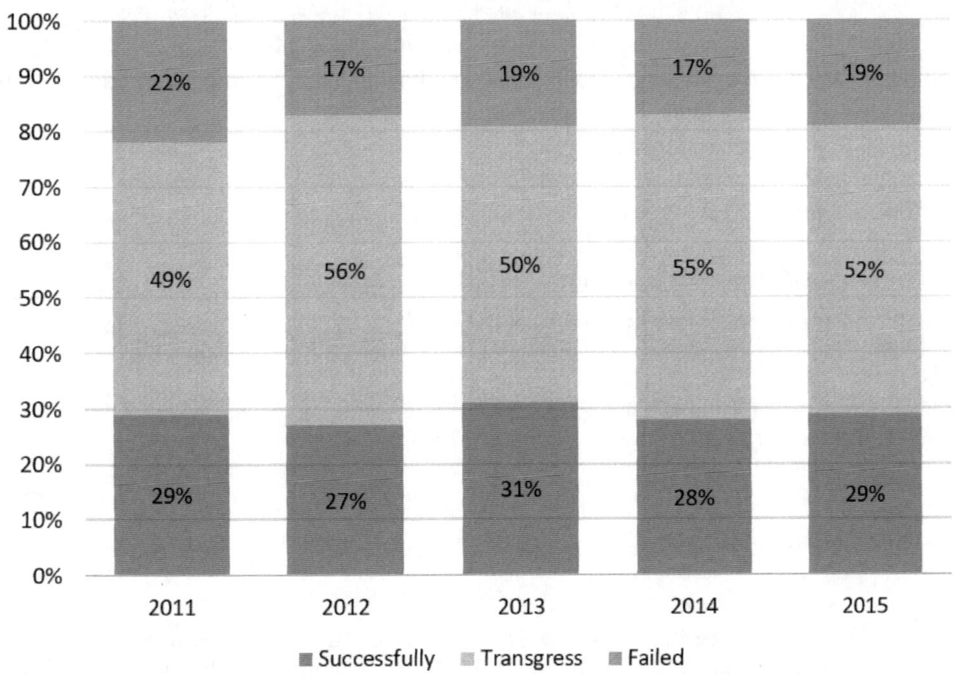

Figure 1. Realization status of 50.000 IT project (2011-2015)

Trend of increase of failed or transgress project is continued from 2013 to 2015 in IT projects. The main identified reasons for this situation are lack of established prioritization process and mismatching of project to long term strategy. According to [3] it is very often situation in which company starts with a project only because project looks like good and popular. Around 68% of companies don't have defined project prioritization process. Purpose of this process is prioritization and creation of list for rank of projects. Only 12,5% of companies take care about align project with long term strategy.

The projects on Figure 1 are different size, scope and complexity. The dominant parts of successfully projects are small projects (62% of all successfully projects). Also, developing methods shows different results – agile methods have significantly more success instead of waterfall method. With increase of scope and complexity goes higher risk of failing.

Positive impact factors on success of projects including PM experts (5%), clear business goals (4%), strong management support (15%), relations between team members (15%), involved users (15%), optimization of results (15%), educated team members (10%), agile skills (7%), resource economy (6%) and software, tools, practices and services (8%) [2].

Also [1] PMI identified 3 positive and 3 negative impact factor success of project:

1) Meet the business goals
2) Project is finished in defined time frame
3) Project is finished in defined budget frame
4) Scope creep – is lack of focus, conditions or purpose. It is about uncontrolled adding of annex or no planed tasks in project.
5) Costs because of surpassing budget frame.
6) Projects that failed before the start – the team or other employers think project is failed before it starts.

All these parameters in last year show undesirable performance. The positive parameters (1, 2 and 3) were moving down, and negative parameters moving up (4, 5 and 6). Analyses of global trends tells that 3 positive parameters move down for last 3 years (2014, 2015, 2016): percent projects that meet goals are 64%, 64% and 62% (respectively), finished on time are 55%, 55% and 53% and finished in budget frame 51%, 50% and 49%. Opposite to this, there are 3 negatives impacts on project success and all of them grow up: scope creep (43, 44 and 45%), costs of surpassing budget frame (30, 30 and 32%) and projects that failed before the start (16, 15 and 16%) [1].

Some of identified causes of failure, according to PMI are:

- Organization priorities in project are changed (41%),
- Projects goals are changed (38%),
- Inadequate set project's goals (37%),
- Chances and risks were not examined enough (31%),
- Inadequate purposes and goals of project (31%).

2.2. CMMI assessment status

According to reports [4], during the last 9 years amount of CMM certified organizations in world is 13.719 in 98 countries. Structure of this amount per maturity level is:

- Initial level organizations (1%),
- Repeating process level (19%),
- Defined process (70%),
- Managed process level (2%) and
- Optimizing process level (8%).

Bosnia and Herzegovina, Serbia, Slovenia, Monte Negro, Albania, Moldova and Estonia are the only countries in Europe that don't require for CMM assessment. The closest CMMI center is in Bulgaria. Spain have the highest number of assessments in Europe. On the global level the highest number of assessments has China, USA and India.

IOP Conf. Series: Materials Science and Engineering **200** (2017) 012044 doi:10.1088/1757-899X/200/1/012044

Table 1. Tools for moving up organizations maturity level [5]

Number	Tool name	Vendor	Web address
1	*Automated Test Designer (ATD)*	*AtYourSide Consulting*	http://atyoursideconsulting.com/
2	*Callis Guide; Callis Reviewer*	*Callis*	http://www.callis.dk/
3	Више софтверских рјешења	*SE-CURE AG*	http://www.se-cure.ch/
4	*DataDrill 5.0*	*Distributive Management*	http://distributive.com/
5	*Polarion QA, Polarion Requirements*	*Siemens*	https://polarion.plm.automation.siemens.com/
6	*Lean Enterprise Self-Assessment Tool (LESAT)*	*MIT Sociotechnical Systems Research Center (SSRC)*	https://dspace.mit.edu
7	*Adobe Analytics Capability-Maturity Assessment Tool*[a]	*Adobe*	http://adobemarketingpro.com/
8	*Knowledge Management Capability Assessment Tool (KM CAT)*	*APQC*	https://www.apqc.org

[a] Tool is used only for moving up marketing function processes or marketing sector maturity level.

The frequent needs of assessment goes to telecommunications, security, safety and recover data, army, defense, government, finance, bank and assurance company. Analyze indicates that the most frequent request for assessment comes from Asia (57%), North America (26%), Europe (10%), South America (6%), Africa (0,97%) and Australia (0,30%).

Software tools for moving up maturity level of organizations and its parts are detailed instructed. We can find only few of them (Table 1).

3. Case study – international manufacturing company

Research of this case study was made in global known company (in Balkan they have factories in Serbia, Romania, Austria, Switzerland and Italy). The study was conducted from 10th to 15th October 2016 in Romania. The company is producing diverse equipment for industrials needs. We have exanimated two parts of maturity project management processes which develops and introduces new products – first one is about the process used in PM and another one is software for moving up maturity level of PM process.

3.1. Project management process in company

Project management process (PMP) of introducing the new products is completely standardized and structured. PMP have:

- Phases – are sets of standardized activities that team implements in company. PMP have 7 different phases that every project goes thru. At the end of every phase is checkpoint.
- Checkpoints – places for quality control and making decision for status of project. Purposes of checkpoints are support good business ideas and projects, and stopping the nonprofit ideas. Decision are made in accordance to risk assessment, reports and check lists.

Result of every checkpoint is status of a project (go to the next phase; conditionally transfer to another phase; repeat some of the parts of the same phase and go again to the checkpoint; temporary stopping of the project; stopping the project). At every checkpoint team must check if project is still aligned with long term strategy of the company.

PMP are: 1) Selection of the projects. Here is necessary to make risk assessment document and strategy aligned documentation. 2) Concept activities are in the second phase and these are solutions for production, market research and earnings possibility. 3) Third phase is about the project goals and tasks for establishing contact with suppliers. Design process is in progress and starts with products introduction at the market. 4) Design and development phase is the plan for optimization the production process and capital costs for project. 5) Validation is the next phase. In this part there is a small test limited edition test product production. This test product will end up on the market for clients test needs, design test and suppliers authorization. 6) Starting with the production. 7) Closing the project.

Quality control is implemented on every checkpoint, after every phase. The most common question raised are: all activities from last phase done? The priority of the project is synchronized with the global portfolio of projects in company. Performed risk assessment, aligned with the project in long term strategy.

3.2. Company software tools for moving up maturity project management level

In this specific company is used Microsoft Project for managing all PMP development and introduction of the new products. All activities and documentation are conducted in Excel Template. Share Point platform ware used for collaboration and access to company documents.

Also, Excel predefined templates are used in every phase of PMP. There is a variety templates that created internally and are the same for every project. Team for quality controls on checkpoints used Share Point platform to access predefined Excel templates and making decision. Besides mentioned software tools, company also uses One Spice design 3D software for designing every new product. Data base of all sample designs is developed and implemented internally for company needs.

Integrated Quality Management System is integrated software for managing and analyzing information in projects. Vendor is Siemens. This solution can be integrated with in information system of the company no matter on size, and also can be customized to the company's needs. QSYS supports periodical inspections – this specific company used this.

In phases 5 and 6 for every product SAP Production Planning module is using assignment code for every product. But, after the introducing the product starting with full production SAP PP module will take the dominant role in production process. Code for every product will contain the part of code of sample design in drawing data base, and also and some other information (production date, where made, from what material, supplier, etc.).

4. Conclusion

Even with a lot of knowledge, best practices and education, there is still a tendency of failed projects on global level. According to data in this research, on every billion of dollars is 122 millions US cost (lost) because of the poor project management (some about 12% lost). This company did not have this kind of issue. From 2009 (after start with EPMO and project management formula which is completely standardized in all processes) to October 2016 company implemented 10 projects and all of them was completely successful. Key of success is in optimal and careful combination of simple software tools (like MS Project, MS Excel, Share Point) and advanced software tools (like QSYS and SAP) with standardized process during the PMP.

First secondary hypothesis - properly selected projects that are in the harmony with strategic goals of company contributes decreasement of costs and performances growth (meet goals, finished on time and in budget frame) is confirmed. Before the selection process started team works on preparation of documentation, risk assessment, capital costs, purpose and alignment of the project with the company portfolio. Also, on every checkpoint alignment with the project portfolio must be checked (long term strategy).

Second secondary hypothesis that use of advanced software has a direct impact on increasement the number of successfully finished projects – is not confirmed. This processes are important more then advanced and expensive software, but combination of advanced software tools with standardized process and simple tools provide the best optimal result.

Main hypothesis confirmed that advanced software tools have impact on increase maturity level of PM process. Therefore, only advanced software tools do not increase maturity level of PM. There must be a standardization, defined processes and carefully combination of software. This would be the key of success.

Future recommendations for researching propose direct comparison software tools for moving up company maturity level and introduction and education of CMMI verification process and his significance in business. On Balkan there is no CMMI center for verification of organization maturity level and there is a need to increase information about its role for global businesses.

References

[1] ***Project Management Institute 2016 *The High Cost of Low Performance, 8th Global Project Management Survey*, Philadelphia, USA
[2] Hastie S and Wojewoda S 2015 *Standish Group 2015 Chaos Report*, available: https://www.infoq.com/articles/standish-chaos-2015
[3] Stanleigh M 2016 *How To Develop a Systematic Approach to Prioritizing All Projects*, availabe: https://bia.ca/how-to-develop-a-systematic-approach-to-prioritizing-all-projects/
[4] ***CMMInstitute 2016 *CMMI Maturity Profile Report*
[5] ***Software Develoopment Tools 2016 *CMM Capability Maturity Model,* available: http://www.softdevtools.com/modules/weblinks/viewcat.php?cid=104

Study of the shielding performances of different materials regarding Electromagnetic Field Interference

Conecici L M, Munteanu C and Purcar I M

Technical University of Cluj-Napoca, Cluj-Napoca – 400020, Romania

E-mail: conecici.madalin@gmail.com

Abstract. This paper presents a computer-aided comparison of some of the most used electromagnetic shielding materials. Computer-aided design (CAD) model and the computer aided engineering (CAE) simulation technologies are used for the analysis of the electromagnetic field shielding performances of each material individually and a comparison will be established. The main topic of this comparison is to establish a proper shielding material for ElectroMagnetic Interference (EMI) sources. A three-dimensional (CAD) model of the circuit breaker coil designed in PTC Creo Elements v.18.1 environment was analyzed in Ansoft Maxwell v.15 environment in order to compute the electromagnetic field distribution. The residual (EMI) values are compared to one another and the best shielding material will be presented for this circumstances.

1. Introduction

Except for the superconductors no other materials is able to block completely a magnetic field without it being attracted to its magnetic force. The most efficient method to protect against magnetic fields is to redirect them, they cannot be removed. In order to redirect magnetic fields, shielding alloys with high-permeability are used. NETIC-AA, NETIC S3-6 and MuMETAL are one of a kind shielding materials with a highly guarded alloy composition, these materials were developed by Magnetic Shield Corp, after years of research and applications [1].

However, shielded enclosures are a poor substitute for good EMC design at the board level. Effective enclosures can add significant cost and weight to a product and a single breach of the enclosure (e.g. an unfiltered cable penetration) can completely eliminate any benefit the enclosure would otherwise provide. In many cases, a product in a poorly designed shielded enclosure will radiate more (or be more susceptible) than the same product without the enclosure [2]. Choosing a proper location, orientation, and material for a shield requires a knowledge of the type of field being shielded and the objectives of the shield. The following sections will describe basic shielding theory and provide several examples of good shielding in various situations.

2. The study domain

The standard European design of a DIN-rail mounted MCB includes the following components as shown in Figure 1:

1. Actuator lever – trips and resets the MCB manually;

2. Mechanism – a tension based system that forces the contacts together or apart;

3. Contacts – moving parts that either allows current to flow, when they are closed or do not allow current to flow when they are separated;

4. Terminals – the contact points of the MCB;

5. Bimetallic strip – used as a tripping system for longer term tripping currents;

6. Calibration screw – set up of the tripping current;

7. Solenoid – used for the short-term, high overcurrent tripping;

8. Extinguish chamber – used to guide the arc flow without damaging the MCB [3].

Figure 1. The MCB components design

As a domain for our study, the coil of the MCB will be used, mainly because the magnetic field reaches a high density in this area. Also, MCB's are often used mounted on rails side by side, the space between the two coils is filled only by the polyurethane casing on both MCB's[4], thus one fault in one MCB can be transmitted to the neighboring MCB via the Electromagnetic interference.

3. 3d model used for the study and considerations

The only reliable way, known at this moment, to reduce EMI generated by parts of the circuit that are influencing other parts of the same or other circuits, is by shielding either the emitter or the receptor against electromagnetic fields [5]. Electromagnetic shielding is composed out of barriers made of conductive or magnetic materials.

The metallic or magnetic shielding can reduce the emitting and receive of radio waves, electromagnetic fields, and electrostatic fields. Also known as a Faraday cage, any enclosure composed of a conductive material is used to block electrostatic fields. The material used in the building of the enclosure severely affects the efficiency of the shielding capacity [6]. Other characteristics that influence the shielding performance are the frequency, size of the shielding enclosure, shape and orientation of the shield.

As a setup used for our study, the coil is situated between the two shields made out of different shielding material, and the study and the graphs are plotted onto two axes situated parallel to the coil, in the same place where the neighboring coil would normally be situated [7], as shown in Figure 2.

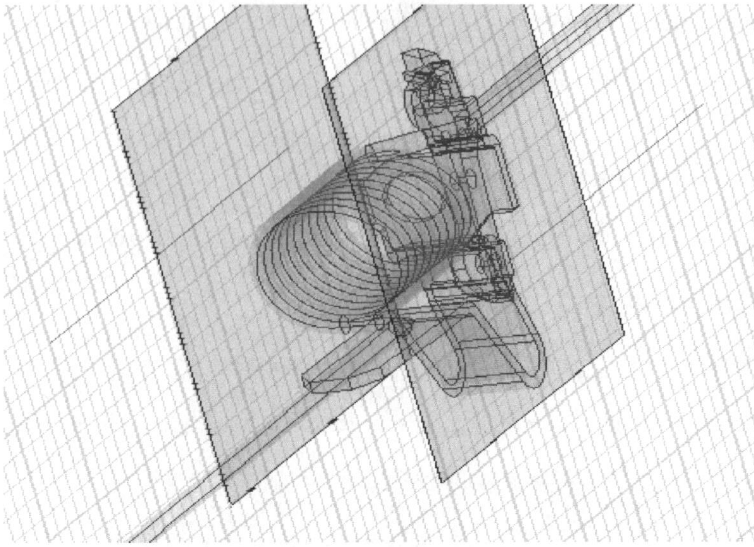

Figure 2. Coil, shielding material and plotting axis

Also as a parallel study, the thicknesses of the shields will differ. The study will take into consideration two thicknesses. One is 0,01mm, a very thin foil, in order to keep costs and space at a minimum. And the second thickness is 0.8 mm plate material which is the maximum width available in the MCB.

4. Materials studied

Electromagnetic Interference (EMI) shielding defines the absorption and/or reflection of electromagnetic radiation due to a material positioned in the way of the electromagnetic interference.

EMI shielding is rapidly becoming a strongly needed resource in the electronics and energetics industry, due to a fast increasing usage of electronics, their reliability and the rapid growth of radiation sources [8].

There is a difference though between Electromagnetic interference (EMI) shielding and Magnetic shielding. For example, magnetic shielding refers to shielding at magnetic fields that have a low frequency (e.g., 60Hz).

On the left side, the materials used for this study are presented with their Electrical proprieties as are the ones introduced in MAXWELL.

- $\boldsymbol{\sigma_r}$ -conductivity relative to copper
- $\boldsymbol{\mu_r}$ - the relative magnetic permeability

Table 1.

Material	σ_r	μ_r	$\sigma_r\mu_r$	σ_r/μ_r
Silver	1.05	1	1.05	1.05
Copper	1	1	1	1
Gold	0.7	1	0.7	0.7
Aluminum	0.61	1	0.61	0.61
Brass	0.26	1	0.26	0.26
Bronze	0.18	1	0.18	0.18
Lead	0.08	1	0.08	0.08
Nickel	0.2	100	20	2x10-3
Stainless steel	0.02	500	10	4x10-5
MUMetal	0.03	20000	600	1.5x10-6

IOP Conf. Series: Materials Science and Engineering **200** (2017) 012045 doi:10.1088/1757-899X/200/1/012045

5. 3d numerical simulation (CAE) and results

5.1. Simulation consideration

Adaptive Mesh Process was used for the simulation in question. This process is working with numerous passes over the model, and refining the mesh with every pass. There are numerous factors that can influence the meshing process, factors like the geometry of the model, field solutions, also a factor is a percent refinement number which in our case was set to 30%. As for Excitation of the solenoid, we used a current of 500A and 60Hz.

- The simulation is preceded by validation of the model. This will find and present any issues with the model and will give a report of the found issues;
- Once all parameters are set, the software is ready to start. Depending on the solution setup (refinement of the solution), and on the hardware used, analyzing the problem can take up to a few hours;
- When the software finishes analyzing the problem the results are represented, as field overlaid plots, graphic plots or numerical values. In this figure, the electromagnetic field generated into the vacuum.

5.2. Graphs and color scheme

- A 2D graph with the values of the magnetic field and the induced current overlaid is presented in the graph from above Figure 3.
- The Green and red graphs represent the intensity of the magnetic field on the left and right side of the MCB's Coil Figure 4.
- For the sake of the experiment, only the maximal values are used.

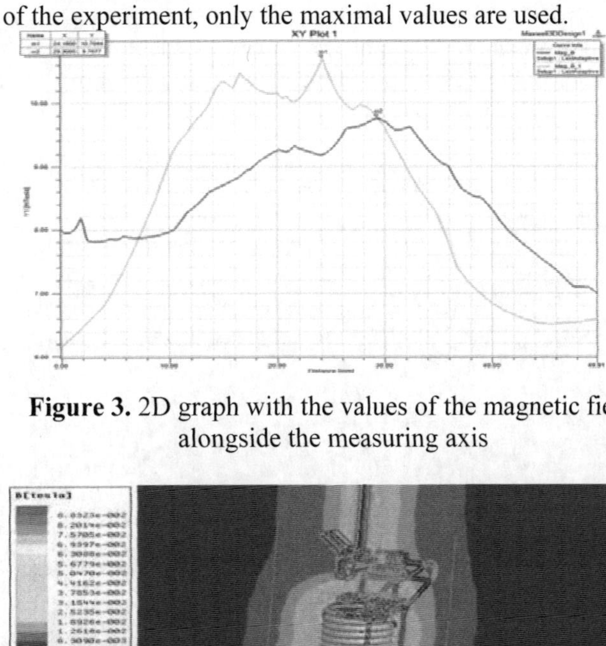

Figure 3. 2D graph with the values of the magnetic field alongside the measuring axis

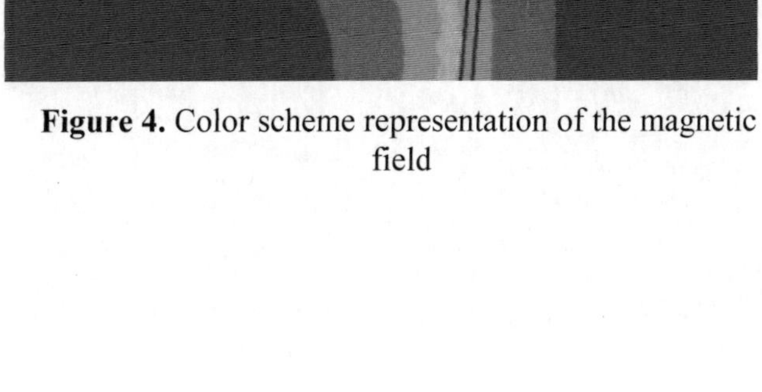

Figure 4. Color scheme representation of the magnetic field

5.3. Comparison of the obtained values

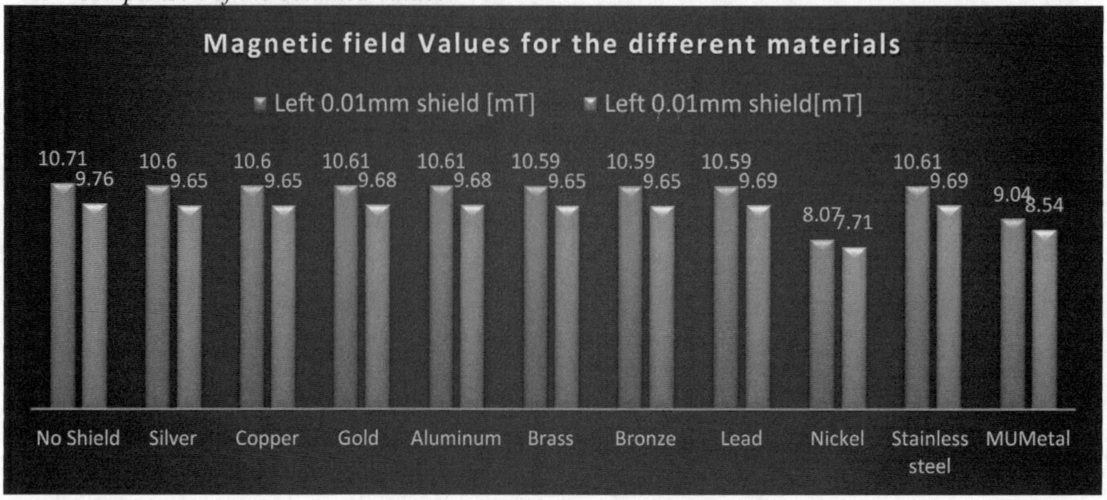

Figure 5. Results of the magnetic field simulation for all different Shielding materials, Values for the 0.01mm shield material

In Figure 5, the results of the simulations are shown, where the shields were made out of 0.01mm thick foil. Immediately it is visible that there are two materials that stand out regarding their shielding abilities. These two materials are Nickel and the Branded MuMetal. Considering these results another simulation is executed but the thickness is 0.8mm, the maximum space available for the application.

In this simulation, though, only Nickel and Mumetal at a thickness of 0.8mm, because all other materials have, as we can see from Figure 5 similar shielding proprieties.

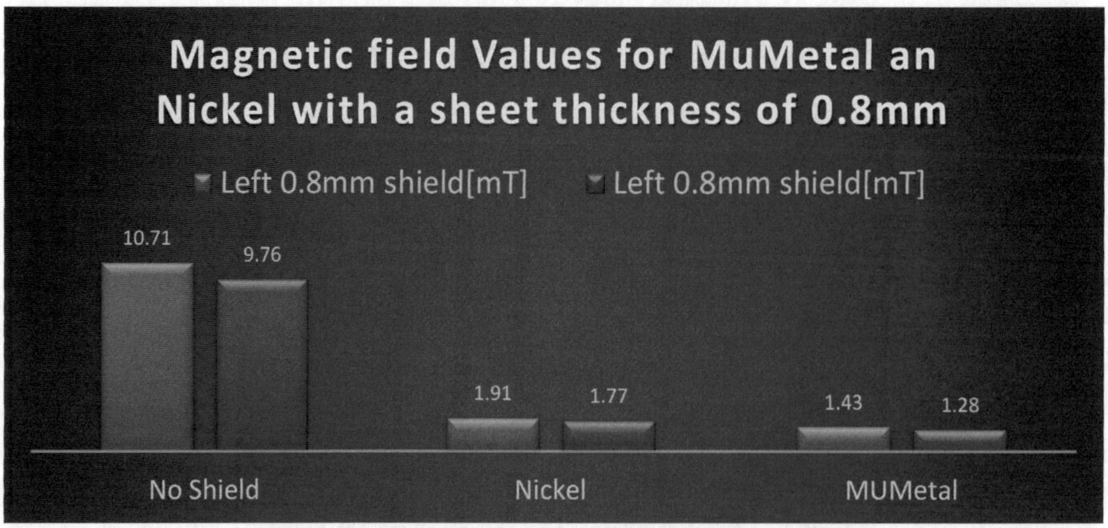

Figure 6. Results of the magnetic field simulation for Nickel and MuMetal, Values for the 0.8mm shield material

IOP Conf. Series: Materials Science and Engineering **200** (2017) 012045 doi:10.1088/1757-899X/200/1/012045

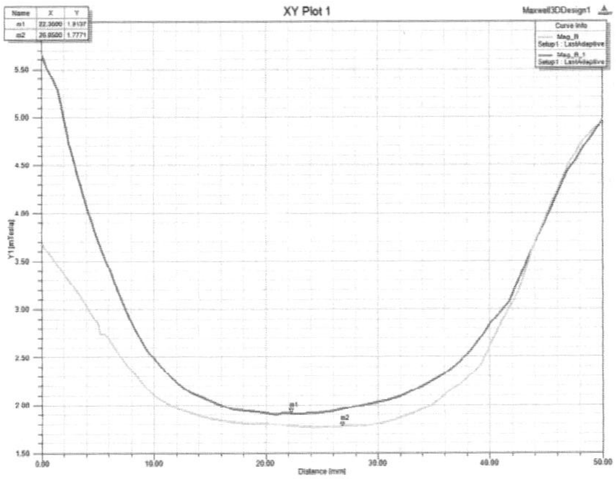

Figure 7. Results of the simulation with proper shielding materials

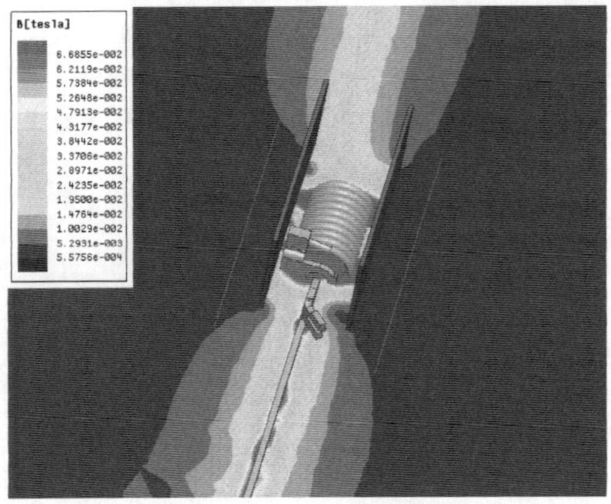

Figure 8. Color scheme representation of the magnetic field with proper shielding

In Figures 6, 7 and 8 The simulation results show the effectiveness of the thicker shielding for the Nickel and MuMetal shields, while Figure 6 shows also a comparison to a no shield situation.

6. Conclusions

Depending on the shape, the size, the composition and the distance to the shielded object, the shielding strategy can be effective or not. For example, if the shields are positioned very close to the shielded object, they will attract the magnetic field. In this case, the MuMetal can get oversaturated and it reacts by increasing the magnetic field surrounding it.

As presented in Figure 5. It is visible that Nickel and Nickel based materials, as the MuMetal is, are the best shielding material for low-frequency applications.

The paper presents a potent solution for the shielding of MCB's, a more reliable shielding solution would be entirely made out of metal, but while the MCB is an Electric protection device it cannot be made out of any conductive material. To bypass this problem, some shielding solutions are implemented at the Distribution Panel level. This shielding solution is designed to protect the devices from the outside interferences, but not from the short circuit faults generated inside MCB's components level.

References

[1] Luo Y, Lu J G and Li Z G 2002 *Study of Reliability Test and Analysis for Miniature Circuit Breakers*, Electrical Apparatus Institute 323# Hebei University of Technology, pp 80-81

[2] Conway J T 2001 Exact Solutions for the Magnetic Fields of Axisymmetric Solenoids and Current Distributions, *IEEE Transactions On Magnetics* **37**(4) 2977-2988

[3] Dwight H B 1931 The magnetic field of a circular cylindrical coil, *Phil. Mag.* **38** 948-957

[4] Conecici L M, Munteanu C and Purcar I M 2015 *Magnetic Field Simulation of a Miniature Circuit Breaker's Coil*, Moden Power Systems MPS 2015, Technical University of Cluj-Napoca

[5] Wu Y, Rong M, Li J and Lou J Y 2006 *Calculation of Electric and Magnetic Fields in Simplified Chambers of Low-Voltage Circuit Breakers*, State Key Laboratory of Electrical Insulation for Power Equipment, Xi'an Jiaotong University, Xi'an, China, 710049

[6] Pop D, Neamt L, Tirnovan R and Sabou D 2013 *3D Finite Element Analysis of a Miniature Circuit Breaker*, Technical University of Cluj-Napoca

[7] Marandi S R 1996 The Magnetic Near-Field Of A Solenoid, *Canadian Space Agency* **1** pp 1-4

[8] ***SIEMENS Primer Miniature Circuit Breakers

[9] ***Metal shields and encryption for US passports, Newscientist.com. Retrieved 18 November 2012
[10] ***MuMETAL Magnetic Shield Corp. 2012 Catalog MU-2. Retrieved 26 June 2016

IOP Conf. Series: Materials Science and Engineering **200** (2017) 012046 doi:10.1088/1757-899X/200/1/012046

Fuzzy control strategy for secondary cooling of continuous steel casting

G O Tirian[1], C A Gheorghiu[2], T Hepuţ[2] and R Rob[1]

[1]Politehnica University of Timisoara, Electrical Engineering and Industrial Informatics Department, 5 Revolution Street, Hunedoara, 331128, Romania
[2]Politehnica University of Timisoara, Department of Engineering and Management, 5 Revolution Street, Hunedoara, 331128, Romania

E-mail: ovidiu.tirian@fih.upt.ro

Abstract. The purpose of this paper is to create an original fuzzy solution on the existing structure of the control system of continuous casting that eliminates fissures in the poured material from the secondary cooling of steel. For this purpose a system was conceived with three fuzzy database decision rules, which by analyzing a series of measurements taken from the process produces adjustments in the rate of flow of the cooling water and the speed of casting and determine the degree of risk of the wire. In the specialized literature on the national plan and the world, there is no intelligent correction in the rate of flow of the cooling water and the speed of casting in the secondary cooling of steel. The database of rules was made using information collected directly from the installation process of continuous casting of the Arcelor Mittal Hunedoara.

1. Introduction

A fuzzy solution is proposed that placed on the existing structure of the management system of continuous casting will reduce fissures in the secondary cooling by generating necessary value adjustments to change the water flow and the velocity of the casting.

There are no current systems in the world that can eliminate cracks if they are detected in the secondary cooling steel. Basically, the database rules were designed for this purpose and they contain measures to mitigate the risk of a fissure.

This principle is original, as the structure of the proposed system. The proposed structure can be applied to any continuous casting plant.

The method of continuous casting consists of introducing the liquid metal with a determined temperature in a shape which has the walls cooled with water named crystallizer, and the evacuation is on the other side where the solidified steel is obtained [1], [2].

A crucial component of the continuous casting installation is the secondary cooling zone. The secondary cooling zone has the role to continue the wire cooling after it has emerged from the crystallizer and to assure the total solidification of the semi –product. It is considered “the heart” of continuous casting and has the benefit of ensuring a better quality of the material. The material surface shape has to ensure a homogeneous cooling and a uniform repartition of the water on the materials surface [3], [4].

2. Design of fuzzy controllers

Fuzzy logic can process vague variables whose values can vary continuously in any numerical range defined by taking decisions depending on the position indicator in the numerical range and predefined rules [5], [6].

Applicability of fuzzy logic is varied, in metallurgical domains fuzzy systems are more commonly used.

For the issue in question a database was built that consists of information gathered from technologist experts. This database was supplemented with information from the mathematical model of solidification process [7].

The fuzzy system receives from the process the following values, depending on the area in which the blank is situated:

- Casting speed v
- Cooling water flow q
- Steel temperature T

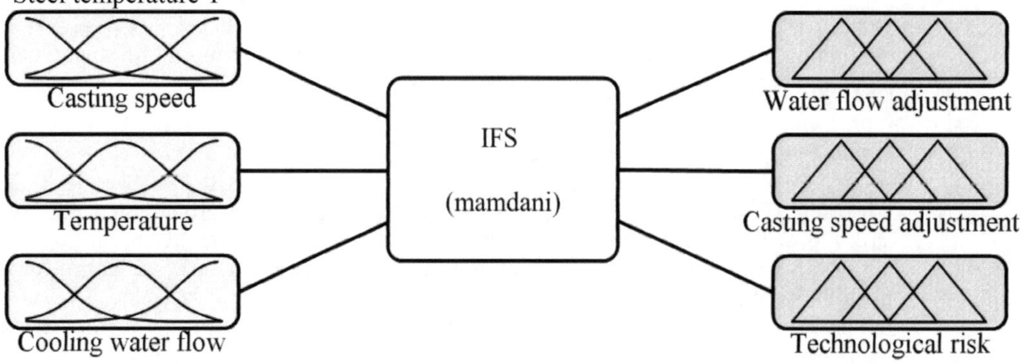

Figure 1. Block diagram of IFS

Laying the foundation of rules required an analysis of all possible outcomes. To this end, a number of expert technologists with extensive experience in the operation of the casting machines were consulted.

Also, for each case was carried out an analysis of the phenomenon of solidification of the crust. In Figure 1 is depicted the block diagram of the intelligent fuzzy system (IFS) [6], [7].

3. IFS projection

IFS actual design was achieved using Matlab toolbox fuzzy. The systems produces 3 outputs namely: water flow adjustment, casting speed adjustment and technological risk. Given the characteristics of the process and that the process is linear were adopted triangular membership functions for both input and output presented in Figure 2.

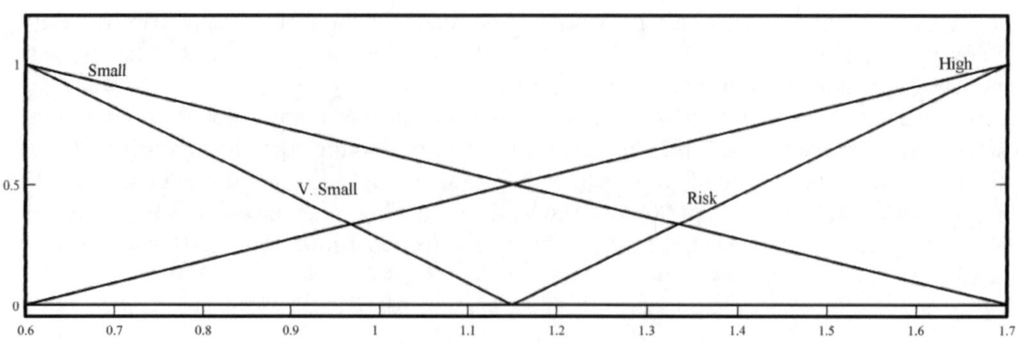

Figure 2. Membership functions

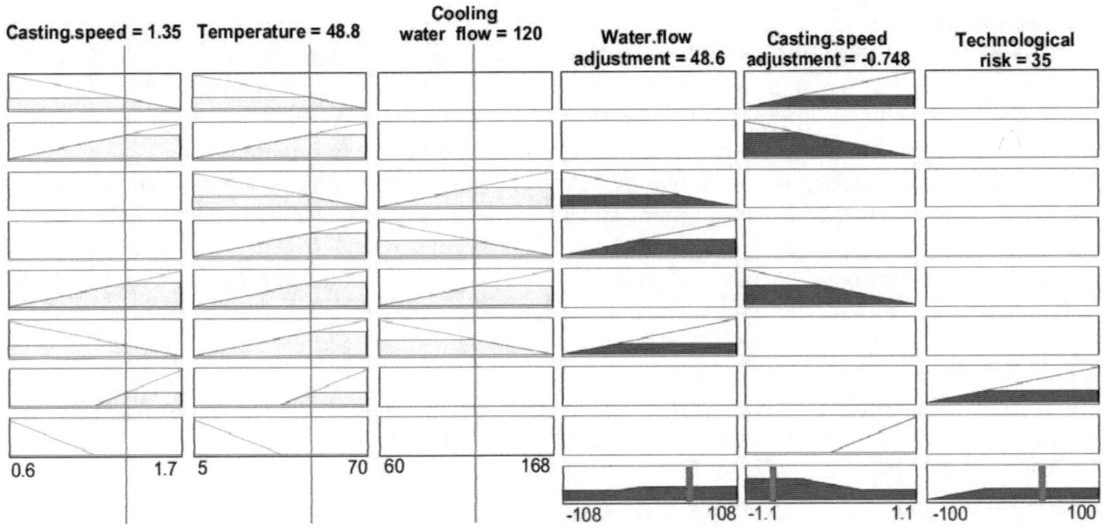

Figure 3. Rule table of IFS

The parameters used for creating the graphs below where: Casting speed 1.35m/sec, temperature difference 48.8 °C and water flow 120 l/s and the generated adjustments +48.6 l/s in the water flow, -0.748 m/sec for the casting speed and an technological risk of 35%.

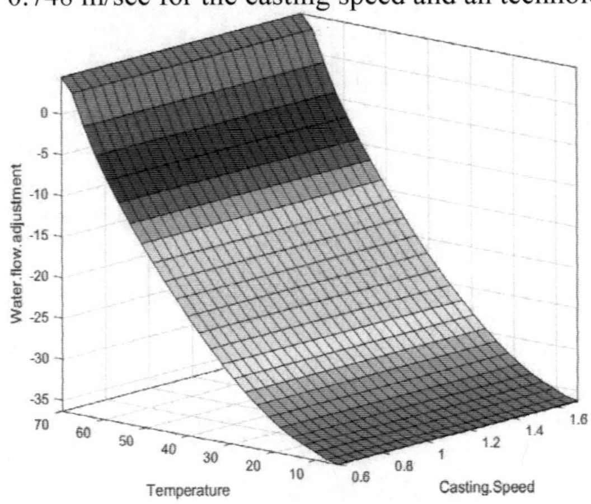

Figure 4. Water flow adjustment

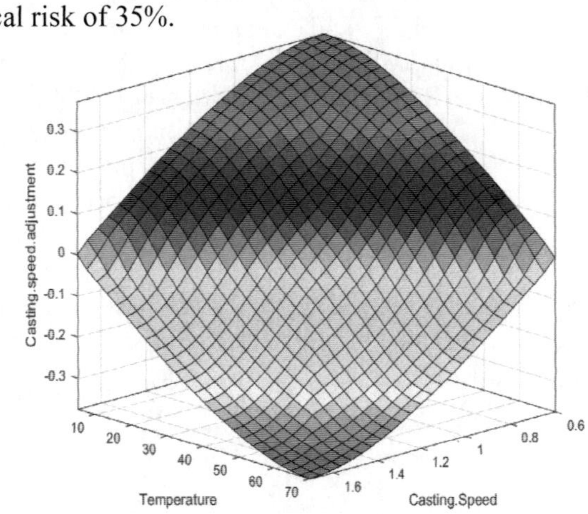

Figure 5. Casting speed adjustment

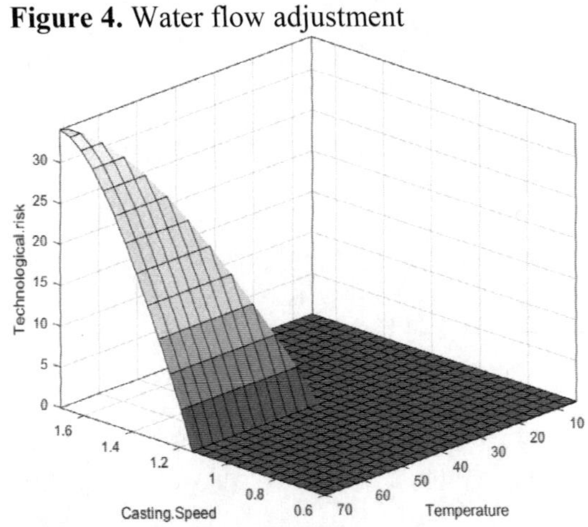

Figure 6. Technological risk

The starting parameters chosen for the graphical representation determined a risk of 35% for crust cracks formation. The parameters were adjusted by the fuzzy rules in such way that after 12 steps the risk dropped down to 0% and was maintained at 0% for the rest of the process. The cooling is made using some valves which adjust the water flow depending on initial measure. The goal of fuzzy system is to improve the classical adjustment system.

IOP Conf. Series: Materials Science and Engineering **200** (2017) 012046 doi:10.1088/1757-899X/200/1/012046

4. Conclusions

Analyzing simulation results shows that regardless of the values of the generated input, IFS develops the necessary corrections casting speed and primary cooling water flow, which confirms the validity of system operation. It is considered that, in terms of quality, use fuzzy decision system is an effective, practical and easy to implement, in order to analyze complex phenomena and nonlinear.

The goal of fuzzy system is to improve this classical adjustment system, the introduction of some adaptive components in the adjustment loops and some overall predictions over the continuous casting machine.

This paperwork introduces a new and original concept of the structure of a control system for the continuous casting.

Acknowledgment

This work was supported by a grant of the Romanian National Authority for Scientific Research and Innovation, CNCS – UEFISCDI, project number PN-II-RU-TE-2014-4-1788.

References

[1] Efimov V A 1986 *Casting and crystallizing of steel*, Technical Publishing House, Bucuresti

[2] Kiflie B and Alemu D 2000 *Thermal Analysis of Continuous Casting Process*, ESME 5th Annual Conference on Manufacturing &Process Industry, Faculty of Technology, Addis Ababa University, Ethiopia

[3] O'conner T and Dantzig J 1994 Modeling the Thin Slab Continuous Casting Mold, *Metallurgical and Materials Transactions* **25B**(4) 443-457

[4] Pinca-Bretotean C and Tirian G O 2006 *The numerical analysis of the asymmetrical thermal tension from hot rolling mill cylinders,* National conference of metallurgy and materials science, Bucuresti, Romania, pp 296-303

[5] Bouhouch S, Lahreche M, Moussaoui M and Bast J 2007 Quality Monitoring Using Principal Component Analysis and Fuzzy Logic. Application in Continuous Casting Process, *American Journal of Applied Science* **4**(9) 637-644

[6] Singh J and Ganesh A 2008 Design and Analysis of GA based Neural/Fuzzy Optimum Adaptive Control, *Transaction on Systems and Control* **3**(5)

[7] Ardelean E, Ardelean M, Socalici A and Heput T 2007 Simulation of continuous cast steel product solidification, *Revista de Metalurgia* **43**(3) 181-187

IOP Conf. Series: Materials Science and Engineering **200** (2017) 012047 doi:10.1088/1757-899X/200/1/012047

Control system of water flow and casting speed in continuous steel casting

G O Tirian[1], C A Gheorghiu[2], T Hepuţ[2] and C Chioncel[3]

[1]Politehnica University of Timisoara, Electrical Engineering and Industrial Informatics Department, 5 Revolution Street, Hunedoara, 331128, Romania
[2]Politehnica University of Timisoara, Department of Engineering an Management, 5 Revolution Street, Hunedoara, 331128, Romania
[3]"Eftimie Murgu" University of Resita, Electrical Engineering and Informatics Department, Street Traian Vuia, 1-4, Resita, Romania

E-mail: ovidiu.tirian@fih.upt.ro

Abstract. This paper presents the results of research based on real data taken from the installation process at Arcelor Mittal Hunedoara. Using Matlab Simulink an intelligent system is made that takes in data from the process and makes real time adjustments in the rate of flow of the cooling water and the speed of casting that eliminates fissures in the poured material from the secondary cooling of steel. Using Matlab Simulink simulation environment allowed for qualitative analysis for various real world situations. Thus, compared to the old method of approach for the problem of cracks forming in the crust of the steel in the continuous casting, this new method, proposed and developed, brings safety and precision in this complex process, thus removing any doubt on the existence or non-existence of cracks and takes the necessary steps to prevent and correct them.

1. Introduction

During the continuous casting of steel, in the secondary cooling stage, the process of crust solidification continues and can develop cracks. In order to detect and remedy cracks that may occur a simulation was made using Matlab Simulink which analyzes the signals from a temperature sensor and provides the necessary corrections. Thus, compared to the old methods of tackling the problem of crust cracks in the continuous casting of steel, such as technological methods (requiring a certain speed, adjusting the composition and quantity of powders, etc.), the new method proposed and developed, brings safety and precision in this complex process.

The output of the proposed system is used as input to a fuzzy decision system, which aims to change the parameters of the technological process to eliminate cracks.

The necessity of manufacturing using continuous casting of some products of high quality and in the same time competitive can be made only by thorough knowledge of the phenomena and the complex processes which take place on the technological development and casting [1], [2].

A very important component of the continuous casting installation is the secondary cooling zone. The secondary cooling zone has the role to continue the wire cooling after it has emerged from the crystallizing and to assure the total solidification of the semi – product. It is considered "the heart" of a continuous casting and has the role of ensuring the quality of the material, the material surface shape and has to ensure a homogeneous cooling and a uniform repartition of the water on the materials surface [3], [4].

In most cases the crust doesn't offer sufficient mechanical resistance to the action of Ferro static pressure. To complete solidification and guidance in good conditions of the wire it is created the secondary cooling zone. This cooling is achieved by direct spraying pressurized water, through nozzles, it is able to cross the steam layer formed by evaporation and ensure continuous water-metal contact [5], [6].

The secondary cooling can be made in different cooling environments. In practice, the water is used especially as cooling agent that is sprayed through nozzles (circular conical, conical ring and slot). In special cases it is added to the water compressed air for optimizing the automation sprayed water [7].

2. Structure of the control system

Figure 1 shows the implementation of intelligent fuzzy system (IFS) in Matlab Simulink. Temperature reading is simulated by using data reads from the continuous casting process. To begin the simulation the starting casting speed and water flow is entered, the program corrects these values using a loop that connects the output adjustment and adds it to the input data. The running time of the simulation is 1000 seconds. Also Figure 1 shows the repartition of water flow for all the cooling zones, 0.24% of the total water pressure for the first zone, 0.34% for zone two and 0.42% on zone three.

Using Matlab Simulink environment IFS allowed for qualitative analysis of its various real world situations, resulting in an IFS that meets all the requirements mentioned above.

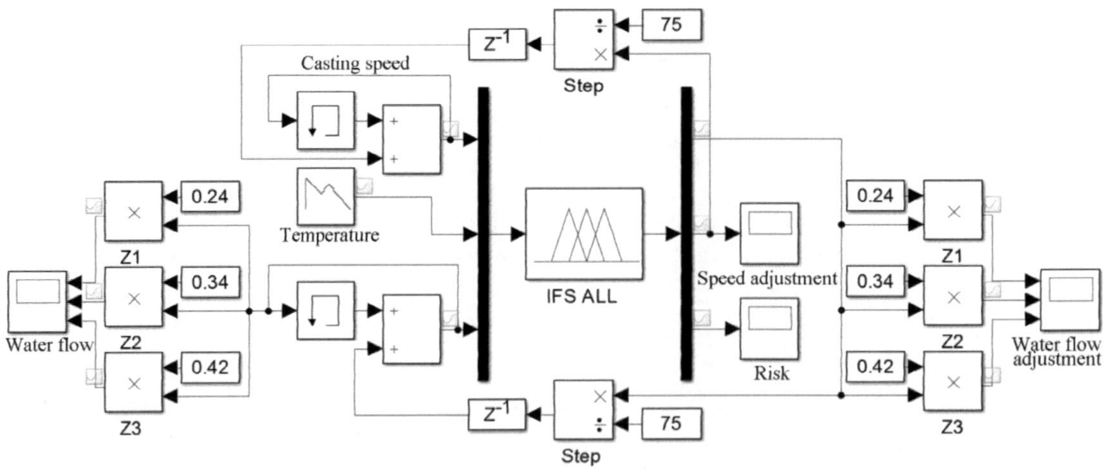

Figure 1. Simulink implementation of IFS-ALL

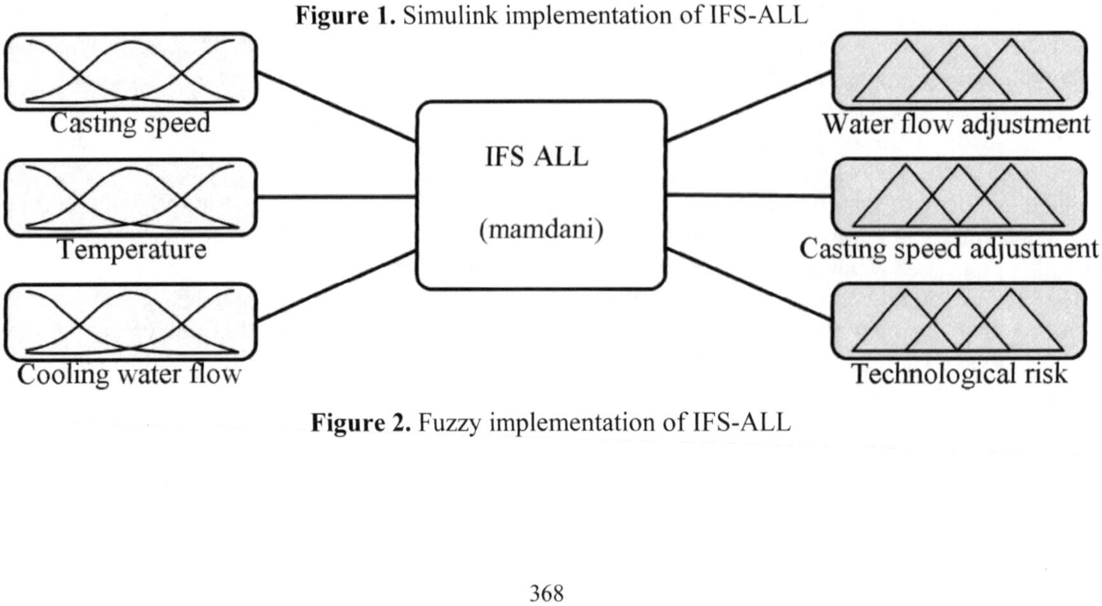

Figure 2. Fuzzy implementation of IFS-ALL

For the IFS seen in Figure 2 a simulation using MATLAB - Simulink environment is used. Figure 2 shows the fuzzy implementation in which we can see the 3 inputs, Casting speed, Temperature, Cooling water flow and the 3 outputs of the system named Water flow adjustment, Casting speed adjustment and Technological risk.

3. Data obtained from the system

A very important component of the continuous casting installation is the secondary cooling zone. The secondary cooling zone has the role to continue the wire cooling after it has emerged from the crystallizing and to assure the total solidification of the product.

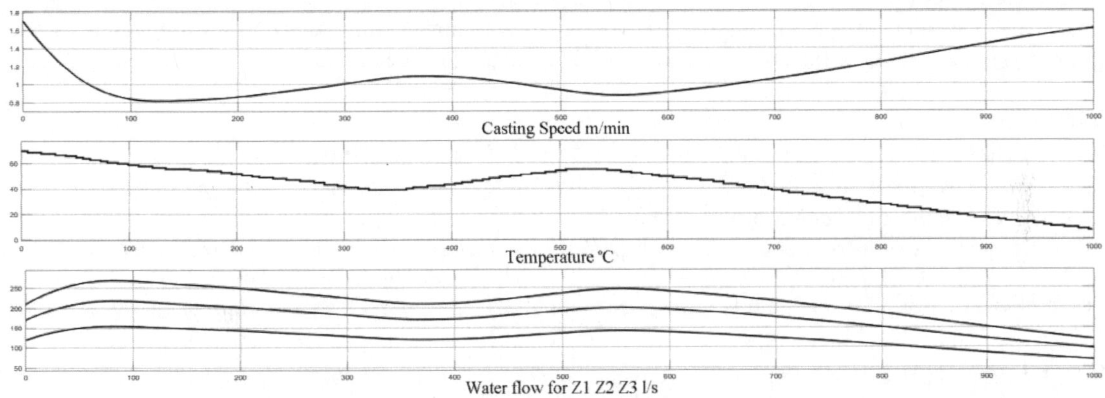

Figure 3. Values of the input variables of IFS

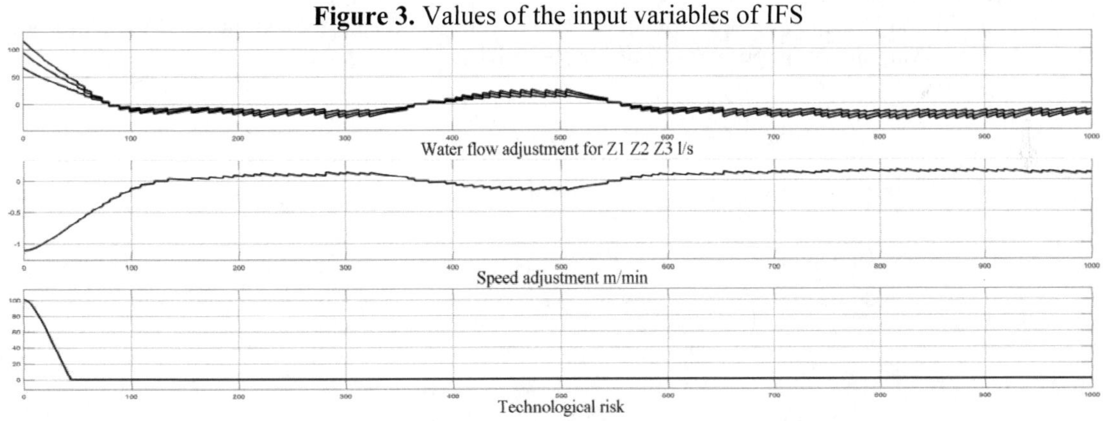

Figure 4. Values of the output variables of IFS

It is considered “the heart” of a continuous casting and has the role of ensuring the quality of the material, the material surface shape and has to ensure a homogeneous cooling and a uniform repartition of the water on the materials surface [3], [4].

Temperature variables are obtained from the process and the starting value for the casting speed is 1.5 m/sec and 168l/min flow of cooling water, we chose these values with high risk factor the risk dropped down to 0% in the first 30 seconds of simulation that can be seen in Figure 4.

Analyzing the simulation results shows that regardless of the values generated the system develops the necessary corrections in casting speed and cooling water flow, which confirms the validity of system operation.

The steels heat is partially removed by heating the cooling water, but the most heat is extracted by water evaporation, even if the evaporation percentage of the sprayed water on the wires surface is in general lower than 20%. All the unevaporated water which flows from the wires surface are collected filtered and reused [5], [7].

4. Conclusions

The goal of the Simulink system is to improve this classical adjustment system.

Technological risk in all cases dropped to 0 after the first few seconds of simulation and was kept constant throughout the simulation.

The simulation has shown accurate results and in future this system can be put into practice and implemented in a programmable logic controller.

A major benefit is that the only input needed is the temperature reading to control the water flow to all areas of the secondary cooling.

Acknowledgment

This work was supported by a grant of the Romanian National Authority for Scientific Research and Innovation, CNCS – UEFISCDI, project number PN-II-RU-TE-2014-4-1788.

References

[1] Beale M 1994 *Neural Network Tolbox for Matlab*, MathWorks Inc., Natick
[2] Cavllo A, Setola R and Vasca F 1994 *Using Matlab, Simulink and Control System Toolbox: A Practical Approach*, Editura Ligouri
[3] Cooper J 1998 *Introduction to Partial Differential Equations with Matlab,* Birkhauser, Boston
[4] ***Math Works Inc. 1993 *Matlab Reference Guide,* Math Works Inc., Natick, Massachussets
[5] ***Math Works Inc. 1997 *Matlab - The Language of Technical Computing,* Math Works Inc., Natick
[6] Wellker H H 1999 *Method and apparatus for controlling the cooling of a strand in a continuous casting installation*, Patent number 5988259
[7] Bordei M and Ciurea A 2004 Influence of the cooling process on the evolution of the solidified crust during the continous casting of the steel slabs, *Metalurgia* **4**

IOP Conf. Series: Materials Science and Engineering **200** (2017) 012048 doi:10.1088/1757-899X/200/1/012048

Some considerations about the use of different sensors, in coordinate measuring of the small parts

L Drăgan

Technical University of Cluj-Napoca, North University Center of Baia Mare, IMTech Department, Dr. Victor Babeş str., no. 62A, Baia Mare, Romania

E-mail: liliana.dragan@cunbm.utcluj.ro

Abstract. The paper presents some particular aspects associated with measuring of the small-size parts with high precision, manufactured by injection procedures. The coordinate measuring machine (CMM) are very used in process of measuring parts with different shapes, dimensions and materials of the most varied. It is studied by experiments, the influence of hygroscopicity on the geometrical properties of polyamide parts, using different types of measuring sensors. We selected a few pieces- cover type, with precision features dimensions and shape tolerances. To measure them was used some sensors which is equipped CMM ScopeCheck S 400 and equipment for dehumidifying. Starting from the need for high precision measurement of geometric characteristics of the parts obtained by injection of plastic, it has been found that the hygroscopicity has a significant influence. To achieve the purpose were used three types of measuring sensors under different conditions of keeping after manufacture. It was observed that the influence of humidity is significantly reduced if the parts are kept in exikator or vacuum dryer.

1. Introduction

The CMM are used to solve of control issues and expertise of industrial products. They operate with numerical data in rectangular or polar coordinates, with tactile or optical probing systems, capable to access all geometrical elements of controlled part. They are used for measuring geometrical characteristics of parts: deviations of form, position and orientation, diameters, angles etc. Measurement results can be displayed both graphically and numerically. Resolution coordinate measuring machines range from 1µm (on the manual control) to 10 nm (on the automatic).

The types of sensors that coordinate measuring machines are using today are:

- optical image: which are based on a high-resolution CCD camera;
- tactile: in which measurement points are recorded by physical palpation of the piece;
- laser: it is recorded and transmitted to individual sets of data captured by the laser fascicle;
- X-rays: when making a 3D scan of the piece, inside and outside, to be completed in measuring program model.

2. The equipment used for experimental verification

For the experiments was used Scope Check S 400, a coordinate measuring machine with dimensions measuring table 400 x 200 x 200 mm and the maximum permissible weight of parts that can be measured 8 kg. The machine has a measurement resolution of 0.1 micrometers, and the uncertainty of measurement up to 2 micrometers. The supply voltage of the machine is 115-230 V (50-60 Hz), maximum power consumption 1.5 kVA and 10A maximum amperage. For the normal functioning of the machine, humidity should be between 40-70% and ambient temperature of 20-21°C.

Scope Check S 400 is equipped with a number of sensors in modular construction for measuring interactive, integrated PLC-controlled production process. To obtain maximum of precision, the main axes of measurement (OX and OY) are separated constructive. On all three axes are set precision guides from steel. Involvement of axes is made with continuous current servomotors and ball screws.

The measuring points are recorded in the machine coordinate system, in conjunction with the data read from the integrated measuring length. The data is processed using a mathematical evaluation software, which determines the geometrical characteristics of measured values and displays them.

WinWerth 8 is a measurement program that has an interface very easy to use with a lot of functions and a variety of measurement modes. The machine is controlled and coordinated the program manually, automatically or interactively. In manual mode, can be inserted values of coordinates and parameters functioning machine that is given; interactive mode, measurements can be made with joystick control, and automatically by using the CAD model or program with interface measurement DMIS (Dimensional Measuring Interface Specification). Scope Check S 400 disposes of 3 sensors: optical (with additional functions such as Autofocus, BestFit, 3D Patch), touch (with interchangeable sensors 3 mm, 1 mm and 0.5 mm) and laser (with the "high speed scanning"). So, the interface of WinWerth 8 program are six modes: screenshot, optical or laser shooting mode, autofocus, stylus, laser mode manually.

2.1. Dimensional measurements technique with multiple sensors

2.1.1. Optical sensor the image processing. The electronic part of matrix camera of the sensor (Figure 1) converts the optical signal into a digital image, which is the used to calculate and measure different points, with appropriate software for processing and measurement images.

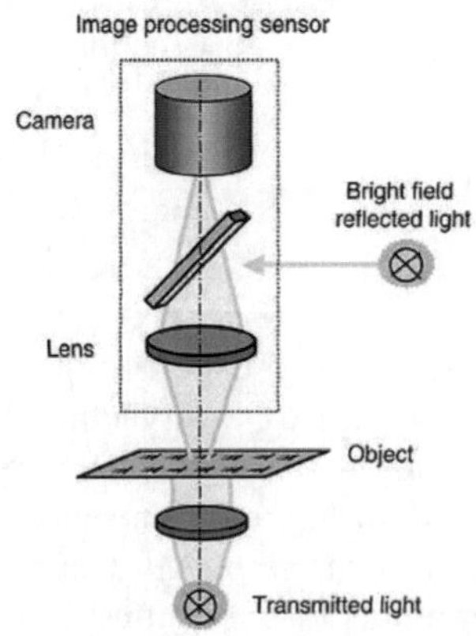

Figure 1. The image processing sensor with transmitted and reflected light

To get the lowest measurement uncertainty the telecentric lens are used. The advantage of this lens is that they do not distort images so even at maximum zoom, they remain unchanged. When are not used this lens, the images of the metered are altered at minimum zoom and appears distortion of image while the measurement uncertainties are high. For a better flexibility is used a multi-lens zoom system (Figure 2a). To increase the accuracy, the measuring machine has a optical system Werth Zoom, which ensures high repeatability (Figure 2b) in repositioning in the same coordinate system and at same values.

There movements of lens are motorized, thus ensuring exactly the same values of zoom and the same accuracy every time. This design is valid for zooms between 1× and 10× at focal length to measured parts between 20 mm and 250 mm. To realize measurements with optical sensor, illumination and contrast have an important role, they must be appropriate.

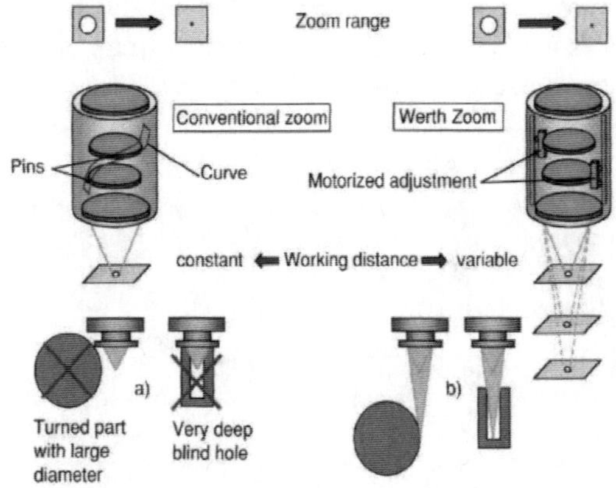

Figure 2. Optical lens system [1]
a) conventional zoom
b) adjustable Werth Zoom and adjustable focus distance

To have an adequate contrast and accentuated of margins, the metered with flat surfaces, measurements can be made with light contour (Figure 3).

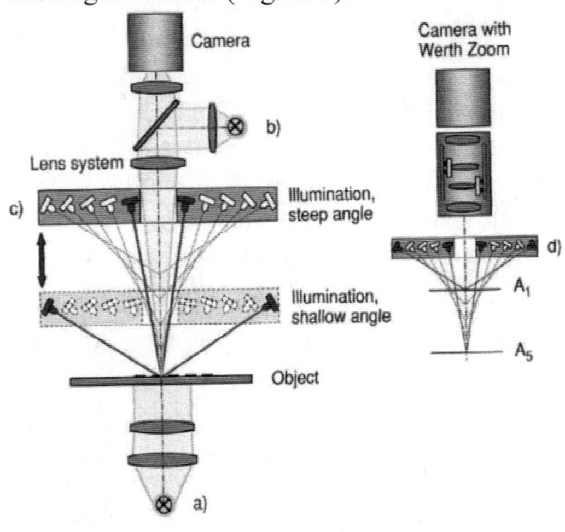

Figure 3. Types of illumination [1]:

a) lower lighting (contour)

b) top illumination (integrated into lens system)

c) d) light side (MultiRing) - adjustable to different heights

d) the light side (MultiRing)

The same hardware is used for autofocus sensor as processing of images. Autofocus is used to measure flat surfaces (parameters, values, tolerances of form). The main disadvantage of autofocus is increased attention that should work for better results and that is very slow.

2.1.2. Point laser sensor. Measuring principle of laser (Figure 4) is based on design of the light ray (red) produced by laser diode, on the object. The reflected ray is absorbed by an optoelectronic sensor. The point is then determined using triangulation technique. The advantage of laser sensor compared to the optical sensor is measuring speed is much higher. In this way can be measured up to thousands of points per second, so this sensor is ideal for scanning surfaces.

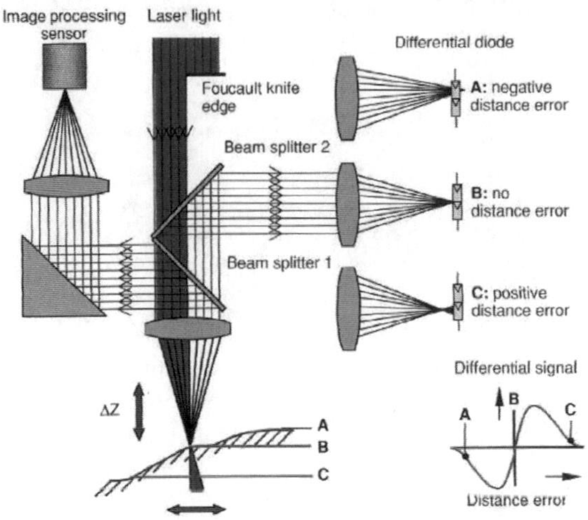

Figure 4. Laser sensor with integrated image processor [2]

2.1.3. Tactile sensor. In principle, all tactile sensors (Figure 5-6) function by mechanical contact with the measured part. The resulting signal is then forwarded for processing. A distinction must be made here between Touch Trigger (connected sensor) and tactile probe (measuring sensor). With tactile sensor geometry (shape and size) of a probe element (scope probes) as the spatial position and geometrical shape of surface of measured object, are contained in the measurements results. To obtain smaller measurements uncertainty, tactile sensor must be calibrated to correct probing errors and contact with surfaces. This is done using the program WinWerth.

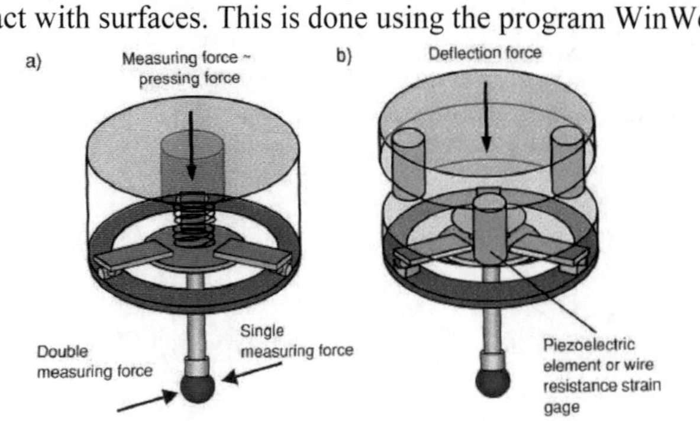

Figure 5. Probes with electromechanical converter (measuring force of pressure/deflection)

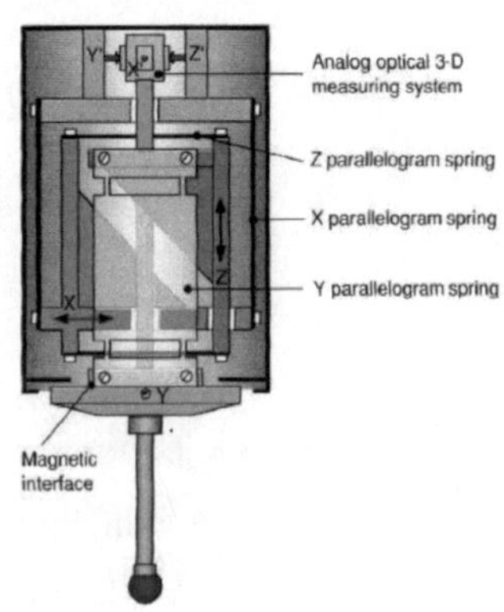

Figure 6. 3D probe with elastic parallelogram mechanisms

The main disadvantage of this tactile sensors is that they are first brought into contact with measured part, all data is saved, then automatically retire and after can be measured the point in the measuring program.

2.2. Comparative study on measurements with different measuring sensors on parts injected from polyamide

2.2.1. Flatness measurement and tolerated dimensions. We selected a few pieces, cover type, made from polyamide, with precision features dimensions and form deviations tolerated. To measure them was used some sensors which is equipped Coordinate Measuring Machine Scope Check S 400 and equipment for dehumidifying parts.

For the beginning, it was measured the part from Figure 7, having deviations from flatness, with tolerances of 0.20 mm on the plain surface. There were used, by turn, autofocus, laser and tactile probe. Then was measured the tolerated dimensions of part from Figure 7b. The dimensions were measured with Scope Check in optical module (autofocus) and with the digital micrometer (thickness).

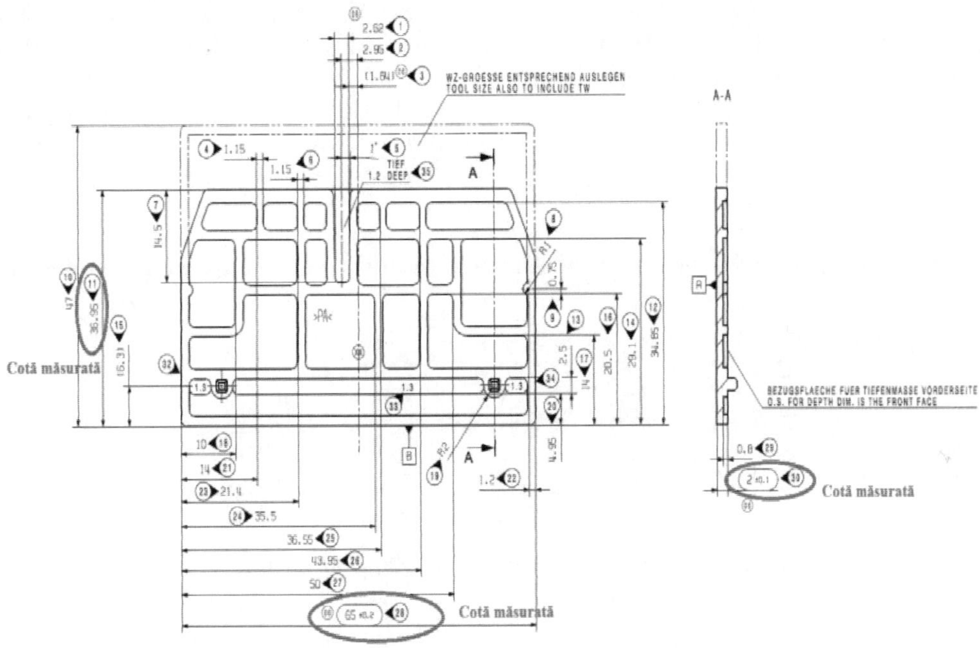

Figure 7. Measured part

For more precise results to measure the flatness with optical sensor autofocus (Figure 8) always is chosen maximum zoom [3], in order to highlight roughness of parts measured.

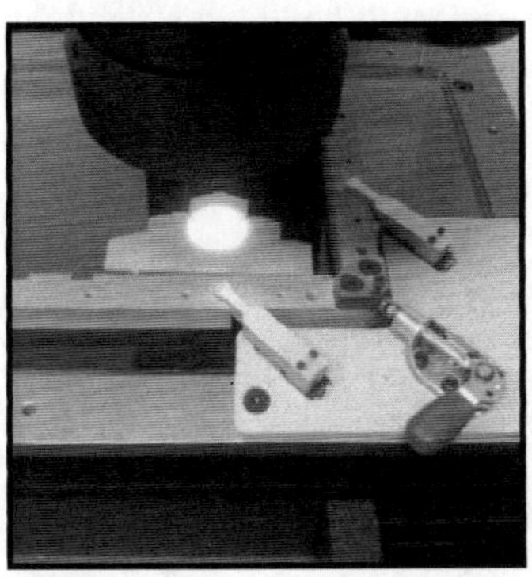

Figure 8. Measurement with Autofocus

Figure 9. Measurement with Laser

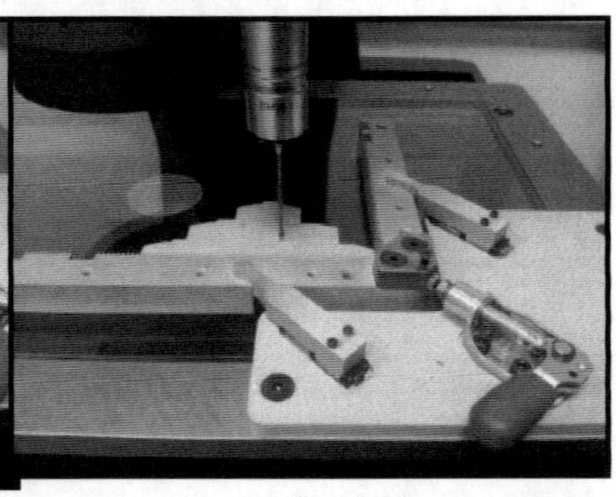

Figure 10. Measurement with Probe

Measurements with laser sensor (Figure 9) are most used because the results are almost as accurate like autofocus sensor and the highest speed of analysis of surfaces. Measurements with tactile sensor (Figure 10) give the most accurate results. Yet if probe is not calibrated, the results will be inaccurate.

2.2.2. Influence of humidity on the dimensions of a part injected from Polyamide. At injection, plastics have some nominal size, but after 2-3 hours they absorb humidity and increase (according to ISO 62 under normal conditions: 23 ° C at normal relative humidity 50%) between 2.6% and 3.2% of their normal that they had at injection. Humidity saturation of the piece injected will be between 8-9%, in other words even if we leave it in the rain will not increase more than 8-9% of the size it had the injection [4], [5].

The big problem is that the parts what are inject [6-8], sometimes come after many hours or even days at specialists in metrology, during which change their some geometrical characteristics because of humidity. For remove the humidity of the pieces, they are placed in exikator for up to 48 hours where the silica gel absorbs humidity. To expedite the process, the pieces can be inserted in vacuum the time preceding measurements with 24 hours. For highlight the influence of humidity on the plastic parts (Polyamide), we conducted a series of measurements on the same piece, kept in different environments.

3. Experimental results

After measuring flatness on the three types of sensors, measuring result were centralized in a diagram (Figure 11).

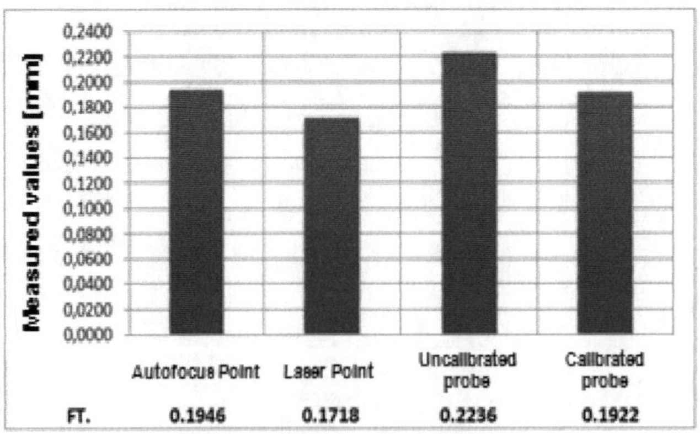

Figure 11. Results of flatness measurements with different sensors

Next, the effect on humidity of injected piece is studied, in different keeping conditions: outdoor (Figure 12), in the exikator (Figure 13-14), in the vacuum dryer (Figure 15-16) and graphs for these situations have been drawn up.

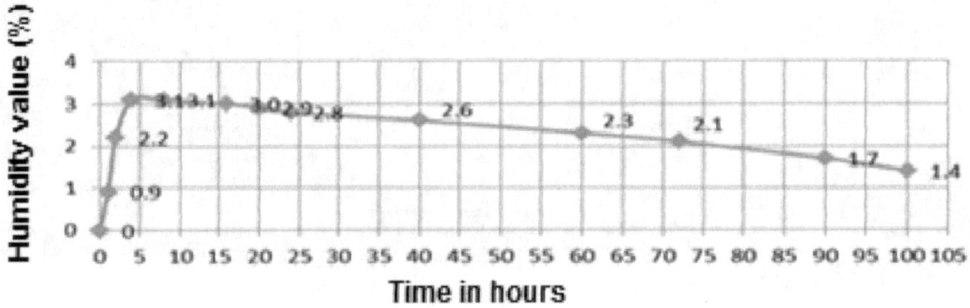

Figure 12. The results of humidity at measured parts, left to outdoor 100 hours

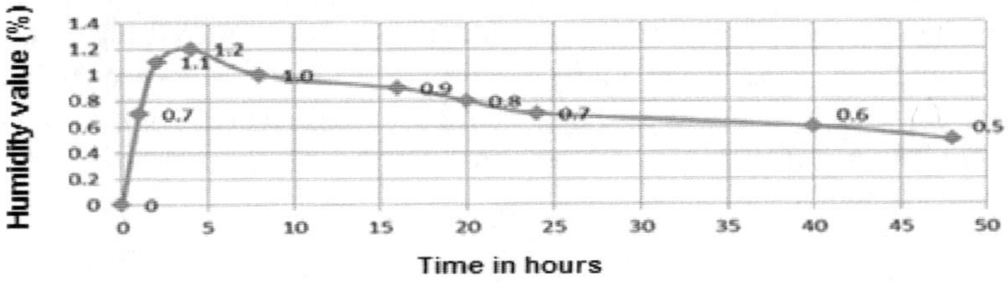

Figure 13. The results obtained for parts placed in exikator immediately after injection

Figure 14. The exikator

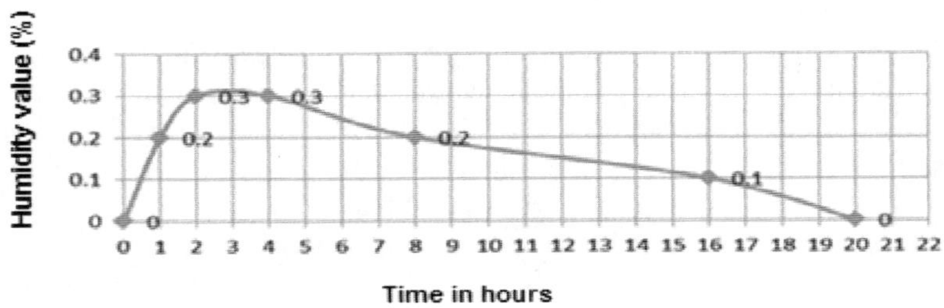

Figure 15. The results obtained for parts kept in the vacuum dryer for 20 hours

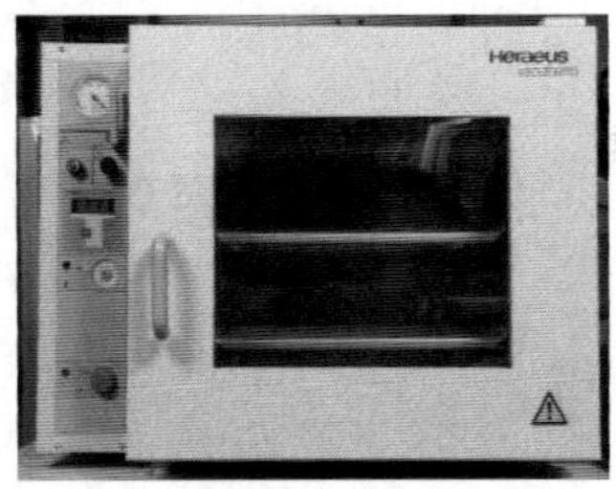

Figure 16. The dryer

As to the effect of humidity on the three tolerated dimensions of a piece measured with Scope Check S 400 and a digital micrometer, kept in different storage conditions, the results can be seen in Table 1.

Table 1. The influence of humidity on tolerated dimensions

	Nominal value +Tolerance (*mm*)	Sample 1 (*mm*)	Sample 2 (*mm*)
AI	36.95±0.28	36.80	36.82
AI	65.00±0.20	64.75	64.80
AI	2.00±0.10	1.97	1.97
AC	36.95±0.28	36.99	36.99
AC	65.00±0.20	65.09	65.10
AC	2.00±0.10	2.08	2.09
EX2.5	36.95±0.28	36.94	36.96
EX2.5	65.00±0.20	65.03	65.05
EX2.5	2.00±0.10	2.04	2.06
EX24	36.95±0.28	36.90	36.92
EX24	65.00±0.20	64.98	65.00
EX24	2.00±0.10	0.02	2.03

AI -The value of dimensions measured immediately after injection.
AC -The value of dimensions measured at 2.5 hours after injection (ambient conditions).
EX2.5 -The value of dimensions measured at 2.5 hours after injection (exikator).
EX24 -The value of dimensions measured at 24 hours after injection (exikator).

IOP Conf. Series: Materials Science and Engineering **200** (2017) 012048 doi:10.1088/1757-899X/200/1/012048

4. Conclusion

Starting from the need for high precision measurement of geometric characteristics of the piece obtained by injection of plastic, it has been found that the hygroscopicity has a significant influence. To achieve the purpose were used three types of coordinate measuring sensors under different conditions of keeping of parts after manufacture. It was also observed that the influence of humidity is significantly reduced if the parts are kept in exikator or vacuum dryer.

References

[1] ***Werth Messtechnik GmbH 2013 *WinWerth 8 User Manual*, Print Edition, November 2013, Germany

[2] ***Neumann R C and Joachim H 2007 *A Multisensor Coordinate Metrology*, Sv Corporate Media, Germany

[3] ***Verlag Moderne Industrie 2012 *X-ray tomography in industrial metrology*, Süddeutscher Verlag Onpact GmbH, Germany

[4] Metin A 2014 *Determining the changes of geometrical properties of plastic materials in different climatic conditions*, Universität Bielefeld North Rhine-Westphalia, Germany, Doctoral Thesis

[5] Barz C, Deaconu S I, Latinovic T, Berdie A, Pop-Vadean A and Horgos M 2015 PLCs used in smart home control, *IOP Conf. Ser.: Mater. Sci. Eng.* **106** 012036

[6] Rabinovich S 2005 *Measurement Errors and Uncertainties: Theory and Practice*, Springer Publication, New York

[7] Lung C, Sabou S and Barz C 2012 *Smart sensor implemented with PicoBlaze multi-processors technology*, International Symposium for Design and Technology of Electronic Packages, SIITME 2012, Alba Iulia, Romania, October 25-28, pp 241-245

[8] Barz C, Todea C, Latinovic T, Preradovic DM, Deaconu S I and Berdie A 2015 Intelligent traffic control system using PLC, *IOP Conf. Ser.: Mater. Sci. Eng.* **144** 012017

IOP Conf. Series: Materials Science and Engineering **200** (2017) 012049 doi:10.1088/1757-899X/200/1/012049

Arts and technology – Mosaic new techniques and procedures

G A Papiu and N Suciu

Technical University of Cluj Napoca, Northern University Center Baia Mare, Faculty of Lettres, Department of Fine Arts – Painting, str. Crişan, no. 10, Baia Mare, Maramureş, Romania

E-mail: gretipapiu@gmail.com

Abstract. The relationship between art and technique has been along the time one that is inseparable and systematic, artists appealing to various technologies, tools and practices that help them stimulate their imagination. Today there is a new category of artists, coming from a technical or scientific field, that are being 'trapped' in this „game of art". The mosaic, even if it is an old technique, responded to the social requirements and it evolved over time, being constantly related to aesthetic and artistic thinking, discoveries of science, assimilating permanent new techniques and technologies, diversifying its artistic forms of expression and methods of transposition. Not being bound any more to a religious institution, which was its birth place, today, she migrated to all public spaces. Works of art in public space have become today an active factor in reshaping the urban aesthetic landscape.

1. History of mosaic

The relationship between art and technology, has been throughout the ages one that is inseparable and systematic, artists turning to various techniques and technologies, tools and different practical processes in their workshops, that were helping to stimulate their imagination. All this is equally an ongoing challenge, the artists fate being to "snatch" materials using tools, different work techniques, their expressive virtues, skillfully discovered and subordinated to our emotional feelings and artistic emotions, giving them life in real artistic creations.

Figure 1. Details with the arrangement of stones in the mosaic

Discoveries, different techniques attributes and technologies in art, were along time a well kept secret in artists and guilds workshops, knowing that working methods have not been explained by artists, they have chosen the power of example through their work, in passing the knowledge, craft and practical exercise witch is visible plainly in recipes, precisely followed codes, hand books of work in achieving the artistic creation. It is more than suggestive the argument of Filippo Brunelleschi (1377-1446) in concealing the knowledge results: "*Do not share your inventions to the many, only share with those few who understand and love science. To reveal too much of your inventions and achievements, is one and the same thing with relinquishing ones fruits of ingenuity. Many are ready when they hear the inventor speaking, to minimize and deny it's accomplishments, so that they do not have credibility in honorable places, but after a few months or a year later, they may use the inventor words in speeches or other projects*" [1].

Figure 2. Detail mosaic from Hagia Sofia Church – Istanbul

Art history confirms and demonstrates that the links, the connections between art and technology are not new concerns. The technological revolution of the last decades, after the '60s that developes in a fast pace and it's becoming in the first place a global political and economic stake, witch constitutes a support for new forms of art. Concerned closely with the evolution of these phenomena, Florence Meredieu mentions the impossibility to set limits for a "*real mutation of a civilization*" - another revolution began recently and is likely to further change our relations with the world: "*the revolution of lights, the virtual image called a synthesis, interactivity, the direct participation of the viewer, creating a world of synthetic sensations, all contributing to creating artificial universes, which makes us believe that in the future, it may even compete in "replacing" our old ways of perception. Our reference to the concept of reality, the relationships we maintained for millennia, with our old planet is likely to falter, the world of images and artificial universes, is becoming prevalent*" [2].

Interference of fine arts is becoming more present; precisely because of the borders mobility that is separating them, be it painting, graphics, sculpture or performing arts, installations, photography, video, new media, that take out of the performance and technology which are always in a perpetual change, "*causing*" thus, new concepts that deliver the artistic message, but also helps to its reception.

It is understood that the invasion of new technologies, materials and processes of their application, have been and are accepted by the other arts: architecture, theatre, music, dance, cinema, so that the result of the artistic product or work of art, is fixed, is implemented in a vast network of communication, in a complex and competitor world and it imposes it's self as the messenger of human sensibility. Artistic mastery perfection of the human spirituality cannot be entirely subordinated to technology, thus the artist is assuming the task to create the artistic and metaphorical universe, witch erases the individual identification of materials and techniques by spiritualizing them, in a distinct message of artistic creation. The german expressionist painter Max Beckman, preoccupied and immersed in his own personal investigation to discover those values and phenomenon that had semantically charged the universe of personal creation, mentions knowingly that : "*What I want to show in my work is the idea that hides behind the so-called reality. I find my self looking for a bridge that leads from visible to invisible [...]. One of my problems is finding my own selft, which has only one form and it is immortal, to find it in animals and people, in heaven and in hell that together makes up the world we live in*" [3].

Works of art in public spaces, in this technique, have become today an active factor in the aesthetic reshaping of urban landscape. In this context, we are mentioning the project from 2005 initiated by American artist Mary Sherman from the Boston College of Art (*Boston - MIT - Program in Art end Technology - The title Project at UNESCO - Trans Cultural Exchange*) where I was invited to participate with a personal creation. The International Project of Art Installations - made from vitrified tiles (burning the ceramic fragment at 1280 degrees C and coloured glazes at over 1180 degrees C) had the end objective of completing and assembling a polychrome mosaics form, with free thematic subjects, in public spaces and public culture institutions, in the following countries: Azerbaijan, Bosnia and Herzegovina, Canada, France, Finland, Germany, India, Israel, Mexico, New Zeeland, Taiwan, Turkey, USA, Slovakia, Vietnam, Wales Country. The only condition at this art project was the imposed size of the tiles witch where 20 by 20 cm.

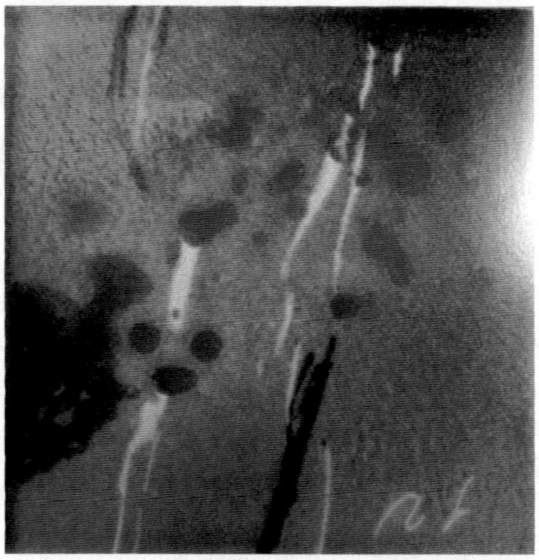

Figure 3. Nicolae SUCIU – Vitrified tile (Detail of personal work)

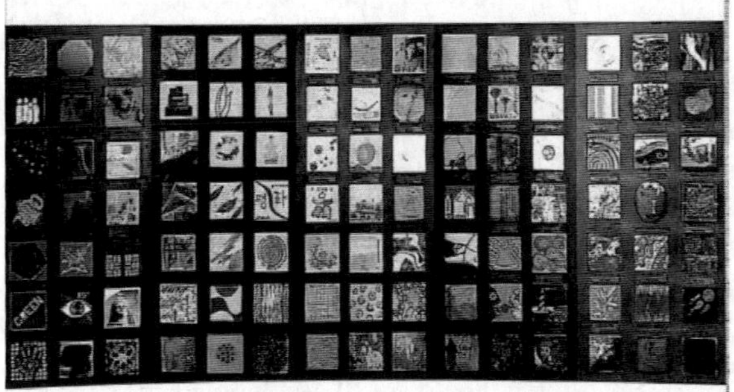

Figure 4. Assembly of Tiles in Ethnological Chinchon Museum in Spain

According to DEX (Romanian Word Dictionary) the meaning of the word mosaic is defined as: "*a decorative work technique, which consists in a artistic assembly of small pieces of marble, ceramic, glass, enamel, and so on; of various colors stuck together with mortar or mastic*." The historical evolution of this art technique has seen many transformations and developments due to the discovery of new materials, and new ways of working with it. From simple pavements stone in two colors, the mosaic has evolved during The Greek Hellenistic Period and later in Ancient Rome, to extensive figurative compositions, with a sober polychromy, but also very charming, knowing that this technique has a decorative character with specific requirements, we simplified the drawing to it's most expresive lines, ennobling with chromatic harmonization, resulting from joining the different colored stone cubes.

Figure 5. Detail from a Roman mosaic

The peak of Mosaic Art is in the Byzantine religious art in an exceptional iconographic register, unsurpassed until today. Cubes of colored stones, where juxtaposed on the principles of harmonic color arrangements were highlighted by the insertion of colored glass cubes or gilded with gold leaf cubes (between two glass plates), conferring thus an immaterial like atmosphere in the scenes with the represented subjects. The triumph of artistic metaphores and visual message of these representations, made invisible the materials witch made up and contributed to the ineffable images. In Islamic art mosaics framed harmoniously laid ornamental combining vegetal and geometrical motifs, joined and completed by calligraphic Persian inscriptions (The Qur'an prohibits isomorphic and anthropomorphic representations). As an artistic technique, alternative to easel painting, mosaic fascinate the imagination, students involved in the learning and assimilation of theoretical knowledge presented in the classroom are excited to practicaly realize this technique, through learning the processes and working methods, using the tools and instruments and materials to complete the projects and the implementing techniques that have always had shortcomings.

Figure 6. Details of tools and instruments used in mosaic techniques

2. Methods and technique use for mosaic

Preparing materials for mosaic technique has always been an ongoing concern for ceramics resistance to variations in temperature, glazes vitrification and mortars application on the wall, whether we talk about the application on both indoor or outdoor spaces. "*I experienced a good period of time with masses of layered tiles, covered by vitrified glazes to be permeabilize the porous area*" [4] frequently using metal oxide, cobalt oxide, for nearly almost all shades of blue, at high-temperature melting and fixing, copper for a variety of green, red iron oxide, for reds, zinc oxide that carries a remarkable flexibility in a high temperature environment.

To get the little cubes covered with precious materials for mosaics we have used a colloidal gold based liquid (9-11% gold and platinum 11%) each time giving to this artistic creation technique a professional standard.

In conclusion, we should mention several technological milestones, which will help to identify the novelty brought by this art techniques, reborned today and used on a increasingly wider scale. In addition to the technological status of this technique, we have pursued a gradual lightening and a more efficient way of transport of the mosaic panels from the workshop where they were created and worked on, to the transposition site from the building, whether it is placed inside or outside the building.

The transport of these mosaic panels is made as safe as possible, their lifting on the scaffolding at the building site is made under effective control, the transposition of the mosaic panel on wet mortar, helps to prevent the falling of the stones and it also makes bonding areas between the panels disappear (both so visible after the „*washing*" of the mosaic) and eliminating the risk of displacement and thus having a perfect matching of the panels, especially in circular or ovoid surfaces.

The steps in this project consist in increasing the drawing at the scale of 1/1 to the final size of the project and establish the local tones, modulation and modelling of the colours on the surfaces for enhanced and intensified effects of light / shadow, the final settlement for the coloured cubes, with the possibility of returning and rearranging them along the way, applying and gluing (Dermacol a rate of 150g / 1 liter water)sheets of wrapping paper over the little cubes surface, and after drying 24 hours later, drawing, cutting and numbering the resulting boardswitch composes the whole image.

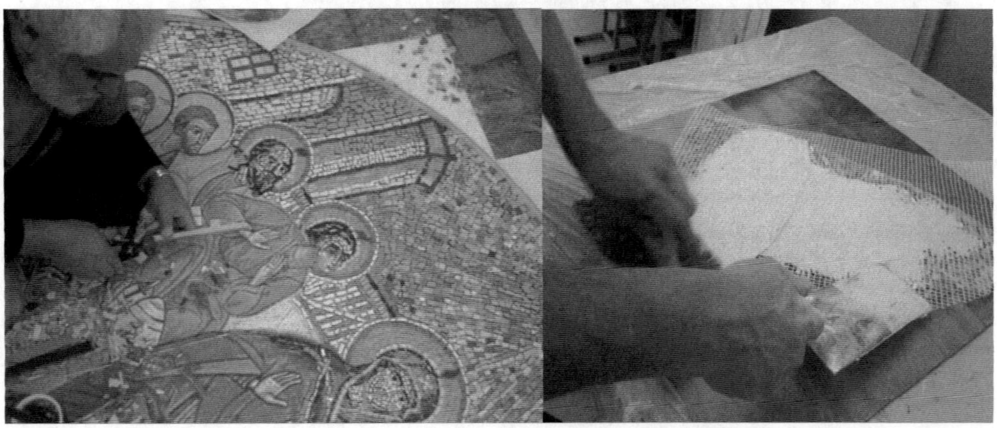

Figure 7. Details of the mosaics execution (Photographs Papiu G)

After cutting to size the contour shapes from the final mosaic work, depending on the to the elements, characters and details of the composition, without putting the accent on the exact dimensions of the plates but on the primary elements of the composition. After turning the boards upside down, we applied on the back a mesh of fiberglass witch is normally used in insulation of facades, using a flexible tile adhesive witch had a 10 mm thickness, thus after drying it was easier to transport them with minimal risks from the workshop to the jobsite. The wall on which the mosaic will be placed, is receiving a layer of mortar 15 - 20 mm thick over which we start placing the component boards of the mosaic.

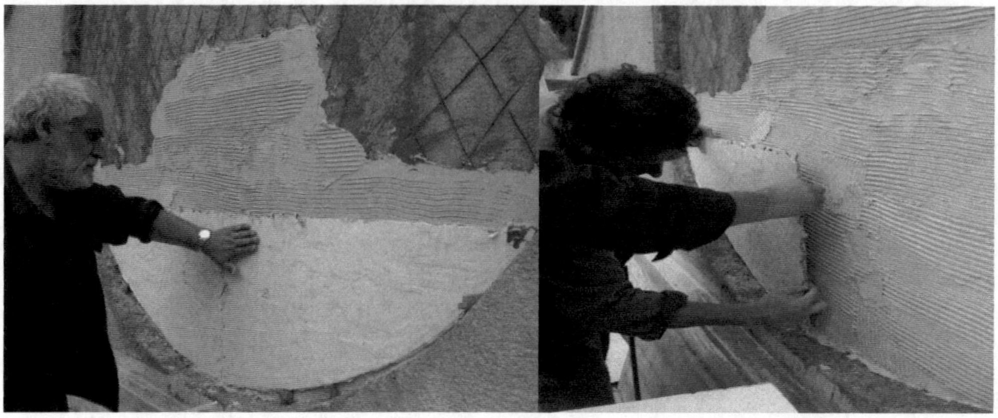

Figure 8. Details of the mosaics execution (Photographs Papiu G)

Attention is focused on setting correctly the essential parts of the image in the original project, and after 48 hours of drying, we eliminate the sheets of paper that kept the boards of the mosaic together, this is called „washing the mosaic" that will be followed by the grouting and refinishing operations.

Figure 9. Detail of *„washing the mosaic"*

In this case also, we will take into account to adapt the colour of the grout to the chromatic and compositional scheme of the mosaic thus completing and continuing the narrative elements. To protect the mosaic, especially when positioned outside, it used to be polished with a layer of beeswax but today weakly a layer or 2 of varnish for stone surfaces.

3. Conclusion

About the purposes and ideals of the artist, **the creator**, always subject to a permanent "review process", subject to permanent improvement and always submitted to compromise, that the contemporaneity demands most of the time, in various justifications. Michelangelo, one of the titans of Italian Renaissance, emphasized the decisive role of **The Creator** in society. "*I would venture to say that a man can not achieve perfection if it aims to please the ignorants and not the real connoisseurs of his own craft, and if it is not" odd "or" peculiar "or however you wish to call it*" [3].

Figure 10. The completed works – **Pentecost and Resurrection scene (dimensions: 2,20 cm diameter)**, St. Elias Orthodox Church, Baia Mare

References

[1] Hockney D 2007 *The Secret science - rediscovering the forgotten techniques of the old masters,* Ed. Enciclopedia RAO, București, Romania
[2] Meredieu F 2005 *Art and new technologies - video art, digital art,* Ed. Enciclopedică RAO, București, Romania
[3] Knobler N 1983 *Visual Dialogue*, Ed. Meridiane, București, Romania
[4] Pamfil M 2016 *Ceramic Techniques and Technologies*, Ed. Performantica, Iași, Romania

IOP Conf. Series: Materials Science and Engineering **200** (2017) 012050 doi:10.1088/1757-899X/200/1/012050

Two fluxes multistage induction coilgun

L Gherman[1], M Pearsica[1], I Circiu[2] and C Rotaru[2]

[1]"Henri Coandă" Air Force Academy, Air Defense Department, Mihai Viteazul str., no. 160, 500183 Brasov, Romania
[2]"Henri Coandă" Air Force Academy, Aviation Department, Mihai Viteazul str., no. 160, 500183 Brasov, Romania

E-mail: lauriang@gmail.com

Abstract. This paper presents a brand new induction electromagnetic launcher, which uses two magnetic fluxes in order to accelerate a projectile. One magnetic flux induce a current in the armature and the second magnetic flux is creating a radial magnetic field. This aproach offer multiple advantages over single flux designs. First we are able to control the induced current in armature because we use the coil just to induce current inside the ring with a great efficiency. Second advantage is the angle of 90^0 between magnetic field density B and the ring. We used the induction to avoid contact between armature and accelerator. In order to create the magnetic field radial we used four coils perpendicular on armature. This approach alove us to control the phase difference between induced current in armature and current in magnetic field coils for a maximum force. The phase difference is obtained by changing the frequency of magnetic field coils power source. We used simulation software to analyze, and simulate a multistage induction coilgun design with two fluxes. The simulation results demonstrated the theoretical results.

1. Introduction

The actual guns to accelerate projectiles based on chemical energy have achieved their limits but the requirements are continuously growing. The electromagnetic launch systems (EMLS) can expand the possibilities of guns [1], [2].

Many EMLS are designed to use the Lorentz force

$$\vec{F} = I \int d\vec{l} \times \vec{B} \tag{1}$$

A well-known design is railgun. The railgun consists of two rails and an armature made by conducting material. The armature is connected with both rails and is accelerated alongside the rails. The armature is accelerated by Lorentz force alongside the rails when the rails are powered by a high current.

In the railgun (Figure 1) the current I is the same current provided by the power system. The rail creates the magnetic field $\vec{B}$.

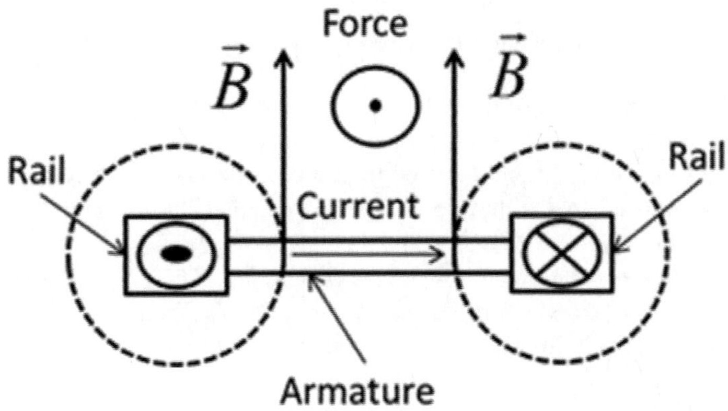

Figure 1. Railgun

A high value of the force is obtained when is used a very high value of current *I* and, therefore, a big power supply system.

Because of this high currents used those electric contacts affect the rails during lunches [3]. In order to reduce the value of current *I* the rails can be replaced by coils.

The current inside armature is obtained by using induction instead of sliding contacts.

$$u_i = -\frac{d\Phi_B}{dt} \tag{2}$$

$$u_i = -\frac{d}{dt}(BA\cos\theta) = -\left(\frac{dB}{dt}\right)A\cos\theta - B\left(\frac{dA}{dt}\right)\cos\theta + BA\sin\theta\left(\frac{d\theta}{dt}\right) \tag{3}$$

where θ is the angle between $\vec{B}$ and $\vec{n}$ (normal unit of surface area A). We assume the magnetic field is uniform distributed in space.

The coilgun design is based on induction where only the magnetic field is variable, the surface *A* and angle *θ* are constant. The Faraday's law can be writhed:

$$u_i = -\left(\frac{dB}{dt}\right)A\cos\theta \tag{4}$$

One approach consists of a barrel of coils which allow a projectile made by aluminium to move inside them [4]. The magnetic flux created by the coils is the only one magnetic flux, which induces current in projectile and provides the radial magnetic field which interacts with the induced current. It is difficult to control in the same time with one coil the rate of change of the axial magnetic density $\overrightarrow{B_a}$ and the radial magnetic flux density $\overrightarrow{B_r}$ [5]. In order to increase the radial magnetic flux density $\overrightarrow{B_r}$ a design with magnetic circuit made by ferromagnetic materials was proposed (Figure 2).

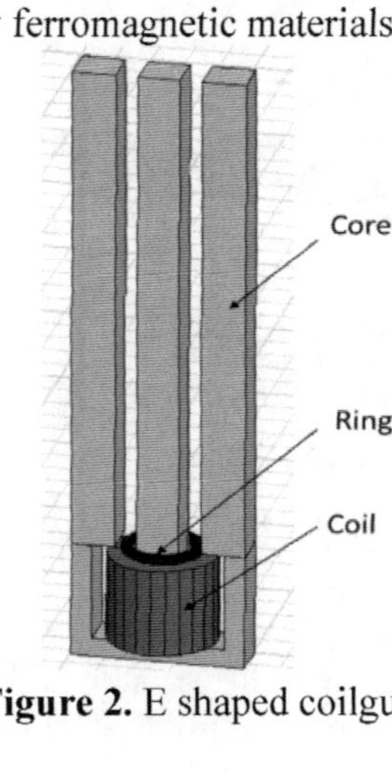

Figure 2. E shaped coilgun [6]

The E shaped design create a magnetic circuit but does not allow to control the difference of phase between induced current in projectile and the phase of the radial magnetic flux density $\overrightarrow{B_r}$.

Based on this observation we can identify the main aspects which should be taken into consideration when an electromagnetic launch system is designed. First the Lorentz force should be used to accelerate the projectile. Second the contact between the projectile and the accelerator should be avoided. Because of this, coils should be used to induce current into projectile. By using induction the value of current I induced into armature can be higher than the current provided by the power system.

Third the design should give us the possibility to change the phase of induced current in armature and the phase of the radial magnetic flux density $\overrightarrow{B_r}$ in order to obtain a high value of the Lorentz force. A coilgun, which obeys all this conditions, is presented in the following section.

2. A multistage induction coilgun design with two fluxes

First we calculate an air core coil which will induce current into a ring placed outside (Figure 3). The aluminium ring is placed outside solenoid because it will act as armature, which will be mechanical coupled with projectile.

$$u_i = -\left(\frac{dB}{dt}\right) A \cos\theta = -\left(\frac{dB}{dt}\right) A \quad (5)$$

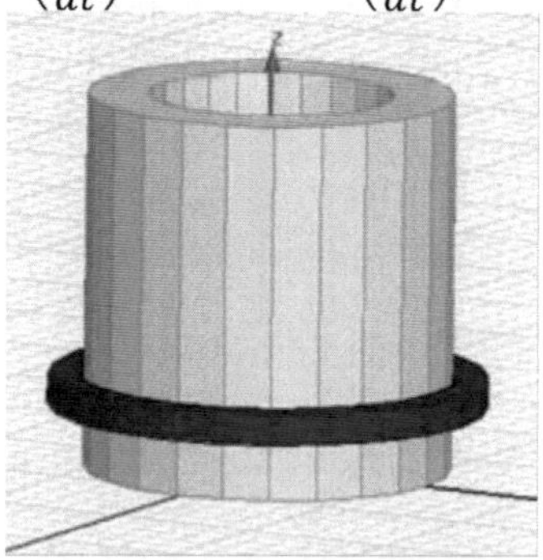

Figure 3. The induction coil

We need another coil, which will create the radial magnetic field [4]. Usually the armature in a coilgun is placed inside the coil and the current I and magnetic field B are produced by the same coil. For our design the magnetic field B is produced by others air core coils placed around ring at $\theta=90^0$ in order to obtain maximum force on the ring (Figure 4).

Figure 4. The magnetic field coils around ring at $\theta=90^0$

All coils (induction and magnetic field) have the same dimensions in order to have the same resistance R and the same inductance L. Because the armature (ring) will be accelerated using multiple stages this design give us the possibility to create multiple stages (Figure 5).

IOP Conf. Series: Materials Science and Engineering **200** (2017) 012050 doi:10.1088/1757-899X/200/1/012050

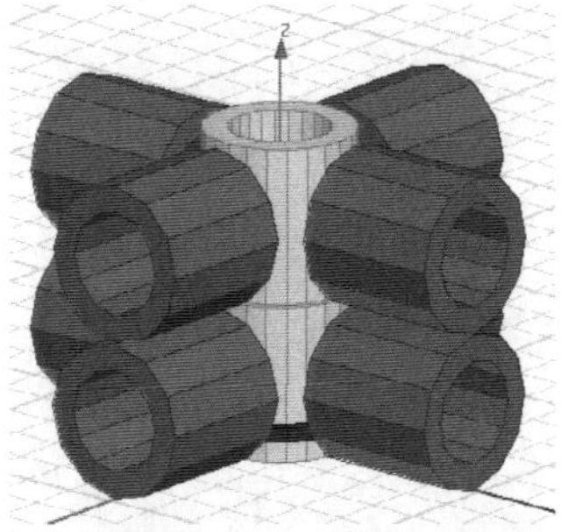

Figure 5. EMLS with 2 stages of acceleration

The simplified drawing of this circuit is displaied in Figure 6. We chose to represent only the induction coil Lb, the ring Li and the coil which produce the magnetic field Lc.

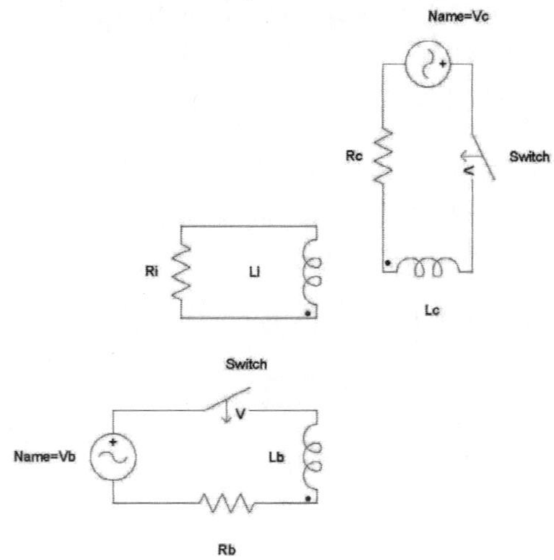

Figure 6. Drawing of electrical circuit

The ring is a short circuit but the coils are powered by two sinusoidal sources with following expression.

$$U_b = U_{bmax}\sin(\omega_1 t + \alpha) \quad (6)$$

$$U_c = U_{cmax}\sin(\omega_2 t + \beta) \quad (7)$$

Because all coils are identical we have the following equality:

$$L_b = L_c \text{ and } R_b = R_c$$

Applying the KVL we obtain the following system of differential equations:

$$\begin{cases} U_b = Ri_b + L\dfrac{di_b}{dt} - M_{bi}\dfrac{di_i}{dt} - M_{bc}\dfrac{di_c}{dt} \\ 0 = R_i i_i + L_i\dfrac{di_i}{dt} - M_{bi}\dfrac{di_b}{dt} - M_{ci}\dfrac{di_c}{dt} \\ U_c = Ri_c + L\dfrac{di_c}{dt} - M_{ci}\dfrac{di_i}{dt} - M_{bc}\dfrac{di_b}{dt} \end{cases} \quad (8)$$

$$\begin{cases} U_b + M_{bi}\dfrac{di_i}{dt} + M_{bc}\dfrac{di_c}{dt} = Ri_b + L\dfrac{di_b}{dt} = V_b \\ M_{bi}\dfrac{di_b}{dt} + M_{ci}\dfrac{di_c}{dt} = R_i i_i + L_i\dfrac{di_i}{dt} = V_i \\ U_c + M_{ci}\dfrac{di_i}{dt} + M_{bc}\dfrac{di_b}{dt} = Ri_c + L\dfrac{di_c}{dt} = V_c \end{cases} \quad (9)$$

$$\begin{cases} V_b = Ri_b + L\dfrac{di_b}{dt} \\ V_i = R_i i_i + L_i\dfrac{di_i}{dt} \\ V_c = Ri_c + L\dfrac{di_c}{dt} \end{cases} \tag{10}$$

The expressions of currents are:

$$\begin{cases} i_b = I_{bmax}\sin(\omega_b t + \gamma) + I_{bmax}\sin(\gamma)e^{-\frac{R}{X}\omega_b t} \\ i_i = I_{imax}\sin(\omega_i t + \lambda) + I_{imax}\sin(\lambda)e^{-\frac{R_i}{X_i}\omega_i t} \\ i_c = I_{cmax}\sin(\omega_c t + \varepsilon) + I_{cmax}\sin(\varepsilon)e^{-\frac{R}{X}\omega_c t} \end{cases} \tag{11}$$

Because the time when the ring is accelerated on a stage is very short (below 10 ms) the Lorentz force acting on ring depend only by the first semi period current oscillation at 50 Hz.

The expression of currents can be approximated:

$$\begin{cases} i_b = I_{bmax}\sin(\omega_b t + \gamma) + I_{bmax}\sin(\gamma) \\ i_i = I_{imax}\sin(\omega_i t + \lambda) + I_{imax}\sin(\lambda) \\ i_c = I_{cmax}\sin(\omega_c t + \varepsilon) + I_{cmax}\sin(\varepsilon) \end{cases} \tag{12}$$

The magnetic field created by the L_c solenoid can be calculated according with Ampere's law:

$$\oint B \cdot dl = \mu_0 i_{enclosed} \tag{13}$$

In our case $i_{enclosed}=i_c$ and the expression of B at the end of solenoid L_c became:

$$B = \mu_0 \frac{N}{2\sqrt{r^2 + h^2}} i_c \tag{14}$$

where:
N=number of turns of solenoid (dimensionless);
r = radius of coil (meters);
h = length of solenoid (meters).
The expression of Lorentz force became:

$$F = l \cdot B_c \cdot i_i = l \cdot \mu_0 \frac{N}{2\sqrt{r^2 + h^2}} i_c \cdot i_i \tag{15}$$

$$F = l \cdot \mu_0 \frac{N}{2\sqrt{r^2 + h^2}}[I_{cmax}\sin(\omega_c t + \varepsilon) + I_{cmax}\sin(\varepsilon)] \cdot [I_{imax}\sin(\omega_i t + \lambda) + I_{imax}\sin(\lambda)]$$

$$F = l \cdot \mu_0 \frac{N}{2\sqrt{r^2+h^2}} I_{cmax} I_{imax}[\sin(\omega_c t + \varepsilon) + \sin(\varepsilon)] \cdot [\sin(\omega_i t + \lambda) + \sin(\lambda)] \tag{16}$$

In order to find the maximum value of Lorentz force F we should find the maximum of expression:

$$max[\sin(\omega_c t + \varepsilon) + \sin(\varepsilon)] \cdot [\sin(\omega_i t + \lambda) + \sin(\lambda)] \tag{17}$$

$$[\sin(\omega_c t + \varepsilon) + \sin(\varepsilon)] \cdot [\sin(\omega_i t + \lambda) + \sin(\lambda)] = \frac{1}{2}[\cos(\lambda - \varepsilon) - \cos(\lambda + \varepsilon)](cos\omega_i t + 1)(cos\omega_c t + 1) + \frac{1}{2}[\cos(\lambda - \varepsilon) + \cos(\lambda + \varepsilon)]\sin(\omega_i t)\sin(\omega_c t) + \frac{1}{2}[\sin(\lambda - \varepsilon) + \sin(\lambda + \varepsilon)](cos\omega_i t + 1)\sin(\omega_c t) + \frac{1}{2}[\sin(\lambda + \varepsilon) - \sin(\lambda - \varepsilon)](cos\omega_c t + 1)\sin(\omega_i t) \tag{18}$$

Because the above expression does not have a global maximum we maximize each expression of the sum. We obtain the following situation:

I. $$max[\cos(\lambda - \varepsilon) - \cos(\lambda + \varepsilon)] = 2 \tag{19}$$

When $(\lambda - \varepsilon) = 0$ and $(\lambda + \varepsilon) = \pi$ where $\lambda = \varepsilon = \frac{\pi}{2}$

The expression becomes:

$$(cos\omega_i t + 1)(cos\omega_c t + 1) \tag{20}$$

And the maximum of:

$$max(cos\omega_i t + 1)(cos\omega_c t + 1) = 4 \quad (21)$$

when $\omega_i = \omega_c = 2\pi f$

I. $$max[\cos(\lambda - \varepsilon) + \cos(\lambda + \varepsilon)] = 2 \quad (22)$$

when $(\lambda - \varepsilon) = 0$ and $(\lambda + \varepsilon) = 0$ where $\lambda = \varepsilon = 0$

The expression becomes:

$$\sin(\omega_i t)\sin(\omega_c t) \quad (23)$$

And the maximum of:

$$max\sin(\omega_i t)\sin(\omega_c t) = 1 \quad (24)$$

when $\omega_i = \omega_c = 2\pi f - \frac{\pi}{2}$

$\omega_i = \omega_c = 2\pi f + \frac{\pi}{2}$

II. $$max[\sin(\lambda - \varepsilon) + \sin(\lambda + \varepsilon)] = 2 \quad (25)$$

when $(\lambda - \varepsilon) = \frac{\pi}{2}$ and $(\lambda + \varepsilon) = \frac{\pi}{2}$

where $\lambda = \frac{\pi}{2}$ $\varepsilon = 0$

The expression becomes:

$$(cos\omega_i t + 1)\sin(\omega_c t) \quad (26)$$

And the maximum of:

$$max(cos\omega_i t + 1)\sin(\omega_c t) = 2 \quad (27)$$

when $\omega_i = 2\pi f + \pi$ $\omega_c = 2\pi f - \frac{\pi}{2}$

IV. $$max[\sin(\lambda + \varepsilon) - \sin(\lambda - \varepsilon)] = 2 \quad (28)$$

when $(\lambda - \varepsilon) = -\frac{\pi}{2}$ and $(\lambda + \varepsilon) = \frac{\pi}{2}$

where $\lambda = 0$ $\varepsilon = \frac{\pi}{2}$

The expression becomes:

$$(cos\omega_c t + 1)\sin(\omega_i t) \quad (29)$$

And the maximum of:

$$max(cos\omega_c t + 1)\sin(\omega_i t) = 2 \quad (30)$$

when $\omega_c = 2\pi f + \pi$ $\omega_i = 2\pi f - \frac{\pi}{2}$

Because λ and ε represent the phase delay of induced current i_i and magnetic field coil current i_c we should analyse all four situations.

First situation is obtained when $\lambda = \varepsilon = \frac{\pi}{2}$ but it is very difficult to obtain the same phase for induced current and current inside magnetic field coils because of mutual induction between coils. The expression of force becomes:

$$F = F_{Imax}(cos\omega_i t + 1)(cos\omega_c t + 1) \quad (31)$$

Second situation is obtained when $\lambda = \varepsilon = 0$ also very difficult to obtain with coils. The expression of force becomes:

$$\mathrm{F} = F_{IImax}\sin(\omega_i t)\sin(\omega_c t) \quad (32)$$

Third situation is obtained when $\lambda = \frac{\pi}{2}$ $\varepsilon = 0$.

The expression of force becomes:

$$F = F_{IIImax}(cos\omega_i t + 1)\sin(\omega_c t) \quad (33)$$

Last situation is obtained when $\lambda = 0$ $\varepsilon = \frac{\pi}{2}$. The expression of force becomes:

$$F = F_{IVmax}(cos\omega_c t + 1)\sin(\omega_i t) \quad (34)$$

In third and fourth case, we create condition to achieve a certain delay between phases of induced current in ring and magnetic field coil current.

Theoretically the maximum of Lorentz force F is obtained in first case when $\lambda = \varepsilon = \frac{\pi}{2}$ and $\omega_i = 2\pi f_i$ and $\omega_c = 2\pi f_c$. Practically the maximum of Lorentz force F is obtained in third and fourth case when $\lambda = \frac{\pi}{2}$ $\varepsilon = 0$ or $\lambda = 0$ $\varepsilon = \frac{\pi}{2}$ and $\omega_i = 2\pi f + \pi$ $\omega_c = 2\pi f - \frac{\pi}{2}$

For these conditions the expression of Lorentz force became:

$$F = 2l \cdot \mu_0 \frac{N}{2\sqrt{r^2 + h^2}} I_{cmax} I_{imax} =$$
$$l \cdot \mu_0 \frac{N}{\sqrt{r^2 + h^2}} I_{cmax} I_{imax} \quad (35)$$

Because we approximated the expression of currents for ideal conditions it is not possible to obtain the same values for phase but we are looking for a maximum value of force between those values.

The following dimensions of this design were calculated [5]. The mass of aluminium ring was calculated at 10 grams. We started from dimensions of device used for E shaped.

Table 1. Coil parameters

	Induction coil	Magnetic field coil	Ring
Inner radius	20 mm	20 mm	31 mm
Outer radius	30 mm	30 mm	36 mm
Length	60 mm	60 mm	5 mm
Number of turns	300	300	1

The Maxwell interactive software package that uses the finite element method (FEM) was used to analyse, and simulate two fluxes multistage induction coilgun design [7], [8].

3. The results of simulation

The coilgun with these dimensions was created using 3D simulation software. The values of voltages and duration of pulses for every stage are presented in Table 2. In this simulation all coils from the same stage were powered with the same voltage.

Table 2. Stages parameters

	Stage 1	Stage 2
Voltage RMS	700V	1400V
Time	6.6 ms	0.7 ms

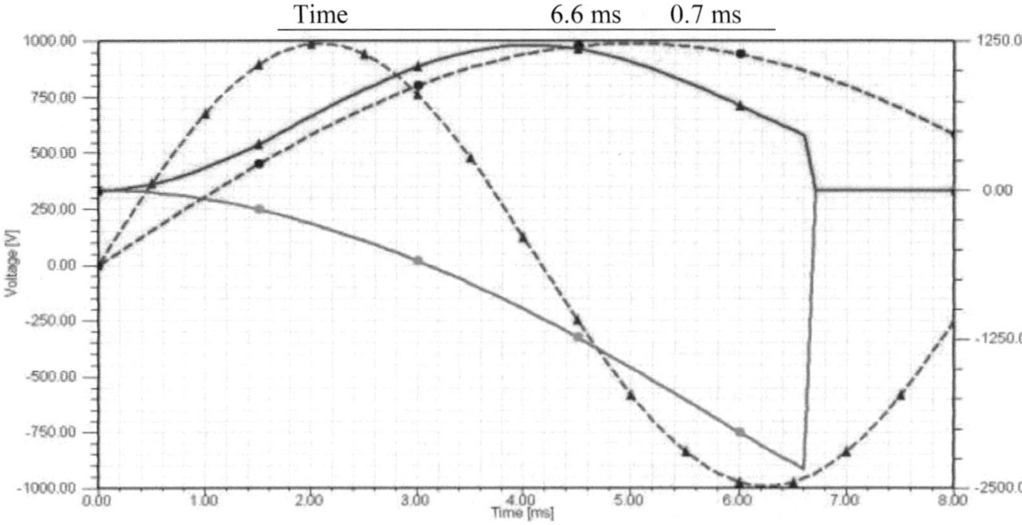

Figure 7. Stage 1 power voltage and current

In Figure 7 is presented the time variation of voltage and current in stage 1 coils. With long dash and circle is presented the time variation of voltage source, which power the induction coil. The frequency is 50 Hz and the phase is zero. The continuous line with circle is the time variation of the current inside the induction coil. With long dash and triangle is presented the time variation of voltage source which power the magnetic field coil. In order to obtain the maximum Lorentz force the frequency of power source of magnetic field coils was changed to 120 Hz with the same phase zero. The resistance of coil was decreased at a very low value in order to obtain the phase 90^0 for current. The continuous line with triangle is the time variation of the current inside the magnetic field coil. At 6.6 ms when the ring left the stage 1 the power line is switched off and the value of currents inside stage 1 coils drop rapidly to zero value.

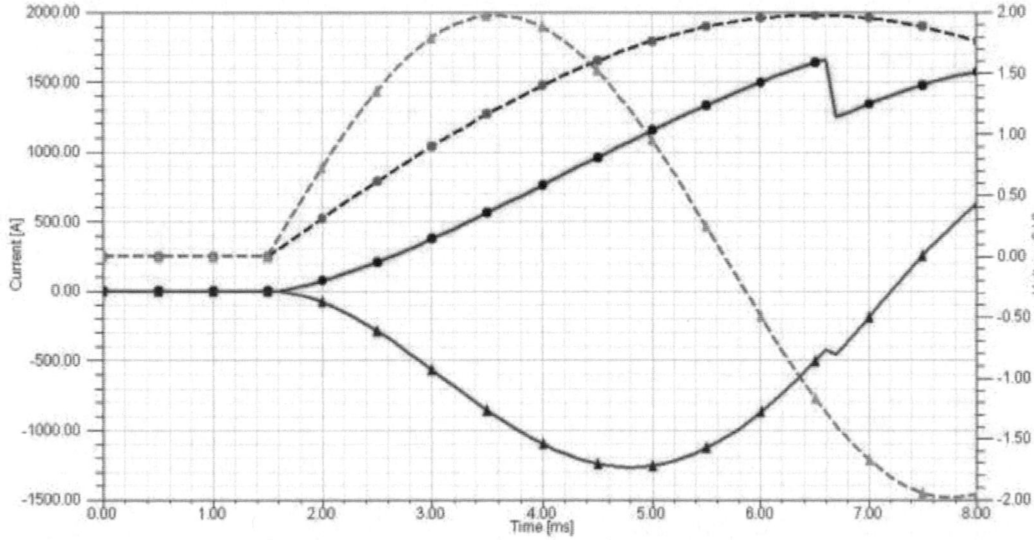

Figure 8. Stage 2 power voltage and current

In Figure 8 is presented the time variation of voltage and current in stage 2 coils. The power for stage 2 coils is switched on at 1.5 ms after first stage is powered in order to create a higher force when the ring arrive al second stage. With long dash and circle is presented the time variation of voltage source, which power the second stage induction coil. The frequency is 50 Hz and the phase is zero. The continuous line with circle is the time variation of the current inside the second stage induction coil. With long dash and triangle is presented the time variation of voltage source, which power the second stage magnetic field coil. The continuous line with triangle is the time variation of the current inside the second stage magnetic field coil. At 6.6 ms when the power line of first stage is switched off the value of currents inside stage 2 coils present a rapid variation. Because after the second stage the ring left the EMLS the power line was not switched off. For multiple stages, we should follow the same pattern.

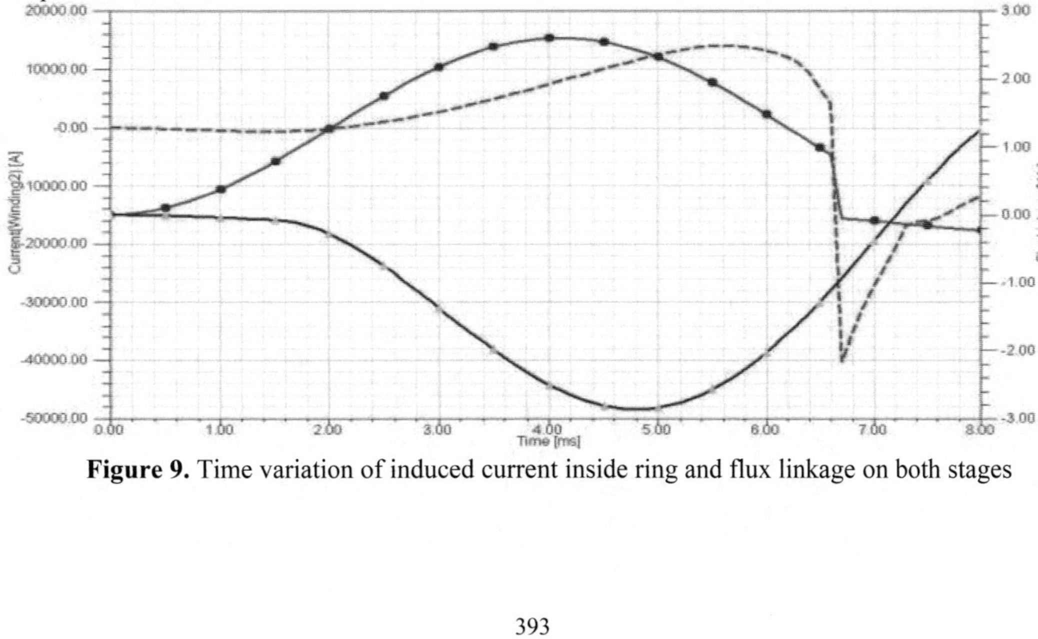

Figure 9. Time variation of induced current inside ring and flux linkage on both stages

The time variation of current induced into the ring is represented by long dash. The maximum value of induced in ring current is 13,973.6 A when the ring is in first stage and 40,236.2 A when the ring is in second stage. The maximum value of current used to induce the current into the ring is 2,332.9 A inside the induction coil from stage 1. All other coils are working with currents below this value. From this point of view, our design is better than the railgun because the value of current inside armature is higher than power supply current. The flux linkage created by stage 1 is presented using a continuous line with circle and for stage 2 is a continuous line with triangles.

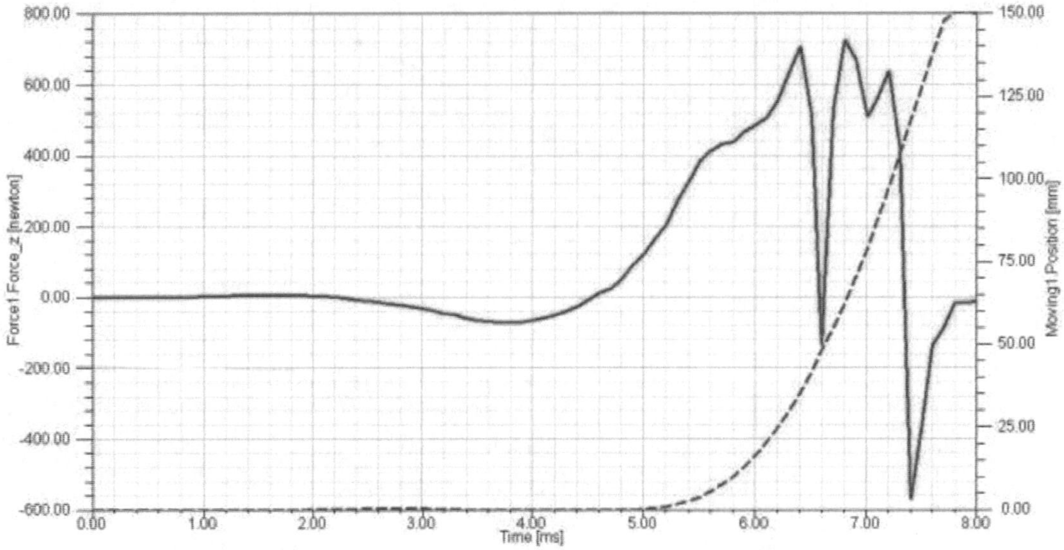

Figure 10. Time variation of Lorentz force and position of ring

In Figure 10 is presented with continuous line the time variation of Lorentz force which is accelerating the ring in Oz direction. The maximum value of force is 724.8 N. The force has a very rapid drop when the ring is passing from stage 1 to stage 2 and the power of stage 1 is switched off and after the ring left the EMLS. With long dash is presented the position of ring during acceleration time. The border between stages is at 40 mm and the limit of EMLS is at 101 mm.

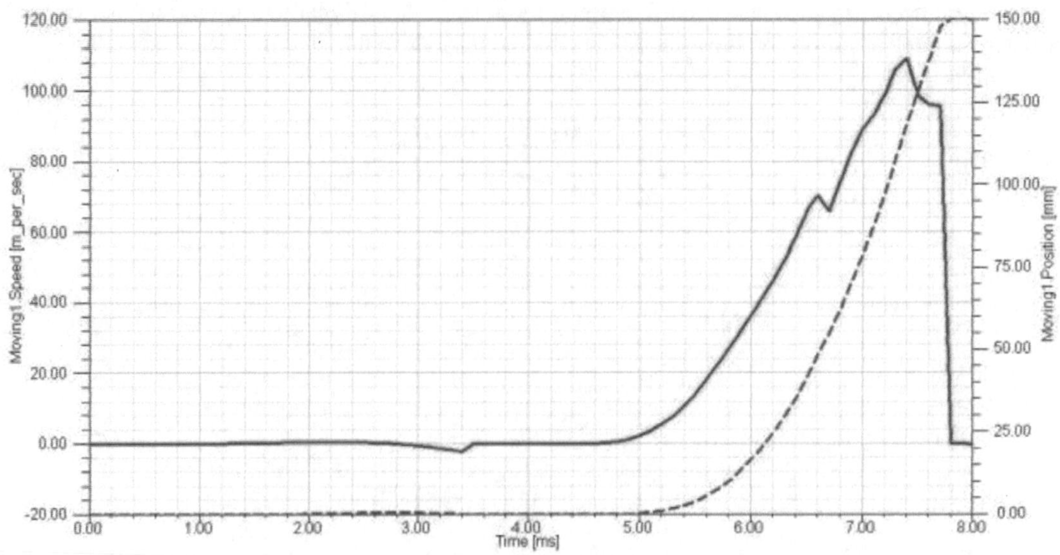

Figure 11. Time variation of speed of ring and position of ring

The time variation of speed of ring is presented with continuous line and with long dash is presented the position of ring during acceleration time. The ring left stage 1 at time 6.49 ms with speed 64.94 m/s and the stage 2 at time 7.23 ms with speed 101.96 m/s. The maximum speed 106.98 m/s is achieved at time 7.33 ms. After that time the ring is no longer accelerated and the speed drop rapidly to zero because the simulation length is 150 mm.

4. Conclusion

The multistage induction coilgun with two fluxes is better than single flux designs. The induced current in armature can be controlled much efficient because we use the coil just to induce current inside the ring with great efficiency. The magnetic field density B inside the air gap is perpendicular on ring and the Lorentz force accelerate the projectile during entire length of a stage.

We used the induction to avoid contact between armature and accelerator and to obtain a higher value of induced current than power current.

In order to create the radial magnetic field we used four coils perpendicular on armature. This approach give us the possibility to control the phase difference between induced current in armature and current in magnetic field coils in order to obtain the maximum force. The phase difference is obtained by changing the frequency of magnetic field coils power source. We do not need less electrical energy because is expensive but because we can use a small power source.

References

[1] McNab I R and Beach F C 2007 Naval railguns, *IEEE Transaction on Magnetics* **43(**1) 463-468

[2] Fair H D 2007 Progress in electromagnetic launch science and technology, *IEEE Transaction on Magnetics* **43(**1) 93-98

[3] Ghassemi M, Molayi Barsi Y and Hamedi M H 2007 Analysis of force distribution acting upon the rails and the armature and prediction of velocity with time in an electromagnetic launcher with new method, *IEEE Transaction on Magnetics* **43**(1) 132-136

[4] Balikci A, Zabar Z, Birenbaum L and Czarkowski D 2007 On the design of coilguns for super-velocity launchers, *IEEE Transaction on Magnetics* **43(**1) 107-110

[5] Wenbo L, Yu W, Zhixing G, Zhongming Y, Weirong C 2013 Connection Pattern Research and Experimental Realization of Single Stage Multipole Field Electromagnetic Launcher, *IEEE Transaction on Plasma Science* **41**(11) 3173-3179

[6] Gherman L, Pearsica M, Strimbu C and Constantinescu C G 2011 Induction coilgun based on "E-shaped" design, *IEEE Transaction on Plasma Science* **39**(2) 725-729

[7] Du Z, Zhan T, Ruan J, Huang G, Zhang Y, Yao Y, Liu K, Huang D and Guan W 2012 *Research on Electromagnetic Performance Affected by Shielding Enclosure of a Coil Launcher*, Electromagnetic Launch Technology (EML),16th International Symposium, ISBN 978-1-4673-0306-4

[8] Polzin K A, Adwar J E and Hallock A K 2013 *Optimization of Electrodynamic Energy Transfer in Coilguns with Multiple, Uncoupled Stages*, IEEE Transaction on Magnetics **49**(4) 1453-1460

IOP Conf. Series: Materials Science and Engineering **200** (2017) 012051 doi:10.1088/1757-899X/200/1/012051

The impact of collaborations between universities and private organizations on cluster development and competitiveness in Romania

D Stoicovici[1], M Bănică[1], M Ungureanu[1] and M Stoicovici[2]

[1]Technical University of Cluj Napoca, Engineering and Technology Management Department, Baia Mare, Romania
[2]Aarhus, Denmark

E-mail: dinus@cunbm.utcluj.ro

Abstract. While the European Union has put a lot of emphasis on cluster development due to their inherent advantages such as lower transaction costs, technological transfer and regional development, little is known about how clusters emerge and what can facilitate their competitiveness. This paper aims to study the impact of public-private cooperation between universities and organizations on cluster development and competitiveness. A literature review is employed to develop the model while 4 qualitative case studies provide the initial test of its validity. The analysis suggests that cooperating with research institutions impacts cluster development first through education of industrial staff, but also by developing innovation processes through the facilitation of the appearance of innovative ideas and also of knowledge sharing among organizations. The research has several implications both for organizations and for government officials. First of all, R&D and top management should actively seek to cooperate with research institutions both for training of their staff but also in seeking new ideas and as a way of collaborating with other organizations within the field without fear of losing competitive advantage. Second, government officials should try to create more incentives both for organizations (through for example tax returns) and for universities (extra funding or salary incentives) that can increase collaboration between these actors. This paper is the first one to asses empirically how cooperation with research institutions affect cluster competitiveness and development, especially within the developing region of Eastern Europe, Romania.

1. Introduction

During the enlargement of the European Union, projects were funded facilitating the integration of all states by creating cross-border areas through cooperation in industry, agriculture, education, culture, technology transfer, etc., in order to develop the component states, to equalize the level of technological advancement and the overall quality of life through sustainable and smart development.

In order to ensure good economical-social-political conditions for development, it is not enough to let natural conditions to act, but it is also necessary to ensure a top-down encouraging system of development by the means of legal and administrative rules. In this context, it was decided to facilitate the creation and development of economic clusters. These clusters lead to the development of geographical areas by the very existence of their operating system. It is generally acknowledged in the specialized literature [1] that the new form of economic organization enables, from the beginning, the rising of the competitiveness and business efficiency of all participating companies. The way that these companies manage to become competitive and to contribute to the rising of the industrial development level of their region depends on the existence of a so called "triple helix" [2].

This is all the more important as economic activities in European Union countries already take place in some economic models that lead to interaction between clusters and innovation systems of processes of generation IV and V. Until now, five generations of models of innovation processes have been defined by literature [3], [4]:

– **Functionally integrated models (generation IV):** the model is built on the integration and simultaneous development of all levels of product innovation model instead of sequentially evolution. These models involve complex business processes, feedback and mutual relations between marketing, research, development and distribution departments of innovative companies with other companies even with competitors.

– **System integrated models (generation V):** a complex overall design integration model of systems and organizations by forming innovative networks, customer units, universities, suppliers and other industrial companies and local communities, a combination of innovative technologies for solving problems of high complexity of new products by working within a network involving the use of CAD/CAM software, performance through simulation modeling, the use of rapid prototyping systems-expert, by the mean of concurrent flows that occur both in the sense of crossing streams and competitive flow between them.

The actual development levels of societies in different countries in EU can be seen as especially the fourth and fifth generations of models of innovation. So, those societies should be in a close relationship with industrial clusters in order to accelerate the development of a certain region. Therefore, problems of regional development should also be solved by increasing the innovation level in that area. A basis study of the development of clusters in the North-West of Romania was made in [5]. In these study some possible potential links between competitiveness of the cluster-type economic agglomerations and systems of innovation process in the North-West of Romania were analyzed. The idea was that, in order to accelerate the development of society, a solution would be to take advantage of the similarities between the two networks: the innovation processes and the one created by the clusters that arise.

In the frame of such similarities of networks, making more complex ties between the universities (as research entities and also as educational sources to ensure the quality of technical and economic education of the specialists working in the field of interest) and businesses (as places where innovation processes are brought to life) should be efficient in the evolution and development of that clusters.
From this point of view, this paper aims to study the impact of public-private cooperation between universities and organizations on cluster development and competitiveness. This paper is the first one to asses empirically how cooperation with research institutions affects cluster competitiveness and development, especially within the developing region of Eastern Europe, Romania.

2. Methodologies

Specific clusters incipient economic relations in North-Western Romania have emerged due to the existence of particular geographical, economic and personnel conditions in the area. To highlight that, four qualitative case studies have been deployed in order to provide the initial theory of its validity.

The study was conducted in the form of recorded interviews face-to-face with decision makers from four companies that operate in the area of Baia Mare.

One of the companies is working in the field of thermal rehabilitation of industrial and civil buildings, but has the ability to carry out activities also in the design of mechanical equipment such as processing of ores and non-ferrous metals. Another company is active in designing mechanical machines in a wide range of industries and is part of a panel of collaboration with other similar units for design and execution of equipment and services in achieving the idea of "turnkey business". The next company is a subsidiary of a European trust working as regards the production of packages. The last company, operating in the IT industry, is part of a specific cluster in the North-West of Romania.

These are enterprises with activities that can be included in the scope of research/technology development, exchange of technologies and introduction of technical progress. These units were selected for study because their areas of activity are very different, both in purpose and in type of industry, but all have preoccupations in innovation/new technologies, while having the tendency to group into the preceding-cluster economic structure type.

Interviews were conducted based on a questionnaire as in Table 1.

Table 1. The questionnaire used on the study

1	How many R&D & technical staff do you employ?
2	What type of learning mechanisms do you use? Do you scan the market for trends, and new technologies? If yes, how?
3	Why have you started cooperating with the university? How did the idea first come up? Did you think of any inherent advantages in starting to cooperate with the university?
4	How was the cooperation? What was good and what was bad?
5	Did you experience any advantages of the cooperation after the fact (more knowledgeable workers, easier to come up with new ideas, easier to get new ideas accepted in the organization)?
6	Do you work with any other firms that cooperate with universities?
7	If yes, how is the cooperation in comparison with other companies?
8	If no, why not?
9	Do you know if any of your competitors are cooperating with universities? Elaborate on what do you know about this?
10	Are you part of any business networks? Do you engage in joint R&D with other firms?
11	Do you see any agglomeration patterns in your industry? Geographically?
12	If yes, why do you think this is happen?
13	If no, what do you think can create it?
14	Do you see any way in which the government/government policies support or impede the creation of clusters in your industry?
15	Do you see any way in which the government/government policies support or impede the collaboration between firms and universities?
16	Do you attend local industry seminars? Why? Do you know of others in your organization that do?
17	Do you hire local graduates? What do you think of them in terms of: on the job knowledge, knowledge of trends in industry, ability to think innovative, ability to come up with new ideas?

3. Results

The responses were generally framed in the following topics.

1. How many R&D & technical staff do you employ?

A weak in R&D and technical staff unable the companies to absorb scientific and technical research results and technological transfer products [6], [7]. Technical staff account from few (6 persons), for about half (46 persons), to even all of the structure of employees. Most employees have technical skills and provides support activity in the field. In one case the R/D component is represented even by a research/development and training department, who ensure the dissemination of research results and the development of technical personnel.

2. *What type of learning mechanisms do you use? Do you scan the market for trends, and new technologies? If yes, how ?*

The current learning activities are summer practice for students, that contribute to diminish the gap between the theoretical knowledge acquired by students in the university and the concrete practical knowledge that companies need of their employees. Other activities are scholarships, lifelong learning, courses based on resources on-line. This can be seen as a learning mechanism due to the capacity of students to bring in new knowledge from the university realm.

Regarding the scanning the market for trends and new technologies, the current main activities are involving departments of R&D that participates at events, internal workshops, seminars, courses, conferences organized by relevant actors/technology leaders, where relevant technologies in areas of interest are presented. Also, there are cases of partnerships with already existing clusters in the region, also e-learning training, job-shops and meetings between students and businesses organized by university, etc.

Other mandatory information sources for technology transfer and introduction of new technologies are the library patent consultation of the State Office for Inventions and Trademarks (OSIM), companies catalogs, web pages (Space-Net).

In one case the research themes are coming from the headquarter of the company from abroad, where the clients are also in concordance with needs of European markets.

This showcases that these companies, even though they don't have very advanced or purposeful mechanisms in place, they do engage in learning especially through networking and hiring students temporarily.

3. Why have you started cooperating with the university? How did the idea first come up? Did you think of any inherent advantages in starting to cooperate with the university?

For one of the companies collaborating with the university as a shareholder was mandatory condition for a project with EU funding, to establish an engineering design unit commercialized for profit, to be an active network of local business units.

Also, the need for qualified personnel helped identify a new option for cooperation - organizing summer internships for students. Throughout time, the relationship with the university developed and became more complex, initiating joint research and development initiatives, with funding under the National Investment Program, and thereafter a framework for continuous cooperation in the field of fund raising for research was established.

This approach, currently in the initial stage, generates optimism because the relationships became more intense by the means of the company itself who was invited to join the Advisory Board of the University, which aims to develop cooperation with business environment and to adapt education to labor market requirements.

This showcases that companies initially decide to cooperate with universities due to a severe lack of qualified personnel and this materializes either by generally employing temporarily (internships) or permanently students or graduates from the university or by directly engaging with the university on projects [8].

4. How was the cooperation? What was good and what was bad?

The respondents described that the university showed openness to the business, and cooperation was very well conducted both at the institutional level (management) and across team activity, in that specific projects (practice, competitions, funding applications) have had support from the university management team and active involvement of academics.

There have been important actions like cross border projects based on EU funds, establishing occupational standards at national level (CAD/CAM operators), distribution of computer software products, training in computer-aided design based also on EU funds.

Cooperation in organizing summer practice sessions, granting scholarships, organizing competitions and events was also very good, and communication was fast, direct and effective.

In many cases the university have had the initiative of starting the cooperation in domains mentioned already.

One respondent insisted on an idea to institutionalize relations between universities and industrial units in order to develop the emerging structures .

One unsuccessful event was described relating to the writing of the grant application within the National Program of Innovation, due to a very late identification of opportunity and the extremely short time in which it was prepared. From this experience resulted an agreement of mechanisms for future cooperation, and to try to materialize this type of collaboration at every opportunity.

5. Did you experience any advantages of the cooperation after the fact?
The respondents agreed on an openness from academics in order to adapt course content to the needs presented by company. Scholarships offered by companies with the cooperation of university had also a positive impact on the company's image among customers and fellow companies community.

The companies also appreciated the higher-skilled are university graduates for knowing the language requirements of advanced CAD / CAM.

6. Do you work with any other firms that cooperate with universities?
7. If yes, how is the cooperation in comparison with other companies?
8. If no, why not?
Generally, the companies are in economic relations with other companies that cooperate with universities.

There is not a significant difference between companies that cooperate with the university and other companies.

9. Do you know if any of your competitors are cooperating with universities? Elaborate on what do you know about this?
Direct competitors collaborates with universities, mainly in the aspects as organizing internships, courses and competitions, obtaining financing for research and development and other projects financed with EU funds, and organizing of joint workshops and conferences .

10. Are you part of any business networks? Do you engage in joint R&D with other firms?
One of the companies interviewed is a member of an industrial cluster specific for its industry. Within the cluster there is openness to perform joint research activities, mainly through the identification of common sources of financing to cover the costs of joint research. It is possible also to attended business meetings, joint training (technical and software skills). A participation was declared also in a project funded by the European Space Agency.

Respondents are aware of the existence of collaborations between university and a variety of industrial units whether they work together or not.

All of the interviewed declare collaborations in the frame of Chamber of Commerce of the region.

11. Do you see any agglomeration patterns in your industry? Geographically?
In general, respondents were quite optimistic regarding spotting the signs of cluster formations in the region. The IT industry was presented as one of the most open structures such as cluster aggregation. In terms of geographic aggregation, today there are two relevant regional entities: *Romanian Association of Electronic and Software Industry*, and a cluster-type structure *iTechTransilvania* based in Cluj Napoca which is trying to expand its coverage. There are also preliminary discussions,

including the interest being expressed by local authorities, to build an association at Baia Mare for the IT industry.

In the design and manufacture of mechanical products industry, *TransMechWorld* was identified as an industry association, with companies from the Noth-West zone of Romania. Collaborations tend to turn into panels collaborative projects providing turnkey approach to customer strategies, etc. Collaboration on mechanical, electrical engineering and CNC machining are also existing among industrial unities in the region.

12. If yes, why do you think this is happen?
The reasons to clustering that the respondents identified were:

- ✓ Funding joint opportunities for this type of structures
- ✓ Sharing access to larger contracts or projects that exceeds the capacity of only one company
- ✓ Identifying common interests themes and support in relation to local or central authorities
- ✓ Reducing the cost of training/life learning of staff, in technical or soft skills
- ✓ Satisfying the need for belonging to the community
- ✓ Image benefits
- ✓ Friendly environment in region
- ✓ The level of development of existing relations and technical activities in the area due to the area's industrial traditions

13. If no, what do you think can create it?
Even though the companies agreed that agglomeration and clustering was taking place, they did identify some potential factors that might accelerate and enhance the process:

- ✓ Enhancing the added value
- ✓ Developing of panels of collaborative projects
- ✓ State financial and legal support to facilitate the structuring of collaborative panels of existing businesses in the area, and between them and the university.

14. Do you see any way in which the government/government policies support or impede the creation of clusters in your industry?
The respondents consider that usually the process of clustering flows naturally, supported by available funding, and does not require government intervention. Also, possible clusters artificially created by government intervention could create a division from the initial purpose, with a dilution effect on the goals.

The respondents consider that the involvement of universities as partners in clusters is beneficial because it brings a distinct vision to that of the economic actors.

One of the interviewees presented the fact that they were engaged in proactive cluster development through concrete stages: „initiating community cluster aggregation, joint training, implementation of common initiatives, research, product development, shared services in advanced stages of maturity". Some of the respondents agreed that the university can act as a catalyst in all stages on the development and evolution of a cluster.

The respondents consider also that a cluster, specifically an innovation cluster, must be oriented towards economic objectives.

Good opportunities to enlarge the players that can contributes to a cluster forming are students start-up companies, technological incubators, financing facilities. In order to ensure those opportunities, the government can play a positive role through its policies.

15. Do you see any way in which the government/government policies support or impede the collaboration between firms and universities?

The triple helix model is a theory that has proved the truth [9]. It proved to be necessary to elaborate a package of legislation to encourage the creation of economic clusters and protecting intellectual property in order to facilitate technology transfer [10].

The main policies through the government can support cooperation between businesses and universities as presented by the respondents are:

✓ Research funding toward universities that can support the needs of business.

✓ Adjustment of tax legislation to facilitate the uptake of research by graduates and businesses by providing incentives to companies who purchase research

✓ Expanding financing lines with joint projects university-companies in the type POCU operational programs at the expense of funding for mass training, which proved to be useless in the 2007-2013 period.

16. Do you attend local industry seminars? Why? Do you know of others in your organization that do?
Companies participate in events organized in the specialized fields and those organized in related fields of business meetings, seminars, workshops. Participation in such events bring benefits both in terms of image and a better understanding of potential partners, competitors, technologies and tools available, etc.

17. Do you hire local graduates? What do you think of them in terms of: on the job knowledge, knowledge of trends in industry, ability to think innovator, ability to come up with new ideas?
The company hires local graduates. Most of them have participated in internships or summer or worked part-time during their studies. Thus, when hiring, graduates have to have a minimum of practical knowledge needed in the workplace. Awareness of trends in the industry, speaking in a highly dynamic industry sectors, depends on manifests of personal interest continuous professional development. Creativity and innovative thinking ability are still visible as personal skills, less trained, practiced or encouraged in schools (universities) technical. Again, as a solution, the companies see the involvement of industries in joint programs with universities.

4. Conclusions

The analysis suggests that cooperating with research institutions/universities impacts cluster development first through education of industrial staff, but also by developing innovation processes through the facilitation of the appearance of innovative ideas and also of knowledge sharing among organizations.

Initial clustering and agglomeration occurred in the region to its geographical and economical advantages for the organizations. However, the cluster cannot survive, let alone develop without proper access to qualified employees, especially in innovation focused industries. The critical role of universities is therefore, to facilitate the access to qualified personnel through various means. From this point of view, in the opinion of interviewees, the presence of an university in the area was an essential factor for the setting up of clusters and positive development by offering a solution to the employment of divers technical and economic and managerial staff. One problem is that all businesses in the area are competing for the same specialists and the demand is greater than supply.

Another factor favoring the emergence of trends towards cooperation between enterprises on clustering is the elaboration of more complex projects, including the mentioned university research contracts, collaboration with university in accessing funding [11].

A general tendency is to consider as primary importance, in order to institutionalize relations between universities and industrial entities that work with transfer of technology, technological and science research results, is for the government to elaborate a package of necessary laws.

5. Implications

The research has several implications both for organizations and for government officials:

Implications for universities as research unit and educational one:

- ✓ Adapt the educational program to the needs of business;
- ✓ Develop a system of collaboration with business in order to access funding from EU and/or National programs for transfer of technology and science and technological research;
- ✓ Promote the innovative ideas, develop innovation processes
- ✓ Encourage knowledge sharing among firms;

Implications for business (companies):

- ✓ To search for relationship which involve financing cooperation contracts with the university
- ✓ Create joint teams to work together with the university to access funds for research and technological transfer;
- ✓ Identifying common interests themes and support in relation with other units on the business environment market.

Implications for government:

- ✓ Adjustment of tax legislation to facilitate the uptake of research by graduates and businesses by providing incentives to companies who purchase research;
- ✓ Research funding towards universities that can support the needs of business;
- ✓ State financial and legal support to facilitate the structuring of collaborative panels of existing businesses in the area, and between them and the university;

Government subsidies for economic research required to favor possible future cluster structure.

References

[1] Porter M 2003 The Economic Performance of Regions, *Regional Studies* **37**(6-7) 549-578

[2] Etzkowitz H 2008 *The triple helix: university-industry-government innovation in action,* Routledge, New York London

[3] *** 2012 *Manual Broker de Tehnologii şi Standard Ocupaţional,* ARoTT, IRECSON, Ediţia I

[4] Stoicovici D, Banica M and Ungureanu M 2008 Consideratii teoretice referitoare la identificarea potentialelor clustere regionale, *Buletin Stiintific, Seria C, Fascicola Management Tehnologic* **IV**(1) 73-78

[5] Banica M, Leganaru B and Stoicovici D 2009 *Identifying potential cluster-type economic agglomerations within Maramures county (Romania),* The 6th International Conference Management of Technological Changes, September 03-05, Vol. 2, pp 5-8

[6] Cohen W M and Levinthal A D 1989 Innovation and Learning: The Two Faces of R&D, *The Economic Journal* **397** 569 - 596

[7] Laursen K and Salter A 2004 Searching high and low: what types of firms use universities as a source of innovation? *Research Policy* **8** 1201-1215

[8] Lobonţiu M, Big R, Lobonţiu G, Coteţiu R and Ungurean N 2008 *Difuzia tehnologică. De la inovare la transferul şi difuzia tehnologiei*, Editura Limes, Cluj Napoca

[9] Klimova N, Kozyrev O and Babkin E 2016 *Innovation in Clusters. Understanding Universities, Special Economic Zones, and Modeling*, Springer

[10] Ungureanu M and Nasui V 2007 *The technological transfer, solution in the development of small and medium sized enterprises in the north-west region of Romania,* 12th International Workshop on Systems and Technologies, DAAAM, Zilina, October 18, pp 155-158

[11] Yuzhuo C 2014 *Implementing the triple helix model in a non-western context: an institutional logic's perspective,* Triple Helix, Springer

Simulation of electrical and thermal fields in a multimode microwave oven using software written in C++

C Abrudean

Politehnica University of Timisoara, Department of Electrical Engineering and Industrial Informatics, Revolutiei str., no. 5, 331128 Hunedoara, Romania

E-mail: cristian.abrudean@upt.ro

Abstract. Due to multiple reflexions on walls, the electromagnetic field in a multimode microwave oven is difficult to estimate analytically. This paper presents a C++ program that calculates the electromagnetic field in a resonating cavity with an absorbing payload, uses the result to calculate heating in the payload taking its properties into account and then repeats. This results in a simulation of microwave heating, including phenomena like thermal runaway. The program is multithreaded to make use of today's common multiprocessor/multicore computers.

1. Introduction

This paper presents the algorithms used and results obtained using a C++ simulation program developed by the author. For experimental validation a thermal imaging (FLIR) camera was used to indirectly measure the electromagnetic field in a multimode microwave oven's payload.

As in any simulation, absolute precision in simulating physical phenomena is unfeasible and some reasonable approximations must be used. For example in our case, electromagnetic field dissipation is considered to take place exclusively in the payload - which is a lossy dielectric - and not inside the cavity which may contain water vapor or cavity walls which are imperfect conductors. Also, since microwave heating is fast and temperature differences relatively small, thermal convection and radiation was ignored. All heat transfer is considered to take place inside the payload and only by conduction.

2. Principle formulation and algorithm descriptions

The electromagnetic field is calculated by the program using a variant of the finite difference time domain method (FDTD), developed by Kane S. Yee [1-4]. This is a method for 3D approximation of Maxwell's equations solutions using centered differences. Electric and magnetic fields intensities are interleaved in both space and time.

There are six field components in each Yee cell (Figure 1). The finite difference equations for the x axis are like:

$$H^{n+1/2}_{x(i,j,k)} = H^{n-1/2}_{x(i,j,k)} + \frac{\Delta t}{\mu \Delta z}\left(E^{n}_{y(i,j,k)} - E^{n}_{y(i,j,k-1)}\right) - \frac{\Delta t}{\mu \Delta y}\left(E^{n}_{z(i,j,k)} - E^{n}_{z(i,j-1,k)}\right) \quad (1)$$

$$E^{n+1}_{x(i,j,k)} = E^{n}_{x(i,j,k)}\left(1 - \frac{\sigma \Delta t}{\varepsilon}\right) + \frac{\Delta t}{\varepsilon \Delta y}\left(H^{n+1/2}_{z(i,j+1,k)} - H^{n+1/2}_{z(i,j,k)}\right) - \frac{\Delta t}{\varepsilon \Delta z}\left(H^{n+1/2}_{y(i,j,k+1)} - H^{n+1/2}_{y(i,j,k)}\right) \quad (2)$$

where μ is the absolute magnetic permeability and ε the absolute real electrical permittivity $\varepsilon = \varepsilon_0 \varepsilon'$. The electrical conductivity σ is specific for conductors, in our case - lossy dielectrics - will consider $\sigma = \omega \varepsilon_0 \varepsilon''$. For the other axes equations are similar.

Exponent notations like n, n+1 and n+1/2 stand for time iteration. Values Δx, Δy, Δz are dimensions of the rectangular parallelepiped and Δt is the time step.

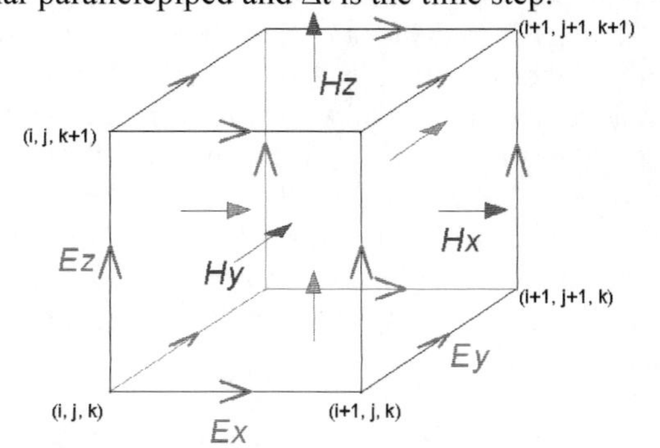

Figure 1. Yee cell with electric and magnetic fields

For model stability spatial dimensions of the cell must be less than a tenth of the wavelength in the considered material, taking into account that this wavelength is smaller in the payload. The time step must be less than 1/10 of microwave's period [3].

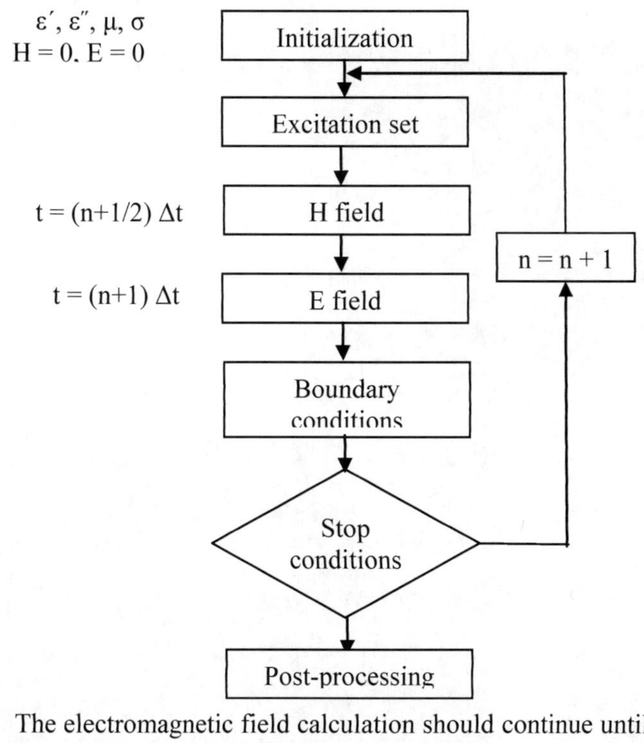

Figure 2. FDTD flowchart for electromagnetic field calculation

The electromagnetic field calculation should continue until a steady state or a maximum number of iterations is reached, in the latter case the conclusion is that the model is not stable and cell dimensions and/or time step were too large and must be reduced.

Because the heating process is many orders of magnitude slower than EM field stabilization, the latter could be safely considered instantaneous. One time iteration for EM field at 2,45 GHz must be less than $4 \cdot 10^{-11}$ seconds; for heating a time step in the order of seconds is normal. But since electrical characteristics of the payload material are dependent on temperature, EM recalculation must be done after each temperature iteration. Thermal processes simulation algorithms are described in [5], [6].

3. Program description, simulation results and experimental validation

The C++ program interface is minimal and most parameters are hard coded, that is written directly into code and changing them implies recompiling the program. This is a CPU intensive application which takes a long time to complete, and such programs are usually designed to run in background, allowing other tasks to take priority. Since most computers today are multiprocessor and/or multicore, calculations must be done in multiple concurrent threads to take advantage of this.

The usual multithreading strategy is to create a number of worker threads that is twice the number of processors/cores and to allow the operating system to make thread-to-processor allocation. This program uses a different technique: the number of worker threads is equal to the number of processors and each is tied to a specific processor using single bit thread affinity masks. A controller thread, which is much less CPU intensive, is feeding small tasks - which could be part of E, H or temperature iterations - to a blocking queue. The worker threads take these tasks from the queue if available, waiting if not. Worker thread priority is set to less than normal, but not idle (minimal).

The 3D graphs resulted in this program are drawn with a software library also designed by the author which relies on OpenGL.

3.1. Simulation results

In Figure 3 is shown the electrical field intensity as color gradient in a transitory state. Blue corresponds to zero and red to maximum with green in between, but it is a logarithmic scale because of wide differences in values. This is a middle section that includes both the wave guide and payload, and both are visible in the image. Because high intensities in the wave guide lead to a "blinding" effect, it will be omitted in the next images.

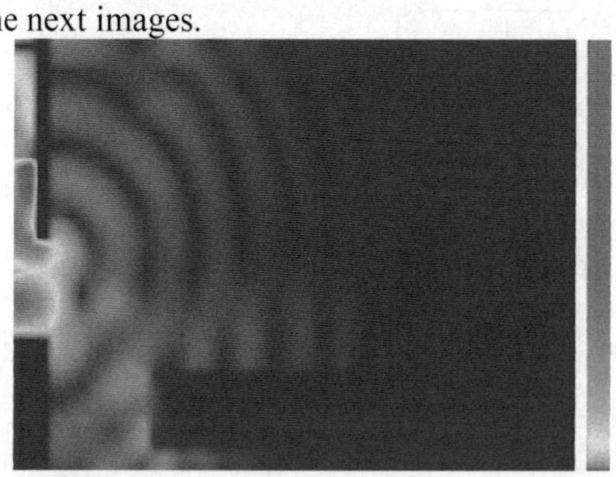

Figure 3. EM transitory state after 500 iterations

Figure 4 shows a steady state without the wave guide.

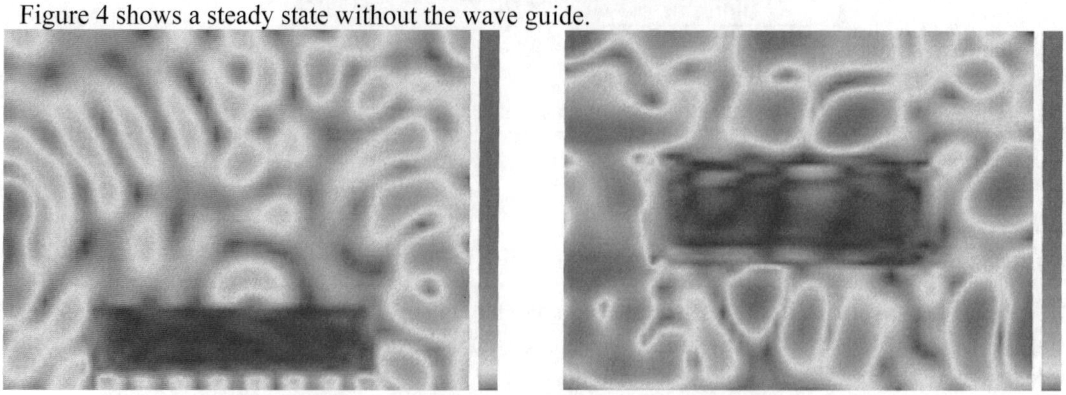

Figure 4. EM steady state from side (left) and above (right)

The following graphs are done for a 200x100x40 mm payload, with ε'=4 and ε"=1 at 0°C, density ρ = 830 Kg/m^3, specific heat capacity 1670 J/KgK, thermal conductivity 1,7 W/mK. Dimensions are scaled to interval [0, 1], the point with coordinates (0, 0, 1) being closest to the cavity feed. The maximum temperature in the payload was 265°C after 120 seconds at 900W microwave power.

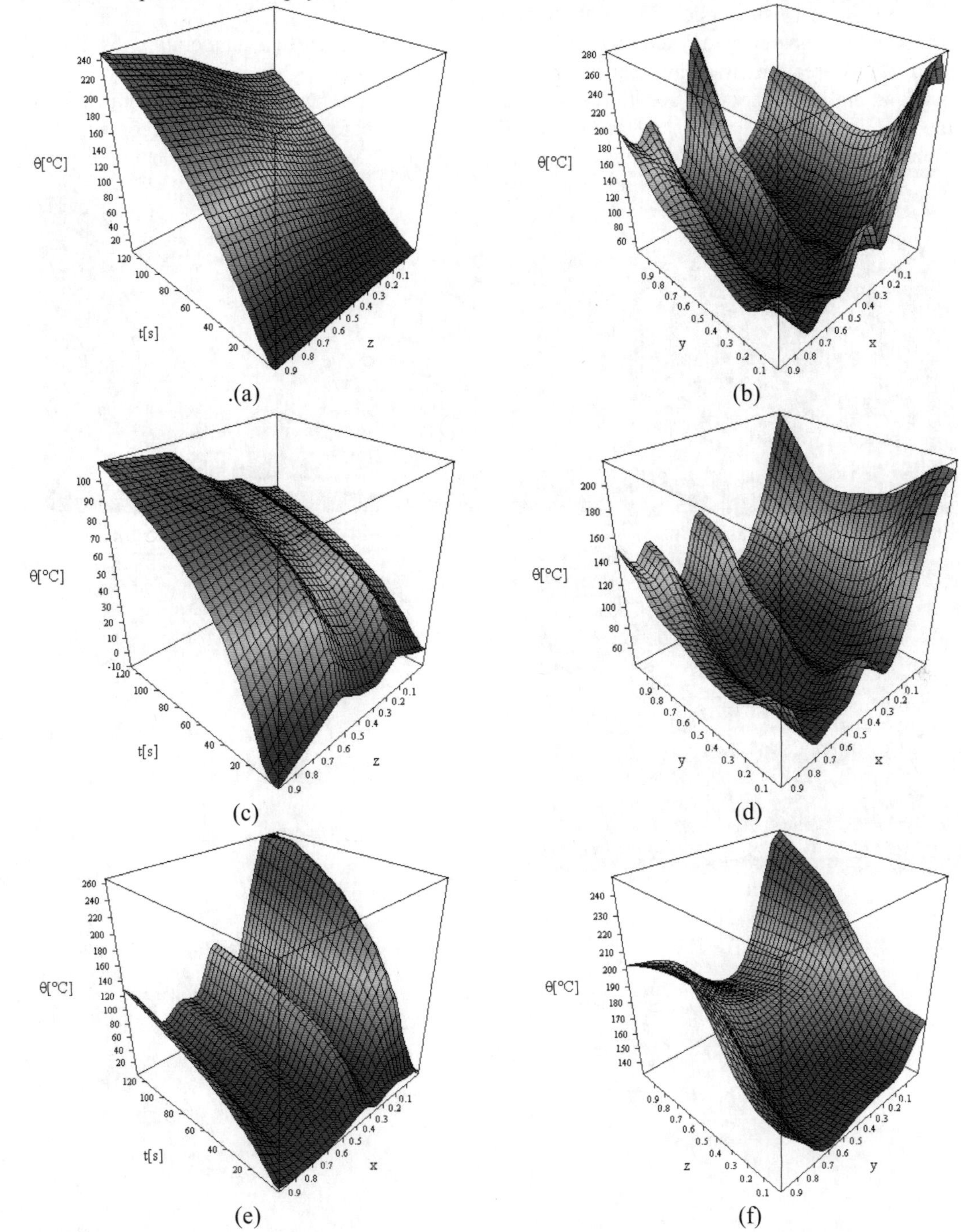

Figure 5. Temperature as function of time (a, c, e) and coordinates at end of simulation (b, d, f)

3.2. Experimental validation

The microwave field is never measured directly in a resonating cavity, such measurements are always done by heating a payload with well known characteristics. To validate the model described above a payload made from dry pinewood was exposed to microwaves for a determined time. The oven used was a commercial Vortex MG8021TP AP with 800W output power. Payload characteristics are: electric permittivity ε' = 2, loss angle tangent tg δ = 0.1, density ρ = 380 kg/m^3, specific heat capacity c = 2200 J/kgK, thermal conductivity λ = 0,16 W/mK.

A thermal imaging camera type Fluke Ti-25 was used to measure surface temperature distributions. Measurement were done by taking snapshots with the oven door open when possible, or by extracting the payload and then shooting quickly enough to avoid cooling. The camera takes pictures in both visible and infrared spectrums and saves specific, proprietary format files which contain all the information. Its software, SmartView, is capable of cropping and saving the relevant data in text files.

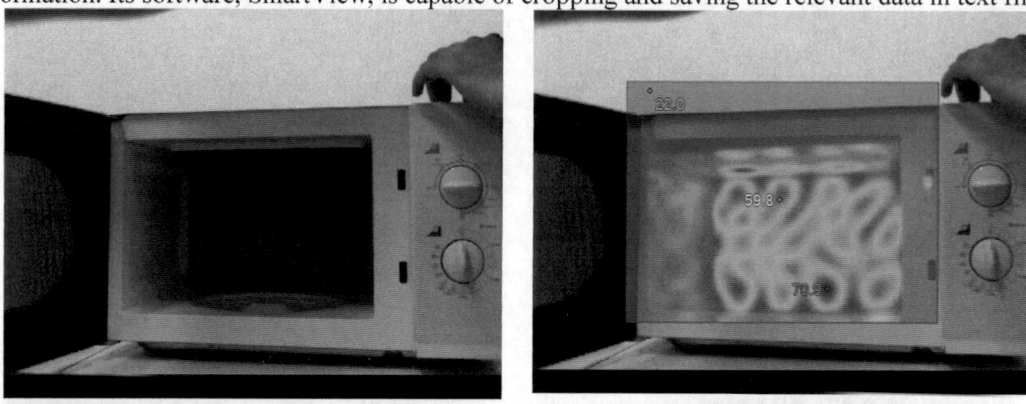

Figure 6. Images taken by Fluke Ti-25 camera, visible spectrum (left) and overlay (right)

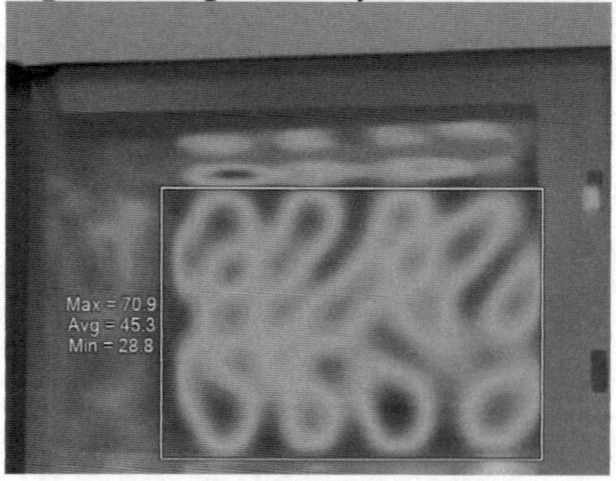

(a)

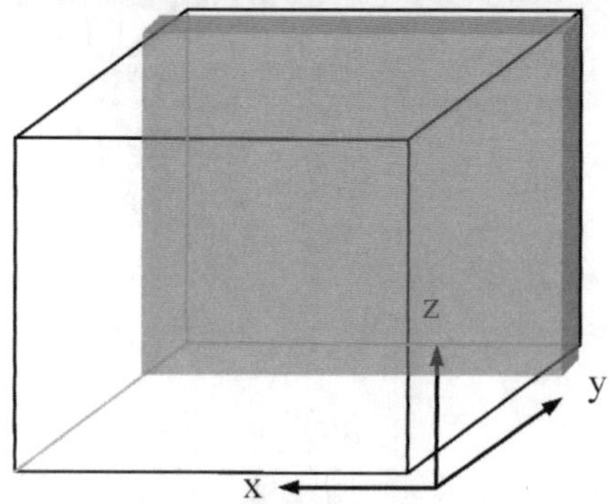

(b)

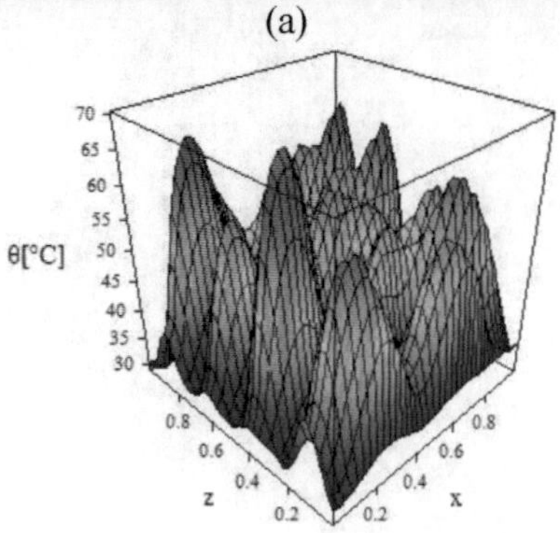

(c)

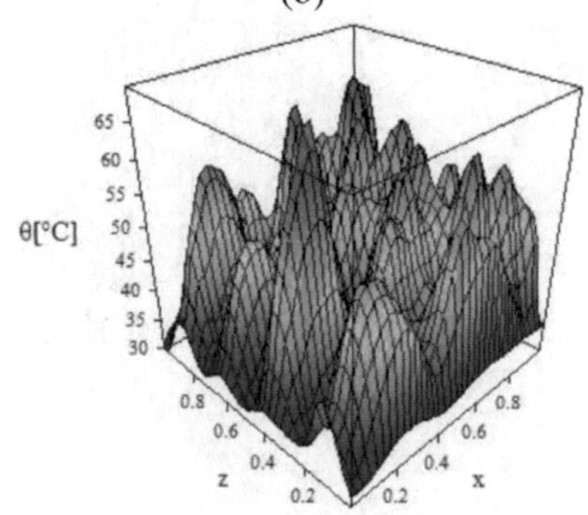

(d)

Figure 7. Thermal image (a), payload position (b), experimental temperature distribution after 60 seconds (c) and simulation temperatures (d)

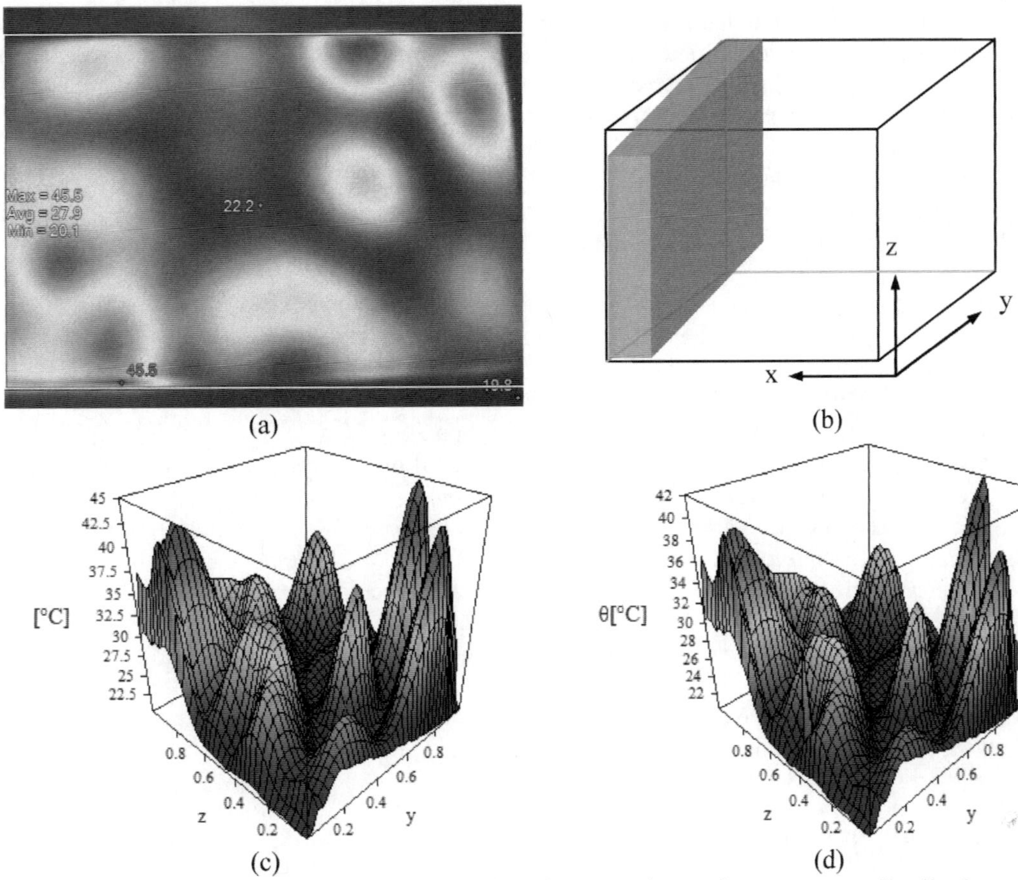

Figure 8. Thermal image (a), payload position (b), experimental temperature distribution after 30 seconds (c) and simulation temperatures (d)

The resonating cavity feed is at x=0 and with y and z coordinates at 50%.

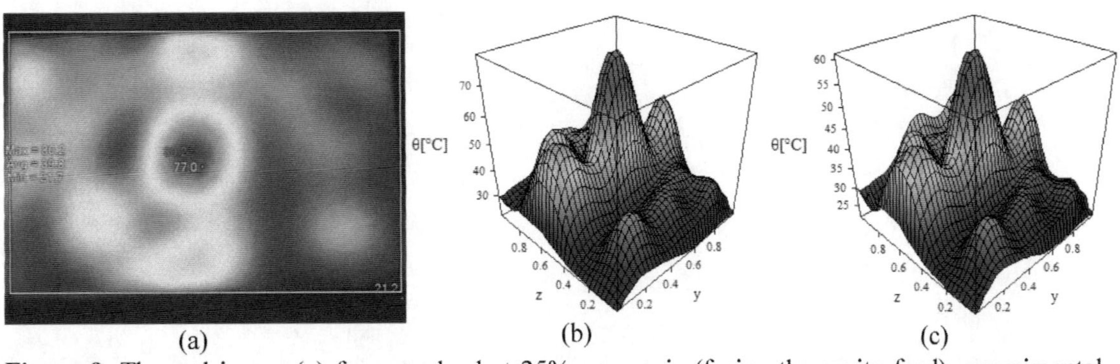

Figure 9. Thermal image (a) for a payload at 25% on x axis (facing the cavity feed), experimental temperature distribution after 60 seconds (b) and simulation temperatures (c)

4. Conclusions

The simulation methods presented here are relatively simple from a mathematical point of view, less sophisticated than frequency domain or finite element methods but easier to implement. Advanced programming techniques led to program execution times on the order of minutes.

From the simulations resulted that the resonating cavity – payload system reaches steady state after a few thousand periods of the excitation feed. With normal sized payloads this is no more than 5000 periods, which means approximately 2 microseconds. Because the magnetron's response time is also very short we can consider that applying anode voltage on an already warmed magnetron leads instantly to a steady state in the cavity. Which means that pulse modulation cannot be used to limit electric field intensity, in order to avoid unwanted phenomena like dielectric breakdown when the payload is too small and/or non-dissipating. To do so would require using pulse frequencies on the order of MHz but that will cause electromagnetic compatibility problems.

References

[1] Metaxas A C 1996 *Foundations of Electroheat: a unified approach*, John Wiley & Sons Ltd., Chichester, England

[2] Metaxas A C 2001 *The use of modeling in RF and microwave heating*, HIS–01, Heating by Internal Sources – Padua 2001, September 12-14, pp 319-328

[3] Dibben D C and Metaxas A C 1996 Time domain finite element analysis of multimode microwave applicators, *IEEE Transactions of Magnetics* **32**(3) 942-945

[4] Abrudean C, Pănoiu M, Pănoiu C and Şora I 2009 *Using Finite Difference Time Domain Method to Simulate the Electromagnetic Field in a Multimode Microwave Oven*, Proceedings of the 13th WSEAS International Conference on Computers, Rhodes, Greece, July 23-25

[5] Abrudean C 2013 Numerical Simulation of Phase Transitions in Heating or Refrigeration Processes, *AIP Conference Proceedings* **1558** 1329-1332

[6] Abrudean C 2014 *Research on efficient electrothermal equipments and processes in microwave field*, Politehnica University of Timisoara, Romania, Doctoral Thesis

IOP Conf. Series: Materials Science and Engineering **200** (2017) 012053 doi:10.1088/1757-899X/200/1/012053

Integration of disabled people in an automated work process

C K Jalba[1], A Muminovic[1], S Epple[2], C Barz[3] and V Nasui[3]

[1]Ulm University of Applied Sciences, Production Engineering and Production Economics, Prittwitzstraße 10, 89075, Ulm, Germany
[2]Kempten University of Applied Sciences, Faculty of Electrical Engineering, Bahnhofstraße 61, 87435 Kempten, Germany
[3]Technical University of Cluj-Napoca, North University Centre of Baia Mare, Victor Babes str., no. 62A, 430083 Baia Mare, Romania

E-mail: klausjalba@t-online.de

Abstract. Automation processes enter more and more into all areas of life and production. Especially people with disabilities can hardly keep step with this change. In sheltered workshops in Germany people with physical and mental disabilities get help with much dedication, to be integrated into the work processes. This work shows that cooperation between disabled people and industrial robots by means of industrial image processing can successfully result in the production of highly complex products. Here is described how high-pressure hydraulic pumps are assembled by people with disabilities in cooperation with industrial robots in a sheltered workshop. After the assembly process, the pumps are checked for leaks at very high pressures in a completely automated process.

1. Introduction

A Lebenshilfe Donau-Iller e. V. is a private welfare organization from Ulm, in Germany with about 1300 members. The organization takes care of more than 1200 handicapped people from the Ulm area.

Figure 1. Wood products made by the life support [1]

IOP Conf. Series: Materials Science and Engineering **200** (2017) 012053 doi:10.1088/1757-899X/200/1/012053

The main objectives of the Lebenshilfe are: life coaching for people with disabilities, the welfare of people with disabilities and their families, human dignity, protection of personality, integration into society [1].

The organization employs around 1000 people at several locations in Ulm and around Ulm, namely Neu-Ulm, Senden, Illertissen, Günzburg, Blaustein and Amstetten.

The life support is financed by donations and membership fees. Another source of income is the production of various products for the free economy. They produce, for example, from simple products such as wood objects (Figure 1), or garden articles to highly complex products for dental technology or industry. Among them are well-known customers such as LIEBHERR (Figure 2 and 3) or BOSCH-REXROTH [1].

Figure 2. Liebherr excavator

Figure 3. Liebherr Hydraulic System

The association needs these financial resources, because on the one hand the care of people with disabilities needs a lot of time and is labor intensive. On the other hand, people with disabilities should be integrated as well as possible into the social life. If necessary, they are provided with accommodation, they are transported between their homes and the workplaces with company-owned vehicles, and they receive food during the working hours. They are also compensated for their work financially.

The workplaces have to be also adapted to the needs of disabled people [2].

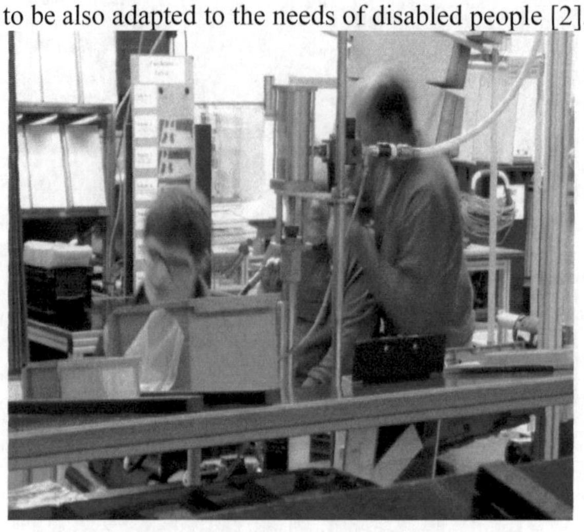

Figure 4. Workplaces for handicapped workers

Here are some examples of the conversion of a workplace into a workplace for disabled people (Figure 4) one-hand operation for an amputated worker, for blind people the equipment of a computer workplace with a Braille line, in the case of visually impaired employees, a large-screen monitor or an extra-large keyboard, in the case of deaf people, picture and writing telephones or light signal systems on machines [2].

Each work process must be meticulously planned and is prepared in detail by the supervisors. Disabled people are also not allowed to be used anywhere in the production facilities and their rest periods are regulated differently than normal workers. For example, they may not work for more than 6 hours a day, and after about half an hour they have to take short breaks.

This way of working requires a great deal of commitment and understanding on the part of the supervisors. Above all, the production technicians have to continually reinvent special tools and ergonomic working equipment in order to prevent accidents, during the products must be produced with normal industrial standards.

2. Production of hydraulic pumps

In this context, the University of Applied Sciences Ulm (Hochschule Ulm) was asked whether it would not be possible to automate some workflows at a production site in the field of life support. It was specifically the production of hydraulic pumps for the company BOSCH – REXROTH AG.

BOSCH-REXROTH AG from Unterelchingen by Neu-Ulm, formerly known as HYDROMATIK, is one of the market-leading manufacturers for high-pressure hydraulic pumps for the construction industry, agriculture and conveyor technology (Figure 5) [3].

Figure 5. Bosch-Rexroth AG in Unterelchingen by Ulm [3]

These pumps (Figure 6) have been produced for Lebenshilfelife e.V. in Neu-Ulm for about 20 years, without errors and other complaints. For this purpose was set up by Lebenshilfe in Neu-Ulm a dedicated production line (Figure 7).

IOP Conf. Series: Materials Science and Engineering **200** (2017) 012053 doi:10.1088/1757-899X/200/1/012053

Figure 6. Hydraulic pumps produced by Bosch Rexroth AG [3]

Figure 7. Production line for hydraulic pumps

The production line is designed in such a way that it can be adapted to the respective requirements (Figure 8) regarding product and production safety, or any changes.

Most of the work steps can be carried out by disabled people. However, there are some work sequences that cannot be executed by them (Figure 9).

Figure 8. Assembly line

Figure 9. Using the torque wrench

In the upcoming project, the main focus was to find out where automation processes are necessary and how they can be integrated into the production flow. For this reason, the production processes were first studied and then the products were analyzed throughout the entire production process. Finally, it was possible to determine where automation processes could be introduced and could be adapted in the production [4], [5].

At the beginning of the assembly components are provided by employees without disabilities (Figure 10 and 11).

Figure 10. Pump housing and shaft

Figure 11. Sealing rings and ball bearings

In the next steps the assembly takes place. This includes steps that require a more complex workload. One of these steps is, for example, the pressing-in of the ball bearings. This work step requires very good physical skills. It can be seen in the picture, eyes, hands and feet must be used together. This work step cannot be performed by people with physical disabilities (Figure 12 and 13).

Figure 12. Pressing-in ball bearings

Figure 13. Pressing ball bearings, assembling housing

During this operation, the almost finished hydraulic pump is placed on the hydraulic press by the assembly line. Because the hydraulic pumps, depending on their size, can be very heavy, it is very difficult to lift them permanently, even for people without disabilities (Figure 14).

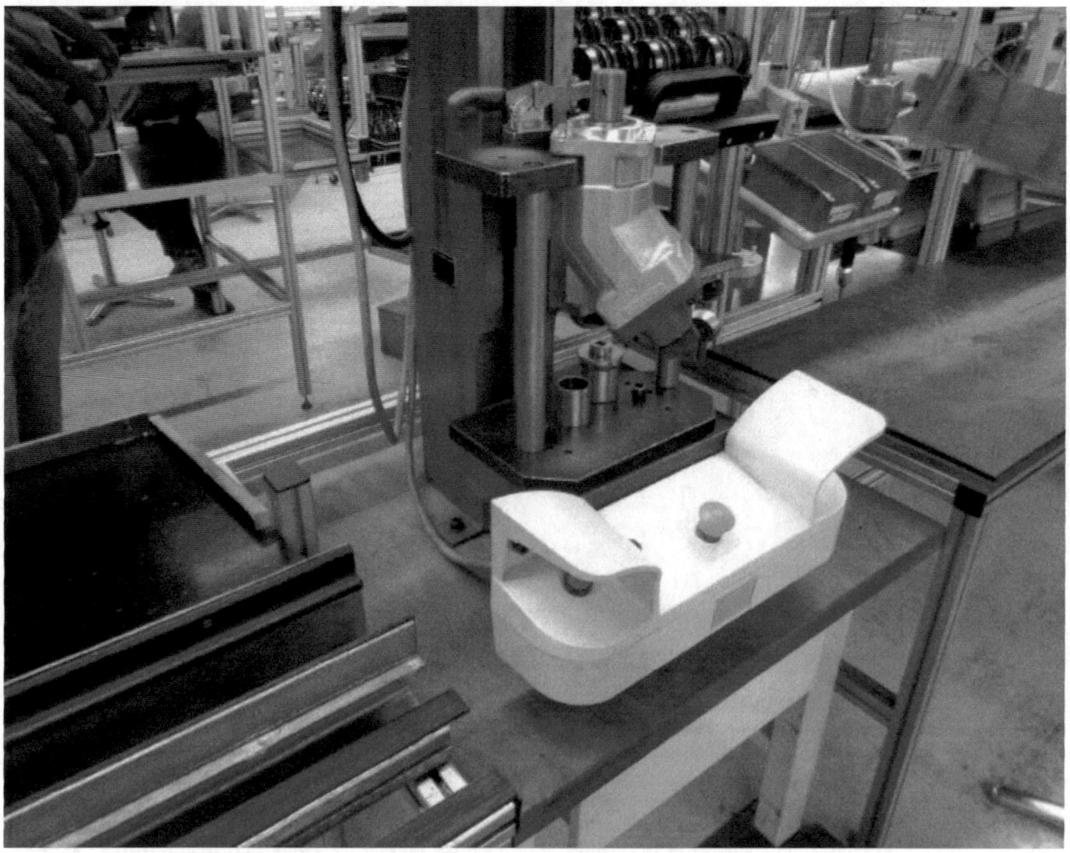

Figure 14. Final assembly

A further challenge was the test bench (Figure 15, 18), where the pumps were checked (Figure 19) under very high pressures (180 – 300 bar) for tightness (Figure 20). In this work section, employees of BOSCH-REXROTH AG were mostly employed because this work could not be trusted to people with disabilities [3].

Figure 15. Test bench

The workplace requires specialist knowledge, physical fitness and craftsmanship. This workplace is not only physically very stressful, but also loud and dangerous [16]. Here, liquids must be handled at very high pressures. At the same time, the final inspection for each product must be recorded (Figure 16).

Figure 16. Monitoring test results

Figure 17. High pressure test

Figure 18. Handling with all senses

Figure 19. Checked hydraulic pumps

Figure 20. Checking under high pressure

If the work on the test bench could be solved differently, the entire value added for the assembly of the hydraulic pumps would remain by Lebenshilfe e.V. and thus the product could be handed over from one hand to the order.

3. Cooperation project

With the help of these examples and the collected data, the experiments were started at the University of Applied Sciences in Ulm in the laboratory for robotics and industrial image processing [6], [7].

One of the first attempts was to classify the individual components for the construction of the hydraulic pumps. Because different pump types are assembled, the individual parts for the respective series have to be identified [8], [9]. This work step was carried out by coding the individual work pieces with barcodes. The encoded components were identified by industrial image processing and the results were recorded in quality assurance files (Figure 21) [10-12].

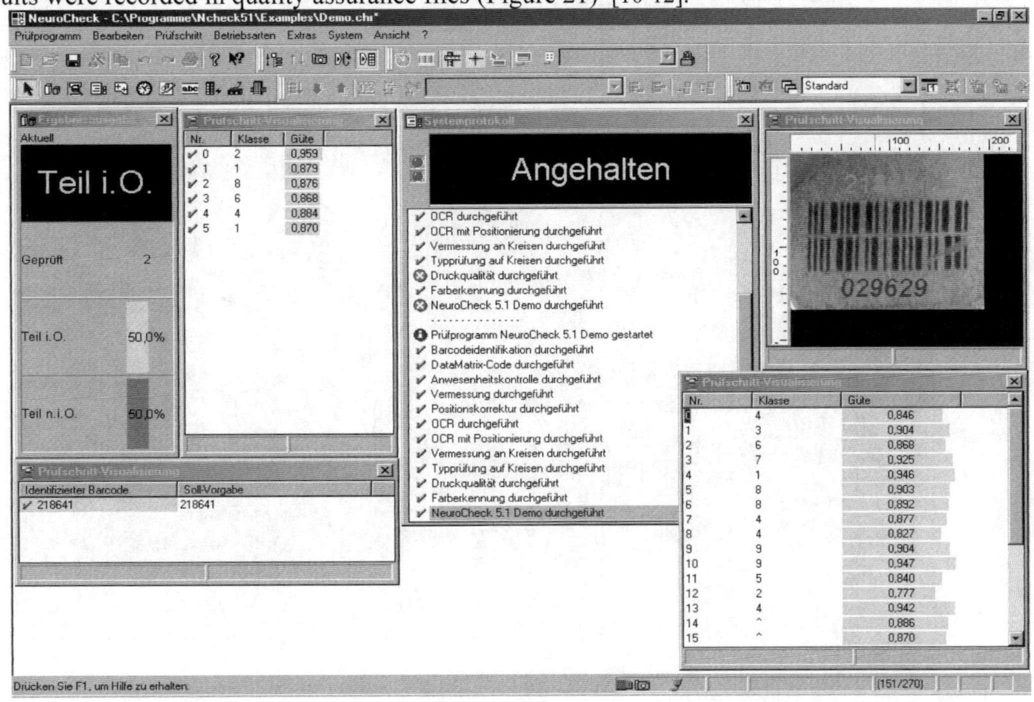

Figure 21. Bar code identification with image recognition

Another step in the process was the recognition of the rotational position (Figure 25) of the multi-shafts (Figure 23), which had to be fitted precisely into the pump bearing. For this purpose, was used a Stäubli robot, [13] which recognized the contours of the waves using a camera system [14], [15] and image processing (Figure 22) [16].

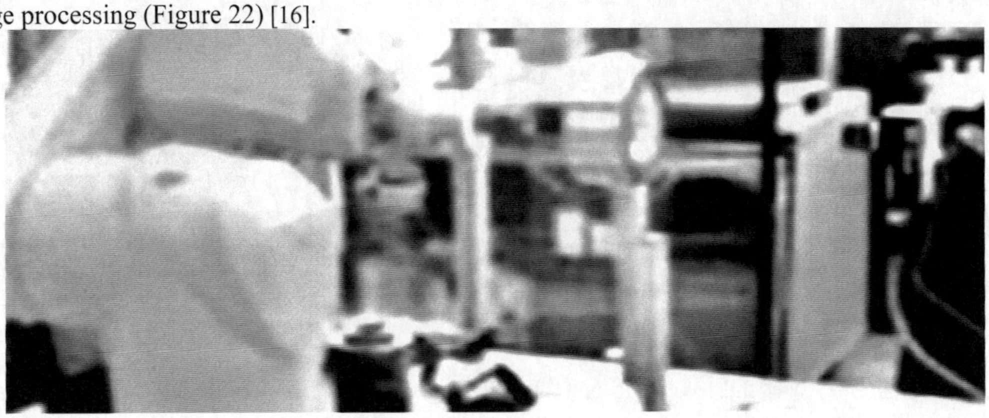

Figure 22. Measuring with robot and camera

The robot had to remove the multi-shafts from a holder and lead to a wing light (Figure 24). A camera took a snapshot and sent a picture to an image recognition program on a connected PC [7], [17], [18].

The image processing had to compare the contours of the wave with a pattern image stored in the working memory. If all features, such as area, number and position of the notches, diameter, etc., coincided with the deposited patterns, the rotational position still had to be detected. Subsequently, the required rotation angle correction was performed on the sixth axis of the robot arm. The shaft could then be guided into the bearing by means of robots.

Figure 23. Bearing and shaft

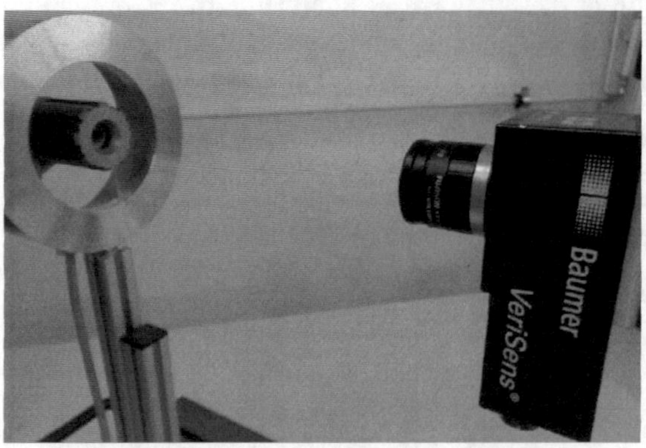

Figure 24. Position recognition

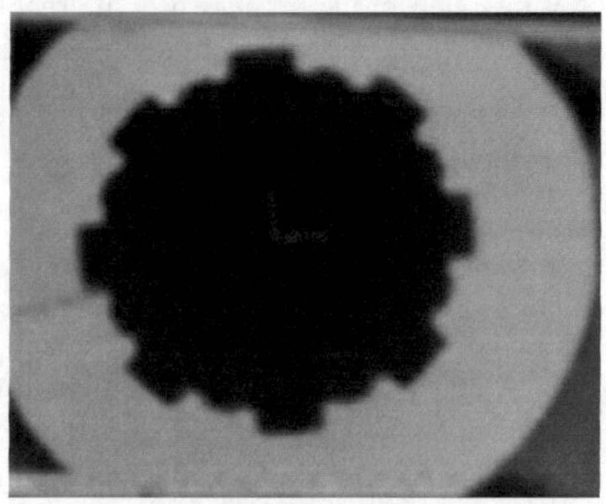

Figure 25. Rotary position

Another challenge was the hydraulic test. At the beginning, we had only one KUKA industrial robot [19] with a load capacity of 15 kg (Figure 26). This load was not sufficient for all pump types. For this reason, some things had to be improvised until the acquisition of a larger industrial robot. The robot arm by KUKA KRC 15 could e.g. under maximum load, cannot be fully extended, or certain tilting movements could not be carried out. After the acquisition of a KUKA KRC 30 (Figure 27) the robot could be finally programmed for all the movements required for the pressure test. With the new robot and with a larger gripper, the larger pump types could also be handled without any problems.

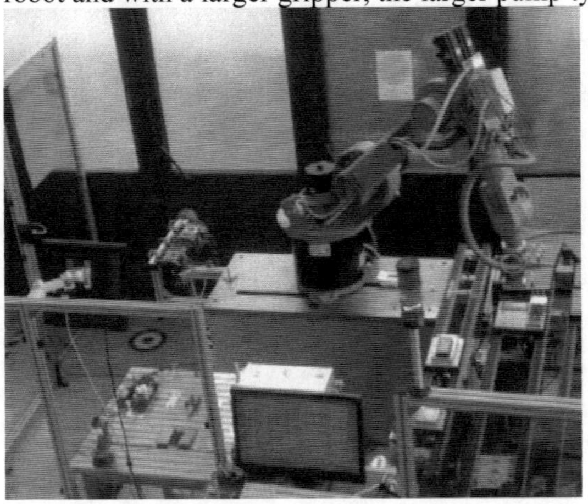

Figure 26. KUKA KRC 15/2

Figure 27. KUKA KRC 30/4

4. Conclusions

Laboratory experiments have shown that a division of labor between disabled people and industrial robots is possible. The cost-effectiveness of this type of cooperation has to be examined on a case-by-case basis.

The costs of acquisition for industrial robots and industrial image processing are relatively high. The investment can be worthwhile if the skilled workers are more likely to be used for the preparation of jobs and to supervise disabled workers rather than for monotonous and dangerous tasks.

With intelligent robot protection devices, it is further possible to allow robots to work together with disabled workers.

References

[1] ***Lebenshilhe Donau - Iller e.V., Eberhard-Finckh-Straße 30, D - 89075 Ulm http://www.lebenshilfe-donau-iller.de

[2] ***Schwerbehindertengesetz (SchwbG) *Gesetz zur Sicherung der Eingliederung Schwerbehinderter in Arbeit*, Beruf und Gesellschaft, http://behinderung.org/gesetze/schwbg.htm

[3] ****Bosch Rexroth AG*, Oberelchingen, Glockeraustraße 2, 89275 Elchingen, Germany, http://www.bvs-cnc.de/produkte/rexrothindramat.html?gclid=CLyt76C6x80CFeUV0wodHWMK6A

[4] Deman C, Streicher-Abel B and Waszkewitz P 2011 *Industrielle Bildverarbeitung Wie optische Qualitätskontrolle wirklich funktioniert*, Springer Verlag Berlin Heidelberg

[5] Bauer N (Hrsg.) 2008 *Handbuch zur Industriellen Bildverarbeitung. Qualitätssicherung in der Praxis*. 2, Auflage, Fraunhofer IRB-Verlag, Stuttgart

[6] ***Hochschule Ulm, Prittwitzstraße 10, 89075 Ulm, Germany, https://studium.hs-ulm.de/de/Seiten/Homepage.aspx

[7] Demant C and Streicher-Abel B 2011 *Industrielle Bildverarbeitung*, 3, Auflage

[8] ***NeuroCheck GmbH, Aldingen, Neckarstraße 76/1, 71686 Remseck am Neckar http://www.neurocheck.de/

[9] Jähne B 2005 *Digitale Bildverarbeitung. 6. überarbeitete und erweiterte Auflage*, Springer, Berlin

[10] ***NeuroCheck GmbH, Engineering Center, Neckarstr. 76/1, D-71686 Remseck, Germany

http://www.neurocheck.de/

[11] ***NEURO CHECK Industrielle Bildverarbeitung, System Schulung, Version 5.1, 2003, NeuroCheck GmbH

[12] ***NEURO CHECK Industrielle Bildverarbeitung, Einsteller-Handbuch, Version 5.1, 2002, NeuroCheck GmbH

[13] ****Stäubli Bayreuth GmbH*, Theodor-Schmidt-Straße 19, 95448 Bayreuth, Germany, http://www.staubli.com/index.php?id=22&L=2

[14] ****Baumer GmbH*, Pfingstweide 28, 61169 Friedberg, Deutschland http://www.kaiser-fototechnik.de/de/produkte/2_1_sortiment.asp?w=381

[15] ***Kaiser Fototechnik GmbH & Co.KG, Im Krötenteich 2,74722 Buchen http://www.kaiser-fototechnik.de/de/produkte/2_1_sortiment.asp?w=381

[16] ***FusionSystems *GmbH, Altchemnitz, Annaberger Str. 240, 09125 Chemnitz* http://www.fusionsystems.de/anwendungsberichte.html

[17] ***NeuroCheck Einführung, Extras Springer

[18] ***NeuroCheck Getting Started, Extras Springer

[19] ****KUKA AG*, Zugspitzstraße 140, 86165 Augsburg, Germany, http://www.kuka-robotics.com/germany/de/company/

Optical Gripper

C K Jalba[1], A Muminovic[1], S Epple[2], C Barz[3] and V Nasui[3]

[1]Ulm University of Applied Sciences, Production Engineering and Production Economics, Prittwitzstraße 10, 89075, Ulm, Germany
[2]Kempten University of Applied Sciences, Faculty of Electrical Engineering, Bahnhofstraße 61, 87435 Kempten, Germany
[3]Technical University of Cluj-Napoca, North University Centre of Baia Mare, Victor Babes str., no. 62A, 430083 Baia Mare, Romania

E-mail: klausjalba@t-online.de

Abstract. With increasing automation, many work processes become more and more complex. Most technical products can no longer be developed and manufactured by a single department. They are often the product of different divisions and require cooperation from different specialist areas. For example, in the Western world, a simple coffee maker is no longer so much in demand. If the buyer has the possibility to choose between a simple coffee maker and a coffee machine with very complex functions, the choice will probably fall to the more complex variant. Technical progress also applies to other technical products, such as grippers and manipulators. In this paper, it is shown how grasping processes can be redefined and developed with interdisciplinary technical approaches. Both conventional and latest developments in mechanical engineering, production technology, mechatronics and sensor technology will be considered.

1. Introduction

Automation processes are used to transfer work from people to machines. As a result of technical progress, with the increasing use of computers and microchips, the applications become more complex, more intelligent and safer for the user. Automation, however, requires more and more peripheral devices, especially in the technical processes. These devices are again becoming more intelligent when they communicate with the environment via sensors or "perceive" the environment.

Also, the application areas become by automation more complex, so today in many technological processes not only machines but systems are used. For example, the highly complex production lines, where the heavy and monotonous human work is largely taken over by robots. The industrial robots are integrated in production in such a way that they communicate with other robots via sensors and control computers.

The control in the automated production processes, where industrial robots are used, is done in most cases via PLC, industrial control computers, camera systems and built-in sensors. An important component of robotics is the gripper technology. The industrial robots in particular need special grippers and gripper systems depending on the application and application area.

1.1. Applications

With the aid of industrial robots (IR), for example, assembly work is carried out or workplaces are machined. It is used in the production as a montage robot, for gluing and sealing, roller folding, for loading machines, for palletizing, stacking, packaging, varnishing, milling, sawing, welding, grinding or the IR is used for measurement tasks (Figure 1, 2, 3, 4, 5).

Robots with special grippers in the automotive industry: [1]

Figure 1. Switching pins

Figure 2. A smaller type of an intercept pendulum

Figure 3. Palletizing robots in the food industry

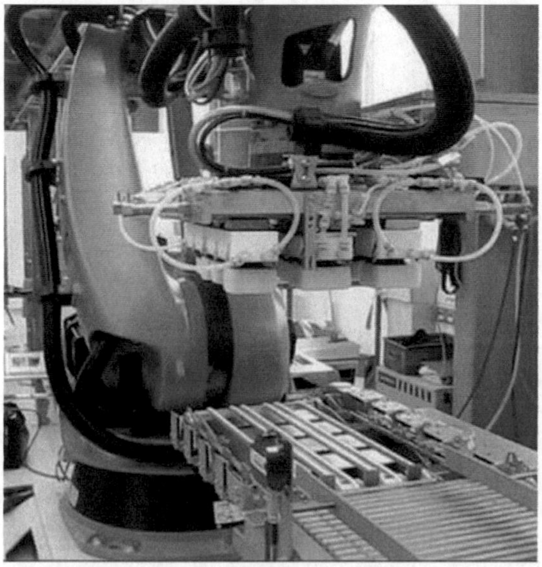

Figure 4. Special gripper of a robot in the pharmaceutical industry

Figure 5. Material handling with hydraulic gripper system

2. Robot grippers

In order for the robot arm to be able to lift objects, it needs at least one gripper, or a gripper system. The gripper is therefore an important component of the robot system and is usually mounted on the 4th axis (Figure 6) on the SCARA robot and on the 6th axis on the industrial robot (Figure 7).

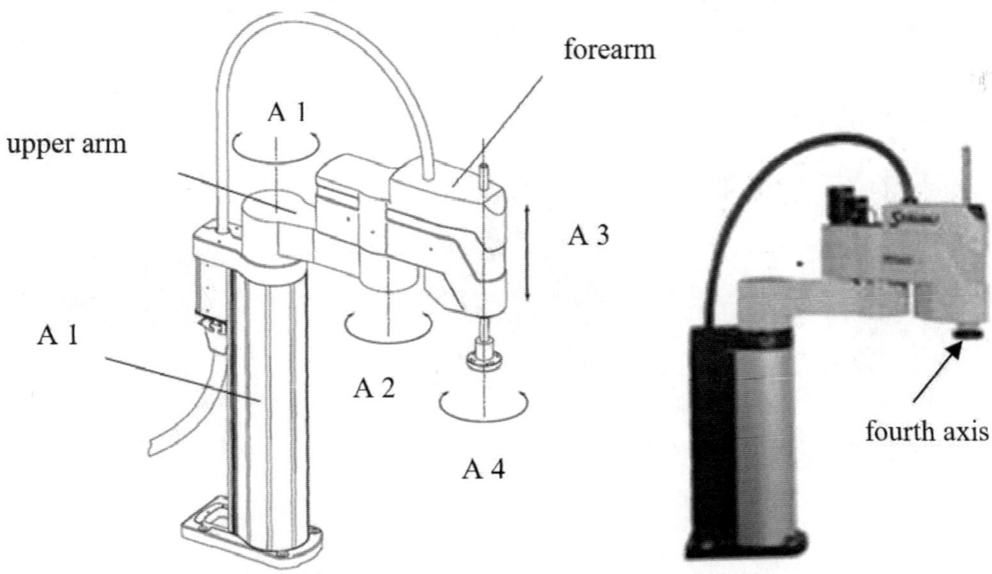

Figure 6. SCARA robot with 4 axis [2]

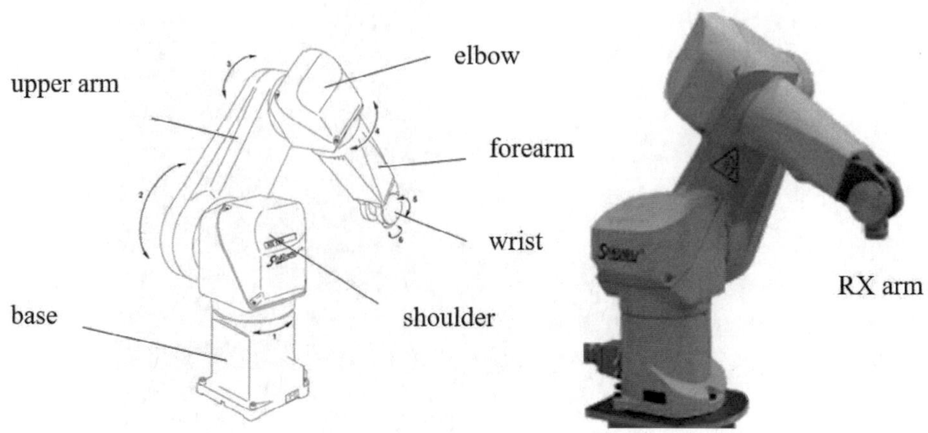

Figure 7. Industrial robot with 6 axes [2]

2.1. Gripper types

Depending on the application and the type of drive, the grippers are divided according to different criteria. Mechanical grippers may, for example, be pneumatically, electrically, hydraulically or in combination, e.g. pneumatically and electrically.

A further division of the grippers takes place via the gripping technique: parallel grippers, two-point grippers, three-point grippers, angle grippers, radial grippers, swivel grippers (Figure 8, 9, 10, 11, 12) or special grippers (e.g. FESTO trunk gripper) [3].

Figure 8. Parallel gripper

Figure 9. Three-point gripper

Figure 10. Angle grippers

Figure 11. Radial gripper

Figure 12. Swing gripping unit

Mechanical grippers are used most for mass production with standardized objects. Since the handling operations take place in the same manner, the objects are recorded in a standardized manner: they are programmed and optimized for the production flow, are adapted for gripping processes, they need gripping devices or gripper changing systems, devices for object acquisition and object storage, and they must work in protected working areas [4].

A robotized production line must be checked and tested before production begins. Only when all prerequisites and requirements are fulfilled, the system is released and production can begin. All of this is bound with a lot of time and costs.

We also came to the same conclusion in various experiments in the robot laboratory at the University of Applied Sciences in Ulm, Germany. Many applications could only be carried out with great time and financial effort [5].

For example, we had to develop a system to assembly fuel cells. For this aim we used two different small grippers, each costing around € 1,000 per unit. For the positioning of the guide pins in the mounting plates we used a parallel gripper. For the assembly of the fuel cell plates we needed a suction gripper. During the fuel cell assembly, these two grippers had to be changed again and again. In addition to the grippers we needed a gripper changing system that had to be purchased for around € 2,000, and an assembly line. The entire investment cost about € 10,000 (Figure 13).

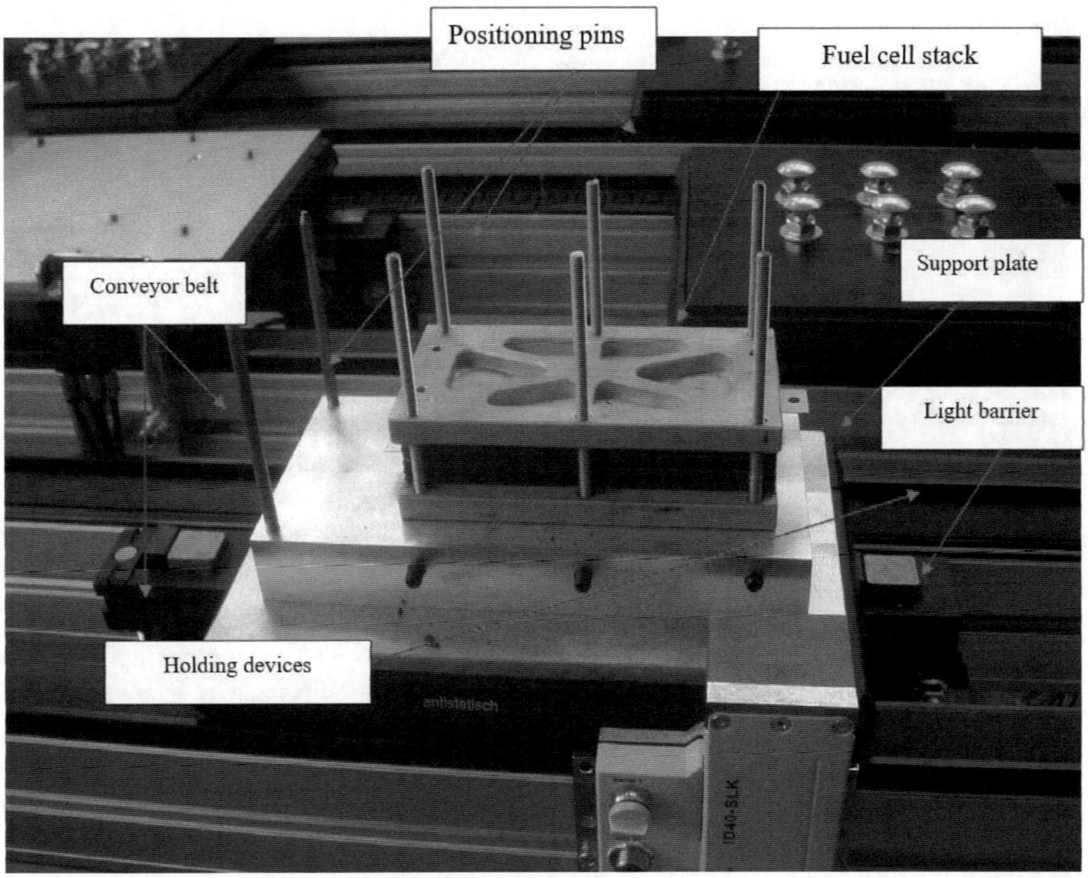

Figure 13. Fuel cell assembly line

This application should serve as a pilot project for the industrial assembling of fuel cells. For this, assembly cycles had to be optimized. The assembly cycles were between 140 and 63 seconds. The shorter assembly time in the range of one minute per cycle could only be achieved if the KUKA industrial robot used operated at 100% maximum speed in automatic mode [1].

In industry, longer assembly cycles are associated with additional costs per assembly cycle. Therefore, the assembly cycles of the fuel cell should be reduced to a minimum.

The higher mounting speed for the above - mentioned mounting could be achieved, for example, if path points were programmed with PTP (point to point) instead of with LIN (linear) movements, or when precision points were over lapped. As a result, the repeatability accuracy was lower, which in turn could lead to damage to the mounting plates and the gripper changing system. The time response in the case of gripper exchange systems was also significant in other laboratory tests.

The use of cameras from industrial image processing proved only to a limited extent as a further solution [6], [7]. The problem was, for example, a dynamic illumination of the objects that should be picked up. Especially in the case of metallic parts, reflections had to be taken because they led to the loss of important information. The object contours could become blurred. This resulted in unintended measurement inaccuracies [8-11].

A further disadvantage was the time required for the programming of the gripping process and the provision of application of specific mounting devices.

All this inconveniences led to the idea of developing a universal intelligent gripper, which even recognizes the objects [12] and adapts and controls the gripping process according to the object contours to be gripped [13].

3. Ideas for a new design

When the gripper has its own sensor system and artificial intelligence, it can independently access any object of different strength without having to be programmed for it.

As an example, from nature served the insect facet eye (Figure 14, 15) [14].

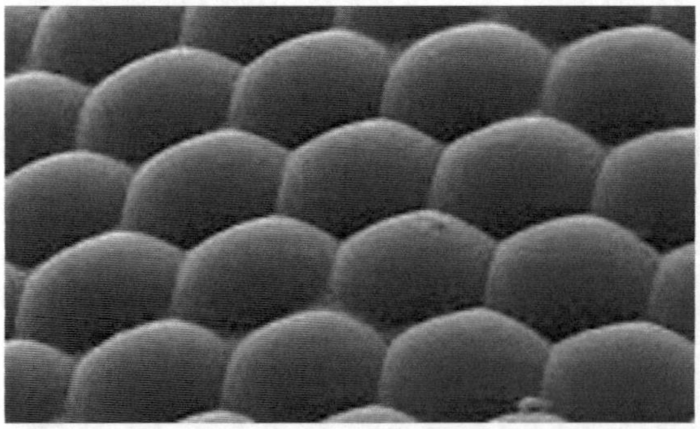

Figure 14. Part of a facet eye

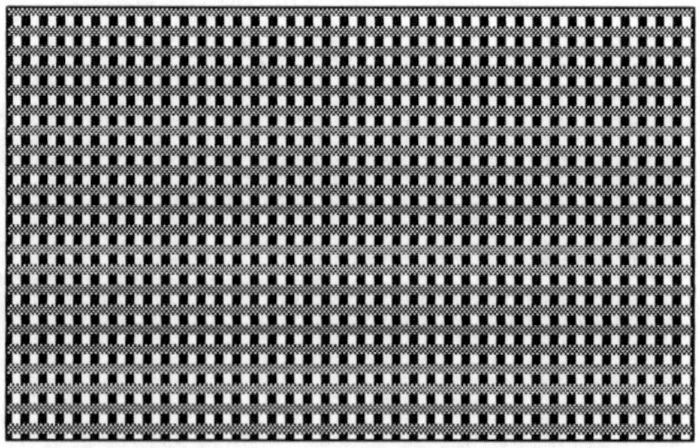

Figure 15. Facet eye as a binary model

If, for example, an insect sees a flower (Figure 16), the optical stimuli are not perceived centrally over an eye and processed further in the brain. The images are instead recognized by the individual facets as light and dark surfaces as partial images and then guided to an overall image (Figure 17).

Figure 16. Analog image

Figure 17. Digitized image

The facet eye can handle up to 300 images per second by fast-flying insects, while the human eye can process approximately 60 to 65 images per second [14], [15].

In this case the light signals can be immediately evaluated by the downstream neurons. In addition, only the change in the light signal can be selectively relayed. This has the advantage that only the most necessary light information is processed (Figure 18), which leads to faster overall processing with less computation effort.

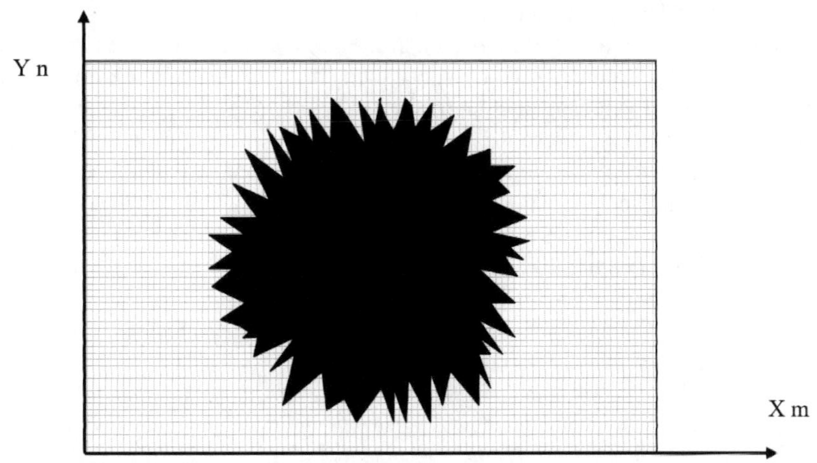

Figure 18. Digitized image in a matrix with relevant information like contour and surface

This perception principle can also be used for the recognition of contours in the case of an intelligent optical gripper. The image is segmented into pixels and is transferred to a projection surface.

3.1. Test setup

To be able to use the principle of the facet eye by an intelligent optical gripper, the object recognition should still take place directly at the object. For this aim was made a test setup with a light source and photocells (Figure 19).

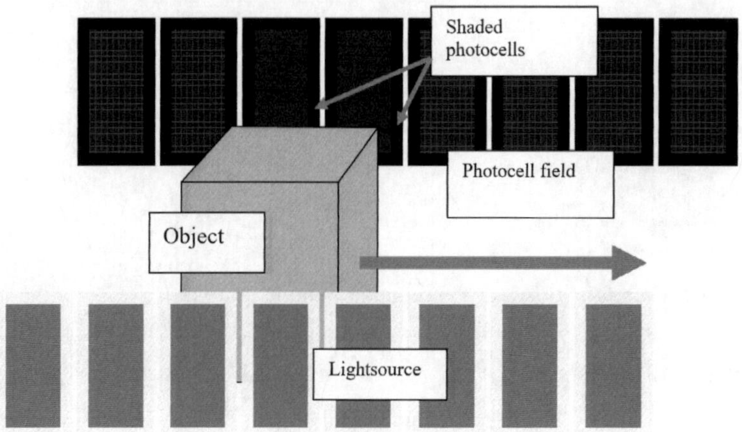

Figure 19. Test setup

In this case, the photocells serve similarly to the facet eyes of the insects for the object recognition. A light source on the opposite side irradiates a parallel number of photo elements. As objects move relative to the photo elements and the light source, they are captured by shading some photocells. The thus shaded photocells provide a dynamic image of the object, or of the objects that are located between the light source and the photoelectric elements.

In this experimental setup, as in the case of industrial image processing, black and white values can hardly be expected for better contrast and the subsequent digitization. The photo sensors only get a similar value, or an attenuated value, against an irradiated reference value (light source).

In the initial state, the photocells were illuminated with light-emitting diodes. The output current at the photocells was assumed to be a reference value with a reasonable tolerance. This value should correspond to the binary threshold on the gray scale of industrial image processing. Values which corresponded to the reference value were interpreted as TRUE values, those which were under it were assumed to be FALSE values (Figure 20).

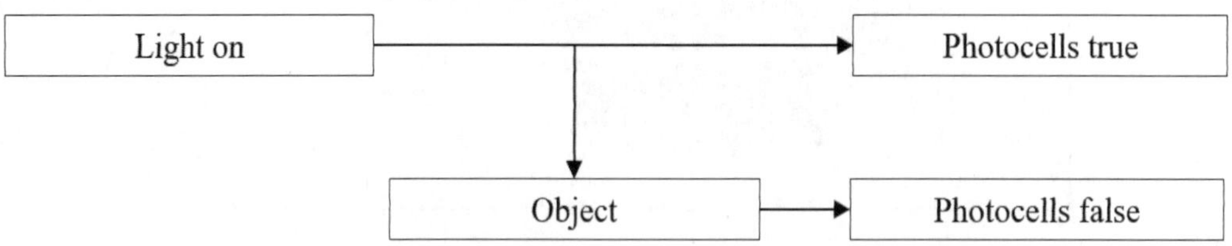

Figure 20. Principle of the switching logic and object detection

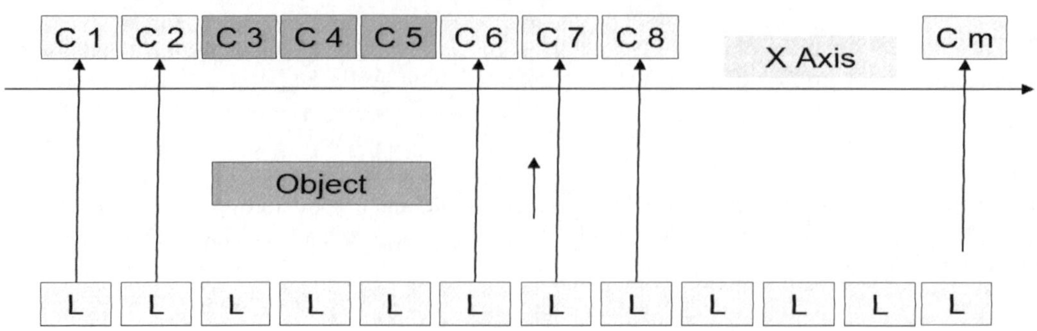

Figure 21. Object recognition

The photocells are represented in a switching logic as a two-dimensional matrix. The rows are numbered from left to right and from top to bottom and are declared as variables (Figure 18, 21).

The light source is represented in the above drawing by individual light emitters (L). The photo cells are light receptors (C1 to Cm). In a static view, the object shadows several photocells, namely the photocells: C3, C4 and C5 (Figure 21). When the photocells are irradiated, the logic state TRUE is accepted in an electronic circuit. If they are shadowed, they are assumed to be FALSE and represent the area of the captured image (Figure 18).

3.2. Object-oriented programming

For the logical processing, simulation and visualization of the switching logic, was used the development environment CoDeSys (Controller Development System) from 3S-Smart Software Solutions GmbH,DE 87439 Kempten, Germany.

CoDeSyss can be configured for a large number of controllers. For example, SIEMENS PLC controls can be easily imported into CoDeSys. Conversely, programs developed in CoDeSys can be exported to SIEMENS controllers. Furthermore, with the simulation and visualization possibilities the programs can be tested for functionality and their functioning can be visualized (Figure 22). A solution to this problem is a customized technical device to significantly increase the lifetime of the clutch. It was accomplished by a precisely controlled activation of the clutch at a point of time when pendulum and motor are at synchronized speed and direction using incremental encoders.

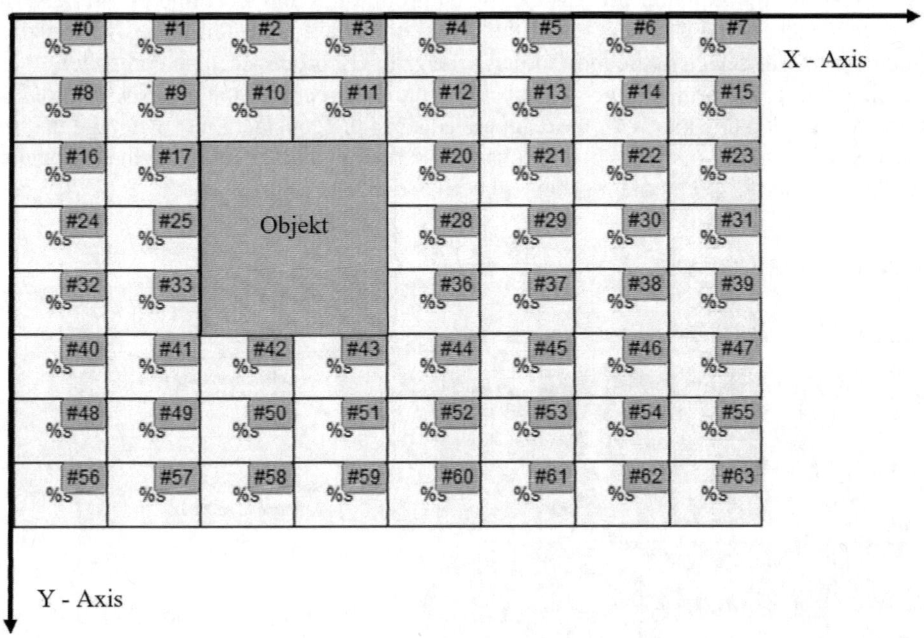

Figure 22. Matrix visualization whit CoDeSys

The fields were programmed and addressed via arrays like in the following figure (Figure 23).

```
PROGRAM PLC_PRG
VAR
   Lichtfeld: ARRAY [1..8, 1..8] OF BOOL:=64 (TRUE);
   Result: ARRAY [0..63, 0..1] OF INT;
   X_Achse: INT;
   Y_Achse: INT;
   counter: INT:= 0;
END_VAR

FOR X_Achse :=1 TO 8 DO
   FOR Y_Achse :=1 TO 8 DO
      IF ( Lichtfeld [ X_Achse, Y_Achse] = FALSE)
         THEN
         Result [counter, 0]: = X_Achse;
         Result [counter, 0]: = Y_Achse;
         counter: = counter + 1;
      END_IF
   END_FOR
END_FOR
```

Figure 23. Arrays Code with CoDeSys

If the objects are located, the gripping process can be controlled through a routine of queries. The precision of the image detection and the accuracy of the gripping process are determined by the size of the arrays. The more pixels are detected, the more precise the gripping process can be controlled.

One way of implementing this principle was to attach the optics directly to the gripper (Figure 24). In this case, the optics is guided together with the gripper. This has the considerable advantage that the gripper can furthermore be made of metal for mechanical reasons. The optics and the associated electronics remain largely unaffected by the mechanical activities of the gripper.

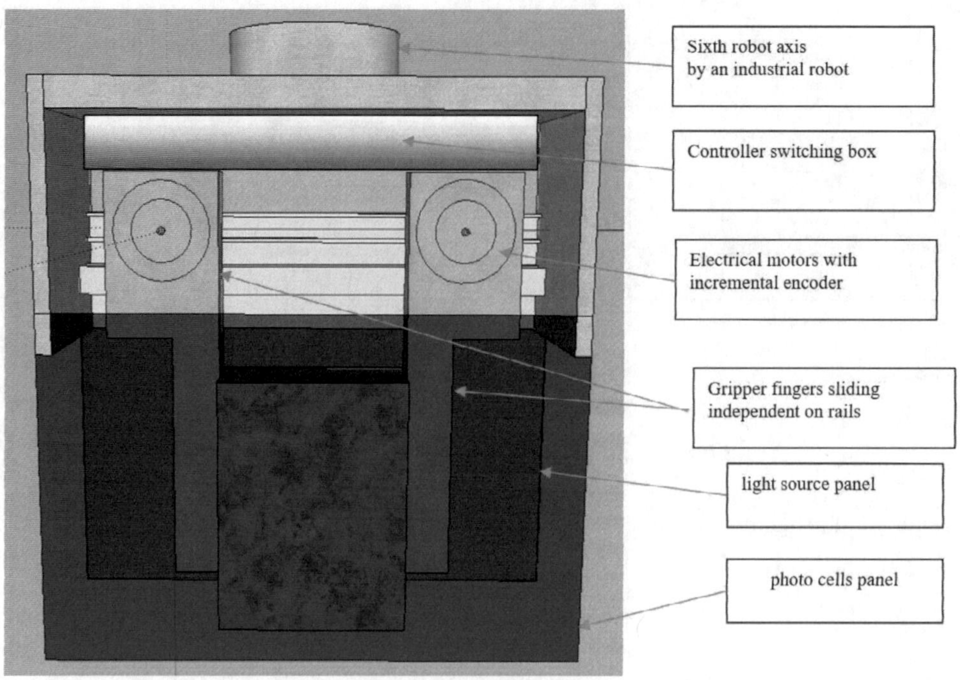

Figure 24. Intelligent optical gripper

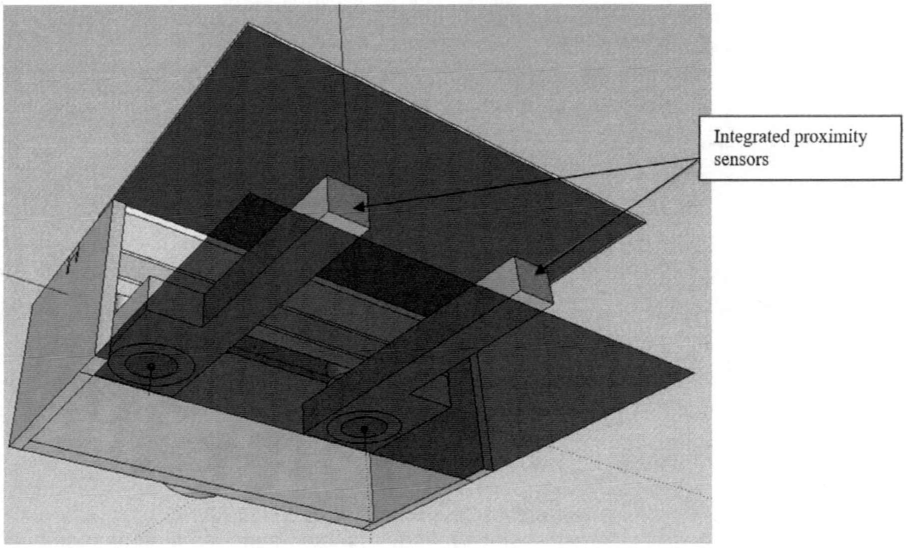

Figure 25. Optical gripper, bottom view

The object recognition can be captured by the attached optics independently of the IR system in real-time and the gripping process can be controlled directly through the integrated gripper controller.

By movements in the Z direction to the object, the gripper can be controlled by means of proximity sensors integrated in the gripper fingers (Figure 25).

4. Conclusions

A major advantage of this gripper concept is the use of conventional and thus affordable optics together with standard grippers, or with adapted mechanics, which in turn can be produced from conventional materials.

With only one intelligent gripper it is possible to grip objects with different sizes and different shapes. The intelligent gripper automatically regulates the gripping force by means of the optics without having to be programmed. All this leads to cost reductions and production acceleration [16].

References

[1] ****KUKA AG*, Zugspitzstraße 140, 86165 Augsburg, Germany http://www.kuka-robotics.com/germany/de/company/
[2] ****Stäubli Bayreuth GmbH*, Theodor-Schmidt-Straße 19, 95448 Bayreuth, Germany http://www.staubli.com/index.php?id=22&L=2
[3] ****Festo AG & Co. KG*, Ruiter Straße 82, 73734 Esslingen, Germany https://www.festo.com/cat/de_de/products_010800
[4] ****SCHUNK* GmbH & Co. KG, Bahnhofstraße 106-134, 74348 Lauffen/Neckar, Germany http://de.schunk.com/de_de/startseite/
[5] ****Hochschule Ulm*, Prittwitzstraße 10, 89075 Ulm, Germany https://studium.hs-ulm.de/de/Seiten/Homepage.aspx
[6] Jähne B 2005 *Digitale Bildverarbeitung. 6. überarbeitete und erweiterte Auflage*, Springer, Berlin, Germany
[7] Demant C and Streicher-Abel B 2011 *Industrielle Bildverarbeitung, 3 Auflage,* Springer Verlag
[8] Demant C, Streicher-Abel B and Waszkewitz P 2011 *Industrielle Bildverarbeitung Wie optische Qualitätskontrolle wirklich funktioniert*, Springer Verlag Berlin Heidelberg
[9] ****NeuroCheck* GmbH, Engineering Center, Neckarstr. 76/1, D-71686 Remseck, Germany http://www.neurocheck.de/

[10] ****NEURO CHECK Industrielle Bildverarbeitung*, Einsteller-Handbuch, Version 5.1, 2002, NeuroCheck GmbH
[11] ****NeuroCheck Einführung*, Extras Springer
[12] Zhang J and Hu J 2008 *Image segmentation based on 2D Otsu method with histogram analysis*, Computer Science and Software Engineering International Conference, Wuhan, China, December 12-14, pp 105-108
[13] ****NeuroCheck Getting Started*, Extras Springer
[14] ****WIKIPEDIA* https://de.wikipedia.org/wiki/Facettenauge
[15] Bredies K and Lorenz D 2011 *Mathematische Bildverarbeitung. Einführung in Grundlagen und moderne Theorie*, Vieweg+Teubner, Wiesbaden, Germany
[16] Bauer N (Hrsg.) 2008 *Handbuch zur Industriellen Bildverarbeitung. Qualitätssicherung in der Praxis,* Fraunhofer IRB-Verlag, Stuttgart, Germany

IOP Conf. Series: Materials Science and Engineering **200** (2017) 012055 doi:10.1088/1757-899X/200/1/012055

Automated assembling of single fuel cell units for use in a fuel cell stack

C K Jalba[1], A Muminovic[1], C Barz[2] and V Nasui[2]

[1]Ulm University of Applied Sciences, Production Engineering and Production Economics, Prittwitzstraße 10, 89075, Ulm, Germany
[2]Technical University of Cluj-Napoca, North University Centre of Baia Mare, Victor Babes str., no. 62A, 430083 Baia Mare, Romania

E-mail: klausjalba@t-online.de

Abstract. The manufacturing of PEMFC stacks (POLYMER ELEKTROLYT MEMBRAN Fuel Cell) is nowadays still done by hand. Over hundreds of identical single components have to be placed accurate together for the construction of a fuel cell stack. Beside logistic problems, higher total costs and disadvantages in weight the high number of components produce a higher statistic interference because of faulty erection or material defects and summation of manufacturing tolerances. The saving of costs is about 20 – 25 %. Furthermore, the total weight of the fuel cells will be reduced because of a new sealing technology. Overall a one minute cycle time has to be aimed per cell at the manufacturing of these single components. The change of the existing sealing concept to a bonded sealing is one of the important requisites to get an automated manufacturing of single cell units. One of the important steps for an automated gluing process is the checking of the glue application by using of an image processing system. After bonding the single fuel cell the sealing and electrical function can be checked, so that only functional and high qualitative cells can get into further manufacturing processes.

1. Introduction

A fuel cell is a galvanic cell which converts the chemical reaction energy of a continuously supplied fuel and an oxidizing agent into electrical energy (Figure 1).

While a lot of heat is generated during the oxyhydrogen reaction (Figure 2), the reaction between oxygen and hydrogen can be controlled in the fuel cell (Figure 3). As a final product, is generated in the fuel cell in addition to water and heat electrical energy.

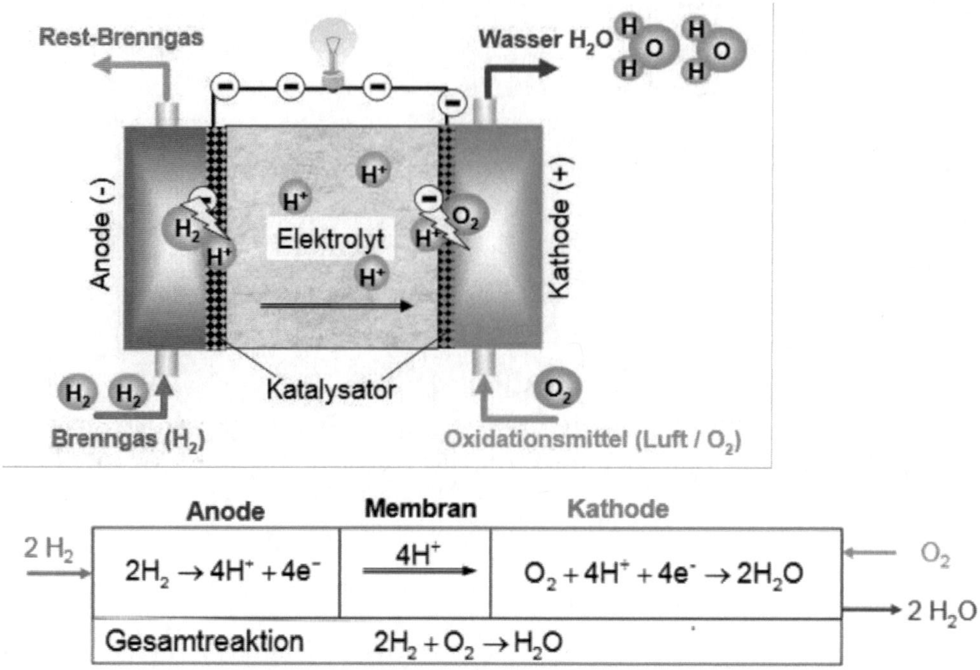

Figure 1. Functional principle of the fuel cell [1]

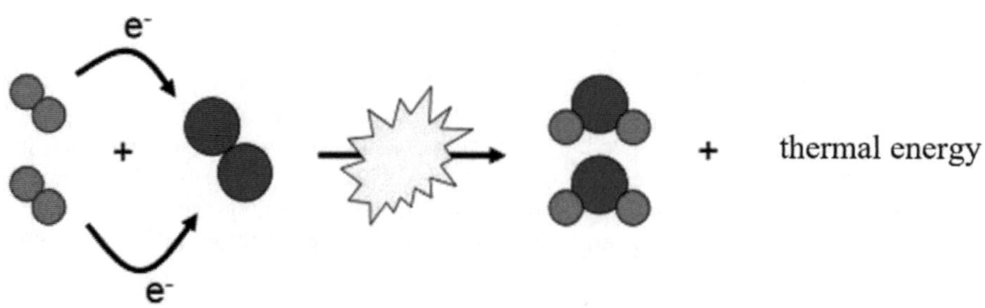

Figure 2. Boiling gas reaction [1]

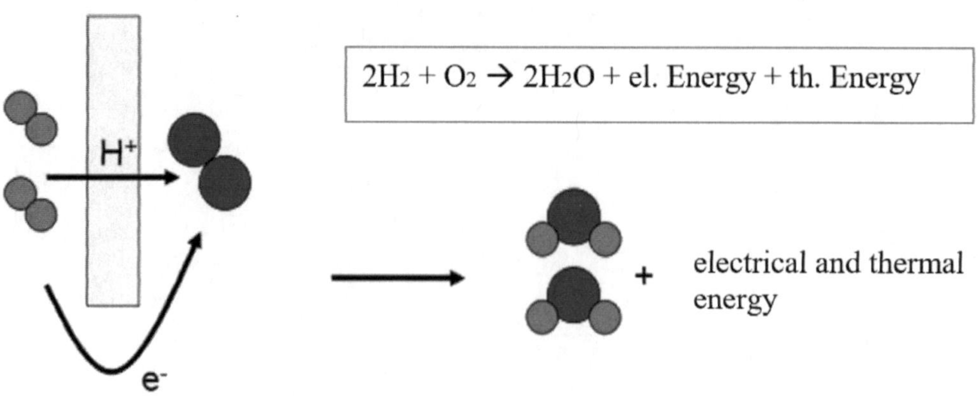

Figure 3. Controlled electrochemical reaction in the fuel cell [1]

IOP Conf. Series: Materials Science and Engineering **200** (2017) 012055 doi:10.1088/1757-899X/200/1/012055

The strength of fuel cells lies in the high-energy density compared to other power storage units. This explains the early interest of the military and space travel in this technology.

Governments are now also interested in the development of fuel cells for ecological reasons. For this reason, projects for the further development of this technology are also being provided in Germany. An institution that is involved in this development is the Center for Hydrogen Technology and Solar-Technology (ZSW) in Ulm, Germany. One of the Daimler Research Centers is also located in Ulm and cooperates intensively with ZSW [2].

Figure 4. Fuel cell vehicle, Mercedes C-class

Daimler developed prototypes of smaller vehicles with fuel cells, which are already available in series; aim to test the size, the weight and the cost of the fuel cell and the appropriate storage of the hydrogen. Daimler presented for example the A-Class and B-Class vehicles with fuel cells (Figure 4). In Hamburg and Stuttgart, hydrogen-powered buses are tested in normal operation.

Fuel cell technology is at a relatively early stage. In order to replace this technology in the long term with other technologies such as the internal combustion engine, fuel cells must become considerably cheaper.

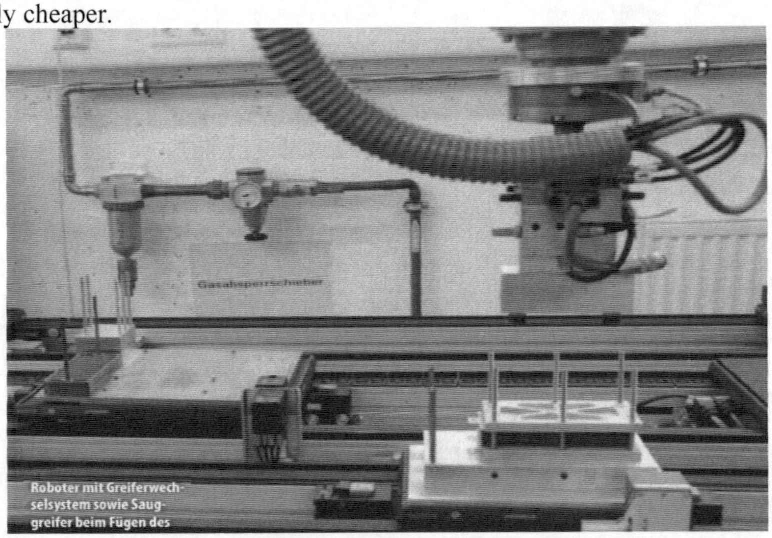

Figure 5. Assembly line for fuel cells [3]

IOP Conf. Series: Materials Science and Engineering **200** (2017) 012055 doi:10.1088/1757-899X/200/1/012055

Because the production of polymer electrolyte membrane (PEM) fuel cell stacks was largely done manually and because it was assumed that by the large number of components, assembly tolerances or material errors had to be expected a higher statistical susceptibility and therefore could cause higher processing costs, one of the aims of this project at the University of Applied Sciences in Ulm, Germany in cooperation with ZSW, was the development of automated manufacturing processes with robots for the medial sealing of membrane electrode assemblies (MEA) with bipolar plates to single cell units. On the other hand, was the next aim the optimization of economic production of fuel cells and quality control by industrial image processing. The individual cells produced and tested in this way should be stored in magazines or provided on assembly carriers (Figure 5) [2].

The first part of the research project describes assembling solutions and the automatic production of individual cells under robotic use. The second part of this paper describes economic advantages and benefits with the new production method.

2. Assembling fuel cells

A fuel cell consists of the following components:

- Electrolyte (membrane), which provides ion transport and separates anode and cathode.
- Electrodes; here take place the electrochemical reaction.
- EME (Electrode Membrane Unit): "heart" of the fuel cell.
- Gas diffusion layers; are necessary for the supply and distribution of the reaction gases.
- Bipolar plates; are fine channels in the plates and ensure the supply and distribution of hydrogen and oxygen. In addition, they serve as "electron-collectors".

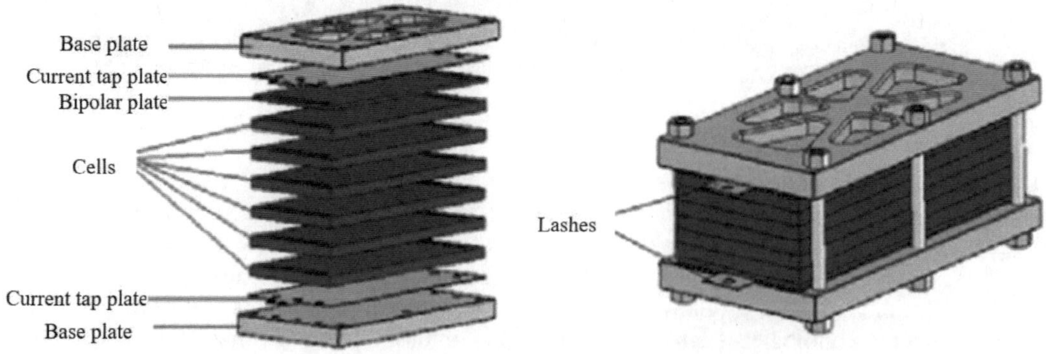

Figure 6. Parts of a 6 V fuel cell stack [3]

Figure 7. Glue bead with robot [4]

In the case of stack mounting, firstly is positioned the base plate. At next is assembled a current tap plate. Depending on the size of the fuel cell, the individual cells are then mounted. Thereafter, a further current take-off plate and an end plate are mounted (Figure 6).

By the laboratory tests were used the weight and cost-saving direct adhesive application (Figure 7).

The individual cells must be separated from each other by adhesive beads. The sealing function of fuel cells can be done either by a metal carrier with sealing beads or by application of the sealing beads directly to the main components.

For this purpose, was used a dispenser system for applying the adhesive seal (Figure 8). The dispenser system was integrated into the robot cell. For quality control was installed an image processing system with camera and PC.

Figure 8. Dispenser system for applying the adhesive seal [3]

For the adhesive hardening, final component inspection and further assembly to finished stacks was implemented a conveyor with carriers and an identification system (Figure 5).

Due to these technical measures were made a lot of modifications: to adapt the grippers, to change the programs from the integrated PLC and the sensor technology and to modify the valve control.

The integrated image processing allows, for example, that boards with a defective glue bead have to be removed before the stack is mounted [5], [6]. The height of the adhesive must not exceed 0.7 mm and must be uniform. Otherwise, problems arise with the joining and the tightness of the cell.

The system allows also through the implementation of the identification unites and the optical recognition system the documentation of every single cell or leakage test result. But also can provide data to create statistics for the general evaluation of the production process: the number of produced cells, required time for each assembly cycle, statistic about failure parts, etc. [7-9].

The complete assembly process from the material supply carrier to the fuel cell stack was led from an industrial robot. It was used a KUKA KRC-15/2 Robot with a gripper change system, a linear gripper and a suction gripper.

The layout of the robot station (Figure 9 and 10) shows the transfer cycle with the material supply and the transport route for mounted fuel cell stacks.

IOP Conf. Series: Materials Science and Engineering **200** (2017) 012055 doi:10.1088/1757-899X/200/1/012055

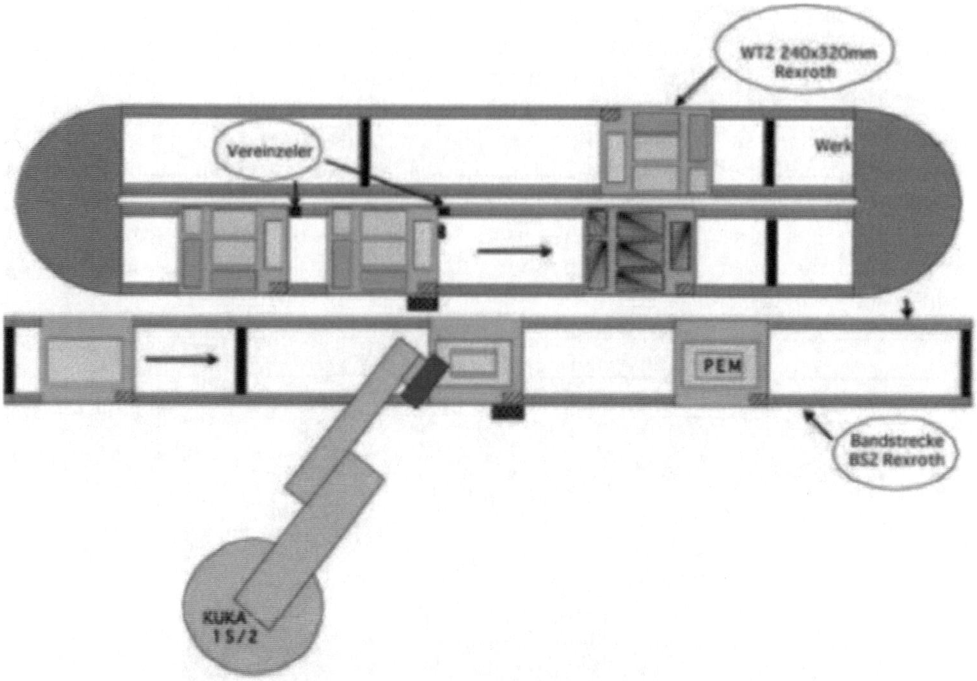

Figure 9. Layout of the assembling line [4]

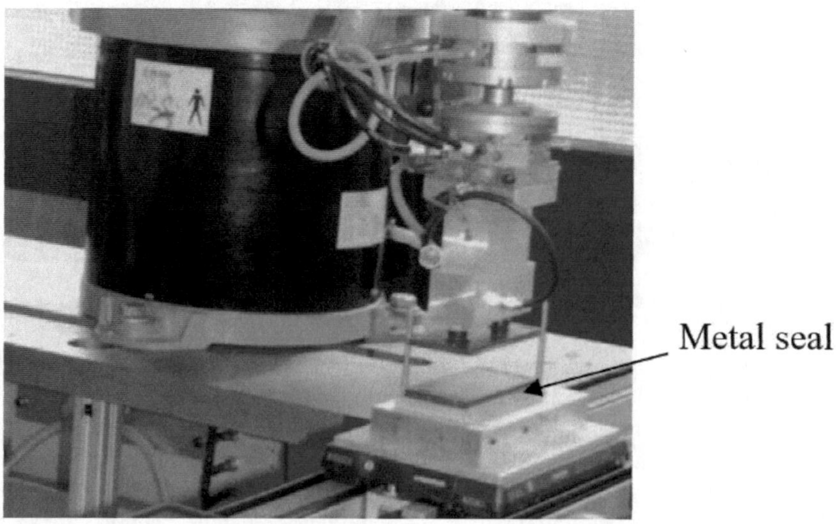

Figure 10. Stack with metal seal [4]

In the tests, it was found that the assembly of the metal seal (Figure 11) is unnecessary when the adhesive bead is applied. The metal gasket does not increase the tightness of the fuel cell. Thus, the production becomes more economical.

IOP Conf. Series: Materials Science and Engineering **200** (2017) 012055 doi:10.1088/1757-899X/200/1/012055

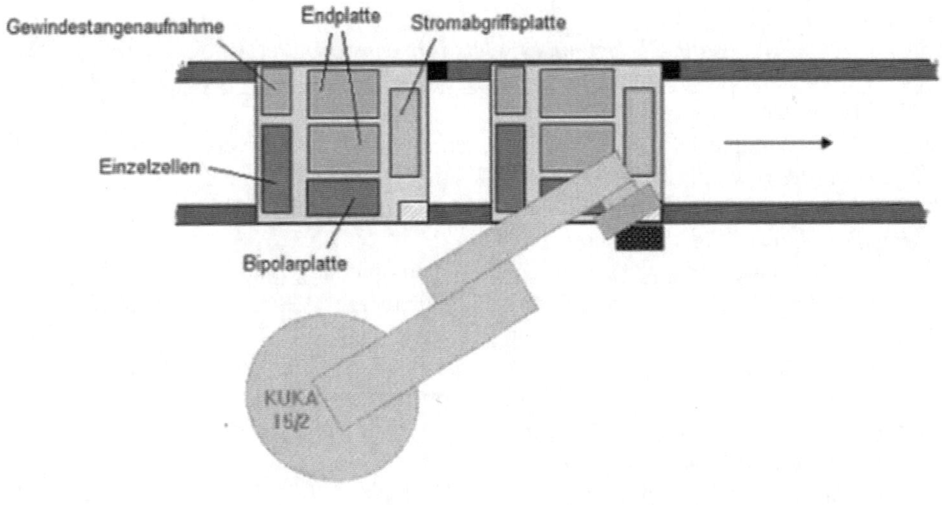

Figure 11. Supplying the robot with fuel cell parts

The following table shows the material cost difference of the mounted cell with and without metal gasket (Figure 12).

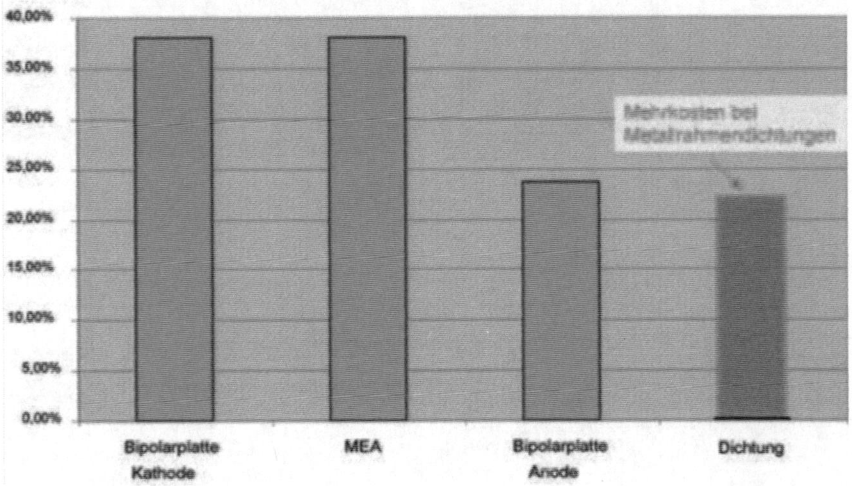

Figure 12. Additional costs for the assembly of metal gasket

Depending on the size of the cell, the material costs of the metal frame cover are up to 25 percent of the other parts of the fuel cell. In addition, there is the separate supplying of the sealing frames in the mountings as well as the required assembly time.

3. Conclusions

The tests performed showed that fuel cells and fuel cell stacks can be mounted with industrial robots. With the developed robotic station, can be assembled even smaller production unites. For larger units is recommended to produce with more assembling robots. For fast movements like peak and place movements can be used 4-axis SCARA Robots.

Through the use of the described automation and savings on the product (removal of the metal support gasket), the production can be more economically. In the case of very large unit numbers, shorter assembly times as well as lower material costs lead to competitive advantages and price reductions of the end products.

References

[1] Craig M 2004 *Brennstoffzelle ist nicht gleich Brennstoffzelle*, In: Telepolis, 17 November https://www.heise.de/tp/features/Brennstoffzelle-ist-nicht-gleich-Brennstoffzelle-3437217.html

[2] Pioch P (WBZU), Aigle T (WBZU) and Jörissen L (ZSW) 2008 *Seminar: Brennstoffzellen- und Wasserstofftechnologie*, January 24,WBZU Ulm

[3] ****A&D Magasin*, Februar 2008, http://www.industr.com/A-und-D-Magazin/de_DE

[4] ****Manager Magazin*, 28.12.2016, http://www.manager-magazin.de/unternehmen/autoindustrie/bild-980873-470666.html

[5] ****FusionSystems GmbH*, Altchemnitz, Annaberger Str. 240, 09125 Chemnitz http://www.fusionsystems.de/anwendungsberichte.html

[6] Demant C and Streicher-Abel B 2011 *Industrielle Bildverarbeitung*, 3 Auflage

[7] ****Baumer GmbH*, Pfingstweide 28, 61169 Friedberg, Deutschland http://www.kaiser-fototechnik.de/de/produkte/2_1_sortiment.asp?w=381

[8] ****NeuroCheck Einführung*, Extras Springer

[9] ****NeuroCheck* GmbH, Engineering Center, Neckarstr. 76/1, D-71686 Remseck, Germany http://www.neurocheck.de/

IOP Conf. Series: Materials Science and Engineering **200** (2017) 012056 doi:10.1088/1757-899X/200/1/012056

Influence of typical faults over the dynamic behavior of pantograph-catenary contact force in electric rail transport

S Rusu-Anghel[1] and A Ene[2]

[1]Politehnica University of Timisoara, Department of Electrical Engineering and Industrial Informatics, 5 Revolution Street, Hunedoara, 331128, Romania
[2]SC Electrificare CFR SA, Bucharest, Romania

E-mail: stela.anghel@fih.upt.ro

Abstract. The quality of electric energy capture and also the equipment operational safety depend essentially of the technical state of the contact line (CL). The present method for determining the technical state of CL based on advance programming is no longer efficient, due to the faults which can occur into the not programmed areas. Therefore, they cannot be remediated. It is expected another management method for the repairing and maintenance of CL based on its real state which must be very well known. In this paper a new method for determining the faults in CL is described. It is based on the analysis of the variation of pantograph-CL contact force in dynamical regime. Using mathematical modelling and also experimental tests, it was established that each type of fault is able to generate 'signatures' into the contact force diagram. The identification of these signatures can be accomplished by an informatics system which will provide the fault location, its type and also in the future, the probable evolution of the CL technical state. The measuring of the contact force is realized in optical manner using a railway inspection trolley which has appropriate equipment. The analysis of the desired parameters can be accomplished in real time by a data acquisition system, based on dedicated software.

1. Introduction

Nowadays, in Romania, each contact line (CL) sector has a maintenance car (Figure 1) for repairs and maintenance activities on 50-60km of CL length. The main goal of this paper consists in proposing for endowment of this maintenance car with a minimal set of measurement elements which permit the identification of the CL real status.

1.1. Establishment of the parameters which influence the energy intake quality

In order to establish the geometrical parameters which must be measured, the operating technical conditions of the catenary suspension (Figure 2) were studied and also the Romanian standards. For an appropriate operating functioning, in maximal safety conditions, the catenary suspension must respect the Instruction no. 353 for technical maintenance and reparation of the electrified railway contact line [1].

As a result of the accomplished study, the following geometrical parameters must be measured:

- **stagger** of the contact wire;
- **height** of the contact wire.

As well, the followings must be accomplished:

- identification of the **dropper absence,** their incorrect tensioning or other **faulty grip** of the CL;
- faults location by **identification the suspension pillars** of the catenary;
- **geographical location** of each measurement set;
- **movement speed of the maintenance car;**
- **temperature** of the environment where the measurements are accomplished.

Beside of the geometrical parameters which will be directly measured, the continuous measuring of the pantograph-catenary contact force is imposed because some faults which are mentioned above (dropper absence, faulty grips) can not be identified directly, but only by analyzing the suspension dynamic regime (a new method presented into this paper).

Figure 1. Maintenance car (VMT 863, SNCFR, București)

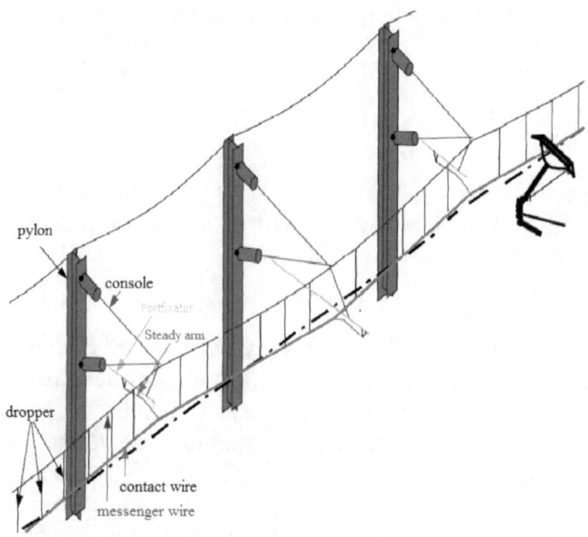

Figure 2. General structure of a catenary suspension

Nowadays, the Instruction no. 353 foresaw two measuring types:

- *measurements accomplished by a „inspection car"* [2] (TMC, Plasser UK etc.) which is unique on the railway network which realizes measurements on large time intervals for determining the exactly technical status of the installation. The accuracy of these measurements is very high, but the costs of the inspection car reach millions of Euro.

The procedure can not be used for permanently knowledge of the technical status of the catenary suspension, therefore, it does not permit the maintenance based on these measurements.

- *periodical measurements* accomplished at predefined intervals and established by Instruction no. 353 which are manually realized by workers teams. The accuracy of these measurements is subjective and they need the supplying interruption. The costs of these measurements are high due to the stuff expenditure.

Neither this procedure does not allow the accomplishing of maintenance operations based on the actual condition of the installations (it is possible that the verified zone has no faults, but these will exist in an unplanned area and will produce effects).

From the exposed reasons, in the present paper a new method of the CL maintenance is proposed based on the exposed status. This status will be determined by automat measurements of the main CL parameters which are established and defined above. The measurements will be realised on short time intervals by daily movement of the maintenance car (existing at each CL sector) for various work.

1.2. Determining of the geometrical parameters influence of the CL over the equipment safety
Stagger failure can lead to „hanging" the catenary suspension, contact wire broking, pantograph destroying and also the suspension on hundreds of meter length until the train stops. It follows considerable material damage, disrupting railway traffic, additional costs for restoring the installation [3].

Failure in contact wire height may cause some electric discharge between the contact wire (supplied at 27.5kV) and adjacent elements (works of art, cars, etc.). This electric discharge can cause fire, electric shock, damages of the contact wire. As well the failure in contact wire height may produce variations in the pantograph-catenary contact force, causing direct effect over the quality of the pantograph-catenary energy intake and over the contact elements wear.

Poor condition of the droppers produces additional vertical displacement of the contact line and by changing the dynamic behaviour of the catenary suspension it will influence the quality of the energy intake. The same effect has other faults of clamping the contact wire.

An improper dynamic behaviour of the catenary-pantograph assembly can cause at high speeds to the cancelling of the contact force and therefore it can lead to the occurrence of the electric arc and adjacent problems (energy faults, premature wear or even melting of the contact wire, electro-magnetic pollution). The problem is much treated in the literature ([4], [5], [6], [7]), especially for trains which travel at high speeds, without receiving a radical solution until now.

Figure 3. Electric arc at pantograph-catenary contact [4]

The essential problem in the measurement of geometrical parameters of CL consists in fact that these measurements must be performed from distance, without direct contact with the elements supplied at 27.5 kV voltage. The only methods that are applicable are optical or ultrasonic.

2. Analysis of dynamic behaviour of contact line in case of faults and their identification based on characteristic variations

The catenary suspension is a mechanical system with cvasiperiodical composition, consisting mainly of tensioned cables. This structure is very flexible, allowing the propagation of waves of large amplitudes, which are reflected and overlap themselves, generating complex movements. However, by design, catenary contains elements that induce a nonlinear behaviour, which accentuates the difficulties of analysing the behaviour of the forces which act over it.

The pantograph, a mechanism with structure composed by articulated bars, presents its own dynamic features and also has a nonlinear behaviour. The dynamic coupling between the two substructures with independent mechanical properties, relative movable to each other, which are linked through a unilateral sliding contact, concentrates a significant number of parameters need to be analysed. As well, the speed of the pantograph is limited up to the values that depend mainly on structural features of the catenary, including technical adjustments necessary for ensuring mechanical tension in the system, and the catenary is considered nowadays as the main limiting factor in the movement of high-speed trains.

The problem of mathematical modelling of the pantograph-catenary assembly is very complex and it has been approached by many authors [8-14]. Due to the fact that the stiffness of the catenary is a variable state parameter, dependent on the position of the reference section and it is much lower in regions located toward the middle of the span, as compared with the areas nearby the support pillars, where the catenary is significantly more rigid, it is clear that the dynamic response of the suspension will take effect as the followings:

- modifying the movement characteristics of the progressive and regressive wave fronts generated by the disturbing action of the pantograph;
- additional vertical movements, quick and relatively frequent, upward and downward of the pantograph, in order to track the fast movements and variable of the contact wire. Catenary stiffness variation and the cvasirandom frequency of this parameter are recognized as determinant factors of energy intake quality and, therefore, through appropriate design and maintenance, suspension stiffness regularization is followed and decreasing the effect of the presence of the supporting pillars.

In terms of the analysis presented and the factors which determine the geometrical and mechanical cvasiperiods in catenary system, they can be classified by various criteria. A classification of interest in the sense issue of this project is the one proposed in [11], which indicates the following two categories: singularities, respectively, faults.

a. *Singularities* - represent catenary areas which not correspond to the nominal intake situation, due to a geometrical and/or elasticity of the catenary.

As examples, the followings are mentioned:

- Road bridges which permit cars circulation at a level situated above the contact line. Changes induced to the catenary include modifications to the suspension geometry and the characteristics of the cable carrier;

- Equipment for mechanical tensioning of the suspension, whose presence requires both major changes to the geometry of the structure overhead, and the existence of an area in which the pantograph must pass on a contact wire on another and, for a certain period of time, it is in contact simultaneously with two or more wires;

- Aerial switches.

b. *Faults* – are the result of the wear caused by operational factors, the most known are the followings:

- Broken dropper, eased or missing, which determine the adjusting the contact wire geometry and also the suspension stiffness. These faults are considered as generating of „hard hits" in catenary system.

- Flaking or deposits on the contact wire, the presence of junction elements mounted permanently to restore the connection between two wires contact, or during the currents maintenance, which have as effect the changes in the system mass emplacement, with implication over its dynamic characteristics

- Breaking or blocking a support arm or other parts or elements, events that have implications over the state parameters, therefore they can greatly affect the dynamic behaviour of the pantograph-catenary assembly.

In conclusion, the analysis of phenomena that influence the dynamic behaviour of the catenary-pantograph in the current electric transport rail involves addressing of complex problems in mechanics of the elastic structures with nonlinear behaviour in connection with the problem in the field of electrical engineering and industrial computer science.

2.1. Determining of the contact force

The pantograph-catenary contact force can be determined with the relation (1):

$$F = K \cdot y \tag{1}$$

where: K is global (variable) stiffness of the catenary suspension;
y is pantograph movement.

The vertical movement of the pantograph can be measured using an ultrasonic distance sensor (Figure 4).

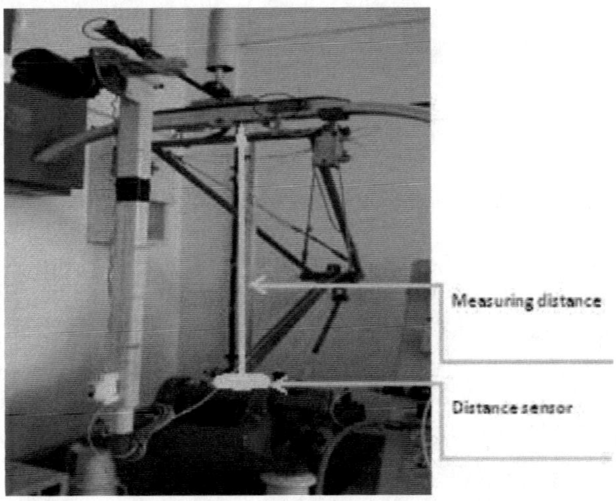

Figure 4. Measuring of the pantograph vertical movement (laboratory equipment)

From the considerations presented in the literature [15] and based on the findings of the practical measurements, it follows that the catenary model can be represented by an equivalent resort, with stiffness, K, usually considered constant. In reality, at higher speed of movement of the pantograph, the catenary behaviour is no longer the same, making this characteristic to be variable, [11] and with the behaviour described by the mathematical expressions (2), (3), (4) and (5):

$$K(x) = K_0\left(1 - \alpha \cdot \cos\frac{2\pi \cdot x}{L}\right) \tag{2}$$

$$K_0 = \frac{K_{max} + K_{min}}{2} \tag{3}$$

$$\alpha = \frac{K_{max} - K_{min}}{K_{max} + K_{min}} \tag{4}$$

$$K(t) = K_0\left(1 - \alpha \cdot \cos\frac{2\pi V}{L}t\right) \tag{5}$$

where: K_{max} – maximal value of the stiffness coefficient between two pillars,
K_{min} – minimal value of the stiffness coefficient between two pillars,
V – pantograph speed of movement,
L – distance between two pillars.

The variation of the catenary stiffness coefficient for distance between two pillars L = 45m is presented in Figure 5.

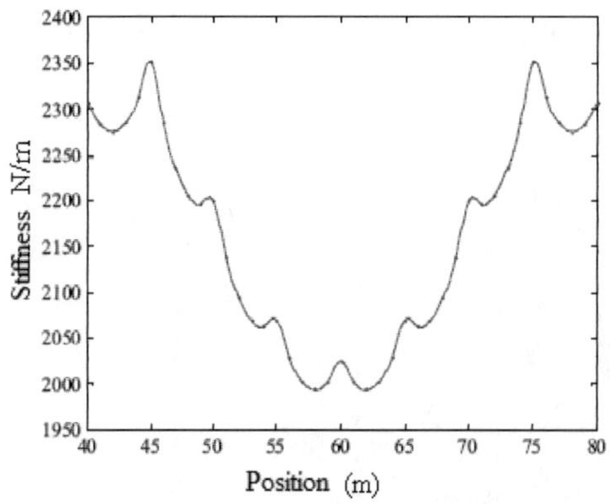

Figure 5. Variation of the catenary stiffness coefficient

Considering the previous relations, the equivalent functions of the catenary stiffness determined for different values of L are presented in Table 1. The expressions presented in Table 1 for K(x) permit a nonlinear mathematical modelling of the pantograph-catenary system.

Table 1. Equivalent functions of the catenary stiffness for different distance between pillars

L (m)	K_0(N/m)	α	K(x)
40.5	2435.6	0.1256	2435,6(1-0,1256cos0,155x)
45	2172	0.0823	2172(1-0,0823cos0,14x)
49.5	2126.6	0.1247	2126,6(1-0,1247cos0,127x)
63	1916.7	0.1776	1916,7(1-0,1776cos0,1x)

In conclusion, the pantograph-catenary contact force can be experimentally determined by measuring the catenary height and its elasticity coefficient which is calculated.

2.2. Simulations

In order to simulate the dynamic behaviour of the catenary suspension-pantograph system, the OSCAR-SAM 5 [16], [17] tool was used. Therefore, a series of faults were express designed for determining of some 'signatures' on the obtained signal.

A waveform of the contact force for a dropper fault, designed by simulation is presented in Figure 6. The broken dropper situated on the horizontally axis at 850m and the weakened droppers situated at 958m and 1065m generate important modifying of the signal (increasing of the contact force).

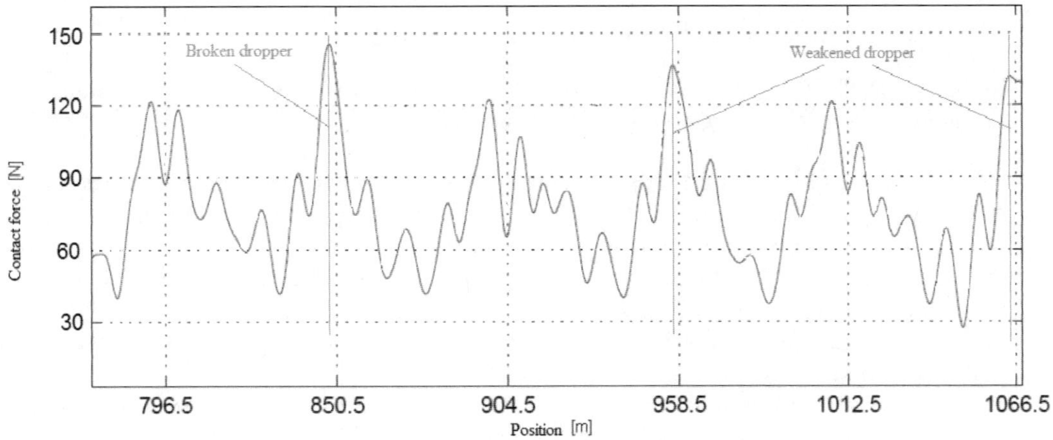

Figure 6. Contact force – OSCAR modelling

These simulations will be experimentally verified. Based on them, the exactly faults locating method will be established which influence the dynamic behaviour of the catenary suspension.

3. Conclusions

In this paper the parameters which influence the energy intake quality were established and will must be measured for ensuring the maintenance based on the real status of contact line, these being the followings: **stagger, contact wire height, contact force and auxiliary parameters: suspension pillars identification, kilometric position, temperature, the maintenance car speed.**

A novelty consists in fact that *the faults determining* is proposed (broken or weakened droppers, faulty grips) using an indirect method by *analysing the dynamic behaviour of the pantograph-catenary system*. The validity of the method was demonstrated by the simulation and in the next step it will be experimentally confirmed.

Acknowledgment

This work was supported by a grant of the Romanian National Authority for Scientific Research and Inovation, CNCS/CCCDI-UEFISCDI, project number 59BG/2016 within PNCDI III.

References

[1] MTT-DCF 1972 *Instrucţia pentru întreţinerea tehnică şi repararea instalaţiilor liniilor de contact ale căii ferate electrificate, Nr.353,* Centrul de documentare şi publicaţii tehnice, Romania

[2] Ministerul Transporturilor, Construcţiilor şi Turismului 2006 *Ordinul nr. 2256/2006 privind aprobarea Instrucţiunilor pentru diagnoza căii şi liniei de contact efectuată cu automotorul TMC*, Romania

[3] Demian D, Rachid A, Gîrniţă I, Alic C I, Rusu-Anghel S, Miklos C C and Roşeanu St 2011 *Contributions to the Development and Improvement of Maintenance Management of Catenary Networks for Railway*, PACIFIC 2011 - International Conference on: Pantograph-Catenary Interaction Framework for Intelligent Control, Conference Proceedings, Amiens, France

[4] Ambrósio J, Pombo J, Pereira M, Antunes P and Mósca A 2012 Recent Developments in Pantograph-Catenary Interaction Modelling and Analysis, *International Journal of Railway*

Technology **1**(1) 249-278
[5] Zhang W, Mei G, Wu X and Shen Z 2002 Hybrid simulation of dynamics for the pantograph-catenary system, *International Journal of Vehicle Mechanics and Mobility* **38**(6) 393-414
[6] Zhou N and Zhang W 2011 Investigation on dynamic performance and parameter optimization design of pantograph and catenary system, *Finite Elements in Analysis and Design* **47**(3) 288-295
[7] Ritzberger D, Talic E and Schirrer Al 2015 *Efficient simulation of railway pantograph/ catenary interaction using pantograph-fixed coordinates*, IFAC-PapersOnLine 48-1 pp 061–066
[8] Alic C, Alic (Miklos) C and Miklos Z I 2009 Upon the Actual Tendencies in Modeling and Simulating the Behavior of the Pantograph-Catenary Pairing Machine Design, *Anniversary Volume of the Faculty of Technical Sciences Novi Sad*
[9] Rob R and Rat C 2013 Mathematical model for studying the variation of the electrical parameters in functioning of nonlinear loads, *AIP Conference Proceedings* **1558** 1345-1348
[10] Collina A and Bruni S 2002 Numerical simulation of pantograph-overhead equipment interaction, *Vehicle System Dynamics* **38** 261-291
[11] Massat J P 2007 *Modélisation du comportement dynamique du couple pantographe-caténaire*, L'École Centrale de Lyon, France, Doctoral Thesis
[12] Rusu N, Averseng J, Miklos C, Alic C and Anghel (Rusu-Anghel) S 2006 *Dynamic modeling of pantograph - catenary system for energy loss control*, IEEE - TTTC, International Conference AQTR 2006, Cluj-Napoca, Romania
[13] Rusu-Anghel S, Gârniţă I, Gavrilă I, Pănoiu C, Pănoiu M and Mezinescu S 2010 *Advanced System for the Control of Work Regime of Railway Electric drive Equipment*, 6th WSEAS International Conference on Energy, Environment, Ecosystems and Sustainable Development (EEESD 10), Timişoara, Romania, pp 115-121
[14] Rusu-Anghel S, Miklos C, Averseng J and Mezinescu S 2010 *Control System for Catenary - Pantograph Dynamic Interaction Force*, Proceedings of 2010 International Joint Conference on Computational Cybernetics and Technical Informatics (ICCC - CONTI), Timişoara, Romania, pp181-186
[15] Lopez-Garcia O, Carnicero A and Marono J L 2007 Influence of Stiffness and contact modelling on catenary - pantograph system dynamics, *Journal of Sound and Vibration* **299(**4–5) 806-821
[16] ***OSCAR SAM 5.0 – User guide
[17] Mitsuo A 2007 Precise Measurement and Estimation Method for Overhead Contact Line Unevennes, *Electrical Engineering in Japan* **160**(2) 77-85

IOP Conf. Series: Materials Science and Engineering **200** (2017) 012057 doi:10.1088/1757-899X/200/1/012057

Factors that increase the electric field of the dielectric barrier ozone generators

I Ganea and R Morar

Technical University, Faculty of Electrical Engineering, Memorandumului str., no. 28, 400114 Cluj-Napoca, Romania

E-mail: ioan.ganea@ethm.utcluj.ro

Abstract. The dielectric barrier discharge cell (DBDC), also called ozone cell, is the main component of the ozone generators widely used. When the cell is powered with a high voltage value (> 5 kV) in the dielectric gas, arises a high intensity electric field that influences ozone generation processes by silent discharge also called dielectric barrier discharge. The intensity of the electric field depends by a number of factors such as the constructive form of the cell, the cell sizes, the value and the waveform of the supply voltage, the type of insulators used as dielectric barrier. The insulators which constituting the dielectric barrier, influence the intensity of the electric field through the dielectric constant of the liquid or solid shows the values of intensity electrical fields from ozone cells, with plane electrodes and ozone cells with cylindrical electrodes. The paper presents a graphic variation of the electric field from the gap and the dielectric barrier of cells with plane electrodes and cylindrical collinear electrodes. Experimental research highlights current-voltage characteristic for several types of dielectrics. Distilled water was highlighted as dielectric with the best results.

1. Introduction

For applications using ozone (O_3) are used generators (ozonizers) which have capabilities very small and small (micro and miniozonizers). These are composed of one or more DBDCs. Ozone cell consists of two electrodes of different shapes, separated by a space, partially occupied by a solid dielectric. Solid dielectric prevents discharge in the gap. A high voltage applied between the electrodes of DBDC, causes ionization processes into the gaseous gap which finally lead to ozone generation.

Generators of ozone are made up of cells with plane parallel and coaxial cylindrical electrodes [1], [2]. Ionization processes in the gaseous gap are heavily depended by the value of the electric field that arises when the cell electrodes are supplied with high voltage. In its turn, the electric field depends by many factors. The increase of the electric field into the gap cell leads to increasing the amount of generated ozone [3].

2. Calculation of intensity electric fields of the ozone cells

Dielectric barrier discharge cell (DBDC) is an enclosure having a form by rectangular parallelepiped or right cylinder bounded by two electrodes. The electrodes are separated to each other by a space occupied partially by a solid dielectric layer. Unoccupied volume between the electrodes and solid dielectric (gaseous or discharge gap), is used for moving the working gas. The electrodes are either two conductive plane parallel or coaxial cylinders [3], [4].

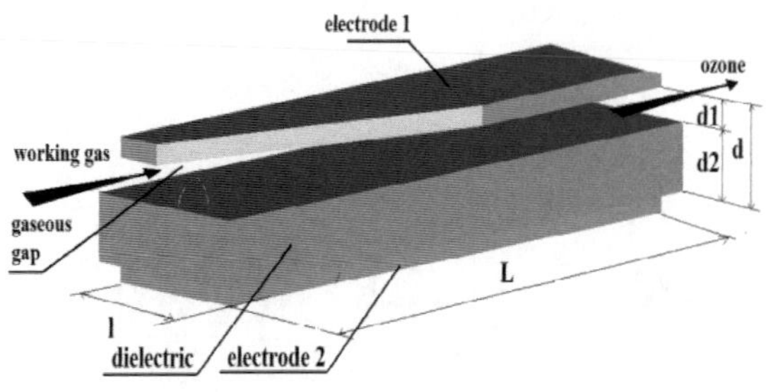

Figure 1a. Ozone cell with plane parallel electrodes

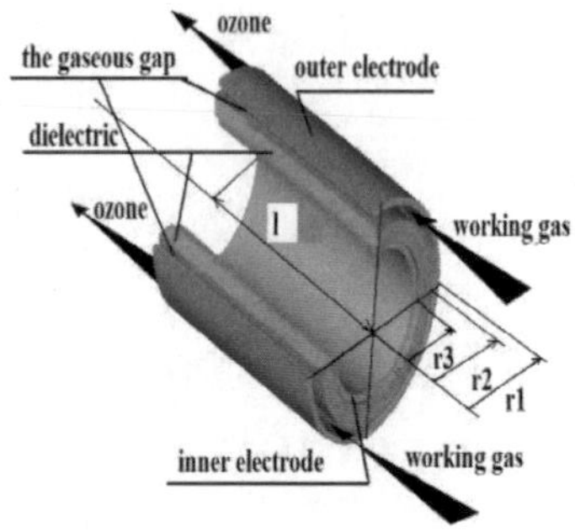

Figure 1b. Ozone cell with coaxial cylindrical electrodes

In Figures 1a, 1b are shown schematic diagrams of a cell with plane parallel electrodes and a cell with coaxial cylindrical electrodes. To develop theoretical considerations, we go from following conditions that allow neglecting edge effects:

- the electric field is uniform inside the cells;
- the value of the electric field intensity, outside the cells is zero;
- charge density, ρ, is constantly and evenly distributed over the entire surface of the electrodes;
- linear dimensions of electrodes (length and width) are much higher compared to the distance between them.

To properly size an ozone cell there is important to know the intensities of the electric fields into the solid dielectric layer and into the gap which depend on the geometrical dimensions of the cell and the voltage value.

2.1. The ozone cell with plane parallel electrodes (Figure 1a)
The intensity of the electric field inside a gaseous gap of a cell with plane parallel electrodes is given by the following relationship [3]:

$$E_1 = \frac{\varepsilon_2 \cdot U}{\varepsilon_1 \cdot d_2 + \varepsilon_2 \cdot d_1} \tag{2.1}$$

E_1 –electric field intensity of the discharge gap,
$\varepsilon 1$ - permittivity of the working gas (air or oxygen),
$\varepsilon 2$ - permittivity of the solid dielectric,
d1 - gaseous gap thickness,
d2 - solid dielectric thickness

$$E_2 = \frac{\varepsilon_1 \cdot U}{\varepsilon_1 \cdot d_2 + \varepsilon_2 \cdot d_1} \tag{2.2}$$

E_2 - electric field strength in the solid dielectric
We can write the following

$$d = d_1 + d_2 \tag{2.3}$$

$$E_1 = \frac{\varepsilon_2 \cdot U}{\varepsilon_1 \cdot d \cdot \left(1 + \left(\frac{\varepsilon_2}{\varepsilon_1} - 1\right) \cdot \frac{d_1}{d}\right)} = \frac{U}{d} \cdot \frac{\varepsilon_2 / \varepsilon_1}{\left(1 + \left(\frac{\varepsilon_2}{\varepsilon_1} - 1\right) \cdot \frac{d_1}{d}\right)} = E_0 \cdot \frac{\varepsilon_2 / \varepsilon_1}{\left(1 + \left(\frac{\varepsilon_2}{\varepsilon_1} - 1\right) \cdot \frac{d_1}{d}\right)} \quad (2.4)$$

$$E_2 = \frac{\varepsilon_1 \cdot U}{\varepsilon_1 \cdot d \cdot \left(1 + \left(\frac{\varepsilon_2}{\varepsilon_1} - 1\right) \cdot \frac{d_1}{d}\right)} = \frac{U}{d} \cdot \frac{1}{\left(1 + \left(\frac{\varepsilon_2}{\varepsilon_1} - 1\right) \cdot \frac{d_1}{d}\right)} = E_0 \cdot \frac{1}{\left(1 + \left(\frac{\varepsilon_2}{\varepsilon_1} - 1\right) \cdot \frac{d_1}{d}\right)} \quad (2.5)$$

$$E_0 = \frac{U}{d} \quad (2.6)$$

In relationships (2.4), (2.5), (2.6) appears expression U/d noted E_0 having the electric field dimension.

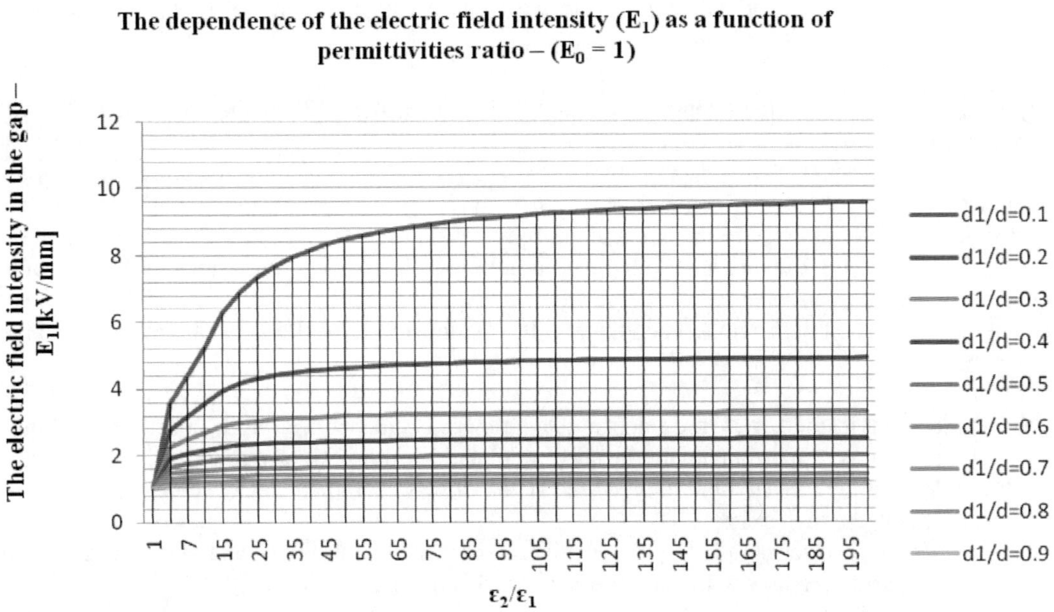

Figure 2a. Graphic representation of the electric field intensity (E1) in the gap of DBDC with planar electrodes (E0 = 1)

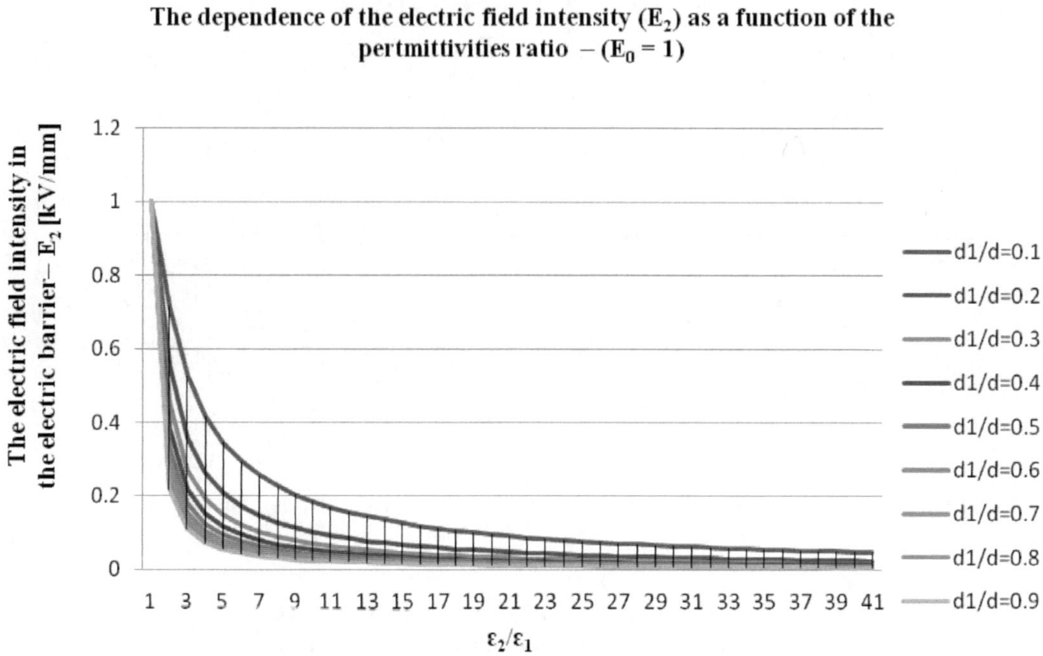

Figure 2b. Graphic representation of the electric field intensity (E2) in the gap of DBDC with planar electrodes (E0 = 1)

The authors of this paper define this report (2.6), the electric field strength inside the DBDC when whole space between the electrodes is occupied by a insulator with relative permittivity equal with unity. In Figures 2a and 2b there are plotted the variation of electric field intensities in the gap, and in the dielectric barrier as a function of $\varepsilon_2/\varepsilon_1$ (relationships (2.4), (2.5) for several values of d_1 (value of d is considered a constant parameter). The value of the relationship $E_0 = U/d$ is equal with unity in these representations.

Conclusions:

- in any point inside DBDC the electric field strength is independent of the coordinates of the point, and it is proportional with the supply voltage U of the cell electrodes;
- the value of the electric field in the gaseous gap differs from the value of the electric field intensity in the dielectric;
- the electric field strength in the gap (E_1) increases with the ratio $\varepsilon_2/\varepsilon_1$ and the electric field int the dielectric decreases with the same ratio (d_1/d is maintained constantly)
- the variations of the electric field strength in the gap and the electric field in the dielectric occur after a nonlinear law;
- the intensity of the electric fields in the gap and in the dielectric decreases when d_1/d increases ($\varepsilon_2/\varepsilon_1$ is maintained constantly).

2.2. The ozone cell with coaxial cylindrical electrodes (Figure 1b)

The ozone cell with coaxial cylindrical electrodes is composed of two coaxial cylindrical metal electrodes between which is a solid dielectric.

The geometrical parameters of the cell (Figure 1b and 3)

r_1- the radius of the outer cylinder

r_2- the radius of the cylindrical surface that separates the two dielectrics

r_3- the radius of the inner cylinder

l - the length of the ozone cell

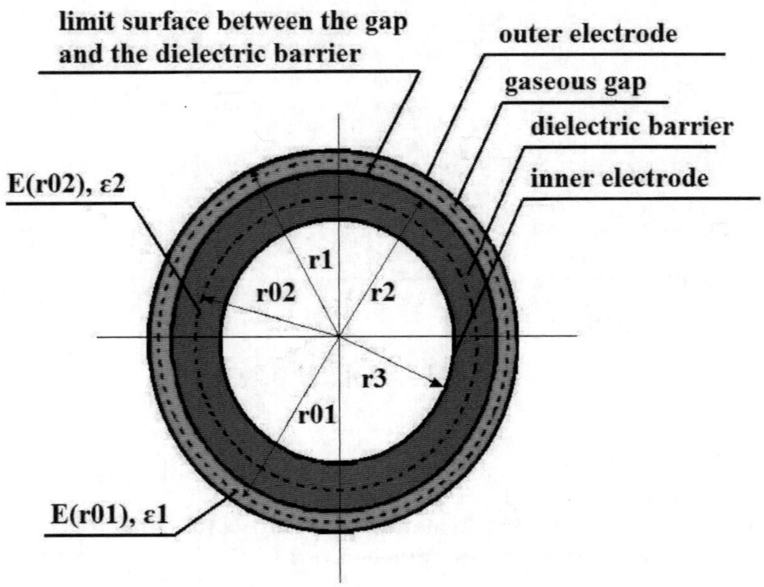

Figure 3. Schematic diagram of a DBDC with collinear cylindrical electrodes

Between r_1, r_2, r_3 there is the relationship:

$$r3 < r2 < r1 \tag{2.7}$$

The intensity of the electric field in the gap (E_1) and the electric field intensity in the dielectric (E_2) are:

$$E1(r01) = \frac{U}{r01} \cdot \frac{\varepsilon 2 \cdot}{\left(\varepsilon 1 \cdot \ln \frac{r2}{r3} + \varepsilon 2 \cdot \ln \frac{r1}{r2}\right)} \quad (r2 < r01 < r1) \text{ [5]} \tag{2.8}$$

ε_1 – the permittivity of the gas (air or oxygen)
ε_2 – the permittivity of the solid insulator
r_{01} – the coordinate of a point in the gap where E_1 is calculated

$$E2(r02) = \frac{U}{r02} \cdot \frac{\varepsilon 1 \cdot}{\left(\varepsilon 1 \cdot \ln \frac{r2}{r3} + \varepsilon 2 \cdot \ln \frac{r1}{r2}\right)} \quad (r3 < r02 < r2) \text{ [5]} \tag{2.9}$$

r_{02} – the coordinate of a point in the solid dielectric where E_2 is calculated
Between the rays r_1, r_2, r_3 we write the equation:

$$\ln \frac{r1}{r3} = \ln \frac{r1}{r2} + \ln \frac{r2}{r3} \tag{2.10}$$

Expressing the values of E_1 and E_2 as functions of the $\varepsilon_2/\varepsilon_1$ and $\ln(r_1/r_2)/\ln(r_1/r_3)$, the relationships (2.8) and (2.9) become:

$$E_1(r_{01}) = \frac{U}{r_{01} \cdot \ln \frac{r_1}{r_3}} \cdot \frac{\varepsilon_2 / \varepsilon_1}{\left(1 + \left(\frac{\varepsilon_2}{\varepsilon_1} - 1 \right) \cdot \ln \frac{\frac{r_1}{r_2}}{\frac{r_1}{r_3}} \right)} \quad (r_2 < r_{01} \leq r_1) \tag{2.11}$$

$$E_2(r_{02}) = \frac{U}{r_{02} \cdot \ln \frac{r_1}{r_3}} \cdot \frac{1}{\left(1 + \left(\frac{\varepsilon_2}{\varepsilon_1} - 1 \right) \cdot \ln \frac{\frac{r_1}{r_2}}{\frac{r_1}{r_3}} \right)} \quad (r_3 < r_{02} \leq r_2) \tag{2.12}$$

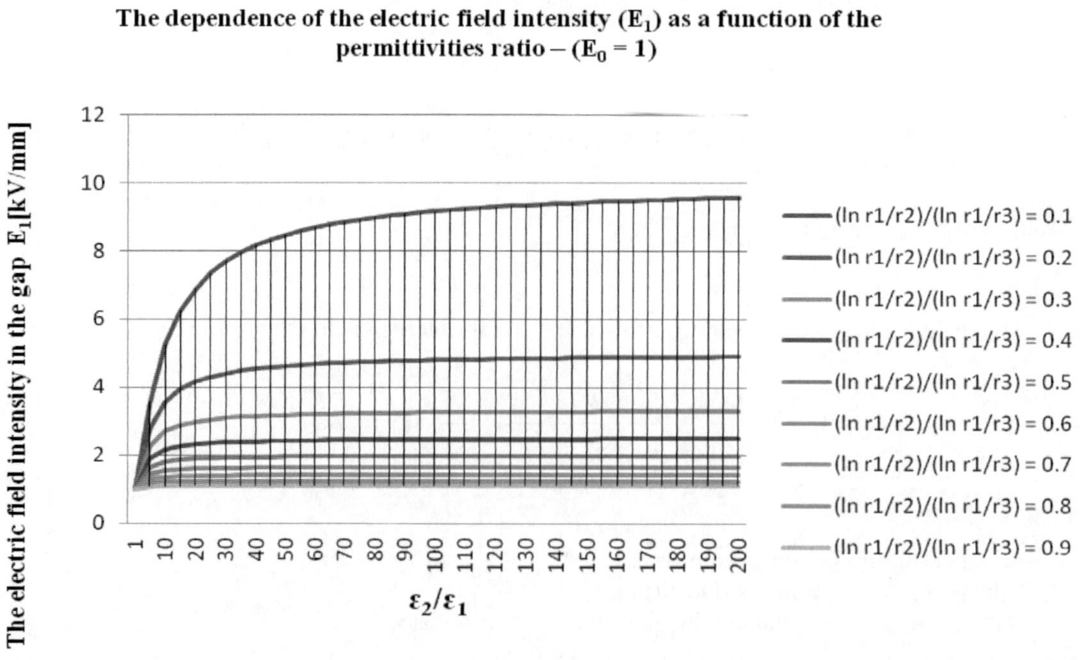

Figure 4a. Graphic representation of the electric field intensity (E1) in the gap of the DBDC with cylindrical electrodes; E01(r01) = 1

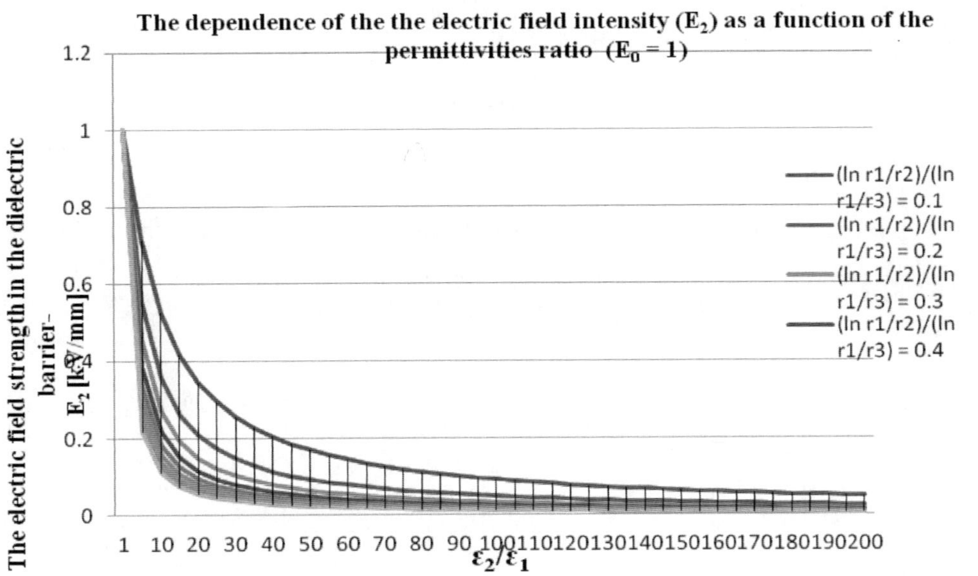

Figure 4b. Graphic representation of the electric field intensity (E2) in the gap of the DBDC with cylindrical electrodes; E02(r01)=1

We denote by:

$$E0(r01)=\frac{U}{r01\cdot\ln(r1/r3)} \tag{2.13}$$

$$E0(r02)=\frac{U}{r02\cdot\ln(r1/r3)} \tag{2.14}$$

Conclusions:

- the value of the electric field in the discharge gap and in the dielectric barrier is dependent by the coordinate where the point is considered;
- the value of the electric intensity in any point on a cylindrical surface of the radius r_{01}, r_{02} (Figure 3) is the same and coincides with the direction of the radius of the surface;
- the value of the electric field in the gap is lower than in dielectric;
- $\varepsilon_2/\varepsilon_1$ increases the electric field intensity in the gap and decreases the dielectric field strength for a given value of the ratio d_1/d;
- the growth of the d_1/d lowers the electric field intensity in the gap and in the dielectric layer when the ratio $\varepsilon_2/\varepsilon_1$ is kept constantly.

3. Similarities between expressions of the electric field intensity of DBDC with plane parallel and coaxial cylindrical electrodes

Between relationships (2.4), (2.5), (2.11), (2.12) we can make a formal resemblance. In the following we write a general relationship of the intensities of the electric fields in the gap and in the dielectric barrier.

These relationships are not depended by the shape of the cell electrodes

IOP Conf. Series: Materials Science and Engineering **200** (2017) 012057 doi:10.1088/1757-899X/200/1/012057

$$E_1 = E_0 \cdot \frac{a}{(1+(a-1)\cdot b)} \tag{3.1}$$

$$E_2 = E_0 \cdot \frac{1}{(1+(a-1)\cdot b)} \tag{3.2}$$

Table 1. The correspondence between relationships of a planar electrodes DBDC and a DBDC with cylindrical electrodes

	The DBDC with plane parallel electrodes	The DBDC with coaxial cylindrical electrodes	Obs.
1	$E_1 = \frac{U}{d} \cdot \frac{\varepsilon_2/\varepsilon_1}{\left(1+\left(\frac{\varepsilon_2}{\varepsilon_1}-1\right)\cdot\frac{d_1}{d}\right)}$	$E_1(r_{01}) = \frac{U}{r_{01}\cdot\ln\frac{r_1}{r_3}} \cdot \frac{\varepsilon_2/\varepsilon_1}{\left(1+\left(\frac{\varepsilon_2}{\varepsilon_1}-1\right)\cdot\frac{\ln\frac{r_1}{r_2}}{\ln\frac{r_1}{r_3}}\right)}$ $(r_2 < r_{01} \le r_1)$	
2	$E_2 = \frac{U}{d} \cdot \frac{1}{\left(1+\left(\frac{\varepsilon_2}{\varepsilon_1}-1\right)\cdot\frac{d_1}{d}\right)}$	$E_2(r_{02}) = \frac{U}{r_{02}\cdot\ln\frac{r_1}{r_3}} \cdot \frac{1}{\left(1+\left(\frac{\varepsilon_2}{\varepsilon_1}-1\right)\cdot\frac{\ln\frac{r_1}{r_2}}{\ln\frac{r_1}{r_3}}\right)}$ $(r_3 < r_{02} \le r_2)$	
3	$E_0 = \frac{U}{d}$	$E_0(r_{0(1,2)}) = \frac{U}{r_{0(1,2)}\cdot\ln(r_1/r_3)}$	
4	d_1	$\ln\frac{r_1}{r_2}$	
5	d_2	$\ln\frac{r_2}{r_3}$	
6	d	$\ln\frac{r_1}{r_3}$	
7	$\varepsilon_2/\varepsilon_1$	$\varepsilon_2/\varepsilon_1$	A
8	d_1/d	$\ln\frac{r_1}{r_2} / \ln\frac{r_1}{r_3}$	B

The terms of this relationship have different expressions as a function of the shape of the electrodes. Table 1 shows the similarities of the formal expressions of the parameters E_1, E_2, E_0, a, b, (3.1), (3.2) regardless of the shape of the electrodes cells. Figures 2a, 2b, 4a, 4b suggests an upper limit of the values of the electric field intensity into the gap (E_1) when the expression $\varepsilon_2/\varepsilon_1$ tends to infinity. This value is given by the expression:

$$E_1 = \frac{E_0}{b} \tag{3.3}$$

Substituting the values of b from Table 1 in the expression (3.3) we obtain the limit values for the electric field strength E_1 in the cell with plane-parallel electrodes:

$$E1 = \frac{U}{d1} \tag{3.4}$$

The limit value of E_l in the gaseous gap of the DBDC with coaxial cylindrical electrodes is:

$$E1(r01) = \frac{U}{r01 \cdot \ln(r1/r2)} \tag{3.5}$$

The limit value of E_2 in the dielectric barrier of the DBDC when the value of the ratio $\varepsilon_2/\varepsilon_1$ tends to infinity is zero regardless of the shape of the electrodes.

4. Experimental results

For the experimental verification of the theoretical results derived in the above presentation the authors used as a dielectric the distilled water, the transformer oil, the acrylic resin, the ordinary glass, and the polyethylene. The experiments were achieved in the laboratory of the intense fields of the Faculty of Electrical Engineering of the Technical University of Cluj-Napoca.

The experimental bench (Figures 5a, 5b) contained a dielectric barrier discharge cell having two plane parallel electrodes with the following geometric dimensions: L = 70 mm (length), l = 25 mm (width), d = 12 mm (distance between the cell electrodes).

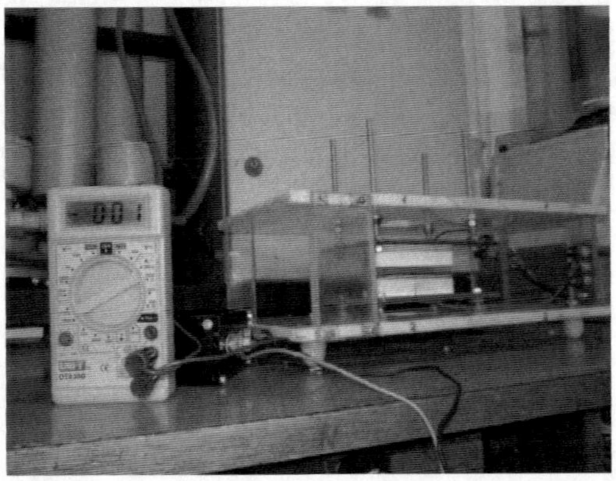

Figure 5a. Experimental bench (front view)

Figure 5b. Frontal view of the DBDC electrodes

The cell employed allowed experiments with solid or liquid dielectrics. In the frame of the experimental procedure values of the electric field into the gap was indirect determined by measuring the voltage value when the ionization processes started.

The voltage current characteristics was plotted for each dielectric employed, by increasing the voltage in the range of 0 – 20 kV and measuring the current through the cell. The gas ionization phenomena were manifested by the appearance of a luminescent violet filaments between electrode and dielectric barrier, accompanied by a specific noise [5].

The light and the noise intensity were intensified when the voltage is progressively increased. The voltage value at which begins the process of the ionization is called ionization voltage (U_{ion}) kV and the value of the electric field into the gap E_{1ion} [kV/mm] is called ionization electric field. Using equation (2.1.4), the values of the electric field in the gaseous gap was calculated considering that the electric field intensity is constantly inside the cell. The air was used as working gas for experimental determinations. Ozone cell was powered with sinusoidal voltage of 50 Hz in range of 0 – 20 kV. For every type of the dielectric was plotted more voltage-current characteristics [6] for several values gap (d_1) maintaining constantly distance value between electrodes.

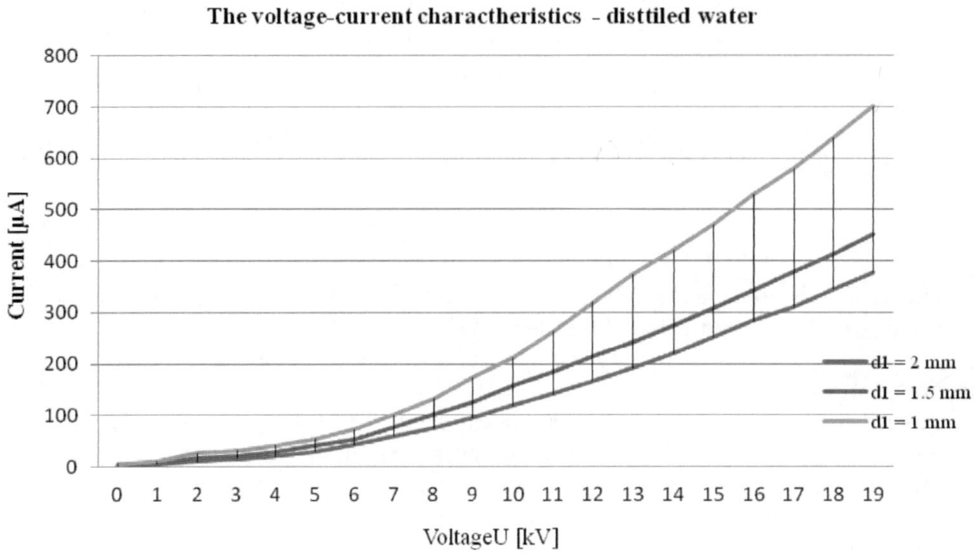

Figure 6. The voltage current characteristics of a DBDC with dielectric barrier distilled water

Figure 6 shows graphic representations of voltage-current characteristics for distilled water ($\mathcal{E} = 81$), and transformer oil ($\mathcal{E} = 2.5$) for several values of d_1 keeping the value d constant at 12 mm.

In the Figure 7 are represented voltage-current characteristics plotted when in DBDC was used distilled water, glass, polyethylene, acrylic resins, transformer oil. Table 2 contains voltage values U_{ion} at which were ionization processes and calculated values of the electric field in the gaseous gap (E_1) for a discharge cell with plane parallel electrodes with the geometrical parameters: L = 70 mm, l = 25 mm, d = 12 mm, d_1 = 1.5 mm.

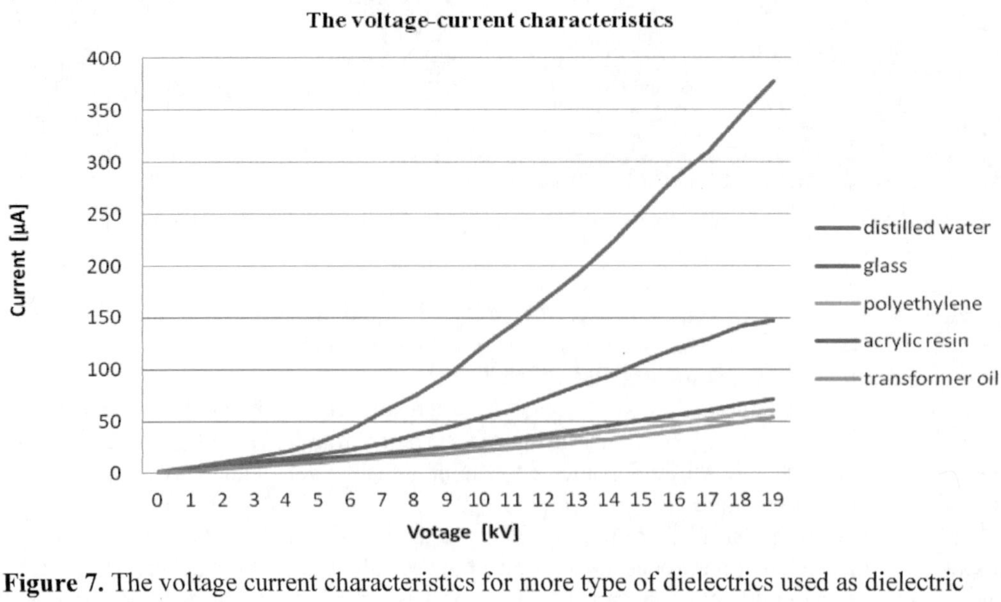

Figure 7. The voltage current characteristics for more type of dielectrics used as dielectric barrier

Electric field intensity E_{1ion} was calculated for dielectric constant of air: $\mathcal{E}_1 = 1.00059$. The calculated values of the dielectric field intensity E_1 in the discharge gap for each type of dielectric tend to 3.5kV/mm (Table 2). These values are influenced by the measurement process.

Table 2 - Ionization voltage of the air into the gap of the DBDC (d = 70 mm, d_1 = 1.5 mm)

	Type of the dielectric	ε_2	U_{ion} [kV]	E_{1ion} [kV/mm]
1	glass	5.5	9	3.45
2	transformer oil	2.5	13	3.33
3	distilled water	81	5.5	3.37
4	acrylic resin	3.5	11	3.38
5	polyethylene	2.9	12	3.39

5. Conclusions

- the expressions of the electric field intensity in the gap and in the dielectric barrier are not the same shape at a dielectric barrier discharge cell;
- the relationships of the electric field in the gap have the same shape but their coefficients are depended of the shape of electrodes (Table 2);
- value of the electric field intensity in the gap of the ozone cell and in the dielectric barrier proportionally increases with the supply voltage (U);
- electric fields values of a DBDC with plane parallel electrodes are independent of the coordinates of the point where they are calculated;
- into a DBDC with coaxial cylindrical electrodes the values of electric field intensity are dependent by coordinates of the point
- rising the value of $\varepsilon_2/\varepsilon_1$ increases the value of the electric field in the gap, and decreases the value of the electric field intensity in the dielectric barrier for given value of d_1/d;
- rising the value ratio d_1/d for given values of $\varepsilon_2/\varepsilon_1$ decreases the electric field intensity in the gap and in the dielectric barrier;
- the distilled water may form a dielectric barrier for an ozone cell because its high value relative permittivity determines an intense electric field into the gap.

References

[1] Suarasan I 2000 *Generarea si utilizarea ozonului*, Editura ETA Cluj-Napoca, Romania

[2] Kogelschatz U 2003 Dielectric – barrier Discharges: Their History Discharge Physics and Industrial Applications, *Plasma Chemistry and Plasma Processing* **23**(1)

[3] Ciupa R 2006 *Bazele electrotehnicii: teorie si aplicatii*, Vol 1 and 2, Casa Cartii de Stiinta, Cluj-Napoca, Romania

[4] Sora C 1982 *Bazele Electrotehnicii* Editura Didactica si Pedagogica, Bucuresti, Romania

[5] Kogelschatz U, Eliasson B and Egli W 1997 Dielectric Barrier Discharges. Principle and Applications, *J. Phys IV France 7 (1997) Colloque C4, Supplement au Journal de Physique III*

[6] Samoilovich V G, Gibalov V I and Kozlov K V 1997 *Physical Chemistry of the Barrier Discharge* (in Russian) Moscow State University (1989), English translation: J. P. F. Conrads, f. Leipold, (Eds), DVS – Verlag GmbH, Dusseldorf (1997)

IOP Conf. Series: Materials Science and Engineering **200** (2017) 012058 doi:10.1088/1757-899X/200/1/012058

Influence of the solid dielectric over the electric field from the ozone cell gap with double dielectric barrier

I Ganea

Technical University, Faculty of electrical Engineering, Memorandumului Str. No. 28 400114 Cluj – Napoca, Romania

E-mail: ioan.ganea@ethm.utcluj.ro

Abstract. The distilled water has the advantage of high value dielectric constant (Ɛ = 81) in relation to ceramic glass materials, currently used for constructing the dielectric barrier. It was necessary to build a thin-walled enclosure of solid insulating material that contain distilled water to achieve a dielectric barrier. This was necessary to avoid exposing the liquid to the direct action of ozone. Dielectric permittivity of the solid dielectric material and the thickness of these walls have diminished the value of the electric field form the gaseous gap of the ozone cell compared to the case with the dielectric barrier from distilled water. The author of this work deduced theoretical relationships that express the values of the electric field intensity in the gap of the cell with two dielectrics and compared them with similar relationships of the intensity of the electric field from the gap of the ozone cell with one dielectric. In this work the author presenting experimental results which confirm the theoretical deducting regarding the use of the solid dielectric as enclosure for the liquid dielectric.

1. Introduction

Generating ozone (O_3) by the action of an intense electric field on air or oxygen is the most common method to produce ozone. Air or oxygen is generic called "working gas". Ionization processes occurring in the working gas cause ozone, as a result of the action of the intense electric field. The process of the generating ozone by the action of an intense electric field over air or oxygen is called dielectric barrier discharge, and the device where the process takes place is called dielectric barrier discharge cell (DBDC) or ozone cell.

Ozone cells (Figures 1a, 1b) are composed of a layer of insulating material (dielectric barrier) with high dielectric strength positioned between two plane parallel electrodes or coaxial cylindrical electrodes. Between the dielectric layer and one of the electrodes is an empty space (gaseous gap) by circulating the working gas during exposure to electric field action. High dielectric strength prevents destructive breakdown to the action of the intense electric field generated between the electrodes of the ozone cell when they are powered with alternative high value voltage (5 – 30 kV).

The insulator used in construction of the ozone cell is a ceramic glass material which has the disadvantage of a low dielectric relative permittivity (ε_2) situated in the range of 8 ÷ 12 but it is resistant to mechanical shocks and high temperatures [1]. Electric field strength in the gap of the ozone cell is dependent by the permittivity of the dielectric barrier.

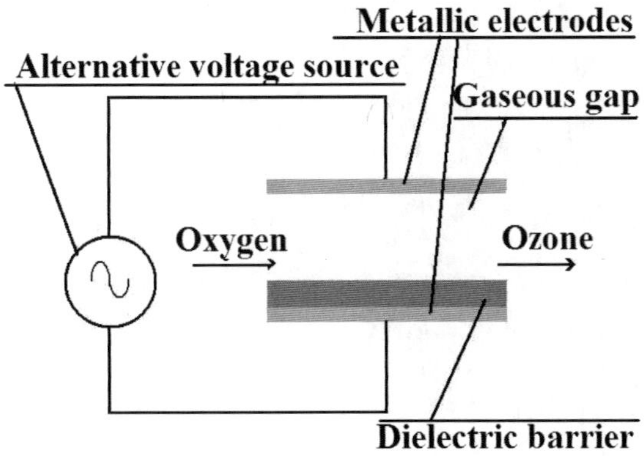

Figure 1a. Ozone cell with plane parallel electrodes

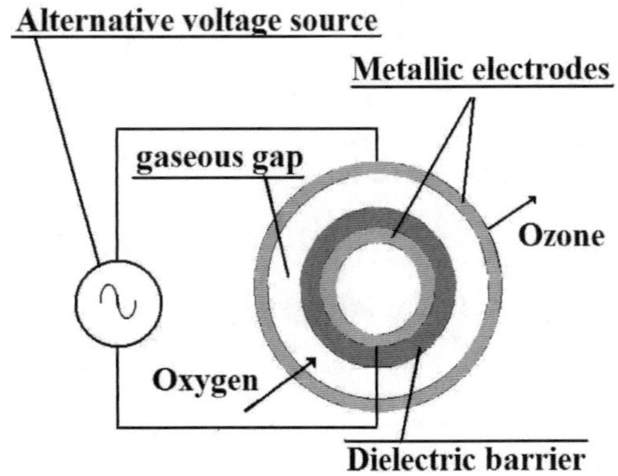

Figure 1b. Ozone cell with coaxial cylindrical electrodes

2. Ozone cells with double dielectric barrier (liquid and solid)
The main disadvantage of the ceramic glass materials, the low dielectric constant, can be removed using a dielectric liquid with high permittivity, distilled water, which has dielectric constant 81.

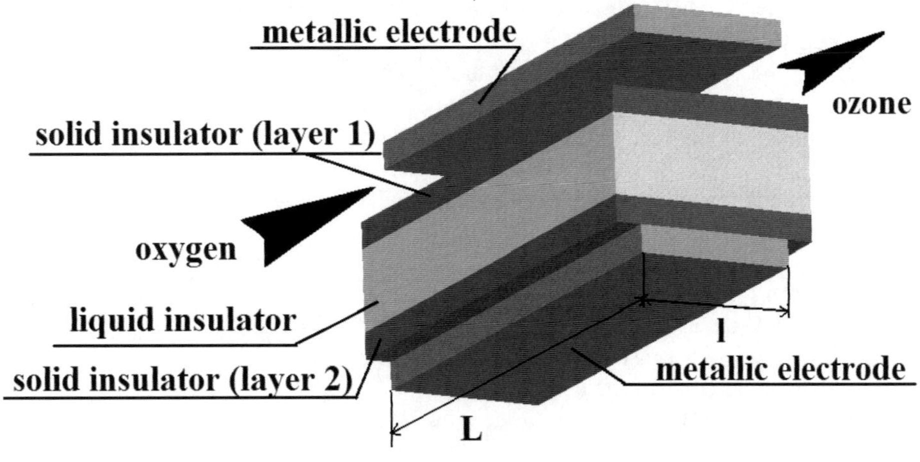

Figure 2. Ozone cell with metallic plane parallel electrodes

In this case dielectric barrier is composed of a liquid material which does not have its own form but can occupy any volume.

Since the dielectric barrier size and its position in relation to the electrodes have a very important role in the functioning of the DBDC, it is required creating a insulating a insulating solid enclosure that contains liquid insulator with high dielectric constant.

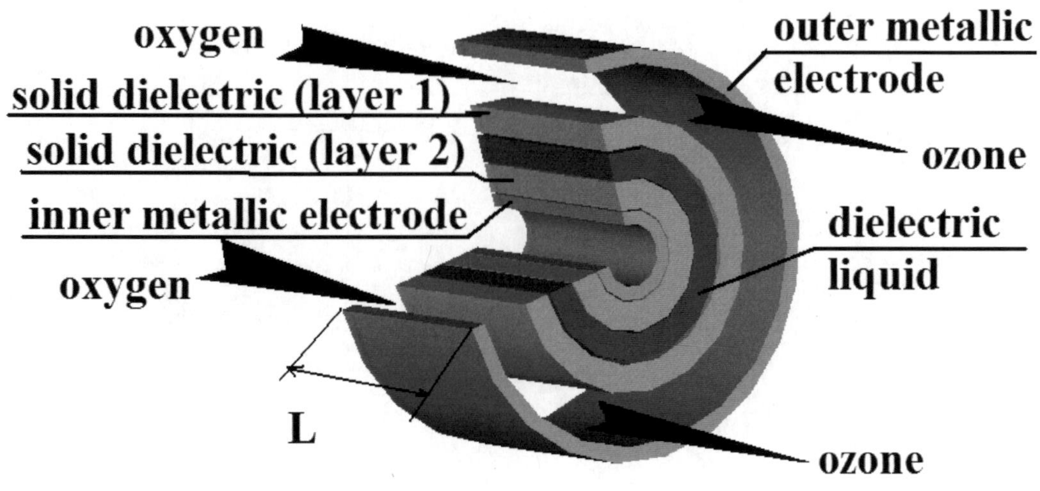

Figure 3. Ozone cell with coaxial cylindrical electrodes

Consequently, the dielectric barrier will consist of three layers: two layers of the solid dielectric which contain a layer of the dielectric liquid.

This type of the barrier is called by the author double dielectric barrier. The solid layers have a technological role, to give an adequate shape to the dielectric liquid layer, and to put this layer in a properly position into the cell.

These layers affect the electric field strength in the gap and in the all others layers of the dielectric barrier. The dielectric barrier shape depends by the type of the cell electrodes. Figures 2 and 3 show the schematic diagrams of the dielectric barrier discharge cells.

In the following the theoretical relationships of the electric field intensity in the gap and in the layers of the dielectric barrier will be deducted for a cell with the plane parallel electrodes and a cell with coaxial cylindrical electrodes.

2.1. Dielectric barrier discharge with plane parallel electrodes (Figure 4)

The cell consists of two metallic plane electrodes with rectangular shape and negligible thickness parallel between them and a plane insulating layer which is the dielectric barrier.

The geometrical characteristics of the cell are:

- the distance between electrodes (d), the length (L) and width (l) of the electrodes (Figures 2 and 4);
- the thicknesses of the insulating layers forming the dielectric barrier the first solid layer (d_{21}), the liquid dielectric layer (d_3), the second solid layer (d_{22}) and thickness of the gap (d_1) (Figure 4);
- dielectric constant of: the gap (ε_1), the solid dielectric layers (ε_2), the liquid dielectric layer (ε_3).

The dielectric barrier is made up of two solid dielectric layers of the same material between these layers there being layer liquid with the permittivity greater than the solid material. Figure 4 shows a schematic diagram of the cell with plane parallel electrodes where are illustrated the main geometrical features of the cell.

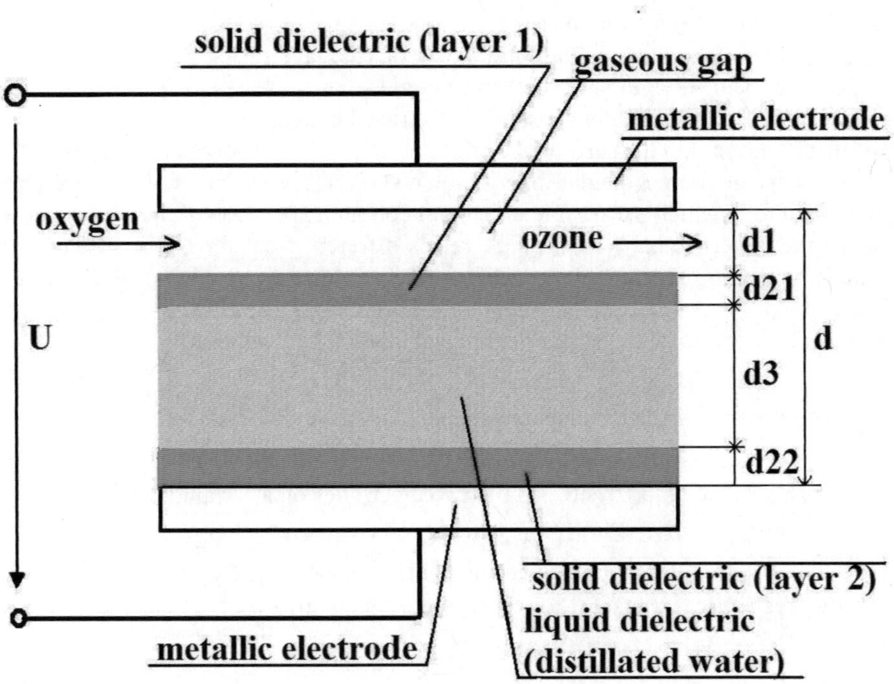

Figure 4. Ozone cell with plane parallel electrodes (double dielectric barrier)

DBDC can be removed electrically as a capacitor with three dielectrics and four layers. One of dielectrics is the working gas flowing through the gap and the others two are constituted of the dielectric barrier (the solid and liquid insulator).

We shall deduce the expressions of the electric field strength in the layers of the cell in the assumption that the value of the electric field strength outside the cell is zero and in a point inside the DBDC is [2]:

$$E = \frac{\rho}{\varepsilon} \tag{2.1}$$

ρ – the surface density on the surface of the electrodes

Ɛ – the dielectric constant of the medium in which the field is calculated

The values of the electric fields intensity into the four dielectric layers are:

$$E1 = \frac{\rho}{\varepsilon 1} \tag{2.2}$$

$$E21 = \frac{\rho}{\varepsilon 2} \tag{2.3}$$

$$E3 = \frac{\rho}{\varepsilon 3} \tag{2.4}$$

$$E22 = \frac{\rho}{\varepsilon 2} \tag{2.5}$$

E_1 – the electric field strength in the gap
E_{21} – the electric field strength in the solid dielectric (the layer 1)
E_3 – the electric field strength in the dielectric liquid
E_{22} – the electric field strength in the solid dielectric (the layer 2)
ρ – the surface density on the surface of a plane electrode is considered constantly

The field lines are perpendicular line segments on the surface of the electrodes but in the simplifying hypothesis which we assume into a real cell the expression of the electric field intensity in a point inside the dielectric barrier discharge cell is different from the expression (2.1) and the field lines are curve segments. The curvature of the field lines toward the edge of the electrodes increases (edge effect).

Into a real cell to minimize the edge effect must be meet the conditions:

$$d \ll l \ , d \ll L \tag{2.6}$$

We can write the equation between parameters d_1, d_{21}, d_3, d_{22} (Figure 4):

$$d = d_1 + d_{21} + d_3 + d_{22} \tag{2.7}$$

The relationship of the voltage between the two electrodes on a certain field line (line segment) is:

$$d = E_1 \cdot d_1 + E_{21} \cdot d_{21} + E_3 \cdot d_3 + E_{22} \cdot d_{22} \tag{2.8}$$

We will express the values of the electric fields intensity E_1, E_{21}, E_3, E_{22}, as a function of the voltage U the dielectric constants ε_1, ε_2, ε_3, the geometrical parameters d_1, d_{21}, d_3, d_{22},using relationships (2.2) ÷ (2.5):

$$E_1 = \frac{U \cdot \varepsilon_2 \cdot \varepsilon_3}{\varepsilon_2 \cdot \varepsilon_3 \cdot d_1 + \varepsilon_1 \cdot \varepsilon_3 \cdot (d_{21} + d_{22}) + \varepsilon_1 \cdot \varepsilon_2 \cdot d_3} \tag{2.9}$$

$$E_{21} = E_{22} = \frac{U \cdot \varepsilon_1 \cdot \varepsilon_3}{\varepsilon_2 \cdot \varepsilon_3 \cdot d_1 + \varepsilon_1 \cdot \varepsilon_3 \cdot (d_{21} + d_{22}) + \varepsilon_1 \cdot \varepsilon_2 \cdot d_3} \tag{2.10}$$

$$E_3 = \frac{U \cdot \varepsilon_1 \cdot \varepsilon_2}{\varepsilon_2 \cdot \varepsilon_3 \cdot d_1 + \varepsilon_1 \cdot \varepsilon_3 \cdot (d_{21} + d_{22}) + \varepsilon_1 \cdot \varepsilon_2 \cdot d_3} \tag{2.11}$$

The expression of the intensity of the electric field E_1 (2.9) using equation (2.7) will be brought to the form:

$$E_1 = \frac{U}{d} \cdot \frac{\frac{\varepsilon_3}{\varepsilon_1}}{\left(1 + \left(\frac{\varepsilon_3}{\varepsilon_1} - 1\right) \cdot \frac{d_1}{d} + \left(\frac{\varepsilon_3}{\varepsilon_2} - 1\right) \cdot \frac{(d_{21} + d_{22})}{d}\right)} \tag{2.12}$$

From relationships (2.10) and (2.11) we obtain relations for the intensity of the electrical fields in both solid and liquid dielectrics (E_{21}, E_{22}, E_3)

$$E_{21} = E_{22} = \frac{U}{d} \cdot \frac{\frac{\varepsilon_3}{\varepsilon_2}}{\left(1 + \left(\frac{\varepsilon_3}{\varepsilon_1} - 1\right) \cdot \frac{d_1}{d} + \left(\frac{\varepsilon_3}{\varepsilon_2} - 1\right) \cdot \frac{(d_{21} + d_{22})}{d}\right)} \tag{2.13}$$

$$E_3 = \frac{U}{d} \cdot \frac{1}{\left(1 + \left(\frac{\varepsilon_3}{\varepsilon_1} - 1\right) \cdot \frac{d_1}{d} + \left(\frac{\varepsilon_3}{\varepsilon_2} - 1\right) \cdot \frac{(d_{21} + d_{22})}{d}\right)} \tag{2.14}$$

2.2. Dielectric barrier discharge cell with coaxial cylindrical electrodes with double dielectric barrier

This type of the cell (Figure 5) has two coaxial cylinders with negligible thickness constituting cell electrodes. Dielectric barrier is composed by three cylindrical layers: a liquid layer between two solid layers. The geometrical characteristics of the cell with coaxial cylindrical electrodes are:

r_1 – the outer electrode radius;

r_{21} – the radius of the first dielectric layer of the solid dielectric with the permittivity ε_2;

r_3 – the radius of the surface of the dielectric layer separating the liquid (dielectric with the permittivity ε_3) and the first solid dielectric layer (dielectric with the permittivity ε_2);

r_{22} – the radius of the surface of separation between the second dielectric layer of the solid and liquid dielectric layer (dielectric with the permittivity ε_2);

r_4 – the radius of the inner electrode of the cell;

L – the length of the cell electrodes (Figure 3).

We deduct mathematical relations expressing intensities of the electric field in the gap and the solid dielectric liquid deducted in the simple assumption that the intensity of the electric field E in a point outside the cell is zero and inside the cell is given by [3], [4]:

$$\vec{E}(r) = \frac{\rho}{2 \cdot \pi \cdot \varepsilon \cdot \vec{r}}, \quad (r21 < r < r1) \tag{2.15}$$

ρ -the surface charge density on the electrodes cell (it is considered constantly in this case)

ε –the dielectric constant of the environment

r – the radius of the cylindrical surface where the point is situated

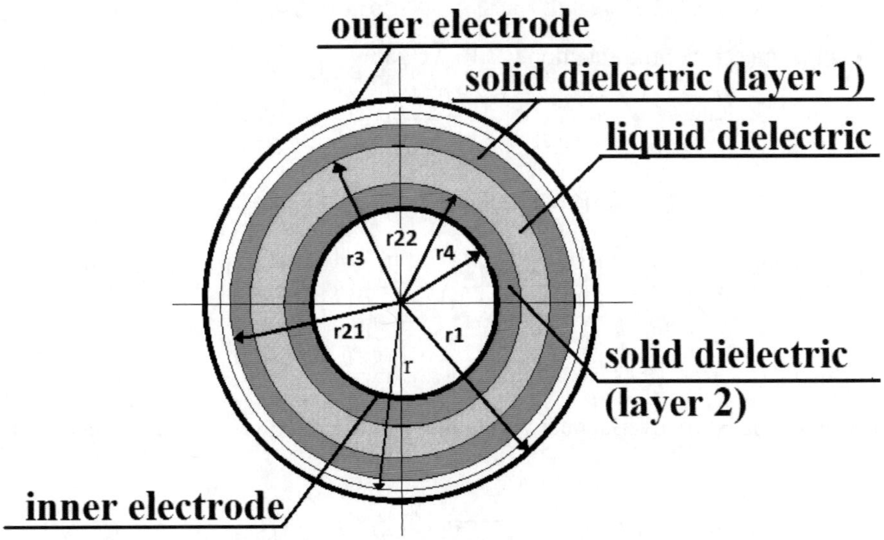

Figure 5. The ozone cell with coaxial cylindrical electrodes (dielectric barrier solid and liquid)

Electric field lines are line segments which overlap with the radius of the cylinders if we idealize the cell. In the real case the field lines are curve segments. The curvature of the field lines increases as approaching the electrode edges (edge effect). To minimize edge effect in a real cell, it must be fulfilled the relationship:

$$(r1 << L) \tag{2.16}$$

Ozone cell can be assimilated to a capacitor with coaxial cylindrical electrodes dielectric having four layers. Voltage U applied to the electrodes of the cell is distributed on each dielectric separately:

$$U = U_1 + U_{21} + U_3 + U_{22} \tag{2.17}$$

U_1 – the voltage drop across the gap
U_{21} – the voltage drop across the first solid dielectric layer
U_3 – the voltage drop across the layer of the dielectric liquid
U_{22} – the voltage drop on the second layer of the solid dielectric
Voltage drop U caused by the dielectric field E_1 in the space between two points a and b is defined by the relationship [3], [4]:

$$U = \int_a^b \vec{E} \cdot d\vec{s} \tag{2.18}$$

Given the relationship (2.2.4) the relationship (2.2.3) becomes:

$$U = \int_{r21}^{r1} \vec{E}_1 \cdot d\vec{s} + \int_{r3}^{r21} \vec{E}_{21} \cdot d\vec{s} + \int_{r22}^{r3} a\, \vec{E}_3 \cdot d\vec{s} + \int_{r4}^{r22} \vec{E}_{22} \cdot d\vec{s} \tag{2.19}$$

E_1 – the electric field intensity of the gap
E_{21} – the electric field intensity of the solid dielectric layer (the layer 1)
E_3 – the electric field intensity in the dielectric liquid
E_4 – the electric field intensity of the solid dielectric (the layer 2)
Given the relationships (2.15) relationship (2.19) becomes:

$$U = \int_{r21}^{r1} \frac{\rho}{2\cdot\pi\cdot\varepsilon_1\cdot r}\cdot ds + \int_{r3}^{r21} \frac{\rho}{2\cdot\pi\cdot\varepsilon_2\cdot r} ds + \int_{r22}^{r3} \frac{\rho}{2\cdot\pi\cdot\varepsilon_3\cdot r}\cdot ds + \int_{r4}^{r22} \frac{\rho}{2\cdot\pi\cdot\varepsilon_2\cdot r} ds \tag{2.20}$$

Calculating the integrals of the equation (2.20) we get:

$$U = \frac{\rho}{2\cdot\pi}\cdot\left(\frac{1}{\varepsilon_1}\cdot\ln\frac{r1}{r21} + \frac{1}{\varepsilon_2}\cdot\ln\frac{r21}{r3} + \frac{1}{\varepsilon_3}\cdot\ln\frac{r3}{r22} + \frac{1}{\varepsilon_2}\cdot\ln\frac{r22}{r4}\right) \tag{2.21}$$

$$\ln\frac{r1}{r4} = \ln\frac{r1}{r21} + \ln\frac{r21}{r3} + \ln\frac{r3}{r22} + \ln\frac{r22}{r4} \tag{2.22}$$

Given the relationship (2.15) the expression of the electric field intensity in a point on the cylindrical surface with the radius r for the gap (E_1) is given by the expression:

$$\vec{E}_1(r) = \frac{\rho}{2\cdot\pi\cdot\varepsilon_1\cdot\vec{r}},\ (r21 < r < r1) \tag{2.23}$$

Since the field lines E_1 are overlapped with radii r_{21}, r_1 the relationship (2.23) can also be written as scalar form:

$$E_1(r) = \frac{\rho}{2\cdot\pi\cdot\varepsilon_1\cdot r},\ (r21 < r < r1) \tag{2.24}$$

We deduce from the relationships (2.21), (2.22), (2.23) the expression:

$$E_1(r) = \frac{U}{r\cdot\ln\frac{r1}{r4}}\cdot\frac{\frac{\varepsilon_3}{\varepsilon_1}}{\left(1+\left(\frac{\varepsilon_3}{\varepsilon_1}-1\right)\cdot\frac{\ln\frac{r1}{r21}}{\ln\frac{r1}{r4}}+\left(\frac{\varepsilon_3}{\varepsilon_2}-1\right)\cdot\frac{\left(\ln\frac{r21}{r3}+\ln\frac{r22}{r4}\right)}{\ln\frac{r1}{r4}}\right)},\quad (r21 < r < r1) \tag{2.25}$$

We denote:

$$E_0(r)=\frac{U}{r\cdot\ln\frac{r1}{r4}} \tag{2.26}$$

The expression (2.26) is the relationship of the electric field inside the two coaxial cylindrical metal electrodes located in a medium with dielectric constant value equals the unity.

The expression (2.25) becomes:

$$E_1(r)=E_0\cdot\frac{\frac{\varepsilon 3}{\varepsilon 1}}{\left(1+\left(\frac{\varepsilon 3}{\varepsilon 1}-1\right)\cdot\frac{\ln\frac{r1}{r21}}{\ln\frac{r1}{r4}}+\left(\frac{\varepsilon 3}{\varepsilon 2}-1\right)\cdot\frac{\left(\ln\frac{r21}{r3}+\ln\frac{r22}{r4}\right)}{\ln\frac{r1}{r4}}\right)}, \tag{2.27}$$

$$(r21<r<r1)$$

The intensities of the electric fields E_{21}, E_3, E_{22} are:

$$E_{21}(r)=\frac{U}{r\cdot\ln\frac{r1}{r4}}\cdot\frac{\frac{\varepsilon 3}{\varepsilon 2}}{\left(1+\left(\frac{\varepsilon 3}{\varepsilon 1}-1\right)\cdot\frac{\ln\frac{r1}{r21}}{\ln\frac{r1}{r4}}+\left(\frac{\varepsilon 3}{\varepsilon 2}-1\right)\cdot\frac{\left(\ln\frac{r21}{r3}+\ln\frac{r22}{r4}\right)}{\ln\frac{r1}{r4}}\right)}, \tag{2.28}$$

$$(r3<r<r21)$$

$$E_3(r)=\frac{U}{r\cdot\ln\frac{r1}{r4}}\cdot\frac{1}{\left(1+\left(\frac{\varepsilon 3}{\varepsilon 1}-1\right)\cdot\frac{\ln\frac{r1}{r21}}{\ln\frac{r1}{r4}}+\left(\frac{\varepsilon 3}{\varepsilon 2}-1\right)\cdot\frac{\left(\ln\frac{r21}{r3}+\ln\frac{r22}{r4}\right)}{\ln\frac{r1}{r4}}\right)}, \tag{2.29}$$

$$(r22<r<r3)$$

$$E_{22}(r)=\frac{U}{r\cdot\ln\frac{r1}{r4}}\cdot\frac{\frac{\varepsilon 3}{\varepsilon 2}}{\left(1+\left(\frac{\varepsilon 3}{\varepsilon 1}-1\right)\cdot\frac{\ln\frac{r1}{r21}}{\ln\frac{r1}{r4}}+\left(\frac{\varepsilon 3}{\varepsilon 2}-1\right)\cdot\frac{\left(\ln\frac{r21}{r3}+\ln\frac{r22}{r4}\right)}{\ln\frac{r1}{r4}}\right)}, \tag{2.30}$$

$$(r4<r<r22)$$

3. Experimental results

Experimental verification of the theoretical relationships deduced above was done in the laboratory of the intense electric fields of Electrical Engineering Faculty of the Technical University of Cluj-Napoca.

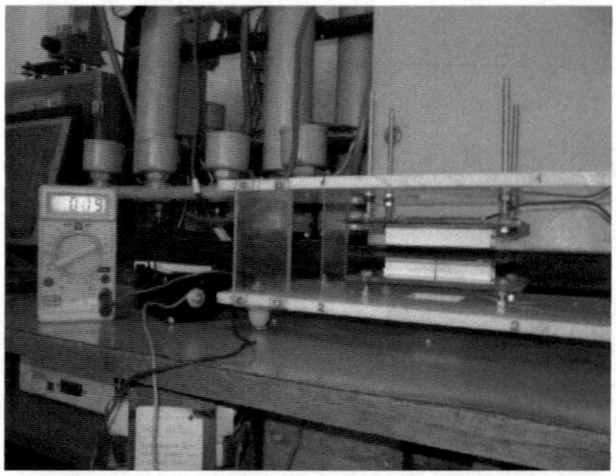

Figure 6a. Experimental bench (front view)

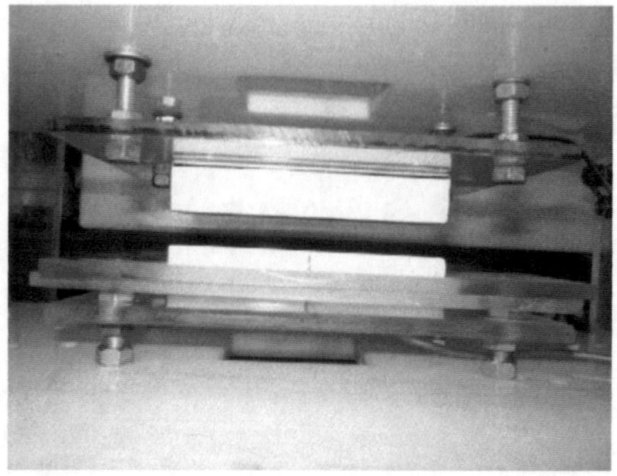

Figure 6b. Frontal view of the DBDC electrodes

The imagines presented in the Figures 6a and 6b show the experimental bench where was mounted the dielectric barrier cell. This cell was powered to a sinusoidal voltage variable frequency of 50 Hz. Determining the values of the electric field intensity was made indirectly by measuring the voltage at which the ionization processes are initiated in the gap (U_{ion}) [5].

The ionization processes is manifested by the appearance of thread-like trails that emit violet light accompanied by a specific low frequency noise [6]. The both phenomena become more intense with increasing voltage (Figures 7a, 7b)

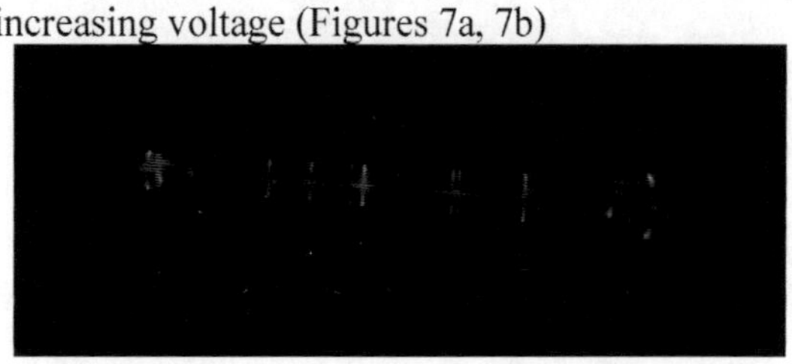

Figure 7a. Experimental bench (front view)

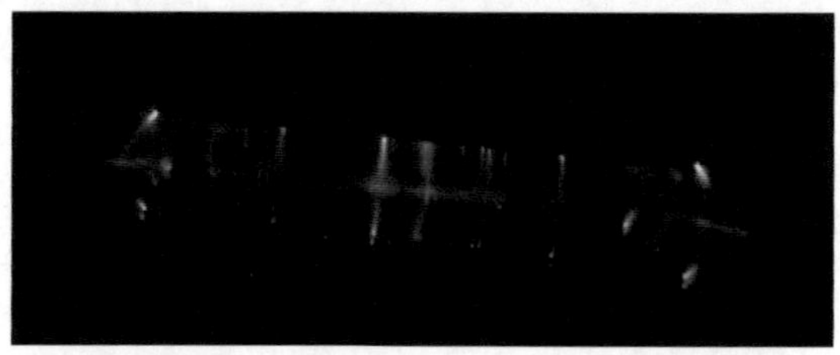

Figure 7b. Frontal view of the DBDC electrodes

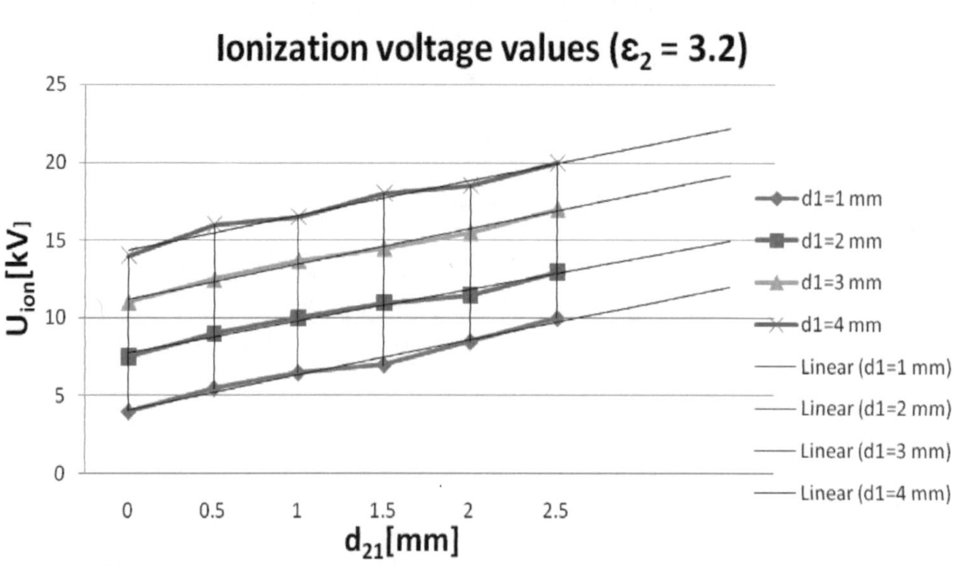

Figure 8. Representation of the experimental results (d = 20 mm)

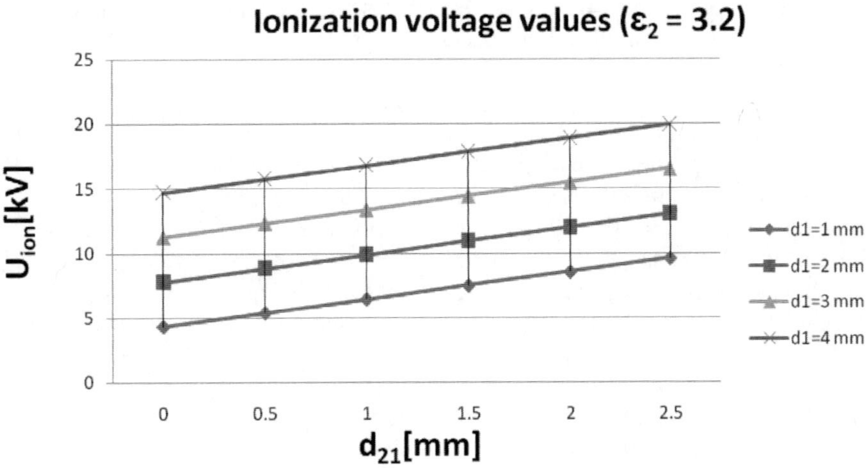

Figure 9. Representation of the ionization voltage values theoretically calculated for E = 3.5 kV/mm (d = 20 mm)

In DBDC the distance between electrodes (d) and the thickness gap (d_1) were maintained constantly. It was changed thickness of the insulating layer (d_{11} and d_{22}) and thus the thickness of the dielectric liquid (d_3). Every measurement was made with identical thickness layers ($d_{21} = d_{22}$). The material used for the solid layer was polyethylene. The supply voltage was gradually increased and the values of starting ionization process was noted. Then using the relationship of the electric field E_1 deduced above (2.26) was calculated for each case, the electric field intensity in the gap of the cell.

The values of the electric field intensity (E_1) were located around value of 3.5kV/mm. Then they were plotted in a chart the experimental measurements of the ionization voltage (Figure 8) and the theoretical values calculated were plotted in the chart shown in Figure 9 calculated for a value of the field E_1 = 3.5kV/mm.

Measurements were made on a cell with plane parallel electrodes having the distance of 20 mm between the electrodes, the dielectric liquid used was distilled water and the enclosure was made by polyethylene with the dielectric constant $\varepsilon_2 = 3.2$.

In the diagram shown in Figure 8 is observed that the experimental data measured at the start of the ionization phenomena of the working gas are grouped around a straight line plotted for each set of the measured values.

The straight line plotted in the Figure 8 (highlighted in the chart with the notation "linear d_1 = 1mm" for example) coincides with the graphical representation of the ionization voltage (U_{ion}) values for the electric field intensity E_1 = 3.5 kV/mm using the equation (2.26). The difference between data measurements and theoretical calculations fall into errors of the measuring apparatus and of the measurement methods.

For experimental verification has been chosen a cell with plane parallel electrodes because it allows a good view of the luminous phenomena taking place in the gaseous gap but also because the theoretical value of the electric field in the gap is constantly in every point.

4. Conclusions:

- the expressions of the electric fields intensity in the cell with plane parallel electrodes and the cell with coaxial cylindrical electrodes have the same form;
- the electric field intensity in the gap of the two types of the cells with double dielectric barrier have maximum values when $\varepsilon_2 = \varepsilon_3$ or d_{21} and d_{22} are zero (for the cell with plane parallel electrodes) and $\varepsilon_2 = \varepsilon_3$ or $r_{21} \cdot r_{22}/r_3 \cdot r_4 = 1$ (for the cell with coaxial cylindrical electrodes);

- it is recommended to reduce to a minimum the thickness of walls of the cell (d_{21} and d_{22} must tend to zero for cell with plane parallel electrodes and the reports r_{21}/r_3 and r_{22}/r_4 must tend to unity for the cell with coaxial cylindrical electrodes).

References

[1] Suarasan I 2000 *Generarea si utilizarea ozonului*, Editura ETA Cluj-Napoca, Romania

[2] Kogelschatz U 2003 Dielectric – barrier Discharges: Their History Discharge Physics and Industrial Applications, *Plasma Chemistry and Plasma Processing* **23**(1)

[3] Ciupa R 2006 *Bazele electrotehnicii: teorie si aplicatii*, Vol 1 and 2, Casa Cartii de Stiinta, Cluj-Napoca, Romania

[4] Sora C 1982 *Bazele Electrotehnicii,* Editura Didactica si Pedagogica, Bucuresti, Romania

[5] Kogelschatz U, Eliasson B and Egli W 1997 Dielectric Barrier Discharges. Principle and Applications, *J. Phys IV France 7 (1997) Colloque C4, Supplement au Journal de Physique III*

[6] Samoilovich V G, Gibalov V I and Kozlov K V 1997 *Physical Chemistry of the Barrier Discharge* (in Russian) Moscow State University (1989), English translation: J. P. F. Conrads, f. Leipold, (Eds), DVS – Verlag GmbH, Dusseldorf, Germany

IOP Conf. Series: Materials Science and Engineering **200** (2017) 012059 doi:10.1088/1757-899X/200/1/012059

The current challenges of teaching ESP

A Fălăuş

Technical University of Cluj-Napoca, North University Centre of Baia Mare, Philology and Cultural Studies Department, 76 Victoriei Street, 430122, Maramureş, Romania

E-mail: anamariafelecan@gmail.com

Abstract. Although the status of *lingua franca* can easily be claimed by English nowadays, there are always plenty of challenges involved in the process of teaching a foreign language. The simple mastering of the four skills (reading, writing, listening and speaking) and the acquisition of general grammar and vocabulary may not be enough in some circumstances. ESP focuses on the specific needs of the learners, concentrating more on language in context and on the students' need to acquire a set of professional skills and particular job-related functions. This paper, consequently, focuses on identifying the current challenges that teachers and students may encounter in the process of teaching and learning English for Specific Purposes.

1. Some general considerations on ESP

As any other kind of language teaching, *English for Specific Purposes* is first and foremost based on the process of learning, a process which nevertheless addresses the needs of certain communities of learners, namely individuals interested in acquiring some professional skills and performing job-related practices. Due to its oriented focus, ESP exhibits some characteristics that differentiate it from ESL (English as a Second Language) or EGP (English for General Purposes). First, it is language in context, this fact requiring real life learning situations, scenarios that tent to replicate the specific working or professional settings the ESP students might be related to or interested in. Instead of focusing on general grammar, vocabulary and language structures, this teaching – learning intercourse stresses the importance of practicing the necessary skills one would mostly employ in their future fields of activity. In comparison with the ESL learning contexts, the ESP students' motivational levels should thus be enhanced by their knowledge of the subject matter, their interest in the field fuelling their active participation in English classes. As Lorenzo Fiorito states, ESP "assesses needs and integrates motivation, subject matter and content for the teaching of relevant skills" [1].

Since its emergence in the late 1960s, ESP has undergone a constant process of development, defining its scope, improving methodology, shaping its objectives and orientations, and enlarging the number of course books designed to serve its purposes. Its emergence, however, was the direct result of a three-directional process [2] which began immediately after the Second World War and had as a starting point a boom in science, technology and economic activities at an international level. This fact triggered off the emergence of a community of people tied by their interest in learning English, not for the mere pleasure of knowing it, but for its utility and efficiency in handling specific job-related practices. The second reason underlining the advent of ESP was a revolution in linguistics which shifted specialists' attitudes and attention towards real communicational language opportunities and contexts, rather than focusing on the old-fashioned formalist view on language. The last step to be taken stressed the role of the learner whose needs and interests played an essential part in the development of future courses and syllabi.

ESP is presently taught in universities all over the world with the express purpose of preparing future specialists in various fields of activity, and it is also promoted by international corporations which organize specializing courses meant to enhance their employees' level of English and competence in various professional areas.

2. Towards a definition of ESP

Various attempts have been made at sketching out, as comprehensively as possible, a definition of this complex subject. In their book *Developments in English for Specific Purposes: A multi-disciplinary approach*, Tony Dudley-Evans and Maggie Jo St John [3] define ESP in terms of absolute and variable characteristics.

Absolute traits or features, as the authors claim, include:

- ESP being designed to meet the specific needs of the learners;
- it making use of the underlying methodology and activities of the disciplines it serves;
- ESP being centered on the language, skills, discourse and genres that are thought to be relevant to the above mentioned activities.

Variable characteristics, in addition to the previously enumerated ones, emphasise the likelihood of:

- ESP being related to or designed for specific disciplines;
- it using a different methodology from that of general English, in specific teaching situations;
- ESP being designed for adult learners, either at a tertiary level institution or in a professional work situation, without excluding the possibility of being used by learners at secondary school levels;
- ESP being aimed at intermediate or advanced students, the basic knowledge of the language system being important but not compulsory, the course having the possibility to be tailored to beginners, too.

This definition suggests that ESP, instead of being a restrictive, narrowing field of study is actually a permissive, flexible one, having influenced the entire process of English teaching. Even EGP teachers nowadays are likely to be influenced by ESP methods and techniques when they choose to design their syllabi starting from learner needs analyses or to use English in real communication contexts, instead of focusing on grammar or vocabulary related activities [4], [5]. The distinction between EGP and ESP offers us a better understanding of the latter's traits. Thus, while EGP has as an ultimate purpose *education* in general, ESP is mostly centred on *training*. The former is usually in the position of predicting with difficulty the very needs of the learners, whereas the latter is meant to be employed in specific professional contexts [6].

In trying to capture the essence of ESP, Hutchinson and Waters give a short but detailed account of it, underlining the fact that it has to be perceived as an *approach*, and not as a *product*. "ESP is not a particular kind of language or methodology, nor does it consist of a particular type of teaching material. Understood properly, it is an approach to language learning, which is based on learner need."[2]

3. Current challenges in teaching ESP

3.1. Course design

One of the greatest challenges of teaching any subject is the course design. Due to the fact that the ESP students have their objectives well-defined from the very beginning, these being directly related to their practical, job-related or professionally oriented needs, the choices the teachers have to make in order to design a course should not, theoretically, be so complicated. The reality, however, is different. According to David Carver [7], [8], an ESP course should be based on three elements; first, it has to offer authentic materials, then it requires a purpose-related orientation, which means that a reasonable simulacrum of reality in which practitioners have the possibility to get involved into communicative tasks that replicate real situations is mandatory, and last but not least, it should be defined by self-direction, i.e. learners are to become active users.

In order to cover all the areas that might play an essential role in the process of course elaboration, the ESP teacher should be ready to ask some questions and gather information in the field so as to create an important database for further developments.

The inquiries to be made are:

- **Why** do the students need to learn?
- **Who** is going to take part in the process (teachers, students, sponsors, experts in the field etc)?
- **Where** is the learning process going to take place? Does the location provide any potential or impose limitations?
- **When** is it set to take place? Is there a time limit to be taken into consideration?
- **What** does the student need to learn? What aspects of the language would be more appropriate under the given circumstances?
- **How** will the learning be achieved, i.e. what theoretical background will be chosen to fuel the methodology that is meant to be used? [2]

The diagram bellow (Figure 1) encompasses all the elements that the above mentioned questions made reference to.

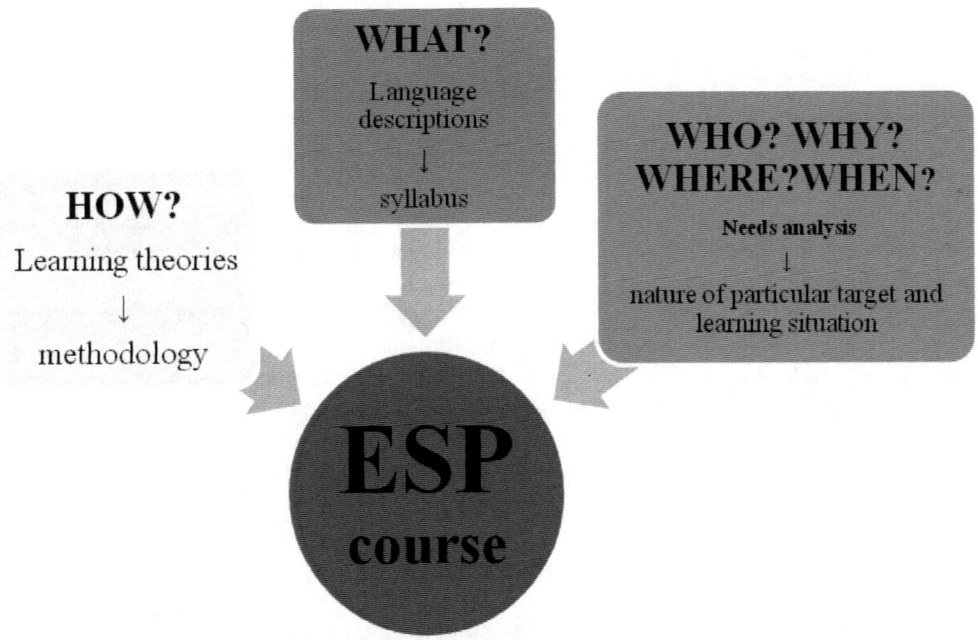

Figure 1. Factors affecting ESP course design [2]

If we are to take these elements apart in order to analyse them, we might reach the conclusion that the challenges and difficulties posed by the ESP teaching process nowadays are not to be ignored.

In what concerns the syllabus design, teachers presently have to face the prospect of being bombarded with a great number of ready-made course books which, however, have been designed with the purpose of easing the teachers of their worries of searching for authentic materials. They offer trainers the possibility to select activities that meet the needs of the learners, but at the same time force them into becoming "slaves" of the published textbooks [4], this plethora of resources reducing "individual instructors' motivation to construct their own course content with a focus on the immediate learners' context and particular needs" [9]. The implication of the above quoted points of view is that ESP practitioners should use textbooks as an alternative option, the suggested procedure being that of teachers collecting "empirical needs-assessment data" in order to create and adapt materials to meet the specific needs identified [10], these materials being selected so as to be able to "equip the students to deal with authentic examples of specialist discourse" [11].

The four main elements that should be taken into account when teachers start writing or designing their own materials are: the input, content focus, language focus and the task [2]. The diagram below (Figure 2) emphasizes the way in which they influence one another and contribute to the creation of a framework which might accommodate various aspects of the language to be taught.

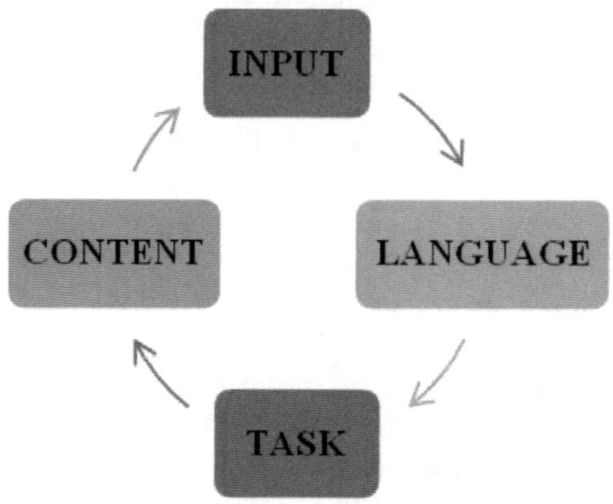

Figure 2. A materials design model [2]

The input may refer to a text, a dialog, a cassette or video-recording. The content presents the information, the subject matter that both teacher and students are dealing with; language is the one that facilitates the transmission of the information and has to be appropriate for the students' level of knowledge and expertise in handling a foreign language, while the task should be designed in such a way to offer the possibility of putting into practice the content meant to be transmitted. In order to create a more comprehensive picture, Hutchinson and Waters [2] expand the above mentioned model with two other elements, namely the student's own knowledge and abilities (the basis on which he/she adds new layers of information and the pace at which it is internalized), and any additional input that the teacher is ready to provide in order to facilitate a better understanding and practice.

Another challenge closely related to the problem of language is the one referring to the grammatical structures and functions that are likely to be identified as specific to certain subject matters. Indeed, when designing an ESP course one usually focuses on specialized vocabulary and on those grammatical structures that seem to have a certain frequency of occurrence. However, one cannot exclusively focus on these categories. For example, in order to have a conversation on a certain technical process, a person should, first and foremost have or gain the ability to get engaged in a general conversation activity. Specialized vocabulary, exclusively related to particular subjects (for example: Biology, Chemistry, Engineering, Business, etc) is of paramount importance, but it cannot be introduced and presented as a list of words that needs to be learnt by heart. They have to be slowly ingested and digested so as to be used without any difficulty, in normal conversational contexts. In what concerns grammar and the grammatical structures to be used, one can easily reckon that certain forms are predominant in ESP, such as Present Simple (used to express general truths or laws of nature), passive constructions, If-clauses (types zero and one) and nominal/adjectival compounds, but in order to create authentic materials teachers cannot design tasks that are solely based on these structures. The proper handling of Passive Voice, for example, requires the ability to use Active Voice and active constructions, as well. The same idea is rendered by Kristen Gatehouse [8], the author claiming that there are three abilities necessary for successful communication in a professional target setting, namely: the ability to use the specialized jargon of a particular professional circle, the ability to employ a set of academic skills appropriate for the occupational setting under discussion and the ability to use the language of everyday informal talk to get engaged in effective communicational activities such as chatting over coffee with a colleague or responding to an informal e-mail message.

Needs analysis is not unique to ESP but it is essential in helping the teacher to define "the specific sets of skills, texts, linguistic forms, and communicative practices that a particular group of learners must acquire" [12]. The students' level of language knowledge is essential as teachers nowadays face the challenge of working with multilevel classrooms, or, if we are to use Ur's terminology [13], "mixed-ability" classes. Students arrive in ESP learning environments with a varied background of linguistic knowledge and this leads to a new hardship, teachers getting engaged in a differential teaching process, so as to satisfy students' needs and necessities. Penny Ur [13] identifies a set of problems that are generally defining for large heterogeneous classes. The following graphic (Figure 3) reunites them all (discipline, correcting written assignments, interest, effective learning for all, materials, individual awareness, participation) under the same umbrella.

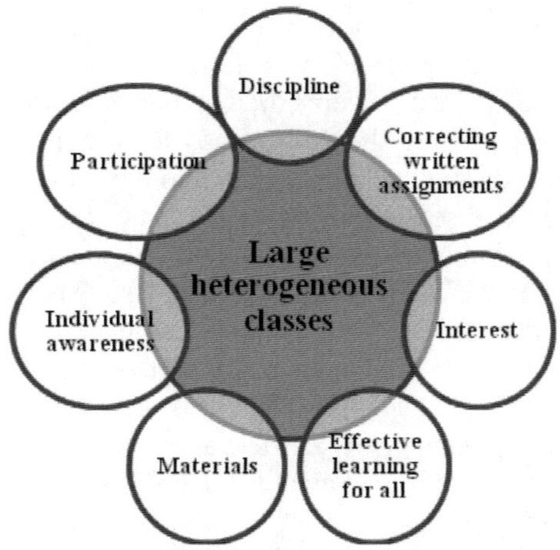

Figure 3. Teaching problems in large heterogeneous classes

The age of the students is also important, classes being presently made up of adults ranging in age from 18 to 50 or more. This fact can bring about serious consequences upon the way teachers organize their instructional process, due to the fact that "people of different ages have different needs, competences and cognitive skills" [14]. For those that have just graduated from high school or secondary school and choose to continue their studies at an academic level, the process of learning would not be a huge challenge. The difficulty appears for the other categories of learners who might have lost the initial ability to study, in time. According to Jeremy Harmer [14], adult learners can exhibit some traits that can easily make learning and teaching problematic. Thus, they might be critical of teaching methods, being sometimes reluctant to engage in role-playing activities, or scenarios that might put them in an uncomfortable position; they may feel anxious or under-confident about learning a language and sometimes worry about their intellectual capacities. All these elements must be thought over carefully when engaging in the process of course elaboration, always keeping in mind the fact that ESP focuses upon *learning* strategies rather than *teaching* techniques.

3.2. The role of the teacher

The ESP teacher plays an essential part in this complex equation due to the fact that the scope of ESP goes beyond the limits of teaching. As far as Tony Dudley-Evans and Maggie Jo St John are concerned [3], the term "practitioner" is preferred to the term "teacher" due to the complexity of the work one is involved in. In their view, the ESP practitioner has to embody five roles, namely: teacher, course designer and materials provider, collaborator, researcher and evaluator [3].

All these defining features are put together to create a composite picture of what the ideal figure of the ESP practitioner should look like. The person under discussion is first and foremost a teacher whose main purpose is to help students to learn. The techniques and methods employed are not too different from those used in the case of teaching General English; the only difference lies in the knowledge of the content to be taught. In this specific case, the teacher is no longer the " 'primary knower' of the carrier content of the material" [3], the students, in many cases, being more accustomed to the subject to be taught, their only flaw residing in their lacking the ability to use English to express their knowledge.

By getting engaged in a needs analysis process, the ESP practitioner becomes aware of all the necessary elements and materials that are meant to be included in the future course, whether he/she chooses to design a syllabus by themselves or make use of the textbooks available, supplementing the given activities with extra materials.

As collaborators, they have to rely on help and guidance coming from various sources. One source is represented by the students. They know exactly what they want to achieve, their goals are very well defined and their knowledge of the subject matter might be an essential asset in the course development. Another source might be the collaboration with specialists in the field, these being able to offer assistance and guidance in what concerns the specialized books and materials to be used.

As researchers, the ESP practitioners should be ready to stay in touch with the latest advances in the subject matters they are teaching (Science, Technology, Economics, Business, etc.) and to incorporate the findings in their course design.

The last role is that of an evaluator. From this position the teachers should always check the attainment of the targets set out at the beginning of the course. They administer tests and evaluation activities "to assess whether students have the requisite language and skills to undertake a particular academic course or career" [3], their final goal being that of quantifying how much learners have gained from a course.

According to Hutchinson and Waters, ESP teachers are "all too often reluctant dwellers in a strange and uncharted land" [2], the difficulties they have to face being grouped in three categories:

a) the lack of an ESP orthodoxy to provide a ready-made guide;
b) the new realms of knowledge the ESP teacher has to cope with, and
c) the change in the status of English Language Teaching.

The very lack of orthodoxy can be translated into the teachers' need to constantly search for and turn to various theses and researches in linguistics and methodology in order to find the proper ways to structure and organize their courses. This hardship they have to endure transforms them into "pioneers who are helping to shape the world of ESP" [2]. The new realms of knowledge make reference to teachers having to master subject matters that are beyond their field of expertise. They need to use texts and materials that focus on concepts they might not be familiar with, or get engaged in activities that they are not accustomed to. Consequently their working effort is double. The last category which focuses on the changes that affect the status of English Language Teaching places the ESP practitioners in the position of being negotiators, too. They have to cooperate with sponsors, specialists or experts in the field and even students who might usually have clear-cut expectations concerning the content of the lessons to be taught, the structures of the activities they are going to engage in, and the achievements of the course in general.

4. Conclusions

In comparison with teaching EGP, teaching ESP usually poses a lot more challenges. Focusing on the specific needs of the learners, concentrating more on language in context and on the students' need to acquire a set of professional skills and particular job-related functions, ESP remains a major testing experience for every teacher in charge of it. Although it is said to be a learning-centred approach, this fact requiring the joint effort of both, teachers and learners, as a practitioner, one needs "an open mind, curiosity, and a degree of scepticism" [2]. The emphasis on learners' wants and interests, and learning autonomy does not diminish the role of the teachers; on the contrary they need to subject themselves to a continuous process of adaptation and evaluation in order to meet the requirements imposed by the subject they are teaching. They need to design courses keeping in mind the nature of the particular target and learning situations they are to deal with, at the same time juggling professionally with the requirements imposed by working with large heterogeneous classes.

References

[1] Lorenzo F 2005 *Teaching English for Specific Purposes*, https://www.usingenglish.com/articles /teaching-english-for-specific-purposes-esp.html

[2] Hutchinson T and Waters A 1987 *English for Specific Purposes. A learning-centred approach,* Cambridge University Press, Cambridge, UK,

[3] Dudley-Evans T and St John M J 1998 *Developments in English for Specific Purposes: A multi-disciplinary approach,* Cambridge University Press, Cambridge, UK

[4] Anthony L 1997 *Defining English for Specific Purposes and the Role of the ESP Practitioner*, http://www.laurenceanthony.net/abstracts/Aizukiyo97.pdf, 20 Nov 2016

[5] Bracaj M 2014 Teaching English for Specific Purposes and Teacher Training, *European Scientific Journal* **10** (2)

[6] Far M 2008 On the Relationship between ESP & EGP: A General Perspective, *ESP World* **1**(17) 7

[7] Bojović M 2006 *Teaching foreign languages for specific purposes: Teacher development*, The proceedings of the 31st Annual Association of Teacher Education in Europe, pp 487-493

[8] Gatehouse K 2001 Key issues in English for Specific purposes (ESP) curriculum development, *Internet TESL Journal* **VII**(10)

[9] Brunton M 2009 An account of ESP – with possible future directions, *English for Specific Purposes* **24**(3)

[10] Belcher D D 2004 Trends in Teaching English for Specific Purposes, *Annual Review of Applied Linguistics*, Cambridge University Press, Cambridge, UK, pp 165-186

[11] Phillips M K and Shettlesworth C C 1978 How to ARM your Students: a consideration of two approaches to providing Materials for ESP, *English for Specific Purposes*, British Council

[12] Hyland K 2006 English for Specific Purposes. Some Influences and Impacts, *The International Handbook of English language education*, Cummins A & Davison C (eds), Vol 1, Norwell, Mass Springer, pp 379-390

[13] Ur P 1991 *A Course in Language Teaching. Practice and Theory*, Cambridge University Press, Cambridge, UK

[14] Harmer J 2007 *The Practice of English Language Teaching*, 4th Edition, Pearson Longman, Harlow, England

Theoretical aspects of the acceptance of new technologies on the smartphone market

N Barkoczi and G Lobontiu

Technical University of Cluj-Napoca - Universitary Center of Baia Mare, 62A Dr. Victor Babes Street, Baia Mare, Romania

E-mail: nadiabarkoczi@gmail.com

Abstract. Adoption of new technologies on mobile phone market in the past 30 years has transformed the industry and had a large impact on the global economy. Technological progress, as the main factor in the adoption of new technologies, has enabled reduce costs and increase equipment performance. The smartphone is perceived by consumers as a sustainable element, to which the price sensitivity is less important. Consumers are willing to pay more in exchange for better quality and features of the product. Smartphones have changed the way people communicate with others, find information and manage their daily life. In addition, recent developments of new operating systems, abundant applications and competition between suppliers have facilitated a marked increase of the users' number.

1. Introduction

Technology diffusion research seeks to understand its spread through modelling the entire life cycle in terms of consumer interaction, too. Numerous marketing studies have focused on how the new technology is perceived by consumers, which is usually tested by the adopters' behaviour and reactions to technological innovation and by the manner it can change over time. Regarding new product development, studies try to discover the stages in the process of technology diffusion or the features of new product developments that are most critical to achieving market success and broad adoption. This paper approaches the manners of acceptance of new technologies in the smartphone market and how they are affected by the characteristics of the society to which the potential users belong.

2. The interaction between technology and the social context in terms of consumer acceptance of new technologies

On the one hand, technology acceptance is the central problem of the transdisciplinarity between technology diffusion and marketing management and on the other hand it relies on the theoretical works of sociology and information management in cognitive psychology [1]. Some researchers believe that understanding the relationship between users may be more important than the factors of the product itself, which highlights the resistance to technology [2]. Barak and Levenberg identified three main factors that may indicate the consumers' dispositional inclination to think flexible in adopting new technologies: the acceptance of new or changing technologies (technology acceptance), openness to others' ideas (open mindedness) and adaptation to changes in learning situations (adapting to new situations) [3]. The transformation of useful information on technology into knowledge leads to competitive advantages [4], communication playing an essential role in this process. Providing to the right people complete, accurate, relevant and timely information is important in order to support the decision-making process regarding the acceptance of new technologies. The quality of information and communication channels affects user satisfaction on the new technology and, consequently, influences their beliefs about its use [5].

2.1. The influence of technology on consumers, in the mobile phone market

The adoption of new technologies in mobile telecommunications, in recent years, has transformed the industry and has had a large impact on the global economy. In particular, mobile subscribers exceeded the population in many countries where there is universal adoption. The mobile revolution began in Europe and United States, but was quickly extended globally and influences the evolution of many developing countries. The empirical literature has confirmed substantial positive welfare effects with the introduction of mobile telecommunications.

Many people and organizations today use mobile wireless technology through mobile WAP, handheld devices, smartphones, wireless laptops, m-commerce, etc. However, despite market growth, research on examining the key factors affecting users' behaviour and adoption are rare. What is missing is an understanding of the motivation resulting in the adoption of new technology by the consumer.

Today, smartphones are no longer interpersonal communication devices via voice and text messaging services. They are a combination of new technologies which are used for browsing the internet, checking e-mails, chat online, time management, entertainment, online / offline games, images and videos sharing, self-expression, establishing identity, online payments and running various mobile applications.

When we talk about smartphones, we talk about custom devices available for use in most contexts and locations. Now they have become ubiquitous and visible to the public. The smartphone pathology is like a double-edged sword: they create and destroy opportunities. The attitude of using a highly customized smartphone allows the fulfillment of pleasures and utilitarian needs of users. Indeed, the favorable attitude towards smartphones could enslave a person, it would influence her to forget the social environment, to spend long hours actively with this device even in inappropriate situations [6].

2.2. Technology acceptance model as a theoretical framework

Technology acceptance model is one of the most widely used theoretical models to examine the factors of adoption and use of technology. Focusing on the associations between perceived usefulness, the perceived ease of use and the intention to use the system effectively, the model attempts to predict the relative importance of perceived usefulness and perceived ease of use, with other variables relevant to research in different contexts. Perceived usefulness is the subjective assessment of a potential user using a particular technology which will improve the performance both at work and in the social life, while the perceived ease of use concerns the extent to which the potential user anticipates the effort to use technology. Both the perceived usefulness and perceived ease of use influence the consumer behaviour and the intention to accept and use the new technology. The model scheme can be followed in Figure 1. The technological complexity, individual differences, social, environmental and trust influences have a positive impact on the usefulness and the ease of use [7].

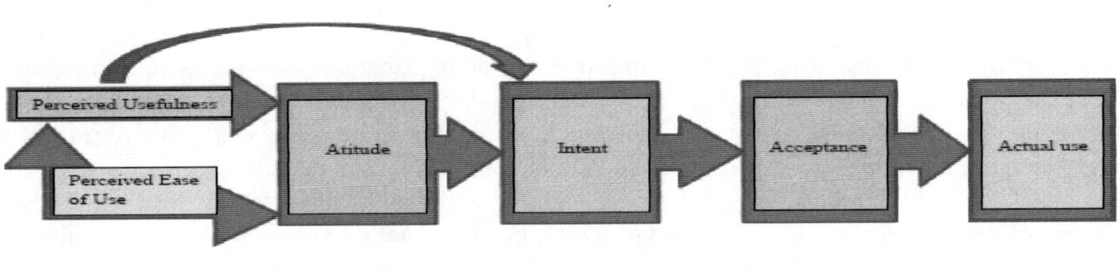

Figure 1. Technology Acceptance Model

There were hundreds of published studies that used the technology acceptance model as a theoretical framework, demonstrating the robustness of the model in areas of personal computers and Internet software applications [8], email [9], word-of-mouth [10], online communities [11], wireless mobile services data [12], etc. In the case of smartphones, their utility with many advanced features and functions has long been observed among users.

2.3. Choosing and using smartphones

There have been several studies on the choice of mobile phones and their use. In his book, Rogers briefly mentioned why a mobile phone has been successfully adopted by consumers. He attributed value to the successful adoption as relative advantages of cordless telephones, mobile phones compatible with existing systems, ease of use, and observability [13]. Although there have been a few studies on the use of mobile phones, there has been no study on the factors affecting a person to choose a smartphone, especially using the economic utility theory and the econometric models.

Many products these days tend to answer to several market segments and to offer consumers the possibility of larger selection than ever before. Consumers also become more aware of their styles and specific preferences and therefore know better what products might suit to their requirements [14]. The preferences for the mobile phone features such as technology, style (color, design, size, etc.) have developed greatly from the launch of the original mobile phone. At first, voice was the dominant feature while today mobile phones include multi-tasking features, such as camera, calendar, mp3 player etc. It is obvious that mobile phones are deeply rooted in every human life everyday [15], and perform many tasks that go beyond traditional communication.

Some researchers developed various methods using Multiple Criteria Decision Making (MCDM) and Quality Function Deployment (QFD) to assess options for mobile phones in terms of user preferences order [16]. The main tool of the QFD is the house of quality represented in Figure 2, through which the voice of the customer is translated as design requirements that become specific target values, which will be operated by the organization in order to be met. Many managers and engineers take into account the house of quality for the main graph in quality planning.

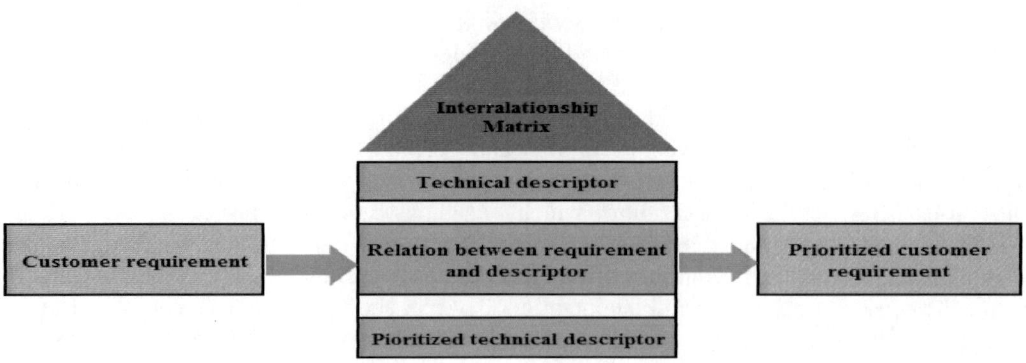

Figure 2. House of quality [16]

Artistic preferences for a product can be improved by adding features to the new technology that the competition does not have yet. The features are the primary sources of benefits received when the product is purchased and thus have a positive impact on product evaluation. This positive impact of the addition of new features through new product technologies is still the subject of research as an interesting phenomenon, robust, even in situations where the new features are irrelevant. Users experience the positive impact of adding new features, however, with a degree of decrease so that the features added to an inferior product have a relatively more positive impact on the assessment of product features added to a product than in the case of a superior product. It is very important if the customer did or did not have experience with the product. Many studies are focused on product features that have an impact on quality assessment by consumers [17] and on integrating concepts used in product development as specialized ontologies [18], [19]. Consumer involvement in the design, creation and development of products is one of the marketers' targets in order to be able to identify their preferences.

3. The influence of marketing strategies on the decision of granting new technologies in the smartphones market

It is important to emphasize the importance of marketing strategies and influences on the consumer purchasing decision in the context of the smartphones market. "The cleverness" of smartphones is still in full operation by consumers [20]. The mobile phone is one of a handful of consumer products that has been accepted globally in a relatively short period of time [21]. The mobile phones market has become an extremely competitive market. If the intensity of competition increases, companies in the industry are more likely to sustain their current level of sales, the profit margin and market share not only by keeping their current customers, but also by acquiring new customers. The interesting feature in the mobile industry is that it evolves rapidly, the products being manufactured for a short lifecycle, to which they add new features, as well as the fierce competition between many companies in the industry. If the client is facing additional functionality and increased complexity of new technologies for the mobile phone it is likely to face problems that can have a negative impact on customer satisfaction, and therefore reduce its intention to repurchase.

3.1. Marketing opportunities brought by new technologies

New technologies create marketing opportunities and challenges. Phones and personal digital assistants increase availability, frequency and speed of communication. Still, the technology associated with these devices, enabling marketers to communicate personally with consumers, continues to evolve. Recognizing the complementary networks World Wide Web and mobile, a content analysis of the Fortune Global 500 websites investigated the adoption of mobile business models and technologies by region and industry. The results suggest geographical differences in the diffusion technology that affects marketing strategies. There are two categories of success factors, messages features and media, influencing three dependent measures of success: consumer attention, the intention of consumption and consumer behavior, as we have seen within the technology acceptance model. The results of these qualitative and quantitative studies give managers insights on mobile marketing and mobile advertising [22]. Many firms vertiginously innovate and create products of successive generations to replace the existing ones, in order to take advantage of the customers' trend to upgrade [23].

The widespread adoption of mobile phones is a huge marketing opportunity to reach and serve customers anytime, anywhere. Paradoxically, while consumers adopt mobile phones to improve their privacy and social life, in marketing mobile phones are seen as a marketing channel. The introduction of smartphones means that they can develop a wider range of services based on engaging a larger set of marketing techniques.

3.2. The importance of technology diffusion and marketing in the adoption of new technologies in the smartphone market

If the concept of marketing was selling the product for profit, now marketing aims at identifying consumer preferences, providing them value through the sold product. This identification is made possible by the abundance of information provided by market research and communication channels and starts from an idea on a technological innovator product. In this direction, the diffusion of technology is based only on knowing the adopter of new technology in order to impress him by the novelty of the product offered by technology.

Technology adoption, which is specific to technology diffusion, is part of marketing, but we cannot ignore the fact that technology diffusion is correlated with R & D management, technology management and strategic management. The new technology is the result of consumer needs and values that are identified through market research and has a specific path before being brought to market and adopted.

With the introduction of new technologies on the smartphone market, marketing and consumers will be exposed to a whole range of marketing innovations that were not possible with the classical mobile phone. For example, applications for smartphones such as Amazon and Google Shopper allow consumers in a retail store to use their smartphone to enter the barcode of a product or take a picture of a product and immediately receive price comparisons, customer reviews, discounts, coupons and other information oblivious of the product from the store. The smartphone has the ability to transform the shopping experiences of consumers and the market value: consumers now can shop quickly and easily across multiple channels (physical, web-based and mobile store) with a significantly higher level of comfort, flexibility, efficiency and customization. The smartphone is seen as another key step in the evolution of technology in the market because it has the ability to seamlessly integrate Bluetooth services, location-based marketing, and other web-based technologies [21].

Research into the mobile market showed that mobile services have increased the adoption rate of mobile phones. Moreover, because the sales of smartphones continued to rise, mobile providers are concerned not only with mobile phones functionality but also with operating systems which are likely to play an important role in the development of smartphones. However, few researchers have focused on competitive technology diffusion among the operating systems of smartphones. Assessing how consumers choose smartphones, in terms of operating systems at the same price level, will allow manufacturers to target consumer segments and produce specialized smartphones [24].

4. Conclusions and discussions

On the smartphone market, and not only, the acceptance of new technologies is conditioned by everything that is involved in promoting and selling the product. It is very important the cooperation of marketing and technology diffusion so that the marketing provides the company the control through advertising, product promotion and confering value to the product. The other side, the diffusion process, on which the company has no control, but which is nevertheless important for market penetration is an essential promotion through use. Fundamentally, one can score three important conditions for the diffusion of technology, for its acceptance by the consumer:

1. to stay informed of the new technology,
2. to accept and adapt to new technology, and
3. to adopt the new technology considering the price and the level of risk.

The diffusion of technology is the cumulative result of a series of calculations that measure the benefits of adopting new technology and decrease costs in an environment of uncertainty (as is the future evolution of technology and benefits) and limited information (about the benefits, cost and technology duration). Although the final decision is given by demand, benefits and costs, it can be noticed the influence of new technologies on customers' decisions. The diffusion rate is thus determined by the sum of these decisions.

In this paper it has been highlighted how technology is perceived by consumers, which is usually tested by the adopters' behaviour and reactions of technological innovation and how they may change over time. An understanding of the smartphones market and trends in terms of consumers is valuable, industry stakeholders can greatly appreciate the information that can be used to build marketing strategies and plans for future research directions.

As noted in the paper, in the smartphone market, an important role is played by developing OS through key factors, such as the performance of consumption, the state of the app stores, software quality, the degree of competition within operating systems, the share of information available for development etc. Companies developing operating systems for smartphones should retain these key factors for decision-making on technology diffusion and therefore, take the best possible marketing decisions.

Technology users who postpone their purchases take into account the difference between the current model and future versions features. Therefore, marketers should target such uncertainties.

Customer satisfaction is one of the most important factors influencing the decision to upgrade, followed by price and the perceived usefulness. Therefore, what makes consumers accept new technologies would be useful marketing strategy for mobile phone manufacturers, making them accept a product with enhanced functionality or a newer model at a reasonable price.

The mobile market is divided into two different markets, namely: mobile phone market and the market for mobile service providers. Choosing a smartphone is often a combination of a phone and services offered. Such a choice can be influenced by social and motivational variables, even if the latter are more relevant in services due to interpersonal relations. These variables affect not only consumers but also companies in launching new products, both in terms of technology and marketing, companies being often able to predict the end of life of the new product in terms of technology.

The extraordinary growth in demand is due to the psychological and social reasons which, combined, have produced specific features both of users and potential users. The interactions between technological advances, changes in consumer and firm strategies influence the evolution of the mobile market. In the mobile industry it is useful to offer new services and to strengthen customer loyalty and customer base rather than acquire more subscribers. In one such industry where product offerings are multiplied through differentiation and versions, this leads, in most cases, to the existence of confused market customers. Therefore, the diffusion of technology and marketing are highly interconnected, as the first changes in the media and branding tools, in addition to developing new opportunities. The latter, in turn, helps companies in differentiating and guiding consumers in their choice.

References

[1] Chiu Y-T H, Fang S C and Tseng C C 2010 Early versus potential adopters: Exploring the antecedents of use intention in the context of retail service innovations, *International Journal of Retail & Distribution Management* **38**(6) 443–459

[2] MacVaugh J and Schiavone F 2010 Limits to the diffusion of innovation: A literature review and integrative model, *European Journal of Innovation Management* **13**(2) 197 – 221

[3] Barak M and Levenberg A 2016 Flexible thinking in learning: An individual differences measure for learning in technology-enhanced environments, *Computers & Education* **99** 39-52

[4] Bach M C, Čeljo A and Zoroja J 2016 Technology Acceptance Model for Business Intelligence Systems: Preliminary Research, *Procedia Computer Science* **100** 995-1001

[5] Wixom B H, Todd P A 2005 A Theoretical Integration of User Satisfaction and Technology Acceptance, *Information Systems Research* **16** 22-38

[6] Saidona J, Musab R, Haruna M H R and Adama A A 2016 The Conceptual Framework of Pathological Smartphone Use (PSU), *Procedia Economics and Finance* **37** 426-431

[7] Kim S H 2008 Moderating Effects of Job Relevance and Experience on Mobile Wireless Technology Acceptance: Adoption of a Smartphone by Individuals, *Information & Management* **45** 387-393

[8] Moon J-W and Kim Y-G 2001 Extending the TAM for the World-Wide-Web context, *Information and Management* **38** 217-230

[9] Gefen D and Straub D 1997 Gender Difference in the Perception and Use of E-Mail: An Extension to the Technology Acceptance Model, *MIS Quarterly* (21:4, December) 389-400

[10] Park D H and Lee J 2008 eWOM overload and its effect on consumer behavioral intention depending on consumer involvement, *Electronic Commerce Research and Applications* **7**(4) 386-398

[11] Chung J E, Park N, Wang H, Fulk J and McLaughlin M 2010 Age differences in perceptions of online community participation among non-users: An extension of the Technology Acceptance Model, *Computers in Human Behavior* **26** 1674-1684

[12] Suki M M 2011 Subscribers' intention towards using 3G mobile services, *Journal of Economics and Behavioral Studies* **2**(2) 67-75

[13] Rogers E M 2003 *Diffusion of Innovations,* 5th edition, Free Press, New York

[14] Sääksjärvi M and Morel K P N 2010 The development of a scale to measure consumer doubt toward new products, *European Journal of Innovation Management* **13**(3) 272-293

[15] Palen l and Salzman M 2002 *Voice-Mail Diary Studies for Naturalistic Data Capture under Mobile Conditions*, in Proc. of CSCW, https://www.researchgate.net/publication/2842528

[16] Saket S, Purbey V, Jagadish and Ray A 2014 Multi Attribute Decision Making for Mobile Phone Selection, *International Journal of Research in Engineering and Technology* **3**(3) 407-501

[17] Haverila M 2011 Mobile phone feature preferences, customer satisfaction and repurchase intent among male users, *Australasian Marketing Journal* **19** 238-246

[18] Lobonţiu M and Petrovan A 2012 Product development ontology (1). Information integration concepts, *Revista de Management şi Inginerie Economică* **4**(46) 43-56

[19] Petrovan A and Lobonțiu M 2011 *Ontology Engineering for Product Development,* 2nd Review of Management and Economic Engineering Management Conference, Cluj-Napoca, Romania, pp 163-168

[20] Osman M A, Talib A Z, Sanusi Z A, Shiang-Yen T and Alwi A S 2012 A Study of the Trend of Smartphone and its Usage Behavior in Malaysia, *International Journal on New Computer Architectures and Their Applications* (IJNCAA) **2**(1) 274 285

[21] Persaud A and Azhar I 2012 Innovative mobile marketing via smartphones: Are consumers ready?, *Marketing Intelligence & Planning* **30**(4) 418-443

[22] Scharl A, Dickinger A and Murphy J 2005 Diffusion and success factors of mobile marketing, *Electronic Commerce Research and Applications* **4** 159-173

[23] Tseng F-M, Liu Y-L and Wu H-H 2014 Market penetration among competitive innovation products: The case of the Smartphone Operating System, *Journal of Engineering and Technology Management* **32** 40-59

[24] Tseng F-M and Chiang H-Y 2013 Exploring consumers to buy innovative products: Mobile phone upgrading intention, *Journal of High Technology Management Research* **24** 77-87

Three lines of defence model and the role of internal audit activities as the response to the global economic crisis

T Dragičević Radičević[1], M Stojanović Trivanović[2] and Lj Stanojević[1]

[1]Faculty of Geoeconomics, str. Bulevar umetnosti 29, 11070 Belgrade, Republic of Serbia
[2]Institute for Scientific Research Work at the Independent University, str. Veljka Mlađenovića 12e, 78000 Banja Luka, Republic of Srpska

E-mail: tanjadragrad@gmail.com

Abstract. The existing framework of corporate governance has shown a number of weaknesses, and the result was a new economic crisis at the global level. The main problems were identified as: increased risk of investors, non-transparency of information, conflict of interest between corporation subjects. European Institute of Internal Auditors in response to the strengthening the trust in information, shareholders activism, better communication, which all will lead to the reduction of risks and restore investors confidence, proposed the Model Three Lines of Defence, where the key role has internal audit.

1. Introduction

The main characteristic of today's market is the financial instability. By looking for the cause of this instability it is recognized that one of the main problems lies in access to information. Access to information is different according to the groups within and outside the company (shareholders, managers, employees, investors, etc.). In such incoherence and conflict of goals and interests there are many problems, such as: relevance, timeliness and transparency of information. In this sense, the more stressed are the necessity of creating functional areas in the company, with basic prefixes on reliability, timeliness and objectivity and with equal access to quantitative and qualitative performances of company.

The introduction of quality as the basic parameter of company performance, it can be concluded that external audit is not a sufficient instrument, especially with reference to its role of unsatisfactory from the perspective of many investors, and in the context of its timeliness (after time conclusions), areas of operation, transparency of activities etc[a] . As the response to the above-set requirements, but also the perceived contradictions in the case of external audit, the European Confederation of the Institute of Internal Auditors (ECIIA)[b], answer finds in internal audit function.

Confederation stresses the importance of truthfulness and relevance of financial statements, from the perspective of market instability, and requires of potential investors, in order to increase confidence in the same [1]. By conducted analysis it was concluded the needs of complex study of the

[a] It is very frequent case where the external auditors appear as a consultant, which will lead to a conflict of interest, especially as the financial compensation in these cases are many times higher than the audit services.
[b] European Confederation of the Institute of Internal Auditors, an association of the Global Institute of Internal Auditors located in 35 countries and employs 35,000 internal auditors.

risk problems, for which financial statements are not sufficient, as they represent the finished action, not a continuum.

In that case ECIIA define the Model of three lines of defense in the function of efficient corporate governance. On the first line of defense processes are focused on risk management by the owners of capital. On the second line of defense it is the focus on compliance, risk management, financial control, etc. On the third line of defense is internal audit, as the "guarantee" that the previous activities and procedures in correlation with efficiency and effectiveness of all interest groups and company goals.

Based on the role of internal audit and issues facing the modern corporate governance, it is clearly an assumption that may withdraw the necessity of building the internal audit function as the catalyst in the creation of corporate social responsibility. Consequently, additional research in this work is aimed to determining the same assumptions, with special emphasis on the situation in the Republic of Serbia and the country in region. In this regard, we are starting from the consideration of the efficiency of existing mechanisms at the defined lines of defense of this model.

2. Internal audit in function of conversion the trust in the value

Conclusions from the Conference of the European Confederation of Internal Auditors in December 2008, emphasized the role of internal audit in terms of consultant and advisor in proactive risk management and dynamic and ongoing reporting, for all interest groups in the corporation. "Its independence in internal frames (always present at all levels) offers shareholders and investors the certainty. The importance and role of internal audit derives from its:

- recognition (expected due care, objectivity, fair, accurate reports);
- reliability (high degree of ethics and integrity, focus on specific areas, respect the object of the audit, providing practical and relevant recommendations);
- value (objective and well-informed correspondent of business risk control and corporate governance, reliable advisor, creates balance advice - guarantee)“ [2].

„The importance of internal audit in corporate governance and risk management is reflected in its analysis of the likelihood and impact of risks, then the impact on the value of risk. This can be achieved by converting the value of trust in the way that is necessary to review all contractual relationships, identify their objectives, assess the risk on the basis of key criteria and their interconnection. In addition, it is necessary to assess the compliance of key control processes as well as their co-ordination, and based on that define appropriate procedures in order to improve efficiency“ [3].

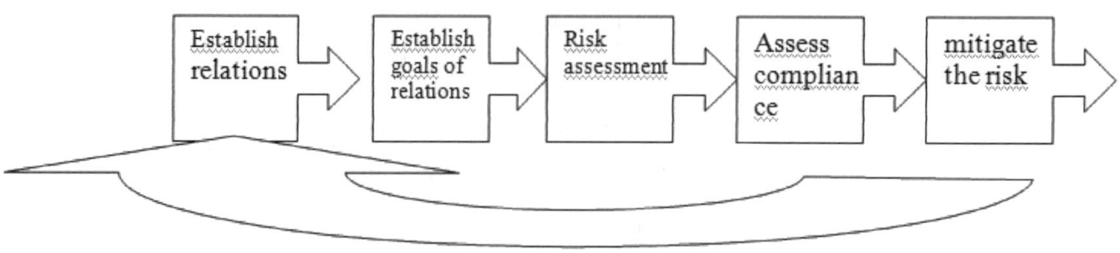

Figure 1. Risk management process [4]

Internal audit can make a vital instrument of corporate control and risk reduction, from the perspective of comprehensiveness and independence in the assessment of risk.

Due to its objectivity and independent functions in the company it might be a good generator of business information from the turbulent environment in terms of globalization. The primary role of internal audit is the function of security for the operational management through a written report and an advisory role on the adequacy of the management control and risk management. The secondary role is to improve governance according to the risks. The ultimate goal of internal audit is to ensure efficiency and effectiveness in dealing with the high level of quality. Internal Audit Standards define the scope of the operation and quality of functioning. It detects potential weaknesses in the internal control system and thereby exclude the possibility of neglecting certain areas. Therefore, the objective of the internal audit can be defined as the instrument by which management and all other employees of the company efficiently and effectively fulfill their obligations. Its mission is reflected in the identification of risk and its avoidance or mitigation.

3. Difference between internal and external audit

Although the internal and external audit are different, the performance of different goals (internal audit is an aid to the management, and external audit verify the reality of performance), they are still partially connected because of similar areas of study, techniques and methods.

External audit is connected to internal audit in two aspects: [5]

a. It can be conducted as an auditor of internal audit departments which control the follows:
   - Degree of independence of internal audit;
   - Areas of auditing and objectivity of performing the internal audit;
   - the expertise and professionalism of internal auditors;
   - technical competence;
   - reports of internal auditors;
   - level of available resources.

b. From the aspects of coordination of internal and external audit.

External audit may not always be a relevant indicator of the contribution of internal audit to successful decision-making, due to conflict of interest. Assessment of the independence and objectivity of the internal audit may be conducted in terms of the assessment of effectiveness, the extent to which external audit may rely on the internal auditing department during the annual audits, not only on the basis of application of generally accepted audit standards.

The link between internal and external audit is in the methods and techniques, and not in the objectives and results (for example, an external auditor will determine the reality expressed in inventories, while the internal auditor has to analyze the effectiveness of the applied method of calculation and stockpiling). So, the basic difference between the internal and external audit is presented in Table 1 [6].

Table 1. Difference between internal and external audit (Source: ECIIA, Response to Green Paper Audit Policy: Lessons from the Crisis, December 2010, www.eciia.eu Download: April 2012., p. 2)

Item	External audit	Internal audit
Recipient of reports	Shareholders or Members	Board members and senior managers
Objective(s)	Add credibility and reliability to reports from the organization to its shareholders by giving an opinion on them	Provide the assurance that members of the board and senior management use to fulfill their duties
Coverage	Financial reports and related disclosures, financial reporting risks and their management[c]	All processes and categories of risks, their management (c) including the flow of information throughout the company, and governance
Timing and frequency	Project(s) tied into financial reporting cycle, focused on objective of audit opinion	Ongoing and pervasive
Focus	Mainly historical	Forward-looking on the basis of "as-is" and recommended "to-be" scenarios
Responsibility for identifying	None – duty to report problems	Responsibility to report

[c] Risk management starts with objectives/purpose, then includes identification, evaluation and assessment of the risk; selection and implementation of the appropriate responses; and monitoring to ensure that the responses are working as required.

Item	External audit	Internal audit
areas for improvement	noted during engagement	recommendations for improvement and promote related action plans, fundamental to the mission of internal auditing
Status and authority	Statutory and regulatory framework and International Auditing Standards (IAS)	International professional standards and Code of Corporate Governance
Independence	Professional ethical standards overseen by audit committee and regulatory framework	Professional ethical standards overseen by audit committee

To what extent will be achieved coordination of internal and external audit, primarily depends on the degree of independence and objectivity of internal audit. The greater degree of independence and objectivity of internal auditing, the greater is the possibility of relying external on internal audit, and vice versa, because by the end of the external audit, the internal auditor will have a more objective and realistic picture of the situation in the department of internal audit and will take the corrective measures and activities that will later contribute to the better functioning of the internal audit, and thus will contribute to more efficient management.

4. Internal audit and new governance framework of corporate governance in EU

The global economic crisis has raised the question of additional insurance in providing a framework of financial stability. External auditors have proved to be insufficient instruments in the creation of this framework. European Confederation of Institutes of Internal Auditors considers that an important role in the creation of this framework should take internal audit, which interacts with other mechanisms of corporate control can provide high-quality frames, which will improve risk management and the framework for future regulatory initiatives, which in the finale results in increased financial stability.

In creating the crisis response, the European Commission in 2010 formed the Reflection Group (Reflection Group), whose task was to draw attention to the current problems in the European legal framework, with a focus on corporate governance domains, ie. to the quality of the Green Paper (Green Paper), of importance for the efficiency of corporate governance assimilated the necessary changes, which will improve and stabilize the financial markets, ultimately.

Reflection Group pointed to the inadequacy of the Sarbanes Oxley act to prevent financial crisis in the US, and that there is an increased need for risk management. At the European level, internal control and risk management is based on only two directives: Directive 2006/43 / EC86, whose Article 41 requires respect for the public interest of the majority, in terms of the formation of the supervisory board, as well as the obligation to establish audit committees to monitor the effectiveness of internal controls, internal audit (if any) and risk management system [7].

Directive 2006/46 / EC88 Article 7 requires from the listed companies, with the disclosure of financial statements, an annual publication, the report on corporate governance, which needs to contain a description of the main features of the internal control of the company, as well as the risk management system in relation to the financial reporting process [7].

The Group emphasizes the importance of European Commission Recommendation 2005/162 on the role of non-executive or supervisory directors, but also emphasized the need for the Commission Audit Committee as well as the assistance in supervising the internal control and risk management as well as instrument of effectiveness of the external audit. This would contribute to the informing process of shareholders and helped the strategic approach to risk management as well as the reduction or elimination of conflicts of interest.

It is clear that the European Commission in the documents related to the new framework of corporate governance highlights two key instruments:

- To increase the security level of investors,
- To improve internal and external communication.

Increasing the level of security of investors includes a true and fair audit report. From this point of view ECIIA sees a solution in supportive role of external and internal audit. Such a role would enable the management of company more efficient approach to managing risks, because of the efficiency of individual perception of risk. In response to such a position was established three lines of defence model, which is increasingly accepted in European companies.

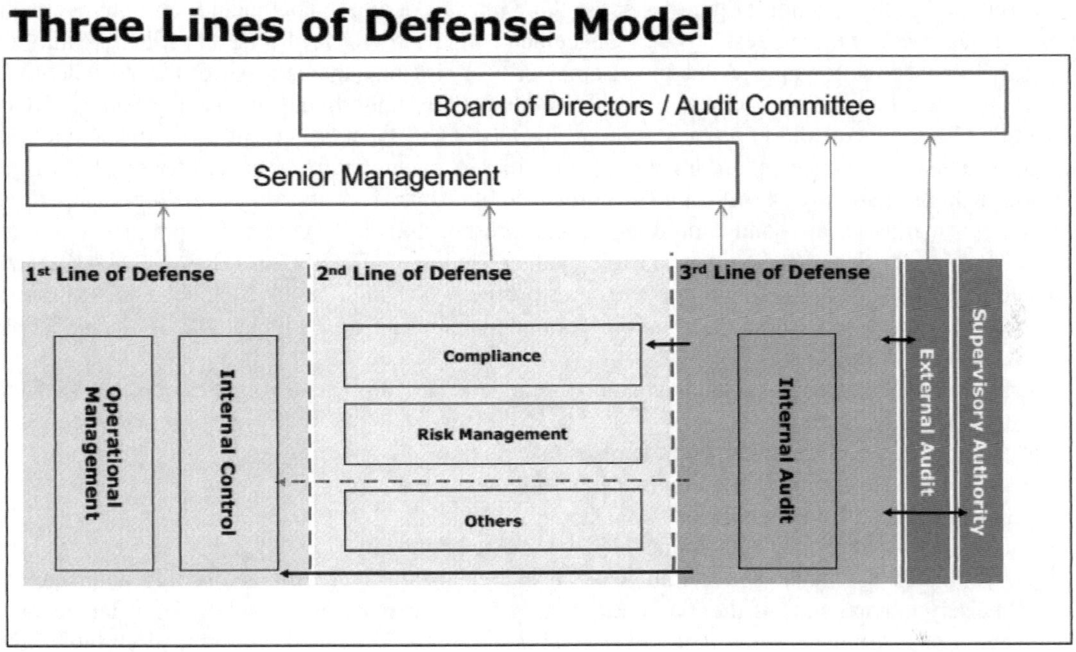

Figure 2. Model three line of defense (Source: ECIIA, Response to Green Paper Audit Policy: Lessons from the Crisis, Decembar 2010, www.eciia.eu, Downloaded: April 2012, p.3)

On the first line of defence function key of responsibilities in the assessment, control and reduction of risk has operational management.

On the second line of defence risk management, compliance, implementation and enforcement of risk management system, is reflected in reporting higher levels of management and owners of the binding information for risk management.

The third line of defence is internal audit, which will provide assurances, by his functional approach, to the higher levels of management the effectiveness of risk management, as well as on the effectiveness of the first two levels of defence. Within this line means all activities related to the elements of risk institutions: risk identification, assessment, communication related to the risk.

This risk management model, which emphasizes the importance of the role of internal audit in risk reduction, ie. improve the corporate governance framework, is largely recognized as an efficient and expeditious finds its application in practice.

Better external and internal communication is a need that arises from the perspective of information provided by external audit. The consensus is that the report of the external auditor provides information related to completed operations (so-called. "Historically focused information"), but also that this information must have value for investors, they have to indicate a potential risk. Research shows that inadequate risk management can mean the loss of 80% of the shareholders [8]. Given that such analytical and scope beyond the scope of the external audit, the solution is to be found in communication with the internal audit. The elements of such claim are arising from its independence and professionalism.

Risk management is one of the key parameters of the new corporate governance framework. Shareholder activism and the value of information for potential investors, created the need to create a model that provides accurate and transparent information, which creates confidence in the reported value. Given the complexity of risk (environmental, social, political, financial, etc.) EKIIR proposed the model of risk assessment, which will contribute to its reduction. This model is useful in risk assessment and in the function of the Model the three lines of defence. This includes proper oversight by the management in the process of risk management, where it will be strategic profiled. Within the EU, it is proposed that companies adopt a framework for risk management (COSO ERM, COSO 2, ISO31000). Benefit from the adoption of this framework is underlined and in a recent report of Reflection Group. Successful implementation of this model of risk management can contribute to: the likelihood and consequences of the materialization of risks, better information in terms of strategic decision-making, reduce cost of capital, more accurate financial reporting, creating competitive advantages, market share and corporate social responsibility. Structured approach for the implementation of risk management is possible by the adoption of Enterprise Risk Management framework (ERM – whose main components are: profile of the internal environment, set goals, risk identification, control activities, information, communication and monitoring.) and it is compatible with ISO 31000.

In this regard, enterprise risk management framework aimed at achieving the objectives of the entity, included in four categories:

• Strategic - high goals, aligned with the mission
• Operations - effective and efficient use of existing resources
• Reporting - reliability of reporting
• Compliance - compliance with applicable laws and regulations

From the listed risk management framework it is a clear the functional relationship with internal audit. Precisely internal audit is the only entity within the company who considers the efficiency and effectiveness of risk management at strategic level, and the assessment of conformity submits to the strategic management. Operational objective comes from its methods, rules and procedures of implementation of activities to evaluate the effectiveness of activities. Reporting by internal audit has a proactive component, in the profiling of potential risk areas, and evaluate the degree of that risk. Compliance comes from its connection with the external audit, but also form his own governing framework regulations internal audit.

Thus, the functional area of internal audit is compatible with the framework to improve corporate governance in the global economic crisis. Creates relevant information, reliable and transparent, which will be for shareholders and potential investors represent a conversion of confidence in the value.

5. The necessity of implementing the new framework of corporate governance

5.1. The efficiency of corporate governance before recommendations of implementing the model of three lines of defence

The Model of three lines of defense, is a response to the efficiency of corporate governance in times of crisis and instability of the global market. While there in 2011 ECIIA is indicative of its importance and the necessity of implementation, electronic research shown that in the Republic of Serbia, this model still largely represent as the unknown. Precisely, with the aim of emphasizing the extent of necessary to implement this model, we used the research about the effectiveness of corporate governance in 2009 [9] (period, which initiated the establishment of this model in Europe). The study surveyed 14 companies in the territory of the Republic of Serbia. The survey was conducted by the survey - a questionnaire Efficiency of corporate governance in the global economic crisis in 2009 year. The survey included 65 questions that were diversified spectrum in terms of establishing a model of corporate governance in the Republic of Serbia. Surveyed enterprise in 47% are in service area, 38% are wholesale companies and 15% have the production activity. In 79% of surveyed companies, number of employees is smaller than 200, 14% have over 500, and 7% have between 200 and 300 employees [9].

Using the results of that research, in order to prove the necessity of the implementation of the Model three lines of defense in the Republic of Serbia, it was made the systematization of certain segments of research, and within defined the lines of defense.

5.1.1. First line of defence – operational management and internal control. The first line of defense in risk management, represents the operational management and internal control system. The task of operational management is based on the identification, assessment, management and control measures to eliminate the risk. Operational management is responsible for taking corrective measures in the functioning of processes and controls [10].

On the basis of the research, it can be concluded that the effectiveness of operational management is based on relevant and timely information in 62% of the respondents (Chart No.1), and that the control procedures in the same percentage assessed as effective. Unfortunately, the percentage is greater than 30% in the negative connotation justified doubt on the effectiveness of implemented first line of defense in the surveyed companies, which indicates the high level of risk at the first line of defense. An additional measure of risk is the fact that 33% of respondents believe that the descriptions of activities only partially regulated, and that 20% of the problems of internal control in just partly or not at all documented.

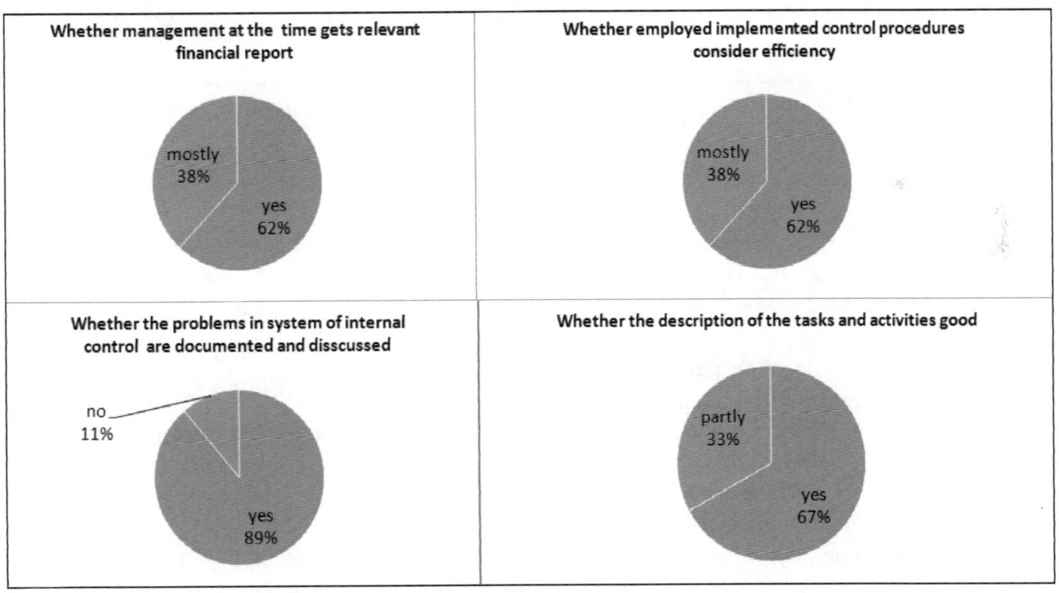

Figure 3. First line of defense

5.1.2. The second line of defence - risk management, compliance, financial control. The fact that high risk is present at the first line of defense, further jeopardizing the effectiveness of the second line of defense, which was confirmed afterwards as well. The functions of risk management, compliance and controlling exhibit a high degree of risk. The lowest risk is on the side of formal management responsibility (7% non formalized). A similar degree of risk present in the control contract (8% partially set), but more in the control of execution of contracting that risk increases (8% did not set up controls and 8% only partially).

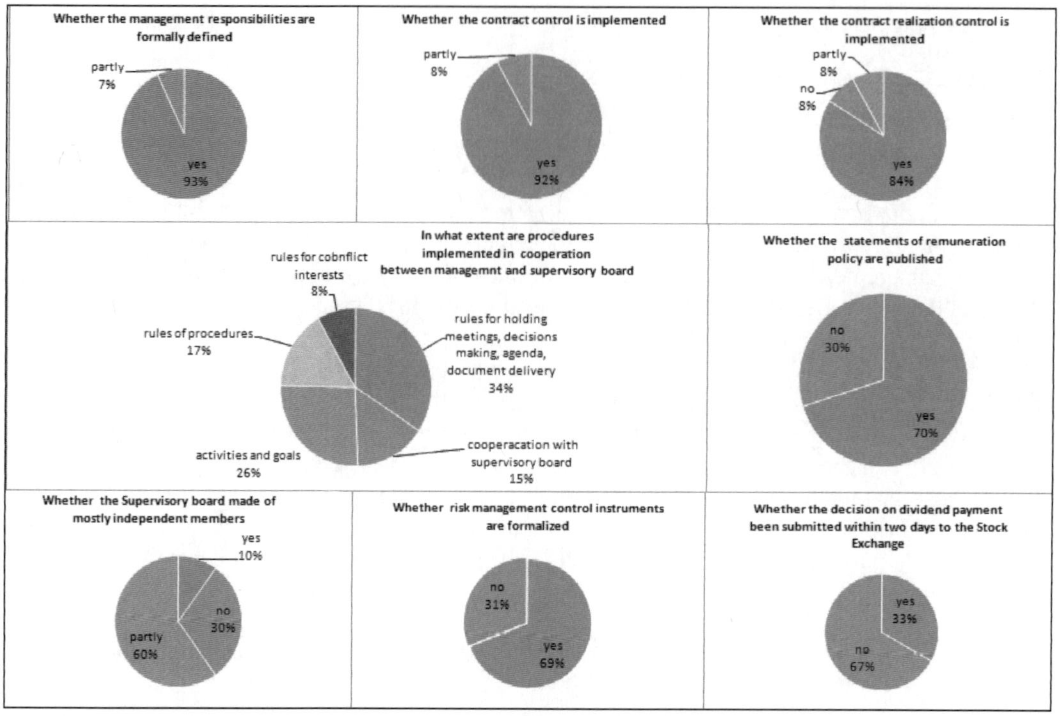

Figure 4. Second line of defense

The highest degree of control operations are related to the formal act (regulations, minutes) as 25% of it, the cooperation with the Supervisory Board is 11% and 19% are the area of operation, rules are 19%, while the conflict of interest is exposed to the greatest risk, because it is regulated by only 6%. The risk of conflict of interest is further substantiated by the fact that the members of the supervisory board in only 10% independent in a given sample. Payment policy management and supervisory boards in 70% are public.

Based on the foregoing, the partial mechanisms justifiably casts doubt on the fact that 92% of respondents implemented risk management mechanisms that are formalized 69%. Even a large degree of deviation from the implementation to the formalization indicates an extremely high degree of risk on the second line of defense, as well as the reasonable doubt to the effectiveness of the implemented mechanisms.

5.1.3. Third line of defence – internal audit. On the third line of defense are the functions that are objective and independent assurance company interest groups that information is timely, relevant, accurate, supportive, precisely those that are arising from internal audit activities.

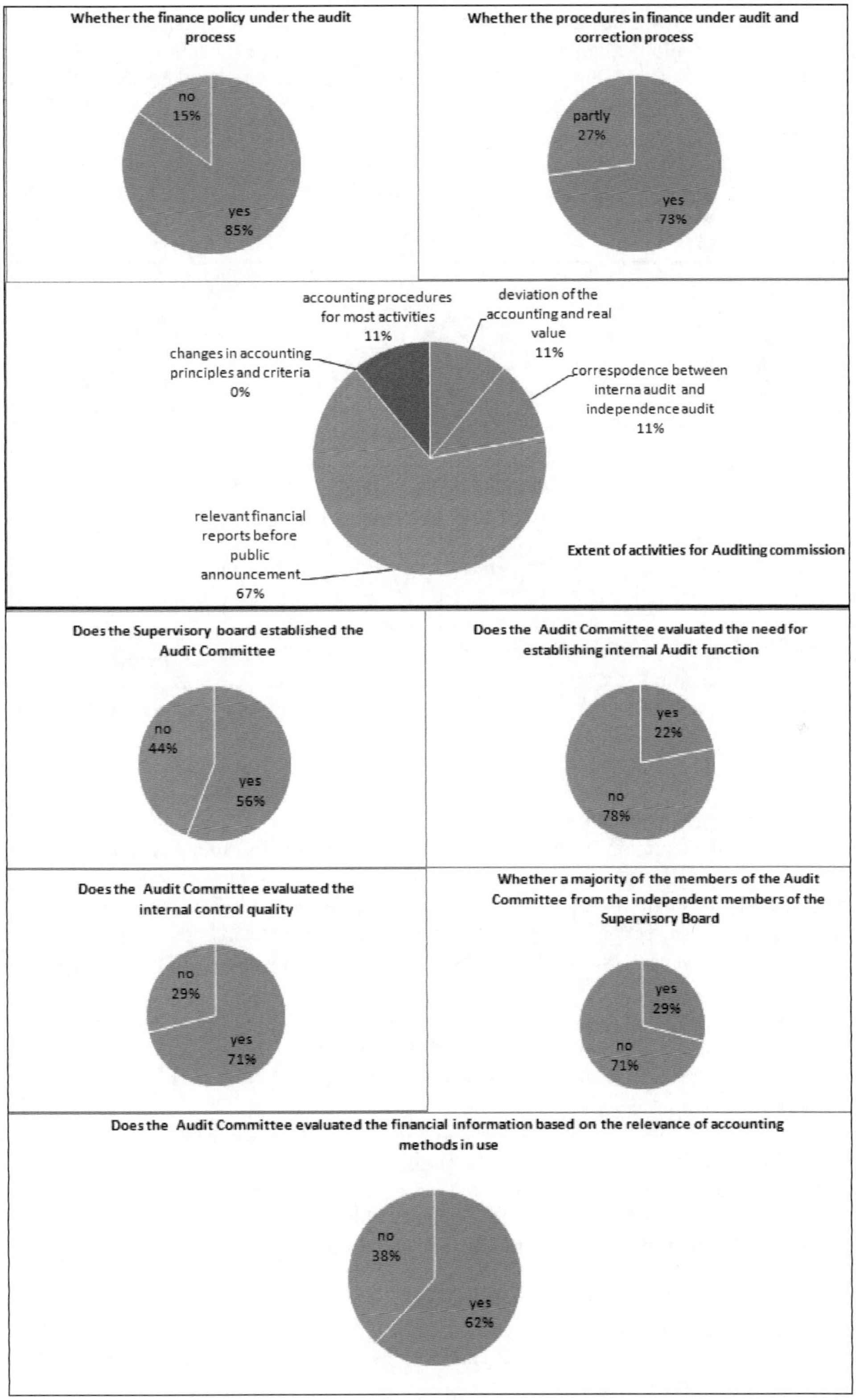

Figure 5. Third line of defence

Analyzing the data of the research can be seen that within the framework of implemented activities in a high percentage is the audit of finance (85%) and adaptation to changes within the finance (73%).

One of the problems in the third line of defense in effective corporate management is the Audit Committee. It was formed in only 56% of surveyed companies, 67% of it is delivered to the financial statements, and even 78% of the assessment was not carried introduction of internal audit by the said Commission. The Commission assesses the internal control system is in 62% of cases. Even 71% of the committee members are not independent members, representing a high degree of vulnerability in terms of objectivity.

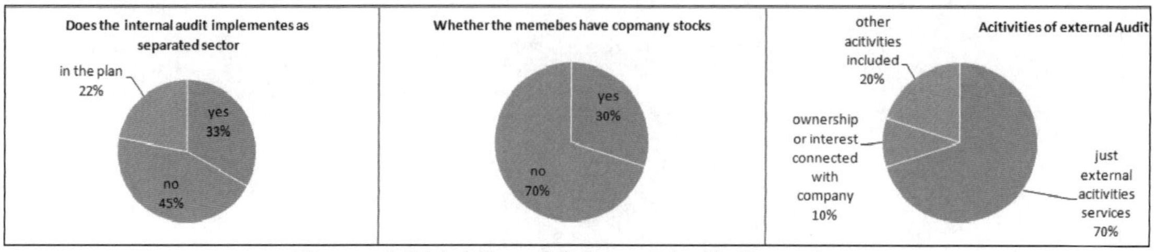

Figure 6. Third line of defence - Audit Committee

When it comes to internal audit, it has been implemented in only 33% of companies, and the objectivity of the external auditor information is further jeopardize because one of the key reasons, and that is that in 30% ownership, interest or through other services, auditor is associated with the company (this cause was one of the principal that has led to the necessity of creating new models of corporate governance, because conflict of interest of auditor is caused the false presentation and biased information which they manipulated shareholders and investors - the case of Enron and Arthur Andersen Company). In addition to previous statements is the fact that the Supervisory Board members have 30% shares of the company.

5.2. Efficiency of corporate governance after recommendations of the model three lines of defence

In order to obtain relevant data, confirm the hypothesis, and given the fact that even after the recommendations of implementation the model of three defence lines in order to reduce the risk of business activities, the model is not implemented, we conducted the short version of the previous survey, and included a small number of companies from the region, because the latest recommendations go in the direction of regional networking. Unfortunately, the results have been devastating. The causes may be founded in contempt of implementing the recommendations of the Model. Risk management is drastically increased.

On the first line of defence (operational management and internal control system) we observed the following results of higher potential risk of defence:

- 62,5% is the efficiency of control procedures (27.5% negative responses)
- 50% activity is only partially defined
- 37,5% problems of internal control are partially or not at all documented.

On the second line of defence (risk management) high potential risk of defence have:

- 37,5 % verify the execution of the contracts
- 25% of members of Supervisory Board, only, are independent members

On the third line of defence (internal audit), is present extremely high risk in the following business segments:

- 75% revision applies only to Financial Statements
- Only 50% have an audit committee
- In 50% case assessment was not carried out by the Audit Commission for the implementation of internal audit
- 50% Commission members are not independent members

Reduced research confirms the urgent need for the implementation of the Model the three lines of defence.

6. Conclusion

Causes and consequences of the global economic crisis caused the need to study the new corporate governance framework and seeking mechanisms and instruments to mitigate the risks which are facing the shareholders, investors, managers, creditors and all other subjects of interest of the corporation. A key question in this context arises, which instruments can lead to the establishment the system of efficiency and effectiveness of corporate governance in a climate of high risk of data inaccuracies, lack of confidence, macroeconomic instability. European Confederation of Institutes of Internal Auditing, the response required in the activities and functioning of the internal audit area. Based on the requirement of creating corporate social responsibility as a response to the need of converting confidence in the value ECIIA defined the Model three lines of defence. The first level of defence is against the risk on the operational management level, while the second level of risk is managed through an enhanced level of communication management and owners of the capital, the third level of defence is internal audit, which is a function of the effectiveness of the risk management and efficiency of the first two levels of defence. Therefore, it is necessary to increase the level of security of investors and establish better internal and external communication. As a result, the set requirements and models, it is emphasized the need for synergic linking of internal and external audit. Internal audit is an instrument that is based on the independence and professionalism and could provide analytical information of risk management, which extend the scope of the external audit, and on the other hand contribute to transparent and comprehensive audit process. All the above would lead to increased safety risk management processes, and ultimately contribute to the construction of corporate social responsibility.

Thus, it is clear proof that the necessity to develop internal audit as a catalyst in the creation of corporate social responsibility. The analysis of the research in the Republic of Serbia, we can conclude that, although the financial and corporate market is still largely undeveloped, the problems facing the global market in these markets even more pronounced and require complex and analytical approach. Also, the data speak in favour of the claim that risk management is largely poor but on the second line of defence, and the fact that the third line of defence is still in the construction phase, which further jeopardizes the market and creates a negative image, which does not attract new investors, or threatens and the interests of the existing ones.

In this sense, the implementation of internal audit, integrated into the company through three lines of defence model, it is necessary and alarming solution in establishing a new framework of corporate governance.

References

[1] ***ECIIA, Response to Green Paper Audit Policy: Lessons from the Crisis, Decembar 2010, www.eciia.eu Downloaded: April 2012

[2] Miller P 2008 *The Global Profession of Internal Auditing and a Vision for the Future,* Conference of ECIIA, December, pp 23 (http://www.ipai.pt/fotos/noticias/miller_gs41_1233773982.pdf)

[3] Živković M and Dragičević Radičević T 2009 *Corporate Management During the Crisis*, 7th International Scientific Conference, Dealing with the Global Economic Crisis by Companies and Economies, Belgrade, November 27

[4] Taken: Patty Miler, op. cit. p 23

[5] Chambers A, Selim G and Vinten G 1987 *Internal Auditing*, Pitman Publishing, London, UK

[6] ***ECIIA, op. cit.

[7] ***Downloaded: Report on a Reflection Group on a Future of EU Company Law, Brussel april 2011.

[8] ***ECIIA, op. cit.

IOP Conf. Series: Materials Science and Engineering **200** (2017) 012061 doi:10.1088/1757-899X/200/1/012061

[9] The survey used in the published paper: Živković M and Dragičević Radičević T *Corporate Management During the Crisis*, 7th International Scientific Conference, Dealing with the Global Economic Crisis by Companies and Economies, Belgrade November 27th, 2009
[10] ***Viewpoint of IIA "Three lines of defense for the effective risk management process and internal control system", IIA, Belgrade, January 2013, p 4, downloaded: march 2015 (https://na.theiia.org)

IOP Conf. Series: Materials Science and Engineering **200** (2017) 012062 doi:10.1088/1757-899X/200/1/012062

Special investigative actions in the criminal procedural law of Bosnia and Herzegovina, and their purposefulness in law

V Rodic

University of Businees Engineering and Management Banja Luka, Republic of Srpska, Bosnia and Herzegovina

E-mail: rodicvesna@yahoo.com

Abstract. Special investigative actions are a novelty in the criminal procedure legislation BiH. In our legislation this has been introduced only in 2003. These are actions that need to assist prosecutors to effectively reveal the perpetrators and the evidence of serious and complex crimes, especially organized crime. There are criminal offenses in whose execution included a larger number of people in different locations and which are carried out with the help of new communication technologies, so that it is not possible to prove the classic means of evidence (documents, physical evidence, witnesses). Unfortunately, criminals are often better organized than the state and in their criminal activities are introduced various technological developments. Therefore, the legislator extended the powers of the prosecutor and gave them into the hands of more effective mechanisms for discovering the perpetrators of these crimes. Special investigative measures are listed in the Penal Code.

1. General Investigations

Special investigative actions are a novelty in the criminal procedure legislation. In our legislation have been introduced in the year 2003 [1].

These are actions that need to assist prosecutors to effectively reveal the perpetrators and the evidence of serious and complex crimes, especially organized crime [2]. There are criminal offenses in whose execution included a larger number of people in different locations and which are carried out with the help of new communication technologies, so that it is not possible to prove the classic the means of proof (documents, material evidence, witnesses).

Unfortunately, criminals are often better organized than the state and in their criminal activities are introduced various technological developments. Therefore, the legislator extended the powers of the prosecutor and gave them into the hands of more effective mechanisms for discovering the perpetrators of these crimes. Special investigative measures are listed in the Penal Code [3].

1.1. Surveillance and technical recording of telecommunications

Surveillance and technical recording of telecommunications called: WIRETAPPING - in practice it is very applicable, useful and fruitful.

This special investigative action is limited to the right to privacy is not only suspicious person, but indirectly, any third party with whom the suspect makes contact by means of communication.

This special investigative action is done by intercepting communications at a time when communication is still going on (so-called, real-time or on-line control).

Surveillance and technical recording of telecommunications is one of the most special investigative actions in combating organized crime for two reasons:

1. Supporting the value of this special investigative action is indisputable person, because words suspects the best proof of his work and his intentions and
2. Electronic control allows action before committing the crime, if the application of these techniques reveal the planning and preparation of its commission.

Under the content of communications include the information regarding the substance, meaning or significance of any wire, oral or electronic communications.

Figure 1. Technical recording of telecommunications

1.2. Access to computer systems and computerized data

Access to computer systems and computerized data apply within the computer crimes - example. double accountancy, correspondence and other.

Access to computer systems means a collection of various technical means and procedures that are used to from one computer system covertly gathered data through interventions conducted from a distance through electronic networks. It is an electronic the opening of a computer system by the state investigative authority and electronic transmission of data computer where the measure is applied using special forensic computer programs [4].

Figure 2. Access to computer systems

IOP Conf. Series: Materials Science and Engineering **200** (2017) 012062 doi:10.1088/1757-899X/200/1/012062

1.3. Surveillance and technical recording facilities

Private and other premises are monitored and recorded. These actions relate to people and their actions, such as communication or action. These actions include optical and acoustic recording of a particular area, or activities in supervised premises.

Figure 3. Monitoring and recording

1.4. Secretly monitor

Secretly monitor and record PERSONS means of transport and objects that are associated with them – this action today is often applied. It can be done as a classic criminal act but but in this case is not evidence. If there is approval judge in the preliminary procedure for undertaking this special investigative action, then the proof.

Figure 4. Record persons means of transport

IOP Conf. Series: Materials Science and Engineering **200** (2017) 012062 doi:10.1088/1757-899X/200/1/012062

1.5. Undercover investigators
Undercover investigator is a police officer who was disguised and inserted into a criminal environment to integrate. The undercover investigator not to carry out a crime, but it can be aider or inciter (provocateur). INFORMANT (INFORMANT) IS THE CITIZEN integrated in the criminal group. It must not be made a criminal offense. Informant long was mounted in special investigative actions. Informant can not be a witness, but remains hidden.,

One of the important specifics of the undercover investigator is reflected in the fact that after his procedural character of the actions taken by the informal. An exception exists in the case of

When undercover investigators heard as a witness, which should be the exception, not the rule. In this situation, the hearing will be conducted undercover investigator so as not to reveal his identity, and that information about his identity are official secrets [5].

Figure 5. Undercover investigator

1.6. Simulated and controlled buy off the subject and simulated bribery
These actions are productive and are very sensitive to provoke offense (Figure 6).

Simulated and controlled buy off the subject - to be used for smuggling drugs, weapons or toxins (buys in / take a smaller amount). In implementing these special investigative actions must not be any provocation of the crime, because that would be overstepping authority (for example, offers the 1 kg of the drug, should not look for another 100 kg).

Simulated bribery - is done to see if the person against whom carried out the special investigative action taking bribes.

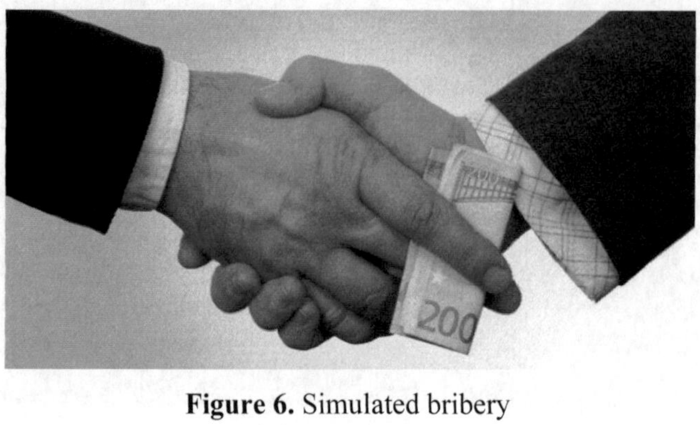

Figure 6. Simulated bribery

1.7. Supervised transport

Supervised transport and delivery objects of criminal offense - is done PASSING with the intent to determine if a specific person has certain types of goods (if the goods are legitimate, whether the goods are smuggled, prices of goods, etc.).

Figure 7. Supervised transport

These investigative activities under the Criminal Code of Bosnia and Herzegovina can be ordered only for criminal offenses against the integrity of Bosnia and Herzegovina, against humanity and values protected by international law, terrorism, and for criminal offenses for which a punishment of imprisonment of at least three years or more.

2. Organized crime

Organized crime is the most severe form of crime. In organized crime Criminal Code of Bosnia and Herzegovina, includes criminal offenses such as:

- terrorism,
- corruption,
- Money Laundering,
- economic delicts,
- drug trafficking,
- human trafficking,
- smuggling of vehicles,
- smuggling of weapons and others.

Organized crime is a negative social phenomenon, but also very complex, and consequently requires the application of special investigative methods such factors effectively fighting organized crime. For these reasons, in addition to well-organized, planned and been realized preventive and repressive activities and apply special investigative measures, as well as measures of covert data collection on the most serious forms of crime. Effectively countering organized crime can not be achieved using the "classic" and traditional methods of evidence because traditional methods of countering crime, especially organized crime are not fully effective and efficient in suppressing this negative social phenomenon.

3. Special investigative measures in the Criminal Procedure Code of Bosnia and Herzegovina

Codes of Criminal Procedure in Bosnia and Herzegovina provide for special investigative action as specific measures for the purpose suppression of complex forms of crime, these measures temporarily restricting basic rights and freedoms [6], [7].

Act regulated the cases when they can be used special investigative measures, given that they represent the contents of which violate human rights and freedoms, especially the right to privacy and protection of family life. Thus, the special investigative actions used under the following conditions:

- classic INVESTIGATION could not be able to reach EVIDENCE,
- criminal offense for which the penalty is over 3 years in prison

- criminal acts that threaten the safety of state, organized crime and others.

Special investigative actions are carried out when there is reasonable doubt in the investigation stage. The proposal to take specific investigative steps gives police the prosecutor, and the prosecutor to the judge for preliminary proceedings, which provides for authorization to take special investigative actions or taking them does not approve. Special investigative actions conducted by the police and their implementation is the responsibility of the Ministry of Interior, and control over their implementation is carried prosecutor.

Special investigative actions must be applied cautiously, and their application in accordance with the Act only provides legitimacy evidence. When conducting special investigative actions, should make an effort to be less damaging human rights and freedoms

Special investigative actions have a limit of up to six months. Their use is controlled and their preservation is required by law. The Law also stipulates the destruction of the material collected during the implementation of special investigative measures and inform the person to whom they were special investigative procedures.

Special investigative actions in Bosnia and Herzegovina mostly used to solve cases of economic crime.

4. Purposefulness of application special investigative actions in Bosnia and Herzegovina

According to data for 2013, the total number of persons against whom was ordered to conduct an investigation is 4.222. Resolved is 2,187 or 52%, while the rest to deal with 2,035 people or 48%.

For the 1,017 people confirmed indictment (of a total of 1,081 defendants).

If data on the number of persons who confirmed the indictment shall relate to the total number of persons against whom an investigation was conducted, it can be concluded that 24% of suspects indictment is confirmed.

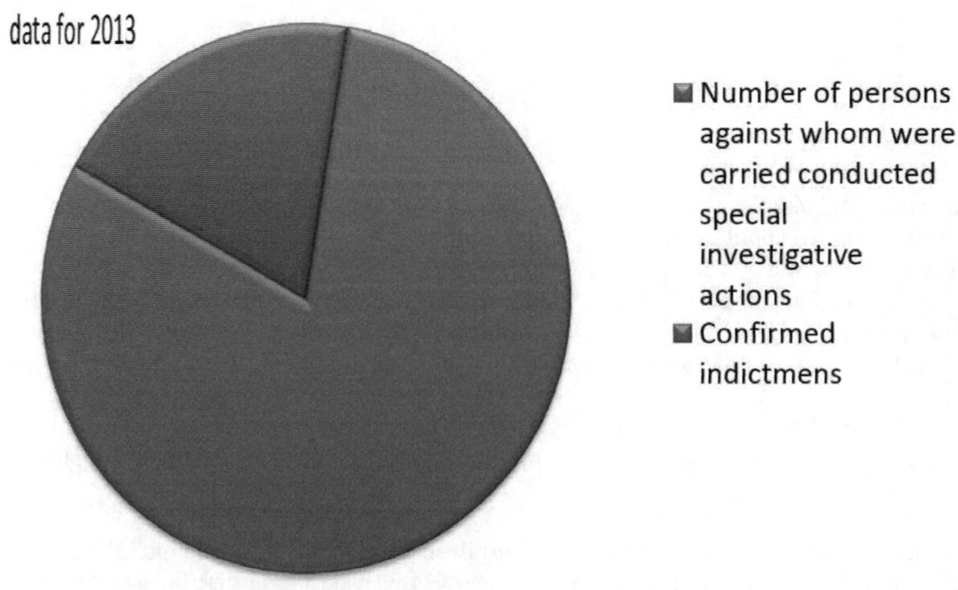

Figure 8. Persons against whom was ordered to conduct an investigation

5. Conclusion

Special investigative measures are very much used in BiH, but do not give satisfactory results. Their use was reduced in the purpose of sale of daily newspapers as well as raising the ratings of television news programs. Very few raised and confirmed indictments and convictions, the best proof of their appropriateness or inappropriateness.

References

[1] ****The Code of Criminal Procedure Bosnia and Herzegovina*, 2003
[2] Kajmakovic D, *The experiences and problems in the application of special investigative actions*
[3] ****Article 116, paragraph 2, The Code of Criminal Procedure* Bosnia and Herzegovina, 2003
[4] Antonic V and Mitrovic D, *Special invastigative action – MODUL*
[5] Bejatović S, *Criminal procedural issues new methods of detecting and investigating criminal acts of organized crime*
[6] ****Art. 116.st 2 items. A) to g) The Code of Criminal Procedure* Bosnia and Herzegovina, 2003
[7] ****Article 33 law on Amendments The Code of Criminal Procedure* Bosnia and Herzegovina, 2003

Rational consumption of nutrients at school-aged children

F Drăgan[1], V V Lupu[2], A Pallag[1], C Barz[3] and K Fodor[1]

[1]University of Oradea, Faculty of Medicine and Pharmacy, Oradea, Romania
[2]"Gr. T. Popa" Faculty of Medicine and Pharmacy, Iaşi, Romania
[3]Technical University of Cluj-Napoca, North University Centre of Baia Mare, Victor Babes str., no. 62A, 430083 Baia Mare, Romania

E-mail: farmafeli@gmail.com

Abstract. Nutrition lies at the basis of life being a factor of permanent action by which the individual, respectively small or large communities have ensured their survival in relation to the environment. Food represents a natural component which provides the energy and the vitality of the body and helps preserve people's health and welfare. Rational nutrition is closely related to the physical and mental development of the young persons, to the adults' work capacity, to the maintaining, as long as possible, of the elderly's health condition, to the prevention of some acute and /or chronic diseases.

1. Introduction

Studies in specialized literature [1] point out that poor nutrition is one of the major causes of the appearance and onset of several chronic degenerative disease [2] such as obesity, diabetes or atherosclerosis. Poor nutrition [3] does not refer only to the abundance or, on the contrary, to the lack of some of the nutritive principles, but also to the food industry technology which has brought a number of changes in human nutrition, thereby creating the premises for nutritional imbalances [4].

One may bring into discussion whether today's young people have a healthy lifestyle, whether today's adults had a truly healthier lifestyle in their childhood and to what extent the abundance of food and fast food products has damaged the quality our children's lives. It is essential to know the principles of a proper nutrition under any life circumstance, depending on age, physiological condition, work, in health and in sickness [5].

Knowing the types of foods and their nature, the nutritive principles and the nutritional ratio, the physiological changes of foods in the body [6], people's health condition and the way foods preserve it as well as setting up a rational and dietetic menu for a healthy person represent essential elements for a healthy lifestyle. All of the above reasons have urged us to set up this exploratory study that we shall present bellow.

2. Material and method

The study was done in several schools from Bihor County, Romania, using the questionnaire method. The aim was to present and highlight the dangers posed by poor nutrition in young people. For this exploratory study, we applied a questionnaire with 26 questions to some school-aged children from Oradea and from the rural area of Bihor County.

The questionnaire, in its form, is divided into two sections. The first section contains questions related to the demographic data used to analyse results as well as more general questions. The second section contains direct questions related to nutrition.

The sample we worked with included 416 children, being an opportunity sampling, a type of non-probability sampling.

3. Results and Discussions

From the total of 316 of subjects, 218 of the respondents were female and 198 males, 241 from urban areas and 175 from rural areas.

The distribution of respondents by age is shown in Figure 1, the distribution of respondents by age groups is shown in Table 1 and the distribution of respondents by gender is shown in Figure 2.

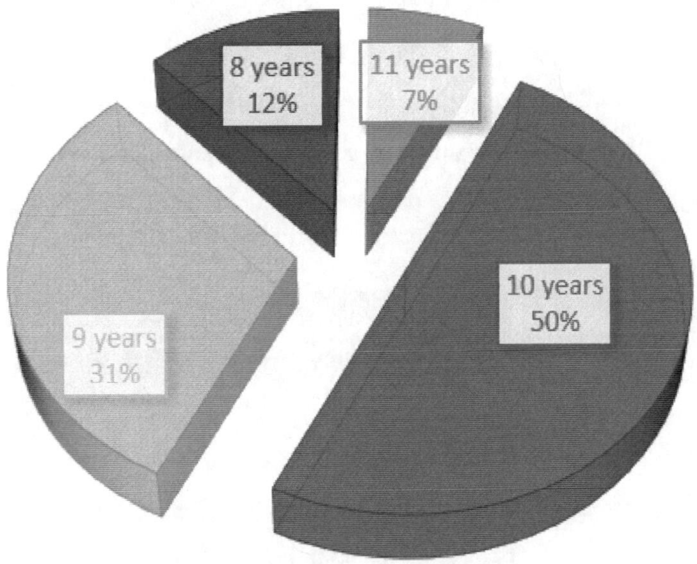

Figure 1. The distribution by age of the respondents from the sample

Table 1. A slightly more complex table with a narrow caption.

Age group (years old)	Male		Female		Total	
	No.	%	No.	%	No.	%
11	14	7.08	15	6.88	**29**	**7**
10	97	48.98	111	50.91	**208**	**50**
9	63	31.82	66	30.27	**129**	**31**
8	24	12.12	26	11.92	**50**	**12**
Total	**198**		**218**		**416**	**100**

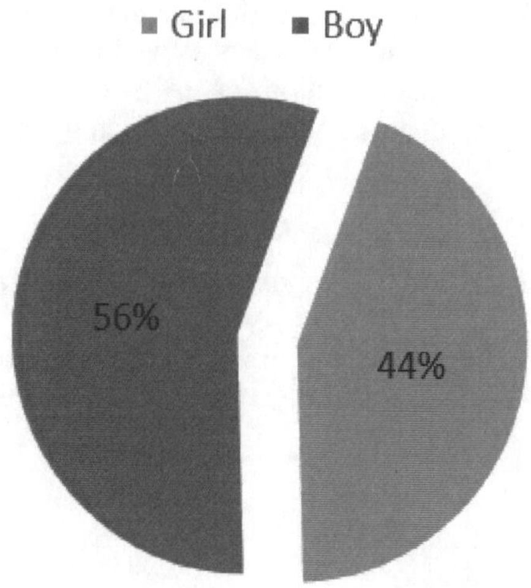

Figure 2. Distribution by gender in the studied sample

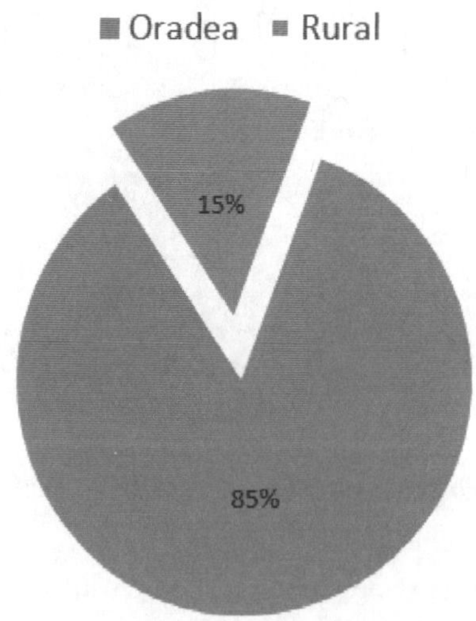

Figure 3. Distribution by background

One can notice that respondents to this questionnaire are children aged between 8 and 11. Our study aimed to clarify an important aspect related to child nutrition in Bihor and to children's adopted, respectively imposed lifestyle.

The child at this age is not aware of the true values of food. This requires proper education on nutrition problems from parents, school and mass media as to ensure a healthy diet. In this questionnaire children were asked whether they eat fruit, vegetables, sweets and of course, fast food products, respectively the frequency of consuming these types of food. Figure 4 shows the consumption of fruit for the questioned children.

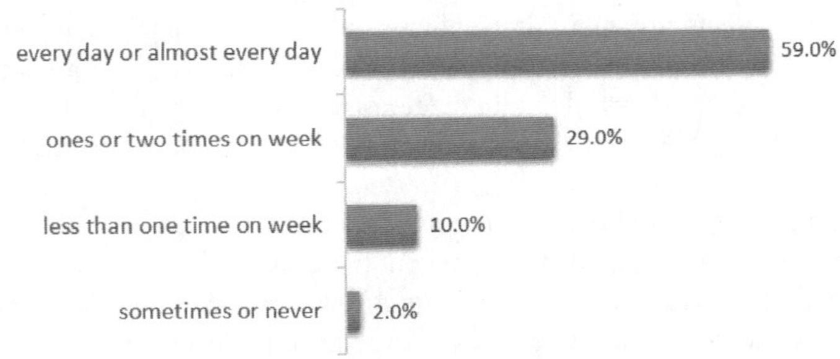

Figure 4. Fruit consumption

One can notice that the majority of children (59%) from the studied sample consume fruit every day or almost every day. The percentage of those who do not consume fruit on a daily basis is very small, only 2%. Fruit are rich in antioxidants and fight the oxidative stress that ages the cell, damages the DNA and predisposes the body to disease.

Fruit also contain fibres, necessary for the proper functioning of the bowel, as well as vitamins, necessary for a harmonious development of the children's body. Figure 5 shows the vegetable consumption of the questioned children.

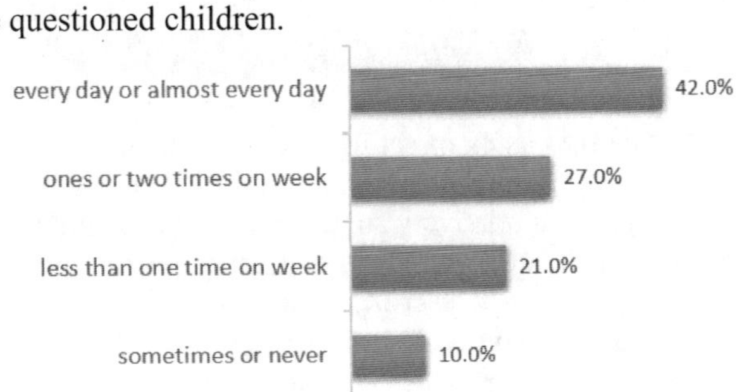

Figure 5. Vegetable consumption

Responses to the questionnaire show that 42% of the children have vegetables in their meals every day or almost every day, but a quite significant percentage of children (10%) has a diet without these very important nutrients. Vegetables play an important role in nutrition, being a source of vitamin A and C, folic acid, minerals, protecting the body and defending it against various diseases.

Vegetables are rich in water that contains concentrates of nutritional elements in the exact amount required by the human body. Figure 6 shows the consumption of sweets.

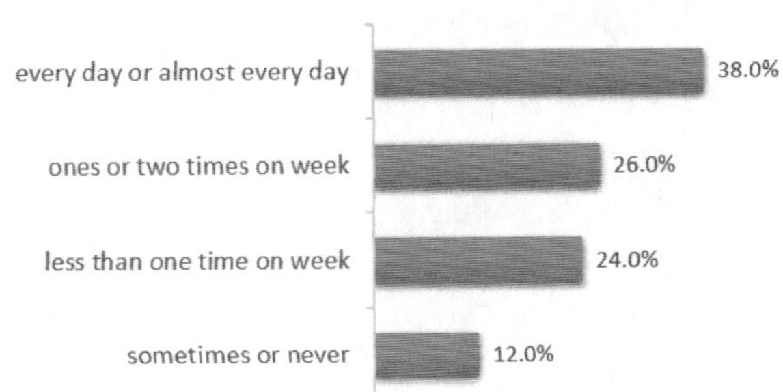

Figure 6. Consumption of sweets

Analysing children's answers one may notice that their nutrition is quite rich in sweets (38%). A small percentage (12%) admits not having introduced sweets in their diet. However, children enjoy sweets; we can say that sweets are their first choice. Sweets are unhealthy primarily because of the contained E numbers and preservatives, but also because of the high sugar content. Sugar-based products lead to an altered lipid profile, atherosclerotic plaques, increased risk of coronary heart disease and cardiovascular dysfunction, rapid rise of the blood sugar concentration, followed by an increased insulin secretion.

The content of caffeine and theobromine in chocolate has stimulant effects on the central nervous system. The beneficial effect of the antioxidants, contained in sufficient quantities in dark chocolate, must also be mentioned here. However, there are sugar-based products such as cakes, ice creams and chocolate that contain milk, butter, dried fruit and oilseeds and that provide some of the nutrients that are necessary for the body. Even here moderation is necessary. Figure 7 shows how frequent the questioned children eat fast food products.

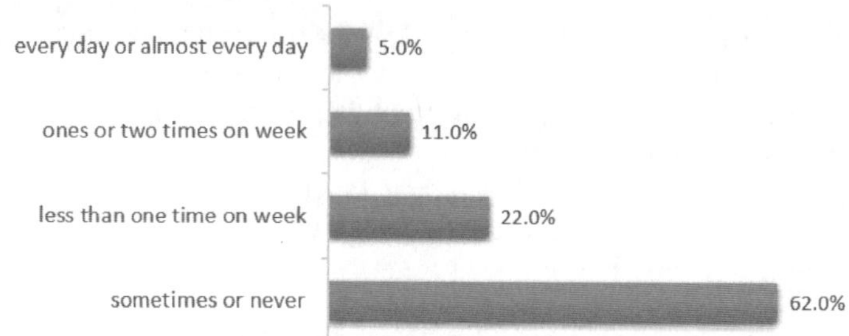

Figure 7. The frequency of eating fast food products

The results of the study reveal that only a relatively small percentage of children (5%) eat fast food products daily, and a high percentage of the questioned children (62%) is not used to consuming these foods. The term fast food or junk food usually describes foods rich in calories, salt, fat, but low in nutrients. This combination can produce, in a short time, changes in appearance, overweight, acceleration of the metabolic processes, insulin resistance, dyslipidemia, diabetes.

If we take into consideration the fact that the main predictors of health through nutrition and, in this case, of the happiness of the children from Bihor "a healthy child is a happy child" seem to be: fruit consumption, sweet consumption, fast food consumption, vegetables consumption, we can say, by extrapolation from the studied sample, that the children from Bihor are healthy-happy.

Figure 8 shows the results of the study in terms of children's health, thanks to nutrition, and their degree of satisfaction regarding food and nutrient consumption in accordance with their age and occupation as pupils in schools from Bihor County.

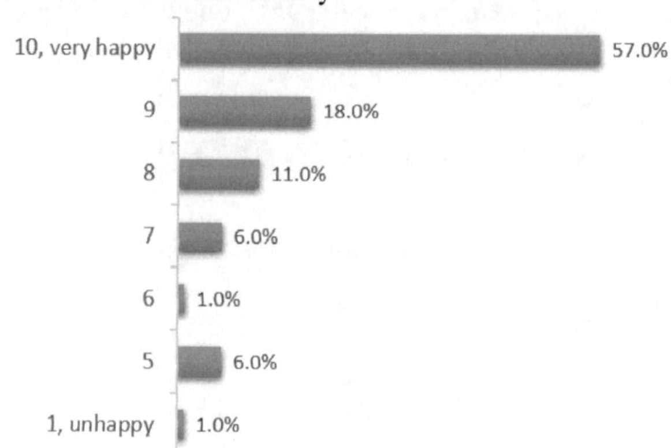

Figure 8. How healthy, thanks to nutrition, and therefore happy are the children from Bihor?

4. Conclusions

In conclusion to the above mentioned issues, one may notice that a large number of children (57%) can be considered healthy and thus happy as far as nutrition is concerned.

One may notice that questioned children have a proper consumption of healthy foods such as fruit and vegetables.

Questioned children admit the pleasure of consuming sweets and this is a good thing as long as sweets are consumed in moderation.

Consumption of fast food products is not yet characteristic to the studied sample, even if advertisements for these products are aggressive and present everywhere.

References

[1] White A et. al 2011 Use of Qualitative Research to Inform Development of Nutrition Messages for Low-income Mothers of Preschool Children, *Journal of Nutrition Education and Behavior* **43**(1)

[2] Roşu S T and Popescu S I 2008 *The therapy of acute pediatric pain. Drug use, abuse and dependency*, Editura „Grigore T Popa", Iași, Romania

[3] Graur M and Societatea de Nutriţie din România 2006 *Ghid pentru Alimentaţia Sănătoasă*, Editura Performantica, Iasi, Romania

[4] Negrişanu G 2005 *Tratat de nutriţie*, Editura Brumar, Timisoara, Romania

[5] Mincu I 2007 *Alimentaţia dietetică a omului sănătos şi a omului bolnav*, Editura Enciclopedică, Bucuresti, Romania

[6] Simu D, Roman G and Syilagyi I 2001 *Ghidul nutriţiei şi alimentaţiei optime*, Editura Dacia, Cluj-Napoca, Romania

IOP Conf. Series: Materials Science and Engineering **200** (2017) 012064 doi:10.1088/1757-899X/200/1/012064

Needs analysis for language course design. A case study for engineering and business students

L Todea and R Demarcsek

Department of Philology and Cultural Studies, Technical University of Cluj Napoca - North University Centre of Baia Mare, Romania

E-mail: ramona.demarcsek@cunbm.utcluj.ro

Abstract. These days, teaching languages – English in particular – has become an ever more complex and challenging task due to the fact that groups of students are eclectic in terms of knowledge of English and level of competence. Teachers try to adapt to this challenging environment by providing a wide variety of activities and materials so as to satisfy the students' demands. This may sometimes prove counterproductive, is usually time consuming, and requires material and financial resources that are not always available. Under the circumstances, we considered that a needs analysis would benefit all parties involved: teachers on the one hand, in that it would somewhat simplify their effort and make it easier to work with the aforementioned groups of students, and the students on the other hand, in that the activities would be better tailored to suit their needs and requirements in terms of knowledge, abilities and skills. To this end, we conducted a survey in which we requested students to state their level of satisfaction with regard to the English course, as well as provide suggestions in terms of topics and language issues they would like to approach during these courses. In this paper we analysed and interpreted the responses in an attempt to adapt the content of the course so as to better meet the expectations of the end beneficiaries, i.e. the students.

1. Designing ESP Courses - Aspects of Theory and Practice

Language courses for students in Business and Engineering should be designed considering their needs, wants and expectations about foreign language communication in a professional setting. The main reason for developing ESP activities should imply a strong purpose of communication while focusing on a specific content and specific language structures. Thus, needs analysis is a useful tool to use in order to assess what language points students need to master, as well as to meet their needs during the learning process. This article discusses the findings of a questionnaire which was administered to students in the second year of Business and Engineering from the North University Centre of Baia Mare, Technical University of Cluj Napoca. The purpose is to assess personal and language information about the selected target group of learners and to establish areas of competence that an ESP learner needs to develop so as to get over the lack of confidence in dealing with specialist types of discourse.

Nowadays, English is considered to be the lingua franca of international technical professions. Romanian technical universities usually demand their students to study at least one course in a foreign language. Most students choose English, French or German. Therefore, technical universities have to re-evaluate the degree to which their students' career needs are being met by the existing language programmes, so that policy and resources should be adjusted accordingly. The question under discussion is whether the courses in foreign languages allow students to communicate effectively in contexts that business or engineering prospective graduates are likely to encounter in their jobs. A user of English as a lingua franca thus has to accommodate to different other speakers from different linguistic and cultural backgrounds with different levels of competence in each speech situation. Real common features have been identified first and foremost regarding discourse style and pragmatics, suggesting that what unifies lingua franca speech is communication strategies rather than the result of any structural convergence [1].

2. ESP and Needs Analysis

It is generally acknowledged that English for Specific Purposes focuses on relating the process of teaching and learning English to the learners' communicative needs. Hutchinson and Waters [2] state that if ESP teachers know why learners need English for, the content of the language course could be tailored accordingly and the teaching process should focus on these specific needs. As a result, curriculum content and developing teaching materials must be based on needs analysis. In Brown's opinion [3], needs should be formulated in terms of goals and objectives which, in turn, could be considered the foundation for developing tests, materials, teaching activities and evaluation strategies.

The process of needs analysis or target situation analysis consists of identifying the target situation and then carrying out a rigorous analysis of the linguistic features of that situation. The needs analysis approach should not be perceived as a procedure delivered only once at the beginning of the seminar; it should rather be an on-going process repeated during the development of a language programme. The factor that distinguishes ESP from GE is the awareness of the need and not the existence of it [2]. The moment learners and teachers establish the context in which English is needed, their awareness will have an influence on what will be acceptable as "reasonable content in the language course" and what "potential can be exploited" [2]. Considering different studies, there is no necessary relationship between necessities as regarded by the ESP teacher and what the learners want or need. Learner motivation is an important aspect in the learning process; hence, the learners' perceived wants cannot be ignored.

Focusing on the situation in which the learners are supposed to work after graduation reveals the target needs. Hutchinson and Waters [2] define target needs as those required by the learners' in the context of use. They identify target needs as necessities and distinguish them from wants which are defined as what the learners believe they need. Although wants are more subjective, both necessities and wants should be considered in the process of needs analysis to establish what the students are like at the start of their language course, in order to investigate their strengths and weaknesses. They also describe the target situation in terms of necessities, lacks and wants. The authors [2] also make a distinction between target needs and learning needs. The target need refers to what the learner needs to do in the target situation and the learning need refers to what the learner needs to do in order to learn [2]. They further subcategorize target need into: (1) necessities, defined as what the learner has to know in order to function effectively in the target situation, (2) lacks, which are identified as the discrepancy between necessity and what the learner already knows, and (3) wants, considered as what the learner actually wants to learn or what they feel they need. The learner's wants may or may not reflect those perceived by the ESP teachers [2].

Furthermore, Berwick [4] defines need as "the gap between what is and what should be". The *what is* could be equated to the current state of the learner's knowledge and skills and the *what should be* could be interpreted as the target situation requirements. The author makes a distinction between felt needs or expressed needs; needs that learners have, and perceived needs; judgment about the educational gaps in other people's experience [4].

In Robinson's view [5] the concept of needs "is established for a particular group of students and will be influenced by the ideological preconceptions of the analyst. A different group of analysts working with the same group of students but with different views on teaching and learning would be highly likely to produce a different set of needs". He also discriminates objective needs, which are defined as all factual information about the learner, for example language proficiency, L1 background, age etc.

Alderson [6], on the other hand, differentiates four types of needs: i. formal needs, which refer to the need to meet the institution requirements such as to pass an exam; ii. actual or obligation needs, which refer to what a student has to do with the language once he has learned it; iii. hypothetical future needs, which is related to the need to become a better professional in the future; iv. wants, which consider what a student feels he/she wants to do or to learn. The first and the forth are types of needs during the process of learning or process-oriented type, while the second and third are types of needs that are future oriented.

According to Altman [7], learners should be placed properly based on their age, level of language proficiency, maturity, time available. Therefore he considers various types of learner needs based on individual differences within the framework of learner-centred language teaching. This requires the institution to make flexible educational arrangements to allow all learners' access to learning that is appropriate to the types of needs they have. The adjustments of the learning resources are made accordingly to meet the individual differences related to time, aims, or expectations of learning.

The most frequent ways of gathering information about target needs are: questionnaires, interviews, observation, data collection and informal discussions with learners, colleagues, others. The questionnaire covers language skills area, out-of-class language experience, aspects of the language course, such as expectations and difficulties they experience during the course, and opinions about their language teachers.

According to Hutchinson and Waters [2] and Robinson [5], needs analysis should be an on-going process carried out during the life of each course since the learners' needs are changing too. This can help teachers take the changes into account in a way that encourages students' progress. Analysis of the target situation can tell us what students do with the chosen foreign language. Teachers also need to know how students learn to do what they do with English language. The role of needs analysis in a skills-centred approach is to provide a basis for discovering the underlying competences that enable students to perform in the target situation and to enable the course designer to discover the potential knowledge and abilities that the learners bring to the ESP class. This approach considers the learner as a user of language rather than as a learner of language.

Hawkey [8] considers needs analysis a tool for course designers. Therefore, it implies a "language training situation with reasonably specific occupational or educational objectives involving a reasonably homogeneous group of learners" [8]. Brindley [9] argues that objective needs should be used as a starting point in course design; thus, "if instruction is to be centred on the learners and relevant to their purposes, then information about their current and desired interaction patterns and their perceived difficulties is clearly helpful in establishing programme goals which in turn can be translated into learning objectives" [8]. He further states that needs analysis is important as a guide in setting broad goals and as a guide in the learning process. Given the information about learner needs a course designer will be able to produce a specification of language skills, functions, and forms as required in the learner needs profile.

Hughes [10] states that "materials for teaching speaking need to synthesize what can be extracted from this immense variety of spoken discourse types and contexts to form the basis of something that can manageably be presented, taught, and assessed". Hughes [10] also suggests that the complexities of the skill itself mean that "materials should generally, and perhaps paradoxically, not be too ambitious in their aims". As the author [10] points out, speaking is a high-risk activity which can never be retracted or erased, making some learners worry about speaking and losing face. Increasing learner speaking time is bound up with understanding both learner and teacher contributions to the quantity and quality of classroom interactions. In the teaching of writing, as in the teaching of speaking, there is tension generated by different views of what teaching should focus on, though these views have major implications for methodology. The main division continues to be between product, process, and genre approaches. In many ESP classrooms, the main approach to writing is still very clearly product oriented. The focus is on models, with little if any thought of the way in which texts function in society; specific genres are provided as prescriptive models intended for copying with minimal adaptation, rather than as sources for generative discussion intended to further an understanding of the genre being explored.

Hutchinson [2] points out that ESP is "an approach to language teaching in which all decisions as to content and methods are based on the learner's reason for learning". The same idea is supported by Strevens [11] because as he says, the "ESP is particular case of general category of special-purpose language training. The same principles apply no matter which language is being learnt and taught". In addition to this definition, Dudley-Evans [12] gives a similar description of ESP. Most researchers seem to agree on the following aspect: ESP is based on a particular context and on the learners' specific needs.

According to Dudley-Evans and St. John [12], there are eight components in today's concept of needs analysis which have been grouped into five broad areas including: (1) target situation analysis and objective needs analysis, such as tasks and activities learners will use English for; (2) linguistic analysis, discourse analysis, genre analysis, knowledge of how language and skills are used in the target situations; (3) subjective needs analysis, including: learners' wants, means, factors that affect the way they learn such as previous learning experiences, reasons for attending the course, expectations; (4) present situation analysis for the purpose of identifying learners' current skills and language use; (5) means analysis, i.e. information about the environment where the course will run.

Furthermore, Stern [13] distinguished four types of ESP teaching objectives: proficiency, knowledge, affective and transfer. Proficiency objectives refer to the mastery of the four language skills: reading, writing, listening and speaking. Knowledge objectives include the acquisition of linguistic and cultural information. Linguistic knowledge objectives refer to language analysis and awareness of the systematic aspects of language, while cultural knowledge objectives refer to the control of socio-cultural rules (mastery of the norms of society, values and orientations and also the ability to recognize culturally significant facts). Affective objectives are about the development of positive feelings toward the subject of study. Transfer objectives concern the ability to generalize from what has been learned in one situation or another.

3. The Subjective Needs Analysis in Practical Application

When teaching students in technical fields, as well as students studying business students, all of the above four objectives are considered. Thus, it is important for students to become proficient in all the four skills, i.e. reading, writing, listening and speaking, while it is equally important for them to be aware of the cultural aspects, i.e. become proficient in the so-called soft-skills, as well. At the same time, having positive feelings about the taught subjects and being able to transfer knowledge and skills from one subject to another are just as essential. In order to reach these objectives teachers need to know what students need in terms of knowledge and skills in the taught language, and to this end, the best option is to run a needs analysis. Obviously teachers need to be flexible and adaptable if they are interested in satisfying the students' needs with regard to the knowledge and the skills they feel are important when learning a foreign language, English in this particular case, and they also have to be willing to use the feedback they get from the students in order to improve the whole teaching-learning process.

After having taught a certain group of students for one academic year, we decided it was an appropriate time to run such a needs analysis. We consider that after two semesters, students can form a relatively objective viewpoint about the teacher, about the subject taught and about what they need in terms of the language they are studying. At the same time, since these students study English for four semesters, it is, from the teacher's viewpoint, an appropriate moment in the development of the course to find out what students think about the teacher's performance, about the course and about how it could be improved to better suit their needs.

Of the four types of needs analyses identified by Dudley-Evans and St. John [12], we chose the third, namely the subjective needs analysis, which focuses on the learners' wants, reasons for attending the course, or expectations. Moreover, we considered that the teacher-student relationship, the classroom environment, and in that line of thought the students' perception of their teacher are all part of this type of needs analysis, i.e. the subjective one, and an integral part of a successful course. To this end, we designed a short questionnaire addressing a group of 179 second-year students from the Faculty of Engineering and from the Faculty of Sciences, studying in domains such as Mechanical Engineering, Electrical Engineering, Business Administration or Management, i.e. either Technical English or Business English. All responses were anonymous.

The questionnaire contained 4 items only: one referring to the students' general impression on the university and their specific field of study, as a quasi-neutral introduction to the actual on-topic questions and which was not analyzed in this paper, being beside the topic; the second item referring to their general feelings or opinion regarding the English course and the teacher; the third referring to the specific aspects that they enjoyed about the English course or those they did not like, if any, and the fourth and most important regarding suggestions that they had for the improvement of the course and the increase of their level of satisfaction related to the English course. These suggestions should all have focused on types of activity the teacher should include in the course in order to increase participation and involvement, improve the quality of the learning experience for the students, as well as their overall level of satisfaction. As expected, there were instances when respondents chose a less formal approach and made irrelevant, non-course related suggestions, yet these were not included in our analysis.

Thus, the first question, which, as mentioned, focused on non-English course related aspects, will be omitted here. The second question, which was meant to assess the students' level of satisfaction with the English course and the teacher, received mainly positive feedback, as one can notice in Figure 1 bellow. One should nevertheless not be mislead by the overall positive feedback, despite the fact that it is encouraging and generates positive feelings for the teachers themselves. A certain level of subjectivity can be expected from the respondents due to the fact that, even though questionnaires are submitted anonymously, generally speaking, students still consider the teacher's feelings, especially since they know that they will interact with the same teacher for the following two semesters. Moreover, the type of needs analysis we decided to implement was the subjective one, as mentioned above, thus subjectivity on the part of the respondents was to be expected.

Thus, on the one hand, a general positive feedback can be generated by the fact that students do not want to offer negative feedback out of consideration for their teacher, and on the other hand, they may be afraid of repercussions in case they provide negative feedback. In certain instances, students may be influenced in their responses by the fact that they think this is what is expected of them, i.e. a positive answer is what the teacher expects. Despite these aspects, it can only be encouraging for the teacher to discover that his/her methods generate such reaction, yet one should not fall pray to it and consider that if the level of satisfaction is high, no need for improvement is required. It should be considered a grater challenge to improve in such a manner as to diminish as much as possible the percentage of dissatisfied students. As shown in the figure below, out of the total number of respondents, 158 offered positive feedback, i.e. 88% of all students taking part in the survey, while 21 (12%) declared a certain level of dissatisfaction. In most cases, reasons were provided: some were related to the choice of activities carried out during the course, others were related to the attitude and behaviour of their peers, while a few were related to such administrative issues as the timetable (i.e. the course being too early in the morning). None had any comments on the teacher's performance, behaviour, knowledge or skill.

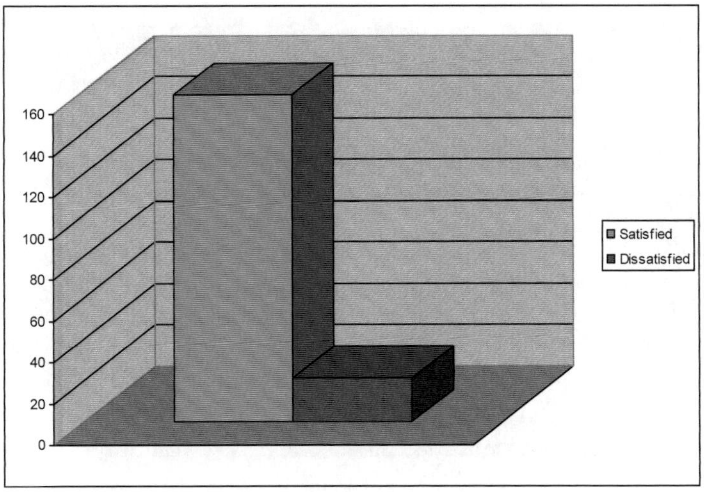

Figure 1. Level of satisfaction

The third question referred to the aspects and activities the students appreciated, or enjoyed, or liked in particular during the English course. Most aspects referred exclusively to the course, while some shifted slightly and included some reference to the teacher in their responses, mentioning the aspects from the teacher's behaviour or attitude that made them enjoy the course (such as the use of humour, for instance). While changing one's behaviour or attitude may be improved to some extent, possessing a sense of humour is an aspect that cannot be controlled by the teacher, unfortunately; nevertheless, it has been proven that humour has always helped in achieving a positive atmosphere in the classroom, so one should strive to include it amongst his/her teaching arsenal.

As regards activities, these are truly controllable aspects in the teaching-learning process, which are exclusively in the hands of the teacher. As one can notice in Figure 2, students have expressed their appreciation of certain types of activities and at the same time a certain level of dislike for others. For some reason, grammar seams to be amongst the least favoured types of activities, probably because it is more rigid in that it requires students to learn and observe rules. When it comes to speaking activities, or presentations which require public speaking and a certain performance in front of an audience, be it even made up of peers (or maybe more so), students have two types of approaches: on the one hand there are those who are really confident and have a good level of English, who are eager to express themselves and share their opinions with the rest of the group, and on the other hand, there are the shy ones or those who are quite self-conscious about their level of English and who prefer to sit quietly and let the former express themselves. This is why activities meant to exercise the ability to communicate orally received such a low score, with only 9 respondents out of 179 favouring these activities. Besides, there is another psychological mechanism which functions in adult learners: the fear of being judged by their peers. In other words, students are aware that they make mistakes when they speak, and choose not to share their ideas and opinion for fear of being mocked at by their colleagues.

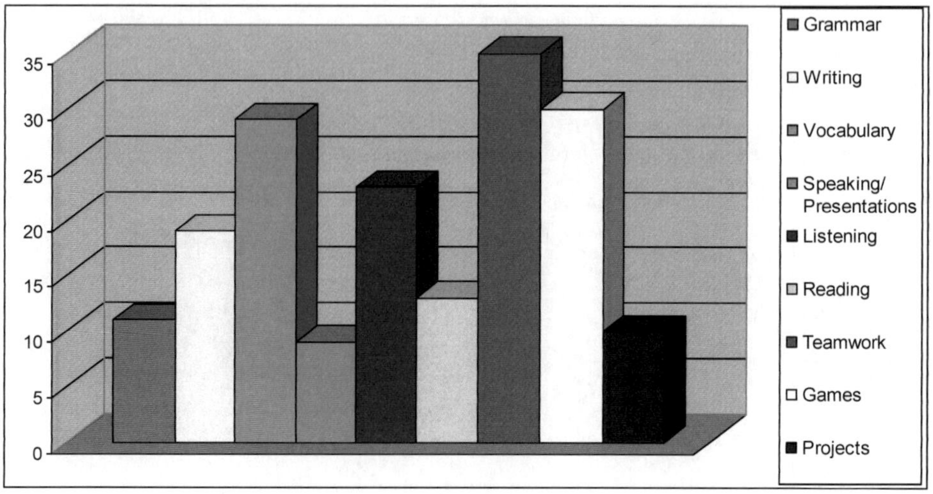

Figure 2. Preferred activities

Reading activities, when it meant reading aloud, were also not very much favoured by respondents (i.e. 13 positive appreciations of these activities), receiving an overall score similar to the speaking activities and being justified basically by the same mechanism: just like speaking, reading exposes the subject to public judgement, and since many are aware that they have poor reading skills or lack confidence, they are afraid of the reaction of the audience. And just like in the case of speaking, students with good reading skills, coupled with a high level of confidence, fuelled specifically by the student's skillfulness, have no problem with this type of activity, and thus will have a positive attitude towards it. Activities which involve writing, not necessarily activities designed to develop the writing skill, also received a low score (19 students mentioned this in their responses), yet this is justified by the students' aversion to individual work (or homework, for that matter). As an argument supporting this final statement is the high score that teamwork received (35 out of 179 students): many students expressed their preference for activities that required them to work cooperatively, especially since a lot of these activities involved using their smart devises [14] in order to look for extra information or for unknown words. The reasons for preferring these types of activities vary from true interest in sharing ideas and dividing the workload, to actually taking advantage of this setting to slack and let others do the work.

The activity coming second in the students' list of preferences was the one that falls under the umbrella of "games". Those games which required competition of any type and which involved team-playing were the most favoured amongst students. The reasons are quite obvious; beside, teamwork and team-play both give the participants the illusion of safety from public judgement (see above statements on speaking and reading). When dealing with vocabulary related activities, students expressed their preference for these because all vocabulary issues were related to their domain of interest, and they were all aware of the importance of studying field-related vocabulary, relevant in their field of study and possibly in their future jobs.

Activities meant to improve their listening skills were mentioned by 23 students in the group of respondents, which is encouraging and proves that they are aware of the need to improve this skill, while projects only by 10 of them. In this context, the term project refers to various tasks that students had to perform either on their own or in a group, and which usually generated a product-like outcome (presentation, essay, report etc.). For example, several students in technical fields had to work in groups in order to prepare a presentation of a science-fiction device, which included the actual design (created by them), the technical details, a description of how it functioned and what it was used for. Some even used short videos in order to illustrate all these aspects. Yet despite the fact that at the moment this type of project-based activity seemed successful from the teacher's perspective, it was appreciated only by few students because these activities are time-consuming, require vision and imagination, and most students are not very willing to invest much of their time in these. Nevertheless, when requested to suggest activities which would improve their knowledge and skills in English, the number of respondents recommending projects was double as compared to the number of those who mentioned these amongst their preferred activities (see specific comments for the fourth question).

The final question was actually the most important because it openly asked for suggestions from the students regarding activities that they considered useful to be approached in order to better meet their needs. This question was in correlation with the previous one, and while one could notice that students who preferred a certain type of activity requested more of the same type, there were situations when students requested others, not included in their list of preferences. The answers can be observed in Figure 3 below.

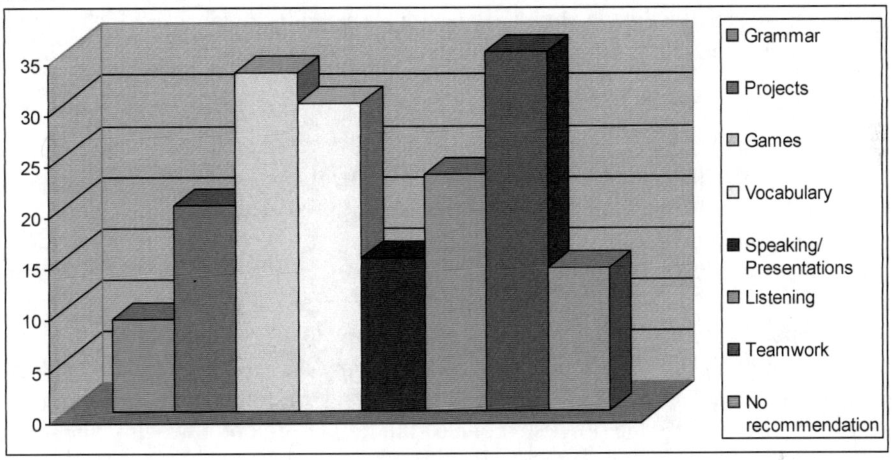

Figure 3. Recommended activities referred activities

Some of the responses might come as a slight surprise, yet one can find explanations for all. For example, despite the fact that grammar is not really appreciated, some of the students, probably those who mentioned it amongst their preferred activities, requested more such activities, acknowledging the fact that no matter how tedious it might seem at a given time, a language needs structure, i.e. grammar, and one cannot properly learn a language without adequately mastering its rules.

Project-like activities were requested by 20 respondents, which represent 12% of participants, and twice the number of students who mentioned these activities as their preferred ones. We have approached this issue at the previous question. Yet one aspect worth mentioning in this context: since the number of students requesting more projects is twice the number of students who like this activity, it proves again the students' maturity and awareness of the usefulness of these activities in developing their skills (not only language related but also cultural and social skills).

Almost a quarter of the respondents (35 students) stated they would like to participate in more activities which require teamwork. Some respondents explained their choices and said that they enjoyed these types of activities because they gave them the opportunity to share ideas, as well as the workload. It comes as no surprise that games were requested by almost a similar number of respondents (33 students), since this type of activity, besides reaching the educational objectives in a less formal manner, is also relaxing and fun. In terms of vocabulary related activities, since for the third question students demonstrated interest for these activities, acknowledging the need for field-specific vocabulary, it made sense that they would request more activities meant to enrich their vocabulary in their field of interest. On the other hand, speaking activities and presentations, despite the fact that these did not enjoy a high level of appreciation, as one can see from the responses provided for the third question, they were requested by 15 respondents, proving that students are aware of the need to exercise their speaking skills in general, as well as their public speaking skills.

Exercising the listening skill was also deemed important and 23 students (13%) requested that this type of activity be used more often during the English course.

Of the whole group of students participating in the needs analysis, 14 had no recommendations or suggestions. This does not necessarily mean a lack of interest; it may simply be a sign of either satisfaction with the activities included in the course, or a lack of inspiration. Some of the students in this group added some extra comments which actually make it easier to understand why they had nothing to say at this point: some said they had no suggestions for improvement because they liked what they had been doing up to that point, others admitted to not knowing what to suggest, while one considered it was not appropriate for a student to make suggestions regarding the activities a teacher should or should not include in his/her course since the latter is more capable of making such appreciations and it is the latter's task to know what activities are useful for the students. We consider that even odd responses like this one offer suggestions, as well as feedback, which the teacher can and should have in mind when designing courses, depending on the circumstances.

4. Conclusions

The questionnaire was both a feedback questionnaire and a needs analysis. Its purpose was to identify the types of activities that students enjoy and consequently use them more frequently during the English course. It was also meant to find out what activities students consider worth including in the English course from the perspective of their own needs as they perceive them, the major goal being to redesign the course in such a way that it best fit the students' needs. Thus, this analysis should result in a change of the design of the English course, and in order to measure its success, another questionnaire needs to be applied at the end of the second academic year to see whether the change was successful. Thus this endeavour will not remain unechoed, some follow-up actions being planned. Needs analysis should be seen as an indispensable aspect of ESP syllabus design not only in order to improve the students' experience during the English course, but rather in order to link the present students' academic needs with their needs in their prospective employment.

After applying the questionnaire we have discovered which activities are more successful and which should be included more often in the English course from the students' viewpoint. The questionnaire has proven to be a useful instrument for the improvement of the teaching-learning process during the English course and will generate change. Yet what we consider probably the most surprising outcome of the questionnaire is not the type of activities the students favour, nor is it the type of activities they would like to focus on more during the English course. The truly surprising aspect, despite the fact that the type of questionnaire used was the so-called subjective one, is that students seem to be quite aware of their needs in terms of English knowledge and skills, i.e. they seem to have quite an objective viewpoint on what activities are useful and generate improvement of their skills and knowledge. Thus we appreciate that such needs analysis should be carried out as frequently as possible and responses should be taken into account when designing ESP courses, since the target group seems to be objectively aware of its needs.

References

[1] Seidlhofer B 2004 Research Perspectives on Teaching English as a Lingua Franca, *Annual Review of Applied Linguistics* **24** 209-39
[2] Hutchinson T and Waters A 1987 *English for Specific Purposes: A Learner-Centred Approach,* Cambridge University Press, UK
[3] Brown J 1995 *The Elements of Language Curriculum: A Systematic Approach to Program Development,* New York: Heinle and Heinle Publishers, USA
[4] Berwick R 1989 *Needs assessment in language programming: from theory to practice,* In R Johnson (Ed) The second language curriculum, Cambridge University Press, UK
[5] Robinson P 1991 *ESP Today,* New York: Cambridge University Press, pp 48-62
[6] Alderson JC 1980 *A Process Approach to Reading at The University of Mexico - Projects in Materials Design,* ELT Documents Special, pp 134-163

[7] Altman H B 1980 *Foreign Language Teaching: Focus on The Learner,* In H.B. Altman & CV James (eds) Foreign Language Teaching: Meeting Individual Needs, New York: Pergamon Press, pp 1-16
[8] Hawkey R. 1980 *Syllabus Design for Specific Purposes, ELT Documents Special, Projects in Materials Design*, London: The British Council
[9] Brindley C 1989 *The Role of Needs Analysis in Adult ESL Programme Design*, In R.K. Johnson (ed) The Second Language Curriculum, New York: Cambridge University Press, pp 63-78
[10] Hughes R 2010 *Materials to develop the speaking skill* in N. Harwood, (ed) English Language Teaching Materials: Theory and Practice, Cambridge: Cambridge University Press
[11] Strevens P 1988 *ESP after twenty years: A re-appraisal*, In M. Tickoo (Ed) ESP: State of the art
[12] Dudley-Evans T and M J St John 2009. *Development in English for Specific Purposes: A Multidisciplinary Approach*, Cambridge: Cambridge University Press
[13] Stern H H 1992 *Issues and Options in Language Teaching*. Oxford: Oxford University Press
[14] Demarcsek R and Todea L 2016 Some of the challenges of teaching multilevel classes and suggestions for how to deal with the situation, *IOP Conf. Ser.: Mater. Sci. Eng.* **144** 012015

IOP Conf. Series: Materials Science and Engineering **200** (2017) 012065 doi:10.1088/1757-899X/200/1/012065

Open source system OpenVPN in a function of Virtual Private Network

A Skendzic[1] and B Kovacic[2]

[1]Politehnica "Nikola Tesla" in Gospic, Bana Ivana Karlovica 16, 53000 Gospic, Croatia
[2]University of Rijeka, Department of Informatics, Radmile Matejcic 5, Rijeka, 51000, Croatia

E-mail: wireless82@gmail.com

Abstract. Using of Virtual Private Networks (VPN) can establish high security level in network communication. VPN technology enables high security networking using distributed or public network infrastructure. VPN uses different security and managing rules inside networks. It can be set up using different communication channels like Internet or separate ISP communication infrastructure. VPN private network makes security communication channel over public network between two endpoints (computers). *OpenVPN* is an *open source* software product under GNU General Public License (GPL) that can be used to establish VPN communication between two computers inside business local network over public communication infrastructure. It uses special security protocols and 256-bit Encryption and it is capable of traversing network address translators (NATs) and firewalls. It allows computers to authenticate each other using a pre-shared secret key, certificates or username and password. This work gives review of VPN technology with a special accent on *OpenVPN*. This paper will also give comparison and financial benefits of using open source VPN software in business environment.

1. Introduction
Virtual Private Networks (VPN) can be used to establish a high level of security in network communication. VPN technology enables high-security networking using distributed or public network infrastructure. VPN uses various security mechanisms and rules within a network. It can be set up using different communication channels, such as Internet or separate ISP communication infrastructure. VPN establishes a security communication channel over public network between two endpoints (computers). OpenVPN is an open source software product under GNU General Public License (GPL) that can be used to establish VPN communication between two computers within business local network using public communication infrastructure [1]. This paper provides a basic overview of the way in which VPN technology works, gives information on GNU GPL for OpenVPN tool and its benefits, demonstrates general benefits of OpenVPN tool compared to other similar software solutions, provides an explanation of security level and finally conclusion.

2. Virtual Private Network
Internet, as a communication platform, is a basic communication system today. In addition to many different services relying on Internet infrastructure, Virtual Private Network (hereinafter:VPN) is a very efficient and economical method of communication, especially when connecting remote (virtual)

offices. VPN technology transmits potentially "sensitive" information, which can be classified as secret or confidential through insecure networks. VPN system is based on setting up of so-called "communication tunnels", previously secured using various cryptographic methods (algorithms).

VPN is basically divided into two types [2]:

1. *Site-to-Site* – an example of such connection would be a company branch that needs to be continuously or occasionally connected to the main location (Figure 2)
2. *Remote Access VPN* – used occasionally for connecting individual users to the company server from mobile phones or in cases where an employee works from home (Figure 1).

Internet is not the only option for establishing a VPN connection, we can use many other technologies such as ATM network or private Internet service provider (ISP) network. However, Internet infrastructure offers several advantages over ATM or private ISP networks [2]:

- more affordable,
- greater mobility,
- transparent communication.

In order to establish a VPN connection, the following requirements need to be met [2]:

- confidentiality - protection from unauthorised access to information,
- data integrity - protection of unauthorised modification of data,
- user authentication - verifying the identity of computers or devices at the ends of the tunnel.

The main advantage of using Internet communication for the VPN service is its price. Available VPN services can be completely free [3] (such as HotSpot Shield, CyberGhost, Tunnel Bear, Spot Flux, with several restrictions), while others are subject to charge and require additional training of persons (administrators) to administer such services.

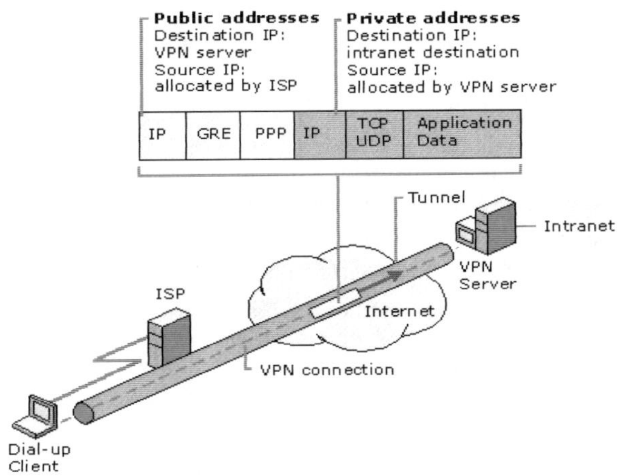

Figure 1. Overview of functioning of a Remote Access VPN network [4]

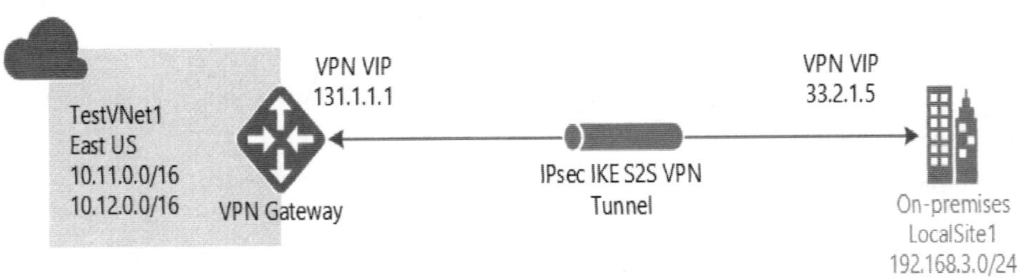

Figure 2. Overview of functioning of Site-To-Site VPN network [5]

3. GNU General Public License

GNU General Public License (hereinafter: GNU GPL) includes sharing and changing of open source software. GNU GPL also introduces certain guarantees that make software free for use [6]. Software under GNU GPL means that it can be copied, shared and distributed. Likewise, there are no limitations regarding the number of copies or the number of computers on which the software can be run. On the other hand, to protect the rights of the ones enabling the use of software under GNU GPL, certain obligations are imposed in case of distribution or modification [5]. When flexibility and limited scopes of software licenses such as EULA are compared to GNU GPL, it is evident that there are clear differences in limiting user rights, assigning user rights and limiting the right to lodge a complaint.

Studies [6] have shown that the following ratios are obtained when EULA[1] license is compared to GNU GPL:

Table 1. EULA – GNU GPL flexibility and limitations

Limitations	EULA	GNU GPL
Limiting user rights	45%	27%
Assigning user rights	15%	51%
iting the right to lodge a complaint	40%	22%

In conclusion, GNU GPL provides users with greater flexibility. User rights are less limited, more rights are assigned, and there are fewer limitations for lodging complaints.

4. OpenVPN

Private network is a new communication model enabling a more flexible and economical communication between employees within a company or within a partnership, that is, a complementary business system [7]. Companies can choose one of the models of setting up a VPN, whether they will take a turnkey service, build a network themselves or, in case of less demanding actions, use free software for setting up a VPN [7].

OpenVPN is also available for other platforms such as MS Windows, even though the benefit of using *OpenVPN* solution comes from the benefit of migrating the business VPN system to Linux platform. Some of the benefits of Linux solution can be [8]:

- robustness,
- flexibility,
- security,
- easy maintenance,
- price.

OpenVPN tool uses SSL/TLS protocol to encrypt data, creating a virtual tunnel between the endpoints. One of the particularities of *OpenVPN* tool is that it does not use a web browser installed on a client computer. Instead, *OpenVPN* package needs to be installed both server-side and client-side.

Advantages of *OpenVPN* tool over other tools can be seen from the following:

- Easy adjustment,
- Possibility of adjustment via script file in order to meet the user's needs,
- Portability, it is available for almost all popular operating systems,

[1]End User License Agreement (EULA) is an agreement between the software developer and the end user. In essence, it significantly differs from GNU GPL. EULA limits the possibilities and rights of users and releases the software developer from any liability.

- Directing network traffic to UDP[2] port 1194 or other. UDP protocol works on the "fire and forget" principle, which means it does not check whether the message reached the receiver, instead it sends short "datagrams" and checks for errors during sending, which results in faster data transmission,
- By using additional software (TUN or TAP), *OpenVPN* orders other applications to communicate only through the created virtual network interface, in a fully automated manner,
- *OpenVPN* enables secure user authentication using a public (static) key, corresponding certificate or user name and password,
- Use of *OpenVPN* also includes membership in a broad user community providing support and suggesting and adding new functionalities.

Table 2. Software for OpenVPN [2]

Software name	Operating system	Price
OpenVPN GUI	Microsoft Windows	Free of charge
Tunnelblick	Mac OS X	Free of charge
Viscosity	Mac OS X	13.95 €
Shimo	Mac OS X	9 $
OpenVPN	DD-WRT	Free of charge
TomatoVPN	Tomato	Free of charge
TunnelDroid	Android	Free of charge

Table 2 provides the overview of software for *OpenVPN*. It should be noted that some network and communication equipment manufacturers integrate *OpenVPN* into their firmware, which means the installation of *OpenVPN* to computers within the network is not necessary.

4.1. OpenVPN and security

As it was pointed out earlier, the protection of data in *OpenVPN* is based on encryption using symmetric and asymmetric algorithms. Hash function is used to protect message integrity, as well as for digital signatures. Symmetric cryptographic systems use the same secret key for both encryption and decryption of data. The problem of using symmetric cryptographic systems lies in the process of agreement, that is, of key exchange, which in this case goes through an insecure communication channel (e.g. Internet). Asymmetric cryptographic systems are a more reliable method of cryptographic protection because they use both public and private keys. Private key is known only to his owner, while public key is distributed to all users communicating within a VPN network. Public and private keys and their owners are part of so-called PKI (Public Key Infrastructure), which also includes digital signatures and digital certificates.
Table 3 provides the comparison between symmetric and asymmetric keys within OpenSSL Library.

Table 3. *OpenVPN* and use of SSL Library. Comparison between the use of symmetric and asymmetric keys [2]

OpenVPN	Use of static key	Use of OpenSSL Library
Encryption algorithm	Symmetric	Asymmetric
Application	Simple	Complicated
Speed	Higher	Lower
CPU usage	Lower	Higher
User authentication	No	Yes
Key exchange problem	Yes	No

[2]**UDP** (**User Datagram Protocol**) is a protocol in the transport layer of OSI model, and in addition to TCP, it is one of the basic Internet protocols.

IOP Conf. Series: Materials Science and Engineering **200** (2017) 012065 doi:10.1088/1757-899X/200/1/012065

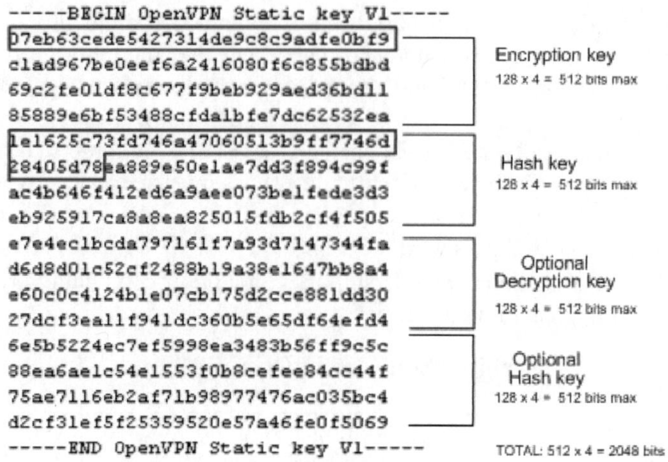

Figure 3. Randomly generated 2048-bit encryption key.
Overview of encryption key structure [9]

4.2. OpenVPN and security

OpenVPN also uses *OpenSSL* program package for its work. In general, SSL (Secure Sockets Layer) is a general security standard, and by using SSL the connection between a web server and a browser is encrypted, which is very important in Internet business environment from the aspect of security [10]. *OpenSSL* provides cryptographic support implemented in form of SSL and TLS security protocols. These are SSL v2 and v3, and TLS (Transport Layer Security) v1. Data encrypted using SSL/TSL technology are not visible, nor can they be intercepted or deciphered within a communication channel (Figure 3). *OpenSSL* can be upgraded within *OpenVPN* in case a new version arrives.

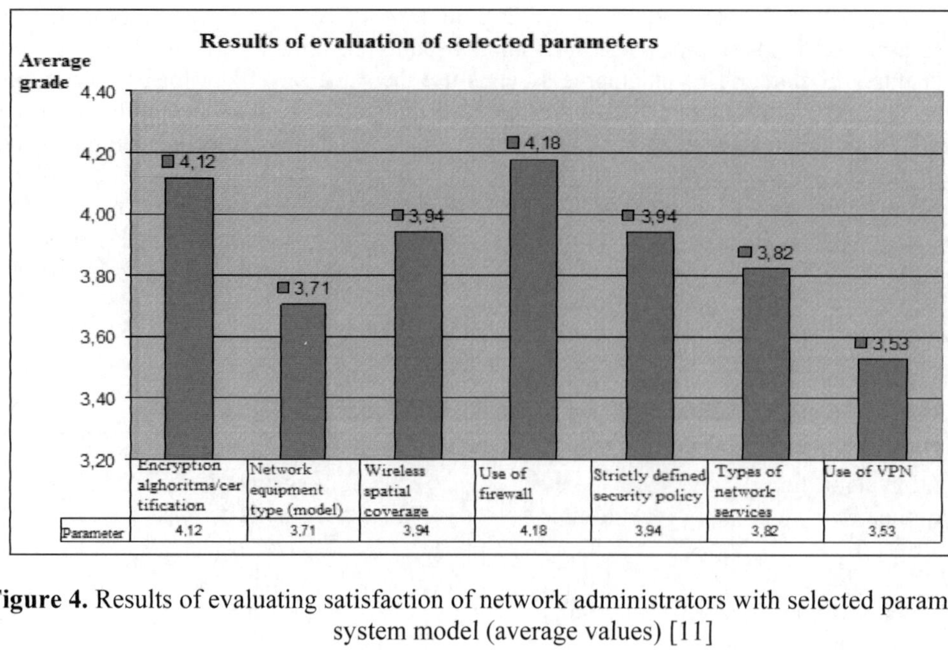

Figure 4. Results of evaluating satisfaction of network administrators with selected parameters in the system model (average values) [11]

Figure 4 shows the results of research[3] on a sample of system engineers (CARNet) who evaluated security parameters that influence the overall security of network communication. Based on the significance expressed for each criterion, weighted values for each criterion (grades 1-5) were established. Since the weighted values obtained were different for each criterion, it was established that different criteria had different influence on the final result of security evaluation. The result achieved by the parameter "Use of VPN" should be singled out, with average result of 3.53. It can be concluded that the sample of subjects on average evaluated the use of virtual private networks in communication as "very good" when it comes to security. This suggests that the use of VPN in combination with a firewall and a clearly defined security policy would achieve an even higher grade regarding security and the use of VPN.

5. Comparison between OpenVPN and SoftEther

Although there are many different software VPN solutions on the market, each solution has its own advantages and disadvantages [12]. It is up to the user to choose the most appropriate solution in line with the user's need, as well as possibilities. SoftEther solution can be given as an example for comparison with VPN software. Differences in their functionalities are given in Table 4.

Table 4. Comparison between OpenVPN and SoftEther tool for setting up a VPN [13]

	OpenVPN	SoftEther
Supported platforms (OS)	Windows, Mac, Solaris, Linux, NetBSD, QNX	Windows, Mac, Solaris, Linux FreeBSD
License	open source	open source
Year	2002	2013
Supported protocols	OpenVPN only	EtherIP, Microsoft SSTP, L2TP/Ipsec
Supported encryption algorithms	AES, Blowfish, 3DES, CAST-128	
Throughput speed	100 Mbs	900 Mbs

There are also various commercial VPN services available on the market. Choice of a high-quality VPN service depends on several factors:

- Supported network protocols,
- Server locations,
- Logs and monitoring,
- Client applications,
- Service price.

Support for modern network protocols is one of the key parameters when choosing a commercial VPN service. The newer and safer the protocol, the faster and safer the communication. The location of VPN server is subject to laws of the country in which servers are located. European countries are more liberal regarding control and privacy than, for example, the USA, where server activities are recorded as so-called logs. Supported client applications enable a simple, fast and secure method of setting up a VPN connection which is fully automated. The service price is proportional to its quality.

In principle, more expensive commercial VPN services offer more.

[3]Research conducted in September 2013 on a sample of 59 CARNet system engineers according to doctoral dissertation: Skendžić A "Model vrednovanja sigurnosti bežičnih lokalnih mreža u obrazovnim ustanovama" (Wireless network security evaluation in educational institutions), Faculty of Humanities and Social Sciences, Zagreb 2014, Croatia. Doctoral Thesis.

OpenVPN, if used on Linux platform, has all the benefits offered by open source licenses. Some of the most commonly used open source licenses are [14]:

- GNU General Public License (GPL),
- Berkeley Software Distribution (BSD) License,
- Artistic License.

Some of the features shared by the licenses above are as follows [14]:

- Software can be installed on an unlimited number of computers,
- Software can be simultaneously used by an unlimited number of people,
- An unlimited number of software copies can be created (free distribution),
- There are no restrictions regarding software modifications,
- There are no restrictions regarding distribution or even sale of software[4].

Since the license allows for free redistribution, once a person receives a copy, they can redistribute it, and even try to sell it. In practice, creation of electronic software copies can actually be done without any costs. Supply and demand will keep the price low. In business environment, the price of product license can be a decisive factor when choosing a product that, according to its specifications, meets the user's requirements. Although commercial VPN services are available through Internet service providers, *OpenVPN* software solution offers an open source license, high level of configurability and, finally, it is free. You only need to ensure a bandwidth that is fast enough depending on the number of VPN network users within a company. The price of VPN service offered by commercial providers can amount to over HRK 20,000[5], depending on the bandwidth (64k – 10G[6]) and the number of users. *OpenVPN* software solution can be a good alternative to the use and adjustment of an own, independently created VPN service.

6. Conclusion

The present-day, modern business environment implies that technology is implemented in the everyday work. Networks and network environments are actual standards in communication, and this will certainly continue in the future since there are over 3 billion Internet users in the world. Mobile office has replaced traditional offices, and employees are now performing their tasks where they are needed in a given moment, always having at their disposal different business materials stored at their own company or office. Fast information flow and mobility are necessary for business success, and in order to achieve this, one must "be connected". In spite of the need for network connection, the issue of security is being raised among users. Communication security within computer networks can be achieved using various technologies. VPN technology certainly belongs to the category of secure remote communication with one's own office, which guarantees data confidentiality, data integrity and secure user authentication. There are many different commercial services offering own VPN solutions, subject to charge. *OpenVPN* does not belong to the category of "chargeable" services, it is part of GNU-GPL community of open source solutions, which is its comparative advantage. Flexibility, upgradeability, support for different platforms, support provided by GNU community are only some of comparative advantages of *OpenVPN*. With respect to disadvantages, it should be noted that *OpenVPN* is not a multi-thread application, which limits its rate of operating network connections (approximately up to 100 connections). In case more connections are needed, additional processes, different ports or IP addresses should be considered. In conclusion, although open source software is not completely free from restrictions, it provides the user with unlimited possibilities, at the same time protecting authors' rights. This is a kind of freedom both authors and users of open source software strive for [14].

[4] The license allows for free redistribution. In practice, creation of electronic software copies can be done without any costs.

[5] According to the current price list of a domestic operator in 2016

[6] According to the current price list of a domestic operator in 2016 Rates 64 kbit/s – 10 Gbit/s)

References

[1] ***https://openvpn.net/index.php/access-server/overview.html (02.11.2016)
[2] ***http://www.cert.hr/sites/default/files/NCERT- PUBDOC-2010-04-298.pdf (02.11.2016)
[3] ***http://onlinetrziste.com/2014/06/predstavljamo-5-besplatnih-vpn-servisa/ (02.11.2016)
[4] ***https://technet.microsoft.com/en-us/library/cc958048.aspx (02.11.2016)
[5] ***https://docs.microsoft.com/en-us/azure/vpn-gateway/vpn-gateway-create-site-to-site-rm-powershell (02.11.2016)
[6] Dražen O 2005 *Ekscentar* **7** December 2005 (url: http://hrcak.srce.hr/9676)
[7] Dražen D and Crnjac M D 2008 *A role of the VPN network modern business*, "Vallis Aurea", Croatia-Austria
[8] Ivan G 2008 *Iskustvo migracije na Linux desktop i server okruženje velikog korisnika.* HrOUGH, Rovinj, Croatia
[9] ***https://openmaniak.com/openvpn_static.php
[10] Skendžić A 2011 *Virtualne private mreže*, VIDI, VIDI-TO, Zagreb, Croatia
[11] Skendžić A 2014 *Wireless network security evaluation in educational institutions*, Faculty of Humanities and Social Sciences, Zagreb, Croatia, Doctoral Thesis.
[12] ***http://onlinetrziste.com/2014/06/predstavljamo-5-besplatnih-vpn-servisa/ (02.11.2016)
[13] ***https://www.limevpn.com/blog/softether-vpn-vs-openvpn-which-one-is-better-for-you-and-why/ (02.11.2016)
[14] ***https://www.carnet.hr/tematski/opensource/licence.html (02.11.2016)

Considerations on stress triaxiality variation for 2P armor steel

V Zichil[1], A Coseru[2], F Nedeff[2] and C Tomozei[3]

[1]"Vasile Alecsandri" University of Bacau, Department of Economical Engineering, Mechatronics, 600115 Bacau, Romania
[2]"Vasile Alecsandri" University of Bacau, PhD student, 600115 Bacau, Romania
[3]"Vasile Alecsandri" University of Bacau, Department of Environmental Engineering and Mechanical Engineering, 600115 Bacau, Romania

E-mail: valentinz@ub.ro

Abstract. Stress triaxiality is considered an invariant of stress, defined as the ratio of hydrostatic stress (hydrostatic pressure by other authors) and the equivalent stress (usually calculated using von Mises criterion). If the values of the main three stresses have comparable sizes, stress triaxiality can be also calculated using the first invariant of the stress tensor. Despite that the stress triaxiality is an invariant, the authors have determined experimentally and analytically its variation with the force at the tensile test, but also with the radius of notches caused in the specimen. 2P armor steel being used in lightweight armor, these notches occur after shocks with foreign objects. Furthermore, the authors have revealed the stress triaxiality variation function of the test type. The tests were performed on tensile specimens loaded for tensile test, pure torsion test, 25% tensile - 75% torsion test, 50% tensile - 50% torsion test, 75% tensile - 25% torsion test. The mathematical model used was designed by Xue.

1. Introduction

2P armor steel is used mainly to achieve coatings for armored military vehicles. There are also some special uses, such as fuel pipes for those vehicles [1]. For this reason, a study for 2P armor steel plate and for a limited length tube, in terms of its behavior in plastic domain has to be done.

1.1. Bridgman's criterion

Bridgman provides a simple analytical solution to calculate the equivalent stress, using von Mises methodology for the minimum section of a revolution specimen [2].

Bridgman's analysis is applied to revolution specimens (with or without the initial unloading) made from a homogeneous isotropic material. The material is considered a plastic one (complying Von Mises plasticity criterion, isotropic hardening and Prandtl-Reuss equations). The specimen is tested to a tensile force in the direction of its *ez* axis (Figure 1).

Bridgman's specific hypothesis reflects the variation of the curvature radius ρ along the *ez* axis. The proposed mathematical relation has the general form:

$$\rho = f\left(\frac{r}{a}, \frac{R}{a}\right) \tag{1}$$

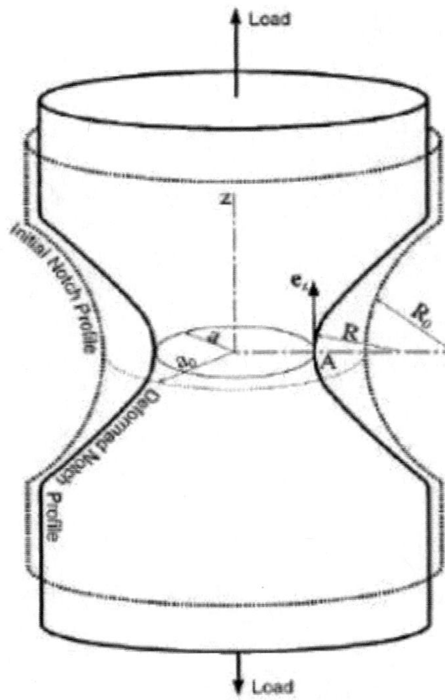

Figure 1. Original and deformed shape profile of a revolution specimen [3]

Bridgman had also proposed the maximum triaxiality ratio of stress τ (the center of the minimum sample section), defined by the equation:

$$\tau = \frac{\sigma_H}{\sigma_{eq,VM}} = \frac{1}{3} + ln\left(1 + \frac{a_0}{2R_0}\right) \tag{2}$$

where σ_H - hydrostatic stress;
$\sigma_{eq,VM}$ - equivalent von Mises stress;
a_0 - the length of the original ligament;
R_0 - initial notch radius

1.2. Barsoum – Faleskog's model

Barsoum and Faleskog have used a tube with slots in their experiments, as it can be seen in Figure 2 [4]. Their results are reported through the Lode parameter, in reference to σ_1, σ_2 and σ_3 main stress as a ratio of the breaking plane, as shown in Figure 3, where Lode's parameter is defined as:

$$\mu = \frac{2\sigma_2 - \sigma_1 - \sigma_3}{\sigma_1 - \sigma_3} \tag{3}$$

1.3. Xue – Wierzbicki criterion

Xue şi Wierzbicki have shown that the ductile fracture is the result of the accumulation of plastic strain and the damage depends both of the Lode's angle and hydrostatic pressure. The model used in analytical calculations is represented by a function of the three tensors of the invariant stress:

J_1 – shows the sensitivity of hydrostatic pressure;
J_2 – shows the equivalent von Mises stress;
J_3 – shows the Lode's angle.

Xue – Wierzbicki criterion suppose that the material is isotope and bears von Mises plasticity criterion, also that the failure is characterized by a deterioration factor formally expressed as (4):

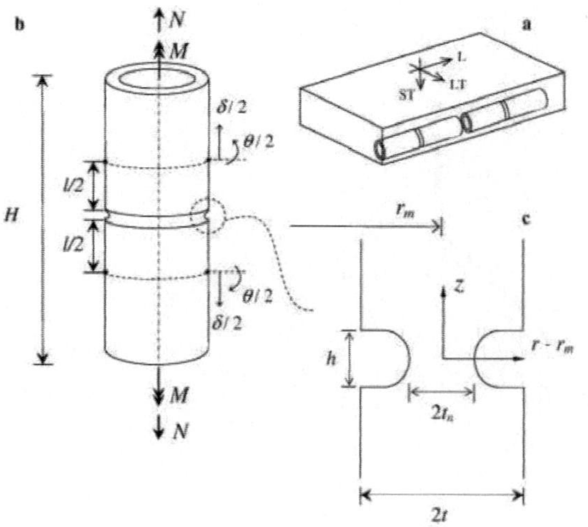

Figure 2. Notched specimen tube used Barsoum and Faleskog [4]

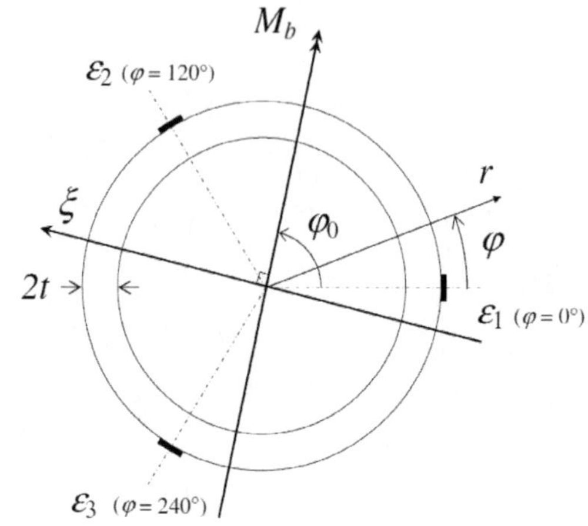

Figure 3. Specific strains along the main stress directions [4]

$$\sigma_{eq,VM} = w(D)\sigma_M \tag{4}$$

where: $\sigma_{eq,VM}$: equivalent von Mises stress ;
$w\ (D)$: the deterioration factor, D = [0,1].

1.4. Lode's angle sensitivity

Lode's angle is represented in a deviator plane, as the angle between the *O'P* and *O'A* directions, where *P* is the stress state and *A* is the condition of pure shear (Figure 4) [5].

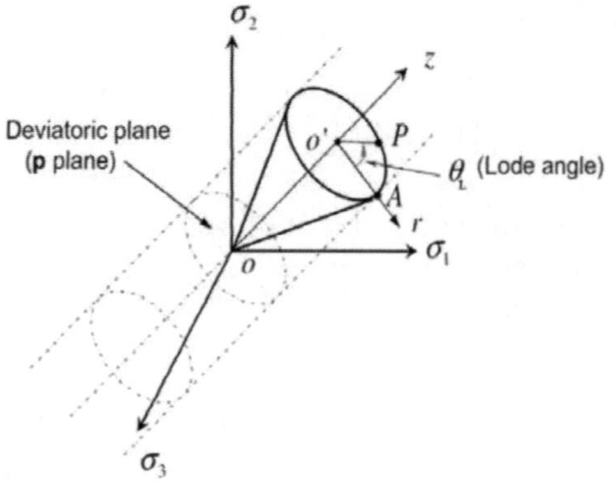

Figure 4. Lode's angle in a deviator plane [5]

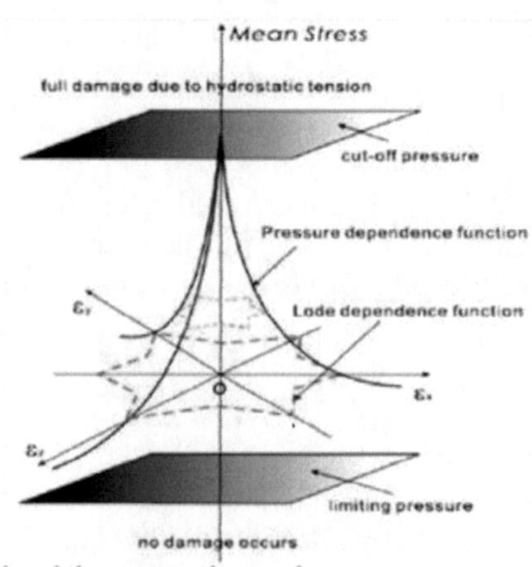

Figure 5. Graphical representation of the Xue - Wierzbicki model [6]

Lode's angle is expressed and closely related to the third stress invariant J_3. Azimuthal dependence is a function on the Lode's angle:

$$X = cos\left(3\left(\theta_L - \frac{\pi}{6}\right)\right) = \frac{27J_3}{2\sigma_{eq,VM}^3} \tag{5}$$

1.5. *Hydrostatic pressure sensitivity*

Stress triaxiality β is defined as the ratio between the hydrostatic stress and the equivalent stress:

$$\beta = \frac{\sigma_m}{\sigma_{eq,VM}} \tag{6}$$

where $\sigma_{eq,VM}$ is the equivalent von-Mises stress expressed by:

$$\sigma_{eq,VM} = \frac{\sqrt{2}}{2}\sqrt{(\sigma_1 - \sigma_2)^2 + (\sigma_2 - \sigma_3)^2 + (\sigma_3 - \sigma_1)^2} \tag{7}$$

and σ_m is the hydrostatic stress :

$$\sigma_m = \frac{\sigma_1+\sigma_2+\sigma_3}{3} = \frac{I_1}{3} \tag{8}$$

Stress triaxiality (Figure 5) is not constant in the nearest notch, the maximum value is determined behind the crack. Critical effective stress triaxiality is defined as the average value from the stress triaxiality distribution on the effective length.

$$\beta_{ef,c} = \frac{1}{X_{ef,c}} \cdot \int_0^{X_{ef,c}} \beta(x)\, dx \tag{9}$$

This model involves identifying six material constants: P_{lim} – material dependent constant; q - pressure dependent constant; γ – Lode's angle dependent constant; ξ_{f0} – crack reference strain; m - damage exponent of accumulation; β – thinning effect.

2. Experimental procedure

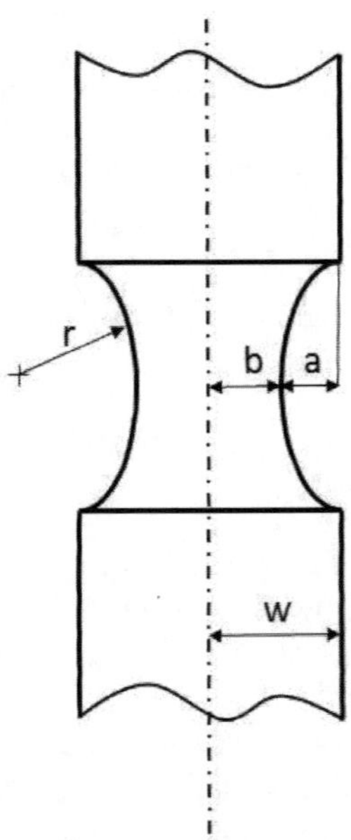

Figure 6. Notch specimen

A tensile test allows measuring the tensile strength of any material. This test involves placing a bar of any material between the jaws of a tensile testing machine to pull the bar until it breaks, according to NF EN ISO 6892-1: 2009 [7]. As a result of the applied axial force upon a specimen by the original length (L_0), a reduction in cross-sectional area is caused, from A_0 to A until the occurrence of breakage. Stress and length variation between two fixed points is recorded and used to determine the stress-strain state.

Torsion test involves twisting a specimen in the absence of other loads (bending or shear) as is specified in ASTM E143: 2013 [8]. Tests were performed on smooth and notched specimens. The tensile strength was carried out by applying a load force between 10 kN and 25 kN. For different notches, the range of the radius was between 0.5 mm and 10 mm (Figure 6). The same type of specimens was used to analyze the influence and the impact of Lode's angle and triaxiality.

To study the effects of Lode's angle, smooth specimens have chosen for different tests (pure traction, pure torsion and combined test: tensile 75% - 25% torque, tensile 50% - 50% torque, tensile 25% - 75% torque).

Lode's effectively parameter ξ_{ef} is evaluated based on actual distance in the middle of the notch. The effective distance X_{ef} is given by the volumetric method [9].

IOP Conf. Series: Materials Science and Engineering **200** (2017) 012066 doi:10.1088/1757-899X/200/1/012066

3. Results

In order to study the effects of hydrostatic stress, selected notched specimens with the radius of the notch between 0.5 mm and 10 mm were studied.

The value of Lode's effectively parameter (ξ_{ef}) at pure tensile test is shown in Figure 7. The values for the radius of the notch and the effective distance (X_{ef}) are reported depending on the length of the notch in Tables 1, 2 and 3.

Table 1. Values for ξ_{ef} and X_{ef} function of notch radius at pure traction test

$\sigma_g/\sigma_N > 1$; 100% Tracțiune		
r [mm]	ξef	Xef [mm]
0,5	0,971068	0,64
2	0,944137	0,7
4	0,955766	0,79
10	0,980475	0,88

Table 2. Values for ξ_{ef} and X_{ef} function of notch radius at pure traction test

$\sigma_g/\sigma_N = 1$; 100% Tracțiune		
r [mm]	ξef	Xef [mm]
0,5	0,909574	0,66
2	0,85741	0,72
4	0,907483	0,86
10	0,966831	0,94

Table 3. Values for ξ_{ef} and X_{ef} function of notch radius at pure traction test

$\sigma_g/\sigma_N < 1$; 100% Traction		
r [mm]	Ξef	Xef [mm]
0,5	0,866405	0,68
2	0,816663	0,74
4	0,854189	0,92
10	0,936259	0,98

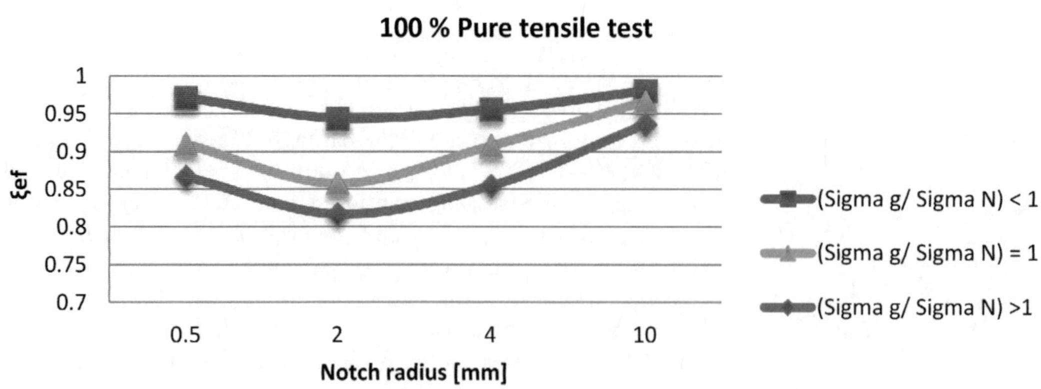

Figure 7. Lode's effective parameter variation at pure tensile test

The results of the tests made for combined loads conditions are presented in Figures 8, 9 and 10. As it may be observed, Lode's parameter values vary between *0* and *1*. A "zero" value corresponds to pure torsion and the "one" value corresponds to pure traction. Values in the range "zero – one" are for combination of tensile and torsion load corresponding to different ratios of these two loads.

Preserving at a constant value the triaxiality (pure tensile load on a smooth specimen), Lode's parameter can take different values (Lode's angle varies between *0°* and *30°*). This variation can cause a significant difference in deformation and also to the material damage, so the material is directly dependent of Lode's parameter. This variation can show the influence of the load level. For a higher tensile load and a reduced torsion load Lode's parameter is close to the value *1*, which means a small influence of this parameter. For a higher torsion and a lower tensile load applied, Lode's parameter is near *0* values, which means a greater influence of Lode's parameter.

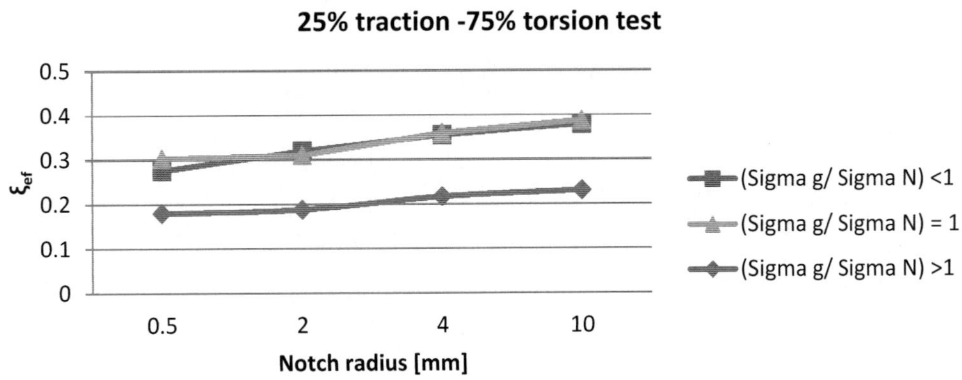

Figure 8. Lode's effective parameter variation at 25% tensile – 75% torsion combined test

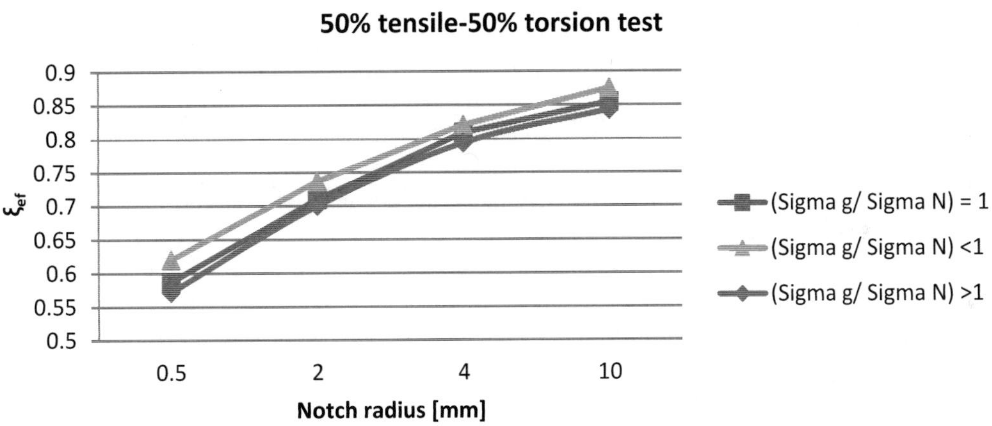

Figure 9. Lode's effective parameter variation at 50% tensile – 50% torsion combined test

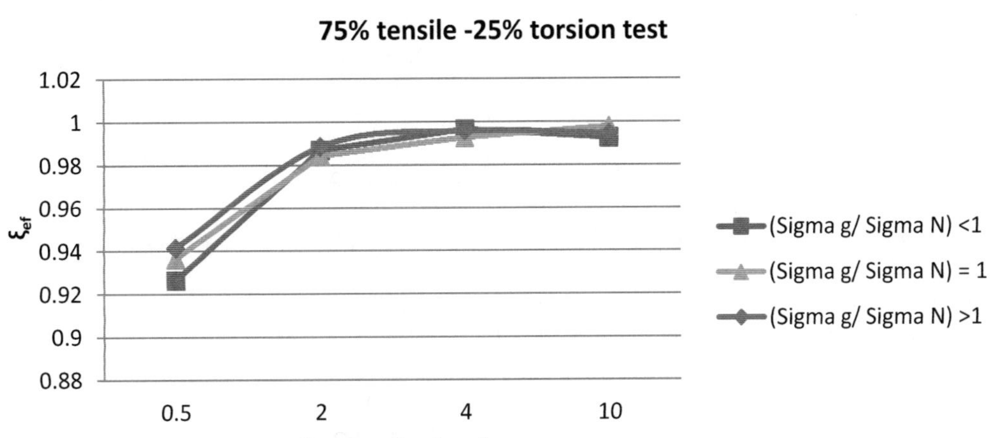

Figure 10. Lode's effective parameter variation at 75% tensile – 25% torsion combined test

Contrariwise, the dependence of the Lode's angle and stress triaxiality (shown in Figure 11 for only one load situation) enables us to make an indirect correlation between the damage risk and the hydrostatic stress.

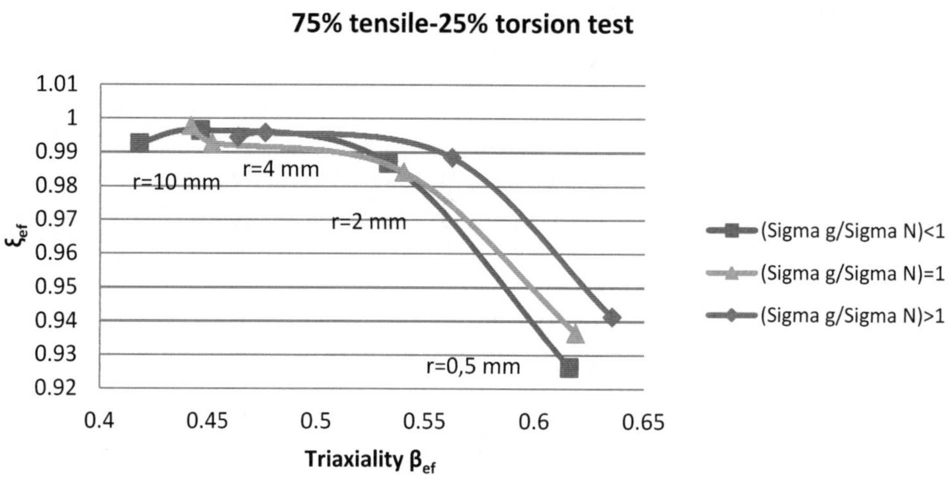

Figure 11. Triaxiality values and Lode's angle for different notch radius in complex load state: 75% tensile - 25% torsion

4. Conclusions

In this paper the authors have proposed to determine stress triaxiality variation on 2P armor steel. Therefor it was determined:

- Stress triaxiality for various types of specimens with different notch radius;
- Lode's angle, using the model Xue-Wierzbicki;
- Effective stress triaxiality and effective Lode's angle;
- The influence of stress triaxiality and the influence of Lode angle upon ductile material breakage.

From the analysis of experimental results is easy to observe that stress triaxiality (via Lode's angle) is a very sensitive indicator for material deformation before a ductile fracture. Thus, for 2P armor steel, only at low levels of stress triaxiality the ductile fracture deformation respect the model described in fracture mechanics. At high levels of stress triaxiality (small parameter values for Lode's) a breakage of the material can occur suddenly. Therefore, the authors believe that the introduction of Lode's parameter as a new indicator in the model of ductile fracture is required. Lode's parameter can be associated with the breakage material characteristics.

References

[1] ***https://www.4tuning.ro/ghidul-soferului/ce-presupune-o-masina-blindata-care-sa-faca-fata-unui-atac-terorist-28474.html

[2] Bridgman P 1944 The Stress Distribution at the Neck of a Tension Specimen, *Trans. ASME* **32** 553-574

[3] Valiente A 2001 On Bridgman's Stress Solution for a Tensile Neck Applied to axisymmetrical Blunt Notched Tension Bars, *Journal of Applied Mechanics* **68** 412-419

[4] Barsoum I and Faleskog J 2007 Rupture mechanisms in combined tension and shear – Experiments, *International Journal of Solids and Structures* **44** 1768-1786

[5] Xue L 2007 *Ductile fracture modeling – theory, experimental investigation and numerical verification*, Massachusetts Institute of Technology, Doctoral Thesis

[6] Xue L and Wierzbicki T 2008 Ductile fracture initiation and propagation modeling using

damage plasticity theory, *Engineering Fracture Mechanics* **75** 3276-3293

[7] ***Norme NF EN ISO 6892-1: 2009, *Matériaux métalliques - Essai de traction - Partie 1* : méthode d'essai à température ambiante

[8] ***ASTM E-143: 2013, *Standard Test Method for Shear Modulus at Ambient Temperature*

[9] Pluvinage G 2007 *General approaches of pipeline defect assessment in Safety Reliability and Risks associated with water, oil and Gas pipelines*, Edited by Pluvinage G and Elwany M , Springer

IOP Conf. Series: Materials Science and Engineering **200** (2017) 012067 doi:10.1088/1757-899X/200/1/012067

Characteristics of the Romanian energy market

M Stet

Technical University of Cluj Napoca North University Centre of Baia Mare, Department of Electric, Electronics and Computer Engineering, Baia Mare, Romania

E-mail: mihaela.stet@cunbm.utcluj.ro

Abstract. This paper highlights the main characteristics of the energy market in Romania. Starting from the mode of organization and operation of the electricity market, there are revealed prices and tariffs for electricity for different categories of customers and their evolution in time. There are pointed also ways of setting electricity prices and tariffs, taking into account the expenditures actually recorded by economic operators.

1. Introduction

Economic development being strongly related to electricity, proper functioning of national electric energy systems is vital for national economies. Given that many national economies have implemented the market model, the concerns have been expanded to the proper operation of electricity market. The peculiarities of this kind of market, its complexity and strategic position in economy requires special attention from authorities.

In this context, there are highlighted in this paper some characteristics of Romanian energy market and its evolution in last years.

2. Literature review

As well as the problems of the electricity market are complex and this type of market has many peculiarities and difference from the other types of market, the scientific literature is not too vast comparative with other markets.

Mookherjee & al. [1] proposed a dynamic electric power model, a day ahead model used by generation firms for planning and bidding purposes. Baldick [2] presents uses of market equilibrium models, like physical model, economic and commercial model. Many studies have been realized in this field [3-6]. Francoa & al. [7] present a system dynamics model that supports analysis of long-term effects of the various policy instruments, focusing on environmental quality, security of supply and economic sustainability.

Glachant and Ruester [8] highlight the steps made to create an EU internal electricity market. Prabavathi & Gnanadass [9] (analyses the state of the art research of bidding strategies in restructured electric power market. Hogan [10] presents problems of virtual transactions and electricity market design.

3. Evolution of Romanian Electricity Market

Even if Romania had not been yet an EU member in 1996, when liberalization of electricity markets was launched by Directive 96/92/EC, since that year she started the creation of a functioning electricity market.

The process started with the restructuring the energy sector and creation the institutional framework for regulating the electricity sector and the model adopted was that of a decentralized markets. But, the opening the energy market started four years later, and in 2005 the openness reached 83.5%, at which time all non-household customers became eligible. For non-household consumers the complete liberalization was produced at the end of 2014.

Full opening occurred after the moment of Romania's integration into the European Union, respectively 2007 [11]. Even though, household consumers have not been able to exercise their eligibility right of the providers. There were a relatively small number of suppliers, who operated each of them on certain distinct geographic areas where, practically, in this way, they exercised their monopoly. No other provider in the country disputed the offer for residential customers on that geographic area.

It was only much later, starting with 2015, when the household consumers have been really able to exercise the right to change their energy supplier, through penetration of new suppliers interested in catching also a part of the market segment of household customers.

From legal point of view in 2007 the degree of market opening was 100%, being achieved the full liberalization. But from a practical perspective, customers could not benefit from the liberalization process. However for households it was even more advantageous to remain on the captive market, with regulated tariffs, due to the state that provide energy for a lower price than the free market.

In 2013 only 10 % from the electricity delivered to household users was at free market price, the difference being billed at regulated price. The market liberalization process is due to close at the end of 2017 (Figure 1), when the consumers will directly negotiate prices with the electricity suppliers.

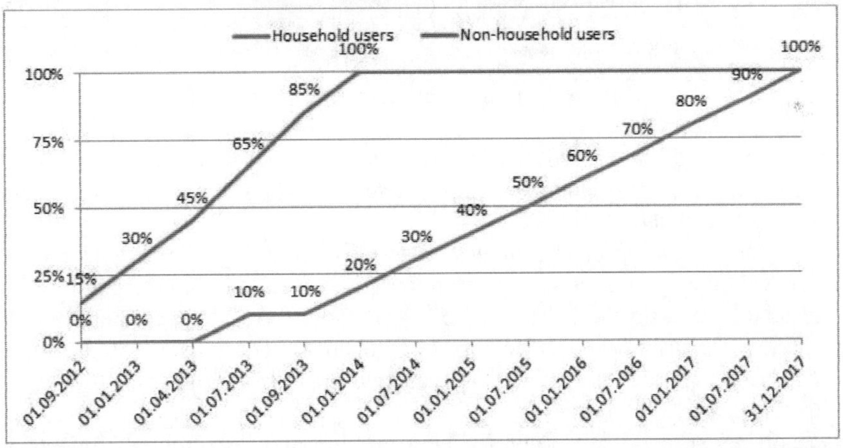

Figure 1. Liberalization calendar of electricity prices [ANRE, www.anre.ro]

The Romanian body that sets the official rules of electricity market and monitors that is ANRE. In 2009 [12] it was considered an autonomous administrative authority, under parliamentary control, fully financed from its own resources, independent decision-making, organizational and functional. Subsequently, it passed under direct coordination of the Vice Prime Minister, thus becoming fully funded from the state budget and the collected revenue flows entirely into the state budget. In 2012 it become again an autonomous authority.

On the regulated market, ANRE may impose public service obligations on producers, transmission and system operator, distribution system operators and suppliers of last resort for certain categories of customers. It also has the role of continuous monitoring of the effect of the regulated market on competitive electricity market, being able take steps to avoid possible distortion.

IOP Conf. Series: Materials Science and Engineering **200** (2017) 012067 doi:10.1088/1757-899X/200/1/012067

The main participants in the electricity market are the producers, suppliers, electricity market operator, transport and system operator OTS, distribution operator, electricity trader, and final customers. Besides them there are, also, other different categories of operators (Figure 2).

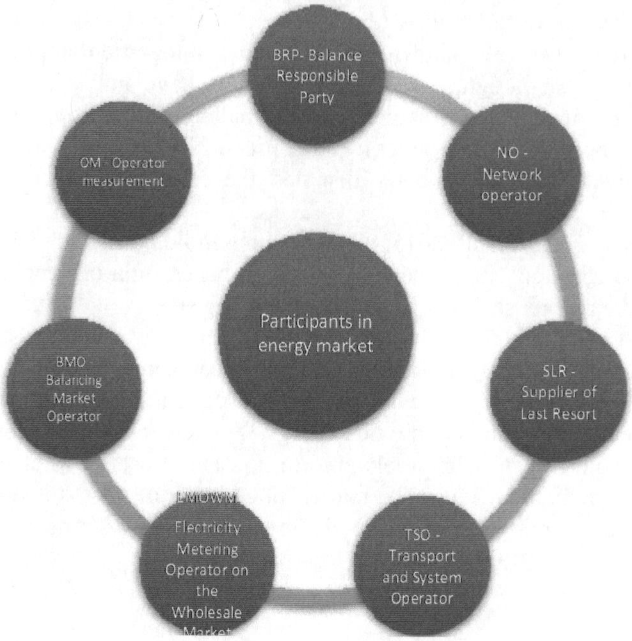

Figure 2. Participants on energy market

OPCOM, as operator of the electricity market, deals with organization and management of centralized markets, in the short, medium and long term, with the exception of the balancing market for wholesale trading electricity.

Distribution operators hold an electric power distribution and they are responsible for operating, maintaining and developing the distribution system in a given area and the interconnections with other systems, as well as for ensuring the long term ability of the network to meet reasonable demand of electricity distribution. Distribution service provided to customers consists of the transmission, in terms of efficiency and safety of electricity between two or more points of distribution network, in compliance with performance standards.

Electric transport network is state property. Transmission and distribution of electricity are fully regulated natural monopolies. Transelectrica, as the transport and system operator, has a transmission electric network and it is responsible for operating, ensuring the maintenance and development of the transmission network in a certain area, its interconnection with other power systems, and for ensuring the long term ability of transport network to cover reasonable requests for electricity transmission. To maintain the level of the reliability into service of the power system, as well as the quality of electricity it performs system services. Also, it provides electricity transmission services between different parts of the transportation network.

There can be metering operators the transmission and system operator, producers and distribution operators.

The supplier of last resort is that provider which is designated by the competent authority to provide universal service for the supply of electricity to end customers who have not secured supply of electricity from any other source. The suppliers of last resort shall be appointed by ANRE from the existing providers on the energy market through competitive mechanisms, based on a regulation which establishes the rules and criteria for their selection for each category of end-customers who they serve.

In the market there can be find horizontally integrated operators performing at least one of the functions of electricity generation for sale, transportation, distribution or supply of electricity, and another activity that is not from the field of electricity.

Economic operators vertically integrated can be traders or groups of traders in the field of electricity who are entitled, directly or indirectly, to exercise control and which perform at least one of the functions of transmission or distribution and at least one of the production or supply of electricity.

In the energy sector there can be identified the market of electricity production and sale, transport and system technological services market, distribution market and electricity supply market.

On the first market of these, there is a quasi-monopoly due to the fact that is dominated by the companies owned by the state, which represents 85-90% of the market. The market has two market segments: futures trading and hourly trading on spot market.

The market of system technological services has for distinctive segments: secondary adjustment, fast tertiary adjustment, slow tertiary adjustment and balancing market to increase or decrease power.

Regarding the distribution market, the energy distributors take over the energy from the transport operator and distribute it to consumers.

On the market of electricity supply market two market segments are distinguished: wholesale trading and supply to the final customers.

Electricity market is composed by regulated market and competitive market and the energy transactions are wholesale and retail [13]. Therefore, electric energy market has two components:

- Wholesale electricity market, attended by producers, the transmission and system operators, distribution operators, electricity market operator and wholesale customers;
- Retail market, in which the transactions are carried between suppliers or producers and consumers.

The growth of competitive market share has been achieved gradually, by ensuring the access on that market for as many as possible participants, producers, suppliers and end customers.

On the wholesale electricity market there are traded electricity and related services and there are as participant electricity generators, the transmission and system operator, distribution operators, the electricity market operator and wholesale customers.

Since 2005, on the wholesale electricity market, there have been introduced new trading platforms:

- Day Ahead Market (PZU) is organized and managed by the operator of the electricity market for buying and selling electricity in the next delivery day following the trading day;
- Centralized Market for Bilateral Contracts where are taking place, through public auction, trading of contracts with physical delivery of electricity based on specific rules approved by the competent authority. Energy producers sell the energy produced on the Centralized Market for Bilateral Contracts PCCB without they could sell them directly;
- Centralized Green Certificates Market:
- Balancing Market, dedicated to offset deviations from the programmed values of the production and consumption of electricity. It is organized and operated by the transmission and system operator, which deals electricity with electricity producers who operates dispatchable production facilities and with the end customers to provide real-time balance between production and consumption.

The functionality of the first three markets is ensured by the wholesale electricity market operator - SC Opcom S.A., while balancing market is managed by Transelectrica, the transmission and system operator. It has the responsibility for the operation, maintenance and development of the transportation network.

The role of Opcom is also that of the settlement operator for centralized markets. Opcom introduced in 2007 also the centralized market for bilateral contracts with continuous negotiation, respectively forward contracts.

There are, also, other centralized markets in the electricity sector:

- ➢ intra daily electricity market PI, organized and managed by the operator of the electricity market, which contributes to improve balancing the portfolio of participants, for a day of delivery, through transactions in sessions held after the completion of transactions in the day-ahead market and before a certain time of delivery start;

- the market for allocation capacities for international interconnection, organized and managed by the transmission system based on specific rules, in order to achieve transactions of the import / export and transit of electricity.
- centralized market for bilateral contracts with continue negotiation PCCB - NC;
- bilateral contracts market for green certificates PCBCV;
- centralized market with double continue negotiation of electricity bilateral contracts PC-OTC;
- electricity market for big final customers PMC.

In the centralized electricity market transactions with electricity occur between various economic operators, brokered by the electricity market operator or transmission and system operator, based on specific rules, approved by the competent authority.

The market participants are required to assume financial responsibility for payment the imbalances they generate on the electricity market, except the imbalances stipulated by legislation [13].

On the electricity retail market the electric energy is purchased by the end customers from suppliers or producers. Suppliers sell electric energy to final customers by bilateral contracts, at negotiated prices or established by standard offers.

For household and non-domestic customers with a number of employees below 50 and a turnover less than EUR 10 million, electricity supply is guaranteed through universal service.

In order to ensure the safety level in operating of the national energy system SEN, as well as the quality of transmitted electric energy at normal parameters, the transport and system operator can ask to producers and dispatchable final customers to ensure the technological system service.

On the competitive market, commercial electricity transactions are carried out transparently, public, centralized and non-discriminatory, as wholesale or retail transactions. On this market, the transport and system operator can acquire technological system services.It is forbidden the cross-subsidy between regulated activities and between the regulated and the unregulated of an economic agent.

Starting to 2015, it is used the trading mechanism by signing bilateral electricity contracts through extended auction PCCB-LE. Participation on the PZU and PCCB-LE markets is voluntary, being allowed for companies that hold participation conventions with OPCOM.

On PZU market transactions are made for trading intervals of delivery day based on the quantity-price offers of participants. After the validation of the offers by the computer system, OPCOM provides physical notifications regarding the transactions to the transmission and system operator OTS and to the parties responsible with balancing.

In 2016. the "Clean Energy for All" package was made public, package which includes action guidelines of the Government represented by the Ministry of Energy. Regarding the energy market, the main objectives are:

- removal of some of the factors distorting the market, such as price ceilings;
- better remuneration of the participation of electricity consumers at the balancing market;
- facilitate the participation in the electricity market in the role of prosumers (producers and consumers);
- guarantee of the right to participate in the balancing market, individual or by the centralized platform;
- use of mechanisms by which the capacity markets can work efficiently;
- guarantee of the free access at the comparison instruments of the suppliers offers which comply with the minimal quality standards.

4. Characteristics of the consumers on the electricity market

Initially, the consumers were differentiated in eligible and captive consumers, but after the full opening of the electricity market, this distinction no longer exists. There is, however, a subcategory of consumers, those of the beneficiaries of universal service.

Suppliers, also, were divided in competitive suppliers for eligible consumers and suppliers for captive consumers. After full market opening, the electricity suppliers are suppliers in competition regime and providers of last resort.

Starting from 2014, the final clients fall into the following categories [14]:

- household customers;
- small non-household customers, i.e. non-household customers for which power by the technical connection approval is less than or equal to 100 kW ;
- large non-household customers, i.e. non-household customers for which power by the technical connection approval exceeds 100 kW.

In Romania, the liberalization of the energy market allows all consumers to benefit from their eligibility right of the provider, based on information related to the price, duration, methods of payment and other contractual conditions. Residential customers who have not exercised their eligibility can opt for any of the regulated electricity tariffs corresponding to the category to which they belong, with the exception of regulated tariff social type CS.

For vulnerable consumers, social protection is insured by maintaining social tariff or other measures of social protection and growing energy efficiency. If household customers have an average income per family member less than or equal to the minimum wage they may require CS tariff to the consume place from their own home [15]. Suppliers are required to keep records of vulnerable consumers in their own area of supply.

On the retail market there can be identified different types of electricity tariffs (Figure 3).

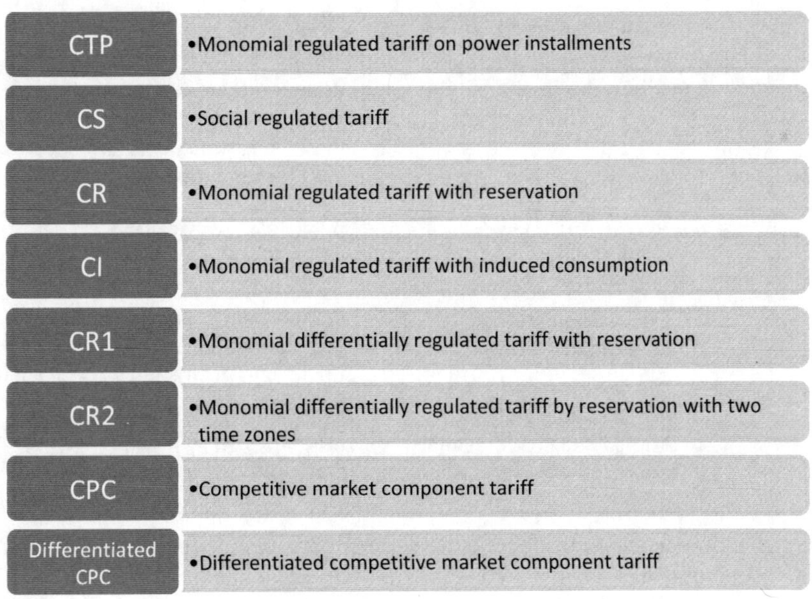

Figure 3. Types of tariffs on electricity market

In the case of the application of the monomial regulated tariff on power installments CTP, the inclusion of residential customers in power installments is based on the maximum contracted power.

In the electricity sector there are used prices resulted from the competitive mechanisms of the electricity market, regulated tariffs for transport, system and distribution of the electric energy, for acquisition of technological system services and for electricity supply to some categories of customers (Table 1), regulated tariffs for connection to the grid, regulated tariffs practiced by the electricity market operator rand regulated prices used by producers or suppliers of electric energy for certain categories of consumers [13].

IOP Conf. Series: Materials Science and Engineering **200** (2017) 012067 doi:10.1088/1757-899X/200/1/012067

Table 1. Pricing methods

Types of tariffs	Pricing methods	
1.Tariffs for post-consumption payment	Social regulated tariffs in three installments CS	In first installment - 2 kWh/day for each day of billing period In second installment - 1 kWh/day for each day of billing period In third installment - the difference between actual consumption and induced consumption in previous installments
	Monomial tariff CR	(lei/kWh)
	Regulated monomial tariff with reservation CR	Price / reservation (lei/day) Price / energy (lei/kWh)
	Regulated differentiated monomial tariff with two hourly zones CR2	Price / reservation (lei/day) Price / energy in day zone (lei/kWh) Price / energy in night day (lei/kWh)
	Regulated differentiated monomial tariff with three hourly zones CR3	Price / reservation (lei/day) Price / energy in peak area (lei/kWh) Price / energy in normal area (lei/kWh) Price / energy in low load area (lei/kWh)
2. Regulated monomial tariff with induced consumption CI	Low voltage (0 - 1 kV inclusive)	subscription (lei/day) Price / energy (lei/kWh)
	Medium voltage (1 - 110 kV exclusive)	subscription (lei/day) Price / energy (lei/kWh)
3. CTP Monomial regulated tariff on power installments - low voltage (0 - 1 kV inclusive), based on the maximum contracted power tranche	Up to 3 kW inclusive	Price / reservation (lei/day) Price / energy (lei/kWh)
	3 - 6 kW	Price / reservation (lei/day) Price / energy (lei/kWh)
	More than 6 kW	Price / reservation (lei/day) Price / energy (lei/kWh)
4. Tariffs for meters with prepayment	Monomial tariff CP	(lei/kWh)
	CP2 monomial tariff differentiated on two time zones	Price / reservation (lei/day) Price / energy in day zone (lei/kWh) Price / energy in night day (lei/kWh)
	CP3 monomial tariff differentiated on three time zones	Price / reservation (lei/day) Price / energy in peak area (lei/kWh) Price / energy in normal area (lei/kWh) Price / energy in low load area (lei/kWh)

In the Romanian electric energy sector there is an area of natural monopoly, specific to transport and distribution for which regulated tariffs for the use of networks shall be based on the ceiling type methodologies. The transport tariff has two components: the price for introducing electricity in the network and the price for energy recovery from the network. Pricing mechanism of the electricity transmission is differentiated on geographical areas, with negative effects on the competition.

On the market of system technological services most of the transactions are realized in a regulated regime.

The support of the electricity production from renewable sources is realized through mandatory quotas for suppliers, proportional to traded energy and the possibility of trading green certificates on the competitive market for green certificates. For example, mandatory quotas in 2017 for financial support through the mechanism of green certificates of the renewable energy is 8.3%.

The promotion system of electricity from renewable energy resources had begun in 2004. In 2006 it was introduced the right of energy suppliers to recover the costs of buying green certificates from the captive consumers, through regulated tariffs and from eligible consumers, increasing the selling electricity price, measure that was canceled in 2011.

5. Conclusions

The process of liberalization of Romanian electricity market have passed through several stages in the last 20 years. The competitive electricity market coexist with the regulated market. In the last few years it has been carried out a process concerning the deregulation of energy prices for household users and non-domestic users. The process had six stages and was completed for companies in 2014, this kind of consumers buying now electric energy only from competitive market.

For household consumers the process includes ten stages and will be finished at the end of 2017. This process has already encounter important effects on the electricity prices.

In these conditions, prices depend mainly on relation between supply and demand, but could be influenced also by some taxes and tariffs practiced on the markets of electricity production and sale, as well as on the markets of electricity transmission or distribution.

Therefore, the regulation authority has to take into consideration a better supervision of registered costs of distribution and transmission companies and a periodic revision and update of the methodologies for distribution and transport tariffs.

References

[1] Mookherje R, Hobbs B H, Friesz T L, Rigdon M A 2010 Dinamic Oligopolistic Competition in an Electric Power Network and Impacts of Infrastructure, Chapter 5, *Economic Market Design and Planning for Electric Power Systems,* Copyright © 2010, Institute of Electrical and Electronics Engineers

[2] Baldick R 2010 Computing the Electricity Market Equilibrium: Uses of Market Equilibrium Models, *Economic Market Design and Planning for Electric Power Systems,* Copyright © 2010, Institute of Electrical and Electronics Engineers

[3] Weidlich A and Veit D 2008 A critical survey of agent-based wholesale electricity market models, *Energy Economics* **30** 1728–1759

[4] Foley A M, Ó Gallachóir B P, Hur J, Baldick R and McKeogh E J 2010 A strategic review of electricity systems models, *Energy* **35** 4522-4530

[5] Chatzigiannis D I, Dourbois G A, Biskas P N and Bakirtzis A G 2016 European day-ahead electricity market clearing model, *Electric Power Systems Research* **140** 225–239

[6] Ventosa M, Baillo A, Ramos A and Rivier M 2005 Electricity market modeling trends, *Energy Policy* **33** 897–913

[7] Francoa C J, Castaneda M and Dyner I 2015 Simulating the new British Electricity-Market Reform, *European Journal of Operational Research* **245** 273–285

[8] Glachant J M and Ruester S 2014 The EU internal electricity market: Done forever?, *Utilities Policy* **31** 221-228

[9] Prabavathi M and Gnanadass R 2015 Energy bidding strategies for restructured electricity market, *Electrical Power and Energy Systems* **64** 956–966

[10] Hogan W W 2016 Virtual bidding and electricity market design, *The Electricity Journal* **29** 33-47

[11] ***GD 638/2007, *Hotarare a guvernului nr. 638 din 20 iunie 2007 privind deschiderea integrala a pietei de energie electrica si de gaze naturale*

[12] ***GD 1428/2009 - *Hotarare privind organizarea si functionarea Autoritatii Nationale de Reglementare in Domeniul Energiei publicata in MOf 847 din 8 dec 2009*

[13] ***Energy Law 2012 *Legea energiei electrice și a gazelor naturale nr. 123/2012, publicată in Monitorul Oficial al Romaniei*

[14] ***Order ANRE 105/2014, *Ordin pentru aprobarea Procedurii privind schimbarea furnizorului de energie electrica de catre clientul final și pentru modificarea anexei la Ordinul președintelui Autorității Naționale de Reglementare în Domeniul Energiei nr.*

35/2010 privind stabilirea unor reguli referitoare la piața de echilibrare a energiei electrice

[15] ***Order 176/2015, *Ordinul nr. 176/2015 pentru aprobarea tarifelor reglementate de energie electrică aplicate de furnizorii de ultimă instanță clienților casnici care nu și-au exercitat dreptul de eligibilitate, precum și a condițiilor de aplicare a tarifelor reglementate și a tarifelor componenta de piață concurențială*

IOP Conf. Series: Materials Science and Engineering **200** (2017) 012068 doi:10.1088/1757-899X/200/1/012068

Research and development regarding the retaining mechanism of lead ions in industrial wastewaters using natural matter with remarkable properties

A Pop and G Iepure

Technical University of Cluj Napoca, North University Center of Baia Mare, Romania, Romania

E-mail: pop aurica2003@yahoo.com

Abstract. The paper shows the studying of the retaining mechanism of lead ions in industrial wastewaters through static and dynamic ion exchange mechanisms. In the experimental determinations of the lead metallic ion retention, metallurgical industry wastewaters have been used on samples of volcanic zeolite tuff (from Barsana, Maramures), samples that show a high concentration of lead ions and an acidic pH. The results showed that both the static and the dynamic ion exchanges ended with good results and they were consistent with other studies conducted on clinoptilolite zeolite tuff. Knowing that the industrial sector is an important source of environment pollution and degradation and being aware of what a serious threat the heavy metal pollution is, due to their high toxicity and stability, the experiment may find applicability in different aspects, both in the Maramures mining basing as well as in the worldwide controlling and directing of the polluting processes.

1. Introduction

The growing tendency of degrading the environment, the irrational management of natural resources as well as the zeolite applicability in various technological areas has led to the anchoring of the present research regarding the technological and economical process of the Sustainable Development Concept.

The natural zeolites are minerals with remarkable ion exchange proprieties, offering numerous application possibilities. The literature data shows concerns about using new inorganic adsorbents (natural zeolites, clay minerals), multiple studies conducted lately having shown a growing tendency in using them.

The ion exchange is a physical-chemical process used especially to eliminate heavy metal ions from watery solutions and it is based on the proprieties of some materials (natural zeolites) of replacing own ions with solution ions (wastewaters containing heavy metals).

The first methods used were the simple chemical precipitation or decomposition which were applied discontinuously, although, the wastewaters volume growth and the diversification of the contained pollutants led to the diversification of the methods and their continuous application.

A study conducted in the northern regions of Brazil which focused on the industrial waste usage when synthesizing the NAA zeolite showed that the temperature was maintained at a constant 110°C throughout the synthesizing process of the NAA zeolite and the durations and ratios of Na/Al were modified. The x-ray diffraction results (XRD) showed an elevated purity for the NAA zeolite, with a high structural order degree; according to the XRD results, which pointed the creation of the KA,

MGA and CAA zeolites, the cation exchange process between the NAA zeolite Na and solution K, Mg and Ca was efficient [1].

Arsenic adsorption on adsorbents like modified natural Ca-clinoptilolite zeolite and MnO2, was investigated in order to explore the feasibility of arsenic removal from underground waters using the adsorbent proprieties of natural zeolites. It was found that the amphoteric proprieties and the arsenic removal efficiency by using the natural Ca-clinoptilolite zeolite were improved significantly after the MnO2 modification. The removal efficiency obtained on the modified zeolite was doubled compared to the one obtained on the unmodified zeolite. The modified-MnO2-Ca-clinoptilolite zeolite seems to be a promising adsorbent for the removal of arsenic from water [2].

The capacity of the natural zeolite tuff for the removal of Pb ions from watery solutions was highlighted by tracking the effects of various parameters: the optimal adsorbent mass, the contact time, environmental pH, the lead concentration (II) and temperature. The maximum removal efficiency of 92% was reached at a 5 pH with a contact time of 25 minutes for 10g/solid-to-liquid ratio and an initial heavy metal concentration of 100mg/L [3].

The cadmium removal efficiency was studied like a function of the contact time, the adsorbent concentration, the pH and the adsorbent dose. The results showed that the cadmium removal was best described by the Langmuir–Freundlich isotherm. The maximum cadmium retention happened at an approximately 6.0 pH for materials.

Bulky organic pollutants, like pesticides and pharmaceutical residues found in used waters, are hard to remove using microporous adsorbents because of their size. The paper investigates zeolites hierarched as potential adsorbents for the removal of watery phase heavy organic pollutants. The obtained materials were fully characterized through X ray diffraction, N_2 physisorption, elementary analysis, pyridine adsorption followed by FTIR. Thus, the collective proves that the hierarched zeolites show a higher adsorption rate and capacity than the microporous ones [5].

Na-Y zeolites made from local Egyptian kaolin were investigated for their adsorption proprieties in order to remove Methylene Blue (MB) from watery solutions. The influence of the chemical composition of the kaolin minerals and the surface proprieties of premade zeolite on the adsorption processes was studied. The effects of various parameters, like the particles surface and dimension, the initial MB concentration, contact time, the adsorbent concentration and subsequent adsorption cycles were examined. The adsorption capacities of the zeolite compared to the MB are influenced, mainly, by the zeolite particle dimension rather than the zeolite content [6].

In this research, Cu^{2+} ion exchange characteristics on natural clinoptilolite zeolite at 2 and 22°C are presented, in order to facilitate de development of a permeable reactive barrier (OEA) for the treatment of heavy metal contaminated waters from Antarctica. The negative effects of cold temperature on the performance of fixed bed will have significant implications for the design of a PRB natural zeolite for the treatment of heavy metal contaminated waters in Antarctica or in other cold regions [7].

Two natural zeolites, clinoptilolite and chabazit, had their mixed heavy metals (Pb, Cd, Cu, Zn, Cr, Ni and Co) contaminated effluents treatment performance evaluated. The effects of relevant parameters, like the chemical treatment, the metal concentration, the pH and the presence of the recurrent ions were examined. Chnoptilolite and chabazite showed different selectivity profiles for all the researched metals, with the exception of lead, for which both used zeolites acted exceptionally well [8].

Zeolite tuff (from Barsana, Maramures) was used in the Pb, Zn static ion retainment experimental determinations, using a ratio of 10g zeolite tuff/100cm^3 of wastewater (taken from the slag-granulating basin of the metallurgical plant S.C. Romplumb S.A. Baia Mare). The wastewater initial concentration of metallic ions was of 3,806 mg Pb^{2+}/l; 129,20 mg Zn^{2+}/l. The evolution of the ion exchange process to retain heavy metal ions on three types of grain size of the zeolite tuff sample was followed: 0.4-0.2mm (G04); 0.2-0.12mm (G02); <0.012-mm (G012).

Clinoptilolite is predominant in the zeolite material used in the experiment and it was emphasized through the electronic microscope quality analysis (SEM) and the quantity analysis through spectral analysis (EDX). Images of the grain surfaces and the chemical components EDX spectrum, with the help of the scanning electronic microscope (SEM) type JEOL JSM – 5600 LV, were made for the G012 S.C. Romplumb S.A. Baia Mare samples set. The zeolite tuff that was in contact with the wastewater got separated through filtration, and the sampling for the determination of the Pb and Zn ion concentration was made at an interval of 12 hours. The Pb and Zn ion concentration was obtained through atomic adsorption spectroscopy at a 5.90 pH value. The efficiency of the ion exchange process was monitored by means of ion exchange efficiency. For lead, ion exchange efficiency values of over 96% were obtained, the maximum value being 99.17% for the G012 sample set. The maximum efficiency value for the Zn ions, for which the concentration in the initial sample was high, was of 52.85% [9].

The usage of Barsana zeolite tuff in the static ion exchange processes for the wastewaters heavy metal ions removal led to good results which were consistent with the ones presented in the specialty literature [10-12].

The structural characteristics of the used zeolite tuff are close to the ones described in literature, like the tuff from the Maramures basin described by [13], [14]; the usage of this tuff for research led to good results regarding the ion exchange process.

The (lot and column) conducted experiments show higher adsorption capacities for ICZ compared to Z and they were generally lower in mixed metal systems than unique metal systems. The column adsorption data was well described by the Thomas model [15].

The heavy metal equilibrium series adsorption, meaning copper, zinc and manganese, from mongolian natural zeolite solution waste samples was made. Natural zeolites are capable of adsorbing and removing heavy metals from watery solutions. In some cases,the metal hydroxide precipitation influenced the adsorption results. The saturated adsorbed quantities of the heavy metals, estimated through the Langmuir equation, were almost the same amongst each other, but they grew along with the pH of the solution and the cationic exchange capacity of the natural zeolite [16].

ZT samples were characterized through XRD, XRF, BET and TGA analysis in order to examine the SO_2 adsorption on the Jordanian zeolite tuff (ZT). The sorption process was realized in a fixed bed column, in various functioning conditions. A single SO_2 measuring technique was used for this examination, which was the UIC sulfuric coulometer. Up until a temperature of 200°C the SO_2 assimilation by the ZT proved to be rising along with the temperature and then it dropped at higher temperatures. The fact that the ZT drying, through conventional and microwave heating, had a considerable effect on the SO_2 adsorption and the breakthrough time was also noted. Experimental data was obtained for the adsorption isotherms and it was noted that it followed the BET (Brunauer-Emmett-Teller) isotherm model. It has been found that the adsorption process is exothermic when it happens in nature. ZT can be regenerated and the initial adsorption capacity was kept after three cycles of regeneration [17].

The results of one research suggest that the adsorption/desorption of any class of compounds can be controlled by temperature operations. Adsorption isotherms were drawn for watery solutions of toluene, cyclohexane, o-xylene, benzyl alcohol, phenol and cyclohexanol in a humus, in combination with zeolite-acid at temperatures of 4, 14, 24 and 34°C and they were used to calculate the isosteric enthalpy ($\Delta_{ads}H_i$) and the isosteric entropy ($\Delta_{ads}S_i$) of the process. For hydrocarbon compounds, toluene, o-xylene, cyclohexane, and o-xylene, $\Delta_{ads}H_i$ as well as $\Delta_{ads}S_i$ were negative, the process being exothermic. Instead, for hydroxyl compounds, benzyl alcohol, phenol and cyclohexanol, $\Delta_{ads}H_i$ și $\Delta_{ads}S_i$ were positive, the entropy growth probably reflecting the water molecules release during the adsorption time.

A study was conducted that aimed to investigate the potential usage of the tuff as a matrix for ammonia removal in an intermittent biofilter flow (IBF), compared to the sand one. The sand filters and the tuff ones were characterized in relation to the respective total microbian biomass biofilms, the number and the diversity of the ammonia oxides and the ammonia removal efficiency. The microbian biomass/1g of tuff was significantly larger than that of the sand. However, the ammonia was removed immediately using the tuff filters, while the sand filter removal started only after ten days, even though the number of the ammonia oxidation bacteria (AOB) was higher in the sand filter. From a practical perspective, the tuff can be used as an effective matrix in IBF for ammonia removal, eventually by creating a more appropriate environment than the sand for bacteria activity. This study suggests an available and inexpensive matrix to produce high quality effluent biofilters IBF [19].

In another study, which aimed to analyses the lead ion removal capacity, from watery solutions in various conditions, of a natural zeolite tuff, various parameters like the optimal adsorbent mass, the contact time, the environmental pH, the lead ions concentration and the temperature have been investigated. On top of that, various adsorption isotherms have been obtained, using lead ions concentrations between 1mg/l and 200mg/l. For a 10g/l-solid-liquid ratio and a heavy metals initial concentration of 100mg/l, the maximum removal efficiency of 92% with a 5 pH at a 25 minutes contact time has been reached [20].

Four geopolymers have been synthesized through a NaOH activation of a mixture made of kaolinite (from Jordan or Ukraine) and filling material (zeolite tuff or silica sand). In order to monitor the reaction and to characterize the geopolymers phases, x-ray powder diffraction (XRD), Fourier transformed infrared spectrometry (FTIR) and nuclear magnetic resonance (MAS-RMN) have been used. The remaining kaolinite in every geopolymer produced sample clearly indicated an incomplete reaction. The ^{29}Si MAS-NMR spectrum of the geopolymers revealed the presence of tetrahedral SiO^4, while the ^{27}Al MAS NMR spectrum revealed the presence of both tetrahedral AlO_4 and octahedral AlO_6. The XRD patterns of the geopolymers showed the formation of a new feldspar mineral. The replacing of the silica sand filling with zeolite tuff significantly improved the corresponding geopolymer's specific surface area [21].

Some studies investigated the cadmium adsorption process from watery solutions, by using Na-zeolite tuff, Fe-zeolite tuff and carbonic material from HCL treated pyrolised sewage sludge. The results show that the adsorption mechanism is a physio-chemical adsorption on heterogeneous materials. In the kinetic studies, the Na and Fe modified zeolite tuffs showed similar adsorption capacities for cadmium but they were higher than the capacities for carbonic material [22].

The functionalization of the Y-zeolite (Faujasite) with the β-cyclodextrin (CD) was achieved by using 3-glycidoxypropyltrimethoxysilane (GPTS) as a coupling agent (linker). Two different strategic courses have been taken: the linker grafting to the β-CD and then carrying out the zeolite reaction or the zeolite linker grafting and then carrying out the β-CD reaction. The obtained materials were characterized in several complimentary ways, including: electronic microscopy with focalized ions scanning beam, spectrometry X-ray dispersive spectrometry, thermogravimetric analysis and differential thermal analysis (TG-DTA), infrared spectroscopy (FTIR) with the Fourier transform and Brunauer-Emmett-Teller analysis (BET). The first strategy led to better adsorption abilities than the second. These new materials can be potential adsorbents for the rehabilitation of the gaseous or aqueous effluents [23].

A study was conducted for the 4A zeolites radiolysis, containing various amounts of water, under ^{137}Cs gamma radiations, focusing on the influence of the water charging ratio. The hydrogen production improvement compared to the bulk water radiolysis was attributed to the energy transfer from the zeolite to the water and to the structural organization of the water inside the zeolite. Both were observed separately with a maximum efficiency for the energy transfer at a loading ratio of about 13%, and with a maximum water structuring impact at a charging ratio of about 4% [24].

The proprieties of the tubular fibers and of the zeolite membranes before and after the heat treatment, are characterized using electronic microscopy, mechanical testing, mercury intrusion method and pervaporation of a watery solution of 90% ethanol weight at 75°C. It has been found that the continuous zeolite layers keep the original crack-free morphologic shapes after the heat treatment. On the other hand, the resistance to compression and the separation performance of the inner side of the NAA zeolite membranes were significantly improved by post-synthesis heat treatment at 240°C for 6 hours [25].

Gas detection is an important environmental issue for this century. Whatever the purpose, be it industrial emissions control or interior air quality supervision, the challenge remains producing efficient detection systems at low costs, adapted to atmospheric conditions. There are studies which focus on the estimation of the microwave transduction technical viability for toluene detection. The sensors are based on zeolite powders, placed on co-planar structures tailored for microwave gamma frequencies [26].

Using advanced electronic transmission microscopy techniques, combined with X-ray powder diffraction and smooth X-ray adsorption structural measurements, the atomic structures of the HEU natural zeolite and two other ion exchange types of zeolites: Ag^{+} (Ag-HEU) and Zn^{2+} (Zn-HEU) have been investigated. In both ion exchange materials, the natural HEU zeolite charging was confirmed. For the Ag-HEU, Ag^{+} ions as well as clusters at the extra frame sites and Ag nanoparticles were confirmed.

A natural zeolite (Z-N), rich in clinoptilolite, was modified (Z-Al) through hydrated aluminum oxide incorporation, with the purpose of simultaneous removal of the ammonia and the phosphate. The incorporation of surface hydroxyl groups (~Al-OH) in the zeolite structure, as active groups for the removal of the phosphate, was characterized by acid titrations (pH PZC = 4.5 ± 0.2). A reduced capacity of regeneration cycles for the zeolite phosphate was observed [28].

Type A Ag- and Zn- liquid ion exchange modified zeolite samples were characterized by physio-chemical and thermal analysis. The resulted materials with improved physio-chemical proprieties are promising candidates for being used as coagulants. X-ray powdered diffraction, as well as the nitrogen adsorption rate show a good zeolite structure and texture conservation after the ion exchange [29].

Interconnected EMT, EMT/FAU and FAU zeolites have been synthesized and applied as additives of FCC catalysts for hydro-enhanced gasoline. The analysis of the physio-chemical proprieties of the synthesis catalysts by various techniques: XRD, N2 adsorption-desorption, HRTEM, SEM, showed that the catalysts with incorporated zeolites have large specific surface areas, pores and suitable acidities.

2. Materials and experimental method

For the determination of the lead ions retaining performances, wastewater from the granulating pool of the lead obtaining metallurgical plant from Baia Mare and a sample of zeolite tuff from Barsana, Maramures were used.

The efficiency of some Barsana tuff in the lead ions (from wastewaters with a lead ion quantity of 5.36 $mgPb^{2+}/L$) retaining process was studied. The static and dynamic ion exchange process was monitored.

The zeolite tuff samples were examined by using the Jeol JSM – 5600 LV electronic microscope and by EDX analysis. The zeolites appear to have the shape of clinoptilolite tabular crystals (Figure 1 & Figure 2), the majority of them having 2-10 μm, some of them reaching 35 μm.

Wastewater was analyzed for the determination of the lead ions concentration, this concentration being determined by mass spectrometry with inductively coupled plasma. The analysis was conducted at the Bucharest Chemical-Pharmaceutical Research and Development Institute.

Chemically untreated zeolite tuff (BN) and chemically treated tuff (Z-Na; noted with BT) with a 0.02 to 0.4 mm grain size and a solid ratio of 1:10 for static exchange were used. The sampling for the lead ions concentration determination was made once every 24 hours until the balance was reached. For dynamic exchanges, the experiments were conducted on a ionic exchange column of 45 mm diameter. 10g of zeolite tuff, a solution flow rate of 0.055 ml/s and gravitational flow were used. The probing was made once every 100 ml of effluent until the ion exchanger exhaustion.

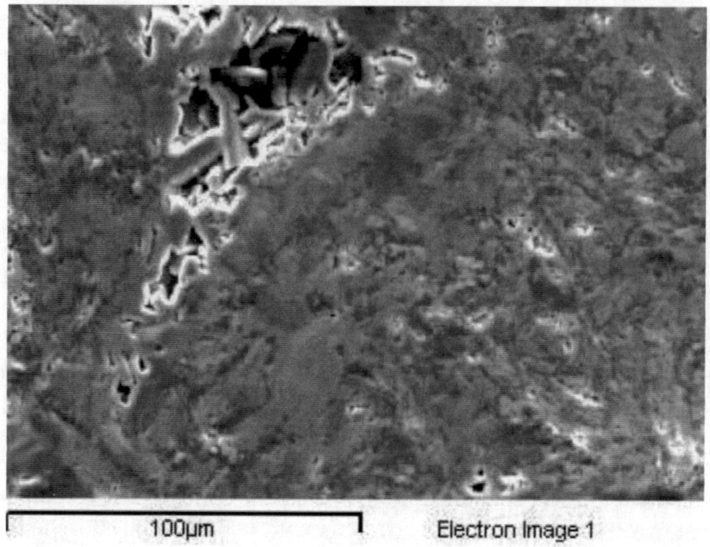

Figure 1. SEM image of the clinoptilolite tabular crystals tuff. (500x)

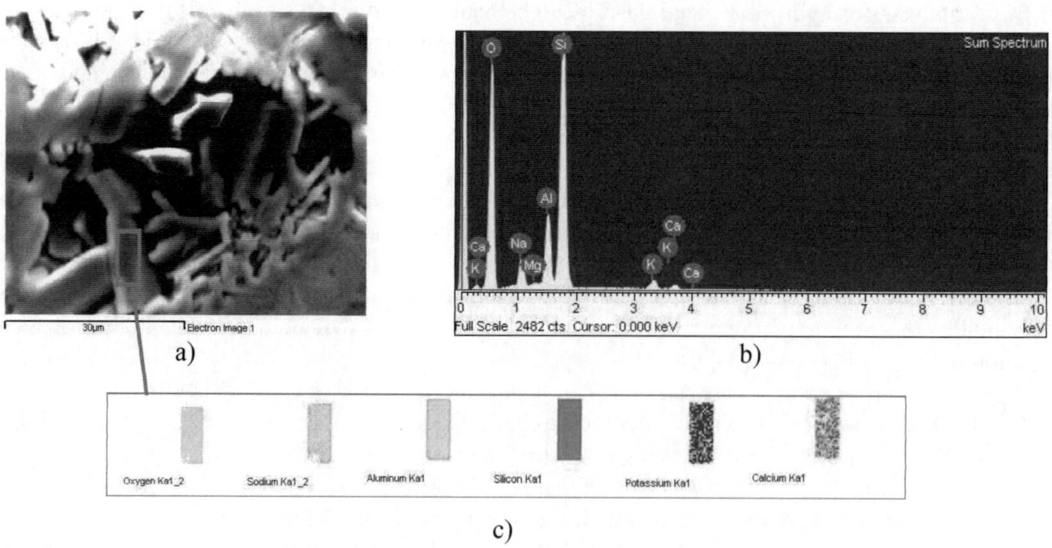

Figure 2. SEM investigation, a) clinoptilolite crystal, b) EDX Spectrum, c) Crystal composition maps [9]

3. Experimental Results

The experiments conducted for static exchange processes in a previous study [11], having the Barsana tuff as zeolite material and wastewater from the metallurgical plant led to the choice as the optimal alternative of a 10g zeolite tuff/100cm3 wastewater ratio for tracking the efficiency of the ion exchange.

The values of the lead ion concentrations were determined by mass spectrometry with inductively coupled plasma for each set of samples and they can be found in table 1. In the Figure 3 the variation of the static ion exchange process for the chemically treated (BT) and untreated (BN) Barsana zeolite tuff sample can be found, and in the Figure 4 the breaking curves for the dynamic exchange process of the two Barsana zeolite tuff can be found. The dynamic exchange experiments were made on column in the conditions mentioned earlier, using wastewater samples containing 5.36 mgPb^{2+}/L, chemically untreated zeolite tuff samples (BN) and chemically treated tuff in alkaline form (BT). The presence of the heavy metals should entail special attention because of their toxicity even in low concentrations.

Table 1. The values of the time depending static ion exchange capacities for the two types of samples BT and BN

Time, [min]	C_i= 5.36 mg Pb^{2+}/L BT	C_i= 5.36 mg Pb^{2+}/L BN
0	0	0
24	0.051	0.038
48	0.052	0.049
72	0.054	0.047
96	0.053	0.05
120	0.053	0.05

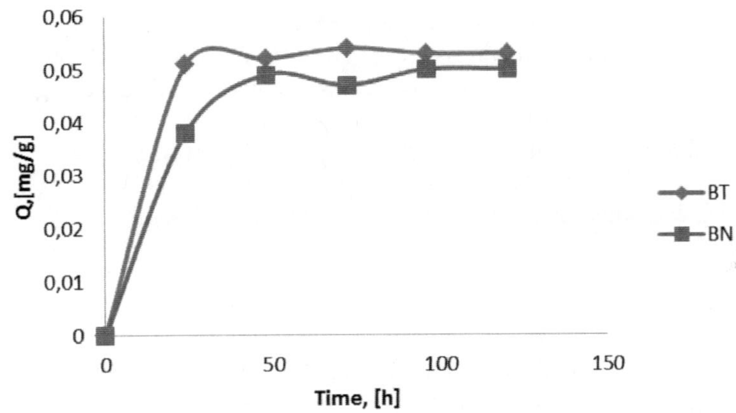

Figure 3. The variation of the static ion exchange capacity obtained for the chemically treated (BT) and untreated (BN) Barsana zeolite tuff

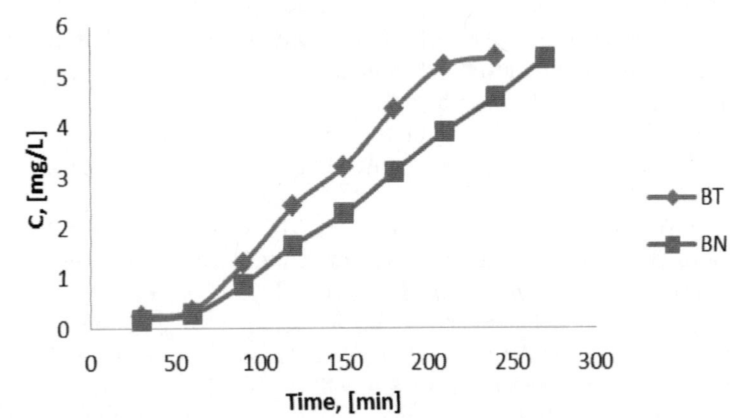

Figure 4. The breaking curves of the chemically treated (BT) and untreated (BN) Barsana zeolite tuff samples in dynamic contact with the wastewater

In Figure 4, the different shape of the breaking curves for the two situations can be seen. Where the chemically untreated zeolite tuff sample was used, the breaking curve slope is a lot smaller, which points to the fact that the lead ions charge of the zeolite tuff takes place at a slower rate and the breaking of the zeolite tuff layer takes place later. Thus the breaking happens after 250 minutes when using chemically treated zeolite tuff compared to 280 minutes when using chemically untreated zeolite tuff.

4. Conclusions

The chemically untreated Barsana natural zeolite tuff used in the static and dynamic ion exchange processes leads to good results regarding the lead ions retention. For the dynamimc exchange we can observe small differences between the two types of samples used for the experiment.

The values obtained from the breaking curves show that the BT zeolite tuff sample is exhausted after 250 minutes, while the BN sample is still active in the ion exchange process, thus more efficient. This is due to the small potential (concentration) difference, the ion exchange process speed being lower, meaning the speed limitating stage becomes the ion exchange reaction.

It is shown that for the static exchange experiments there are no significant differences between the two chosen samples (BT and BN).

The Barsana Maramures natural zeolite with remarkable proprieties due to its clinoptilolite content, was pointed out by various studies conducted by [13], [14] and considered to be a natural material used for wastewater heavy metals ion retention by [9-12] and it led to very good results for ion exchange processes.

Lead can be found in the environment on natural ways, but most of it comes as a result of human activities which lead to unwanted polluting processes that can have a serious impact on the health, the various effects of the lead intoxication on the human body being well known.

References

[1] Aurea A, Maia B, Neves R F, Omulo S, Angelica and Pollmann H 2015 Synthesis, optimisation and characterisation of the zeolite NaA using kaolin waste from the Amazon Region. Production of Zeolites KA, MgA and CaA, *Applied Clay Science* **108** 55-60

[2] Camacho L M, Parra R R and Deng S 2011 Arsenic removal from groundwater by MnO2-modified natural clinoptilolite zeolite: Effects of pH and initial feed concentration, *Journal of Hazardous Materials* **189**(1-2) 286-293

[3] Karatas M 2012 Removal of Pb(II) from water by natural zeolitic tuff: Kinetics and thermodynamics, *Journal of Hazardous Materials* **199-200** 383-389

[4] Gutierez-Segura E, Solache-Rios M, Colin-Cruz A and Fall C 2012 Adsorption of cadmium by Na and Fe modified zeolitic tuffs and carbonaceous material from pyrolyzed sewage sludge, *Journal of Environmental Management* **97** 6-13

[5] Rasamimanana S, Mignard S and Batonneau-Gener I 2016 Hierarchical zeolites as adsorbents for mesosulfuron-methyl removal in aqueous phase, *Microporous and Mesoporous Materials* **226** 153-161

[6] EL-Mekkawi D M, Ibrahim F A and Selim M M 2016 Removal of methylene blue from water using zeolites prepared from Egyptian kaolins collected from different sources, *Journal of Environmental Chemical Engineering* **4**(2) 1417-1422

[7] Woinarski A Z, Stevens G W and Snape I 2006 A Natural Zeolite Permeable Reactive Barrier to Treat Heavy - Metal Contaminated Waters in Antarctica: Kinetic and Fixed-bed Studies, *Process Safety and Environmental Protection* **84**(2) 109-116

[8] Ouki S K and Kavannagh M 1999 Treatment of metals-contaminated wastewaters by use of natural zeolites, *Water Science and Technology* **39**(10-11) 115-122

[9] Pop A, Vida-Simiti I, Damian Gh and Iepure Gh 2012 Removal of heavy metals from wastewater by using zeolitic tuff, *Carpathian Journal of Earth and Environmental Sciences* **7**(1) 239-248

[10] Pop A 2011 Experimental studies on the influence of pH in wasrwwater's lead retention processees using volcanic zeolite tuff, *Buletin Ştiinţific al Univ. De Nord din Baia Mare, seria D* **XXV**(2) 117-123
[11] Pop A, Iepure Gh, Pop E, Hotea V and Juhas J 2011 Experimental studies on the influence of the ratio of amount of zeolite volcanic tuff/wasterwater volume on the ionic exchange process, *The Annals of "Dunarea de Jos" University of Galati, Fascicle IX. Metallurgy and Materials Science* 254-258
[12] Pop A, Vida-Simiti I, Iepure Gh, Pop E, Hotea V and Juhasz J 2011 *Experimental studies on the usage of zeolite volcanic tuffs for the reduction of concentration of heavy metal ions from wastewaters*, 43rd International October Conference on Mining and Metallurgy, pp 242-245
[13] Damian Gh, Damian F, Macovei Gh, Constantina C and Iepure Gh 2007 Zeolitic tuffs from Costiui zone, Maramures Bazin, *Carpathian Journal of Earth and Environmental Sciences* **2**(1) 59-74
[14] Damian Gh, Pop N, Kovacs and Peter P 1991 *La pétrologie des tufs volcaniques badénienne du Bassin Maramures*, in: The Volcanic Tuffs from the Transylvanian Basin (ed. Mârza I), Universitatea din Cluj-Napoca, România, pp 233-243
[15] Nguyen T C, Loganathan P, Tien Vinh Nguyen T V, Vigneswaran S, Kandasamy J and Naidu R 2014 Simultaneous adsorption of Cd, Cr, Cu, Pb, and Zn by an iron-coated Australian zeolite in batch and fixed-bed column studies, *Chemical Engineering Journal* **270** 393-404
[16] Egashira R, Tanabe S and Habaki H 2012 Adsorption of heavy metals in mine wastewater by Mongolian natural zeolite, *Procedia Engineering* **42** 49-57
[17] Al-Harahsheh M, Shawabkeh R, Batiha M, Al-Harahsheh A and Al-Zboon K 2014 Sulfur Dioxide Removal using Natural Zeolitic Tuff, *Fuel Processing Technology* **126** 249-258
[18] Leone V, Iovino P, Salvestrini S and Capasso S 2013 Sorption of non-ionic organic pollutants onto a humic acids-zeolitic tuff adduct: Thermodynamic aspects, *Chemosphere* **95** 75-80
[19] Sabbah I, Baransi K, Massalha N, Dawas A, Saadi I and Nejidat A 2013 Efficient ammonia removal from wastewater by a microbial biofilm in tuff-based intermittent biofilters, *Ecological Engineering* **53** 354-360
[20] Karatas M 2011 Removal of Pb(II) from water by natural zeolitic tuff: Kinetics and thermodynamics, *Journal of Hazardous Materials* **199-200** 383-389
[21] Yousef R I, El-Eswed B, Alshaaer M, Khalili F, Rahier H 2011 Degree of reactivity of two kaolinitic minerals in alkali solution using zeolitic tuff or silica sand filler, *Ceramics International* **38** (6) 5061-5067
[22] Gutiérrez-Segura E, Solache-Ríos M, Colín-Cruz A and Fall C 2010 Adsorption of cadmium by Na and Fe modified zeolitic tuffs and carbonaceous material from pyrolyzed sewage sludge, *Journal of Environmental Management* **97** 6-13
[23] Mallard I, Städe L W, Ruellan S, Jacobsen P A L, Larsen K L and Fourmentin S 2015 Synthesis, characterization and sorption capacities toward organic pollutants of new β-cyclodextrin modifiedzeolite derivatives, *Colloids and Surfaces A: Physicochemical and Engineering Aspects* **482** 50-57
[24] Frances L, Grivet M, Renault J P, Groetz J E and Ducret D 2014 Hydrogen radiolytic release from zeolite 4A/water systems under γ irradiations, *Radiation Physics and Chemistry* **110** 6-11
[25] Zhan Z, Ma N, Yan H, Peng Y, Wang Z and Yan Y 2014 Heat treatment for improving performance of inner-side zeolite NaA membranes on composite hollow fibers, *Journal of Membrane Science* **485** 94-102
[26] de Fonseca B, Rossignol J, Bezverkhyy I, Bellat J P, Stuerga D and Pribetich P 2015 Detection of VOCs by microwave transduction using dealuminated faujasite DAY zeolites as gas sensitive materials, *Sensors and Actuators B: Chemical* **213** 558-565
[27] Filippousi M, Turner S, Katsikini M, Pinakidou F, Zamboulis D, Pavlidou E and Van Tendeloo G 2014 Direct observation and structural characterization of natural and metal ion-

exchanged HEU-type zeolites, *Microporous and Mesoporous Materials* **210** 185-193
[28] Guaya D, Valderrama C, Farran A, Armijos C and Cortina J L 2015 Simultaneous phosphate and ammonium removal from aqueous solution by a hydrated aluminum oxide modified natural zeolite, *Chemical Engineering Journal* **271** 204-213
[29] Benaliouche F, Hidous N, Guerza M, Zouad Y and Boucheffa Y 2014 Characterization and water adsorption properties of Ag- and Zn-exchanged A zeolites, *Microporous and Mesoporous Materials* **209** 184-188

IOP Conf. Series: Materials Science and Engineering **200** (2017) 012069 doi:10.1088/1757-899X/200/1/012069

Simulation of defects on contactor coil

Z Erdei, M Horgos and O Zetea

Technical University of Cluj-Napoca, North University Centre of Baia Mare, Electrical Engineering, Electronics and Computers Department, Baia Mare, Romania

E-mail: erdeiz@cunbm.utcluj.ro

Abstract. By definition, a coil is an electrical passive device, which have two terminals, use in electrical circuits to keep the power in magnetic field or to detect the magnetic fields. In winding process, is possible to appear different defects, or issues which can cause in time problems in functionality of products. In this paper, we will analyze two types of defectives what were observed in winding process. In first type of defect, some wires of beginning of winding remain out of normal winding, and the wires are visible from outside, and in second type of defect, same beginning of winding remain inside of coil, under the normal winding, not in correct position. For simulation, we will used an assembly compose by anchor, electromagnet and coil. Those are parts of contactor.

1. Introduction

A coil, which is part of contactor, is compose by: coil body (plastic part), contacts (copper with silver surface), and copper wire. The coil what we will used for studying the defects, have wire by 0.112 mm diameter, covered with isolation lake. Together with it, the wire has 0.125 mm. On coil body were winded 4900 windings, divide in 27 levels. In each level are winded 181 windings. The defects what will be described in these papers, appeared when one coil is finished, and winding machine did not cut the wire correctly. These wires will be out of normal winding as in Figure 1 [1], [2].

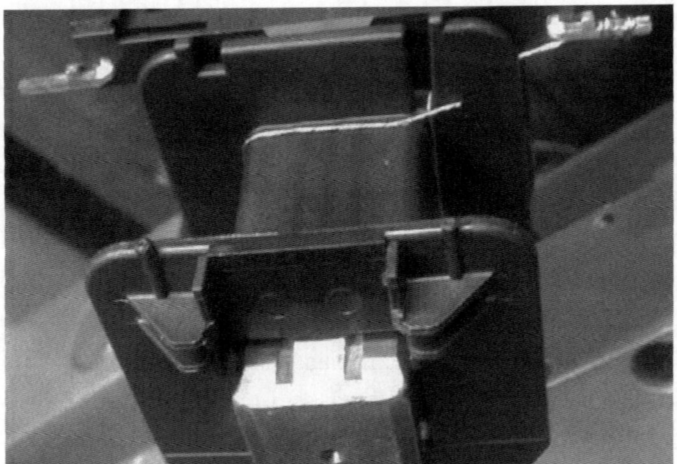

Figure 1. Beginning of defect winding

After finished the winding process, these wires can be hide inside of normal winding, and cannot be observe from outside (Figure 2), or can remain outside of normal winding (Figure 3), and can be observe from outside and scrap.

Figure 2. Wires will be hide by normal windings

Figure 3. Wires remain outside of normal windings

2. Parameters of coil

In circuit, this type of coil is charge with 27 V c.c., and have resistance 764 Ω. This mean, according with Ohm law, the current through coil is 0.035 A [3-8].

The inductance is calculating with equation (1):

$$L = \frac{0.008 \cdot N^2 \cdot d^2}{3 \cdot d + 9 \cdot l + 10 \cdot h}\ [\mu H] \tag{1}$$

Where:
N - number of turns
d - middle diameter of coil
l – length of coil
h – high of coil

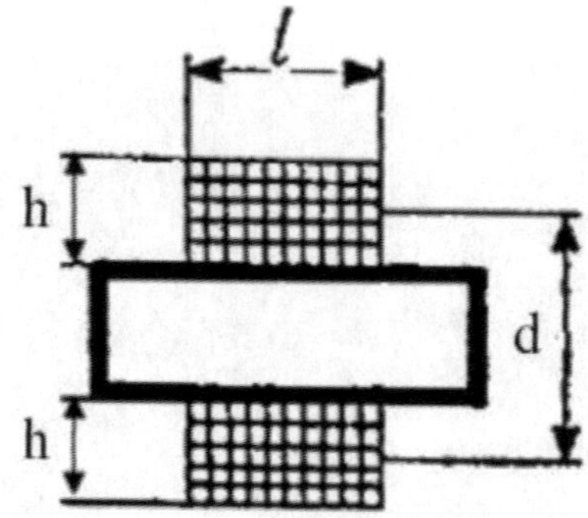

Figure 4. Parameters for inductance calculation

If we replace all values in equation (1), we have:

$$L = \frac{0.008 \cdot 4900^2 \cdot 21.8^2}{3 \cdot 21.8 + 9 \cdot 22.4 + 10 \cdot 3.3} = \frac{192080 \cdot 475.24}{65.4 + 201.6 + 33} = \frac{91284099.2}{300} = \\ = 304280.33\ [\mu H] = 0.3\ [H] \tag{2}$$

The density of current can be calculated with equation (3):

$$J = \frac{I}{A} \text{ [A/mm}^3\text{]} \tag{3}$$

Where:
I – the current through wire
A – section of wire
For wire with diameter 0.125 mm2:

$$A = \pi R^2 = 3.14 \cdot \left(\frac{0.125}{2}\right)^2 = 3.14 \cdot 0.0625^2 = 3.14 \cdot 0.004 = 0.01\, mm^2 \tag{4}$$

If we use equation (3), we obtain:

$$J = \frac{0.035}{0.01} = 3.5\, A/mm^2 \tag{5}$$

3. Simulation of coil

For simulation of coil we used FEMM software. We simulate from magnetic point of view, and also from electric point of view. In Figure 5, is a section of coil, design in FEMM, before charging, when the anchor is up [9-12].

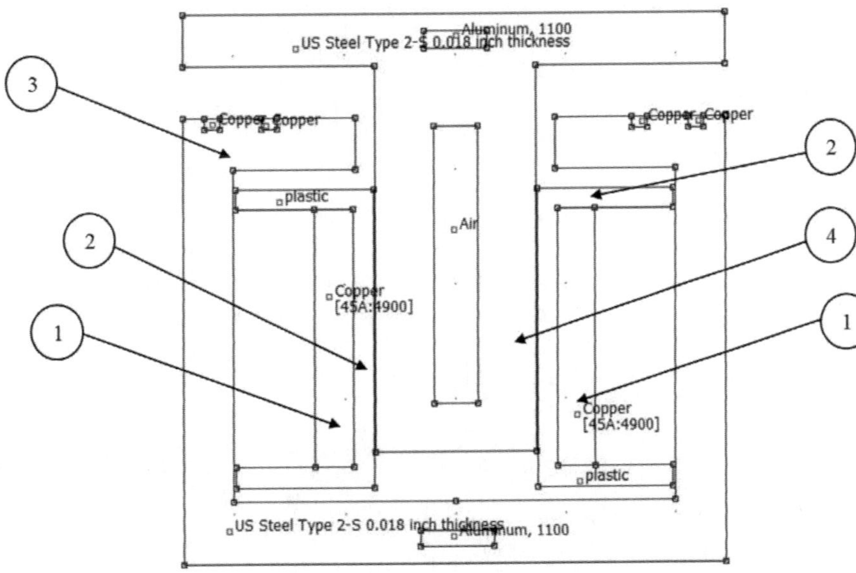

Figure 5. Section of coil: 1 – winding of coil; 2 – coil body; 3 – electromagnet; 4 – anchor

3.1. Magnetic simulation

The magnetic field of correct winding coil can be saw in Figure 6. This simulation is made in function position, when coil is charge with 27 V cc., and has 35 mA thought wire. If we simulate the two defects, in same conditions, the magnetic field has another form in anchor. This could be saw in Figure 7 and Figure 8.

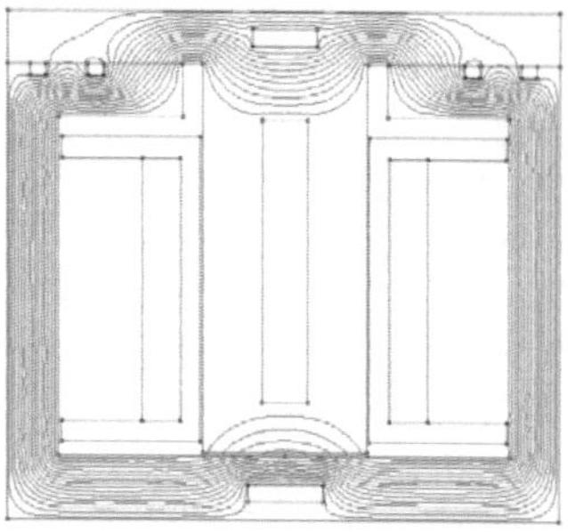

Figure 6. Normal winding coil

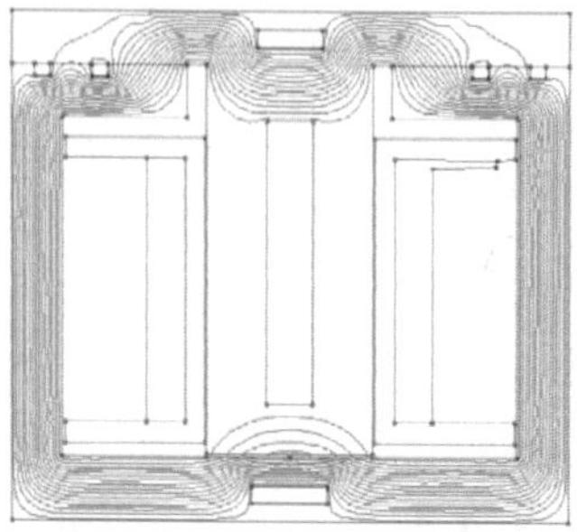

Figure 7. Wires outside

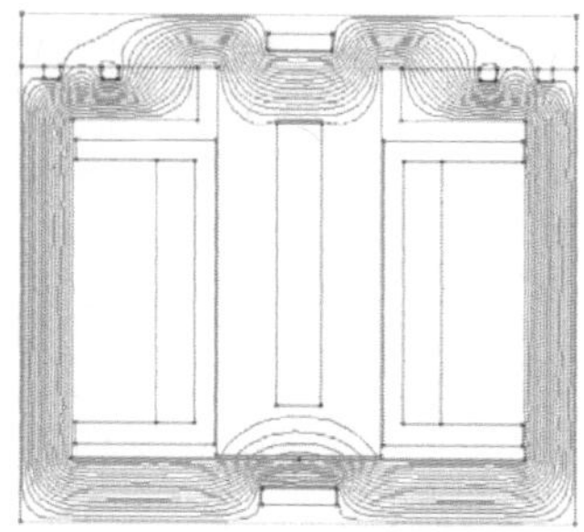

Figure 8. Wires inside

The magnetic field is modified by additional wires, because all parameters of coil are modified in right side of section. The defect is composed by 5 wires, winding together. These wires have 10 mm length [4], [5], [11].

The resistance of these 5 wires can be calculated with equation (6):

$$R = \delta \cdot \frac{l}{A} \ [\Omega] \tag{6}$$

According to FEMM simulation, the average of magnetic inductance value of anchor increase. This mean the forces applied in anchor increase also, because:

$$B = \frac{F}{I \cdot l} \ [T] \tag{7}$$

$$F = \frac{B}{I \cdot l} \ [\mathrm{N}] \tag{8}$$

The value of is around of 0.06 N. In Figures 9, 10, 11, could be see the evolution of magnetic inductance on entire surface of assembly. The difference could be easily saw on anchor surface. We have areas where values of inductance drop down from 1 T to 0, but also surfaces where values increase from 0.7 to 1.2 T. all those values influence the functionality of coil. In points where wires are out of normal winding, the forces are bigger than in normal cases. These forces produce bigger vibration and in time can destroy the lake of wire [13-15].

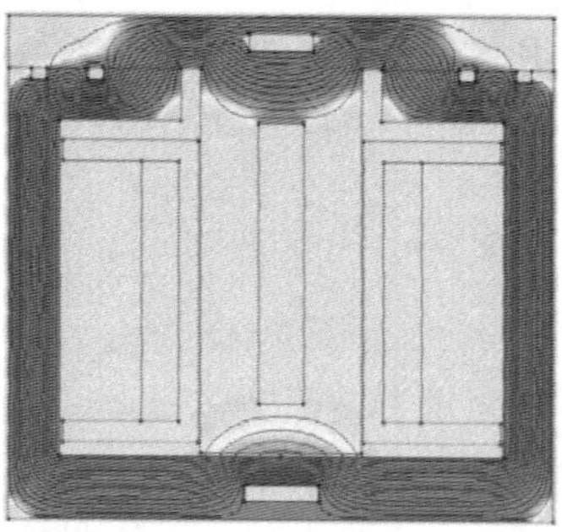

Figure 9. Normal winding coil

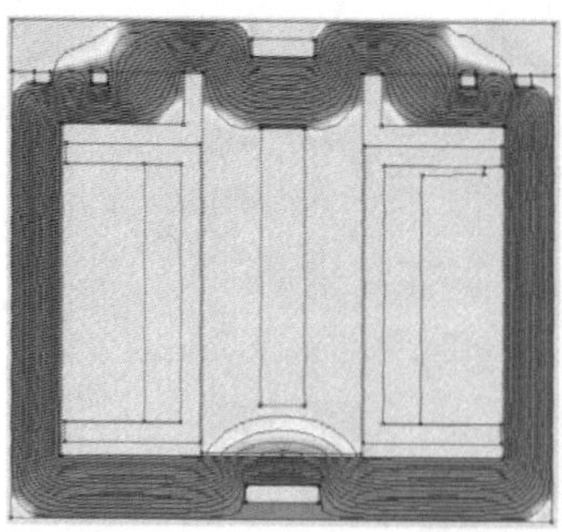

Figure 10. Wires outside

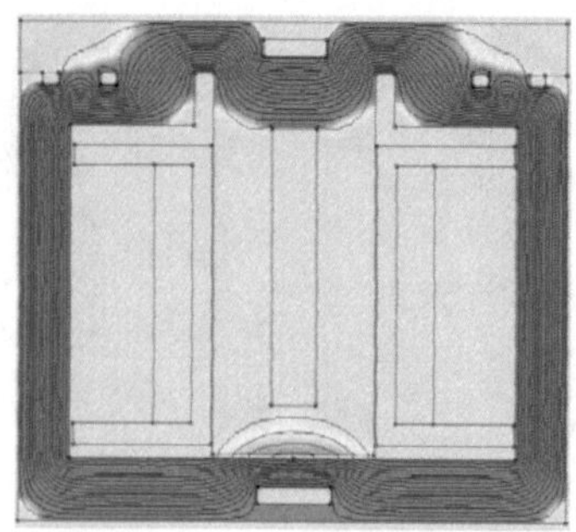

Figure 11. Wires inside

The meaning of each color from figures can be seen in Figure 12.

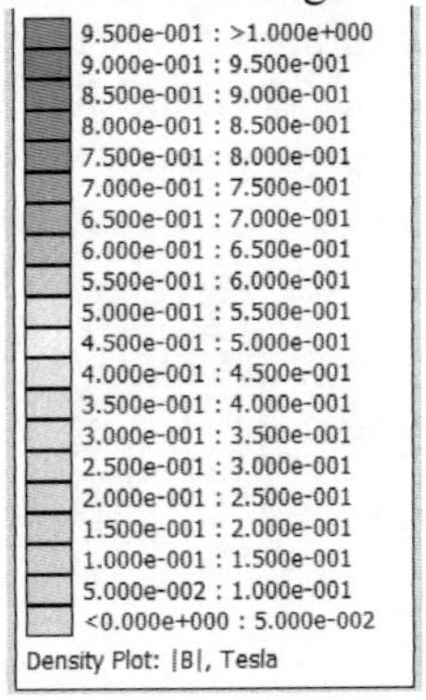

Figure 12. Meaning of color

3.2. Electrical simulation

The electrical simulation is made also in FEMM, in same conditions as magnetic simulation. If we look in Figure 13, we can see that in upside and in down side of coil, is bigger density of voltage, than in the middle of coil. Also, in Figure 14, because in upside we have the other 5 wires, which are remaining out because of bad winding.

If we analyze each case, and we compare the density of voltage from entire surface of coil, we can say that, where density of voltage is bigger, the temperature is bigger. The defects wires, have other direction of electrical field. This is opposite with the direction of electrical field of normal winding.

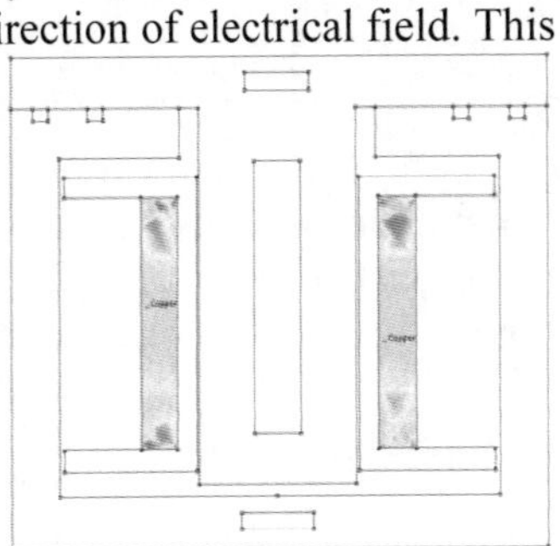

Figure 13. Normal winding coil

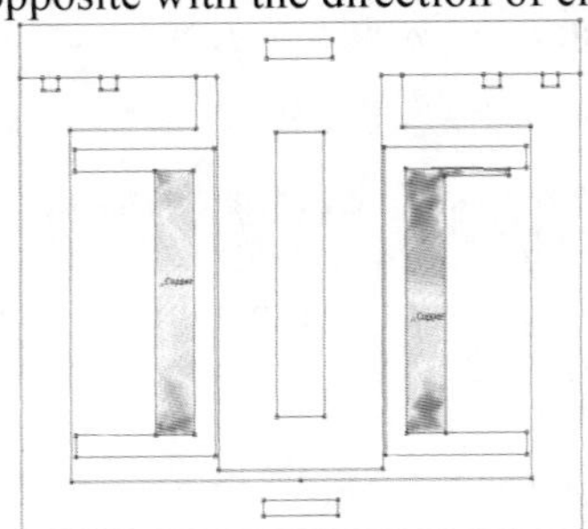

Figure 14. Wires outside

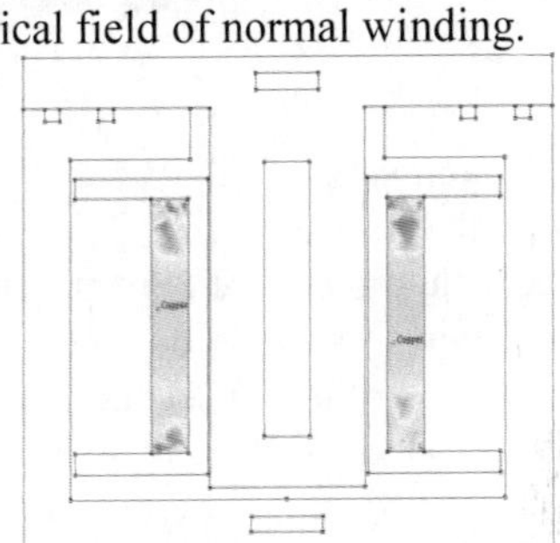

Figure 15. Wires inside

The graphics for voltage density on surface of coil, show us the instability of voltage in defect area. In Figure 16 is presented the voltage density for correct coil, and in Figure 17 is density of voltage for defect coil [16], [17]. The real component of voltage density from Figure 16 is constantly, but in Figure 17, in defect coil we can saw that drop of voltage in defect area. This voltage drop can produce inside of coil some shocks, which can affect the functionality and the life time of coil. Especially because, according with bellow graphic, this drop is on short distance of wire.

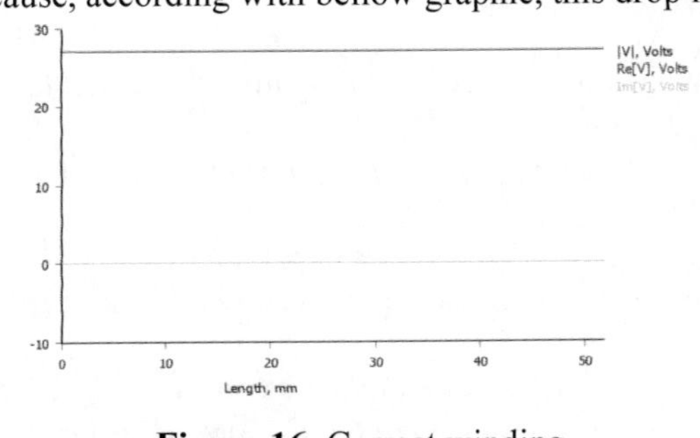

Figure 16. Correct winding

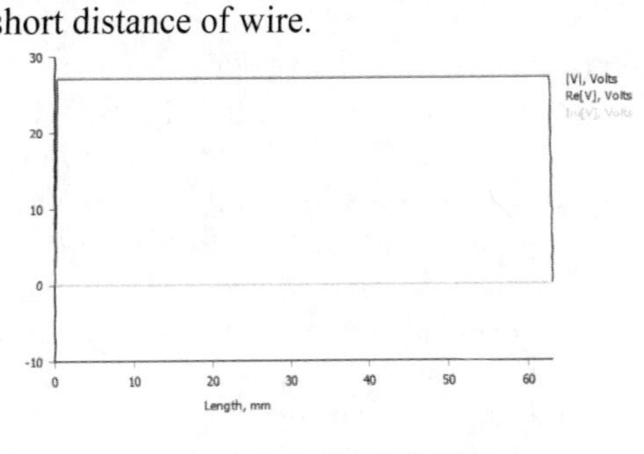

Figure 17. Bad winding

In electrical simulation, we can study also the moving of electrons inside of electrical field. In Figure 18, we can see this moving in coil with wires outside of normal windings. This moving should produce a normal electrical field, but because the wire has opposite orientation, we have two electrical fields. One produce by normal windings, and one produce by 5 wires which are out of normal windings. This mean, there we have an intersection of two different electrical fluxes, with different directions.

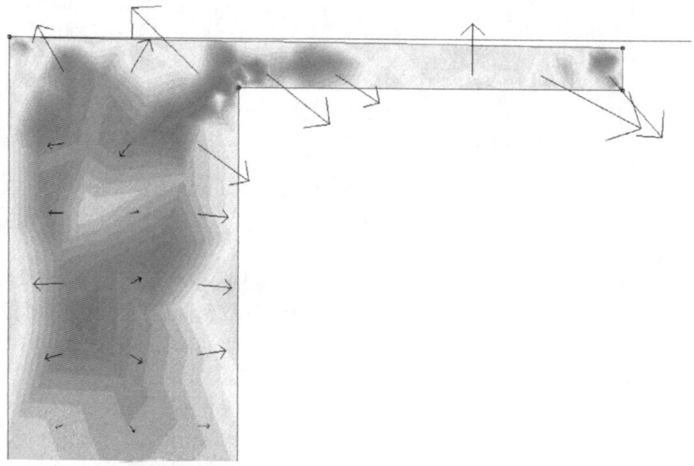

Figure 18. Moving of electrons

4. Conclusions

After all those simulation, we can affirm that each little deviation from normal process can affect more or less the functionality of products. The products must respect some parameters, and, as in our cases, some wires can take out the coil from normal parameters. Because those wires what have different direction, produce a second electrical field, and inside of coil, result a difference of potentials. At firs look those values are not so bigger than normal, but in time can cause damages. The lake of wires can be destroyed, and a short circuit can appear.

Those differences can grow up the forces and the temperature. According with Joule Lentz law, the temperature of conductor is proportionally with current values, time and resistance. So in this case, few Ohm will increase the temperature in coil.

$$Q = I^2 \cdot R \cdot t \tag{9}$$

If we charge three contactors with those three studied types of coils, the correct one will have a longer life. The two defective coils have bigger resistance. This difference, together with voltage drop saw in Figure 17, can reduce the life of coils. This thing cannot be anticipated exactly, because these two types of defect can be different for each coil.

References

[1] Erdei Z, Horgos M, Grib A, Preradović D M and Rodic V 2016 MCCB warm adjustment testing concept, *IOP Conf. Ser.: Mater. Sci.* **144** 012014

[2] Erdei Z, Horgos M, Lung C, Pop Vadean A and Muresan R 2017 Frequency behavior of the residual current devices, *IOP Conf. Ser.: Mater. Sci.* **163** 012053

[3] Dumitriu L and Iordache M 1998 *Teoria moderna a circuitelor electrice - Vol. I – Fundamentare teoretica, Aplicatii, Algoritmi si Programe de calcul*, Editura All Educational S.A., Bucuresti, Romania

[4] Iordache M and Dumitriu L 2000 *Teoria moderna a circuitelor electrice - Vol. II – Fundamentare teoretica, Aplicatii, Algoritmi si Programe de calcul*, Editura All Educational S.A., Bucuresti, Romania

[5] Iordache M and Dumitriu L 2002 *Simularea asistata de calculator a circuitelor analogice*, Editura Politehnica Press, Bucuresti, Romania
[6] Iordache M, Dumitriu L and Matei I 1999 *ASINOM – Analiza Simbolica bazata pe Metoda Nodala Modificata, Manual de utilizare*, Catedra de Electrotehnica, U.P.B., Bucuresti
[7] Iordache M and Dumitriu L 2000 PACER – Program de Analiza a Circuitelor Electronice Rezistive
[8] Erdei Z 2002 *Analiza toleranţelor circuitelor electrice – proiect de diplomă*, Facultatea de electrotehnică, Universitatea Politehnica Bucuresti
[9] Barz C, Oprea C, Chiver O, Erdei Z, Neamt L, Horgos M and Pop Vadean A 2014 *The Advantages of Numerical Analysis for Claw Pole Alternator*, International Conference and Exposition on Electrical and Power Engineering EPE 2014, Iasi, Romania, October 16-18, pp 353-356
[10] Chiver O, Micu E and Barz C 2008 *Stator winding leakage inductances determination using finite elements method*, 11 th International Conference on Optimization of Electrical and Electronic Equipment OPTIM 2008, Braşov, Romania, May 22-23, pp 69-74
[11] Schwarz A E 1987 *Computer-aided design of microelectronic circuits and systems*, Academic Press, London, UK
[12] Orchard H J and Temes G C 1968 Filter design using transformed variables, *IEEE Transactions on Circuit Theory* **15** (4) 385-408
[13] Chiver O, Neamt L, Barz C and Costea C 2014 *Frequency domain numerical analysis of rotor cage induction motor*, International Conference and Exposition on Electrical and Power Engineering EPE 2014, Iasi, Romania, October 16-18, pp 327-331
[14] Barz C and Oprea C 2011 Study of electromagnetic field in claw-poles alternator, *Annals of Faculty Engineeering Hunedoara-International Journal of Engineering* **IX** 359-362
[15] Boldea I, Topor M, Marignetti F, Deaconu S I and Tutelea L N 2010 *A Novel Single Stator Dual PM Rotor, Synchronous Machine: topology, circuit model, controlled dynamics simulation and 3D FEM Analysis of Torque Production*, 12th International Conference on Optimization of Electrical and Electronic Equipment OPTIM 2010, Brasov, Romania, May 20-22, pp 343-351
[16] Barz C, Deaconu S I, Latinovic T, Berdie A, Pop-Vadean A and Horgos M 2016 PLCs used in smart home control, *IOP Conf. Ser.: Mater. Sci.* **106** 012036
[17] Barz C, Oprea C, Erdei Z, Pop Vadean A and Petrovan F 2014 *The control of an industrial process with PLC*, International Conference on Applied and Theoretical Electricity ICATE, Craiova, Romania, October 23-25, pp 1-4